THE ELECTRONIC DESIGN AUTOMATION HANDBOOK

The Electronic Design Automation Handbook

Edited by

Dirk Jansen et al.

Director of ASIC Design Center,
University of Applied Sciences, Offenburg, Germany

KLUWER ACADEMIC PUBLISHERS
BOSTON / DORDRECHT / LONDON

A C.I.P. Catalogue record for this book is available from the Library of Congress.

ISBN 1-4020-7502-2

Published by Kluwer Academic Publishers,
P.O. Box 17, 3300 AA Dordrecht, The Netherlands.

Sold and distributed in North, Central and South America
by Kluwer Academic Publishers,
101 Philip Drive, Norwell, MA 02061, U.S.A.

In all other countries, sold and distributed
by Kluwer Academic Publishers,
P.O. Box 322, 3300 AH Dordrecht, The Netherlands.

Printed on acid-free paper

Printed in the Netherlands.

Foreword

When I attended college we studied vacuum tubes in our junior year. At that time an average radio had five vacuum tubes and better ones even seven. Then transistors appeared in 1960s. A good radio was judged to be one with more then ten transistors. Later good radios had 15–20 transistors and after that everyone stopped counting transistors. Today modern processors runing personal computers have over 10 million transistors and more millions will be added every year.

Prof. Dr. Gajski demonstrates the Y-chart

The difference between 20 and 20M is in complexity, methodology and business models. Designs with 20 transistors are easily generated by design engineers without any tools, whilst designs with 20 M transistors can not be done by humans in reasonable time without the help of automation. This difference in complexity introduced a paradigm shift which required sophisticated methods and tools, and introduced design automation into design practice.

By the decomposition of the design process into many tasks and abstraction levels the methodology of designing chips or systems has also evolved. Similarly, the business model has changed from vertical integration, in which one company did all the tasks from product specification to manufacturing, to globally distributed, client server production in which most of the design and manufacturing tasks are outsourced.

In general, the product creation process can be divided into several tasks including requirements gathering, specification generation, component selection, architectural exploration, component synthesis, physical design, verification, simulation, prototyping, and manufacturing. Furthermore, each product can be represented on different levels of abstraction defined by the complexity of components used in the design, such as transistors, gates, registers, processors, or general types of cores, or any other intellectual property. For each task and each level we have created many different methods and different tools, and thus have created the Electronic Design Automation (EDA) industry.

The set of tasks and tools which lead from product specification to manufacturing is called design methodology. Each system company uses a slightly different methodology in the creation of different products, depending on market needs, product requirements, quality, cost, and business model.

This application specific methodology combined with the advances in chip fabrication allow system companies to build complete systems on a single piece of silicon. That has introduced many changes in the way we used to think about building systems.

In this new design environment the system companies are defining products in their application domain whilst semiconductor companies are building systems on silicon. Design activities are usually outsourced to the third party vendors. The design activity is mostly focused on specification, architecture exploration, software, and verification. The cost of most SOCs is in software and not in hardware. Software and hardware are designed together, and we are forced to think about the whole product and not just about one block of design. Therefore designers must be aware of the whole design process and not just of one task on one level of abstraction.

This insight is also leading our university education to focus on system engineering in order to generate graduates with system knowledge and not just programmers or circuit designers.

In such a changing environment this Handbook on Electronic Design Automation represents a welcome help for the practitioners of the system design as well students taking courses in the same area. The handbook surveys the main tasks of system design methodology, explains different methods and tools for design specification, synthesis, simulation, and verification, introduces hardware description languages and modeling practices, and introduces techniques for the design of circuits, modules and systems on different levels of abstraction. The main advantage of this Handbook is that readers quickly obtain a good overview of the design methodologies and of the design automation field as a whole, instead of only one aspect of it.

For this reason I sincerely welcome this book and recommend it highly to all practitioners in the field of designing and building electronic systems

Irvine, California

Daniel D. Gajski

Editor's Preface

This book was jointly developed in a co-operation of colleagues lasting many years from the Universities of Applied Sciences, Germany, who are engaged in the design of integrated electronics in education and research, and which form the MPC Group of the Universities of Applied Sciences of Baden-Wuerttemberg/Germany. MPC stands for 'Multi Project Chip', the idea of placing on a wafer several projects at the same time in order to distribute the high costs of one project-run over several projects. The realization of this idea requires teamwork, mutual agreement, and knowledge about the current state of the art. Although after more than 10 years the MPW service has changed over to professional brokers such as Europractice and CMP, the solidarity of the group, which covers in the meantime 13 engineering schools from the southwest, preserved nevertheless remained. MPC works today also as a network of partners with industry and is able, owing to the widely varying experiences of the institutes involved, to cover the entire range of modern circuit design. Each year more than 600 students are educated in the laboratories of MPC members. Their own experience from student projects and industry projects ensures authenticity – by the close relationship with teaching and research. There is plenty of practical experience in entirely normal, unspectacular, as well as exotic projects. This experience has been attempted to be laid down here in book form.

The book therefore addresses itself to the user of EDA, the practical working engineer, and the student. The specialists and researchers working in this field may forgive me some simplifications and abstractions. We tried to gather, to the best of own kowledge, what one **must know** today to design modern electronics with computer programs effectively. This knowledge must be interdisciplinary and covers the areas of classical computer science as well as digital design, semiconductor physics, electronic circuit technology up to the manufacturing processes of ICs and printed circuit boards. Does one have to really know all that? I am asked this question again and again by students and answer them with "**that's the minimum**!". Whoever wants to penetrate these subjects more deeply, there are numerous good special books available today. We attached these references, therefore, in the respective chapters.

There are, admittedly, better books for each section. This book can not replace those special books. But the most important content from these books (and perhaps still a little more) is written into this **Handbook of Electronic Design Automation**. All that **has to be known** is packed handily, is didactically prepared, and is easily consumable by a stressed industrial engineer too, just as one expects it from a handbook. This book therefore belongs on the table of every developer of electronics!

Whoever expects to find in this book a guide how to use software programs, and perhaps in the appendix also a CD with demo versions; he has understood nothing. With the present innovation speed of 2 updates/year such software would already be soon outdated just after the printing of this book. In order to receive current software descriptions and also downloads from demonstration software, a detailed table with current web addresses can be found in the references section. Likewise there is some information on the research landscape of EDA in Europe and the US and a list of active federations and organizations, as well as a list of websites of software providers.

Here I would like to thank again all the authors for their contributions, for the support by the publisher and the publishing company, and my coworkers in the ASIC Design Center of the University of Applied Sciences, Offenburg, Germany, who all contributed in various way to the success of this work.

The Editor

Prof. Dr.-Ing. Dirk Jansen

Acknowledgements

This book is the result of a cooperation between several members of the MPC group working as authors, co-authors, translators, and provider of examples, pictures, or whatever makes this book readable. I have to thank these friends, for without their knowledge, patience, and support the German version of this book, and, of course this new and updated English version, would never have been written. I am very grateful to my friends Horst Nielinger and Wolfgang Ritzert, and Wolfgang Ludescher, all member of MPC from the beginning, who helped significantly in translating the chapters and supported the authors in many ways. Thanks also to all of the authors who spent their precious time on the chapters again, reading corrections and delivering all the material needed.

I again have to be grateful to all the students and assistants, who helped to prepare 741 pictures, 176 tables, and uncounted examples, and to Wolfgang Vollmer, Jürgen Hauser, and several others, working in the laboratories of the authors and giving many suggestions for examples and presentation of the material.

I would like to thank Daniel Gajski, who arranged a stay for me at his CECS at the UCI, giving me support, recommendations, and, not least, convincing me to accept the challenge of translating the original local German version of the book and of publishing it on the international EDA scene.

My further thanks have to go to Giesela Hagedorn for helping with translation, Dr. Naake at Chemnitz, for doing all the formatting and typesetting, and Mark de Jongh from Kluwer Academic Publishers for arranging everything necessary for the publication of this book, and of course Michael Cole who advised on changing German style of many chapters to something sounding more like the English language, and lastly, but not least, the staff at Kluwer for their highly professional production of this book.

Thanks are also due to my wife Rosi for her unquestioned support and patience during half a year of evenings full of work.

Contents

Overview EDA

1, 2

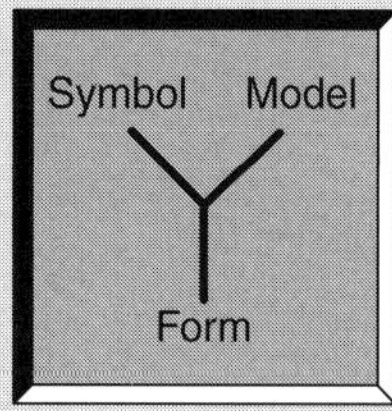

Symbolic Design

3

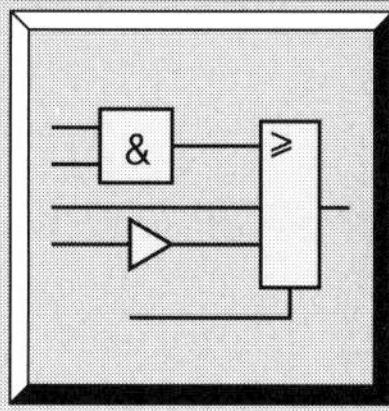

High Level Language Design

4, 5, 6, 7, 8

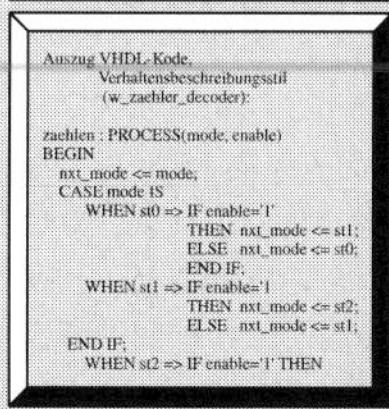

Modelling and Verifications

9, 10, 11, 12, 13, 14, 15

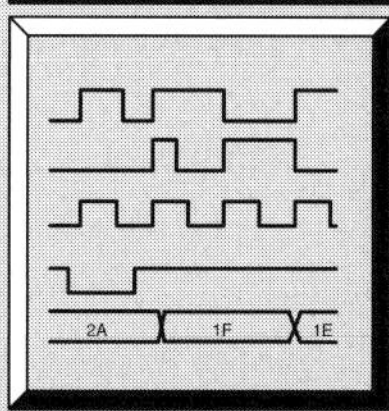

Implementation

16, 17, 18, 19, 20, 21, 22, 23, 24, 25

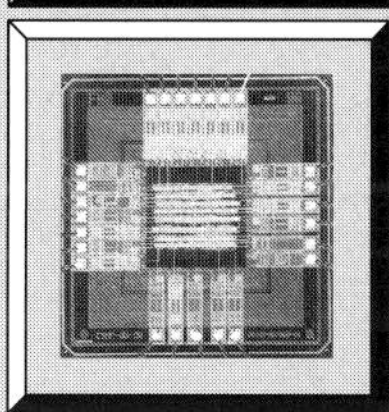

Tutorial

26

Appendix

A, B, C, D, E

1 Introduction

Dirk Jansen

In 2000 the Semiconductor Industry Association (SIA), an American trade association of the semiconductor industry, announced a world-wide electronic industry market volume in excess of $200 billion. The average growth is anticipated to remain at a rate of 17 %/year [1.10]. Market volumes for the last 20 years can be seen in fig. 1.1. This figure shows the exponential growth of the curve; the first billion dollar was reached in 1961. By 2010 the SIA expects a trillion dollar market volume for an industry that was not in existence 50 years ago. No other industry in history has experienced this phenomenal growth rate.

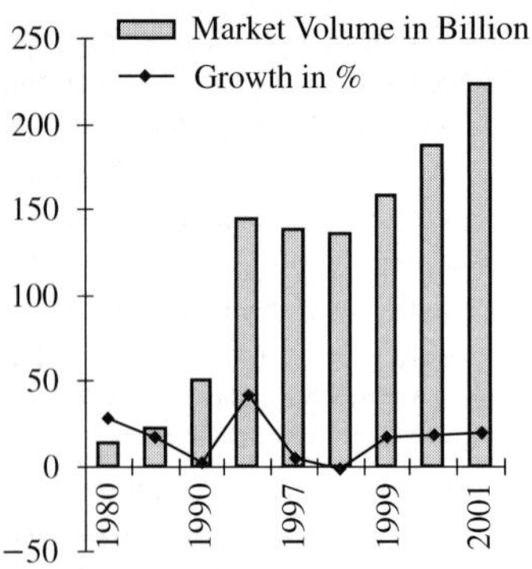

Fig. 1.1 Worldwide market volume and growth of the semiconductor industry [1.10]

The largest sales market is in the US. Japan, the second largest market, has sales on a par, with the rest of the European countries combined. The Asiatic-pacific nations of Taiwan and South Korea also have a great importance table 1.1. Europe has a subordinate role in the market; Europe is more or less a producer of semiconductors and electronics. Only one German company, Siemens, is amongst the 60 largest semiconductor manufacturers. Siemens has a ranking of 16 [1.9]. The consumption of German made semiconductors within Germany's own industries is infinitesimally small. The same is true within the PC-production sector, nevertheless a 100 million PC/YEAR production rate was observed in 1998. This is more than television production world-wide; a sector where European countries have been unable to conquer a considerable market share.

Table 1.1 Electronics market shares (consumption) according to data of the Semiconductor Industry Association SIA [1.10]

Market	1998	2001
USA	32.7 %	33.1 %
Japan	21.6 %	18.7 %
Asien/Pazifik	23.0 %	25.0 %
Europa	22.7 %	23.2 %

How it started

The transistor was originally invented by J. Bardeen and W. Brattain in 1947 at Bell Laboratories. Advance were quickly made by W. Shockley and the production of the first integrated circuits by Kilby at Texas Instruments began in 1958, and subsequently by Noyce at Fairchild Inc. [1.9]. The field of microelectronics had begun its triumphant advance. Before the creation of the transistor, electronic circuits were limited to simple designs relying heavily on vacuum tubes. Electronics were used almost exclusively in the area of radio and television broadcast and reception. Vacuum tubes were used on a large scale in these fields until the 60s. Even before transistors breached this vacuum tube stronghold, it was used as a discrete device or in simple integrated circuits in niche technological markets (e. g., medical electronics (hearing aids), in military equipment and space technology). This provided the means of funding the first useful transistors. Later Japan began to use the first semiconductors in consumer applications such as portable radios, tape recorders

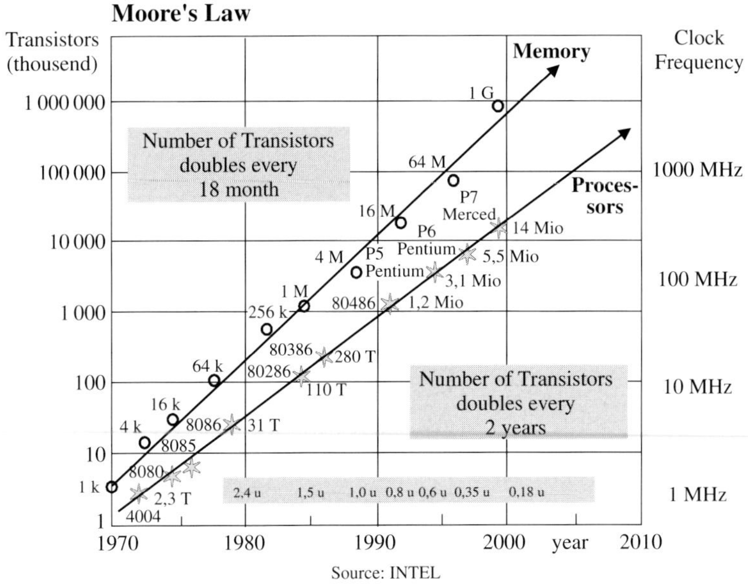

Fig. 1.2 Moore's Law: Trend in numbers of transistors per integrated circuit for memory IC and microprocessor IC, taken from INTEL [1.1], [1.4]

and entertainment electronics. Mass production of transistors began, and a correspondingly drastic decline in prices for electronic goods was observed.

Moore's law

The first affordable hand-held calculator was made around 1972. An invasion of electronics into offices and administrations soon followed. Environments where rudimentary centralized data processing was being performed by large central computers were especially open to the new electronic gadgets. Computer technology and associated digital apparata had played a minor role compared to radio and television, but that rapidly changed with the invention and wide spread application of MOS, and subsequently CMOS technology. This technology forms the basis of highly integrated circuits (LSI, Large Scale Integration). One of the first major accomplishments in 1972 was the invention of the microprocessor at Intel. Intel had fewer than a dozen employees at that time. GORDON MOORE, director of the R&D department of Fairchild Semiconductor Corp. and later president and CEO of Intel, prophesied at that time (1965!), what today is known as the so called 'Moore's Law' (fig. 1.2): "*The number of transistors on a chip will double each year!*" [1.4]. In creating his now famous relation, MOORE had only extrapolated 5 points on the new curve and said his statement should be valid for the next 10 years. In 1975 MOORE again repeated his assertion, but with more precision: After the semiconductor technology had reached a certain maturity: "**The number of transistors on each chip will double every 18 months!**" (Moore's Law), which still remarkably well describes the situation today. It is not obvious whether this exponential growth will continue to persist; it may persist longer or it may soon begin to falter. In either case the official roadmap of SEMATECH [1.12] shows that in the next few years there will not be a waning in the trend, at least the physical limitations will not yet be reached in the near future.

No limits can be seen

The usable of wavelength of the light was thought to be the limiting factor size for device fabrications. Since light is used to expose masks in the photolithography step, it was regarded as an absolute physical limit. However, with the transition to even shorter wavelengths (currently 248 nm Deep

UV is state of the art) structures down to 0.13 mm can be produced. Even now, next generation photolithography with a 193 nm exposure wavelength is going into production. This wavelength allows structures down to 90 nm, and possibly down to 65 nm. The next available wavelength 157 nm will follow for geometries below 45nm; EUV will go even further down. With EUV and some more exotic technologies like proximity-X-rays, electron beam exposure (the SCALPEL process), and ion beam projection systems a further decrease of geometry will be possible, although the limit of the classical UV-lithography cannot be seen. In the laboratories there are already exposure equipment which are able to produce gate lengths down to 15 nm [1.11], [1.15].

Using current state of the art processes about 500 million transistors/chip in a 0.13 mm technology are manufacturable with a good yield. Figure 1.3 shows a scanning electron microscope image of a chip cutout with 6 layers of copper metallization (IBM CMOS 7S-Process), similar processes are offered by Motorola Corp., INTEL and others in the meantime too and may be considered as state of the art.

Fig. 1.3 Scanning beam electron microscope image of a cutout of a chip, made by the IBM-CMOS-7S process with 6 layers of copper metallization and smallest features of 0.12 mm after [1.4]

Exponentially Increasing Complexity

The complexity producible today is already unimaginable. The PENTIUM 3 chip in 0.35 μm technology had about 7.5 million transistors (fig. 1.4), the contemporary Pentium 4 chip increases this number to approx. 42 million transistors in a 0.18 mm technology. The ITANIUM, the actual flagship development by INTEL in co-operation with HEWLETT PACKARD, has more than 200 million transistors integrated, and even larger numbers in 0.13 μm technology are on the way. The current designs can no longer exhaust the production potential.

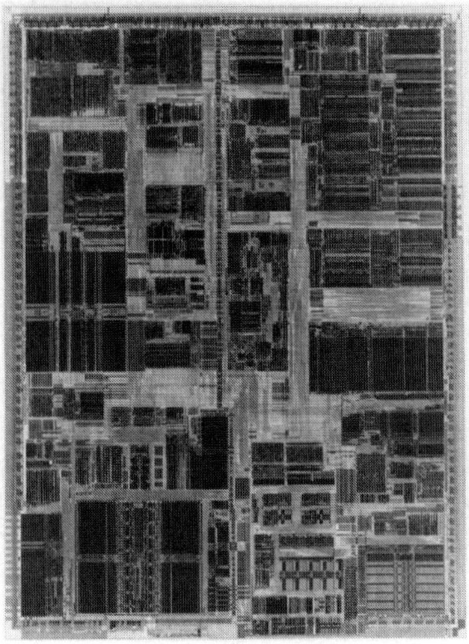

Fig. 1.4 Chip Photograph of the Pentium-P6 Microprocessor with about 7.5 million transistors, about 196 mm² area, 0.35 μm Technology, after INTEL Corp. [1.14]

Within this area there is an unsaturable need: semiconductor memories. Figure 1.5 shows the progression of generations of dynamic DRAM-Memory modules [1.5]. Components with 256 Mbit, which corresponds to a density of over 256 million transistors/chip with 1 transistor/memory cell, were introduced to the market in 1998. The gigabit memory chip has existed since 1999 [1.13]. Integrated circuits for mass applications such as memory are produced on a 300 mm Wafer, so called 'Pizza Wafers', whereby lots (batches) with more than 200 Wafers are processed in one run.

There is an Enormous Market for Electronics in the Future

The main driver of this trend is, amongst other things, the triumphant career of the personal com-

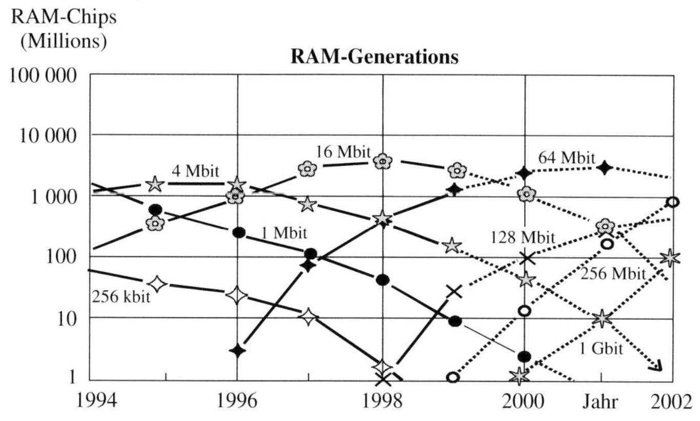

Fig. 1.5 Sold memory chips (RAM-IC), ordered in generations (source: IEEE-Spectrum, Technology Forecast 1998 [1.4])

puter (PC). Today more than 100 million are produced each year. The success of the PC made the company INTEL the largest semiconductor company of the world with an annual revenue of $26.1 billion in 2001 after $33.7 billion in 2000. Huge revenues in electronics were also found in telecommunications, which is today fully digitally and computer controlled, and within the mobile telephone market with device generations of less than 2 years. Expanding markets are found in the classical consumer goods for radio broadcast and television, which are on the way to digital concepts. Special decoders, 'set top boxes' are designed in connection with private television companies, containing a huge amount of digital encryption and decryption technology. As well as there is the performance hungry computer games market, which increasingly operates with its own dedicated hardware. Many markets are in start up condition or still not yet opened. The use of digital electronics in the automobile industry is still at the beginning, traffic and road systems are still nearly completely without electronics. Trade using digital accounting systems for a long time, and smart cards are used more and more for paying goods and deliveries. But huge parts of daily life and administration still run conventionally without electronics support.

Computer Aided Engineering

How is it possible to design circuits with up to 200 million transistors? The first microprocessors were designed by hand with the help of large cut tables and a great deal of drafting effort. A circuit design with hundreds of thousands of transistors without the aid of a computer is inconceivable. Fast semiconductor electronic development was only possible because computers built from semiconductors became more efficient and faster, which again allowed the development of still more efficient integrated circuits. Semiconductor technology and Electronic Design Automation (EDA) are closely interlocked and depend on one another. Progress in EDA leads to progress in semiconductor technology and vice versa. From process improvements to circuit design technology, the development of the EDA taken from SEMATECH Roadmaps [1.13] impacts all areas of semiconductor technology.

In the beginning there was Computer Aided Design (CAD), a common term in mechanical engineering. CAD provided substantial support in mask design tasks through the use of suitable polygon editors. Soon the term *Computer Aided Engineering* (CAE) was implemented to more accurately describe the use of simulation programs such as SPICE. This term expressed that not just design drawing was being performed, but rather that more complex tasks were incorporated into the computer calculations.

Currently one can speak of CAX, where X may stand for Q such as *Quality Assurance*, P for *Planning*, S for *Simulation* or M for *Manufacturing*. Probably no area in an industrial environment can now get along without substantial aid of computers.

The Usefulness of Computers

The progress provided by the acceleration of activity is not the sole contributor to rapid microelectronic development. In particular, there are the *demand for accuracy* in the obtained results, and the *quality of documentation*, more or less produced automatically and without additional effort, as well as cost savings for a given product over its lifetime owing to reduced servicing and maintenance. These driving forces have encouraged the usage of CAX; in some cases no alternative even exists, because a product may be too complex to be developed by any other means.

Typically development in industry takes place one step after the other; a completely new design or substantial redesign is more the exception than the rule. With gradual development or *incremental engineering* the use of well known, established components plays a crucial role. For example, the IC design flow is a well known concept and EDA systems can readily be used to design ICs. In fact, in EDA libraries substantial information exists, which can be recombined to make designs for new applications. A completely new design of an integrated circuit is estimated to be below 30 %. This value also tends to be decreasing, whilst discrete electronic systems may be an even smaller portion. In other words, complicated design may not be performed without the help of a computer.

Description of Devices by Data Models

CAX has three important areas of applications:

- Economically oriented applications;
- Device- or function-oriented applications;
- Integration and networking of a completely integrated system.

In all three areas of application *data models* play a central role. *Data models* are kept in relational databases and mutually interact with one another. Through the use of *data models*, one understands the relevant objects, attributes, and integrity relationships, this describes an *entity*. An entity may be described by several different aspects or *views*, see fig. 1.6. For example, the *wrist watch* entity can be described as follows:

- mechanical components such as housing, bracelet, circuit board, display, etc. (*mechanical view*);
- Schematic description of the connection diagram (*electrical view*);
- Geometrical drawing, instruction for assembly (*integration view*);
- Description of functions, possibly of embedded software, operating instructions, testing instruction, etc. (*functional view* or *behavioral view*);
- Assembly effort, bill of material and purchase list, trading and selling procedures, prices, market models (*commercial view*).

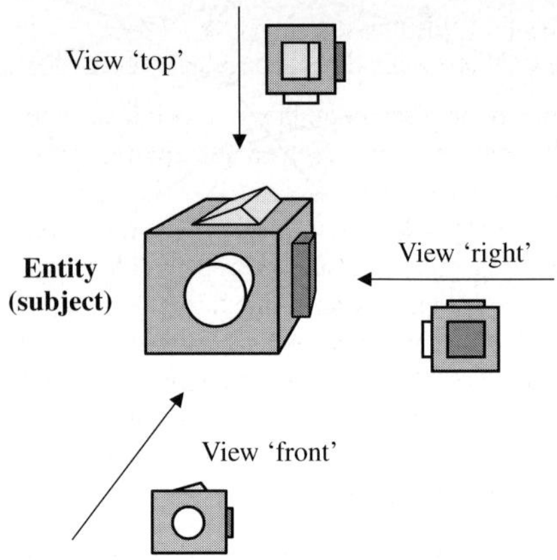

Fig. 1.6 Views of an entity

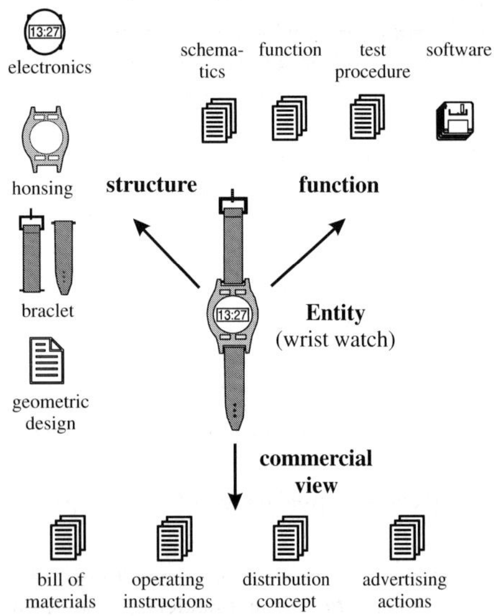

Fig. 1.7 Views of the example wrist-watch

These views, traditionally coming from different sources and maintained in different departments within a company (see fig. 1.7), are now regarded as an integrated common information space. This is a new concept in computer science. The individual views, when regarded as *views of an entity*, describe the same thing from different points of view, therefore each description does not have to be complete. The data model of the entity consistently maps the different views and information in a common model of a relational data base. It should be noted that this procedure is not only a collection of information about the object, but the individual entries are directly related to each other.

Imagine a product containing a printed circuit board; however, the purchasing department obtained a less expensive board from the Far East. The board is too big to fit into the housing; therefore the housing will also need to be made bigger. The *geometry view* is an immediate concern; assembly can no longer be performed. Management may decide to obtain a larger housing (*economic view*) with a penalty owed to a lead time, but it does not sell on the market because the shape has been altered (*marketing view*).

Consistency between Views of the Data Model

The wrist watch scenario gives an understanding of the term *consistency*, the model of the entity must show different views which must be correlated to each other by their relations. The consistency of the model no longer can be controlled by humans alone, it is controlled by the data processing in a CAX System which is able to handle and edit the respective views of the entity. A further goal is the creative application of data processing. By definition or modification of one of the several views of an entity, the remaining views can be derived or generated from other views automatically in a consistent manner. This has yet to be obtained in many areas of CAX system design.

From Variant Construction to Design Automation

In the wrist watch example the desires of the marketing department (at the customer's request), "*We need a clock with a 'poppy' appearance and with special functions*", automatically leads to a suitable product design. From the specification of the desired target product, a qualifying 'poppy' housing is selected. The list of housings and suitable electronic modules are supplied by vendors. Software modules which understand the demand for special functions, parts lists, delivery schedules, assembly plans, testing instructions, etc. (i.e., the complete manufacturing documents for the new clock) automatically generate all necessary documents. If a suitable market model and selling volume exists, and the modeled procurement, assembly, selling, advertisement, etc., also exist, then costs and long term company profit may be predicted. Such extensive application of data processing is understood today as **Design Automation (DA)** and goes beyond what was called *variant construction* in former times.

Top Down Design or Bottom Up Design?

Going down in the design flow from the customer's request to the details of the design is generally called *top down design style* in contrast to the *bottom up design style*, in which first the detailed blocks are designed and then integrated at a higher system level. The names *top* and *down* is owed to the usual traditional representation on hierarchy plans, where the super ordinate building blocks are drawn on *top* and detailed assembly units are drawn further *down* on the schematic. Companies which use the *top down design style* call themselves proudly *system companies*, as companies which manufacture devices in the typical bottom up design style are known as *device supplier*.

Interface Problems

Design Automation at the product level still has yet to be realized with most manufactured goods. One of the most important reasons for this is the *interface problem*, poor modeling, and the problem of undefined relations between the different views of the entity. Because of the traditional origin of the work areas, these views are normally not compatible and data may not be exchanged between them without any difficulties. The exchange of geometrical information between CAD systems of different manufacturers is a problem, despite various interface standards. The same is also true for the exchange of data between CAE programs in the electronics area [1.8].

Standardisation

Standardization is the keyword. Efforts of international organizations such as the International Standards Organization ISO in the STEP-Initiative (STEP stands for **ST**andard for the **E**xchange of **P**roduct model data) led to the group of standards ISO 10 303. These standards demand a well defined, consistent description of product data within an *integrated product model* [1.7]. The ISO 10 303 standards contain:

- Overview and fundamental principles;
- Description methods;
- Implementation methods;
- Conformance testing methodology and framework;
- General basis models and application resources;
- Application protocols;
- Abstract test suites.

For the description of products using the standard, a uniform model specification language EXPRESS is defined. This language describes information units both textually and graphically. The application of these standards and acceptance in the industrial environment needs still more time, even if subareas are yet applied in the meantime in individual disciplines. A more precise explanation of the STEP Initiative would magnify the framework of this depiction, see references for that in [1.3].

Development Process According to ISO 9000 ff.

Design Automation is a process that documents the gradual and stepwise development procedures using a pre-defined procedural model. A substantial demand of the ISO 9000 ff., standardization is in compliance with the *Design Automation* process, and it is the basis for quality assurance in development. The transformation of the product data model usually happens in several steps which are characterized by reviews and evaluations of the development stage. The fully automatic transformation of the design process remains a task for the future in many areas; various creative engineers must devote a great deal of effort in this area; nevertheless, *Designs Automation* systems will supply extensive documentation for reviews and, in the long run, the results. The advantages are important:

- Control of systems with high complexity;
- Forecast of the behavior;
- Design productivity;
- Quality of the product;
- Shortening of design time;
- Avoiding of design- and documentation errors;
- Improvement of the documentation quality;
- Maintenance and support during the product lifetime.

Electronic Design Automation

Electronic Design Automation EDA belongs to the few technical disciplines, which utilize *Design Automation* in some way. The design of integrated circuits follows a persistently mutual *transformation between 3 views: the structural view; the behavioral view; and the geometrical view* [1.2], described by GAJSKI in the so called Gajski-Y. Many tasks in design flow are done automatically today; the design engineer makes some specifications, accompanies the flow, makes design decisions and controls the results. In former times this was called a *Silicon Compiler*, an over-simplified design system not adequate for many of the designs made today. EDA covers more than simply actual chip design, even if this is the focus of the design task. EDA means more: the design of the *entire system* fitting a written and well defined specification, down to a series production device; this is called *top down design style*. The printed circuit board and the embedded system device driver software are included in EDA design. Although there may be dependences on other components, the mechanical components like the housing are excluded here, and are left to mechanical engineers to optimize.

Increase of Developer Productivity by EDA

Why is productivity so important? Since technology progresses quickly, the market life span of electronic products has been reduced to only a few years. In former times a radio was used for in excess of 15 years, televisions had to be replaced after 10 years, recorders after 5 years and computer technology products (e. g., PCs, printers and scanners) after 3 years. These items are replaced, not because they are defective or worn, but because they are inefficient, and insufficient in performance. Additionally, there are new products which are interspersed in strong international competition. Today there is only one chance for profit: A company has to be fast and, whenever possible, one of the first with a new product to market. Competitive companies in a new market earn half as much as the first to market. This price purge is owed to a learning curve and the following mass production lag that the first to market has already overcome.

Time to market becomes the central motivation and the demand driver for faster development and significant increase of the developer productivity.

Concurrent Work on all Views of the Product

The traditional work flow (e. g., building laboratory breadboards, prototypes, and pre-production models) requires too much time because of the sequential character of the development steps. This work flow is completely unsuitable for modern IC-design. Soldering irons and oscilloscopes have, to a large extent, been relegated to retirement in a modern electronics laboratory.

In contrast, a concurrent work flow is maintained, working in parallel on all aspects of the design (*concurrent engineering*). By intensive application of simulation and design tools, a realization of the final product at the first throw is tried (*first time success*). Even if this is not always a direct success, the entire period of development is, nonetheless, drastically shortened. Japanese successes in the consumer goods area may be attributed a good deal to this concurrent design concept. For example, after the decision to build a *Walkman*, simultaneous development, marketing, construction of a manufacturing plant, the sales force, etc., will be initiated in parallel. This, of course, needs a far sighted commercial decision; European CEOs still seem to have a problem with far sighted commercial decisions.

The Demand for Zero Errors

In the design of integrated circuits, a substantial component of many of today's new products, the demand for *first time success* is even stronger. With only a 2 to 3 month turn-around time for chip manufacturing, there may be only a single chance to redesign; incremental improvement is not realistic in this environment. Here a comprehensive and accurate simulation is required and promotes application of EDA on a large scale. Inclusion of software development into the development cycle of hardware (so called embedded systems or *system on a chip SOC* designs) is often implemented. *Zero Error request* and *first time success* is an inevitable demand for IC development and a strong motivation to use EDA tools wherever possible in a concurrent design environment.

EDA even for Small Electronics Enterprises

Today's PC technology provides enormous computing performance; a standard PC exceeds the computing performance of a CRAY-supercomputer of the 80s. Using PCs, inexpensive and efficient EDA tools are open for normal electronic engineers. Design of printed circuit boards are exclusively made with EDA tools today, filters with filter design programs, discrete circuits are analyzed and optimized with SPICE-like simulators, digital circuits with PLD or FPGA design systems. Large IC design systems from CADENCE, MENTOR or SYNOPSYS, which ran on high speed and costly UNIX workstations in the past, are now available on Linux or even on the native NT operating system of PCs. Prices for licenses permit an entrance also for the middle and smaller electronic enterprises. Because:

Electronics Design is IC-design!

The products of the future will contain complete systems on ICs, putting the creation of value thereby into the hand of the manufacturer again. Copying of products by cheap asiatic companies is much more difficult now. *System design capability* is the keyword. Since multiple components are combined into single chip designs, the simple combination of standard devices which are available on the market is not efficient, and cannot be the future of IC design; rather, development and optimization of user specific integrated circuits (ASICs) will become more the norm. A few example applications where ASICs have found their use: quartz wrist watch, bicycle computer, mobile telephones, GPS receivers, pagers, and many things we use daily in our lives.

Organization and Equipment Allotment for EDA

The decision for application of EDA tools in the development department of an enterprise has important business, organizational, and financial aspects. Some of them are drafted here:

Application areas

- Schematic entry and documentation;
- Printed circuit board design, interactively or automatically;
- design of PLDs/FPGAs;
- logic design and synthesis;
- circuit simulation, digitally and analog;

- ASIC design, only digital design;
- ASIC design, mixed signal with analog library cells;
- design of full-custom cells;
- integration of IPs, SOC designs.

Each additional point in the list doubles the requisite investment expenditure for equipment and software (fig. 1.8). This estimate does not include costs for investment in personnel and training. ASIC design usually requires a team of specialized engineers, which may only be financable in larger enterprises. Simpler design scenery (e. g., FPGA-design) is found in small and medium enterprises (SME). If SMEs need user-specific circuits they then depend on *Design Houses*, specialized enterprises for IC design, or universities which have appropriate equipment and have the appropriate knowledge. With the availability of acceptable low-cost EDA systems for ASIC design on inexpensive PCs, ASIC development of SMEs will become economically feasible in the future. System knowledge is of great importance for SMEs, it has to be completely kept in-house. By performing all design work in-house, product secrecy is preserved; this is an important factor in protecting the company against the competition. In addition, if all work is done within the company the employees themselves are the designers, and a protection against fading knowledge can be enjoyed.

Equipment allotment:
- PC-Network with server, operating system Windows NT or LINUX;
- Workstation-Network with UNIX-operating system (e. g., SUN SOLARIS).

More and more systems installed today rely on personal computers PC or X86-architecture based servers, with the classical workstation clusters only installed in large enterprises with high technical demands. Unix systems are expensive in investment, maintenance and support and replaced by LINUX-based computers with central servers or *server farms* for application software and storage management. The newest ITANIUM 64 bit servers are able to run operating systems such as UNIX, LINUX, or NT on the same hardware, and can address one terabyte of main memory. These systems are highly scalable to each required performance level without breaking the architecturale concept.

Software License Procurement:
- Licenses for schematic editor, e. g. ORCAD;
- Licenses for printed circuit board design e. g. PROTEL, ORCAD, MENTOR BOARD STATION;
- Licenses for PLD/FPGA design, simulation and programming, e. g. from ALTERA, XILINX, ACTEL etc.;
- Licenses for analog/digital-simulation, e. g. PSPICE;
- Licenses for ASIC-design, e. g. TANNER, MENTOR, CADENCE, SYNOPSYS.

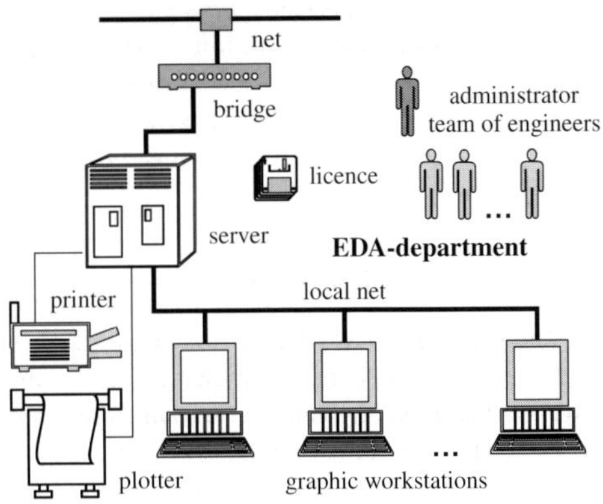

Fig. 1.8 Typical configuration of an Electronic Design Automation Department

License procurement may follow different business models, e. g., purchase with update support and maintenance, annual royalty after low single down payment, or royalties only in the case of use. Licenses take the largest part of investments in the EDA area; they may easily cover a multiple of the computer equipment investment. PC licenses normally cost only a fraction of the UNIX licenses with frequently only small performance differences. This is caused by the completely different market with an enormous installation basis (approx. 100 million PCs per year!). UNIX licenses usually cost more, maintenance costs too and higher administration expenditures. PC licenses with good performance are available with some \$10000/seat, commercial volume licenses in the UNIX area easily exceeds several \$100k/seat. A list of EDA software providers is to be found in the appendix.

Libraries:

Libraries keep the material from which the designs are formed. Basic libraries of standard cells are usually provided by the manufacturing companies gratis or for small costs. Company independent libraries already cost substantial royalties, with specialties (A/D cells, operational amplifier, etc.) much more. Larger function modules such as processor cores, interface blocks, etc, so called IPs, may easily cost several $10k. There is a market today for IPs and building blocks with decreasing prices for well established modules. The collection of company owned libraries from earlier designs or for the purpose of internal standardization makes sense, but the expenditure for documentation and support may be severe. Hard macro-libraries suffer from fast technology changes with each CMOS generation and need a great deal of maintenance effort.

Administration, Maintenance, and Support:

EDA departments with several seats require significant administration and maintenance costs. Beside the administration of the domain, the questions of data security, the backup service, and the license administration, a great deal of work is involved in the installation of updates, installation of patches (e. g., bug correction) and there is the continuous effort of library administration. Highly qualified personnel are needed to administer the network; the personnel also remove pervasive inadequacies, and correct errors in the software or those caused by an operator. Installation and testing of new software packages are also a substantial expenditure. New software is typically evaluated in small pilot projects, where the results can be surveyed. Here scripts are written which contain control instructions for standard procedures. Larger companies maintain their own personnel infrastructure, who organize and establish complete design flows for semi-automatically script driven procedures. Since software is heterogeneous (i.e., supplied by different software companies), problems with software interface and data exchange are frequently encountered. Such heterogeneity is difficult to avoid within the fast advancement in the area of EDA. Much time is devoted to writing interface programs and performing data transformation. An all encompassing integral data model is still far away from existing in reality and remains an important task for the future.

Training and Personnel Management:

Efficient use of the high investment tools, both hardware and software, is possible only with trained and highly qualified personnel. Constant training of personnel must be carried on, which must be relevant to the configuration of installed software. Further training is also necessary in content related subjects. Usually several courses per year are required for each engineer, whilst new employees require an even greater training effort. Investments for personnel are secured by a suitable cooperative and informative leadership style and appropriate salaries. The employees have a substantial knowledge of the company's system; therefore they are a constant target of acquisition from outside people. Head hunters in the industrial culture are notorious for their attempts at acquiring vital employees. Some employees may leave and form their own competing business (e. g., Silicon-Valley/US large companies), which later become a substantial competitor.

Support:

With the complexity of the software used and the multiplicity of possible faulty operations, regular disturbances have to be expected. So that productivity does not drop to zero organizational support must be provided. Support is usually organized in three levels:

- In the *first level support* an individual, a particularly experienced engineer, e. g., the system administrator, is responsible for the many small problems. He may be trained specifically in giving basic computer help. This kind of support takes place fast and unbureaucratically.
- In the *second level support* people from the central computer center are to be provided for solving problems concerning the network. They should be able to solve central server administration problems of the operating system.
- In the *third level support* there is a fall back to external companies, e. g., over so called hot lines, e. g., if problems with the application software exist. A call on the hot line is, however, always expensive after an initial phase of software run-in. Larger disturbances, which require a local support, can become a serious financial

problem both for the software license provider as well as for the application user.

Outsourcing of subtasks to external companies belongs to support (i.e., library maintenance and extension, writing of adaptation software and the establishing of automation methods and procedures for often used tasks), this is known as *customizing*. Internal company software developments are losing importance as the usage of commercial software increases, or through the purchase of additional customized software modules. In the US there is an efficient cooperation with universities in this field.

What Does One has to Know in Order to Use EDA?

EDA is an area in which computers do a substantial part of the engineering process. EDA cannot replace the experience of the engineer, but they can be used to increase an engineer's productivity. A meaningful application of EDA needs knowledge of applied methods (in models and in design languages), as well as a fundamental knowledge of the underlying electronics. The style of the final design is strongly affected by EDA. In the following chapters a symbiosis between Electronic Design Automation (belonging more to the field of computer science) and a basic knowledge of electronics and VLSI design (belonging more to hardware engineering science) is discussed.

The stated goal of the authors, all working in the EDA application development area, is the concentration on *information that the designer really needs* in his job. Actual trends, handling of program surfaces, as well as characteristics of software products are excluded because developmental progress is fast in this area. By the time this book is published some of the functions will have changed, companies will have merged, have been renamed, or simply dissolved, and some software programs will already have been revised. However, we think it is important to deliver a book to an electronics engineer which will remain relevant and contains information, useful for the next decade.

1.1 References

[1.1] *Craig, R. Barrel*, President INTEL Corp.: Keynote Presentation on the Advanced Technological Education in Semiconductor Manufacturing. Third Annual conference, 1997

[1.2] *Gajski, D. D.; Dutt, N. D.; Wu, A. C.*: High-Level Synthesis: Introduction to Chip and System Design. – Kluwer Academic Publishers, 1992

[1.3] *Grabowski, H.; Erb, J.; Polly, A.; Anderl, R.*: STEP – Grundlage der Produktdatentechnologie. Aufbau und Entwicklungsmethodik. CIM Management 10 (1994) H. 4, S. 45–51, H. 5, S. 36–43, H. 6, S. 45–49

[1.4] *Geppert, L.; Sweet, W.*: Technology 1998, Analysis and Forecast. IEEE Spectrum, January 1998

[1.5] IEEE Spectrum, Jan. 1998, S. 27

[1.6] Internet page of EDA Consortium, USA, with exhaustive statistical data. http://www.edac.org

[1.7] ISO 10 303: Standard for the Exchange of Product Model Data.

[1.8] *Jaeger, K. W. (Hrsg.)*: Schnittstellen bei CAD/CAE-Systemen. – VDI-Verlag, 1991

[1.9] *Mellar-Smith, M.; et al.*: The Transistor: An Invention Becomes a Big Business. Proceedings of the IEEE, Vol. 86 No. 1, January 1998, P. 86 ff.

[1.10] Proceedings of the Semiconductor Industry Association, http://www.semichips.org/

[1.11] *Schulz, W.*: Lithographie, Verfahren der nächsten Generation. ELEKTRONIK 4, 1999, S. 24

[1.12] SEMATECH: National Technology Roadmap for Semiconductors. March 1995, http://www.sematech.org

[1.13] SEMATECH: Technology Computer-Aided Design (TCAD) Roadmap: A Supplement to the National Technology Roadmap for Semiconductors. March 1995, http://www.sematech.org

[1.14] Website of INTEL Corp.: http://www.intel.com

[1.15] INTEL Corp.: Technology Journal, Voll.6 Issue 2, May 2002

2 The Concept of Electronic Design Automation

ALFRED SCHÜTZ

2.1 Design Methodology

The design of an integrated circuit (IC, chip) can be an extremely complicated task and requires a structured approach. In short, this endeavor is nothing else than the transformation of a mixed electric/logic specification into geometrical patterns suited for mass production. Today such patterns are usually printed onto a set of masks (also called reticles) shaping an image which will be optically projected onto a photographic film layer deposited at the surface of the chip. For a typical CMOS technology the number of masks may range from 12 up to 30 (the main reason for more mask layers required being the number of wiring layers). The geometrical patterns on those masks are represented by a large number of polygons each characterized by its corner points. With the broad scope of ASICs today, the total number of the points can vary widely. For instance, an ASIC with 1,000,000 gates (which can be considered a commodity in these days rather than advanced) comprises some 200 million points. Needless to say, the coordinates of each point are meaningful, the proper function of the chip is in question if a particular point is offset.

A chip with a million gates would nowadays not span over more than 20 mm^2 (including area occupied by the peripheral circuits needed to interface with the outside world). Much larger chips are technically feasible and the cited complexity numbers can easily go up by two orders of magnitude.

The geometric representation

As indicated above, the ultimate goal of an ASIC development process is the definition of all polygons. The development process will be completed once a set of polygons has been found ready for transfer to the production division, enabling them to manufacture chips showing the desired functionality.

An integrated circuit is completely described by its set of polygons. Owing to its nature, this kind of viewing is being referred as *geometric representation*.

The behavioral representation

The development of an integrated circuit starts with its specification; it is a description of what the chip is supposed to do. The specification describes the behavior of the chip under all operating conditions, in particular the logical (functional) behavior, the timing behavior and the response of the circuit to input signals (stimuli). The desired functional behavior can be captured by formulae, state diagrams, and formal languages similar to programming languages. The chip may be part of an electronic system (printer, digital voice transmission) and has to perform a certain function, thus determining the specification.

The specification describes the functional behavior of the chip. It can be considered *complete* if every chip meeting all formal descriptions would function properly within the target system. While such specification does not define how the chip is to be built, it has to clearly describe how it shall behave. A comprehensive specification can be considered the *functional* or *behavioral representation*.

The structural representation

A direct transformation of the behavioral representation into the set of polygons is not feasible. The main reason is that the behavioral representation does not uniquely correspond to the geometric representation. In other words, given a certain

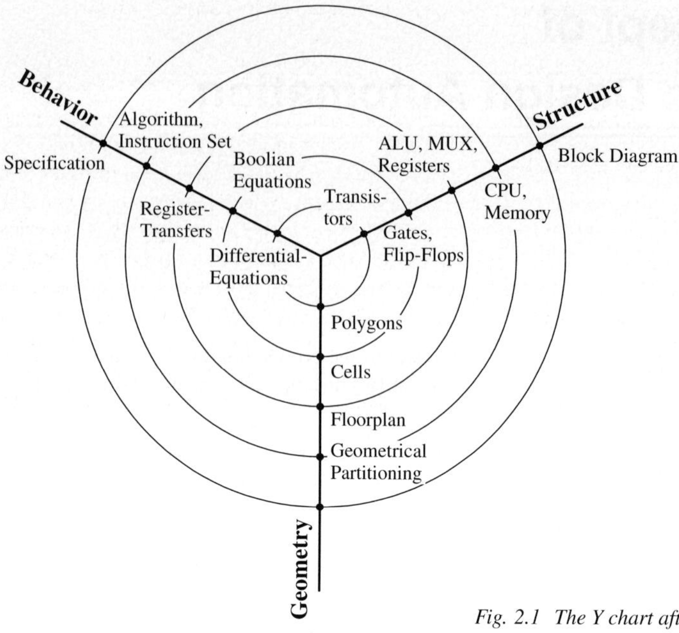

Fig. 2.1 The Y chart after D. Gajski et al. [2.4]

formula (a fragment of the behavioral representation) there are many possible translations into the geometric domain. Whilst all of them would be valid if just the given formula is concerned, most of them can be ruled out if the formula is seen within the overall context. This will be corroborated later in this chapter.

To solve this dilemma a new type of description is being introduced, the *structural representation*. It can be seen as a construction plan of the circuit. It describes how the chip is built up from basic elements and how those are to be interconnected. This representation can be captured by a schematic or a netlist.

The structural representation does not describe how the circuit behaves; however, the final behavior can be forecasted by means of simulation. The basic elements (gates, transistors, resistors, etc.) are the building blocks and have a well defined (usually simple) behavior. Once mathematical or numerical models exist for the elements the overall behavior can be calculated.

The value of the structural representation is also established by practical work: one part of the design process is partitioning a complex system into blocks which in turn may also contain further blocks. The partitioning process will be repeated until we arrive at components small enough for translation into the geometric domain. This procedure is also being referred to as 'divide and conquer' meaning repetitive partitioning until only well known components remain.

Within the context of the structural representation, the terms 'hierarchy', 'modularization' and 'chip architecture' are also frequently used.

The meaning of these three views – *behavioral*, *structural*, and *geometrical* – had originally been recognized by D. Gajski et al. [2.4], [2.7] when he proposed the Y chart in 1982 (Figure 2.1), also referred to as the 'Gajski-chart'. That diagram is composed of three arms representing the three views, and the distance from the center of the Y represents the level of abstraction. Concentric circles around the center of the Y connect corresponding abstraction levels within the three views. The outer circle corresponds to overall specification, top level schematic, and geometric partitioning. Going inward we arrive at lower levels of hierarchy, processor cores, arithmetic logic units (ALU), registers, gates, and, finally, transistors. If we walk along the behavioral axis we first see the instruction set, algorithms, then register transfer instructions, Boolean equations, and, at the lowest level, differential equations. Further out, an even higher level, not shown here, would describe the interaction of several chips within a circuit board. This would go beyond the scope of this.

A compact description of the meaning of the three arms would read as:

What? How? Where?

The behavioral representation tells us **what** the chip does, the structural representation describes **how** this behavior is going to be implemented, and the geometry captures **where** its basic elements are located.

2.2 Development steps

In this section we will work out what particular tasks are required for the development of an application specific integrated circuit (ASIC) and what a reasonable flow (or sequence) would look like. Common terms are design steps, and design flow.

Within the Y chart we can identify the start and end points of the design process. We start at the overall specification and we want to reach the center of the Y, – more precisely, find the lower level of the geometric representation which is the set of polygons – as those represent the data needed for manufacturing.

We do have a certain amount of freedom in how to traverse through the Y chart. As mentioned above, we have seen that a direct translation from behavior into geometry will be difficult – if not impossible – at the outer levels. It is more appropriate to partition and structure the circuit first. A few examples will show us that effort and data volume will grow if we walk down the behavioral axis. EDA can help by turning over to a computer the time consuming tasks and creating the vast amount of data. EDA tools will automatically generate the lower abstraction levels. Within the Y chart the migration to the structural and geometrical views will happen earlier (at an outer level) compared to a manual flow.

The design steps mentioned below are typical for a design flow of an integrated circuit. In this section we will not differentiate between manual and automated processes.

2.2.1 Creation of the specification

The specification is a description of how the chip will behave at its pins or interfaces. This description must capture the logical behavior, electrical parameters, and environmental conditions. The latter typically being ranges of temperature and supply voltages within the chip are supposed to work. Electrical parameters are clock frequencies, timing behavior (minimum and maximum delays), and latences (number of clock cycles that may lapse until the chip must show a response to an input signal). Priorities may vary, depending on the segment of the application: handheld devices, in particular, have a tight power budget, as they may drain power from a battery, and may also have size and weight constraints, usually not critical for fixed devices with access to the power grid. A good specification completely captures the necessary behavior of the circuit (i.e., 'every chip showing this behavior will work in the target application'), but does not impose constraints or give hints of how this functionality can be implemented.

Today there is no standardized language for system specification. We often see a combination of timing diagrams, free text, and figures. Usage of classical programming languages such as 'C' enjoys growing acceptance, but a behavioral style of VHDL or Verilog is also common. Analog behavior is more difficult to capture because of non-standardized terminology and measurement techniques.

Within the context of this book the chip specification is the interface between IC development and system development. It can be seen as a duty book and captures what the design process has to aim at. After first samples of the chip have been manufactured they will be compared (and qualified) against the target specification, and potential mismatches (ideally there are none) must be understood and justified. The target specification will then turns into the final specification and become the *defacto* documentation for all potential users who aim to use the particular chip in an electronic system.

A comprehensive specification will also describe the measurement setup needed to verify a certain behavior. A common source for project delays and exploding costs is incomplete specification. Unfortunately, complete capture of the specification

C <= A + B;

```
C(0) <= A(0) xor B(0);
C(1) <= A(1) xor B(1) xor (A(0) and B(0));
C(2) <= A(2) xor B(2) xor (A(1) and B(1)
            or (A(1) or B(1)) and A(0) and B(0)));
  .
  .
  .
C(N-1) <= A(N-1) xor B(N-1) xor (A(N-2) and B(N-2) or ....);
```

RTL, Algorithm **Logic (behavior)**

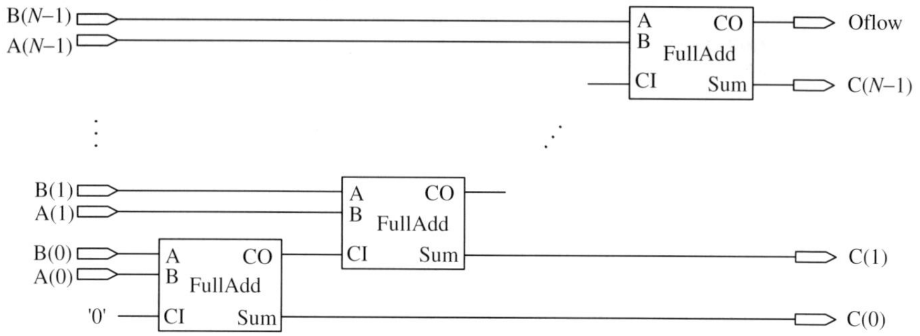

Logic (structure)

Fig. 2.3 Transformation of an RTL representation into a logic description

2.2.5 Transistor Level Circuit Design

The mission of circuit design is the transformation of a logic description into a transistor level netlist and the sizing of the transistors. Figure 2.4 shows the transformation of a logic symbol into the transistor level representation. A CMOS technology provides two types of transistors: n-channel (NMOS) and p-channel transistors (PMOS). Their primary parameters are *channel length, channel width*, and *oxide thickness*, and those mainly influence the I V diagram, the relationship between the *drain current* and terminal voltages applied.

One of the goals at this level is to determine the transistor parameters such that the desired electrical properties – primarily performance – can be met. The target technology (the manufacturing process) sets certain constraints:

- The *oxide thickness is fixed* in a particular technology; it cannot be changed by the designer.
- *Channel length and width* must be chosen above a certain minimum boundary.

For digital circuits it is advisable to set the channel length for all transistors to the *minimum value* supported by the target technology. This will both maximize performance and minimize the area of the chip (and hence cost). The optimization space is, therefore, limited to selecting of a value for the *channel width.* Narrow channels cause currents to decrease but also delays to go up, whereas wider channels will lead to higher performance at the expense of chip area, and power consumption, of course.

Normally **ASICs** are built from *gate array* or *standard cell* libraries according to a *semi-custom* design methodology. Rather than optimizing at transistor level, pre-designed, *standardized* cells will be (re-)used, and, placed within a schematic diagram or instantiated within a gate level netlist. To allow for the speed vers area trade off modern cell libraries offer several speed versions for the same logic function. They differ in the size (or strength) of the internal transistors[1)]. For instance, a NAND function can be offered with strength 1, 2, 4, 8, 12, ... Whilst the designer, or the EDA tool cannot pick the individual optimum size, he can choose a cell that comes closest. Therefore a semi-custom design flow will lead to a circuit

[1)] Advance CMOS technologies may offer several flavors for each of these transistor types. This may include a choice of oxide thickness and threshold voltages. For the sake of simplicity this will not be taken into consideration here.

which is slightly larger than the smallest possible chip. This is the price you have to pay for design time and effort reduction.

Transistor level design is still being done for analog circuits and the design of standard cell libraries. For the latter, optimization is less of an issue – the library is to enable a number of chips, therefore no particular performance goals can be set. *Characterization* of all cells is the major task: A very good delay model must be provided for every cell, sufficiently accurate within a wide range of operating conditions. These are input slew rates, output loads, temperature, supply voltage, and manufacturing tolerances. Careful characterization drives the cost of the development of a new standard cell library. Nevertheless, cells in a library will be reused so many times such that the effort is well justified.

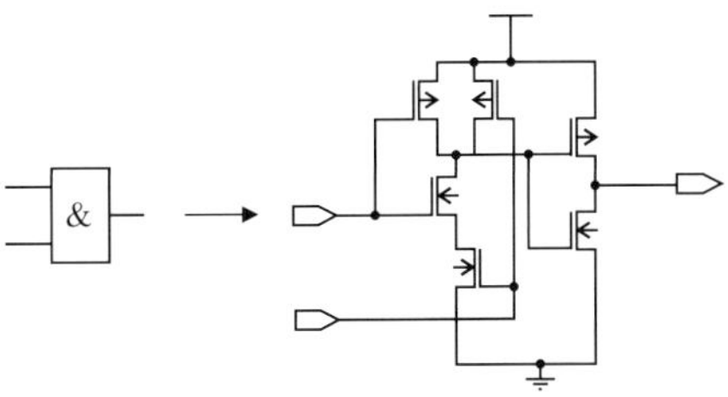

Fig. 2.4 Transformation of an AND gate from logic symbol to transistor schematic

2.2.6 Polygon Pushing (Layout)

Once the size of the transistors and their interconnects have been fixed, the transformation into the geometric domain can start. Here the mission is to find a geometric placement for the polygons that build the transistors or cells and their interconnects in such a way that the silicon area is minimal. The result is commonly referred to as *layout* or *physical design*. Figure 2.5 shows the transformation of an inverter schematic into its corresponding layout.

Each semiconductor technology has its own set of ground rules that a physical design has to comply with. The most important rules constrain the *width of and the distance between polygons*, and they have the primary impact on chip area. Each technology has its own recipe of how to implement physically the structures at the silicon surface, and, depending on the resolution, polygons which are too close may melt together, and vice versa polygons which are too narrow may disappear completely.

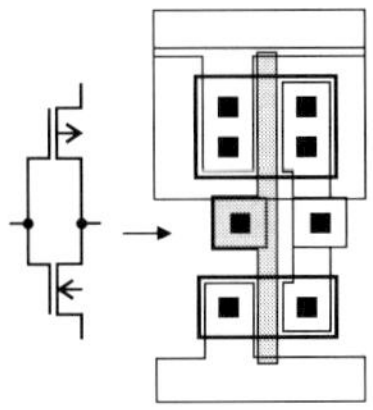

Fig. 2.5 Transformation of transistor level schematic of a CMOS inverter into the geometrical representation

In semi-custom methodology layouts of cells and chips are to be seen under different aspects: cell layouts are crafted by semiconductor manufacturers and represent their intellectual property. Usually customers only receive the abstracts with a simplified layout containing only the outlines and the locations of the input and output terminals. The wires which interconnect the terminals are custom for the particular chip.

2.2.7 Design for Test

The topic *test* has not been recognized as part of an ASIC design for very long, and it is not mentioned within the Y chart. Designers were reluctant to spend effort on testing development, as that had not been considered a constructive step. However, test is key for *quality of a chip* and it is extremely critical because of high drop out rates typical for semiconductor manufacturing. Nowadays, the importance of test design and *design for test* is no longer doubted – in many cases a result of painful experience. It has been seen that:

- The cost of test grows faster than complexity (gate count) of a chip.
- Unless appropriate measures are taken during chip design, test of a state of the art ASIC can no longer be afforded both owing to the cost of test program development and recurring cost in production.

The answer to this crisis is *design for test*. This methodology requests the addition of circuitry that is of use only during production test. The additional elements provide access for the test equipment to the internal wires of the chip.

Apart from the design of the auxiliary test circuitry, test vectors are to be generated. These are a set of stimuli which will be applied to the circuit during test as well as the correct responses of the circuit.

The request for testability and the desired test method should also be part of the specification, together with a target for test coverage.

2.3 Implementation and Verification

Every step of an ASIC design flow can be categorized as either implementation or verification. In this section we will work out a better understanding of those terms.

Implementation is always a creative process. Within the context of ASIC development, the result of an implementation step will be one of the representations listed in section 2.2. This can be the creation of a specification, an RTL model, or a gate level netlist as well as the creation of a test program. In general, implementation is the transformation of a more abstract level into a lower, more detailed level.

An implementation task will always be ambiguous. The number of solutions for modeling the multiplier of Section 2.2.3 in RTL is virtually unlimited. The reverse operation, the transformation into a higher, more abstract level, must be unique. In our example the RTL model has to show a multiplier behavior. Assume we had initialized the loop counter in Figure 2.3 to zero and have it increment in each clock cycle. The result of the whole sequence (16 cycles) would be:

```
C <= A * B*, with
B*(15 DOWNTO 0) <= B(0 TO 15)
```

(reversed bit order)

and would not meet the requirements.

Now we bring up the *definition* for verification:

Verification is the task to verify a previous implementation step.

Whilst the human genius can efficiently carry out the implementation tasks, we tend to be blind to mistakes we have made before. Data volumes in ASIC design have grown beyond any limit that could be overseen; therefore powerful tools are needed for verification. Traditionally simulators have been used for verification tasks. In the previous example we can use a VHDL simulator and choose arbitrary values for A and B and find out whether the expected result of $A \cdot B$ can be found in result register C after the 16th clock cycle. However, even if we find a match would we know whether the circuit works for other input values as well? To be very sure we would have to repeat the simulation for all possible input combinations. This requires us to simulate $65\,536 \cdot 65\,536$ different multiplication operations. Since each multiply operation requires 16 clock cycles, no less than $66 \cdot 10^9$ clock cycles would be necessary. A powerful VHDL simulator may simulate 1000 cycles per second and would then need approximately 2 years for an exhaustive verification. Very frustrating, indeed.

The previous conclusion tells us that we cannot achieve exhaustive verification by means of simulation. Therefore we need to select a set of vectors which is capable of detecting all potential errors. The selection of the vectors and the evaluation of simulation results can be very time consuming. For a typical ASIC project more time and resources will be spent on verification than implementation.

An ideal design flow provides for a verification step after each implementation.

It is also meaningful to perform small – local – verification runs during an implementation task to uncover obvious mistakes. It is of utmost importance, however, to perform a comprehensive verification after each of the implementation steps mentioned in sections 2.2.2 up to 2.2.6. Usually verification is a comparison relative to the previous, more abstract level. If an error introduced at an implementation step is not immediately detected at the subsequent verification step, the implementation result will become the reference for the next implementation and verification steps. This would eventually lead to faulty masks, and the resulting chip would not work properly. A redesign step would become necessary and generate a huge cost overhead. However, real life has shown that things can become even worse:

In 1995 it became known that the core of the Pentium processor – in those days the working horse of more than 80 % of all PCs shipped – under very rare circumstances would generate incorrect results. Because the error happens so infrequently the manufacturer of the chip, Intel Corporation, had not discovered it. Once that flaw had been uncovered, the message was spread among PC users. Finally, Intel offered to replace every affected chip. The cost of that operation is said to have been in the order of 100 million US$.

2.4 Top Down or Bottom Up?

We shall now corroborate a deeper understanding of how the vast amount of geometric pattern for a chip can be created. Before starting a chip design we have to understand what the chip is supposed to do. As discussed before, the ultimate goal of the design process is to arrive at with a set of polygons ready to print on masks and submit to manufacturing. The development sequence outlined in section 2.2 is based on recursive refinement. We start at a coarse, very abstracted representation, and systematically bring in more details. Such a methodology is referred to as a *top down* process. When proceeding along the behavioral axis in the Y chart, top down is the only meaningful direction. There may be only one exception: during an implementation step we might discover that there is no solution, e.g. it is impossible to meet the performance requirements. Let us give a practical example: after selecting a *ripple carry* architecture for an adder we work out that we cannot meet the target clock frequency because the carry delay would be too long. We would then have to go back to logic level and change the architecture to *carry look ahead*.

Now let us consider the structural axis in the Y chart. Going down (which means towards the center of the Y) the chip is being partitioned into its components. Partitioning is a *divide and conquer* approach unfolding a more complex entity into several smaller entities. No details are known yet about the components, they remain black boxes and their content will be defined at the next level. Speaking more practically, when drawing the schematic of some higher level entity we introduce symbols of the next level entities and we connect them together. In the next step we would design their schematics. Recursive application of that procedure will finally bring us down to a level where only transistors, gates, or other known elements are placed. Structural design would then have been completed.

Going bottom up along the structural axis may also be meaningful: we would first design lower level modules by placing and connecting together transistors or gates, and then use the newly designed modules to build higher level modules until we are able to draw the schematic of the top level of the ASIC. This approach avoids the need to work with black boxes. If functional representations exist for all levels then it is possible to follow that approach. Analog chips are commonly designed that way.

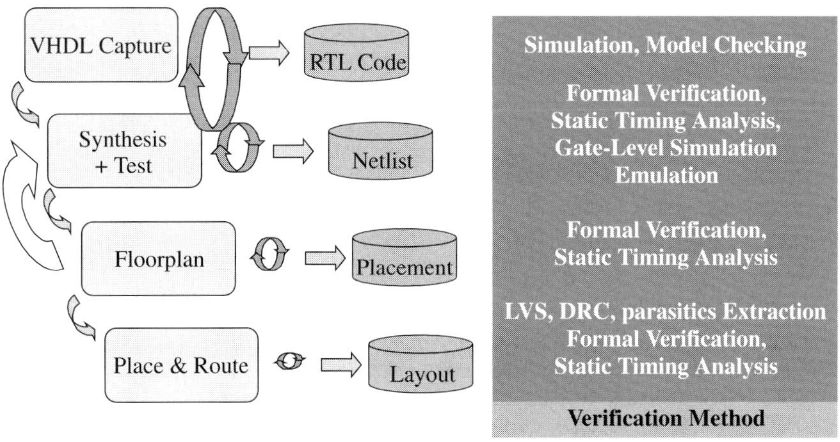

Fig. 2.6 A modern top down design flow

In the next section we will see that the introduction of EDA removes the need for manual design work in the structural and geometrical domains. Design work is now moved into the functional domain, and, as mentioned above, top down is more appropriate here. Modern design flows work primarily top down, as demonstrated in Figure 2.6. The left hand side shows the implementation steps that capture their result in a certain *design exchange format* for transfer to other EDA tools. The result of the implementation will be verified before hand over to the next implementation step. The method for verification is shown at the right hand side. Usually several iterations (this includes going backwards) will be needed between RTL (VHDL) capture and synthesis, because at RTL we cannot reasonably forecast the signal delays, as they strongly depend on the choice of cell library and semiconductor technology. Quite frequently synthesis cannot meet the performance requirements, and then RTL architecture must be revised, e.g. by introducing a pipelining concept.

At the lower levels delay predictions are more accurate, and fewer iterations are required, as indicated by smaller circles. This is very fortunate because the amount of data and computational work grow when going down to the lower levels.

2.5 Short History of EDA

2.5.1 The first Generation

The advent of EDA came in the early 70s. It was a few years after the invention of the integrated circuit by Jack Kilby, and chips with 20 to 100 transistors were state of the art in those days. Popular ICs were multiple logic gates and flip flops (e.g. the 7400 series), and operational amplifiers. Development of those circuits was hand work and design was based upon empirical methods, simplified formulas, estimations, and assumptions. Layout was manually and polygons were cut out of rubilith foils for mask making.

In the light of modern design flows, that process sounds primitive and time consuming. The biggest design problem, however, was that the intrinsic non-linearity of the transistors did not allow one to accurately predict the behavior of the circuit. With increasing complexity of the chips it became more likely that first prototypes did not work as desired. Several prototype runs were needed to allow for learning and iterative refinement of the circuit (redesign) until an implementation had been found meeting the specification. The **computer program SPICE** developed at the university of California in Berkeley [2.7] laid the foundation for EDA. That program, after many revisions still in use today, can simulate electric networks even if they contain non-linear elements like transistors and allows one to accurately predict either time or frequency behavior [1)]. Circuit optimization now no longer had to rely on trial and error. Productivity received a boost because simulation on a computer is orders of magnitude faster and cheaper compared to the physical implementation of a prototype.

Computer Aided Design (CAD) arrived on the horizon once computers became more handsome and affordable. Initially CAD targeted mechanical and construction engineering, but it became clear that those tools could be used for any kind of geometric design work. CAD systems substantially improved productivity of layout design. Polygon data are now being entered into a computer via a mouse, and can easily be modified and maintained. Once polygon capture is complete, a mechanical optical system (old fashioned, not used any more) or an electron beam writer transfers the geometric structures into a physical image (masks).

Design errors were a burden for chip design from the early days. Simulation allowed the prediction of the electrical behavior, but there was no proof that the manual transformation into the geometric domain was correct. Of course, manual checks were applied before the design data were released for production, but these checks became increasingly inefficient as complexity grew. More time had to be spent, and chances of an error being missed in the check grew.

Layout verification tools were developed and solved that problem. Two kinds of layout verification tool exist today:

- **Geometric checks** (DRC: Design Rule Check). This is software that uncovers a violation of geometrical rules such as minimum width of

[1)] SPICE numerically solves mesh and node equations of a network that represent a system of non-linear differential equations with several unknowns.

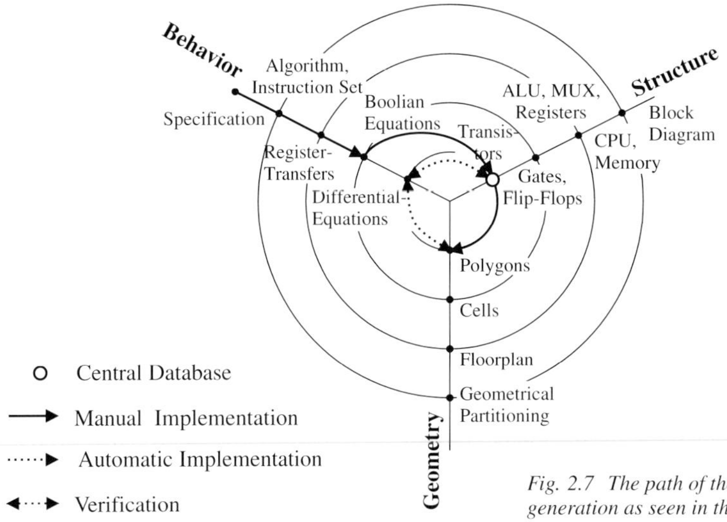

Fig. 2.7 The path of the first EDA generation as seen in the Y chart

polygons and minimum distance between polygons.

- **Layout extraction** is the recognition of active and parasitic elements and their interconnects. This is a requisite for checking the match between layout (physical design) and the symbolic/structural (schematic) design. The latter task is commonly referred to as *layout versus schematic check*. Extraction of parasitic elements such as wire capacitance is needed for estimating their impact on delays.

To summarise, the first generation of EDA comprised tools for verification of the electric behavior (simulation) and tools for correlating the electrical and geometrical representations. It came along with a substantial reduction of time consuming and costly redesigns. A layout editor was another member of the first EDA generation valued as an efficient capture tool for mask data. Layout editors are still used today for the design of analog circuits and cell libraries.

The first generation of EDA did not contain a tool for automation of an implementation step (hence the term EDA is not fully correct). Layout editors may offer some very limited automation capability, but this can only be applied to repetitive structures typical for memories.

Figure 2.7 shows the trace through the Y chart if using the tools of the first EDA generation. We walk down the behavioral axis and break the functional behavior down into Boolean equations. Then we create structure by crafting a transistor level netlist matching the Boolean equations. Circuit simulation (SPICE) will then solve the differential equations to verify that transformation. Once the transistor netlist has been found to be correct it will be translated into polygons. That translation is to be verified by DRC and LVS.

2.5.2 The Second Generation of EDA

In the early 80s semiconductor technology had advanced enough to allow for the integration of up to ten thousand gates. Layout creation by means of a layout editor and circuit simulation using SPICE started to consume intolerable amounts of time and cost. Additionally, computer requirements skyrocketed, in line with the amount of data. A new approach was needed.

Automatic Place & Route tools arrived on the horizon and rendered manual layout entry obsolete. Those tools create chip layout by instantiating basic cells and interconnecting their terminals.

Logic simulators were introduced and they boosted the productivity of electrical and logic verification. Contrary to simulation with a circuit simulator, a signal is considered to have only three discrete states $(0, 1, X = \text{unknown})$ rather than solving for the analog voltage level. Although not implied by the term 'logic simulator', delay calcu-

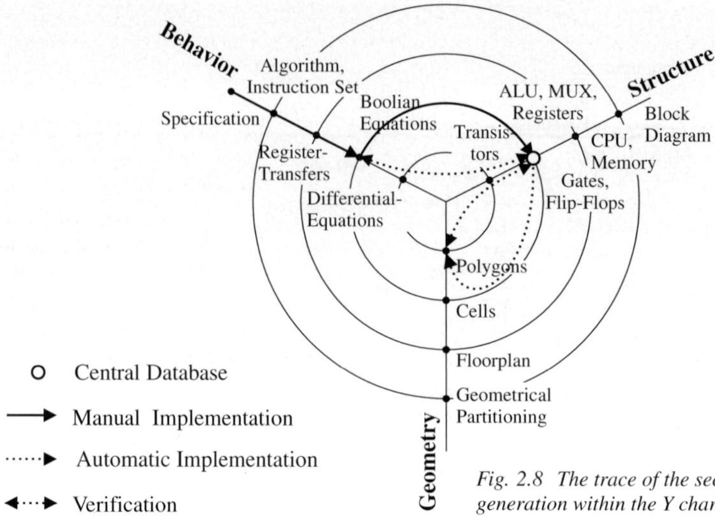

Fig. 2.8 The trace of the second EDA generation within the Y chart

lation or at least delay estimation is performed to ensure that the design will meet the performance targets. For that purpose logic simulators have built in a kind of a delay model that utilizes data obtained from the characterization of the cell library.

Initially circuits were captured in the form of a *netlist* – a tabular representation listing every occurrence of the basic cells including all wires (nets) needed for interconnect. Netlists are efficient in terms of data storage, but they are not well adepted for human understanding. Electrical engineers 'think' in schematics which graphically show all interconnects between circuit elements. Schematics had been used for documentation purposes since the early days of electrical engineering. Once *schematic editors* became available they quickly became the state of the art for circuit capture. These tools allow for graphical entry, capture, and output (plot) of the circuit schematic. However, the graphics information is only needed for human understanding – it is redundant for the circuit itself. The graphics information will be removed before design data are sent to other EDA tools. Typically EDA tools use standardized netlist formats (e.g. EDIF or Verilog) for the exchange of structural circuit information.

The second EDA generation allowed for the moving of the electric design work to the gate level, i.e., no manual work at lower levels was needed any longer. The structural gate level schematic or netlist becomes the most detailed design description created manually and can be considered the reference database for all further computer aided (or automated) design work (see figure 2.8). The logic simulator is a tool for the verification of the last manual transformation process and allows for the correlation of the schematic or netlist with Boolean equations and timing behavior.

2.5.3 The Third Generation of EDA

The early eighties saw a few universities starting to work on algorithms for logic design automation. The goal was to read in e.g. register transfer design or Boolean equations and automatically perform all design steps down to the polygon level. The term *silicon compiler* had been invented. The early experimental tools demonstrated nice examples, but failed completely in real life. Looking back, we can say that it is inappropriate to perform several transformation steps (as seen, e.g. in the Y chart) without human interaction. Every ASIC has custom requirements which mandate a high degree of customization in the design flow. Early silicon compilers simply did not provide that level of flexibility.

An important requisite for the later success of design automation was the standardization of hardware description languages, e.g. VHDL in 1987 by the IEEE and, later, Verilog. They had initially been intended for functional design documentation and cell level modeling at the RT level[1]. Later they also enjoyed acceptance in use for capturing the functionality of complex digital systems. Based upon these languages, logic synthesis tools were developed which allowed for automatic translation.

After early versions of logic synthesis tools had provided the world of engineering with limited success, the company Synopsys introduced their tool *DesignCompiler*. This tool performs logic synthesis and is closely tied to a companion tool for RTL synthesis called *HDL Compiler*. *DesignCompiler* and *HDL Compiler* can virtually perform two tasks in a single step. This tool suite saw some early adopters, but it took several years until it became the defacto standard for RTL synthesis. The reasons for slow acceptance were:

- The price tag for those tools was considerably high;
- RTL synthesis revolutionized the design methodology. There was no place for schematic diagrams widely used in those days and engineers had to change their 'way of thinking';
- The quality of results of early synthesis runs did not meet that of manual designs.

Early weaknesses of that product were solved and the VHDL and Verilog languages were finally taught at universities. Soon young engineers 'thinking' RTL rather than schematic arrived on the working floor. In addition, by the early 90s ASICs started to become so complex that the traditional design methodology started to break down.

Register transfer level synthesis automates the implementation steps below the RT level. The RTL description becomes the *reference database* (see Figure 2.9) and is typically expressed in VHDL or Verilog. Now simulation tools are needed to predict the logic behavior of the design: *VHDL and Verilog simulators* work similarly to a gate simulator, but they can handle the high level constructs available in those languages. This additional abstraction compared to the gate level opens the door for computer run time improvements because of similarities between VHDL or Verilog and computer programming languages. A '+' operation can be executed in a single machine instruction, while at gate level, the operator is decomposed into individual bits and a number of full adder cells, each having a logic complexity of approximately 10 gates.

After verifying the correct behavior of an RTL design with the aid of VDHL or Verilog simulation, RTL synthesis will translate the design into a gate level netlist. How can we know that the translation was error-free? At first glance one could argue that a computer program does not make mistakes and the results would be *correct by construction*. Real life had shown many cases where synthesized gate level netlists did not show the desired behavior. Tool flaws and ambiguities of RTL languages, especially with clock trees, are the main source of wrong synthesis results. Because of the high incremental cost of a redesign and the associated project delay, it is mandatory to verify the netlist even if it is machine-generated.

The conventional approach to verifying the generated netlist is to refer to the second generation EDA tools. A logic simulator can use the same set of stimuli as an RTL simulator. Fortunately, VHDL and Verilog simulators are also capable of gate level simulation, therefore, the stimuli (the 'test bench') do not have to be re-written in another format. However, with very complex circuits (e.g. more than a few hundred thousand gates) gate level simulation requires excessive run times and becomes the bottleneck for design productivity.

Formal verification and **static timing analysis** are the resort. A formal verification tool checks whether two circuits are equivalent, i.e., have an identical logic behavior. The circuits can be represented at different levels of abstraction, e.g. one is given by its RTL model and the other is described at the gate level. The principle of formal verification is quite simple: The behavior of logic gates can be modeled by very simple Boolean expressions. The behavior of a circuit represented by its gate level netlist can be derived from the gate instances and their interconnects. Using Boolean algebra the tool can then see whether the gate level behavior is equivalent to that captured in the RTL design. Of course, formal verification tools are very complex (and expensive), but they are very fast, indeed.

[1] The intent of VHDL and Verilog was to capture RT level behavior. For practical reasons it was necessary to introduce the concept of hierarchy similar to subprograms in software programming languages. Hierarchy provides a link to structural domain, and VHDL and Verilog can, therefore, be used to capture gate level netlists.

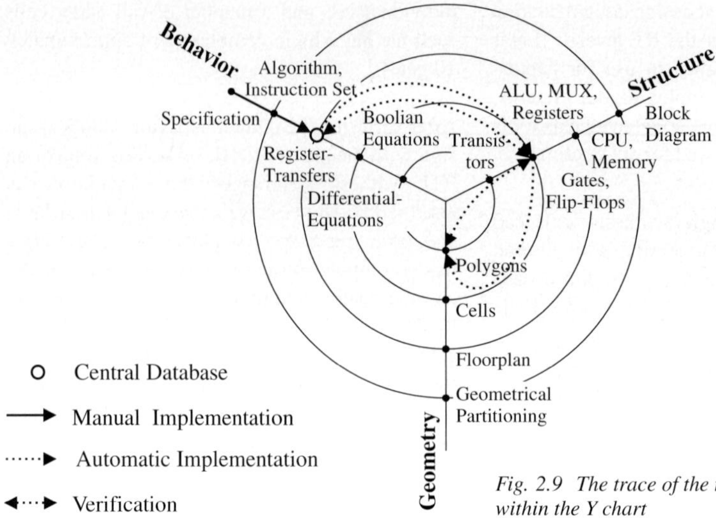

Fig. 2.9 The trace of the third EDA generation within the Y chart

Static timing analysis is the method of calculating and adding up gate delays along a signal path. After considering all structurally possible paths between a given start and end point, those with minimum and maximum delays are considered for further analysis.

Formal verification and static timing analysis are referred to as 'static' tools because they do not require stimuli (test vectors). Static tools have been accompanied by a tremendous improvement in computer run times: A circuit requiring days for logic simulation can be checked in several minutes with static tools. The use of static tools mandates a *synchronous design style*. Ideally this implies that the whole circuit changes it state only at the active edge of a single clock. In practice, this implies that all clock pins of all registers be connected to the same clock signal.

To summarise,

- Hardware description languages;
- RTL simulation;
- Logic synthesis;
- Formal verification; and
- Static timing analysis;

all build the tool suite of the third EDA generation. They represent a complete development system for design at the RT level and its transformation into the gate level. An additional tool is also considered a member of this generation:

Automatic Test Pattern Generation (ATPG)

ATPG cannot be seen as beeing detached from logic synthesis: as mentioned in section 2.2.7, a plain functional test is not feasible and structural test is the only resort. This methodology calls for creating test vectors needed to identify physical wire defects (*stuck at faults*). This task cannot start before the netlist has been frozen, and, similarly to any work below the RT level, will be automated according the charter of the third EDA generation.

Figure 2.9 sketches the traces within the Y chart if the third EDA generation is applied. A deeper understanding can be found in [2.5] and [2.8].

2.5.4 Design Productivity

The amount of time needed for the development of an ASIC has a major impact on the commercial success of an electronic product. ASICs are primarily used within new products with short life cycles. If the development of the ASIC is late then also the market introduction of the target product will be delayed and the life cycle will be shortened. Less time left to sell the product also means less revenue and less profit [1)].

[1)] Within the context of product life cycle the market window is an important parameter. This is the duration between product acceptance in the market place and its replacement by a more powerful product. For typical high tech products life times are in the range of one to two years. If development is late and delays the introduction of production by just 2 months then potential revenue would decay by 8 %.

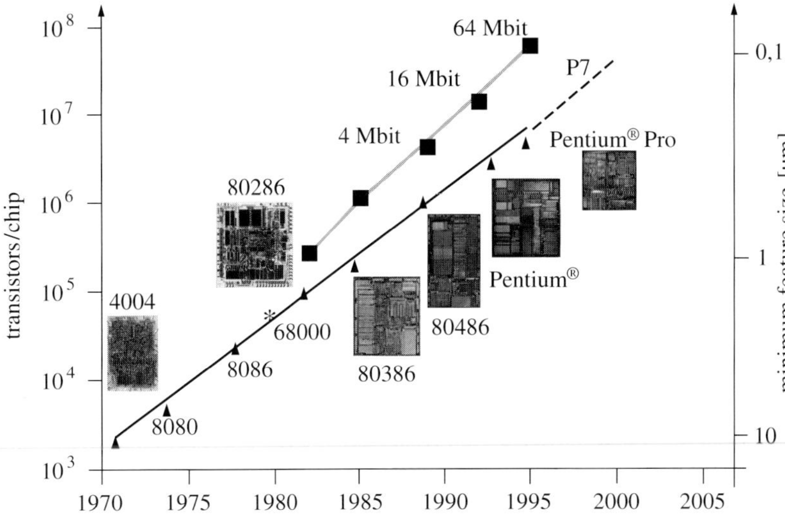

Fig. 2.10 Evolution of complexity of Integrated Circuits (Provided by Infineon Technologies)

Figure 2.10 demonstrates how the complexity of microprocessors and DRAMs – expressed in terms of transistor count – have evolved over the last 30 years. ASICs enjoyed a very similar progress. According to Moore's law every 8 years the complexity will increase by a factor of ten, on average. Let us now define design productivity as the number of logic gates a designer can implement in an ASIC within one year,

$$P_{\mathrm{E}} = \frac{\text{total gate count}}{\text{dev. time [years]} \cdot \text{number of designers}}$$

If productivity does not change over time then ASIC development time would also grow tenfold every 8 years – if the size of the design team remains constant. Alternatively, we could pump up the design team tenfold to keep the development time constant. Neither approach is viable, nor do we see that design issues would soon invalidate any progress in semiconductor manufacturing technology. A continuous increase in design productivity is the only resort. Fortunately the evolution of EDA does provide the required productivity improvements.

In table 2.1 we have summarized typical figures for design productivity when using the tool suite available in the EDA generations covered in the previous sections. The author interviewed several ASIC design teams and managers to compile that overview. Of course, design productivity is not a simple constant, but strongly depends on the type of an ASIC, especially on the amount of regularity within the circuitry. The ratio between the values in table 2.1, however, can be considered typical.

Table 2.1 Evolution of design productivity

EDA generation	Gates/man year
I	500
II	8 000
III	50 000

By migrating from EDA generation I to generation III, design productivity grew by a factor of 100. From that we can conclude that a typical ASIC design in 1979 using EDA tools of generation I required approximately the same amount of human resources like an ASIC typical for 1995 when using the tool suite of generation III.

2.5.5 Outlook on the Fourth EDA Generation

For another ten or fifteen year we can expect to see the complexity of ASICs grow as a result of advances in semiconductor technology. EDA needs to accommodate that trend, and design productivity will have to increase continuously, and new tools will be needed. Currently the following trends can be seen:

Design Re-use (multiple usage of a single design).

Design re-use had been widely practiced at a lower level (in the context of the Y chart) for a while: usage of pre-designed standard cells is a way for design re-use. We can easily boost design productivity if we raise design re-use to a higher level, e.g. by reusing microprocessor cores rather than simple gates. However, there arrives one of the challenges of design re-use: specification of logic gates is quite easy because of the simple logic behavior. Semiconductor manufacturers not only provide their customers with a detailed data book of their cell libraries, but also provide the characterization data in various formats (views) needed by EDA tools. For a more complex circuit, however, the size of and effort in documentation grows rapidly.

Historically circuit ideas had not been considered marketable products and nobody was willing to spend the cost of user-ready documentation. More recently design re-use has been seen as essential key to system integration (*system on a chip*), and the *Virtual Socket Initiative*, an international forum, [2.12] pushes, promotes, and develops standards for design re-use. A new market place has developed and several new firms, but also existing companies, sell or broker circuit designs, frequently referred to as intellectual property, e.g. [2.10].

The significance of design re-use can be demonstrated within the Y chart. By inserting an existing circuit in a new design we can immediately migrate to the structural axis rather than having to traverse further down the functional arm, and there is no need to describe behavior at lower levels.

Graphic State Chart Editors

During the era of the second EDA generation, schematic editors had been introduced to allow electronic designers to think in schematics. At the RT level some members of the designer community also desire to design graphically. However, RTL descriptions are usually viewed more like a computer program rather than as a structured electronic circuit. Flow charts describing behavior would benefit most from graphical entry, but they are applicable only to a fraction of a typical circuit. It depends on the designers‘ preference whether a mixed graphical language oriented entry is preferred against a pure language oriented one. Several EDA companies offer state chart editor today, but these have not yet gained much ground.

Behavioral Synthesis

Behavioral synthesis is the continued evolution of the concept of EDA: raise manual design work to higher levels of abstraction. It automates the transformation of an algorithmic description to the register transfer level. C level synthesis is a certain type of behavioral synthesis and those tools can read ‘C’ language rather than VHDL or Verilog. These tools are gaining ground [2.4], [2.5], [2.6] but they appear still to be in their infancy.

Hardware Software Co-design

Most electronic systems contain a microprocessor. In such systems the software executed by the microprocessor controls all peripheral hardware. When developing such systems the designers have the choice of implementing a certain function in either hardware or software. A software solution, in general, is cheaper, but whenever highest performance is required a hardware oriented solution may be required. The partitioning between hardware and software impacts the overall cost of a system. Hardware software co-design improves the productivity of design of a hybrid system by allowing one to simulate both hardware and software. Tools are available today, but again they are in a state of infancy.

In general, the emerging fourth generation of EDA systems will support the design of *systems on a chip* (SOC), increasingly calling for support of mixed analog digital (*mixed signal*) circuitry. The extension of VHDL to VHDL-AMS allows for capturing behavior of mixed signal design and is a step towards that direction.

2.6 References

[2.1] *Gajski, D. D.; Kuhn, R. H.*: ‘Guest Editors Introduction - New VLSI Tools’, IEEE Computer

[2.2] *Gajski, D. D., Padua, A., Kuck, D. J., and Kuhn, R. H.*: ‘A second opinion on data flow machines and languages’. *IEEE Computer*, 15:58-69, February 1982

[2.3] *Gajski, D. D.*: 'Silicon Compilation', Addison-Wesley, 1988

[2.4] *Gajski, D. D.; Dutt, N. D.; Wu, A. C.*: 'High-Level Synthesis: Introduction to Chip and System Design', Kluwer Academic Publishers, 1992

[2.5] *Gajski, D. D.; Vahid, F.; et al.*: Specification and Design of Embedded Systems. – Prentice Hall: Englewood Cliffs, 1994

[2.6] *Gajski, D. D.; Iianwen Zhu; Dömer, R.; Verstlauer, A.; Shuquing Zhao*: SpeL: Specification Language and Methodology. – Kluwer Academic Publishers, 2000

[2.7] *Nagel, L. W.*: 'SPICE2: A Computer Program to Simulate Semiconductor Circuits', PhD Thesis, University of California, Berkely, 1975

[2.8] *Siegl, J.; Eichele, H.*: 'Hardwareentwicklung mit ASIC', Mikroelektronik Bd. 8, Hüthig Buch Verlag, 1990

[2.9] *Weste, N.; Eshraghian, K.*: 'Principles of CMOS VLSI design', Addison-Wesley, 1993

[2.10] http://www.mentorg.com/inventra

[2.11] http://www.synopsys.com

[2.12] http://www.vsia.org

2

Overview EDA

1, 2

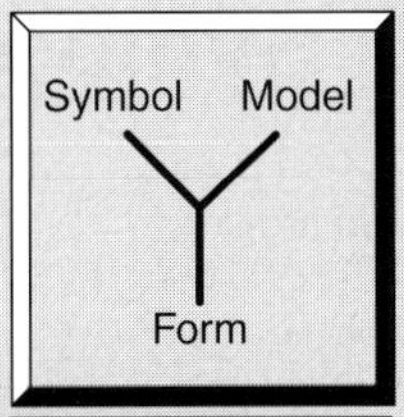

Symbolic Design

3

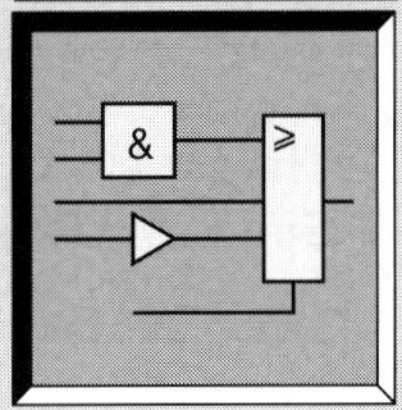

High Level Language Design

4, 5, 6, 7, 8

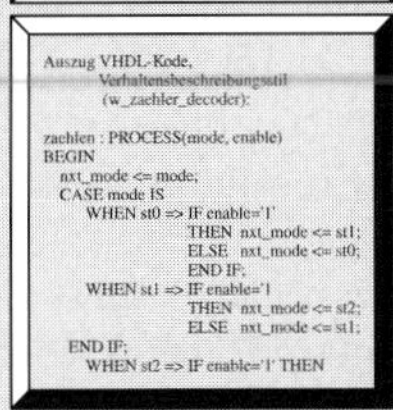

Modelling and Verifications

9, 10, 11, 12, 13, 14, 15

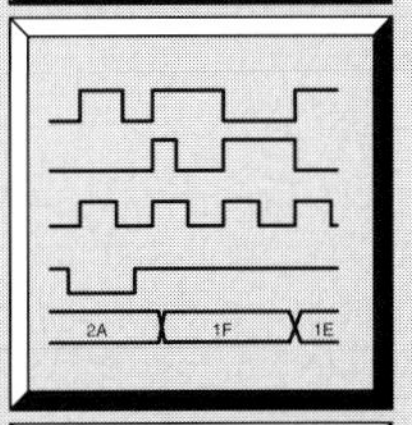

Implementation

16, 17, 18, 19, 20, 21, 22, 23, 24, 25

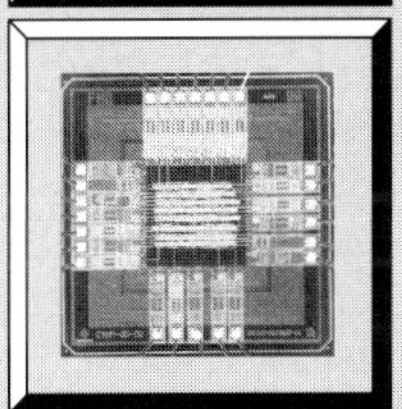

Tutorial

26

Appendix

A, B, C, D, E

3 Symbolic Design Entry

Gert Voland

3.1 The Role of Symbolic Design Entry

3.1.1 The First Steps in Electronic Design

Symbolic design entry, also called schematic entry, describes the process of entering the circuit schematics of an electronic design into a computer. This is done with the aid of a special graphics program, the schematic editor. The schematic entry usually represents the first step of a design process. At this point there are two major goals to be reached: firstly, the graphical documentation; secondly, the starting point for further, automatic design steps.

True to the motto 'A picture tells a thousand words' the use of circuit editors has been developing steadily since the early 1980s and they are now available on any type of computer and under any operating system. The symbols of the electronic devices to be used are positioned in an editor and their pins connected by lines which represent the wires of the implementation. The topological arrangement of the symbols normally has no influence upon or relationship with the physical placement of the devices in the final design.

The graphics will then be converted into a netlist which basically has exactly the same information content as the circuit diagram; that is to say, devices and their connections. Netlists contain the data for the consecutive design steps. In principle there are two basic types: a verification step by analog, digital, or even mixed signal simulation or the implementation step with the possible target of a PLD (Programmable Logic Device), a PCB (Printed Circuit Board) or an IC (Integrated Circuit). This situation is explained in fig. 3.1.

3.1.2 Structural and Behavioral Description

Historically speaking, the schematic editors have replaced the process of designing circuits with pencil on paper. The strongest motivating forces for that development have already been mentioned: documentation in machine readable form for the subsequent automated processing steps. The possibility of introducing much more information into screen schematics than can be contained in a simple pencil drawing has, of course, also been widely employed.

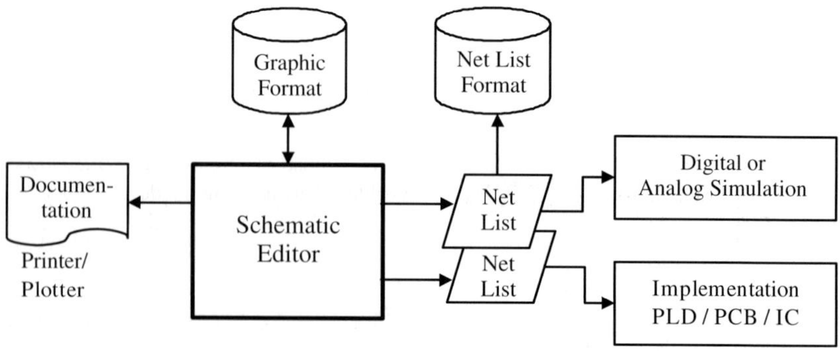

Fig. 3.1 The role of the schematic editor as an entry tool

The design process of circuit entry produces a structural description. It implies that the designer has already constructed the circuit and knows exactly which devices are to be used and how they will be connected. This chapter deals mainly with this type of structural design entry. A more abstract method, the graphical behavior description, is presented in chapter 4.

Schematic entry is still the preferred method for small to medium designs. In contrast, the technique of microelectronics allows the integration of millions of devices on a single die. For that purpose methods of high level design languages and circuit synthesis have been developed (see chapters 5 and 6). They describe not so much the final use of single devices as the entry of abstract behavioral descriptions.

In order to cope with the increasing complexity of circuits the concept of hierarchy is extensively used. Higher levels of the design contain block symbols, which represent design modules of lower levels, as well as elementary symbols, so called primitives. Such a hierarchy mechanism allows the coexistence of structural design entry next to behavioral high level entry. A block symbol is used for graphical schematics as well as textual descriptions. Both forms of design representation may be mixed and have to be fed into a synthesis program which will then interpret and optimize the design and map it into a structural netlist (see fig. 3.2). Thus the synthesis program may even be used to re-map the symbols of a structural schematic into the cells of a different implementation library, as long as the functionality of entry and the physical design are convertible.

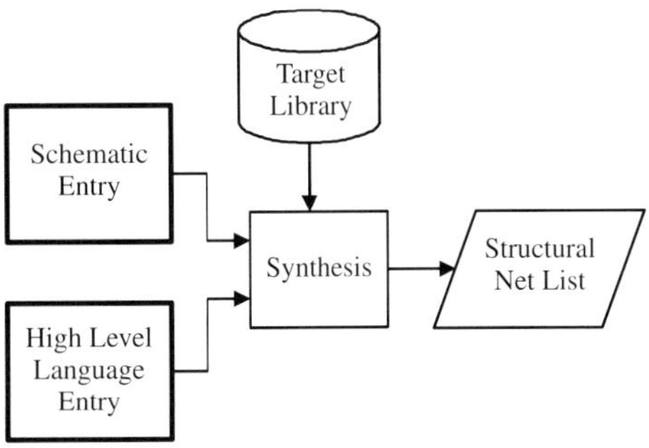

Fig. 3.2 Parallel entry of schematic and high level design language

3.1.3 Standardization

Ideally, the design of a circuit should be completely independent of the manufacturer or the production foundry. In practice, there are two principle difficulties which usually cannot be overcome.

On the one hand, there is no compatibility between different schematic editors. There are only very few cases where the format of one graphical tool can be read in, or converted into, the format of another CAD tool, even if the symbol libraries represent the same physical devices. Altera Corporation, the supplier of FPGAs (Field Programmable Gate Arrays) and CPLDs (Complex Programmable Logic Devices) allows, for example, reading in OrCAD STD schematics. Naturally, one may only use symbols in the primary editor which can be understood by the reading editor.

There have been attempts to represent the circuit diagram in standardized EDIF (Electronic Design Interchange Format, see chapter 8). Whilst EDIF is widely used for netlist transfer, the use for symbol and schematic exchange has not been very successful. The main reason for failure seems to lie in the various incompatible editor features and that EDIF standardizes the syntax but not the content. As a result every tool manufacturer may create his own EDIF dialect.

On the other hand, there is no standardization of the symbol libraries. This is true for symbol names, shapes, functionality, and symbol pins. Even on the same tool one cannot easily translate the schematics based on symbols for semiconductor foundry A into schematics for foundry B. The D-flipflop of library A may switch on the rising clock edge and the flipflop of library B on the falling clock edge or the reset input may be of active high or active low type. A relatively costly attempt to translate an EDIF library is described in [3.5].

Only the use of circuit synthesis has solved this problem and has thus made circuits portable. A circuit may be drawn with the symbols of one foundry and can then be translated into the netlist of a second foundry.

Although the basic structure of different editors is very similar, the characteristics of three different design tools will be explained:

- **Graphic Editor of Altera Corporation**, which allows the schematic entry for ALTERA CPLDs. The ALTERA system is optimized for easy entry of hierarchical schematics in parallel with textual entry [3.1].
  The Windows version of MAX2PLUS V9.23 is used (short form: ALTERA).
- **Schematic Editor by Cadence/OrCAD** (ex MicroSim), which is used for digital, analog, mixed and behavioral simulation with PSPICE and for the design of PLDs and PCBs [3.4].
  The Windows version 9.1 is used (short form: PSPICE).
- **Design Architect by Mentor Graphics Corporation** with the programs NETED and SYMED. This system is the most universal one of the three [3.3].
  Version C1 on HP Unix V10.20 is used (short form MENTOR).

3.2 Schematic Editors

3.2.1 Graphical Elements of Schematic Editors

The following elements of schematic editors will be described:

- Symbols;
- Instance Names;
- Important Elements of Symbols;
- Electrical Connections;
- Buses;
- Back Annotation;
- Page Frames and Title Blocks;
- Additional Graphical Objects.

Symbols

Symbols may represent a lot of different design objects of a circuit. The most important one is the device symbol. A device may be an electronic object that either will be mounted on a PCB by soldering or a cell which will be placed on an integrated chip. Besides such primitive symbols, modules of primitives may be combined to subdesigns with a corresponding block symbol. Both types of symbols are handled in the same way.

Figure 3.3 shows various types of symbols for different design systems. The decision as to which kind of symbol to use depends strongly on the target application.

- The logic symbols for PLD designs represent only the logic behavior. This is because after the optimization process which follows, mapping to the chosen PLD structure will take place.
- For the development of full custom digital or analog IC cells, mainly symbols of single transistors and some other analog elements are needed.
- A design for CBICs (Cell Based Integrated Circuits) must only contain cells from a standard cell library which will be supplied by the chosen semiconductor foundry.
- Symbols for printed circuit boards have to identify the devices and thus specify exactly the shape and size of the target packages. These devices could be standard parts, PLDs or even ASICs (Application Specific Integrated Circuits).

Instance Names

The data of a single symbol are stored in a symbol library. When a symbol is placed in a schematic on the screen the data set of that symbol does not usually get copied into the design. Only a pointer is stored in the design. Since the same symbol may be instanced any number of times in various places in the design, the single placements have to be distinguished by their placement or instance names.

Instance names are a special symbol attribute in the object oriented organization of symbol data (see section 3.2.3). Most editors use an automatic numbering mechanism for the instance names.

For example, the library symbol could use the string 'U?' as a generic instance name. When placing the symbol, the '?' gets replaced by a running integer number starting at 1. The user may re-edit the instance name and replace it with a unique character string. Thus a risk of multiple instances with the same name may appear but can easily be detected by an Electrical Rules Checker. The method of distinguishing symbols in hierarchical structures is described in section 3.2.2.

Figure 3.4 shows an IEC symbol, mostly used in Europe, and an American ANSI symbol of NAND3 (7410) which is taken from a 74 library. The most important elements are: Symbol Icon, Symbol Name, Instance Name, Input, Output and Supply Pins, Bounding Box, and Origin.

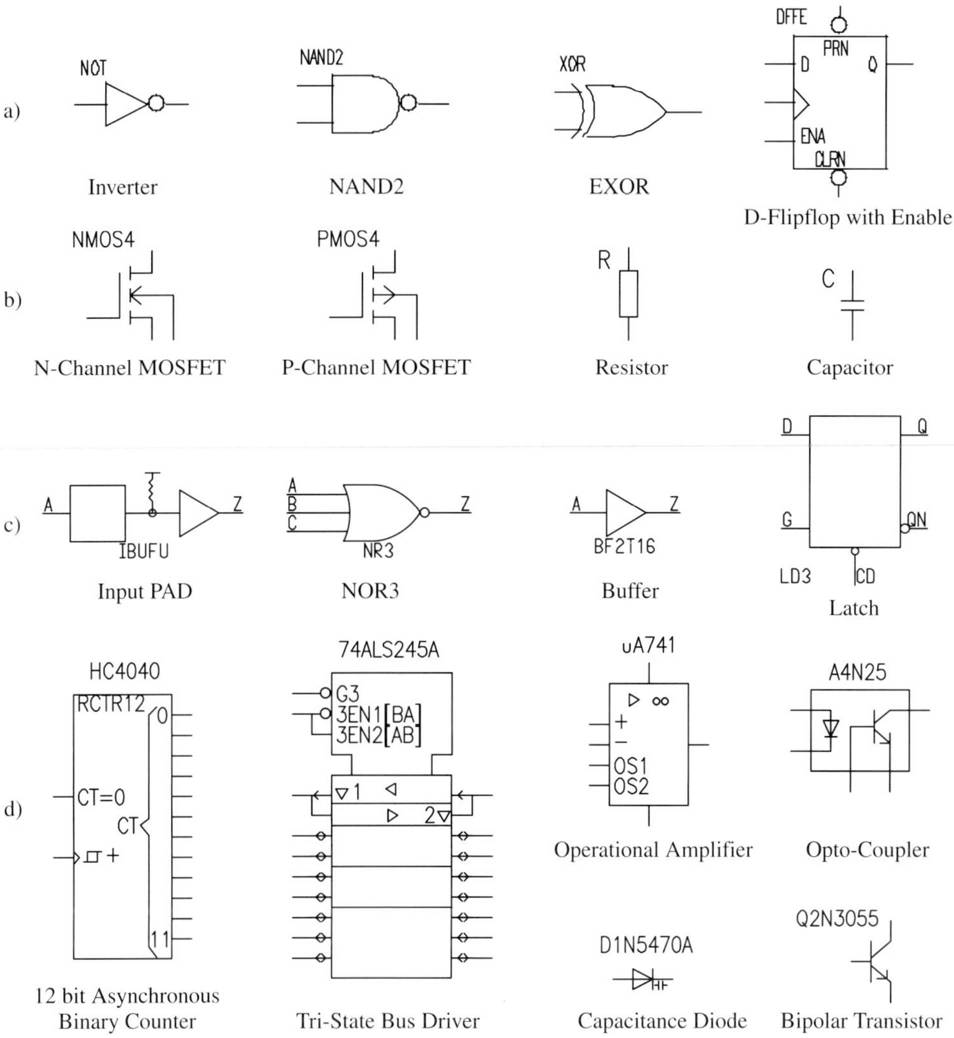

Fig. 3.3 Examples of device symbols for various schematic editors:
a) Logic symbols for CPLD designs on ALTERA;
b) IEC[1] symbols for the design with analog circuits on PSPICE;
c) symbols of ASIC standard cells by the semiconductor foundry MIETEC on MENTOR;
d) IEC[1] symbols for printed circuit boards with PSPICE

PCB design systems usually call the placement name a 'reference designator'. These designators also carry the information of which instance shall be used if there are multiple occurrences of the same function in one package; for example, in 74 device families.

Important Elements of Symbols

Owing to their graphical character, the position of the symbol pins is extremely important because only the pins establish the connectivity. In fig. 3.4 the end point of the pins is marked with a little

[1] EN 60 617-5

cross (x) to show the point of connection. In the final schematic these crosses will not be visible when they are properly hooked up. The rest of a symbol only has the character of an icon for better visual recognition.

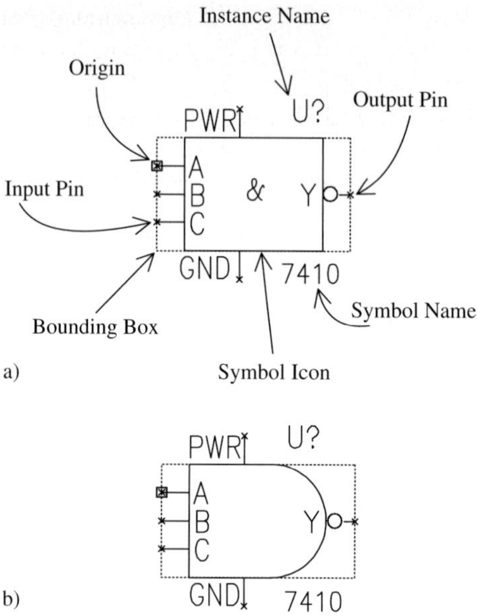

Fig. 3.4 7410 symbol in two different standards: a) IEC; and b) ANSI

The power supply pins PWR and GND will also not be visible in the schematic since they are declared as 'hidden pins'. The designer does not have to wire them up because they will be connected to global nodes during netlist creation. Figure 3.4 also shows the bounding box, which is a rectangular outline to highlight the symbol when the mouse cursor is clicked inside this box.

This now leads to the subject of standardization. Basically, two internationally accepted standards exist for electronic symbols: IEC/DIN/EN and ANSI/IEEE standards. The latter are usually used for ASIC designs. It depends solely on the designer's experience and taste which symbols lead to better readability. For most of the following examples the IEC type of symbols are used, but they can easily be 'translated' using the explanations in Appendix A.

In an editor the replacement of one symbol type by another is only possible when the symbols carry the same device names and if they have the same pin coordinates as well as the same origins. Given that those are the same, it is only a question of the library configuration as to which symbol will be displayed. Since the crosses of the visible pins A, B and C are the same for both symbols of the device 7410 in fig. 3.4, they may be replaced by each other.

Depending on the specific application, further symbols may be necessary which do not always represent electronic devices:

- Symbols for voltage supply and ground unless they are connected implicitly via hidden pins. They are needed mainly in analog designs and PCB schematics;
- Input and output ports (I/O-ports) for subdesigns which connect the module to other levels in the hierarchy through block symbols;
- Symbols for stimuli which are only used for simulation and not for the implementation;
- Symbols for global signals, as long as they are not generated during netlist conversion (for example, clock signals);
- Non-electrical information, such as markers to set attributes for a net to appear in the waveform display or to set initial conditions.

Electrical Connections

In order to connect devices wires are drawn from pin to pin. Graphically the wires are represented by lines which may branch off but always represent the same electrical potential of a signal node. For better documentation and orientation a wire may be labeled with a signal name. Crossing of wires may lead to electrical connection, in which case the system must provide a graphical element of a junction dot. Crossing wires represent two different signal nodes as long as there is no junction dot connecting them. In some cases junction dots are forbidden so that intersections have to be drawn with an offset if a connection is wanted (see fig. 3.5d)).

Any non-labeled signal nodes will automatically be numbered by the editor. As a consequence the signal names are no longer self-explanatory, which does not ease circuit verification or debugging after simulation. The property 'connect by name' bears a high risk, causing all nodes with the same label to be automatically shorted. If two wires are equally named by coincidence they will be shorted even without any graphical contact in the schematic. In PSPICE this property can be set by an extra switch and then all signals through any level of hierarchy are shorted. With the ALTERA and MENTOR system all equally named signals of one sub-design level are always shorted.

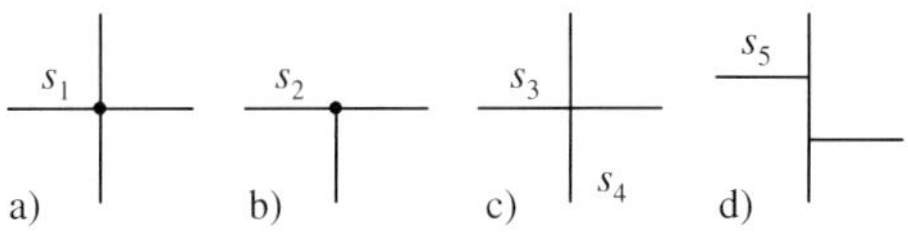

Fig. 3.5 Different possibilities of intersection and crossing of wires
a) Crossing with junction dot: signal s_1;
b) T-junction with dot: signal s_2;
c) Crossing without connectivity: signals s_3 and s_4;
d) Offset T-junctions without dots: signal s_5

PSPICE and ALTERA do not provide a 'renaming symbol' to change the name of one signal without losing the electrical property of a common potential. MENTOR provides such a symbol 'NetCon' to short two differently labeled nets without introducing any electrical functionality. Further possibilities are described in 'Global Signals' of section 3.2.2.

Buses

In datapath designs the same circuit function often has to be used for different signals. For example, all bits of an A/D converter have to be inverted. The editors provide different mechanisms for compacting repetitive parts of the displayed circuits:

- Signal buses; and
- Hierarchical designs (see section 3.2.2).

Buses are the combination of several wires to a line that is usually drawn in a wider style. In principle, the same rules are used for connecting buses as are used for single wires. A bus pin on a symbol has to become connected to a bus of equal width. The identification of the signals contained in a bus bears some risks. The designer has to carefully label the bus according to the number and the sequence of the signals in the source and the target notation.

There are two basic methods for naming a bus. On the one hand, there are implicitly indexed signals. A bus label is followed by a range of numbers within square brackets: data[3–0] represents the four signals data3, data2, data1 and data0 in exactly that sequence. On the other hand, any sequence of net names may be used when they are separated by commas without any spaces. Another four-signal bus could be labeled a,b,c,d. Depending on the design system, one may find differing syntactic and semantic rules for the same mechanisms.

In order to branch off from a bus the easiest way is to use a T-junction or dot-crossing, which will produce the same width and signal order as the source bus. A more delicate task is forking off partial buses or single signals. The branches have to be labeled with a true subset of the source bus. Figure 3.6 demonstrates the situation: four signals, bit3 to bit0, are combined to a single bus bit[3–0]; a partial bus, bit2,bit3, branches off (with different sequence!); bit0 connects to a single wire and the internal ports D3, D2, D1, D0 of DAC4 are connected.

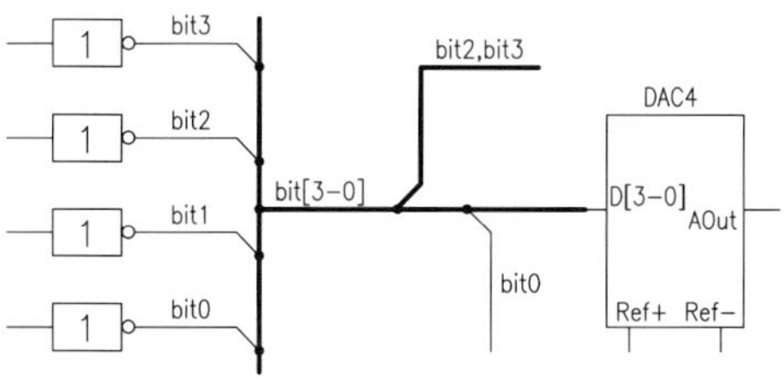

Fig. 3.6 Wiring with buses; all branch signals have to be labeled

Back Annotation

Back annotation describes the process of writing values of subsequent calculations back into the schematic. Examples are the values of a DC analysis, capacitance values on output wires, logic levels, or the placement in multi-part packages such as the 74 series. Dedicated symbol attributes are prepared for such a mechanism and the resulting values get assigned to those attributes to be displayed in the schematic.

Page Frames and Title Blocks

Each page automatically has a rectangular border or frame usually containing grid information for easy location of elements on the page. The grid partitions could use numbers in the horizontal direction (1, 2, 3, ...) and alphabetic characters vertically (A, B, C, ...). The title block serves to register information such as the design house, designer's name, project title, date, and version. In addition, all editors allow placement of any kind of text for documentation. Such text will then not be relevant for any further processing step.

Additional Graphical Objects

Some editors provide placement of further graphical objects such as vector graphic elements (lines, circles, rectangles, arrows, etc.) or bitmaps for documentation purposes.

3.2.2 Structure and Organization of Graphical Designs

The basic structure of a graphical design is a page or sheet surrounded by a border and containing a title block. The user has the optional choice of the page size. The decision for a certain size depends on the design size and the printing or plotting possibilities. Figure 3.7 demonstrates possible settings of PSPICE.

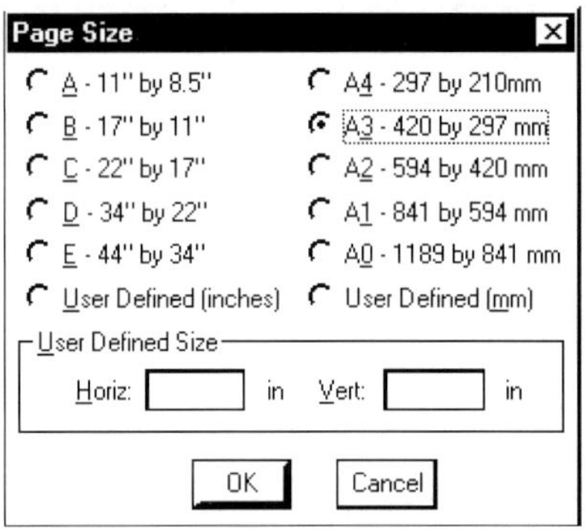

Fig. 3.7 Settings of page size as offered by PSPICE

In order to keep the overview of large designs, two mechanisms are supported:

- Multi-Page Designs; and
- Hierarchical Designs.

Multi-Page Designs

Multi-page designs are usually organized by design managers, i.e., programs that keep record of all documents and files belonging to one design. Off-page symbols define the interface signals between all pages that constitute an extended design. Signals connect by name and establish a single electrical node over multiple pages as long as they are connected to off-page symbols. Figure 3.8 displays two different methods of handling off-page symbols. The instance names must be unique for one level of hierarchy even if it extends over several pages.

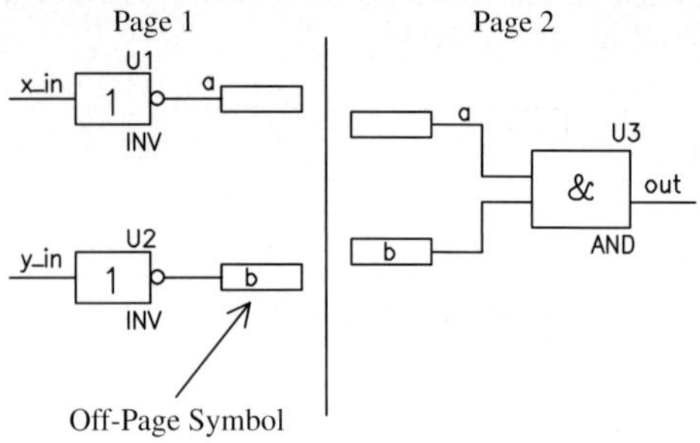

Fig. 3.8 Connectivity of two design pages. The upper off-page symbols connect the two signals labeled 'a' together; the lower symbols must both be labeled 'b' because the wires carry no user label

Hierarchical Designs

The characteristic of hierarchical designs is the assignment of sub-designs or sub-modules to a block symbol. Such a block symbol may be instanced any number of times on higher hierarchy level (but of course not recursively). The sub-design may contain primitive symbols as well as other block symbols. The lowest hierarchy level consists only of primitives. Resolving the hierarchy (flattening) thus leads to a flat structural netlist containing solely primitive elements. Each element of the flat netlist represents a different distinguishable object for either simulation or physical implementation. In principle the sub-designs build up a tree structure (see fig. 3.9). The schematic representation of each level is usually stored as a separate file.

Connections in Hierarchical Designs

Sub-designs are most easily connected through the block symbols to a higher level by port symbols. In the underlying module the input and output signals are connected to input and output port symbols. Just like the off-page symbols, such I/O ports need distinguishable, unique names. Schematic editors usually have built-in symbol generators for such block symbols. Thus the I/O port types and names are converted to the corresponding pins of the

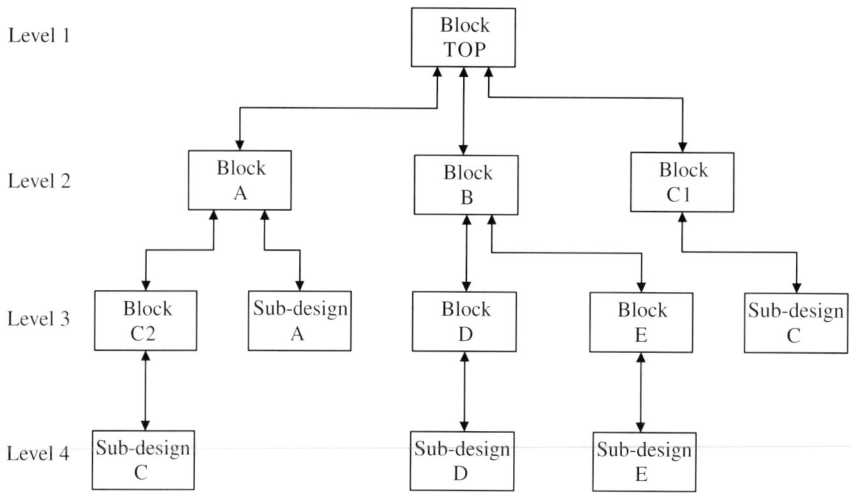

Fig. 3.9 Tree structure of a hierarchical design. The sub-designs contain only primitives. Instances of block symbols appear in various branches (e.g., block C). The sub-design C itself exists only once.

block symbol. When the sub-design contains a bus the block pin will be of bus type with a pin name that is equal to the bus label.

Numbering of Devices and Signals

At the lowest level of hierarchy only primitive symbols are found; these have to be distinguishable. An inverter in sub-design C will switch at different times when it is instantiated in TOP_BlockC1 than in TOP_BlockA_BlockC2. In order to differentiate between physical devices, the instance name of an individual device gets extended by the hierarchical path name. Different editors may use different separation characters. PSPICE uses an underscore '_' like this:

```
TOP_BlockA_BlockC_Inv1
```

ALTERA separates the branch names by '|' and adds the instance name (a number) at the end after a colon. The top level name 'TOP' is left out:

```
|blockA:1|blockC:1|invert:3|
```

MENTOR uses the same scheme with '/'.

The signal names are handled in a very similar way. A signal s1 in Sub-design C will become TOP_BlockC1_s1 in the first branch while it is called TOP_BlockA_BlockC2_s1 in the second one.

Global Signals

Next to the local signals, global signals are found which will be connected directly through any level of hierarchy. Besides supply nodes, there are also clock and reset signals belonging to that group. PSPICE provides symbols, such as 'Bubble' and 'Global' for that purpose. ALTERA uses an implicit 'connect by name' mechanism for every level. Signals have to be connected by explicit I/O-symbols between any level. MENTOR provides a symbol, 'Global', for clock and reset signals. Connect-by-name is also restricted to one page or one level only.

Views

The view of a block is a mechanism for representing different types of description behind a symbol. A Schematic View is used to define the circuit behind a symbol as a graphical schematic drawing. A VHDL View would describe the content as VHDL text. Depending on the type of VHDL description, there could be behavioral, RTL, or structural text. In any case, a synthesis compiler is then needed. The text has to be converted into a structural netlist which may be combined with other netlist parts from schematic entries. Such a mixed entry can be seen in fig. 3.10. It shows the hierarchy tree of the ALTERA system where graphical files (.gdf) are mixed with textual files (.tdf). Such a scheme

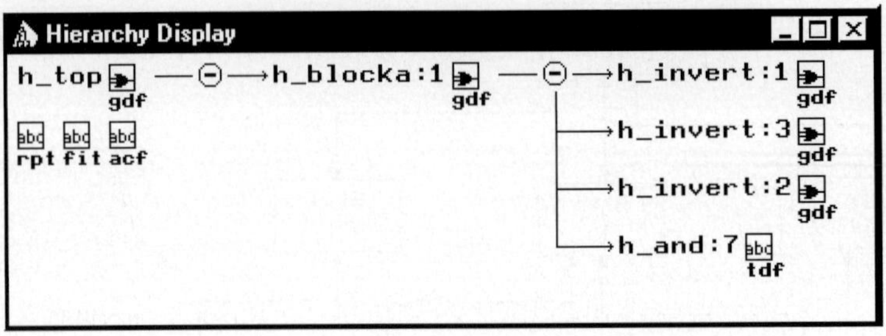

Fig. 3.10 Hierarchy display in ALTERA: The top level design (project) is called h_top and is a graphical file (.gdf) while the sub-block h_and is a text file (.tdf). The numbers following the colon are placement names; h_invert is instantiated three times.

allows a high degree of complexity. Any graphical or textual sub-design may be referenced in any higher level graphical or textual block. Therefore, tools are provided for creating symbols from either .gdf or .tdf files and also a mechanism to create so called include files to be created from either .gdf or .tdf. Thus a view is an abstract form of modeling a design block.

Soft Macros

Soft macros are pre-designed flat or hierarchical circuits with a corresponding block symbol. The underlying design is usually drawn as a schematic of primitives from a specific library on a specific editor. As a consequence, soft macros are very dependent on technology and design tools. They may be ported to other design systems if the tool supplier offers the exchange of schematics and symbols as a defined product feature.

Many systems provide the functionality of the 74 device families as soft macros. In such a case the schematics behind 74 device symbols are often drawn with the primitive INV, AND, NAND, OR, NOR, and DFF symbols of the implementation library. For simulation and implementation purposes the soft macro will be 'flattened' into a netlist of primitive functions. The structure of the soft macro is usually completely lost in such a flat netlist.

This also shows the weakness of soft macros. The final structure and the timing behavior depend only on the available primitives and the mapping mechanisms. In any case, the function of the soft macro has to be carefully verified in the final design.

Design Style

The two methods, hierarchy and views, allow complex designs to be created with relatively little effort, which is especially necessary for FPGA and ASIC designs. A preferred method is to use a graphical top level in which one block symbol is placed next to all input and output pads. Such a style guarantees the overview over all the physical I/O-pins.

The second level may represent the partitioning structure of the design as a block diagram. The blocks are then single units which may be designed separately by different people. A top down style requires each function to be refined down to the final structural level. A high level function may thus be described first in a behavioral manner and verified by simulation before the refinement into the final structure.

On the other hand, a bottom up approach combines functional groups to more abstract blocks after thorough structural verification. The method gets repeated until the top level is reached. In practice, a combination of both approaches may be used which usually get applied alternately.

3.2.3 Assignments, Properties, Attributes

All modern editors build up the designs in an object oriented manner. The basic concept will then be applied systematically to create possibly very complex designs. One of those basic concepts is the mechanism of assigning a well defined set of properties to the design objects. The names which are used may be 'property', 'assignment'

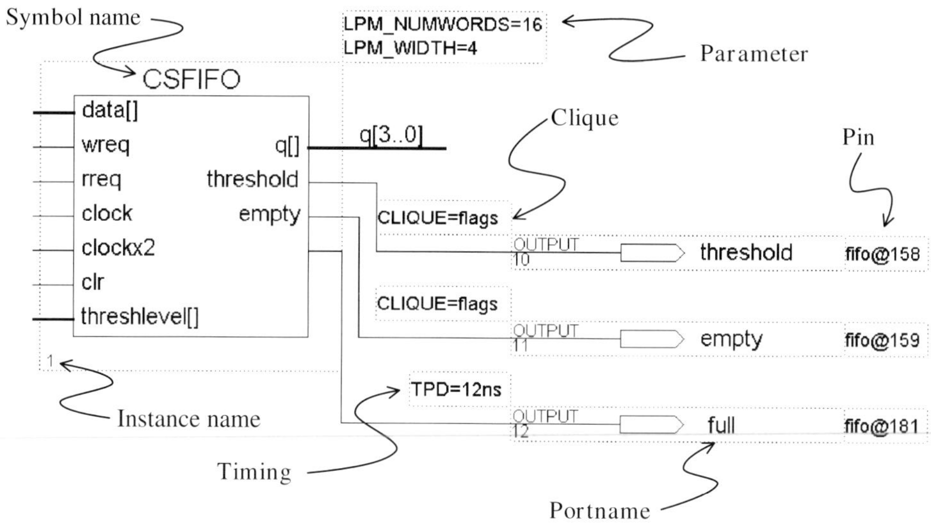

Fig. 3.11 Visible assignments and parameters of an ALTERA schematic

or 'attribute', but they basically all use the same methods: attributes with names and value assignments. A very strict implementation is found with MENTOR. The types of attributes and the range of values are always checked automatically.

Some attributes are used in exactly the same way with all editors: the symbol name is the value of an attribute 'Symbolname'. In a similar way, the attributes 'Instancename' and 'Pinname' are used for instance names and pin names. Even if the basic mechanism is very similar, editors may differ immensely in some detail of handling.

ALTERA places the assignments for further processing in the following way. Figure 3.11 displays a partial zoom of a schematic with a symbol CSFIFO and some attributes which have been switched to 'visible'. The assignments are displayed in little dotted boxes next to the symbols, wires or I/O symbols. Besides the already known attributes 'Symbolname', 'Instancename' and 'Portname', others are found, such as:

- **Timing**: assignment of a timing constraint on a specific wire (signal);
- **Clique**: grouping of certain logic blocks in order to keep a short critical path;
- **Pin**: assignment of an I/O port to the pin of the chosen package.

Further possible attributes could be:

- **Probe**: alias name for a signal hidden deeper down in the hierarchy;
- **Device**: placement of the logic in a certain device if the complete design has to be partitioned over several physical CPLDs;.
- **Logic Options**: choice of optimization switches for logic synthesis.

Another class of attributes are parameters of a block symbol. They will be passed down to the underlying HDL (High Level Design Language) text. In Figure 3.11 it can be seen that the number of words of the FIFO is set to 16 (LPM_NUMWORDS) and the data width is set to 4 (LPM_WIDTH). This method applies to all ALTERA LPMs (Library of Parameterized Modules). A user may apply the same method of passing parameters to self-written HDL modules.

In PSPICE the user may configure the attributes more freely. Figure 3.12 shows the dialog box for editing the attributes' names and their values used by the 7410 NAND3 symbol. The standard attributes of this symbol are:

- **PART**: symbol name;
- **MODEL**: pointer to the logic and timing model of this device. Alternatively, this could also be a pointer to a textual subcircuit description;

- **REFDES**: short for reference designator. It is used to assemble the placement name. In the library symbol REFDES has the value 'U?'. When the symbol is placed, the '?' is replaced by a running number (a '2' in fig. 3.12). Additionally, the assignment to a physical place in the final package can be seen (here it is U2A with the pins 1, 2, 13, and 12). REFDES and the pin numbers are thus used for back annotation;
- **TEMPLATE**: the value of this attribute contains the formatting commands for the netlist line of this part;
- **IO_LEVEL**: assignment of the analog interface when a mixed signal simulation is started;
- **MNTYMXDLY**: setting of method for the calculation of minimum, typical and maximum propagation delay times;
- **ipin(PWR)**: assignment of the global hidden digital supply node $G_DPWR;
- **ipin(GND)**: assignment of the global hidden digital ground node $G_DGND.

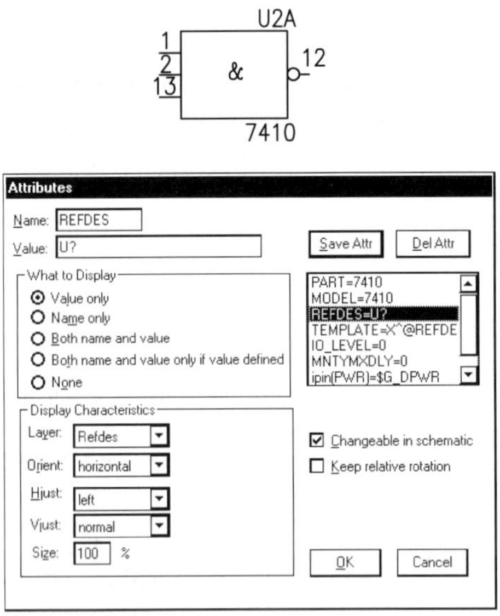

Fig. 3.12 Symbol and editing window for the attributes of the 7410 device of a PSPICE library

For each attribute the type of display (What to Display) may be set, as well as display formatting (Display Characteristics). A very important feature is whether or not the attribute may be changed by the user when placed in a design (Changeable in schematic).

For the selected attribute REFDES only the value (Value only) will be visible and the user may change it in the schematic. The attributes PART, TEMPLATE and the two ipins cannot be changed in the schematic since they represent basic data of this library element. The user may add attributes with new names and values to a symbol. With those he/she may, for example, pass more data to a netlist or to models. For that purpose the value of an attribute in the formatting TEMPLATE string can be used with the prefix @. PSPICE neither checks the type of attributes nor a violation of the value range.

In PSPICE the pins represent a sub-class of restricted symbols with pre-defined properties. They have a name, number, type, and display property such as orientation and visibility. The attribute pin number is used for back annotating the part assignment in a package. A further attribute 'ERC-Type' may be set to control subsequent ERC checking (Electrical Rules Check).

A pin may be of one of the following types: input, output, don't care, highZ, bidir, open collector, open emitter or power. According to the type, the permitted method of connection can be checked by the editor or netlist converter. For example, two pins of type output are not allowed to be connected. Two inputs do not need that restriction. This mechanism is valid not only for digital designs but also for analog and mixed signal designs. Netlisting is done according to the rules of the SPICE netlist which has the character of an industry standard. For analog cell designs the width and length of MOS (Metal Oxide Semiconductor) transistors may be entered in the instances allowing individual sizing of the devices. Thus, the W- and L-attributes can be set in the schematic and are passed through the TEMPLATE down to the MOS-models.

In the MENTOR system, the strictest type-oriented methodology of attributes is found; they are called properties. Only the following objects may be associated with properties:

- Symbol icons;
- Instances;
- Nets (nodes);
- Pins;
- Comments;
- Frames.

Each property may be of one of the following types:

- Text string;
- Number;
- AMPLE expression;
- Triplet (min, typ, max timing values).

Figure 3.13 shows the so called Component Interface of the MENTOR system. It displays the properties belonging to a symbol (see also the MENTOR symbol editor in fig. 3.16).

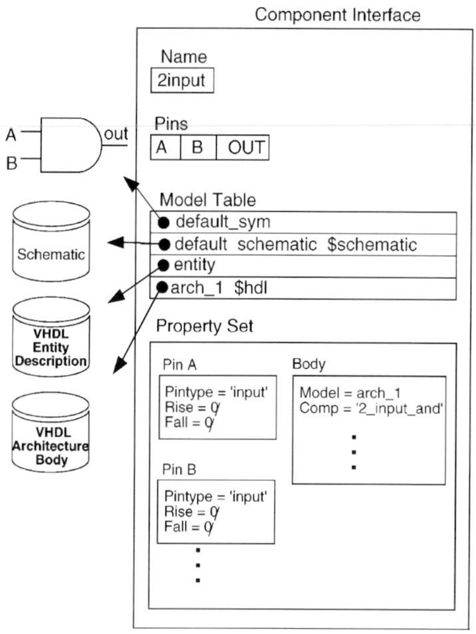

Fig. 3.13 Component Interface of MENTOR with the properties: Name, Pins, Model Table, and the Property Set for Pins

3.2.4 Symbol Libraries

Symbols are organized in symbol libraries. Usually one library is stored as one file. More details concerning the content and usage of ASIC libraries is described in chapter 17. Library files have to be configured for an editor before symbols are ready to be used in schematics. Each editor then has distinct mechanisms of handling symbols.

In PSPICE the configuration is accomplished by

- entering the library name in an options menu including the complete path; or by
- entering the library name relatively to a separate path entry (Library Path).

Using the second method, libraries can easily be exchanged by simply changing the path name.

Symbols with the same name may exist in different libraries. Thus, the sequence of library name entries determines which symbol shall be used: it is always the first appearing symbol.

This control mechanism allows the exchange of, for example, ANSI and IEC symbols by just reversing the order of library entrances.

PSPICE has an easy to use library browser for selecting the symbols. All symbols of the configured libraries are displayed in a list sorted in alphanumerical order. Next to the symbol name and a textual description of the function the user can also visually check the symbol's icon in a graphical preview. By providing three further mechanisms the searching process is made easier when large numbers of symbols are configured:

- Search by name with replacement character '*';
- Search by keywords in the textual description;
- Search in the single libraries.

In PSPICE there is no method of freezing library elements. A symbol placed in any schematic may be edited by any user with the symbol editor. Such openness is potentially dangerous because the position and attributes of pins are extremely sensitive. A wrongly moved pin can seriously disturb the function of the circuit.

ALTERA uses a similar mechanism, although library elements exist in the same directory as design files of a project. Symbols are also prioritized by sequence. The following files are the basis for symbol placement:

- The current project directory is searched first;
- Then a configured list of 'userlibs' is used;
- At the end the search continues through the preconfigured system libraries 'max2lib'.

The selection is only done by scanning the symbol lists in those directories. ALTERA does not allow editing of primitive symbols for security reasons.

In the MENTOR system the libraries are configured by a script containing all symbol names. Then a library palette is available in the graphics editor. An example with a MIETEC ASIC library

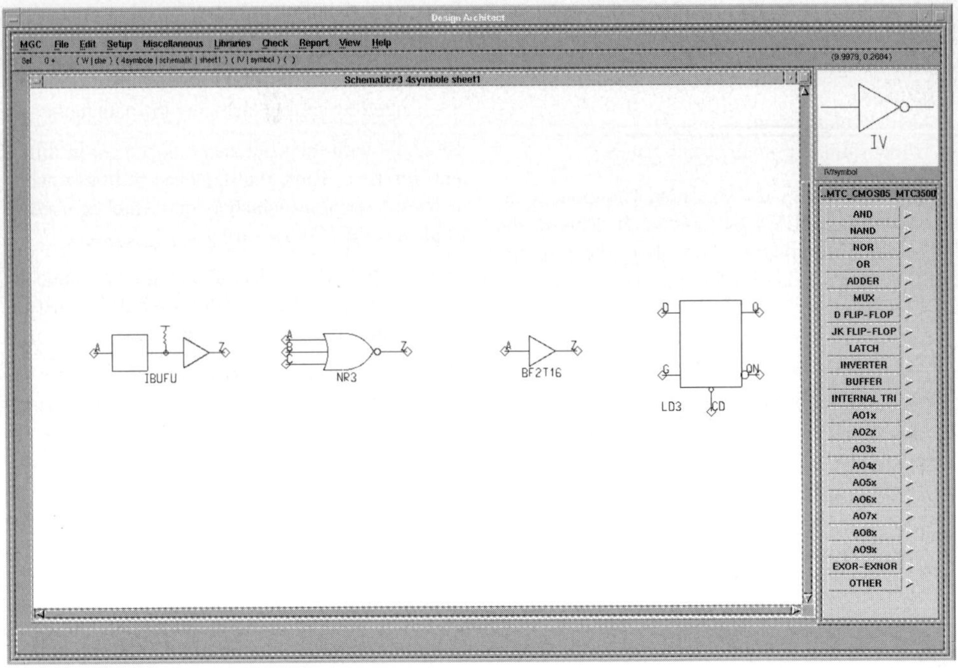

Fig. 3.14 Selection of symbols in MENTOR's Design Architect

is shown in fig. 3.14. Four symbols have been selected and placed.

3.2.5 Symbol Editors

The use of a symbol editor has two goals:

- to design new symbols; and
- to update existing symbols, especially those generated automatically.

A symbol editor is a program which works in a way very similar to the standard editor except that the design objects are symbols and not electrical schematics. Creating a symbol can be done from scratch or by copying and renaming existing symbols.

Each system provides a mechanism for automatically generating symbols from circuit modules. The size of such a normally rectangular symbol is usually determined by the number of input pins (usually on the left edge) and the number of output pins (on the right edge). In any case, the symbol has to be placed in a library in the library path. In order to adapt the symbol to specific needs it may then be changed in the symbol editor. The same restrictions regarding the sensitivity of pin positions still apply as to any other symbol.

The important drawing elements of a symbol editor are:

- Elements for the icon, such as lines, rectangles and (partial) circles;
- Pin symbols of normal, inverted (with inverter bubble) or clock (with clock triangle) type;
- Text for labeling;
- Bounding Box, which determines the selection area;
- Origin, which becomes referenced in the schematic. All the symbol coordinates are only relative to the symbol origin.

Figure 3.15 shows the window of the PSPICE symbol editor. In the drawing area elements may only be placed on visible grid points. Any element can be selected and edited by cut, copy and paste. The NOR4 symbol has normal input pins whilst the output uses an inverting pin. In any case, the little cross is the point of contact for a wire in the schematic. The symbol icon has only a

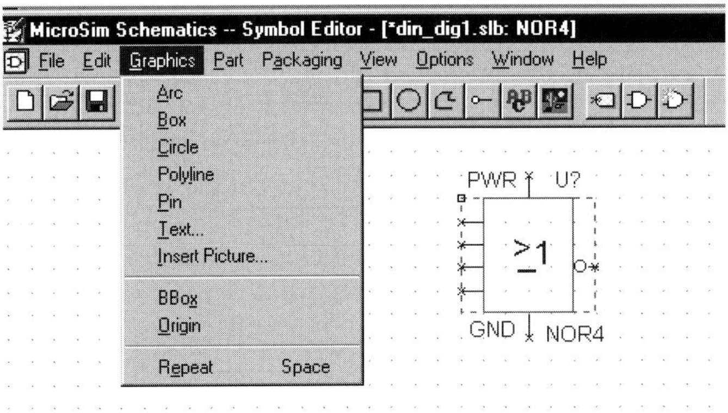

Fig. 3.15 Drawing elements of the PSPICE symbol editor

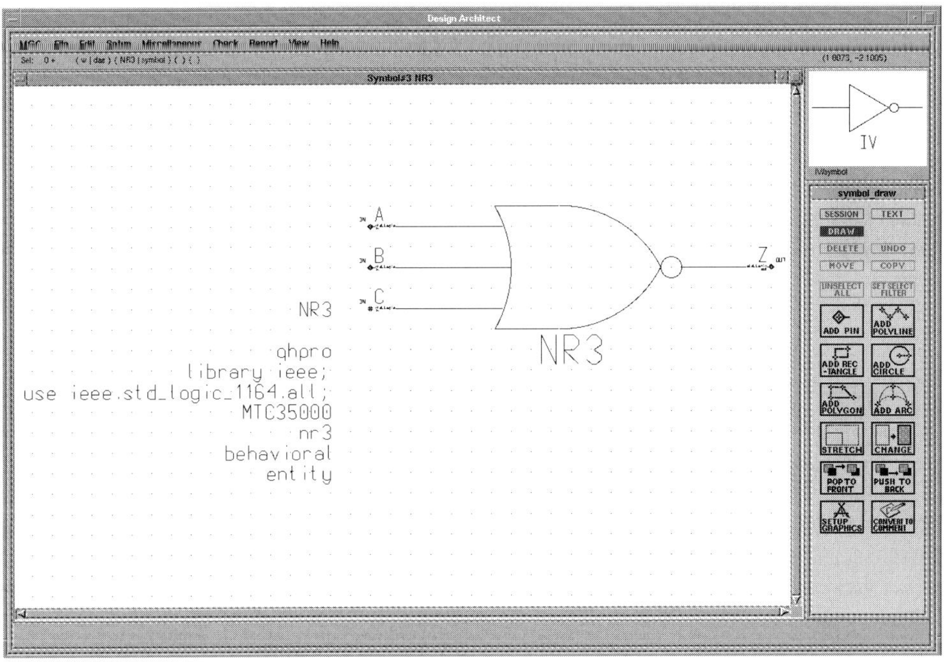

Fig. 3.16 Functions of the MENTOR symbol editor with a cell NR3 (NOR3)

descriptive character so that a user can recognize or identify the symbol. The netlist generator only needs the wire connection to a pin. An attribute editor for all the symbol properties is also built into the symbolic editor.

The ALTERA symbol editor is much simpler. It is limited to updating automatically generated symbols. The user may change the size of the rectangle, the bounding box, and the position of the pins. Of course, type and size of the selected text font may also be adapted.

MENTOR also provides a symbolic editor (SYMED) for changing appearance and properties of symbols (see fig. 3.16).

3.2.6 Edit Functions

This part describes the generation of a schematic with its fundamental steps. The concepts used have been described so far in chapter 3. As a result some examples are presented in section 3.4. Basically the on-line help systems are well enough established so that the use of printed manuals may become almost obsolete.

Configuration of Libraries

At the beginning of a project the target symbols have to be specified. Since most systems already have pre-installed primitive libraries, the user has to configure his/her specific project libraries in the library search path.

Choice of Page Size and Editor Configuration

The choice of the page size basically depends on the target printer or plotter. Hierarchical designs should be drawn to fit easily onto a letter sized or A4 page. Other editor settings are grid spacing, snapping to grid, and measurement units 'inch' or 'mm'. The grid spacing of the editor should never be larger then the minimal pin spacing of symbols because otherwise not all pins can be connected. It should also not be much smaller to prevent placing a wire only in the vicinity of pins. For details see fig. 3.17.

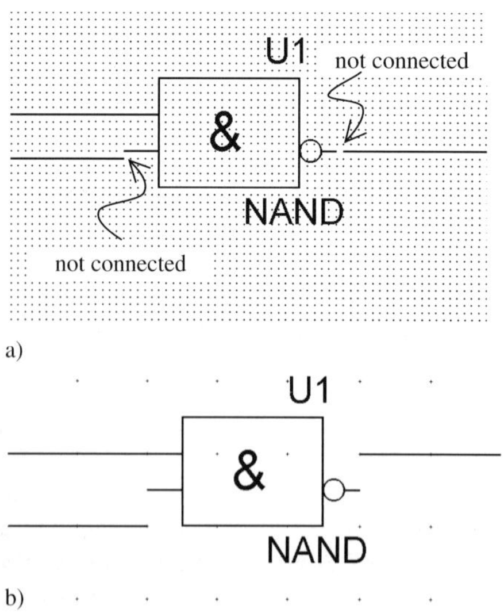

Fig. 3.17 Grid setting of a schematic editor:
a) grid spacing too fine; wires may be placed only close to pins;
b) grid spacing too coarse; connections not possible because not all pins are positioned on grid points

Any other settings, such as visibility of grid points (sometimes grid lines), non-orthogonal wiring, or visibility of certain attributes, may be set or changed at a later point in time. Those settings do not change the positions of symbols or the connecting wires.

Placement of Symbols

The chosen symbols are selected in the library browser and deposited on the page in a meaningful way. Normally the signal flow goes from left to right. All systems use a 'drag&drop' mechanism for symbols except ALTERA. A selected symbol is shown by its outline and gets dragged across the screen while holding down the left mouse key; upon release the symbol is dropped on a grid point. ALTERA, though, needs an insertion point defined prior to the selection of the symbol. Directly after the placement the symbol remains selected for easy rotation, flipping, or alignment operations. With a function 'Repeat Placement', some editors allow easy arrangement of symbol arrays.

Undo, Redo

As with text editors, the last commands may be corrected by undo or redo functions. Modern editors allow setting of the undo depth.

Wiring of Symbols

In PSPICE and MENTOR the drawing mode has to be set explicitly by a command (Draw Wire). In the ALTERA system the mode switches automatically when the mouse cursor moves over a pin or wire end. The wire can then be drawn by pressing the mouse button and dragging to the target point. Wire junction points are automatically added when a wire begins or ends on an existing one.

PSPICE uses two different objects for a wire and a bus. ALTERA and MENTOR differentiate them only by the line width.

Signal Labels

Normally the edit mode of a signal label is entered after double clicking on a wire or a bus.

Setting of Attributes

The appearance of a schematic may be influenced by editing some properties and attributes after all symbols have been placed and wired up. PSPICE allows the visibility of basically any attribute to be set. The instance names may also be changed instead of using the automatic numbering system. Often it may be necessary to move part names, instance names, or signal names to more adequate places.

Pan and Zoom

Pan and zoom are graphical commands to place the desired selection on the screen. Usually there is also at least one function to fill the screen with the complete schematic in an optimal way.

Correcting the Entry

In order to change errors during design entry, there has to be a set of graphical edit commands as they are known from text editing:

- Selection (highlighting) of one or several elements (wires, symbols or text);
- Moving selected elements;
- Copy, cut, and paste;
- Delete.

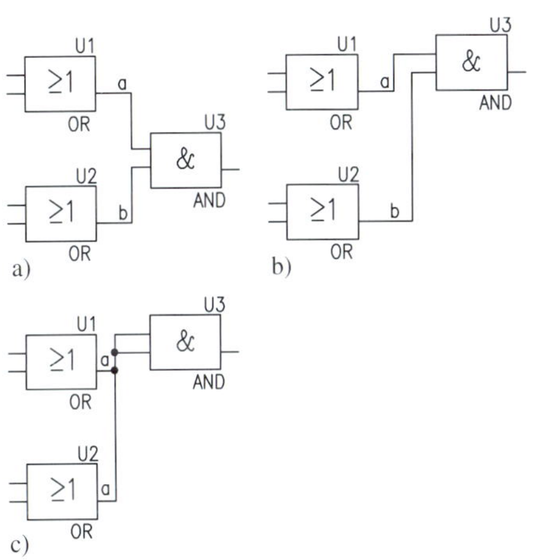

Fig. 3.18 Problem caused by 'rubber banding';
a) Original drawing with two signals 'a' and 'b';
b) U3 has been moved upwards and to the right; there was enough room to re-arrange the jogs;
c) U3 moved upwards without enough space; the two signals 'a' and 'b' coincide, are shorted and renamed to 'a'

Problems may arise during correction when the function 'rubber banding' is switched on. With that function, connections remain in place even when symbols and wires are moved about. In those cases, unwanted shorts may incidentally occur which should be repaired immediately by an undo function. An example is shown in fig. 3.18.

MENTOR treats signals shorted in such a way during an edit process by a very special method. Those signals are not shorted (and not renamed) but a special warning symbol is placed at the crossing (similar to a 'No Parking' sign).

Schematic Checking

Most topological and electrical errors of a schematic can be found by an ERC (Electrical Rules Check). In the ALTERA system a special switch, Design Doctor, has to be turned on during the compilation run. A netlist may only be created after all errors have been removed. An extremely helpful function is localizing an error in the schematic by back annotation. For that purpose an error message is double clicked and the schematic page is brought forward with the cursor placed at the error.

Printing and Plotting

The pages of a design should be printed or plotted for documentation purposes. The possible options may have to be combined carefully to obtain satisfying results:

- Print or plot 'print/plot only selected area' or 'print full page';
- Print/plot on a single page or divide the circuit into segments for several pages.

Placing Schematics into other Documents

In order to place schematic drawings in documentation systems, one may 'print' them to a file in a PostScript format (*.eps for Encapsulated PostScript). The result is a graphic file in vector format. Vector graphic files may then be imported into programs such as CorelDRAW!, Micrografx Designer or Visio. Subsequent operations may change line thickness or colors. The resulting file would be of *.eps format again to be read into word processing programs. The screen content may also be grabbed as a pixel bitmap. In that case a high resolution screen must be used in order not to lose any lines due to interference. A very suitable format is .png (portable network graphics) which stores such bit maps in very small files. Important is a reduction of the number of colors to the absolute minimum. Schematics can usually be well displayed with two (black and white) or 8 colors.

3.2.7 Special Characteristics of Schematic Editors

Even if the basic tasks of schematic editors from different suppliers are all the same, each program is equipped with some individual properties which are only found in a few or even only one of the programs. Some of the characteristics worth mentioning are described here.

Library Manager

A library manager is a set of tools which allows the administration of graphical and non-graphical information of the involved libraries. The necessary functions are the export and import of symbols in an ASCII format. Such a mechanism may be necessary to convert libraries to other versions of an editor, for example when the binary symbol format has been changed.

The PSPICE Schematic Editor product of OrCAD offers such functions to the user. ORCAD's Capture product even allows access to libraries through the internet.

ALTERA and MENTOR do not make these tools available to the end user.

Parameterizing

The original Berkley version of SPICE had already provided the keyword PARAM, which allows the value of device parameters such as size, resistance, or capacitance to be controlled. This mechanism has been carried over into the graphical editor of PSPICE. Parameters may be given global assignments which are evaluated at each instance of a graphical object.

For example, instead of assigning a fixed value to a resistor an expression may be used which contains a symbolic name of a parameter. Two resistors may have the assignment {3 * Rbase} and {12 * Rbase}. The global parameter Rbase may have a value of 10 k leading to final values of 30 k and 120 k for those two resistors.

In a similar way MENTOR provides passing of parameters in its Falcon Framework programs by using the scripting language AMPLE. ALTERA uses such a mechanism only for passing information down to text files, such as AHDL (ALTERA High level Design Language), VHDL or Verilog but not to graphical elements.

Scripting Languages

On the basis of the AMPLE language MENTOR allows access to almost any part of the design data base. AMPLE scripts may be used to configure programs and data files. In a similar way AMPLE is also used to automate the generation of symbols, simulation stimuli or layout elements. For the future a C based language has been announced. Starting with versions PSPICE V9, ORCAD wants to offer a scripting language which is similar to Visual Basic.

Engineering Change Order (ECO)

The ECO method provides a mechanism to follow up design changes from the schematic to other consecutive programs in the design flow. During PCB design the Forward ECO method is well established. Incremental changes in a schematic have to be marked in the final layout. Small changes may still be done by hand without re-running the time consuming complete Place&Route procedures.

The Backward ECO method works in a similar way. Changes to the connectivity, to names or devices on the layout level are fed back into the schematics. Thus a consistent set of design data can be kept through all design steps even if the changes happen with long intervals of time between them.

Stroke

The MENTOR design systems provide a method of assisting the user when entering commands. Instead of using key strokes or menu clicks, mouse movements are interpreted as commands. While pressing the middle mouse button, the total mouse movement is evaluated by dividing the area covered into nine sections. The relative sequence of movement is then interpreted as a command. Figure 3.19 shows the sequence of the most important commands.

Machine-generated Schematics

When using synthesis tools a behavioral, an RTL (Register Transfer Level) or a structural text description is optimized and mapped to the library

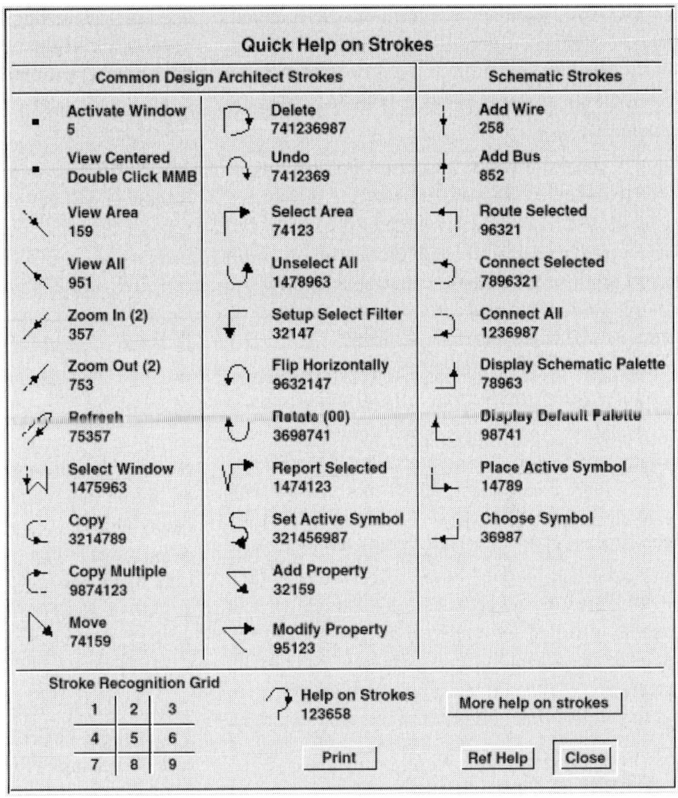

Fig. 3.19 MENTOR's stroke method for entering commands by mouse movements

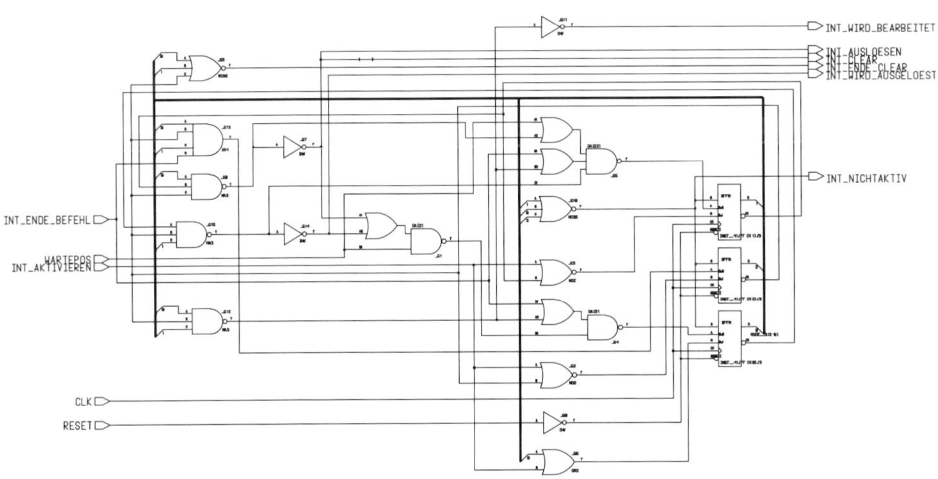

Fig. 3.20 Machine-generated circuit schematic of a synthesized VHDL description (interrupt circuit)

elements of an FPGA or an ASIC library. In any of those cases the result is a form of ASCII netlist. Under certain conditions it may be desirable to visually check the resulting logic. For that purpose, some design systems provide built in schematic generators.

Such a generator needs access to the symbols of exactly those library elements that are referenced in the netlist. The logical content of the library has to be reproduced exactly. In addition, the generator has to produce the topological arrangement of the circuit. Primarily a placement of non-overlapping symbols and wires has to be achieved.

A manually drawn schematic may use the neighboring relationship of symbols in order to increase the readability for humans. Such a characteristic is usually not found in generated schematics. They may be logically correct but are not always well readable. Figure 3.20 shows an example from the MENTOR system. A VHDL description has been optimized and mapped to the cells of a MIETEC ASIC library by a synthesis run. The reader can easily see which logic elements have been used, such as gates and flipflops. He also obtains an impression of the number of wires and buses used, but the logical function is not necessarily apparent.

3.3 Netlist Generation

Netlists establish the connection between the graphical or textual design entry and the subsequent design steps, such as simulation and implementation. Before generating a netlist, the entry has to be checked for errors and inconsistencies. In an ERC (Electrical Rules Check) run, for example, multiple instance names, unconnected inputs or shorted outputs may be detected.

The structure of different types of netlists is described in chapter 8. Normally a flat netlist is generated. In hierarchical netlists the hierarchy information has to be preserved for device names and node names by the use of path structures.

Any digital or analog circuit gets translated into a SPICE netlist in the PSPICE system. It is a device oriented format with positional signal assignments (see section 8.1). The SPICE netlist contains control information in addition to the circuit information. Model definitions for simulation can be found directly in the final SPICE list. Such a combined control list may of course also be written by hand and fed into the simulator directly. In that sense the graphical entry is one way of generating a structural netlist. Examples, partly with parameters and control statements, can be found in lists 3.3 and 3.5 of section 3.4.

The PSPICE system also supports the design of printed circuit boards. For that purpose a different type of flat, non-hierarchical netlist is used. The format uses a qualifier .nfl and is basically node oriented. An example is shown in list 3.4. This format also has to carry the socket names and foot prints to be used for the drilling mask.

The ALTERA system feeds any design entry to the synthesis tool. As a result the user sees a netlist using the syntax and semantics of ALTERA's proprietary design language AHDL. In theory such an output may be used directly as an input again. Nevertheless, it is a structural list using only the available primitives of the target CPLD. Basically only the following elements are used:

- INPUT, OUTPUT, NODE;
- NOT, AND, OR, EXOR;
- FLIPFLOP.

The AND/OR terms do not directly represent single logic devices but they get mapped to the PAL (Programmable Array Logic) or SRAM based logic cells in the ALTERA CPLDs.

An example is shown in list 3.1 (see section 3.4) where the following short forms have been used:

!	NOT
$	EXOR
&	AND
#	OR
LCELL	Logic Cell
EXP	Expander Cell

LCELL and EXP do not represent logical functions. They control the placement in the final device. The double hyphen '–' is used to separate comments.

The MENTOR system generates the netlists during the 'Viewpoint Generation' design step. The netlists are an integral part of the Framework and are not meant to be accessible by the user. For a gate level simulation with a VHDL simulator a

Table 3.1 Examples for schematic entry

Example	Design Style	Implementation	System
4 bit Adder	Primitives, Hierarchy, Macros, Digital Simulation Variation: LPM-Function (Macro) Variation: AHDL Behavioral Description	ALTERA FPGA/CPLD	ALTERA
Transistor Amplifier	Discrete Devices, Flat Netlist, Analog Simulation	PCB Layout	PSPICE
Adder Cell	Flat MOSFET Circuit, Analog Simulation for Characterization	Full Custom Layout, LVS	PSPICE
Shift Register	Hierarchical Standard Cell Design, Digital Simulation with VHDL	Standard Cell IC	MENTOR

truly structural, hierarchical netlist description is created. In this case it does not matter if the source was a schematic with VHDL views behind the symbols or a synthesis run from a RTL description. Lists 3.6a to 3.6d in section 3.4.4 shows such a VHDL netlist.

3.4 Examples of Schematic Entry

This section describes examples of various design styles on different design systems. The examples demonstrate the concept of design entry introduced in the previous sections. The symbols described in Appendix A may be used to translate between IEEE and IEC symbol icons. A list of the different types of examples is given in table 3.1.

3.4.1 Example of an FPGA/CPLD Design

Design with Logic Symbols

This example describes a hierarchical circuit of a 4 bit adder on the ALTERA design system (see fig. 3.21). It only consists of primitive symbols XOR and AND2 at the lowest level. This circuit is represented by a block symbol 'half_add'. Two 'half_add' block symbols and an OR2 primitive are placed on the second level. It leads to the next level 'full_add'. On the top level this 'full_add' symbol is placed four times and the input and output symbols define which nodes are to be connected externally through the device pins.

At the time of compilation, the ALTERA system passes pin names to the Waveform Editor where stimuli for a simulation have to be supplied to the inputs. The sequence of output signals gets displayed after the simulation run.

The following details should be noted:

- The buses A_i, B_i and Sum_i consist of 4 bits each. The 'Connect-by-Name' mechanism produces the necessary netlist;
- The supply pins VDD and GND are not drawn. They get implicitly generated and connected to the corresponding pins of the FPGA/CPLD package;
- The three graphic files are stored as .gdf files. The top level file name 'add4_log' is also the project name of the design;
- The netlist of this circuit has an internal binary format which will be used by the synthesis program of the compiler. The result is an optimized structural list in ALTERA's AHDL format. It references only primitives that can be implemented in ALTERA chips (see list 3.1).

Variation: LPM Function (Macro)

A 4 bit adder may also be created by using a predefined LPM (Library of Parameterized Modules) function. A parameterized AHDL description can be found behind the symbol LPM_ADD_SUB. The parameters are set in such a way that the same function as 'add4_log' results (see fig. 3.22). The AHDL macro text gets converted into a structural AHDL netlist in a similar way as the symbolic entry before. Because the adder is now processed based on the adder function of the macro, the circuit may become 30 % smaller but also a bit slower. The I/O pins and the simulation stimuli are the same as for 'add4_log'

Variation: AHDL Behavioral Description

The language AHDL also allows behavioral constructs, for example the addition in the form of a

```
- Node name is '|full_add:18|half_add:2|~1~1' = '|full_add:18|half_add:2|SUM~1'
 _LC115 = LCELL( _EQ009 $ GND);
 _EQ009 = A0 & A1 & A2 & CIN
        # A1 & A2 & B0 & CIN
        # A0 & A1 & A2 & B0
        # A0 & A2 & B0 & B1
        # A2 & B0 & B1 & CIN;

- Node name is '|full_add:18|half_add:2|~1~2' = '|full_add:18|half_add:2|SUM~2'
 _LC113 = LCELL( _EQ010 $ GND);
 _EQ010 = A0 & A2 & B1 & CIN
        # A0 & B1 & B2 & CIN
        # B0 & B1 & B2 & CIN
        # A0 & B0 & B1 & B2
        # A0 & A1 & B0 & B2;

- Node name is '|full_add:18|half_add:2|~1~3' = '|full_add:18|half_add:2|SUM~3'
 _LC114 = LCELL( _EQ011 $ GND);
 _EQ011 = A1 & B0 & B2 & CIN
        # A0 & A1 & B2 & CIN
        # A1 & A2 & B1
        # A1 & B1 & B2;
END;
```

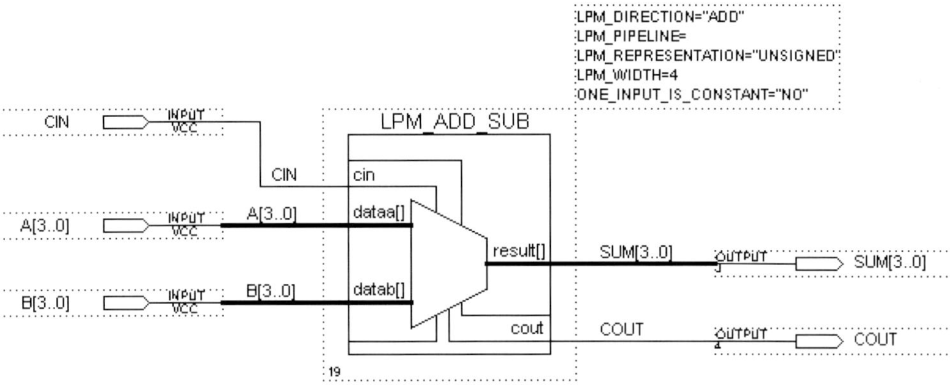

Fig. 3.22 The adder circuit implemented with an LPM Macro LPM_ADD_SUB

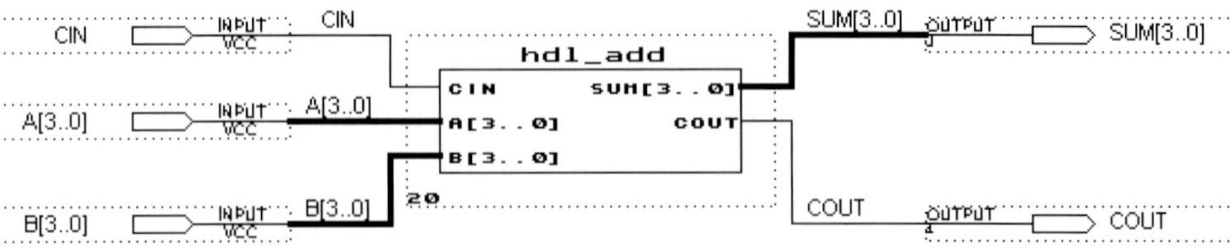

Fig. 3.23 The adder circuit with a behavioral description behind the block symbol

List 3.2 AHDL description of the 4 bit adder

```
SUBDESIGN hdl_add
(  CIN, A[3..0], B[3..0] :INPUT;
   SUM[3..0], COUT       :OUTPUT; )

BEGIN
   (COUT, SUM[]) = (0, 0, 0, 0, CIN) + (0, A[]) + (0, B[]);
END;
```

3.4.2 Example of a Printed Circuit Board Design

A relatively simple one-transistor amplifier stage has been chosen for this example. The DC operation point is stabilized with the resistor RE and still has the full AC gain owing to the capacitor CE (see fig. 3.24a). Using the PSPICE schematic editor, a core circuit can be drawn to be used for analog simulation as well as for a printed circuit board layout. For that purpose, a block symbol 'tr_amp' is created and stored in a symbol library. The block symbol of fig. 3.24 has been edited for its final appearance.

Analog Simulation

In order to verify the function of a circuit with an AC or transient analysis the following elements have to be placed around the core circuit (fig. 3.24b):

- voltage supply (VCC and GND);
- signal source (including source resistance); and
- load impedance (only resistive in this case).

These elements should not appear on the PCB since they become connected through the interface plug (fig. 3.24c). The marker symbols in the core schematic only serve to make signals visible in simulation. They are ignored for the implementation. List 3.3 displays the corresponding netlist for the analog simulation. All circuit elements have been automatically extended by a prefix (see also chapter 8). The supply voltage VCC thus becomes V_VCC.

The node names of automatically named signals such as '$N_0002' can easily be recognized. The netlist has become flat, although the hierarchical path information is still present: the base node of transistor BC550 is now called HS1_VB. The name of its instance, 'HS1', has been added in front of the node. The device values have been entered directly (not parameterized). The resistor R1 has a value of 120 kOhm, for example. The signal generator Vgen is a sine wave source with an amplitude of 20 mV at 20 kHz. An AC simulation reveals a gain of 45 dB in a frequency range from 50 Hz to 10 MHz.

List 3.3 Netlist in (P)SPICE format

```
* Schematics Netlist *
V_VCC          $N_0002 0 DC 15V
R_HS1_RE       0 HS1_UE 220
R_HS1_R1       HS1_UB $N_0002 120k
R_HS1_RC       HS1_UC $N_0002 1.5k
R_HS1_R2       0 HS1_UB 15k
Q_HS1_Q1       HS1_UC HS1_UB HS1_UE BC550C
C_HS1_Caus     HS1_UC Uaus 1uF
C_HS1_CE       0 HS1_UE 470uF
C_HS1_Cein     Uein HS1_UB 68uF
R_Rload        0 Uaus 25k
R_Rgen         $N_0001 Uein 50
V_Vgen         $N_0001 0 AC 20mV
               SIN 0 20mV 20kHz 0 0 0
```

PCB Layout

The core circuit only consists of discrete devices, each with a specific foot print for the PCB layout. The pin numbers, as they are visible in the schematic, represent those of the foot print. The top level PCB diagram (fig. 3.24c) contains the additional connector so that the PCB may be supplied. A special PCB netlist has to be generated for the subsequent Place&Route program.

List 3.4 shows the corresponding PCB netlist for the program PCBOARDS. It is a node oriented format (see chapter 8). All capacitors use the same foot print CK05, which of course is not generally the case. The resistors use a foot print RC05. Similar formats exist for many other PCB programs.

List 3.4 Amplifier netlist for PCBOARDS

```
*PCBoards Netlist Version 8.0: Format 1.0
*timestamp:  netlist 920302938
*component CONN5 DIN5 Bu1
*component C CK05 Caus
@attribute VALUE=1uF
```

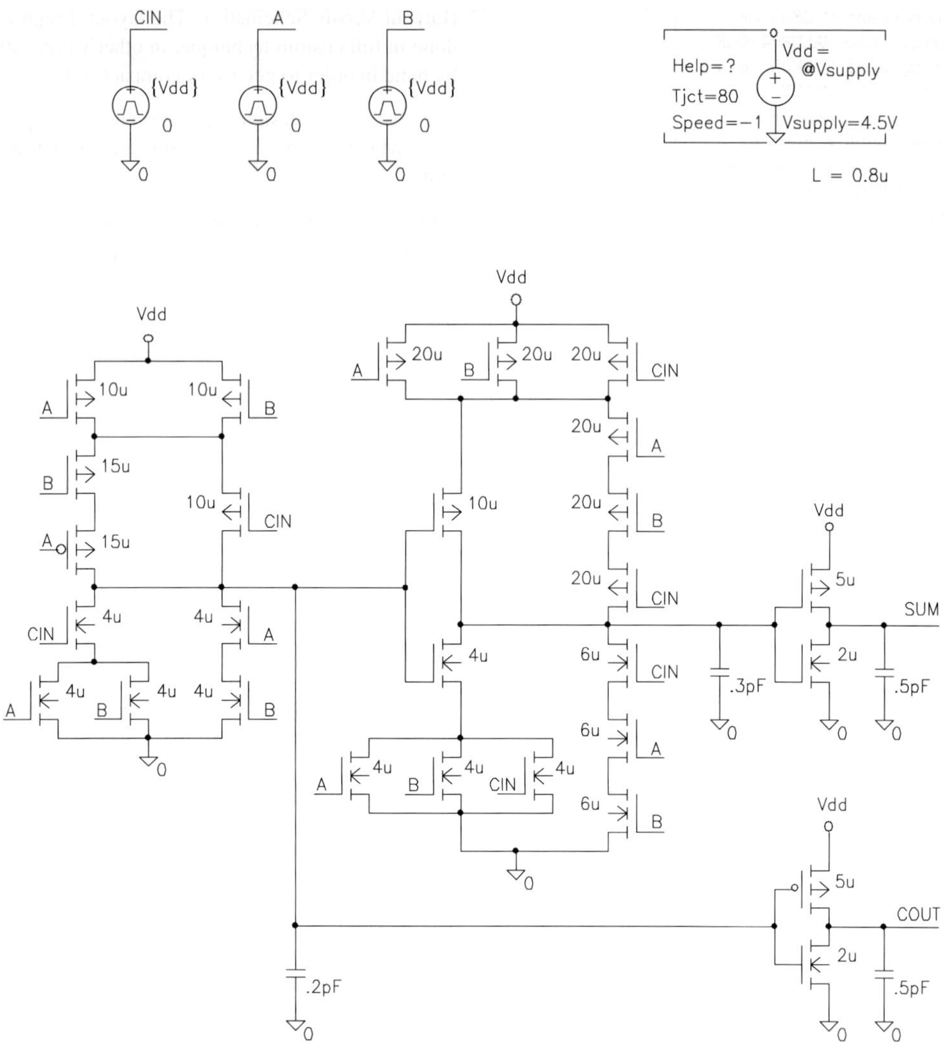

Fig. 3.25 Schematic for characterization of a 1-bit CMOS adder cell

- A parameter 'speed' to vary the strength of the CMOS transistors through their model parameters. The range of values for speed is: $-1 = slow$, $0 = typical$ and $+1 = fast$.

In the example the values have been set to 80 °C, 4.5 V and −1. Thus, the longest delay times will result.

List 3.5 shows a part of the generated PSPICE netlist. It is important to note that a parameter Vdd exists (4.5 V) and a node Vdd.

For each transistor type a 'Level 3' model with the model names 'NMOS1u' and 'PMOS1u' is used. In those models the parameter 'speed' is also used to introduce the production tolerances by varying model parameters, such as K0 and VTH.

List 3.5 Partial PSPICE Netlist of the 1-bit adder in CMOS transistor logic

```
.PARAM        Speed=-1
.PARAM        Vdd=4.5V
.TEMP         80
.tran 20ns 300ns
.OP

V_V3          Vdd 0 4.5V
V_V4          CIN 0 PULSE 0 {Vdd} 20ns 2ns 2ns 98ns 200ns 30ns
V_V5          A 0 PULSE 0 {Vdd} 20ns 2ns 2ns 23ns 50ns
V_V6          B 0 PULSE 0 {Vdd} 20ns 2ns 2ns 48ns 100ns

C_C1          0 $N_0001 .2pF
C_C2          0 COUT .5pF
C_C3          0 SUM .5pF
C_C4          0 $N_0003 .3pF
M_M34         $N_0002 A 0 0 NMOS1u L={0.8u-0.1u*speed} W={4u+0.2u*speed}
M_M37         $N_0003 $N_0001 $N_0002 0 NMOS1u L={0.8u-0.1u*speed} +W={4u+0.2u*speed}
M_M38         $N_0003 $N_0001 $N_0004 Vdd PMOS1u L={0.8u-0.1u*speed} +W={10u+0.2u*speed}
M_M39         $N_0004 A Vdd Vdd PMOS1u L={0.8u-0.1u*speed} W={20u+0.2u*speed}
M_M24         $N_0001 A $N_0005 Vdd PMOS1u L={0.8u-0.1u*speed} +W={15u+0.2u*speed}
M_M10         $N_0006 A 0 0 NMOS1u L={0.8u-0.1u*speed} W={4u+0.2u*speed}
M_M12         $N_0001 CIN $N_0006 0 NMOS1u L={0.8u-0.1u*speed} +W={4u+0.2u*speed}
M_M23         $N_0007 A Vdd Vdd PMOS1u L={0.8u-0.1u*speed} W={15u+0.2u*speed}
M_M22         $N_0005 B $N_0007 Vdd PMOS1u L={0.8u-0.1u*speed} +W={15u+0.2u*speed}
M_M11         $N_0006 B 0 0 NMOS1u L={0.8u-0.1u*speed} W={4u+0.2u*speed}
M_M13         $N_0008 B 0 0 NMOS1u L={0.8u-0.1u*speed} W={4u+0.2u*speed}
M_M21         $N_0001 CIN $N_0007 Vdd PMOS1u L={0.8u-0.1u*speed} +W={10u+0.2u*speed}
M_M20         $N_0007 B Vdd Vdd PMOS1u L={0.8u-0.1u*speed} W={10u+0.2u*speed}
M_M14         $N_0001 A $N_0008 0 NMOS1u L={0.8u-0.1u*speed} W={4u+0.2u*speed}
...
```

3.4.4 Example of a Standard Cell IC Design

This example shows the shift register schematic for an IC design with standard cells supplied by the semiconductor manufacturer MIETEC [3.6]. The top level 'shift4' contains all the necessary I/O and supply pad symbols next to the block symbols 'shift1'. The basic schematic of 'shift1' contains a flipflop FD2 and a multiplexer MUX21 for switching from serial to parallel data entry. The schematic of fig. 3.26 can be used in this form for logic simulation as well as for the IC Place&Route program.

Since a VHDL simulator is used for logic simulation (in this case, ModelSim), the schematic gets transferred into a structural VHDL netlist. List 3.6a shows the Entities and the Architectures of this hierarchical design.

List 3.6a Entity of the VHDL netlist of the basic circuit 'shift1' corresponding to fig. 3.26

```
----------------------------------------
-- VHDL object: Entity "shift1" (component
   interface "$MY_HOME/shift1:shift1")
-- Generated on: Fri Mar 19 14:13:36 1999
-- Generated by: student1
-- Source from:  $MY_HOME/shift1/part
-- Program:      VHDLwrite  v8.6_1.3  Sun
   Nov 30 23:24:17 PST 1997
----------------------------------------
-- LIBRARY STATEMENT
LIBRARY ieee;
-- PACKAGE STATEMENT
USE ieee.std_logic_1164.ALL;

entity shift1 is
   -- GENERIC LIST
   -- PORT LIST
   port(
      Clk : in std_logic;
      D_par : in std_logic;
      D_ser : in std_logic;
```

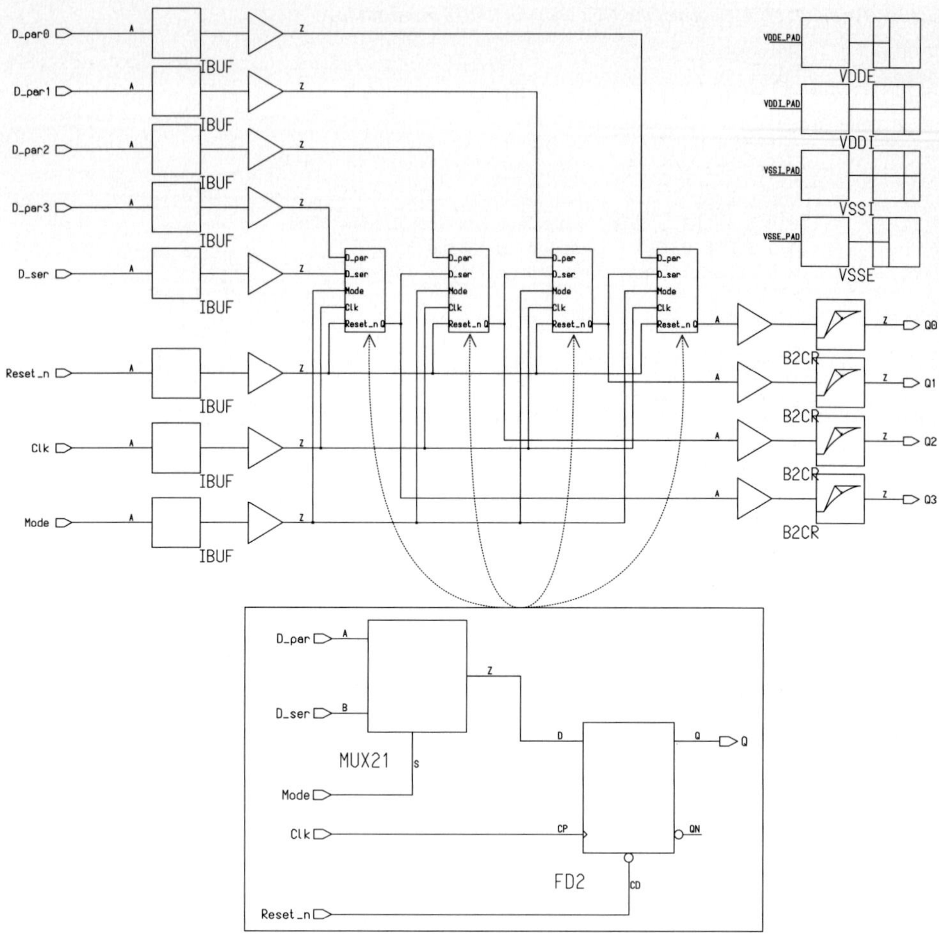

Fig. 3.26 Schematic of a shift register with symbols for an IC standard cell design

```
      Mode : in std_logic;
      Reset_n : in std_logic;
      Q : out std_logic
   );
   attribute original_name : string;
end shift1;
```

List 3.6b Architecture of the VHDL netlist of basic circuit 'shift1'

```
------------------------------------------
-- VHDL object:  Architecture "structure"
   (schematic "$MY_HOME/shift1/schematic")
   of entity "shift1" (component interface
   "$MY_HOME/shift1:shift1")
-- Generated on: Fri Mar 19 14:13:36 1999
-- Generated by: student1
-- Source from:  $MY_HOME/shift1/schematic
-- Program:      VHDLwrite  v8.6_1.3  Sun
   Nov 30 23:24:17 PST 1997
------------------------------------------
library MTC35000;
library ieee;
use ieee.std_logic_1164.all;
architecture structure of shift1 is
   -- TYPE DECLARATIONS
   -- SIGNAL DECLARATIONS
      signal local_Clk : std_logic;

      signal local_D_par : std_logic;
      signal local_D_ser : std_logic;
      signal local_Mode : std_logic;
      signal local_Q : std_logic;
      signal local_Reset_n : std_logic;
```

```
      signal N_0364 : std_logic;
      signal N_0368 : std_logic;
   -- COMPONENT_DECLARATIONS
      component fd2
         port(
            CD : in std_logic;
            CP : in std_logic;
            D : in std_logic;
            Q : out std_logic;
            QN : out std_logic
         );
      end component;
```

List 3.6c Entity of the VHDL netlist of the top level 'shift4'

```
-----------------------------------------
-- VHDL object: Entity "shift4" (component
   interface "$MY_HOME/shift4:shift4")
-- Generated on: Fri Mar 19 14:13:36 1999
-- Generated by: student1
-- Source from:  $MY_HOME/shift4/part
-- Program:      VHDLwrite  v8.6_1.3  Sun
   Nov 30 23:24:17 PST 1997
-----------------------------------------
-- LIBRARY STATEMENT
LIBRARY ieee;
-- PACKAGE STATEMENT
USE ieee.std_logic_1164.ALL;
entity shift4 is
   -- GENERIC LIST
   -- PORT LIST

   port(
      Clk : inout std_logic;
      D_par0 : inout std_logic;
      D_par1 : inout std_logic;
      D_par2 : inout std_logic;
      D_par3 : inout std_logic;
      D_ser : inout std_logic;
      Mode : inout std_logic;
      Q0 : inout std_logic;
      Q1 : inout std_logic;
      Q2 : inout std_logic;
      Q3 : inout std_logic;
      Reset_n : inout std_logic
   );
   attribute original_name : string;
end shift4;
```

List 3.6d Architecture of the VHDL netlist of the top level 'shift4'

```
-----------------------------------------
-- VHDL object:  Architecture "structure"
(schematic "$MY_HOME/shift4/schematic")
of entity "shift4" (component interface
"$MY_HOME/shift4:shift4")
-- Generated on: Fri Mar 19 14:13:36 1999
-- Generated by: student1
-- Source from:  $MY_HOME/shift4/schematic
-- Program:      VHDLwrite  v8.6_1.3  Sun
   Nov 30 23:24:17 PST 1997
-----------------------------------------
library MTC35400;
library MTC35400_ETC;
library ieee;
library work;
use ieee.std_logic_1164.ALL;
architecture structure of shift4 is
   -- TYPE DECLARATIONS
   -- SIGNAL DECLARATIONS
      signal N_0361 : std_logic;
      signal N_0362 : std_logic;
      signal N_036205 : std_logic;
      signal N_036206 : std_logic;
      signal N_036207 : std_logic;
      signal N_036208 : std_logic;
      signal N_036209 : std_logic;
      signal N_036213 : std_logic;
      signal N_036216 : std_logic;
      signal N_036218 : std_logic;
      signal N_036221 : std_logic;
      signal N_036222 : std_logic;
      signal N_036223 : std_logic;
      signal N_036224 : std_logic;
      signal N_036225 : std_logic;
      signal N_036226 : std_logic;
   -- COMPONENT_DECLARATIONS
      component b2cr
         port(
            A : in std_logic;
            Z : out std_logic
         );
      end component;
      component ibuf
         port(
            A : in std_logic;
            Z : out std_logic
         );
      end component;
      component shift1
         port(
            Clk : in std_logic;
            D_par : in std_logic;
            D_ser : in std_logic;
            Mode : in std_logic;
            Reset_n : in std_logic;
            Q : out std_logic
         );
      end component;
      component vdde
         port(
            VDDE_PAD : out std_logic
         );
      end component;
```

3

```
    component vddi
      port(
        VDDI_PAD : out std_logic
      );
    end component;
    component vsse
      port(
        VSSE_PAD : out std_logic
      );
    end component;
    component vssi
      port(
        VSSI_PAD : out std_logic
      );
    end component;
 -- INLINE CONFIGURATIONS
    for I_0361 : shift1 use entity
    work.shift1(structure);
    for I_03610 : ibuf use entity
MTC35400.ibuf(behavioral);
    for I_03611 : ibuf use entity
MTC35400.ibuf(behavioral);
    for I_03612 : ibuf use entity
MTC35400.ibuf(behavioral);
    for I_03613 : ibuf use entity
MTC35400.ibuf(behavioral);
    for I_03614 : ibuf use entity
MTC35400.ibuf(behavioral);
    for I_03615 : ibuf use entity
MTC35400.ibuf(behavioral);
    for I_03616 : ibuf use entity
MTC35400.ibuf(behavioral);
    for I_03617 : b2cr use entity
MTC35400.b2cr(behavioral);
    for I_03618 : b2cr use entity
MTC35400.b2cr(behavioral);
    for I_03619 : b2cr use entity
MTC35400.b2cr(behavioral);
    for I_0362 : shift1 use entity
work.shift1(structure);
    for I_03620 : b2cr use entity
MTC35400.b2cr(behavioral);
    for I_0363 : shift1 use entity
work.shift1(structure);
    for I_03633 : vdde use entity
MTC35400_ETC.vdde(behavioral);
    for I_03634 : vddi use entity
MTC35400_ETC.vddi(behavioral);
    for I_03635 : vssi use entity
MTC35400_ETC.vssi(behavioral);
    for I_03636 : vsse use entity
MTC35400_ETC.vsse(behavioral);
    for I_0364 : shift1 use entity
work.shift1(structure);
    for I_0369 : ibuf use entity
MTC35400.ibuf(behavioral);

 begin

-- COMPONENT INSTANTIATIONS
  I_0361 : shift1
    port map(
      Clk => N_036213,
      D_par => N_036218,
      D_ser => N_036207,
      Mode => N_036208,
      Q => N_036222,
      Reset_n => N_036205
    );
  I_03610 : ibuf
    port map(
      A => Reset_n,
      Z => N_036205
    );
  I_03611 : ibuf
    port map(
      A => Clk,
      Z => N_036213
    );
  I_03612 : ibuf
    port map(
      A => Mode,
      Z => N_036208
    );
  I_03613 : ibuf
    port map(
      A => D_par3,
      Z => N_036216
    );
  I_03614 : ibuf
    port map(
      A => D_par2,
      Z => N_036218
    );
  I_03615 : ibuf
    port map(
      A => D_par1,
      Z => N_036209
    );
  I_03616 : ibuf
    port map(
      A => D_par0,
      Z => N_036221
    );
  I_03617 : b2cr
    port map(
      A => N_0362,
      Z => Q0
    );
  I_03618 : b2cr
    port map(
      A => N_0361,
      Z => Q1
    );
  I_03619 : b2cr
    port map(
      A => N_036222,
```

```
      Z => Q2
   );
I_0362 : shift1
   port map(
      Clk => N_036213,
      D_par => N_036216,
      D_ser => N_036206,
      Mode => N_036208,
      Q => N_036207,
      Reset_n => N_036205
   );
I_03620 : b2cr
   port map(
      A => N_036207,
      Z => Q3
   );
I_0363 : shift1
   port map(
      Clk => N_036213,
      D_par => N_036209,
      D_ser => N_036222,
      Mode => N_036208,
      Q => N_0361,
      Reset_n => N_036205
   );
I_03633 : vdde
   port map(
      VDDE_PAD => N_036223
   );
I_03634 : vddi
   port map(
      VDDI_PAD => N_036224
   );
I_03635 : vssi
   port map(
      VSSI_PAD => N_036225
   );
I_03636 : vsse
   port map(
      VSSE_PAD => N_036226
   );
I_0364 : shift1
   port map(
      Clk => N_036213,
      D_par => N_036221,
      D_ser => N_0361,
      Mode => N_036208,
      Q => N_0362,
      Reset_n => N_036205
   );
I_0369 : ibuf
   port map(
      A => D_ser,
      Z => N_036206
   );
end structure;
```

3.5 References

[3.1] 'MAX+PLUS II, Programmable Logic Development System, Getting Started'. – San Jose: ALTERA Corporation, 1997

[3.2] 'MAX+PLUS II, Programmable Logic Development System, AHDL'. – San Jose: ALTERA Corporation, 1997

[3.3] 'Design Architect User's Manual'. – Wilsonville: Mentor Graphics Corporation, 1998

[3.4] 'MicroSim Schematics, User's Guide'. – Irvine: MicroSim, 1997

[3.5] *Smith, M. J. S.*: 'Application Specific Integrated Circuits'. Addison-Wesley, 1997

[3.6] 'Standard Cell Library 0.5 μm'. MIETEC MTC 35, EUROPRACTICE, 1998

4 Design using Standard Description Languages

FRIEDEMANN STOCKMAYER, HANS KREUTZER

4.1 Introduction

VHDL as a hardware description language is becoming even more popular in the field of the development of digital hardware and it offers a wide range of possibilities for dealing with ever increasing requirements. The following sections, along with several examples, give a basis for the practical interpretation and present an approach to problem solving for frequently asked questions. The sections follow on systematically, bringing together the different aspects to give an overall picture. The aim therefore is not only to present a complete syntactical language description, but to emphasize the important elements used in the practical implementation of a synthesizable design.

Points of main topics:
- Structure of VHDL designs;
- Concurrent und sequential statements; application in synchronous designs;
- Structure of a simulation environment.

4.1.1 The Foundation of VHDL

VHDL is a **description language for digital circuits**. It was developed by an initiative within the VHSIC programme (*Very **High** Speed **I**ntegrated Circuits*) of the American Ministry of Defence. The driving force was the difficult situation that suppliers used different description languages for the development of their systems. This made it almost impossible to exchange designs or to be able to work together on larger projects. Also there was no guarantee that the individual languages would last the lifetime of the systems.

The solution was a standard language with a guaranteed future. At the beginning of the 80s the first versions of this language were mentioned, which gave it the name VHDL (**V**HSIC **H**ardware **D**escription **L**anguage). Following on from that the Institute of Electrical and Electronics Engineers (IEEE) developed this as a standard language which was ratified in 1987 with the title IEEE 1076 (VHDL'87). This opened the way for the development of the next generation of integrated circuits.

In comparison with the classical method of design at the gate level, the new method is more suitable for larger and highly complex designs. Above all, the various possibilities of the early detection of conceptional and logical error during the simulation increases the quality of the design and reduces the period of development.

VHDL is mainly a **technology independent description**, therefore it is possible to create designs for a chosen technology using synthesizing tools. This characteristic is a requirement which allows one to keep up with the fast development of semiconductor technology. Not only is the aspect of the shifting of existing designs important, but also the re-use of already tested and proved components. The **standardization of VHDL** is also not completed. Owing to the increasing usage of this language it is necessary to make regular extensions and amendments. The last extension took place in 1993 (VHDL'93). This status is recorded in the IEEE standard 'VHDL Language Reference Manual' [4.3] and is also the basis for this introduction to the description language VHDL.

4.1.2 VHDL Design Cycle

VHDL supports all **phases of circuit development**.

Typical steps are:
- System modelling (specification);
- Description on the register transfer level (design);
- Netlist (Implementation).

Because of its **high versatility**, access to the language appears to be complicated. First steps seem to be very difficult. However, VHDL is a multi-layered language. It offers distinctive features for each of the three phases mentioned above. For a practical application it is only necessary to use a subset of the entire vocabulary. This makes it easy to keep track of things, and simplifies learning.

The Phase of Specification

The system model contains algorithms regardless of the possibility of the model being able to be synthezised. Using a formal specification a first model serves to investigate the system's response and hence its functionality. This allows the orderer and customer to find out if the problem has been understood and if all requirements have been met. The main effort focuses on algorithms which describe the principle behavior of a circuit.

The Phase of Design

During the next step the formal specification is converted to a description which can be synthesized. To meet requirements, e.g., chip area, gate count, and clock frequency, the circuit developer has the difficult task of mapping program language statements (IF, ELSIF, CASE, …) in an optimal way to the register transfer level. In detail this describes the type of combinatorial logic and the way signals are passed from one flip flop (register) to the next. Figure 4.1 shows the structure of a fully synchronous design. In this phase the primary interest is focused on the way in which the circuit interacts with other parts of the design.

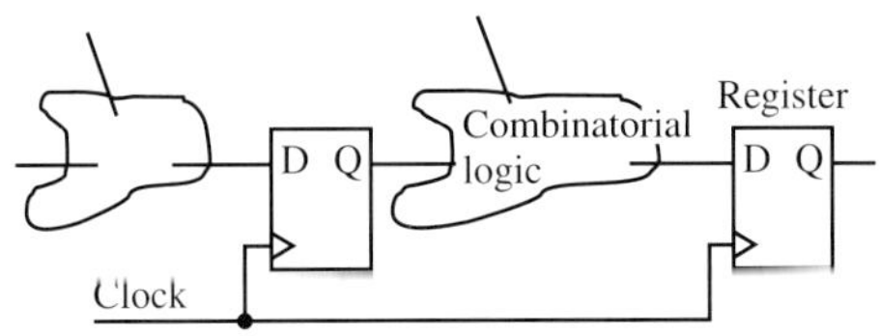

Fig. 4.1 Circuit description on a register transfer level

The **description can be in text or in schematics** such as block diagrams, flow charts, state diagrams, and truth tables, followed by an automatic VHDL code generation. Figure 4.2 shows a typical example.

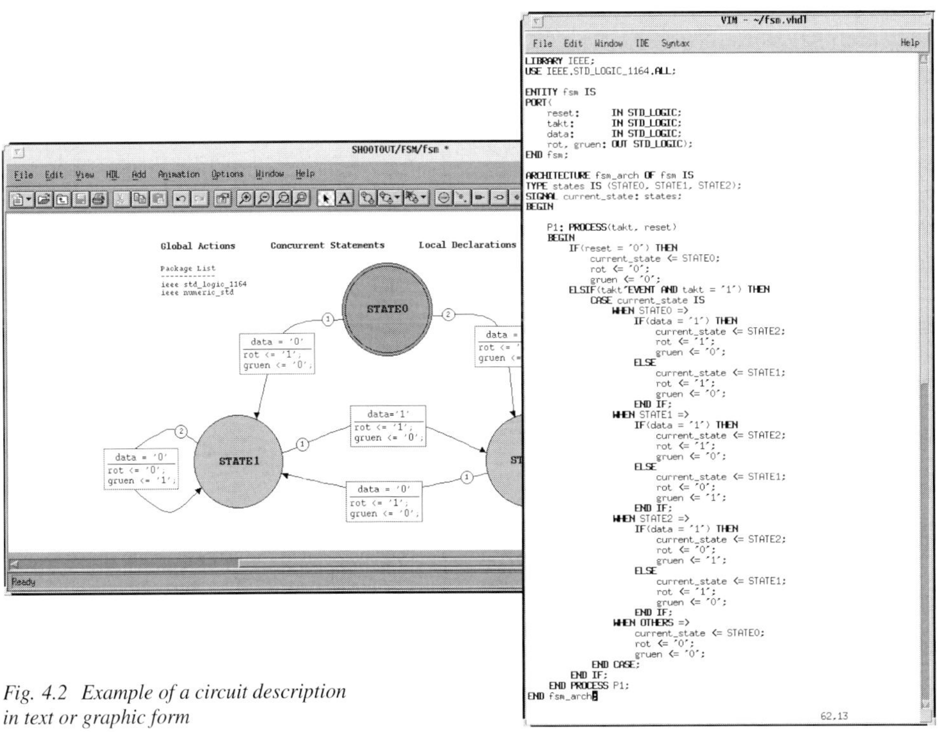

Fig. 4.2 Example of a circuit description in text or graphic form

One of the big advantages of VHDL is the possibility of designing a test environment, which is called a test bench. The **test bench** allows the stimulation of the inputs of the circuit under test with realistic signal patterns. Simulation shows whether the outputs produce the expected responses. Verification of the simulation results can be done by the test bench itself. Figure 4.3 shows the simulation results of a circuit model which enters all states and performs all state transitions once. Simulation and verification of results take a huge amount of time in a design cycle. It is necessary to make sure that there are no functional errors, which means that the design is ready for the next step: the synthesis.

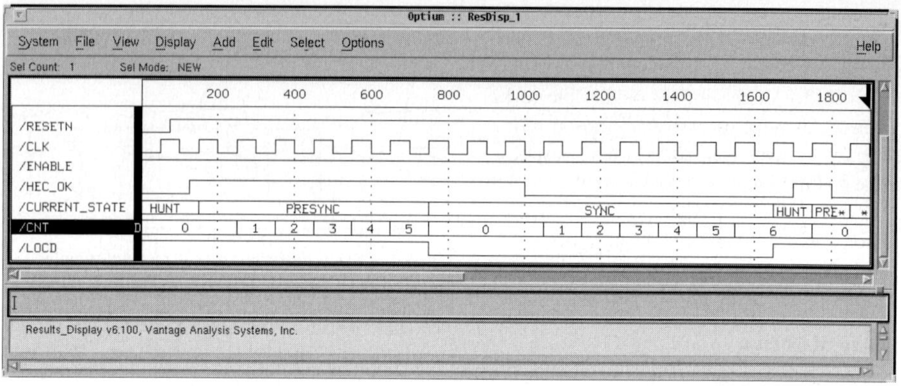

Fig. 4.3 Result of a VHDL simulation

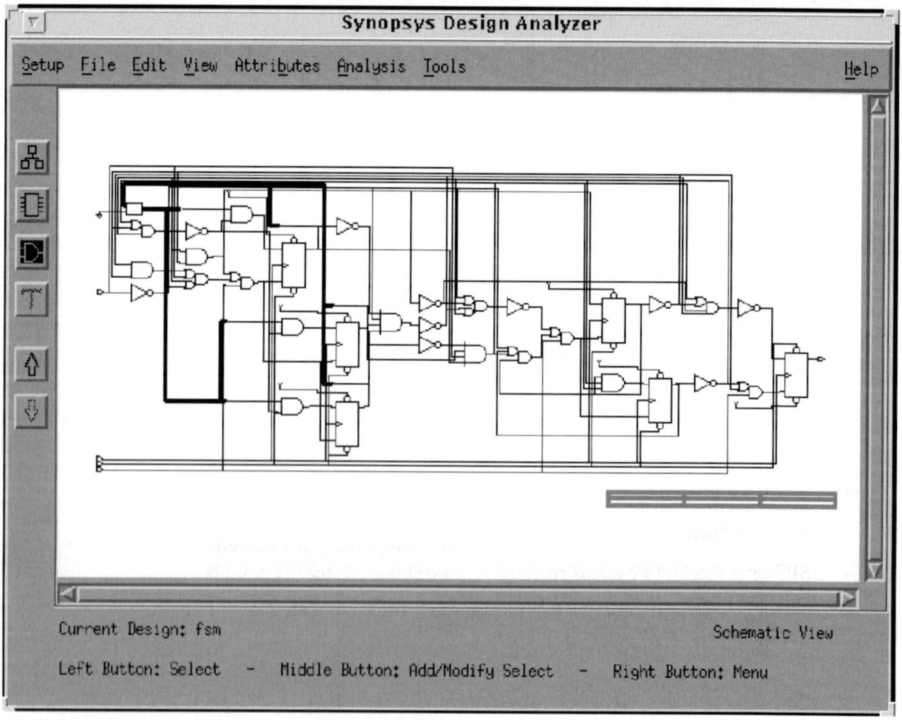

Fig. 4.4 The schematic of the synthesized circuit

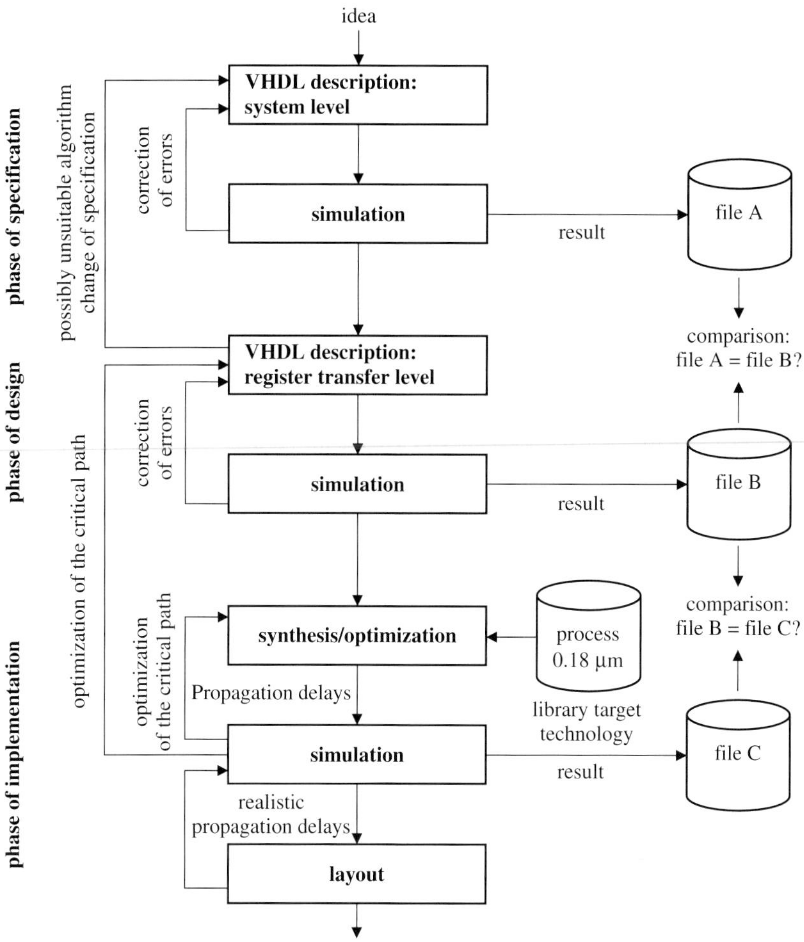

Fig. 4.5 Design flow of a circuit with VHDL

At times a critical path can appear after synthesis. Sometimes it can only be removed by a change of the VHDL description. In the worst case this could imply a change of specification. As a consequence there could be several iteration loops on the way to a useable netlist (fig. 4.5).

The Phase of Implementation

Is it to be an ASIC or a FPGA? Physical realization only depends on the library attached during synthesis. Assuming the synthesis tool has access to all relevant data, the generated circuit is a true picture of its textual description made in the first design step. However, it is very difficult to understand the functionality of all the gates in detail. Requirements regarding area and speed are met by the tool in an optimal way. A simulation of the synthesized circuit at the gate level, compared with the simulation results of the design phase, must produce identical results.

The first two phases (specification and design) can be performed in one step if it is a small and straightforward design. That means that the modelling of the system is carried out using a description which can be synthesized at the register transfer level. Synthesizing the design, normally carried out in the third phase, runs automatically. Thus describing a circuit at the register transfer level using VHDL will be the main issue and the main point of interest in this chapter.

4.2 Structure of a VHDL Design

This chapter deals with some basic **aspects of how to structure a VHDL design**:

- Signals and multi-level logic;
- Interface signals and ports;
- Entity; how a module is presented to the outside world;
- Architecture; how to describe the functionality of a module;
- Configuration; how to join parts to a system.

4.2.1 Signals

Signals are very important in VHDL. Similarly to variables in any computer language, they carry data. A signal assignment specifies a logic value of a signal. After a signal assignment it can be tested or be used in other language constructs. Just as does a piece of wire, a signal connects an output and an input of two parts of a circuit (e.g., gate, flip flop, ...). However, after synthesis, not every signal will turn into a piece of connecting wire. During synthesis signals can appear or disappear as a result of optimization.

Basic signals

Prior to their use signals must be declared in a VHDL description.

```
-- Comments start with:  --
-- Signal declaration:
SIGNAL a,b,c:  BIT;
```

Such a **signal declaration** consists of three parts. The first part is the so-called key word (reserved word). In this case **SIGNAL**. The second part assigns a name to individual signals, namely a, b, and c. The third part determines the type of the signals, which is BIT in this case. The type of the signal defines the logic values which the signal can take. A signal of type BIT can show the logic values '0' or '1' (character literal). The assignment of the logic values 0 and 1 is denoted by a character framed by two apostrophes. As already mentioned, individual signals can receive values assigned, also it is possible to combine and evaluate signals in expressions.

```
-- Signal assignment
a <= '1';
-- Signal assignment
a <= b AND c;
-- Signal evaluation
IF(a = '0') THEN ...
```

Bus Signals

Next to individual signals we have signal buses of different width.

```
SIGNAL bus_a:  BIT_VECTOR(7 DOWNTO 0);
SIGNAL bus_b:  BIT_VECTOR(0 TO 7);
```

Both declarations define an eight bit wide signal bus. However they are different in terms of indexing the bus signals, which affects the position of the most significant bit (MSB). As regards bus_a, the MSB is carried by the signal bus_a(7), the signal with index seven. The least significant bit (LSB) is carried by the signal bus_a(0), the signal with index zero. Looking at bus_b, it is the exact opposite.

```
bus_a <= "10000000" -- MSB: bus_a(7)
-- Index:  76543210
bus_b <= "10000000" -- MSB: bus_b(0)
-- Index:  01234567
```

The value of a bus is always the same, independent of indexing. It corresponds with the well known way of reading bus values having the MSB in the first place. The bus value is a string of characters enclosed in quotation marks.

To prevent confusion and remove a possible source of error in the development team, a unique and consistent way of declaring bus signals within the design and its interfaces is highly recommended. This means: all buses are declared using ascending (TO) or descending (DOWNTO) indexing. The DOWNTO version has the advantage that the index of the individual signal corresponds with its equivalent of the binary value.

Multi-value logic

WHEN it comes to defining a bi-directional databus of a processor interface, it is obvious that a two-valued logic with the logic values '0' and '1' is not sufficient to describe the behavior correctly. In a system which contains multiple sources it must be guaranteed that only one source actively drives the data bus. All other sources have to set their bus driver to a high impedance value. If this were not the case bus contention would occur. The worst case would be that a signal would be driven with opposite logic values (0 and 1) at the same time, which causes higher power consumption and an increase in the probability of failure.

The IEEE library contains a so called package which defines a nine-valued logic system (Std_logic_1164, see app. C) as an extension of the Standard but not of the language itself. Including this package into the design and using type definitions std_ulogic or std_logic instead of the type definition BIT makes it possible to describe additional values such as 'Z' for high-impedance or 'X' for bus contention. Section 4.7.2 shows the declaration of std_logic as an ENUMERATOR.

```
TYPE std_ulogic IS(
 'U' -- Uninitialized
 'X' -- Forcing Unknown
 '0' -- Forcing 0
 '1' -- Forcing 1
 'Z' -- High Impedance
 'W' -- Weak Unknown
 'L' -- Weak 0
 'H' -- Weak 1
 '-' -- Don't Care
);
```

'U' stands for 'uninitialized', which means that no logic value has yet been assigned to the signal. The logic values 'L' and 'H' describe soft values, e.g., caused by pull up or pull down resistors. These logic values can be overdriven by a forcing '0' or '1' value. If the logic value can be chosen freely, the character('-') symbolizes a 'don't care' condition which provides additional degrees of freedom during synthesis.

What is the difference between the logic types **std_ulogic** and **std_logic**. In the form: The u in std_logic stands for unresolved. Derived from std_ulogic, std_logic provides a so called resolution function. This function will be automatically called when more than one value should be assigned to a signal at the same time (e.g., fig. 4.6). The resolution function takes the place of a referee and decides which logic value will be assigned to the multi-driven signals. This feature allows one to describe signals having a so called tri-state condition.

Today std_logic is the de facto standard. All CAE tools support the IEEE library. Compatibility of interfaces is an important condition for the re-use of VHDL components or for its purchase to increase synergy.

```
-- Attaching the IEEE Library
LIBRARY IEEE;
USE IEEE.STD_LOGIC_1164.ALL;
...
-- Signal types of the IEEE library
SIGNAL d,e,f:  STD_LOGIC;
SIGNAL bus_c:  STD_LOGIC_VECTOR(15 DOWNTO 0);
```

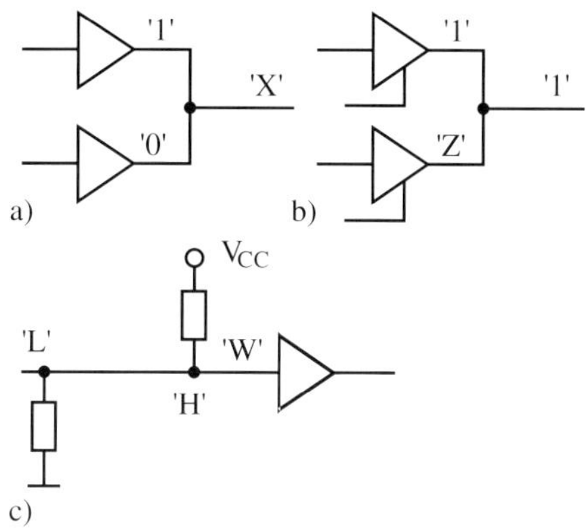

Fig. 4.6 Example of resolving a bus contention

Initialization of Signals

A signal of type std_logic which has not been initialized correctly, or has not been assigned a logic value at the start of the simulation, takes the logic value defined in the type declaration at the first position which is 'U'. Types, representing a range of values (e.g., INTEGER) rather than an ENUMERATION are set to the value at the left boundary. To set up a user defined starting point at the beginning of the simulation, signal declaration allows an explicit initialization of the signal. This possibility must be used with great care, because signal initializations will be ignored during synthesis. Because of a random starting point the realized circuit can reach an unpredicted and, even worse, not simulated state. The explicit definition of a reset function avoids that kind of surprise. Furthermore, it is highly recommended that the circuit can change state from a random to a regular state of operation. When designing state machines this precaution is necessary and causes a small additional circuit complexity.

```
-- Initialization of a signal
-- at its declaration
SIGNAL data:  STD_LOGIC := '1';
SIGNAL data_bus:
STD_LOGIC_VECTOR(7 DOWNTO 0):= "10100101";
```

4.2.2 Ports

The **interfaces** of a VHDL component are signals with additional properties. Apart from its type (e.g., std_logic), such a port is characterized by information of its signal direction: the so called mode (IN, OUT, INOUT, BUFFER, LINKAGE).

```
SIGNAL i1:   IN STD_LOGIC;
SIGNAL o1:  OUT STD_LOGIC;
```

Within a port declaration only signals (no constants or variables) are allowed as an object class. Therefore the key word SIGNAL is optional and is omitted most of the time.

```
i1:   IN STD_LOGIC;
o1:  OUT STD_LOGIC;
```

The meaning of the signal modes in detail:

`IN` This port is an input and only readable within the component;

`OUT` Output port, only writeable within the component;

`INOUT` Bidirectional port: Readable and writeable within the component. Will be a tri-state signal when being synthesized;

`BUFFER` Like an output, readable within components;

`LINKAGE` Connection between components, signal direction unknown.

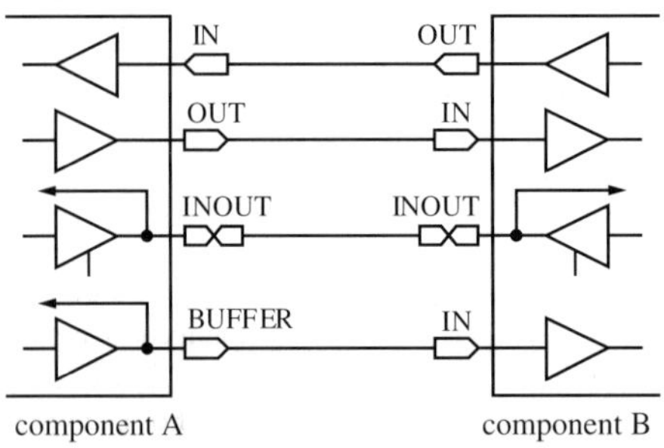

Fig. 4.7 Example of Ports

4.2.3 Entities

The **entity** defines the name of a functional unit and its interfaces to the outside world. The function of the component is not defined (fig. 4.8) so far. The completion of the empty box with a detailed plan (so called ARCHITECTURE) describing its internal structure and its behavior produces a model which can be simulated and changes the black box into a component with well designed properties.

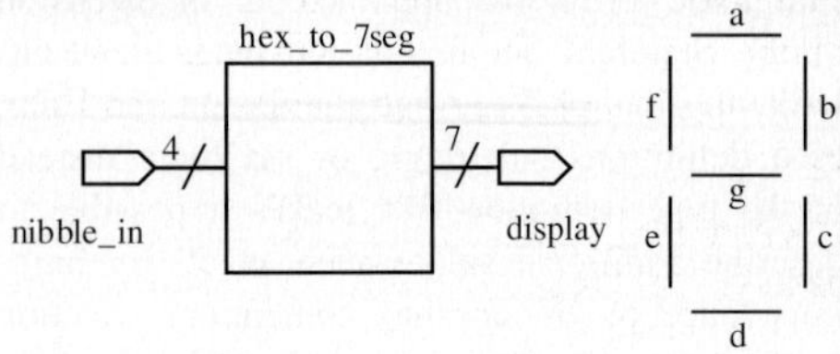

Fig. 4.8 Interface of a Hex to 7segment decoder

The following example shows a complete declaration of an entity of the symbol shown in figure 4.8. It is important to link the IEEE library with a reference to the package std_logic_1164 when using type std_logic or std_logic_vector.

```
LIBRARY IEEE;
USE IEEE.STD_LOGIC_1164.ALL;

ENTITY hex_to_7seg IS
PORT (
    nibble_in:
        IN STD_LOGIC_VECTOR(3 DOWNTO 0);
    display:
        OUT STD_LOGIC_VECTOR(6 DOWNTO 0)
    );
END hex_to_7seg;
```

The key word ENTITY is followed by the component name. In this case the component has two inputs, a 4-bit input bus and a 7-bit output bus, both listed in the port list PORT(...).

Design units

An entity is a design unit. Written to a file, that example can be compiled and tested for syntax. It is not possible to distribute design units such as architecture, package, and configuration to different files. However, they can be part of one file. The compiler treats every unit in a way as if written to separate files. If several files describe a project the required sequence of compilation is defined by the structure of the design and its dependencies. A typical sequence would start compiling packages (if present) followed by the entity, its architectures, and an optional configuration.

4.2.4 The library WORK

WORK is a VHDL reserved word and identifies the actual working library. Independent of symbolic and physical names of a library, WORK is the location where a compiler stores its analyzed modules. Modules will be retrieved there during simulation at a later point in time. To access a library coming from another simulation environment, its symbolic name (e.g., SIM_LIB) will be used.

```
LIBRARY SIM_LIB;
USE SIM_LIB.ALL;
```

4.2.5 Architectures

Because of the separate external and internal point of view, namely interfaces and functions of a component, VHDL supports a hierarchical top down and bottom up design, which partitions, during a first step, the overall system into small modules (components) with simple interfaces. At a starting point of a design the key issue is the structure of the design rather than detailed thoughts about the module behavior and attempts to implement circuit functionality. The architecture supports both: it can define a circuit structure using individually connected components or describe its behavior by means of suitable statements. The entity declaration and its architecture represent a model of a piece of hardware, which can be simulated: the so called design entity.

Structure

A hardware structure is defined by its components as well as its interfaces and its signals. In this case (see fig. 4.9) the architecture looks more like a netlist.

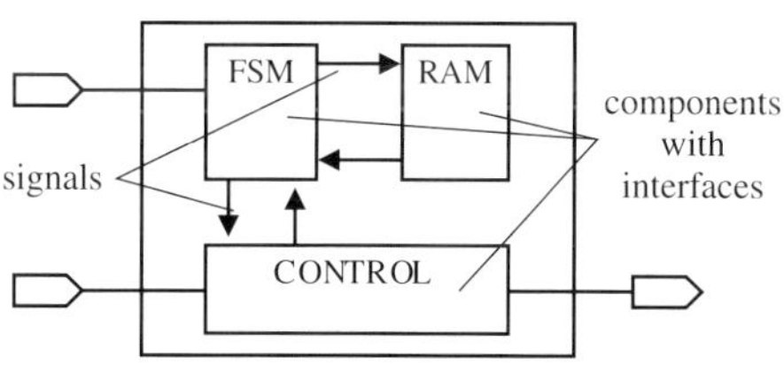

Fig. 4.9 The components define the structure of a design

Behave

Circuit behavior is determined by the processing of data between input and output, influenced by logical and physical timing to take signal and gate delays into account (fig. 4.10).

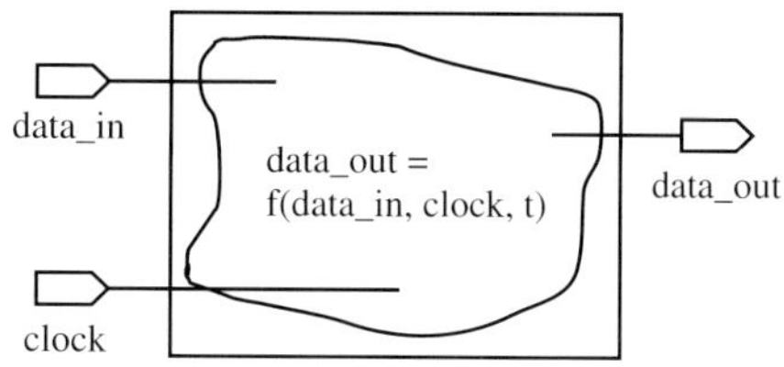

Fig. 4.10 Behave of a component, defined by processing of data and timing

```
-- Example of an architecture
ARCHITECTURE behave OF hex_to_7seg IS
-- declaration of components,
-- signals, types etc.
BEGIN
  -- declarations
  -- abcdefg
display <=
    "1111110" WHEN nibble_in = "0000" ELSE
    "0110000" WHEN nibble_in = "0001" ELSE
    "1101101" WHEN nibble_in = "0010" ELSE
    "1111001" WHEN nibble_in = "0011" ELSE
    "0110011" WHEN nibble_in = "0100" ELSE
    "1011011" WHEN nibble_in = "0101" ELSE
    "1011111" WHEN nibble_in = "0110" ELSE
    "1110000" WHEN nibble_in = "0111" ELSE
    "1111111" WHEN nibble_in = "1000" ELSE
    "1111011" WHEN nibble_in = "1001" ELSE
    "1110111" WHEN nibble_in = "1010" ELSE
    "0011111" WHEN nibble_in = "1011" ELSE
    "1001110" WHEN nibble_in = "1100" ELSE
    "0111101" WHEN nibble_in = "1101" ELSE
    "1001111" WHEN nibble_in = "1110" ELSE
    "1000111" WHEN nibble_in = "1111" ELSE
    "0000000";
END behave;
```

In the first line the architecture is given a unique name. Furthermore, this is a connection to the entity declaration, which makes it possible to provide several versions for simulation. For example, a first version can describe system behavior in a way which can not be synthesized. The same design, ported to the register-transfer- level, would refine to a description which can be synthesized. Normally this kind of description does not show any timing information of the circuit. The simulation of the decoder circuit is shown in figure 4.11. A purely structural architecture would be a result after synthesis. This third version of an architecture would contain all gate elements connected in a netlist. This means there are three alternative descriptions for one component, producing three

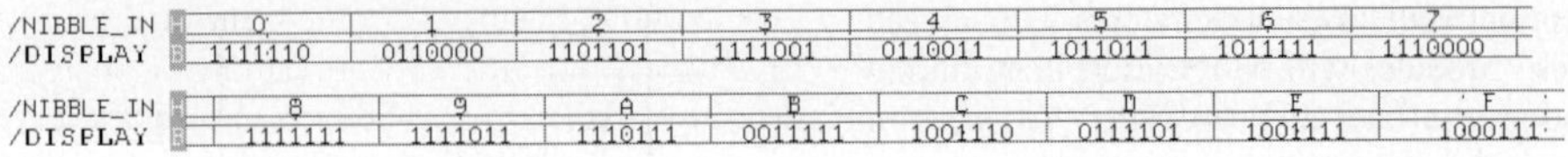

Fig. 4.11 Simulation of the decoder in VHDL without timing

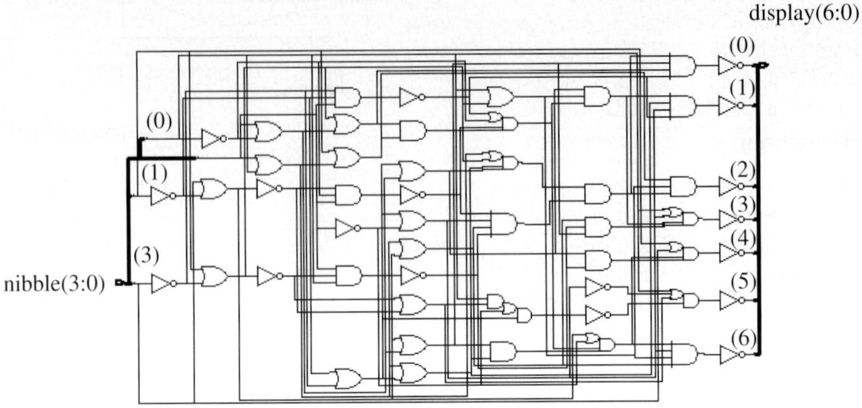

Fig. 4.12 Synthesized circuit of the decoder example

simulations on different levels of abstraction. The simulation results must be functionally identical.

The architecture is divided into a part containing declarations (that is, before the reserved word BEGIN) and a part between BEGIN and END containing statements. All signals, constants, types, and components used within the architecture are defined in the declaration part. They are only valid within this architecture. Interfacing signals, present in the entity declaration, must not be declared again. The statements behind BEGIN contain the circuit description, e.g., by the placing and wiring of existing components or by using concurrent statements. Each statement would represent a small part of the overall circuit.

4.2.6 Components

A hierarchical circuit design is formed by connecting individual design entities (entity architecture pairs). Design entities of lower hierarchical levels turn into components of a structural description in the level above. This is interesting to note that: a component is the substitute for a design entity. A configuration will define which particular design entity will replace the substitute. A working example applying components, their declaration, and instantiation is shown in section 4.2.8.

Component Declaration

A component to be used in an up to date description must be made known (declared). Normally, the interface description in the component declaration, namely signal names connected to the individual type and their sequence, correspond exactly to the equivalent entity declaration. This information enables the compiler to find out whether the component is connected in the right way. On the other hand, the declaration does not provide information about where to find the equivalent design entity. Perhaps it is not available at all within the actual design hierarchy at compile time.

A component declaration is made within an architecture or, in a global way, within a package.

Component Instantiation

If a component declaration is compared to the labelling of a drawer to mark its content, taking out and wiring a component can be seen as an instantiation (fig. 4.13). The instance is only an instance of the component declaration, not of the design entity. The connection to the design entity is made by the configuration. An individual name must be assigned to every instance, because a component might be instantiated more than once.

Wiring of components can be achieved in two ways. The quickest way to wire up signals is to list signals in the PORT MAP according to the sequence of the component declaration. Based on the declarations, the compiler tests only whether signals of the same type are connected to each other. If the sequence of signals is not maintained, it could mean that signals (e.g., a clock and a reset signal) are swapped. This could produce errors eventually detected at a late point in time of the design cycle.

A safe method is to assign every connection of the component to the associated signal in an unmistakable way. In the first place, the PORT MAP shows the interface description as defined in the declaration followed by the sign '=>' pointing to the signal to be connected.

```
-- component instantiation
U1:  hex_to_7seg
  PORT MAP(
    nibble_in => byte(7 DOWNTO 4),
    display => display_a);
```

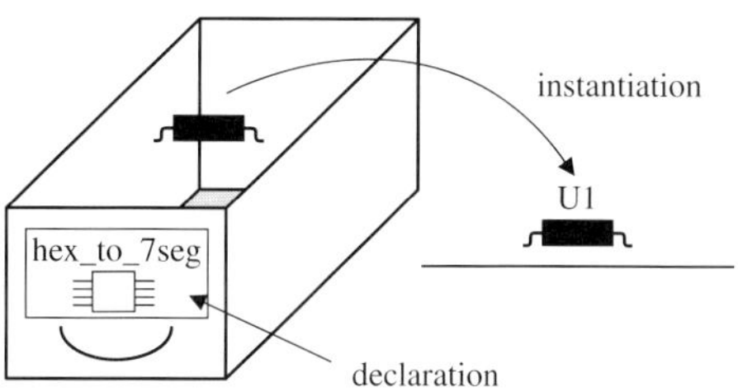

Fig. 4.13 Difference between declaration and instantiation of a component

4.2.7 Configurations

A configuration statement is meaningful and necessary for several reasons. Firstly, it chooses from several versions of implementation (architectures) exactly one for a simulation and closes the gap between the entity declaration and the architecture. Secondly, it matches component instance and design entity. Both declarations do not match exactly at times. A different name of the component and its entity, different notation of interfaces, its sequence, types, and numbers can be taken into account in the configuration. Flexibility and readability increase because of that. The parts of a hierarchical design form a circuit description which can be simulated.

Assuming that names of components and entities as well as names and modes of interfaces, are identical, a configuration is not necessary. In this case the simulator takes the architecture most recently compiled. According to these rules synthesis is performed. Here configurations are ignored as well.

Configuration Declarations

The configuration declaration is a separate compile unit and identifies, for each component, an entity architecture pair explicitly. The overall model can be configurated without modifying individual modules. A change in the design kit of the system only causes a new compilation of the configuration, provided all model versions are available. The skeleton of a typical configuration looks as follows:

```
-- example of a configuration declaration
CONFIGURATION conf_name OF entity_top IS
  FOR architecture_name_top
    FOR instance_name_comp1:
        component_name1
      USE ENTITY WORK. entity_name_comp1
               (architecture_name_kompl);
    END FOR;

    FOR instance_name_comp2:
        component_name2
      USE ENTITY WORK. entity_name_comp2
               (architecture_name_comp2);
    END FOR;
  END FOR;
END conf_name;
```

The configuration receives a name (conf_name) related to the top level entity (entity_top). After that a corresponding architecture for this entity will be selected (FOR architecture_name_top).

If this architecture is hierarchically designed and if it contains components, the configuration will be two-dimensional. In a first step each component will be identified using instance_name:component_name to decide which library should provide a design entity to be connected to the component. This is because the standard library WORK must always be used. If there is an existing configuration for a lower hierarchy level, instead of using the statement USE ENTITY …, it can be attached using

```
USE CONFIGURATION WORK.conf_name
```

Configuration Specification

There are some limits to using the configuration specification as a second possibility to configure. In this case it is not a separate design unit any longer. Being placed in the declaration part of the architecture it causes a loss of flexibility.

```
-- when declaring the architecture:
FOR U1, U2:  hex_to_7seg
  USE ENTITY WORK.hex_to_7seg(behave);
```

To match different naming conventions in declarations:

```
FOR ALL: hex_to_7seg
-- Having different port identifiers:
-- nibble_in in the entity,
-- nib in the component
  USE ENTITY WORK.hex_to_7seg(behave)
  PORT MAP
                    (nibble_in => nib, display => disp);
```

4.2.8 Dual Digit Conversion 'Hex to 7Segment'

❑ *Example* (list 4.1):

This example shows declaration and instantiation of components and the design of a suitable configuration. A structural description will be used.

List 4.1 Dual Digit Conversion 'Hex to 7Segment'

```
LIBRARY IEEE;
USE IEEE.STD_LOGIC_1164.ALL;

ENTITY dualhex IS
PORT (
    byte_in :    IN  STD_LOGIC_VECTOR(7 DOWNTO 0);
    display_a :  OUT STD_LOGIC_VECTOR(6 DOWNTO 0);
    display_b :  OUT STD_LOGIC_VECTOR(6 DOWNTO 0));
END dualhex;

ARCHITECTURE structure OF dualhex IS
-- Component declaration
COMPONENT hex_to_7seg
PORT (
    nibble_in :  IN  STD_LOGIC_VECTOR(3 DOWNTO 0);
    display :    OUT STD_LOGIC_VECTOR(6 DOWNTO 0));
END COMPONENT;

BEGIN
-- Component instantiation with wiring
U1:  hex_to_7seg
PORT MAP(
    nibble_in    => byte(7 DOWNTO 4),
    display      => display_a);
U2:  hex_to_7seg
PORT MAP(
    nibble_in    => byte(3 DOWNTO 0),
    display      => display_b);
END structure;

-- optional
CONFIGURATION dualhex_conf OF dualhex IS
    FOR structure
        FOR ALL: hex_to_7seg USE ENTITY WORK.hex_to_7seg
                                                  (behave);
        END FOR;
    END FOR;
END dualhex_conf;
```

4.3 Concurrent Statements

The hardware description language VHDL differs from other programming languages because concurrent and sequential statements are possible.

All statements within an architecture describe activities taking place concurrently. Part of those are the ordinary signal assignment, conditional and selective assignments, component instantiation, and the entire process. The sequence of concurrent statements in a VHDL text is not important. This reflects the general behavior of digital hardware, in which signal transitions occur in parallel.

Within a process the execution of statements is performed sequentially. This corresponds to the understanding of a classic programming language.

❑ *Example*: full adder (fig. 4.15)

```
ENTITY adder IS
PORT (
    op_a:   IN STD_LOGIC;
    op_b:   IN STD_LOGIC;
    carry_in:   IN STD_LOGIC;
    sum:  OUT STD_LOGIC;
    carry_out:  OUT STD_LOGIC);
END adder;
ARCHITECTURE behavior OF adder IS
BEGIN
    sum <= op_a XOR op_b XOR
             carry_in;
    carry_out <= (op_a AND op_b) OR
                 (op_a AND carry_in) OR
                 (op_b AND carry_in);
END behavior;
```

4.3.1 Concurrent Signal Assignment

Apart from their task of connecting components, signals can be combined with each other and form a part of combinatorial logic in this way. Simple signal assignments are:

```
a <= '1';
z <= a XOR b;
```

(to be read: 'signal z is assigned the logic value of a XOR b')

The signal type on the right side must correspond to the signal type on the left side.

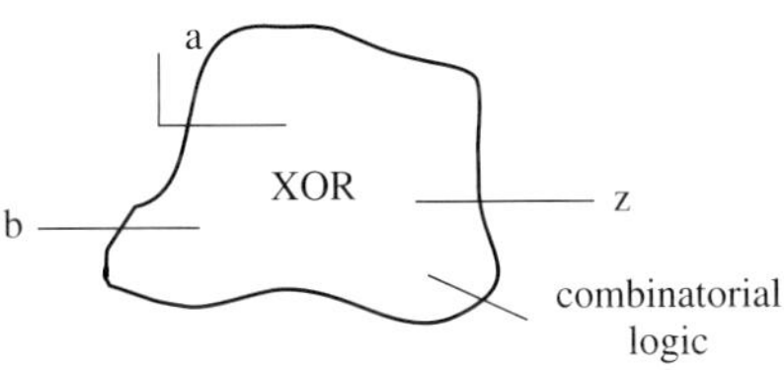

Fig. 4.14 A signal assignment produces combinatorial logic

Next to the VHDL description of the full adder, it would not be a big problem to draw an equivalent circuit containing XOR, AND, and OR gates. However, this circuit would not be a result of an automatic synthesis. Synthesis has to meet a set of constraints and find a compromise. The result depends on the target technology and requirements concerning area and speed as well as periphery. Requirements such as fan out, driving capabilities, capacitive loads, or the availability of, e.g., XOR components, must all be taken into account. The answers to these questions decide what the circuit looks like after synthesis.

4.3.2 Conditional Signal Assignment

The simplest way to assign a signal conditionally describes a multiplexer (fig. 4.16):

```
z <= input_a WHEN sel = '0'
             ELSE input_b;
```

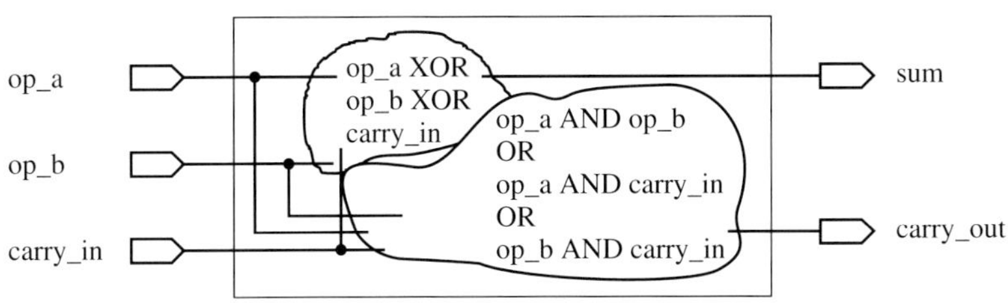

Fig. 4.15 Combinatorial logic of the full adder

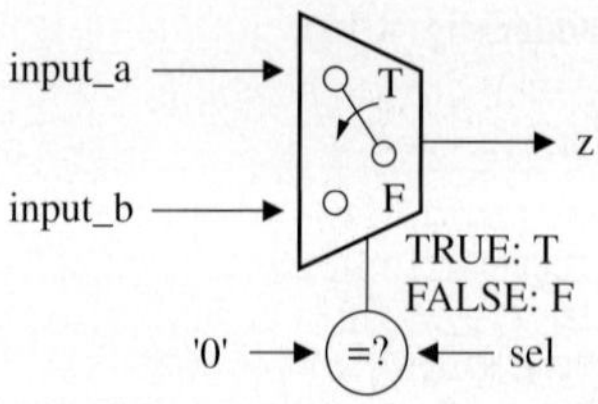

Fig. 4.16 Multiplexer with a conditional signal assignment

Depending on the condition sel='0', which can be TRUE or FALSE, one of the two sources input_a or input_b is selected and connected to the target, the output z. In this case source and target must be of the same type (e.g., std_logic). In most cases the condition is formed by operators such as =, >, <, >=, <=. The assignment can contain boolean equations.

```
sum <= op_a XOR op_b
           WHEN carry_in = '0'
           ELSE NOT (op_a XOR op_b);
```

Conditional signal assignments can be expanded to realize complex hierarchical multiplexers (fig. 4.17).

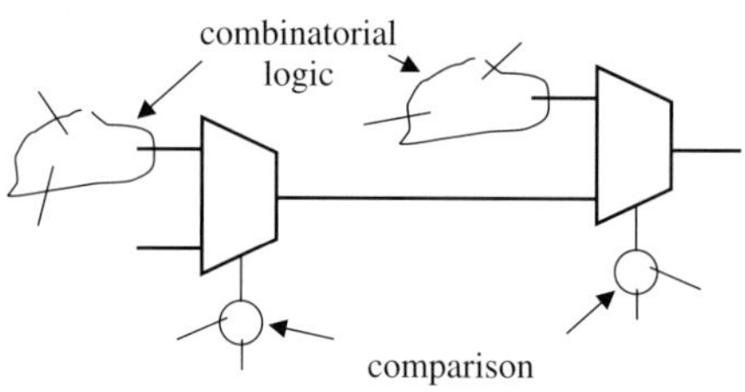

Fig. 4.17 Structure of a hierarchical multiplexer

It can not be assumed that mutual dependences of conditions will be detected by the synthesizer. Having dependences, it is recommended that a selected signal assignment is inserted in order to avoid redundant logic.

The last ELSE statement carrying no condition should handle all conditions not covered so far. Otherwise a situation could occur in which a new value would not be assigned to the output. This means that during synthesis feedback paths or asynchronous storage elements would be inserted which is not desirable at all.

4.3.3 Selected Signal Assignment

This type of signal assignment allows only one condition to be specified. The result of this test determines the source and target being connected.

```
SIGNAL choice:  NATURAL RANGE 0 TO 3;
...
WITH choice SELECT
    z <= a WHEN 0,
         b WHEN 1,
         c WHEN 2,
         d WHEN 3;
```

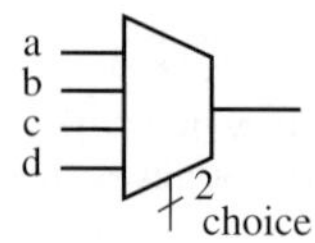

Fig. 4.18 Multiplexer with selected signal assignment

Contrary to a conditional signal assignment, individual tests (WHEN 0, WHEN 1, ...) describe no hierarchy. They have the same priority and exclude each other mutually. All elements of the set of values the condition can take (see examples 0...3) must be enumerated.

The last test (WHEN OTHERS) can combine all remaining possibilities.

```
WITH choice SELECT
    z <= input_a WHEN '0',
         input_b WHEN OTHERS;
```

4.3.4 Coder with Prioritizing of Inputs

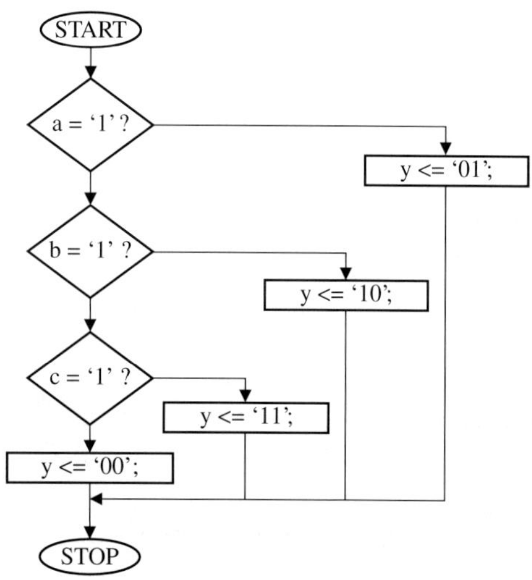

Fig. 4.19 Flow chart of coder

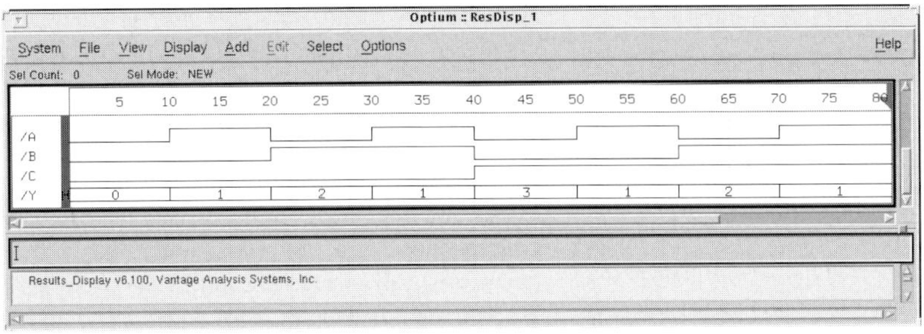

Fig. 4.20 Simulation of the architecture coder_arch

The following example shows two different versions of implementing the functional description in figure 4.19. Both solutions produce the same results in simulation and synthesis (figures 4.20 and 4.21).

List 4.2 Coder with Prioritizing of Inputs

```
LIBRARY IEEE;
USE IEEE.STD_LOGIC_1164.ALL;

ENTITY coder IS
PORT(
    a,b,c:  IN STD_LOGIC;
    y :  OUT STD_LOGIC_VECTOR(1 DOWNTO 0));
END coder;

ARCHITECTURE coder_arch OF coder IS
-- constants to improve readability
CONSTANT one:
    STD_LOGIC_VECTOR(1 DOWNTO 0) := "01";
CONSTANT two:
    STD_LOGIC_VECTOR(1 DOWNTO 0) := "10";
CONSTANT three:
    STD_LOGIC_VECTOR(1 DOWNTO 0) := "11";
CONSTANT zero:
    STD_LOGIC_VECTOR(1 DOWNTO 0) := "00";

BEGIN
    y <= one WHEN a = '1' ELSE
         two WHEN b = '1' ELSE
         three WHEN c = '1' ELSE
         zero;
END coder_arch;

ARCHITECTURE complicated OF coder IS
TYPE mychoice IS (0, 1, 2, 3);
SIGNAL conditional:  mychoice;
SIGNAL selected:
    STD_LOGIC_VECTOR(1 DOWNTO 0);

BEGIN
    -- concurrent signal assignment
    y <= selected;
    -- concurrent conditional signal
    -- assignment
    conditional <= 0 WHEN a = '1' ELSE
                   1 WHEN b = '1' ELSE
                   2 WHEN c = '1' ELSE
                   3;
    -- concurrent selected signal
    -- assignment
    WITH conditional SELECT
    selected <=    "01" WHEN 1;
                   "10" WHEN 2;
                   "11" WHEN 3;
                   "00" WHEN OTHERS;
END complicated;
```

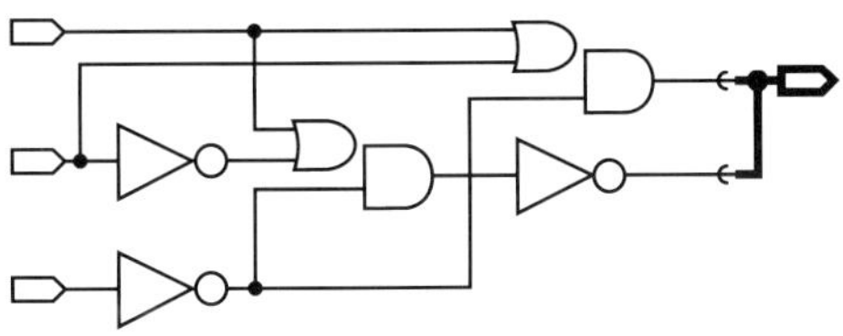

Fig. 4.21 Synthesized Example: The same result for both architectures

4.4 The simulation model in VHDL

After a look at concurrent signal statements in chapter 4.3, basic mechanisms of the language can be discussed in detail. Originally the language VHDL was developed for simulation only. A discussion about the theory of operation of a VHDL simulator helps to understand VHDL modelling and simplifies the choice of suitable language constructs to describe hardware constructs. The topic simulation model has an emphasis on:

- Driving signals;
- Delta delay mechanisms;
- Modelling of delay times.

4.4.1 Drivers

Every concurrent statement, including processes, causes an event responsible for the progress of a signal. Within the simulator there is a list containing points in time to schedule signal events. Such an entry in that list is called a transaction. This list is the basic information for scheduling signal values during simulation (fig. 4.22). A change of a signal value caused by an entry in that list, is called an event in that signal. If a new entry is the same as the actual signal value this signal assignment has no effect.

❑ *Example*:

```
-- construction of a list of drivers
y <= '0' AFTER 0 ns,
     '1' AFTER 20 ns,
     '0' AFTER 50 ns;
```

time	value
0 ns	'0'
20 ns	'1'
50 ns	'0'

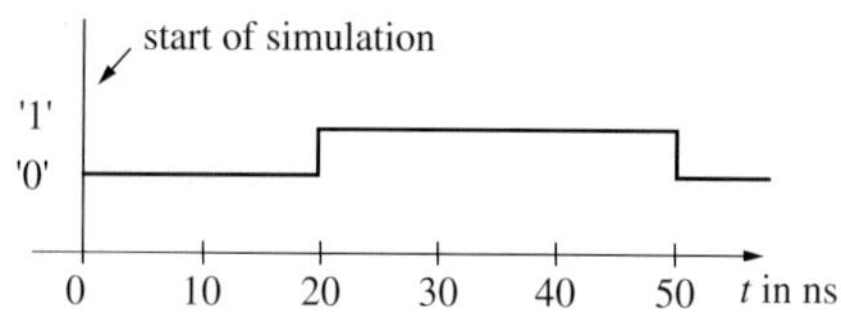

Fig. 4.22 Driver list with signal timing diagram

4.4.2 Delta Delay Mechanism

To answer the question of how the simulator finds an unequivocal result despite several concurrent signal assignments, it is helpful to look at a simulation cycle in detail. A simulation cycle with a step forward in time (e.g., 1 ns) and a cycle without progress in time ($\Delta t = 0$ ns) will be distinguished.

VHDL simulators are driven by events. The simulator advances the simulation time and searches all driver lists to find out if an event, that is a change of state of a signal, is due to take place. Signals associated with a driving list containing an entry for the time being, will be updated actual time to the corresponding value of the driver list. The new signal value is presented by the graphical output of the simulator now. All other signals without a scheduled transaction keep the values they have.

A change of state of a signal now activates the simulator. The simulator examines all concurrent statements to find out if any of these statements have an effect on that signal. If this is the case, those statements would be executed as well. After all concurrent statements have been examined, new values and new entries to the driving list can be a consequence. If a signal assignment has a zero delay time, all concurrent statements associated with the modified signal must be reviewed again. There must be no increase in time, zero delay cycles will apply.

A stable simulation result is achieved when there are no new entries in the driver list at this point in time. This means all signals remain stable for at least one delta time. After that the simulation time step can increase again. In this case the step in time does not have a constant increment, it can be dynamically enlarged according to the next event.

This sequence of events maps parallel activities within a VHDL description to sequential steps of a simulator program. The sequence of concurrent statements in a VHDL text makes no difference to the results of the simulation . Events in signals only determine the work flow. This property makes VHDL simulation very efficient, because computer time will be only needed when there is a change of state in the simulation model.

❑ *Example*: Zero delay RS flip flop (table 4.1)

```
x <= NOT (y AND lset); -- (1)
y <= NOT (x AND reset); -- (2)
```

Table 4.1 Simulation cycle without signal delay

Simulation time	lset	x	y	reset	Evaluation of statement
20 ns	↓	0	1	1	(1)
20 ns + 1Δ	0	↑	1	1	(2)
20 ns + 2Δ	0	1	↓	1	(1)
20 ns + 3Δ	0	1	0	1	—
30 ns	↑	1	0	1	(1)
30 ns + 1Δ	1	1	0	1	—
40 ns	1	1	0	↓	(2)
40 ns + 1Δ	1	1	↑	0	(1)
40 ns + 2Δ	1	↓	1	0	(1)
40 ns + 3Δ	1	0	1	0	—

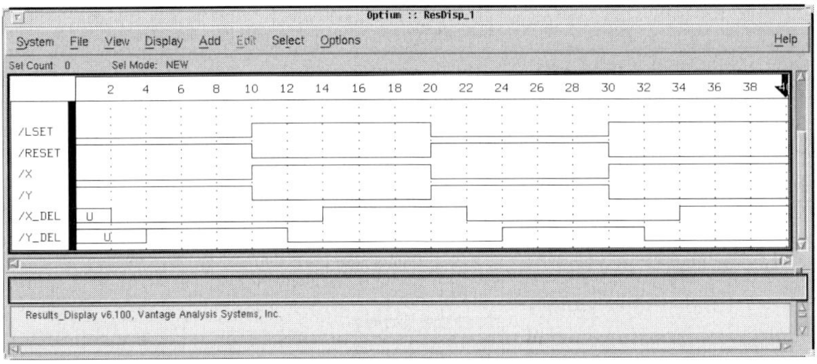

Fig. 4.23 Simulation result of the RS FF for the zero delayed (X, Y) and delayed (X_DEL, Y_DEL) model

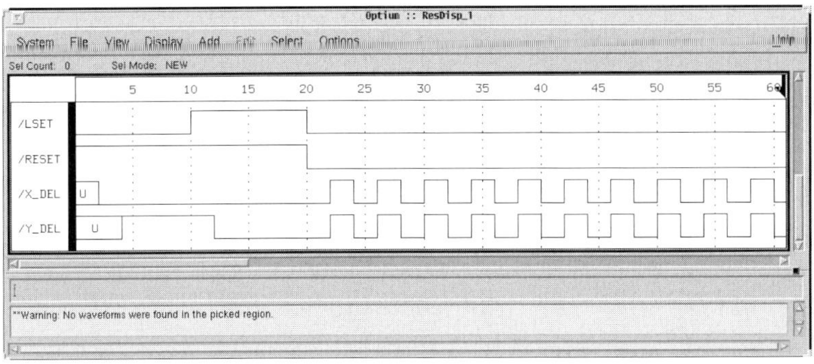

Fig. 4.24 Oscillating RS flip flop

❑ *Example*: RS flip flop with Signal Delay (table 4.2)

```
x_del <= NOT (y_del AND lset)
   AFTER 2 ns; -- (1)
y_del <= NOT (x_del AND reset)
   AFTER 2 ns; -- (2)
```

Table 4.2 Simulation cycle with signal delay

Simulation time	lset	x_del	y_del	reset	Evaluation of statement
20 ns	↓	0	1	1	(1)
22 ns	0	↑	1	1	(2)
24 ns	0	1	↓	1	(1)
24 ns + 1Δ	0	1	0	1	—
30 ns	↑	1	0	1	(1)
30 ns + 1Δ	1	1	0	1	—
40 ns	1	1	0	↓	(2)
42 ns	1	1	↑	0	(1)
44 ns	1	↓	1	0	(1)
44 ns + 1Δ	1	0	1	0	—

There is an amazing situation in figure 4.24. Despite both inputs lset and reset carrying a static low logic level, the outputs x_del and y_del change state periodically. The circuit oscillates. The period of the oscillation corresponds exactly to the delay of the two signal assignments, x_del and y_del. If the delay is decreased the change of state of the signals increase. The extreme case, a delay of zero seconds, causes infinite delta steps. The simulator does not find a stable result and eventually exits. Table 4.3 shows the simulation run.

Table 4.3 Simulation of the oscillating RS flip flop

Simulation time	lset	x_del	y_del	reset	Evaluation of statement
20 ns	↓	0	0	↓	(1), (2)
20 ns + 1Δ	0	↑	↑	0	(1), (2)
20 ns + 2Δ	0	↓	↓	0	(1), (2)
20 ns + 3Δ	0	↑	↑	0	(1), (2)
etc.	0	...	...	0	(1), (2)

4.4.3 Modelling of Delay Times

A VHDL description (RTL) to be synthesized usually does not contain delay times in signal assignments. Delay times caused by delays at the gate level (e.g., combinatorial logic) or by propagation delays (flip flops) depend on the target technology, which will be defined at a later point in time of the design cycle. However, the specification of delay can describe the timing behavior after synthesis or provide a tool for generating complex input stimuli in a simulation environment.

```
clk <= NOT clk AFTER 100ns;
```

There are two different types of **delays**:

- A transport delay for modelling delays on wires;
- Inertial delay, causing suppression of very short pulses.

Transport Delay

The effect of a transport delay: the new value is added to the list of the events at the scheduled point in time. If there are other entries in the list, scheduling events at a later point in time, these entries will be cancelled now (fig. 4.25).

Assuming $y(t)$ was not a delayed signal, then $y(t - T)$ is a signal delayed by T.

❑ *Example*: Multiplexer with transport delay (fig. 4.25)

```
z <= TRANSPORT input_a AFTER 15ns
                    WHEN sel = '0' ELSE
      input_b AFTER 10ns;
```

Inertial Delay

Although the inertial delay is the default value, this version is a bit more difficult to understand.

A pulse shorter than the setup time will be suppressed.

```
y <= input_a AFTER 10 ns;
-- A set up time of 10 ns is assumed
y <= REJECT 10 ns INERTIAL input_a
AFTER 200 ns;
-- Definition of minimum pulse width 10 ns
y <= REJECT 0 INERTIAL input_a
AFTER 200 ns;
-- Equivalent to TRANSPORT input_a
```

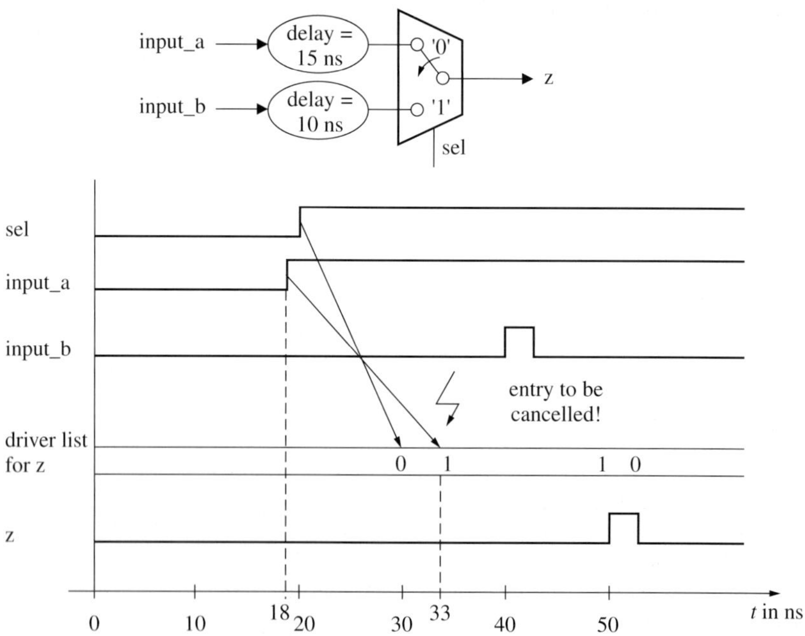

Fig. 4.25 Example of Transport Delay

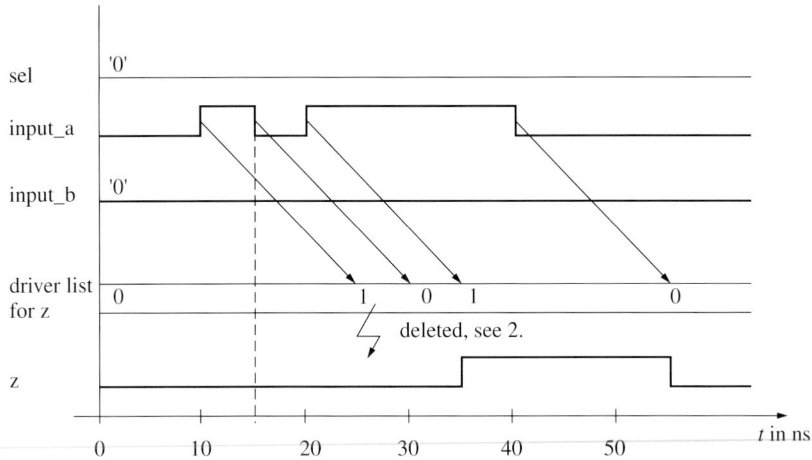

Fig. 4.26 Example of Inertial Delay

The main difference from a transport delay is that the history of the signal plays a part as well. Delayed signals are built according to the following rules:

- Label the new entry to the driving list;
- Label that entry prior to the new entry which has the same value;
- Label the entry which is responsible for the present signal value;
- Remove all entries with no labels.

❑ *Example*: Multiplexer with an inertial delay (fig. 4.26)

```
z <= input_a AFTER 15 ns
     WHEN sel = '0' ELSE
     input_b AFTER 10 ns;
```

Explanations regarding the sequence in figure 4.26:

1. Point in time: $t = 10$ ns:
   Entry of '1' at $t = 25$ ns.
2. Point in time: $t = 15$ ns:
   Entry of '0' at $t = 30$ ns. Labelling of entries at 0 ns and 30 ns, entry at 25 ns deleted.
3. Point in time: $t = 20$ ns:
   Entry of '1' at $t = 35$ ns, entry at 30 ns deleted.
4. Point in time: $t = 40$ ns:
   Entry of '1' at $t = 35$ ns already active. No entries between $t = 40$ ns and $t = 55$ ns. No entries to be deleted.

4.4.4 Comparison of Transport and Inertial Delay

❑ *Example*: Two stage delayed multiplexer (fig. 4.27)

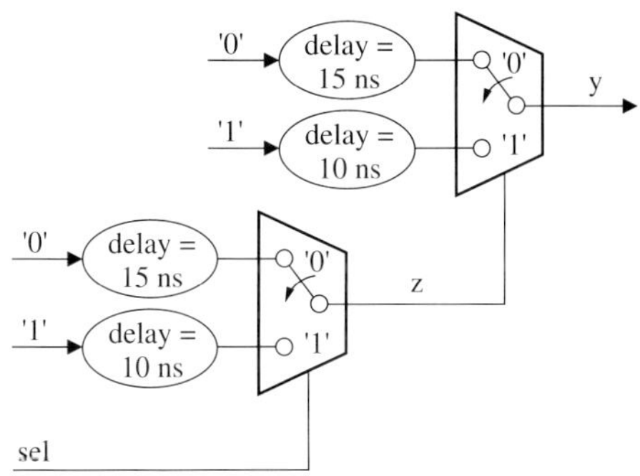

Fig. 4.27 Two stage delayed multiplexer

The following question must be answered: What is the minimum pulse width of the signal sel in figure 4.27 which will produce a visible signal change on output y? Is there a difference of simulation results using inertial or transport delays?

```
LIBRARY IEEE;
USE IEEE.STD_LOGIC_1164.ALL;

ENTITY comparison IS
PORT (
  sel:  IN STD_LOGIC;
  y_inertial:  OUT STD_LOGIC;
```

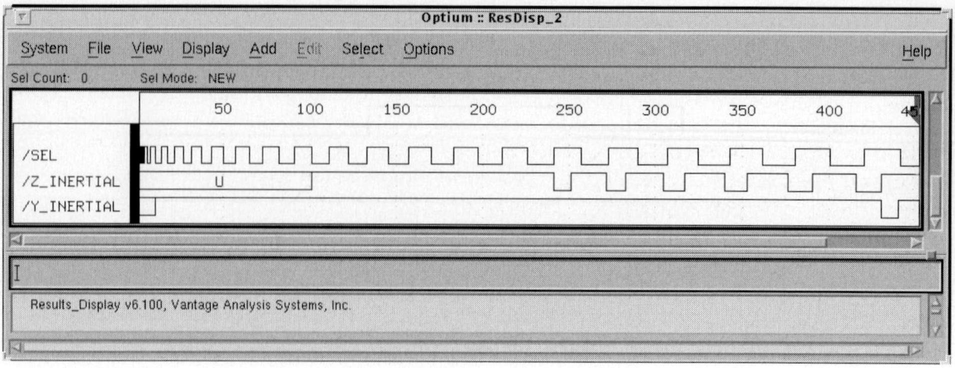

Fig. 4.28 Simulation with inertial delay: pulse width of sel minimum 20 ns

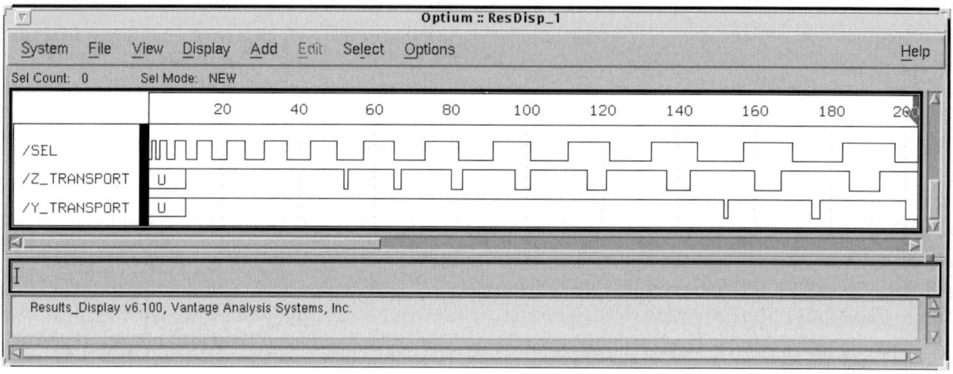

Fig. 4.29 Simulation with transport delay: pulse width of sel minimum 12 ns

```
 y_transport:  OUT STD_LOGIC);
END comparison;
ARCHITECTURE comparison_arch
OF comparison IS
SIGNAL z_inertial:  STD_LOGIC;
SIGNAL z_transport:  STD_LOGIC;

BEGIN
 z_inertial <=  '0' AFTER 15ns
                WHEN sel = '0' ELSE
                '1' AFTER 10ns;

 y_inertial <=  '0' AFTER 15ns
                WHEN z_inertial = '0' ELSE
                '1' AFTER 10ns;

 z_transport <= TRANSPORT '0' AFTER 15ns
                WHEN sel='0' ELSE
                '1' AFTER 10ns;

 y_transport <= TRANSPORT '0' AFTER 15ns
                WHEN z_transport = '0' ELSE
                '1' AFTER 10ns;
END comparison_arch;
```

The answer is shown in the simulation results of figure 4.28 and figure 4.29.

4.5 Process

As shown in the previous chapters, concurrent statements are only legal within an architecture. Concurrency is an important requirement for describing digital circuits. However, complex sequences of events are difficult to describe using the language constructs known so far, such as simple, conditioned, or selective signal assignments. Each modern programming language provides constructs for supporting a structured programming style. Because of that, sequential language constructs such as IF ELSIF, CASE or loops must be available in VHDL as well. So far there is no answer to how to describe a combinatorial or synchronous design with a higher complexity. Within VHDL the process is the answer to this question.

Being a part of the architecture, a process acts concurrently from an external point of view. Internally it provides space for sequential statements, hence combining the requirements for a hardware description language and the demand for structured programming ideally. These properties turn the process to the most used statement, which explains its central role in VHDL. At a later point in time it will be obvious that the signal assignments mentioned above represent a simplified form of a process.

4.5.1 Properties

A process contains sequential statements. The program execution is defined by this sequence. The behavior of other, non-concurrent, languages and VHDL can only be compared partly. The main difference between a process and a classical program is: the process, under normal circumstances, will be repeatedly activated. An obvious example are processes used in synthesizable code.

Structure of Processes

Processes may be located anywhere in the statement region of an architecture. A process always contains a sensitivity list, a declarative part, and a statement part.

```
PROCESS (sensitivity list)
-- declarations
BEGIN
  -- sequential statements
  -- IF, CASE, LOOP
END PROCESS;
```

Sensitivity List

The process will be activated when a signal in the sensitivity list changes state. A process is either in the state 'executed' or 'suspended'. An event in any of the signals in the sensitivity list causes the process to be executed. After the last statement the process is put into the suspended state and waits for a new event.

Process Declarative Part

This part is used to declare types, functions and procedures. Its main purpose is to declare signals and variables (see section 4.2). All these statements are used locally. They are only known and valid within this process.

4

Process Statement Part

The process statement region contains sequential statements executed as a consequence of an event in a signal in the sensitivity list. The following constructs are sequential statements:

- IF, CASE, LOOP, WAIT;
- Assignments of signals und variables.

The reserved words BEGIN and END define the statement part within the process and mark a part of the circuit which belongs together. This could be a counter, adder, or a state machine.

❑ *Example*: Adder (fig. 4.30)

```
LIBRARY IEEE;
USE IEEE.STD_LOGIC_1164.ALL;

ENTITY adder IS
PORT (
  in_a,in_b:  IN NATURAL RANGE 15 DOWNTO 0;
  sum:        OUT NATURAL RANGE 15 DOWNTO 0);
END adder;

ARCHITECTURE behave OF adder IS
BEGIN
  P1:  PROCESS(in_a, in_b)
  BEGIN
    sum <= in_a + in_b;
  END PROCESS P1;
END behave;
```

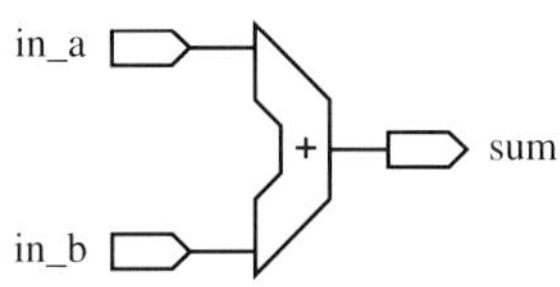

Fig. 4.30 Adder

The example presents a VHDL process containing combinatorial logic only and no storage elements or registers.

> For combinatorial processes the rule applies that all signals being read must be mentioned in the sensitivity list.

That process, describing an addition, can only produce a correct result in the signal sum when it is sensitive to any change of the input signals in_a and in_b. An incomplete sensitivity list causes unexpected results or models that can not be synthesized. The worst case is that the simulation prior to and past synthesis differ. Some synthesis tools ignore the sensitivity list because they assume a process is to be re-executed if there was a change in a relevant signal. Assuming the signal in_b is not mentioned in the sensitivity list, the VHDL simulator will only calculate a new result at a signal change of in_a. This means: if only signal in_b changes its value the signal sum will keep the present value, which is wrong by now. If the synthesis tool assumes the same, it will 'try to keep the actual value', inserting a storage element (latch). Normally this is not wanted. Even worse, it is not desirable at all.

As a supplement, the use of the type NATURAL – a subtype of INTEGER – is mentioned to reduce the value range to natural numbers. An additional reduction to the range 0 to 15 is caused by the statement RANGE 15 DOWNTO 0.

4.5.2 Signals and Variables

Signals

Signal assignments within a process act completely differently from the concurrent statements presented so far. Without knowing some details, processes hide some potential sources of error. Valid in any case is:

> A process as a whole is a concurrent statement. To be processed, it takes exactly one delta of simulation.

Signals within a process do not change their value during process execution. Signal assignments show, as long as they do not have delays, results not before the end of the process. This is a simulation delta later. Signals being read in a process produce values coming from preceding cycles of simulation. This assures concurrency related to other processes. Within a process several assignments, following each other, to the same signal, are valid. However, only the last mentioned assignment stays in effect. Signals carrying a delay time will have the assigned value written to the driver list at the proper point in time, according to the rules of inertial or transport delay (see section 4.4.3).

The example of section 4.5.1 should be changed in a way that, if it exceeds the value 10, the result would be reset to zero (fig. 4.31). At the first glance it seems that the following implementation describes that behavior. However, a more detailed analysis shows that an erroneous function can occur. That would be the case when both input signals take a value larger than 5 at any point in time (e.g., $t = 20$ ns). What is happening is then shown in table 4.4 in detail.

```
ENTITY adder IS
PORT ( in_a, in_b:
       IN NATURAL RANGE 15 DOWNTO 0;
       sum:
       BUFFER NATURAL RANGE 15 DOWNTO 0);
END adder;

ARCHITECTURE bold OF adder IS
BEGIN
   V1:  PROCESS(in_a, in_b, sum)
       BEGIN
         sum <= in_a + in_b; -- 1.  line
         IF(sum > 10) THEN   -- 2.  line
           sum <= 0;         -- 3.  line
         END IF;
       END PROCESS V1;
END bold;
```

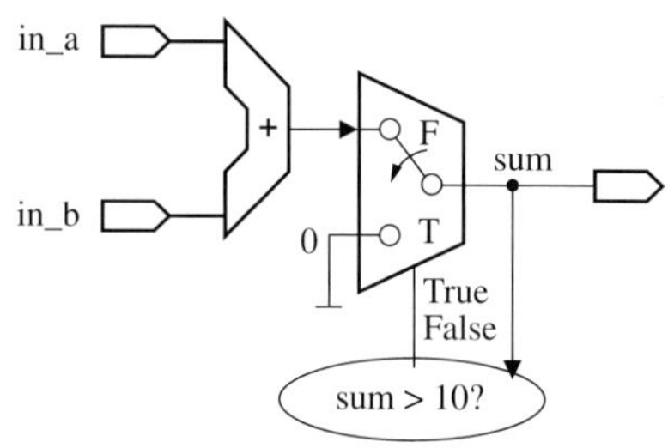

Fig. 4.31 Adder with comparison

The workflow within a process is sequential. The first line calculates the sum and passes the result to the signal sum. The second line finds out whether the legal range of values was exceeded and resets; in the third line, if necessary; the signal sum to 0. Because the execution of signal assignments within a process is performed sequentially, an assignment performed later can overwrite a prior assignment. The execution of assignments is performed sequentially; however, the assignment of new values to signals is made at the end of the process or at a break caused by a wait statement. A change of state in a signal is visible not before the next simulation delta. The example above shows a comparison using values from the prior simulation delta. Hence, depending on the history, the signal sum can take values larger than 10.

Because the process is sensitive to a change of the signal sum, which has now just changed its value, it will be resumed. The following simulation delta will produce a comparison with the true condition. The statement in the third line will then be executed. Continuing this analysis shows that the result can not be stable. The value of signal sum changes between 0 and 12, every change of the signal adds another simulation delta (see section 4.4.2).

Table 4.4 Simulation progress of process V1

Simulation time	in_a	in_b	sum
	2	3	5
20 ns	6	6	5
20 ns + 1Δ	6	6	12
20 ns + 2Δ	6	6	0
20 ns + 3Δ	6	6	12
...	...	...	...

Variables

Signals do not change their values during the execution of a process. Therefore it is not possible to store intermediate results on a signal. That is the reason for useing variables as an important supplement for describing sequential activities in VHDL. Variables obtain their value immediately after the assignment. The value is available within the actual simulation delta and can be used at once for further operations in the process. An assignment to a variable is always performed immediately. This implies that it is impossible to add a time delay (AFTER ...) to a variable assignment. To emphasize that fact variable assignment has a different symbol.

When declaring a variable there is no restriction on its type. However, the range of legal values of the variable is limited to the process where it has been declared.

```
VARIABLE a,b:  STD_LOGIC;
VARIABLE c:  STD_LOGIC_VECTOR(15 DOWNTO 0);
```

Variables, like signals, should be initialized. This is done explicitly by a declaration or, by default, it takes the left value presented by the type declaration, which is a 'U' for STD_LOGIC. Initialization is performed only once at the start of the simulation ($t = 0$ ns) and not, as sometimes assumed, when a process is re-entered. Care must be taken when initializing signals, because hardware normally can not perform that operation, and, as a consequence, synthesis would ignore initial values of signals. The following example shows a correct behavior according to specification and is a typical application of a variable.

```
ARCHITECTURE correct OF adder IS
BEGIN
  V2: PROCESS(in_a, in_b)
  VARIABLE temp: NATURAL RANGE 15 DOWNTO 0;
  BEGIN
    temp := in_a + in_b;
    IF(temp > 10) THEN
      sum <= 0;
    ELSE
        sum <= temp;
    END IF;
  END PROCESS V2;
END correct;
```

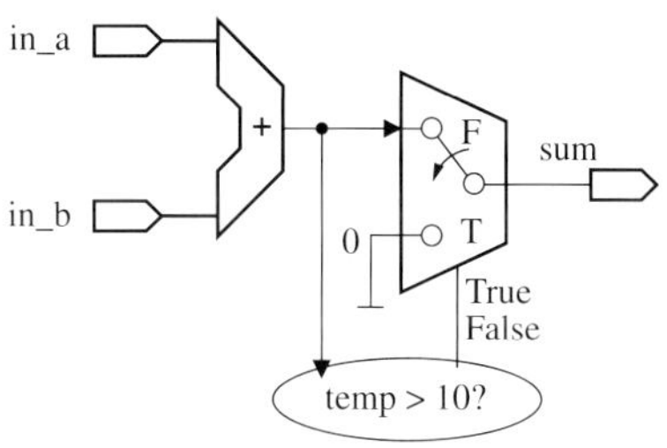

Fig. 4.32 Usage of variable

Table 4.5 Simulation progress of process V2

Simulation time	in_a	in_b	temp	sum
	2	3	5	5
20 ns	6	6	12	5
20 ns + 1Δ	6	6	12	0
20 ns + 2Δ	6	6	12	0

Shared Variables

Normally signals are responsible for process to process communication. VHDL'93 defines a global variable by using the add-on SHARED in its declaration. This global variable can be

In the following example a typical application for the IF statement is shown.

❑ *Example*: Address decoder

An address decoder distributes the address space available amongst peripheral units. The process 'decodeA' decodes the full address space, whereas 'processB' only prepares the selection to a maximum of 16 pages of 4k each. To address a piece of peripheral equipment, the corresponding chip select signal must be active. Thus the second version is independent of the physical location of the memory map (fig. 4.35).

```
LIBRARY IEEE;
USE IEEE.STD_LOGIC_1164.ALL;

ENTITY addr_decode IS
PORT (
  address:  IN STD_LOGIC_VECTOR(15 DOWNTO 0);
  cs1, cs2, cs3, cs4:  OUT STD_LOGIC);
END addr_decode;

ARCHITECTURE addr_decode_arch OF
             addr_decode IS
BEGIN
  decodeA: PROCESS(address)
  BEGIN
    cs1 <= '0';
    cs2 <= '0';
    cs3 <= '0';
    cs4 <= '0';
    IF(address >= X"0000" AND
      address < X"2000") THEN
    cs1 <= '1';
    ELSIF(address >= X"2000" AND
      address < X"3000") THEN
    cs2 <= '1';
    ELSIF(address >= X"3000" AND
      address < X"7000") THEN
    cs3 <= '1';
    ELSIF(address >= X"7000") THEN
    cs4 <= '1';
    END IF;
  END PROCESS decodeA;
END addr_decode_arch;

decodeB: PROCESS(address)
VARIABLE addr :  STD_LOGIC_VECTOR(3 DOWNTO 0);
BEGIN
  addr := address(15 DOWNTO 12);
  cs1 <= '0';
  cs2 <= '0';
  cs3 <= '0';
  cs4 <= '0';
  IF(addr = "0000" ) THEN
    cs1 <= '1';
  ELSIF(addr = "0001") THEN
    cs2 <= '1';
  ELSIF(addr = "0010") THEN
    cs3 <= '1';
  ELSIF(addr = "0011") THEN
    cs4 <= '1';
  END IF;
END PROCESS decodeB;
```

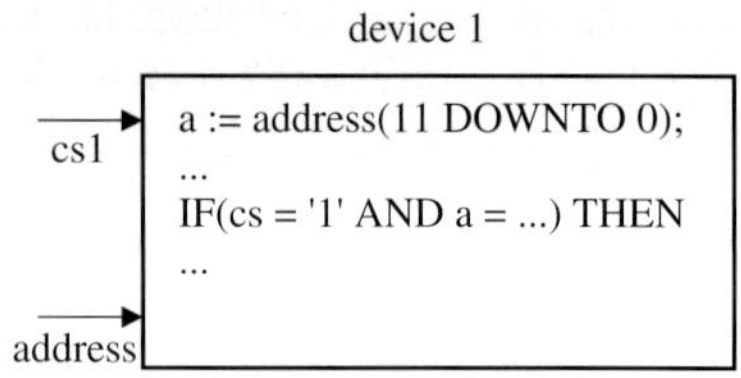

Fig. 4.35 Address decoder

CASE WHEN

Similarly to an IF statement, CASE statements allow a branch of program flow depending on a signal's or variable's value or on the result of a logical expression.

To control the program flow only a discrete data type (such as INTEGER or ENUMERATOR) is allowed. However, this offers a wide range of possibilities, because most of the signal types are declared to be of the type ENUMERATOR. The CASE statement can be compared with a table. 'expression' indicates the line in which the instruction to be executed can be found.

```
CASE expression IS
  WHEN value1 =>
    do something;
  WHEN value2 =>
    do something else;
  WHEN OTHERS =>
    do the normal;
END CASE;
```

A clear example is the VHDL code to describe the output of a state machine.

```
TYPE states IS
(STATE1, STATE2, STATE3, STATE4);
TYPE color IS
(RED, GREEN, BLUE, YELLOW);
SIGNAL actual_state :  states;
SIGNAL output:  color;
...
PROCESS(actual_state)
BEGIN
```

```
  CASE actual_state IS
    WHEN STATE1 => output <= GREEN;
    WHEN STATE2 => output <= YELLOW;
    WHEN STATE3 => output <= RED;
    WHEN STATE4 => output <= BLUE;
  END CASE;
END PROCESS;
```

Every branch can contain several instructions, including additional CASE or IF statements. The list of choice (WHEN ...) must contain all values *expression* can take. If it is not desirable to list all remaining cases, they can be combined in the last choice (WHEN OTHERS). Apart from that, groups or ranges are possible:

```
-- STATE1 or STATE2:
WHEN STATE1 | STATE2 =>

-- All states, of STATE1 up to STATE4
-- with attribute
  'LEFT and 'RIGHT:
WHEN states'LEFT TO states'RIGHT =>
```

After synthesis has been carried out, once again, we will find multiplexers in the hardware. Branching alternatives always depend on only one input. They are mutually exclusive. Because of that, at times the CASE statement is the better choice, because multiple branches in IF statements assume independent controlling conditions.

LOOP

VHDL distinguishes three different variations of a loop:

- The *simple LOOP*, which is an infinite loop. It can be terminated with EXIT.

```
LOOP
  do something repeatedly;
  EXIT WHEN (condition=TRUE)
END LOOP
```

- The *WHILE loop* runs as long as the loop condition is true.

```
WHILE (condition=TRUE) LOOP
  while condition is TRUE; END LOOP;
```

- The *FOR loop* repeats the body of the loop *n* times. *n* is specified in the head of the loop.

```
FOR i IN 0 TO 15 LOOP
  16 times to repeat;
END LOOP;
```

The loop allows the repetition of a part of a VHDL program several times, for example to process each element of an array in the same way. If the program is to be synthesized it is important to know how often the loop will be processed, because this determines the complexity of the circuit. In a FOR loop the loop parameter defines how often the body of the loop will be repeated.

4

❑ *Example*: Parity generator with a FOR loop

This example shows a FOR loop to perform a XOR operation to all individual signals of a vector. The result adds an even checksum to the data word. The checksum is attached to the data word using the concatenation operator '&'.

```
ENTITY parity_gen IS
PORT ( data_in:
       IN STD_LOGIC_VECTOR(15 DOWNTO 0);
    data_out:
       OUT STD_LOGIC_VECTOR(16 DOWNTO 0));
END parity_gen;

ARCHITECTURE parity_arch OF parity_gen IS
BEGIN
  gen:  PROCESS(data_in)
  VARIABLE parity:  STD_LOGIC;
  BEGIN
    parity := '0';
    FOR i IN 15 DOWNTO 0 LOOP
      parity := parity XOR data_in(i);
    END LOOP
    data_out <= parity & data_in;
  END PROCESS gen;
END parity_arch;
```

It is worthwhile mentioning some special **features**:

- The loop parameter 'i' is only visible within the loop;
- The loop parameter is like a constant;
- `15 DOWNTO 0`: first loop iteration starts with `i=15`;
- `0 TO 15`: first loop iteration starts with `i=0`;
- A loop parameter can be of type ENUMERATOR: `FOR i IN states'RANGE LOOP`, 'i' subsequently takes all values as defined in the type declaration of states;
- The use of a signal 'parity' instead of the variable is not possible, because the result would be the very last assignment to the signal with the parity calculated in the preceding simulation delta. `parity <= parity XOR data_in(0)`.

Use of the EXIT command

❑ *Example*: Counting leading '1's in a WHILE loop

A WHILE loop is a big challenge for a synthesis tool. A WHILE loop can not be synthesized if the compiler can not detect the maximum number of iterations or if the maximum number is a condition dynamically evaluated at run time.

In the following example a WHILE loop is used to count the number of leading ones in a data word. The first zero being detected causes a break of the loop. Assuming the vector only contains ones, the loop is terminated with the EXIT command because the index of the vector 'data' must not exceed the value 15 (run time error). The result *cnt* serves as an index to address individual bits of the vector at the same time.

The sample VHDL code can not be synthesized, because the index *cnt* is changed within the loop. It would be better to realize that task using a FOR loop (process countB).

```
ENTITY number_leading_ones IS
PORT (
  data:   IN STD_LOGIC_VECTOR(0 TO 15);
  number:  OUT NATURAL RANGE 0 TO 16);
END number_leading_ones;

ARCHITECTURE number_arch OF
  number_leading_ones IS
BEGIN
  countA: PROCESS(data)
  VARIABLE cnt:  NATURAL RANGE 0 TO 16;
  BEGIN
    cnt := 0;
    WHILE data(cnt) = '1' LOOP
      cnt := cnt + 1;
      EXIT WHEN cnt = 16;
    END LOOP;
    number <= cnt;
  END PROCESS countA;
END number_arch;

  countB: PROCESS(data)
  VARIABLE cnt:  NATURAL RANGE 0 TO 16;
  BEGIN
    cnt := 0;
    FOR i IN 0 TO 15 LOOP
      EXIT WHEN data(i) = '0';
      cnt := cnt + 1;
    END LOOP;
    number <= cnt;
  END PROCESS countB;
```

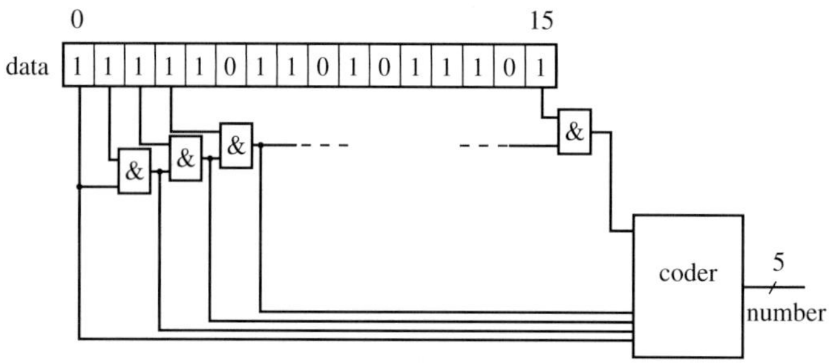

Fig. 4.36 Interpretation of the loop within the process countB

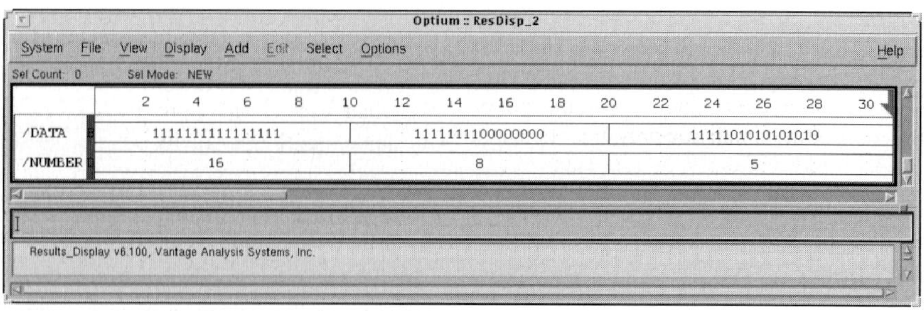

Fig. 4.37 Simulation result: counting leading ones in a data word

Example without loop:

```
...
cnt := 0;
IF(data0) = '1') THEN cnt := 1;
  IF(data(1) = '1') THEN cnt := 2;
    IF(data(2) = '1') THEN cnt := 3;
      ...
    END IF;
  END IF;
END IF;
```

Use of the NEXT command

❑ *Example*: Counting of all '1's in a data word

```
ENTITY number_of_all_ones IS
PORT (
  data:  IN STD_LOGIC_VECTOR(0 TO 15);
  number:  OUT NATURAL RANGE 0 TO 16);
END number_of_all_ones;

ARCHITECTURE number_arch OF
  number_of_all_ones IS
BEGIN
  count:  PROCESS(data)
  VARIABLE cnt:  NATURAL RANGE 0
    TO (data'HIGH+1);
  BEGIN
    cnt := 0;
    FOR i IN data'RANGE LOOP
      NEXT WHEN data(i) = '0';
      cnt := cnt + 1;
    END LOOP
    number <= cnt;
  END PROCESS count;
END number_arch;
```

WAIT

So far the sensitivity list of a process exclusively decides when the process is to be executed. An alternative path of a process employs a wait instruction, instead of the sensitivity list, to put the process into a suspended state. For synthesis, however, there are substantial limitations when using the wait statement. There must be only one wait statement, either at the beginning or at the end of the process. For the simulation only, this statement offers quite a lot of possibilities to handle process flow.

```
PROCESS
BEGIN
  y <= a AND b;
  -- complete replacement of the
  -- sensitivity list:
  -- Wait on an event in signal a or b
  WAIT ON a, b;
END;
```

Processes which have a sensitivity list are executed once, during the so called elaboration phase for initialization, at the start of the simulation. The WAIT statement at the end of the process is an equally good replacement, because all statements placed in front of it are at least executed once at the start of the simulation. The initialization of a process in VHDL has no counterpart in hardware. That is why initialization can be switched off in some simulators. This prevents the simulation starting in a well defined state, whereas the synthesized circuit could enter an undefined state. That is the reason for placing the wait statement at the beginning of a process.

Additional forms of a **WAIT assignment** are:

- WAIT FOR time_expression;

```
PROCESS
BEGIN
  reset <= '0';
  WAIT FOR 100ns;
  reset <= '1';
  WAIT; -- for ever
END PROCESS;
```

- WAIT UNTIL (boolean expression);

```
PROCESS
BEGIN
  WAIT UNTIL (clk'EVENT AND clk = '1');
  q <= d;
END PROCESS;
```

ASSERT

The assert statement is a suitable tool for trapping an erroneous behavior of the VHDL model. The statement knows, based on a boolean expression, the correct situation and announces every discrepancy during simulation. The assert statement is, depending on its location within VHDL code, concurrent when placed outside a process, or sequential when placed inside a process. Its form remains the same, but the sensitivity list will be different. The concurrent statement is sensitive to all signal changes of the logical expression, whereas the sequential statement is tied to the sensitivity list of the associated process.

```
ASSERT boolean_expression
REPORT "output of a message"
```

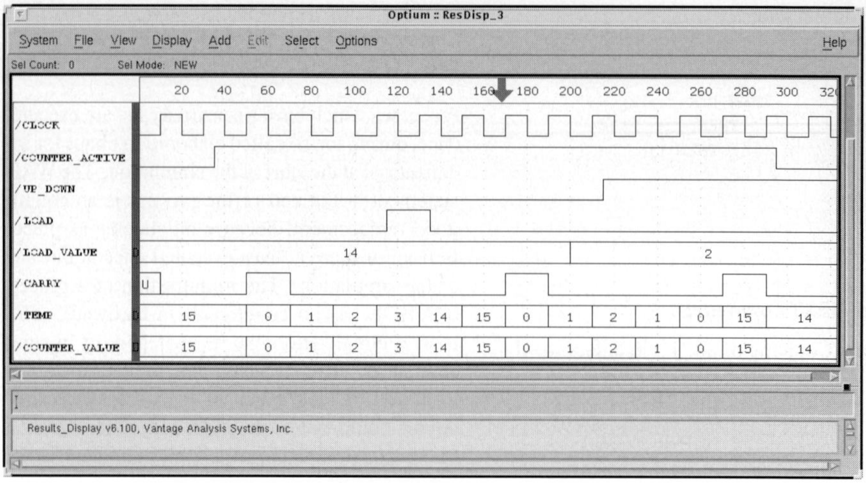

Fig. 4.38 Simulation of the synchronous counter

```
SEVERITY severity_level: NOTE, WARNING,
ERROR, FAILURE;
```

Example:

```
ASSERT (COUNT < 255)
REPORT "Invalid Counter value"
SEVERITY FAILURE;
```

If the logical expression returns the result FALSE a message is issued. Without a REPORT statement the message simply is 'assertion violation' plus stating the module, which causes the error. The level of the error status decides if the simulation is to be continued.

NOTE: This error status denotes insignificant output, such as general comments regarding the state or progress of the simulation or a notification of time consuming computing activities.

WARNING: A warning indicates a situation which is exceptional but not necessarily a fault. The simulation can continue.

ERROR: Replacement value when no error states are specified. In this case often a condition is not met, which causes things to go wrong. Often simulation will be terminated.

FAILURE: A severe error condition occurred, e.g., division by zero. Simulation will be terminated.

NULL

Sometimes it is useful to have a statement which does nothing at all. The NULL statement could, e.g., be situated in a branch of an IF or CASE statement. VHDL requires the explicit use of a NULL statement, the opposite of the programming language C in which an additional semicolon produces an empty instruction.

At the end of this chapter an example of a typical application will be presented. Some of the components discussed so far will be used.

❑ *Example*: Synchronous counter

```
LIBRARY IEEE;
USE IEEE.STD_LOGIC_1164.ALL;

ENTITY counter IS
PORT(
  clk:  IN STD_LOGIC;
  counter_active:  IN STD_LOGIC;
  up_down:  IN STD_LOGIC;
  load:  IN STD_LOGIC;
  load_value:
    IN NATURAL RANGE 15 DOWNTO 0;
  counter_value:
    OUT NATURAL RANGE 15 DOWNTO 0;
  carry:  OUT STD_LOGIC);
END counter;

ARCHITECTURE counter_arch OF counter IS
SIGNAL temp:  NATURAL RANGE 15 DOWNTO 0;
BEGIN

  counter_value <= temp;

  count:  PROCESS
```

```
  BEGIN
  WAIT UNTIL clk'EVENT AND clk = '1';

  carry <= '0';
  -- might be overloaded later
  ASSERT (load AND counter_active) = '0'
  REPORT "Counting and loading at same
  time?!"
  SEVERITY WARNING;

  IF(load = '1') THEN
    temp <= load_value;
  ELSIF(counter_active = '1') THEN
    CASE up_down IS
      WHEN '0' =>
        IF(temp = 15) THEN
          temp <= 0;
          carry <= '1';
        ELSE
          temp <= temp + 1;
        END IF;
      WHEN '1' =>
        IF(temp = 0) THEN
          temp <= 15;
          carry <= '1';
        ELSE
          temp <= temp - 1;
        END IF;
      WHEN OTHERS => NULL;
    END CASE;
  END IF;
  END PROCESS count;
END counter_arch;
```

4.6 Sequential clock synchronous logic

Digital systems such as processes, finite state machines, or communication equipment need a clock. The clock determines their operating speed and is responsible for partial results being present at the right point in time. Without a clock pulse the circuit remains in its actual state. In CMOS technologies this halt of activity causes a small, often insignificant, quiescent current. If the circuit consists of several parts or if a result depends on the present state of the circuit, a clock ensures a regular and well defined flow of events. The whole circuit only changes state when a special event occurs, for example a rising edge of the clock. Consequently the storage elements (registers or flip flops) change their values at the same time, synchronously with the clock. The principle can be compared to a bucket chain, working only when all participants take the bucket coming from one side and pass it over to the other side.

The design of a clock synchronous circuit is called a 'description at the register transfer level', having transfer functions to describe the data flow between registers. In hardware, flip flops often make up the registers, whereas combinatorial operations implement the transfer function (fig. 4.39).

Partitioning a design into individual registers and transfer functions is an important step in the design cycle of a digital circuit. The synthesis tools expect that structure in a VHDL description too. This chapter shows how to describe synchronous state machines using VHDL and points out when to take care if a synthesis is to follow.

4.6.1 Combinatorial Logic

VHDL offers a lot of possibilities for describing combinatorial logic. Some of them have already been covered in the chapters before. This includes:

- concurrent statements and the use of the logical operators such as AND, NAND, OR, NOR, XOR, XNOR and NOT;
  Example: `y <= (a AND b) AND NOT z;`

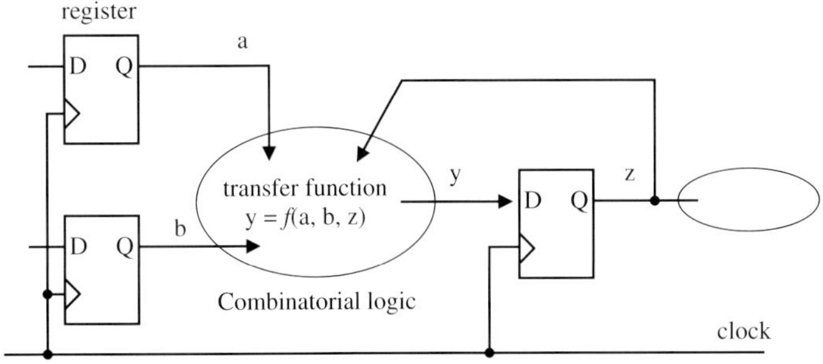

Fig. 4.39 Principle of a sequential and synchronous logic (state machine)

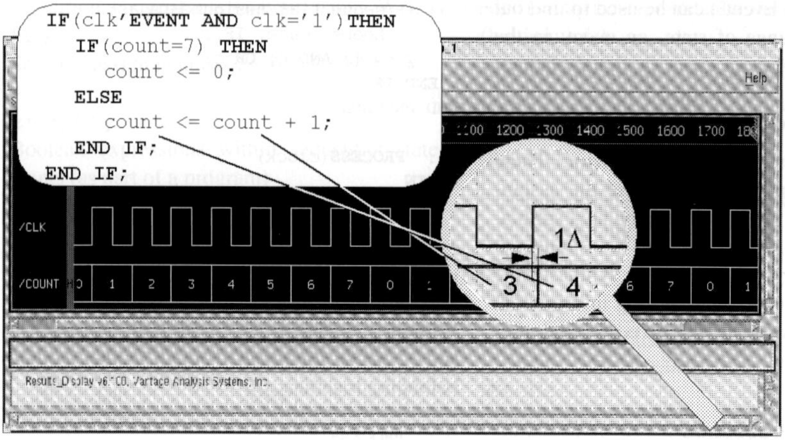

Fig. 4.43 Simulation of clocked processes

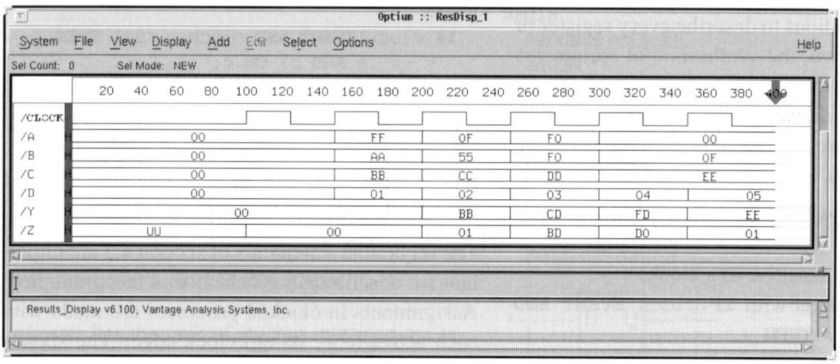

Fig. 4.44 Simulation result of the VHDL code without timing

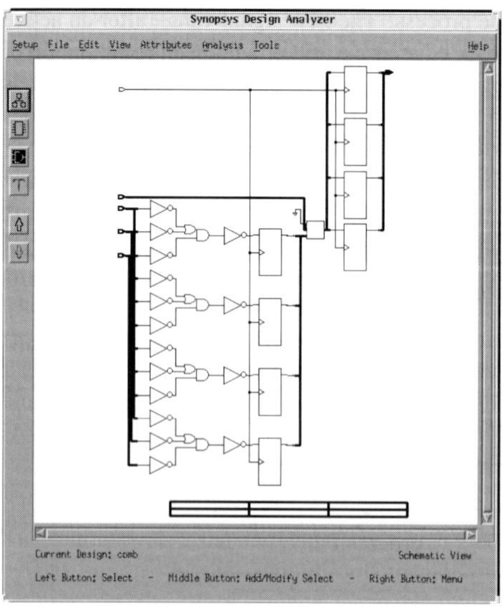

Fig. 4.45 Synthesized process with combinatorial logic and registers

The synthesis tools convert the process description directly. Within a statement IF(clock'EVENT AND clock = '1') all signals located on the left side of an assignment turn into register outputs. The operation on the right side of the assignment forms the combinatorial part of the register. Figure 4.43 and fig. 4.44 show the simulation result of the VHDL code describing a clocked process. Figure 4.45 shows the result of the synthesis, and fig. 4.46 is an example to display the critical path in the timing analysis of a synthesized circuit.

Reset in Clocked Processes

Registers are storage elements and take a random value after the supply voltage has been applied. The correct operation of the circuit often depends on a defined starting value at the output of the register. Initialization of registers is performed with a special signal (reset) which is connected to all registers. This (hardware) initialization must not be confused with initialization in a signal dec-

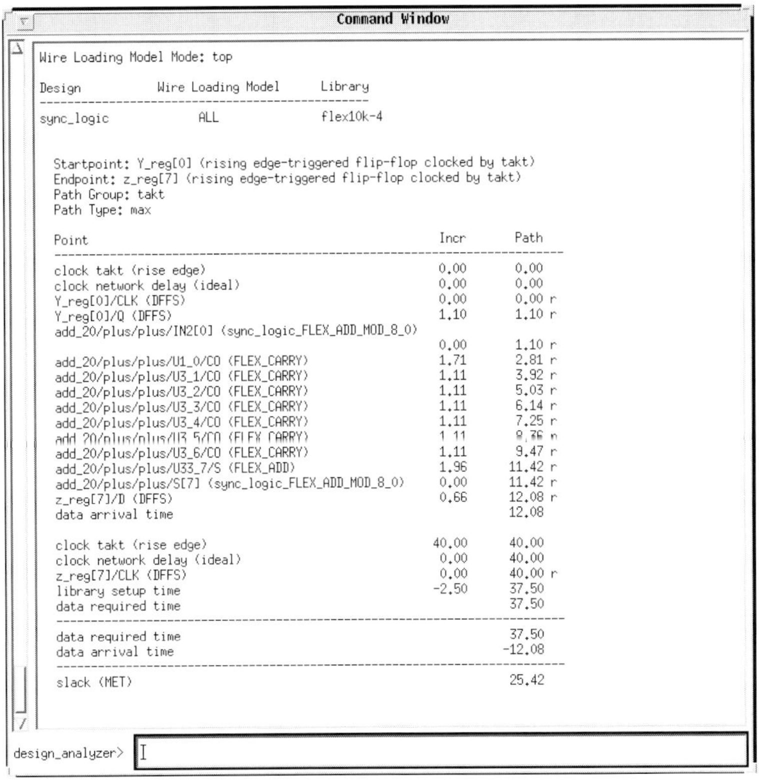

```
Command Window

Wire Loading Model Mode: top

Design              Wire Loading Model      Library
------------------------------------------------
sync_logic               ALL                flex10k-4


  Startpoint: Y_reg[0] (rising edge-triggered flip-flop clocked by takt)
  Endpoint: z_reg[7] (rising edge-triggered flip-flop clocked by takt)
  Path Group: takt
  Path Type: max

  Point                                                    Incr       Path
  --------------------------------------------------------------------------
  clock takt (rise edge)                                   0.00       0.00
  clock network delay (ideal)                              0.00       0.00
  Y_reg[0]/CLK (DFFS)                                      0.00       0.00 r
  Y_reg[0]/Q (DFFS)                                        1.10       1.10 r
  add_20/plus/plus/IN2[0] (sync_logic_FLEX_ADD_MOD_8_0)
                                                           0.00       1.10 r
  add_20/plus/plus/U1_0/CO (FLEX_CARRY)                    1.71       2.81 r
  add_20/plus/plus/U3_1/CO (FLEX_CARRY)                    1.11       3.92 r
  add_20/plus/plus/U3_2/CO (FLEX_CARRY)                    1.11       5.03 r
  add_20/plus/plus/U3_3/CO (FLEX_CARRY)                    1.11       6.14 r
  add_20/plus/plus/U3_4/CO (FLEX_CARRY)                    1.11       7.25 r
  add_20/plus/plus/U3_5/CO (FLEX_CARRY)                    1.11       8.36 r
  add_20/plus/plus/U3_6/CO (FLEX_CARRY)                    1.11       9.47 r
  add_20/plus/plus/U33_7/S (FLEX_ADD)                      1.96      11.42 r
  add_20/plus/plus/S[7] (sync_logic_FLEX_ADD_MOD_8_0)      0.00      11.42 r
  z_reg[7]/D (DFFS)                                        0.66      12.08 r
  data arrival time                                                  12.08

  clock takt (rise edge)                                  40.00      40.00
  clock network delay (ideal)                              0.00      40.00
  z_reg[7]/CLK (DFFS)                                      0.00      40.00 r
  library setup time                                      -2.50      37.50
  data required time                                                 37.50
  --------------------------------------------------------------------------
  data required time                                                 37.50
  data arrival time                                                 -12.08
  --------------------------------------------------------------------------
  slack (MET)                                                        25.42

design_analyzer>
```

Fig. 4.46 Identification of the 'critical path' (example of the FPGA Compiler Synopsys)

laration! A signal explicitly designated as reset can often not be avoided in a VHDL simulation, because at the simulation's start signals of the type std_logic take the value U (undefined). Structures with a feedback do not have a chance of going past the undefined state, because a boolean combination of undefined U and 0 or 1 results in U again. However, in the real world, after power up, a circuit must not enter a state which needs a reset.

There are two ways in which to realize a reset, the synchronous or the asynchronous way. Having a reset working asynchronously, the register will be reset independently of the clock signal. The reset signal has a priority in this case. On the other hand, a synchronous reset is realized by a boolean operation in the data path (fig. 4.47).

Process with asynchronous reset

```
-- Process is sensitive on the clock
-- and reset signal
PROCESS(clock, reset)
BEGIN
  -- reset has priority before the clock
  IF(reset = '0') THEN
    signal_out <= '0';
  ELSIF(clock'EVENT AND clock = '1') THEN
    signal_out <= signal_in;
  END IF;
END PROCESS;
```

Process with synchronous reset

```
-- process is only sensitive
-- on the clock
PROCESS(clock)
BEGIN
  IF(clock'EVENT AND clock = '1) THEN
    IF(reset = '0') THEN
      signal_out <= '0';
    ELSE
      signal_out <= signal_in;
    END IF;
  END IF;
END PROCESS;
```

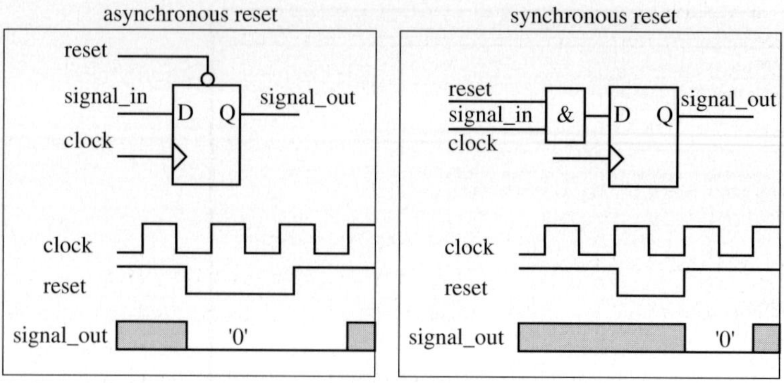

Fig. 4.47 Synchronous und asynchronous reset

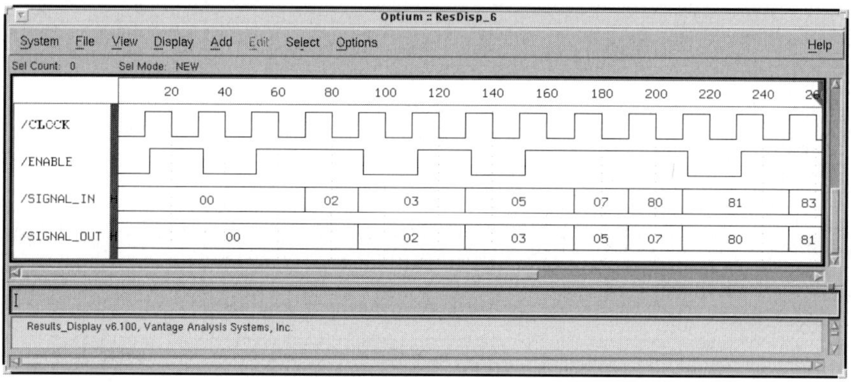

Fig. 4.48 Simulation of a process with an enable signal

Clocked Process with an Enable Function

In many cases it is desirable to stop activities in clocked processes during one or several clock periods, or to make signal assignments only when a valid input signal is present. A dedicated signal (enable) takes control and makes sure that the register retains its actual value if there is an interruption. This is carried out by a feedback from the output to the input (figures 4.48 and 4.49).

❑ *Example*: Process with an Enable Function

```
PROCESS(clock)
BEGIN
  IF(clock'EVENT AND clock = '1') THEN
    IF(enable = '1') THEN
      signal_out <= signal_in;
    END IF;
  END IF;
END PROCESS;
```

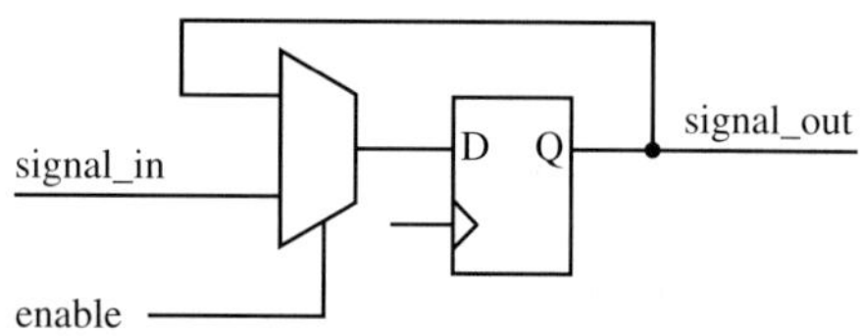

Fig. 4.49 Principle of clocked processes with an enable function: processed only when enable = '1'

Variables in Clocked Processes

As shown before, variables take their values immediately after an assignment. That is why they have a special status in a clocked process. The answer to the question of whether a register for a variable is to be synthesized or not depends on the location of the variable within the process. Assuming the assignment of the variable occurs prior to its evaluation, no register will be synthesized to store its value. In that case the variable is a substitute for a complex expression. On the other hand, if the variable were be read prior to its assignment, a register will be synthesized to memorize the value from the preceding clock cycle.

```
VAR1:  PROCESS(clock)
VARIABLE temp:  STD_LOGIC;
BEGIN
  IF(clock'EVENT AND clock = '1') THEN
    -- write to variable,
    -- if storage is not wanted
    IF(a = '1') THEN
      temp := b AND c;
    ELSE
      temp := b OR c;
    END IF;
      read variable after assignment
    signal_out <= temp;
  END IF;
END PROCESS VAR1;
```

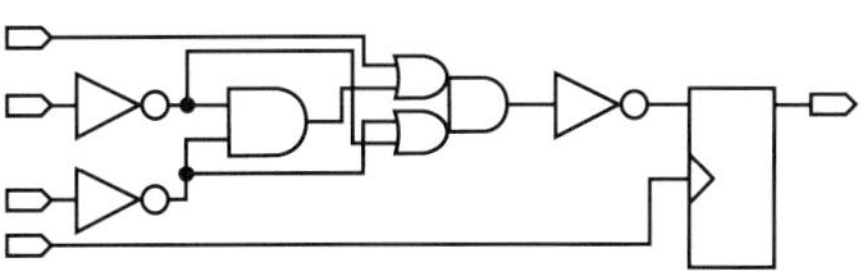

a) process VAR1

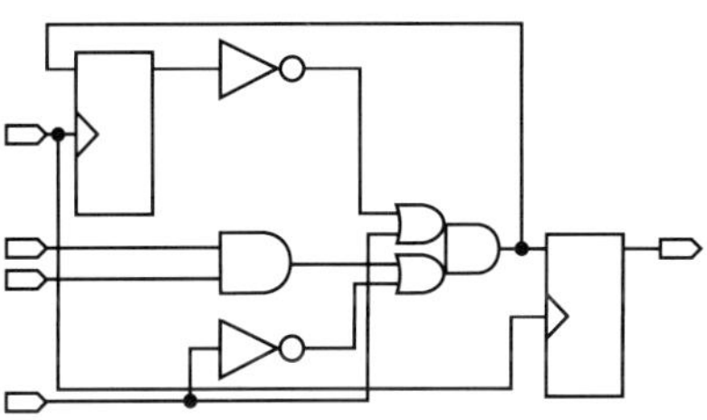

b) process VAR2

Fig. 4.50 The handling of the variables in synthesis

```
VAR2:  PROCESS(clock)
VARIABLE temp:  STD_LOGIC;
BEGIN
  IF(clock'EVENT AND clock = '1') THEN
    IF(a = '1') THEN
      temp := b AND c;
    ELSE
    -- If a='0', read temp first
    -- (value of the registers 'temp')
    -- and then assign value
      temp := NOT temp;
    END IF;
    -- second evaluation of temp
    -- corresponds to the value
    -- of the register input 'temp'
    signal_out <= temp;
  END IF;
END PROCESS VAR2;
```

Figure 4.50 Shows the handling of the variables in synthesis.

4.6.4 Processing of Asynchronous Bus Signals

In clock synchronous circuits, input signals with no reference to the clock signal must be looked at more closely. If it is a single signal (e.g., STD_LOGIC) the asynchronous signal can be 'synchronized' using two or three cascaded flip flops. A change of state on the asynchronous signal will be detected with a resolution of a clock period. This principle can not be applied when asynchronous bus signals (e.g., STD_LOGIC_VECTOR or INTEGER) are concerned as each signal of the bus can carry a displacement of one clock period.

❑ *Example*: the transfer of a bus signal (count) between two systems having two different operation clocks (clock1 and clock2) will be examined.

It is not the bus (named count in the example) that is going to be synchronized, but two control signals (request and acknowledge) which request and acknowledge a transfer. Inverting the signal request initiates the transfer, inverting acknowledge indicates the value to be transferred is available in an intermediate register.

```
ARCHITECTURE SYNCH_ARCH OF SYNCH IS
SIGNAL count, temp, read_count :
  NATURAL RANGE 65535 DOWNTO 0;
SIGNAL request, request1, request2,
  request3:  STD_LOGIC;
SIGNAL acknowledge, acknowledge1,
  acknowledge2, acknowledge3:  STD_LOGIC;

BEGIN

SYS1:  PROCESS(clock1)
BEGIN
  IF(clock1'EVENT AND clock1 = '1') THEN
    request1 <= request;
    request2 <= request1;
    request3 <= request2;

    -- count is to be past
```

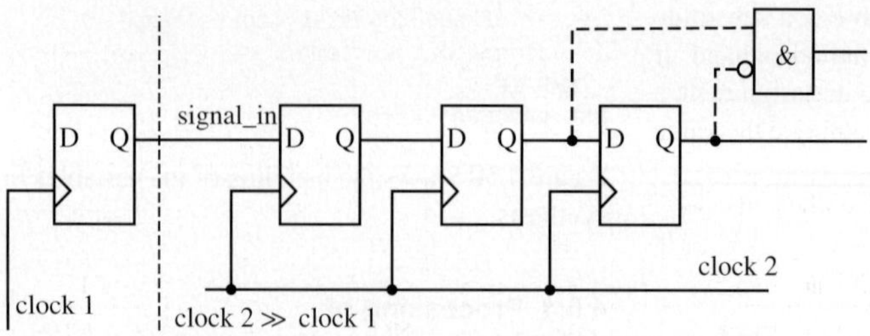

Fig. 4.51 Synchronization of an asynchronous signal (signal_in)

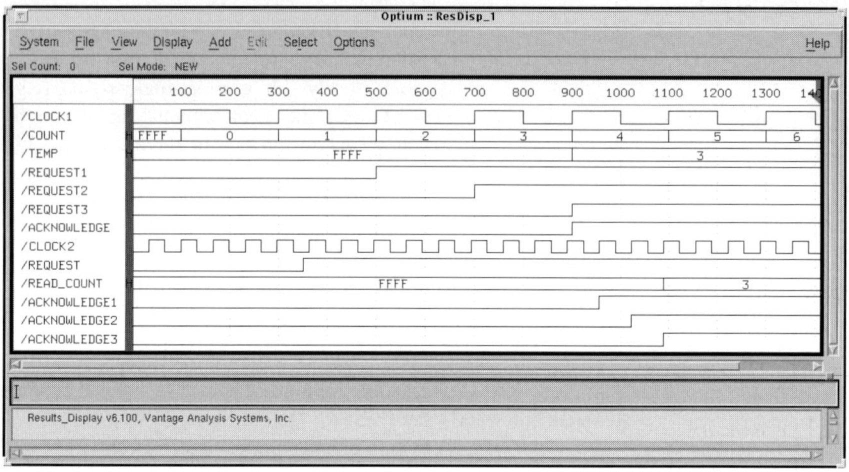

Fig. 4.52 Processing asynchronous bus signals; a simple protocol controls the data flow

```
    count <= count + 1;

    -- request to transfer the value?
    IF((request3 XOR request2) = '1') THEN
      temp <= count;
      -- show intermediate result
      acknowledge <= NOT acknowledge;
    END IF;
  END IF;
END PROCESS SYS1;

SYS2:  PROCESS(clock2)
BEGIN
  IF(clock2'EVENT AND clock2 = '1') THEN
    acknowledge1 <= acknowledge;
    acknowledge2 <= acknowledge1;
    acknowledge3 <= acknowledge2;

    -- request value
    request <= NOT request;

    -- accept intermediate result
    -- after acknowledgement
    IF((acknowledge3 XOR
         acknowledge2) = '1') THEN
      read_count <= temp;
    END IF;
  END IF;
END PROCESS SYS2;
```

4.7 Types

Each of the objects signals, constants, or variables has a specific type assigned in VHDL. Object types inform us about the characteristics of data flow and provide a specification telling how the data is to be processed. VHDL is said to be a strongly type oriented language. Apart from the already available standard types, one's own type definitions can be added easily. Looking at existing types, new definitions (so called sub-types) can be derived. Table 4.6 mentions the possible type classes and the tables 4.7 up to 4.15 describe pre-defined standard types of the package STANDARD as well as the operators applicable to these types.

Table 4.6 Class of Types

Class of types	Usage
Enumerator	List of identifiers and character literals
Integer	Positive or negative whole number with zero
Floating Point	numbers in scientific notation
Physical	Physical units
Array	One- or multi-dimensional fields
Record	Composition of elements
Access	Access with addresses (pointer)
File	A collection of data

4.7.1 Standard Types

Predefined Types in the Package STANDARD

Table 4.7 Type BOOLEAN

BOOLEAN	
Type declaration	**TYPE** BOOLEAN **IS** (FALSE, TRUE);
Type	Enumerator
Description	Used for relation operation
Logical Operators Relational Operators	NOT, AND, OR, NAND, NOR, XOR, XNOR =, /=, <, <=, >, >=
Example	**SIGNAL** a, b: **INTEGER** ... **IF**((a > b) **AND** (c < d)) **THEN** ... **IF**((a > b) = TRUE) **THEN** ...

Table 4.8 Type BIT

BIT	
Type declaration	**TYPE** BIT **IS** ('0', '1');
Type	Enumerator
Description	Used for the logical states 0 and 1
Logical Operators Relational Operators	NOT, AND, OR, NAND, NOR, XOR, XNOR =, /=, <, <=, >, >=
Example	**SIGNAL** a,b,c: **BIT**; ... **IF**(a = '0') **THEN** b <= **NOT** c;

Table 4.9 Type BIT_VECTOR

BIT_VECTOR	
Type declaration	**TYPE** BIT_VECTOR **IS ARRAY** (NATURAL RANGE <>) **OF BIT**;
Type	Array
Description	one-dimensional Array of BIT
Logical Operators Relational Operators Shift Operators Concatenation Operator	NOT, AND, OR, NAND, NOR, XOR, XNOR =, /=, <, <=, >, >= SLL, SRL, SLA, SRA, ROL, ROR &
Example	**SIGNAL** data: **BIT_VECTOR** (7 **DOWNTO** 0) ; ... data <= "10100101";

Table 4.10 Type CHARACTER

CHARACTER	
Type declaration	**TYPE** CHARACTER **IS** (...); see Appendix C
Type	Enumerator
Description	ASCII characters
Relational Operators	=, /=, <, <=, >, >=
Example	**SIGNAL** c: **CHARACTER**; ... c <= 'A';

Table 4.11 Type INTEGER

INTEGER	
Type declaration	**TYPE** INTEGER **IS RANGE** −2147483648 **TO** 2147483647;
Type	Integer
Description	32 Bit two's complement description of whole numbers
Relational Operators Arithmetic Operators	=, /=, <, <=, >, >= Sign +/−, ABS, +, −, *, /, MOD, REM, **
Example	**SIGNAL** x,y: **INTEGER**; ... y <= x + 4;

Table 4.12 Type NATURAL

NATURAL	
Type declaration	**SUBTYPE** NATURAL **IS INTEGER RANGE** 0 **TO** INTEGER'**HIGH**;
Type	Subtype of Integer
Description	Subtype derived of Integer. Restricted to natural numbers. Calculation of provisional results are taken with the type integer. In an assignment the range of the subtype must be observed.
Relational Operators Arithmetic Operators	=, /=, <, <=, >, >= Sign +/−, ABS, +, −, *, /, MOD, REM, **
Example	**SIGNAL** x,y: NATURAL; ... y <= x + 4;

Table 4.13 Type REAL

REAL	
Type declaration	**TYPE** REAL **IS RANGE** −1.70141l0E+38 **TO** 1.70141l0E+38;
Type	Floating Point
Description	Representation of floating point numbers (not synthesizable)
Relational Operators Arithmetic Operators	=, /=, <, <=, >, >= Sign +/−, ABS, +, −, *, /, **
Example	**CONSTANT** pi: **REAL** := 3.14

Table 4.14 Type TIME

TIME	
Type declaration	**TYPE** TIME **IS RANGE** −9223372036854775807 **TO** 9223372036854775807 **UNITS** fs;
Type	Physical
Description	Representation of time: delay time, simulation time
Relational Operators Arithmetic Operators	=, /=, <, <=, >, >= Sign +/−, ABS, +, −, *, /

Example	**SIGNAL** simulation_time: **TIME**;
	...
	simulation_time <= **NOW**+10 ns;
	CONSTANT delay: TIME:=5 ns;
	...
	d_out <= d_in **AFTER** delay;

Table 4.15 Type STRING

STRING	
Type declaration	**TYPE** STRING **IS ARRAY (POSITIVE RANGE <>) OF CHARACTER;**
Type	Array
Description	One-dimensional Array of CHARACTER
Relational Operators Concatenation Operator	=, /=, <, <=, >, >= &
Example	**SIGNAL** w1, w2: **STRING**(4 **DOWNTO** 1); **SIGNAL** w3: **STRING**(8 **DOWNTO** 1); ... w1 <= "this"; w2 <= "+that"; w3 <= w1 & w2;

Predefined Types of the IEEE library: Package IEEE.STD_LOGIC_1164

Table 4.16 Type STD_ULOGIC

STD_ULOGIC	
Type declaration	**TYPE** STD_ULOGIC **IS** (...); see Appendix C
Type	Enumerator
Description	Nine-level logic without resolution function
Logical Operators Relational Operators	NOT, AND, OR, NAND, NOR, XOR, XNOR =, /=, <, <=, >, >=

Example	**SIGNAL** d_out, d_in:
	STD_ULOGIC;
	...
	IF(ena = '1') **THEN**
	    d_out <= din;
	ELSE
	    d_out <= 'Z';

Table 4.17 Type STD_LOGIC

STD_LOGIC	
Type declaration	**SUBTYPE** STD_LOGIC **IS RESOLVED** STD_ULOGIC;
Type	Sub-type of STD_ULOGIC
Description	Nine-level logic with resolution function
Logical Operators	NOT, AND, OR, NAND, NOR, XOR, XNOR
Relational Operators	=, /=, <, <=, >, >=
Example	**SIGNAL** d_out, d_in: STD_LOGIC; ... d_out <= 'Z'; d_out <= d_in;

4.7.2 Type Class Enumerator

Enumerators are versatile and provide a readable and clear VHDL text. The type definition of an ENUMERATOR assigns a symbolic name to all values in a list. Very often the signal representing the state of a state machine is an enumerator. This has the advantage that the symbolic names of the states are visible during simulation. They will not be replaced by tangled bit combinations. Apart from reserved words, every name is legal in VHDL.

```
TYPE current_state IS
(RED, GREEN, YELLOW, BLUE);
```

This type owns four literals: Red, Green, Yellow, Blue. Corresponding to the position in the list each literal receives a number assigned, starting with position 0. Looking at the number of this position, a comparison using the operators (=, /=, <, <=, >, >=) is possible. In addition to the symbolic names the list of an enumerator can contain character literals as well. Character literals are single letters, being part of the ASCII character set, framed by apostrophes:

```
TYPE BIT IS ('0', '1');
```

The advantage of character literals is the way they describe arrays as a string. They are suitable for designing a logic system containing single signals and bus signals.

```
single_signal <= '0';
one_bus <= "101010101";
```

4.7.3 Physical Types

The physical types are a speciality of VHDL. A unit must be specified here. Physical types not only define the legal range of values, but also identify the absolute unit and provide a conversion table for deducted units. Because of the conversion table, units in expressions can be mixed, e.g., the addition of minutes and nanoseconds.

```
TYPE TIME IS RANGE -9223372036854775807
TO 9223372036854775807
  UNITS
    -- Basic unit:  femto-seconds
    fs;
    ps = 1000 fs;
    ns = 1000 ps;
    us = 1000 ns;
    ms = 1000 us;
    sec = 1000 ms;
    min = 60 sec;
    hr = 60 min;
  END UNITS;
...

SIGNAL start :  TIME := 1 min + 10 ns;
```

Physical types have no meaning for the synthesis, but they are meaningful for the design of simulation environments (test bench, section 4.10). There are, e.g., objects of the type TIME, responsible for the correct progress of simulation.

4.7.4 Type Class Record

A record is similar to the structure of the programming language C, a collection of elements into one object. The only operators usable on records are tests on equality or non-equality (=, /=). Elements are compared on a one by one basis, followed by a boolean AND operation performed on all prior results.

```
TYPE interface IS RECORD
  data:  STD_LOGIC_VECTOR(1 DOWNTO 0);
  parity:  STD_LOGIC;
END RECORD;

SIGNAL d_in, d_out:  interface;
...
  d_out <= d_in WHEN (d_in.parity =
```

```
    (d_in.data(1) XOR
     d_in.data(0))) ELSE
...
```

Additional forms of assignments to records or to elements in records are:

```
d_out <= (data => "10", parity => '1');
d_out <= ("10", '1');

d_out.  data <= "10";
d_out.parity <= '1';
```

4.7.5 Type Class Array

An array is a combination of elements of the same type. Individual elements are addressed using an index, which can be of the type integer or enumerator. Before a signal can be declared to be of the type array it must be specified which elements the array can take and how the indexing of the array is to be done (index type and range of value). There are **two different kinds of type definitions**:

- Constrained: The number of array elements is already specified in the type definition. All signals and variables of that type have the same size.

  Example:
  ```
  TYPE word IS ARRAY
    (NATURAL RANGE 15 DOWNTO 0) OF
  STD_LOGIC;
  SIGNAL a, b:  word;
  ```

- Unconstrained: The size of an array is not specified in the type definition. The fill-in 'RANGE <>' permits specification at a later point of time, e.g., when declaring the signal. Only the index type is specified, e.g., NATURAL. Because of that a family of similar arrays, varying only in size, can be defined; this provides a universal use of the type array.

  Example:
  ```
  TYPE STD_LOGIC_VECTOR IS ARRAY
    (NATURAL RANGE <>) OF
  STD_LOGIC;
  ...
  SIGNAL data:
    STD_LOGIC_VECTOR(7 DOWNTO 0);
  SIGNAL address:
    STD_LOGIC_VECTOR(9 DOWNTO 0);
  ```

The example declares a signal 'data' to be a one-dimensional array having 8 elements with an index in the range 7 DOWNTO 0 to address then. The first element has the index 7, the second element the index 6, and the last element has the index 0. This descending sequence of indices corresponds to the usual presentation of bus signals. The index of an addressed bit is equal to the power, when calculating its binary value. The access to an individual element can be done either by a static index, e.g.,

```
data(0) <= '1';
```

or dynamically using a variable, a signal, or a loop parameter:

```
FOR i IN 7 DOWNTO 0 LOOP
data(i) <= '0';
END LOOP;
```

Other similar assignments are:

```
data <= ('1', '0', '1', '1', '0', '0',
         '0', '0');
daten <= "10110000";
daten <= ('1', '0', '1', '1',
  OTHERS => '0');
data(7) <= '1'; data(6) <= '0';
data(5) <= '1'; data(4) <= '1';
data(3 DOWNTO 0) <= "0000";
```

Two-dimensional Arrays

The following example shows the structure of two-dimensional arrays. The first type definition declares a one-dimensional array (rom_data) in the first instance. The second type definition takes four members and combines them to an 'array made of arrays'. A suitable application of two-dimensional arrays would be the structuring of small storage blocks, e.g., RAM or ROM.

Example:

```
TYPE rom_data IS ARRAY
  (NATURAL RANGE 7 DOWNTO 0) OF
STD_LOGIC;
TYPE simple_rom IS ARRAY
  (NATURAL RANGE 0 TO 3) OF
rom_data;

CONSTANT rom:  simple_rom := (
('0', '0', '0', '0', '0', '0', '0'),
('0', '0', '0', '0', '0', '0', '1'),
('0', '0', '0', '0', '0', '1', '0'),
('0', '0', '0', '0', '0', '1', '1'));
SIGNAL data:  rom_data;
SIGNAL address:  NATURAL RANGE 7 DOWNTO 0;
...
```

```
FOR addr IN 0 TO 3 LOOP
  data <= rom(addr);
END LOOP;
```

4.7.6 Access

Access types are comparable to addresses or pointers in other programming languages. Being not synthesizable and most probably being a data type used very seldom, they allow the detection of complex functions. A typical application of pointers can be seen when modelling FIFOs (first in, first out) or building chained lists within a simulation environment (test bench, section 4.10).

Furthermore, access types are only legal for variables within a process.

Dealing with access types, two pre-defined functions are prominent: NEW and DEALLOCATE. The function NEW allocates storage for the size of the object and returns a pointer to the memory area. From that moment the object can be accessed using the pointer. If the object is not be needed during simulation, the allocated memory can be returned. To achieve that the function DEALLOCATE deletes the object addressed by the pointer and returns the memory to the system. Section 4.7.8 shows an example of a chained list using access types.

4.7.7 Type Class File

File objects allow a direct access to the file system. Within a VHDL description files are very suitable for reading or storing sequences of test vectors (section 4.10) or to initialize blocks of storage, such as RAM or ROM. The use of files provides a high degree of flexibility in a VHDL description, because files can be modified or exchanged without a new compilation step. The interfaces to files are procedures for transferring data (read and write) or functions to detect the end of a file (end of file).

The first step in setting up a file object is the definition of its content. Possible entries are scalar types, records, or arrays. The second step generates an object, links the object and the file name (including path information), and specifies the direction of the data transfer (IN or OUT).

```
TYPE integer_file IS FILE OF INTEGER;
TYPE text_file IS FILE OF STRING;
-- VHDL87:
FILE my_integerfile:  integer_file IS IN
"int.txt";
FILE my_textfile:  text_file IS OUT
"/home/example/text.txt";
-- VHDL93:
FILE my_integerfile:  integer_file;
FILE my_textfile:  text_file;
file_open(my_integerfile,
"int.txt", READ_MODE);
file_open(my_textfile,
"/home/example/text.txt", WRITE_MODE);
file_open(status, my_integerfile,
"int.txt", READ_MODE);
```

Whereas a file according to the VHDL87 standard has to be explicitly opened during the elaboration phase, the VHDL93 standard provides procedures solely for opening and closing files. Hence these two versions of the standard are not compatible. A second form of the procedure 'file open' additionally returns a status to indicate the successful opening of a file (OPEN_OK) or notifies when there is a problem:

- STATUS_ERROR: File is already in use;
- NAME_ERROR: File does not exist;
- MODE_ERROR: File can not be opened using this mode.

To read, write, or close a file the corresponding procedure is called with the name of the file object. Prior to each read operation a test for end of file should be carried out.

```
IF(NOT endfile(my_integerfile)) THEN
read(my_integerfile, integer_data);

write(my_textfile, string_data);
file_close(my_integerfile);
```

4.7.8 Modelling of a stack using access types

Example:

```
PACKAGE list_types IS
  TYPE data IS ARRAY (7 DOWNTO 0)
  OF STD_LOGIC;
  TYPE list_element;
  TYPE ptr_to_list_element
  IS ACCESS list_element;
  TYPE list_element IS RECORD
    list_entry:  data;
```

```
    wheretocontinue:  ptr_to_list_element;
  END RECORD;
END list_types;

USE WORK.list_types.ALL;

ENTITY list IS
PORT (
  clock:  IN STD_LOGIC;
  cs:  IN STD_LOGIC;
  rd_wr:  IN STD_LOGIC;
  din:  IN data;
  dout:  OUT data);
END list;

ARCHITECTURE list_arch OF list IS
BEGIN
  LAST_IN_FIRST_OUT: PROCESS(clock)
  VARIABLE list_start:
    ptr_to_list_element := NULL;
  VARIABLE temp:
    ptr_to_list_element := NULL;
  BEGIN
    IF(clock = '1' AND clock'EVENT) THEN
      -- writemode
      IF(rd_wr = '0' AND cs = '0') THEN
        -- allocate new storage
        temp := NEW list_element;
        -- enter data
        temp.list_entry := din;
        -- new element at start of list
        temp.wheretocontinue := list_start;
        list_start := temp;
      -- readmode
      ELSIF (list_start /= NULL AND
      rd_wr= '1' AND cs = '0') THEN
        -- goto start of list
        temp := list_start;
        -- read entry
        dout <= temp.list_entry;
        -- point to following entry
        list_start := temp.wheretocontinue;
        -- deallocate memory
        DEALLOCATE(temp);
      END IF;
    END IF;
  END PROCESS LAST_IN_FIRST_OUT;
END list_arch;
```

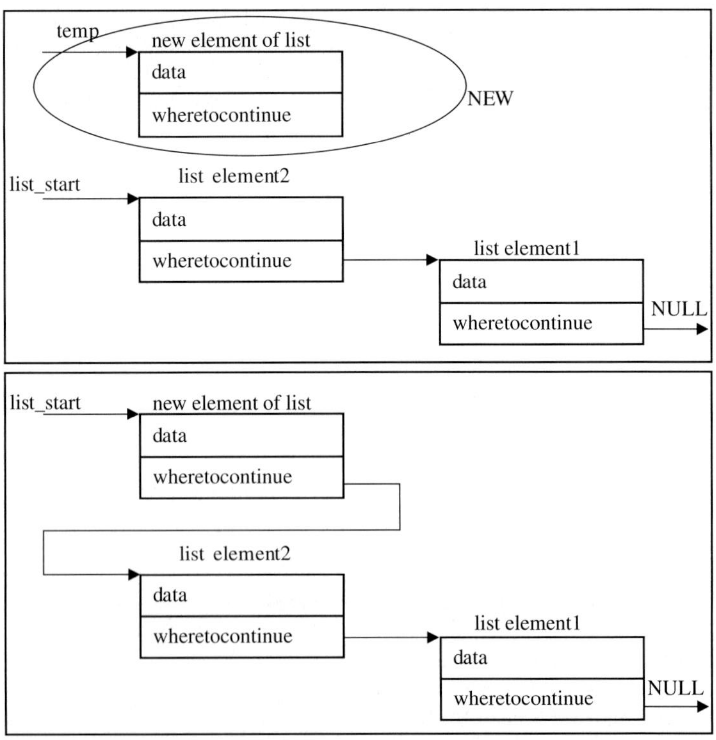

Fig. 4.53 New element to be positioned on the top end of the list

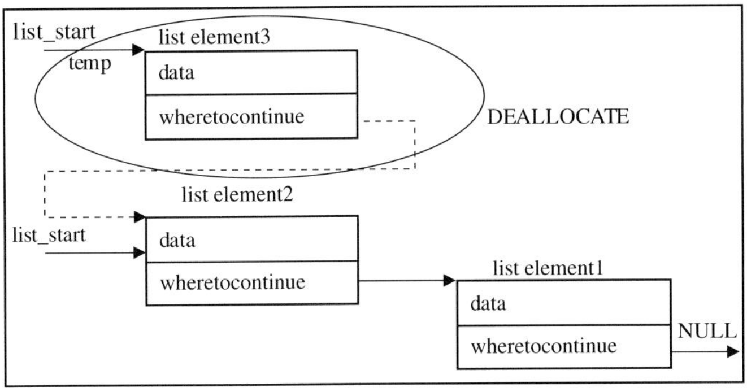

Fig. 4.54 Delete the top element of the list

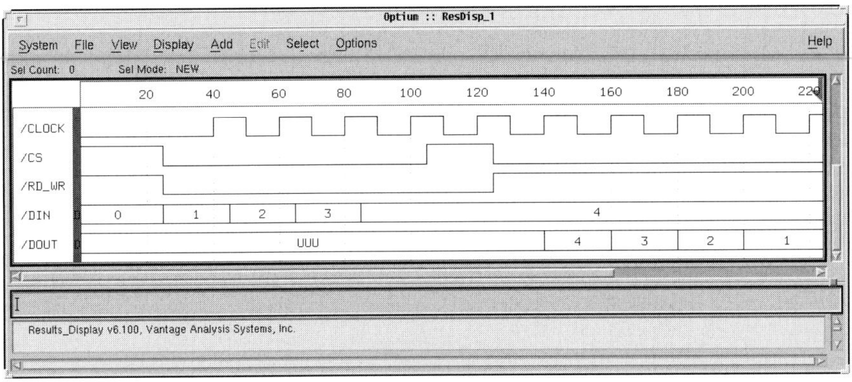

Fig. 4.55 Chained list in simulation

4.8 Operators

VHDL offers a large set of different operators for defining comparisons, boolean equations, or arithmetic in RTL models. Operators are always connected with objects (signals, variables or arrays) of a specific type (BIT, BOOLEAN, INTEGER). At the definition of a type it is necessary to supply a set of operators. Looking at the predefined types BIT and BOOLEAN, as well as their arrays, this is carried out automatically. If there are no implicit operators, either one's own operators must be added or the behavior of built in operators must be modified. The simple enumerator STD_LOGIC is the most popular type for which a package in the IEEE library implements operators. As a rule, a test to 'equal' or 'not equal' values is implemented for all types, because a test bit by bit is sufficient to perform that test.

Quite on the contrary, there is a difference in performing an addition with, e.g., integer objects or enumerators. The enumerators need specific instructions in order to provide a right result. An overview of available operators is shown in table 4.18. Some predefined operators for the different types are shown by table 4.19.

Table 4.18 Operators in VHDL

Class of operators	Operator
Logic	NOT, AND, OR, NAND, NOR, XOR, XNOR
Relational	=, /=, <, <=, >, >=
Shift/Rotate	SLL, SRL, SLA, SRA, ROL, ROR
Addition	+, −, & (Concatenation)
Sign	+, −
Multiplication	*, /, MOD, REM
Miscellaneous	**, ABS

Table 4.19 Predefined Types and their standard operators

Type	Operator		
	Logic	Compare	Arithmetic
Bit Boolean	✓	✓	
Enumerator		✓	
Integer		✓	✓
Arrays		✓	✓
Arrays of Bit or Boolean	✓	✓	✓

4.8.1 Standard Operators

The standard separates operators into individual classes. Different priorities (fig. 4.20) define the sequence of execution, especially when there are several operators in one expression. E.g., the result of the example

$2 + 3 * 4$

is 14, not 20, because multiplication has a higher priority than addition. If an expression has several operators of the same priority, processing is performed according to the arrangement in the expression from left to the right: $2 - 3 + 4$ will be interpreted as $(2 - 3) + 4$.

Table 4.20 Priority of the Standard Operators

Priority	Operator
low	AND,NAND, OR, NOR, XOR, XNOR
	=, /=, <, <=, >, >=
	SLL, SRL, SLA, SRA, ROL, ROR
	+, −, &
	+, − (sign)
	*, /, MOD, REM
↓ high	**, ABS, NOT

4.8.2 Logical Operators

The following logical operations are available: NOT, AND, NAND, OR, NOR, XOR, and XNOR. A boolean combination returns TRUE or FALSE. Opposite to other programming languages, in VHDL all logical operators except the NOT operator have the same priority. The **use of brackets** is necessary under all circumstances when having mixed types of operators:

```
carry <= (a AND b) OR (a AND c)
         OR (b AND c);
z <= NOT a AND b;
```

is equivalent to

```
z <= (NOT a) AND b;
```

because of the priority of the NOT operator.

In VHDL brackets must be used, because inverting operators (NAND, NOR, XNOR) are not associative.

```
y<= a NAND b NAND c;
-- is not allowed, because
-- it returns different results
y <= (a NAND b) NAND c;
y <= a NAND (b NAND c);
```

Logical operators can be handled easily by a synthesis tool which uses a circuit model using boolean equations.

It is possible to find the VHDL operators in the internal description which has sum or product terms. However, the result is influenced by many factors. Depending on constraints, a re-structuring of the circuit occurs during synthesis. For example, optimizing area or speed decides about the use of operators. The same goes for the choice of a target technology. Having no XOR gate available would cause a transformation to sum (OR) and product terms(AND) such as

```
XOR = (NOT a AND b) OR (a AND NOT b)
```

Having look up tables in the design, it is only a question of what data is assigned to the RAM cell, in order to define which operator is represented that way.

❑ *Example*: XOR operator using a look up table (see table 4.21)

```
y <= a XOR b;
```

Table 4.21 XOR Operator using a look up table

Input (RAM address)		Output (RAM content)
a	b	y
0	0	0
0	1	1
1	0	1
1	1	0

4.8.3 Relational Operators

A most frequent use of relational operators is the test of numerical types (see table 4.22). Such a test has the format, e.g., $(a < b)$ or $(a > 123)$. Since the result is of the type BOOLEAN (which is TRUE or FALSE) a mixture of boolean and relational operators is possible:

```
IF((a XOR b) = '1') THEN ...;
```

Table 4.22 Relational Operators

Operator	Meaning	Result
=, /=	equality, inequality	TRUE or FALSE
>, >=	greater than, greater than or equal	
<, <=	less than, less than or equal	

4.8.4 Overloaded Operators

> An operator is a function, its name corresponds to the symbol of the function.

Because of that there is no difference between overloaded functions or operators. To overload functions means that functions have the same name but are different in their parameters and return values. Because of the parameters and the return value (number and type) the compiler can determine the right function. The same applies for operators, in which case the type of the operand selects the desired operator.

However, there is a restriction namely that only predefined operators can be overloaded. Furthermore, the number of operands of the overloaded and the predefined operator must be the same. The most important application of overloading is to provide a newly introduced type with a set of suitable operators. Certainly the most needed operators are comparing operators (table 4.22) and arithmetic operators.

Operator		*Example*
+, −	Addition, Subtraction	`y <= a+b;`
+, −	Sign: Identify, Negation	`y = -a;`
*, /	Multiplication, Division	`y <= a * b;`
MOD	Modulo Arithmetic	`y <= a MOD b;`
REM	Remainder	`y <= a REM b;`
**	Exponentiation	`y <= a ** 2;`
ABS	Absolute value	`y <= ABS(a);`

The following examples can be used as a sample to overload operators. Type is a fill in for the newly defined type for which the operator is to be implemented for:

```
-- unary operators
FUNCTION "NOT"  (r:  type) RETURN type;
FUNCTION "-"    (r:  type) RETURN type;
FUNCTION "+"    (r:  type) RETURN type;
FUNCTION "ABS"  (r:  type) RETURN type;

-- binary operators
FUNCTION "and"  (l,r:  type) RETURN type;
FUNCTION "or"   (l,r:  type) RETURN type;
FUNCTION "nand" (l,r:  type) RETURN type;
FUNCTION "xor"  (l,r:  type) RETURN type;
FUNCTION "xnor" (l,r:  type) RETURN type;
FUNCTION "="    (l,r:  type) RETURN BOOLEAN;
FUNCTION "/="   (l,r:  type) RETURN BOOLEAN;
FUNCTION "<"    (l,r:  type) RETURN BOOLEAN;
FUNCTION ">"    (l,r:  type) RETURN BOOLEAN;
FUNCTION "<="   (l,r:  type) RETURN BOOLEAN;
FUNCTION ">="   (l,r:  type) RETURN BOOLEAN;
```

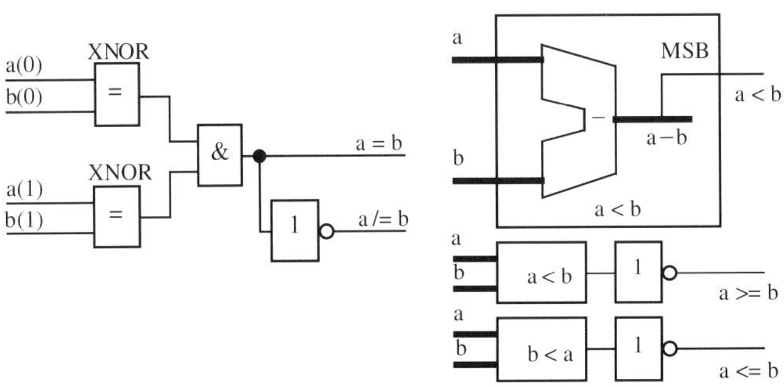

Fig. 4.56 A two bit comparator after synthesis

```
FUNCTION "**"   (l,r:  type) RETURN type;
FUNCTION "*"    (l,r:  type) RETURN type;
FUNCTION "/"    (l,r:  type) RETURN type;
FUNCTION "+"    (l,r:  type) RETURN type;
FUNCTION "-"    (l,r:  type) RETURN type;
```

4.8.5 Addition of Enumerators with an Overload '+' Operator

Example:

The '+' operator is to be applied to a self-defined type (here: COLOR) in such a way as to return the resulting mixed color (COLORMIX) in the case of addition. The type, function declaration and implementation of the function are placed in a package (my_package, see section 4.11). Thus the new operator is visible after attaching the package by the statement USE WORK.my_package.ALL. The specialized function (for performing addition) is realized by a two-dimensional table (mix tab). It returns the correct result for all input combinations. It should be noted that a signal of the enumerator type COLOR can be used as an index for the table, because the sequence of enumeration in the list (UNKNOWN, YELLOW, BLUE, etc.) is only another name for the integer values 0, 1, 2, 3 etc. (see list 4.3).

List 4.3 Addition of Enumerators with an Overload '+' Operator

```
LIBRARY IEEE;
    USE IEEE.STD_LOGIC_1164.ALL;

PACKAGE my_package IS -- declaration part of the package
    TYPE COLOR IS (UNKNOWN, YELLOW, BLUE, RED, GREEN, VIOLET, ORANGE);
    FUNCTION "+" (l,r:  COLOR) RETURN COLOR;
    TYPE COLOR_ARRAY IS ARRAY (COLOR'LOW TO COLOR'HIGH) OF COLOR;
    TYPE COLOR_TAB IS ARRAY (COLOR'LOW TO COLOR'HIGH) OF COLOR_ARRAY;
END my_package;

PACKAGE BODY my_package IS -- implementation part of the package
    CONSTANT mix_tab:  COLOR_TAB :=(
    (UNKNOWN,UNKNOWN,UNKNOWN,UNKNOWN,UNKNOWN,UNKNOWN,UNKNOWN),
    (UNKNOWN,YELLOW, GREEN,  ORANGE, UNKNOWN,UNKNOWN,UNKNOWN),
    (UNKNOWN,GREEN,  BLUE,   VIOLET, UNKNOWN,UNKNOWN,UNKNOWN),
    (UNKNOWN,ORANGE, VIOLET, RED,    UNKNOWN,UNKNOWN,UNKNOWN),
    (UNKNOWN,UNKNOWN,UNKNOWN,UNKNOWN,GREEN,  UNKNOWN,UNKNOWN),
    (UNKNOWN,UNKNOWN,UNKNOWN,UNKNOWN,UNKNOWN,VIOLET, UNKNOWN),
    (UNKNOWN,UNKNOWN,UNKNOWN,UNKNOWN,UNKNOWN,UNKNOWN,ORANGE));

    FUNCTION "+" (l, r:  COLOR) RETURN COLOR IS
    BEGIN
        RETURN mix_tab(l)(r);
    END "+";
END my_package;

LIBRARY IEEE;
USE IEEE.STD_LOGIC_1164.ALL;
USE WORK.my_package.ALL;
ENTITY mix IS
PORT (
    color1:  IN COLOR;
    color2:  IN COLOR;
    colormix:  OUT COLOR);
END mix;

ARCHITECTURE mix_arch OF mix IS
BEGIN
    -- Application of the overloaded operator
    colormix <= color1 + color2;
END mix_arch;
```

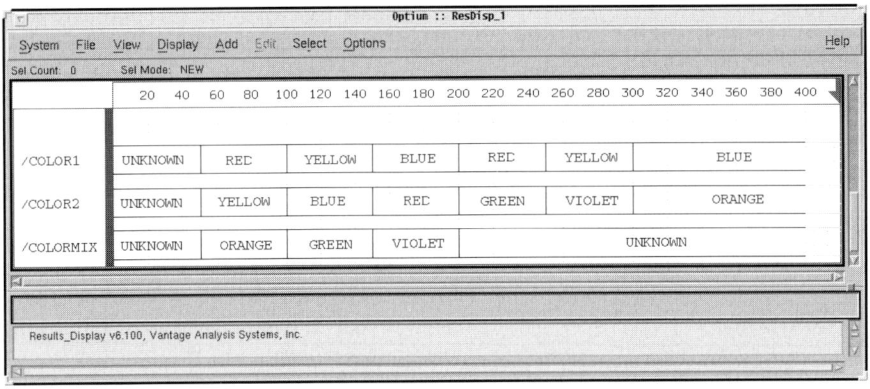

Fig. 4.57 Effect of the overloaded operator

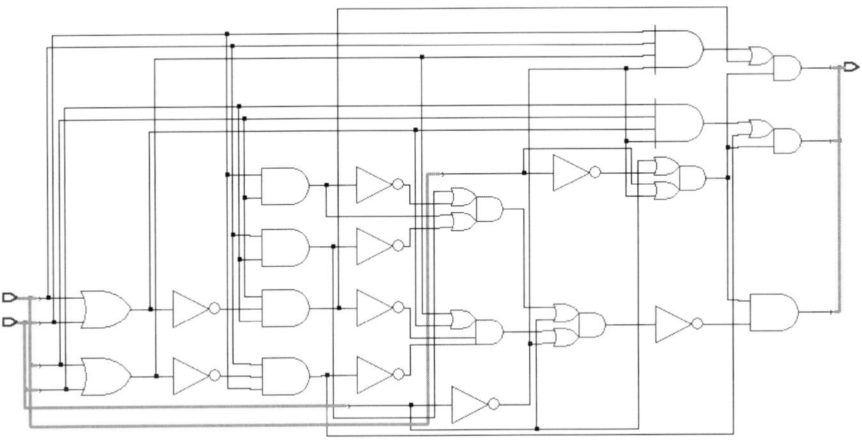

Fig. 4.58 Synthesized Operator

4.9 Sub-programs

In classical programming languages (e.g., C), functions and procedures are elements building hierarchy. The work flow is divided and each individual task is coded in a sub-program. Calling the sub-program forms the program. In VHDL a hierarchy is not created by sub-programs, but by entity-architecture pairs or components. Functions and procedures are preferably used for collecting frequently used operations (macro), the type conversions at interfaces or the supply of operators for newly defined types.

Functions in signal definitions are mainly located on the right hand side or embedded in conditions of IF or CASE statements, whereas procedures are autonomous statements. Depending on its location within or outside a process, they represent a sequential or a concurrent statement. Internally, similarly to processes, functions and procedures use sequential statements (fig. 4.59).

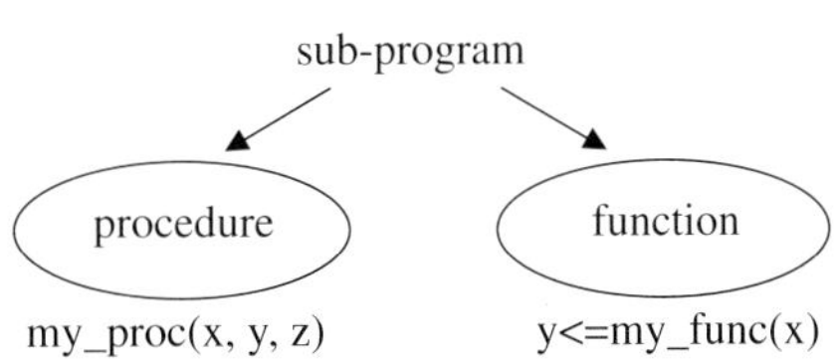

Fig. 4.59 Sub-programs: Functions und procedures

4.9.1 Functions

The interface of a function consists of submitted parameters and a return value. The return value can be assigned to a signal, to a variable, or can be evaluated in a comparison. The following example shows various applications. The most significant bit has either a value of '0' or '1', with the purpose that a checksum (XOR operation of all bits in the vector) returns the value '0'.

❑ *Example*: Calculation and evaluation of a checksum (even parity)

```
-- Application of the function parity
-- in a concurrent statement
d_out(8) <= parity(d_in(7 DOWNTO 0));

PROCESS(d_in)
BEGIN
  -- Test the return value
  -- of the function parity
  IF(parity(d_in(7 DOWNTO 0)) /= d_in(8))
    THEN
    -- Usage within a sequential assignment
    d_out <= parity(d_in(7 DOWNTO 0)) &
    d_in(7 DOWNTO 0);
  END IF;
  -- Alternatively to the IF Statement:
  -- d_out <= parity(d_in(7 DOWNTO 0)) &
  -- d_in(7 DOWNTO 0);
END PROCESS;
```

It is assumed in the example presented that the function has already been declared. There are several possibilities of how it can be done:

- In the declaration part of the process, in front of BEGIN;
- In the declaration part of the architecture;
- In the declaration part of a package (section 4.11);
- Very seldom in a package body, as it would be only visible there.

Structure of a Function

Using the example presented previously, the structure of a function can be shown. Its structure is similar to a process; however, looking at variables, the function has a different behavior. The function declaration consists of the function name and details of submitted parameters and the return value.

```
FUNCTION parity (d:  IN STD_LOGIC_VECTOR)
RETURN STD_LOGIC IS ...
```

As submitted parameters are, as a rule, only readable within the function, they must show the mode IN. IN is the default setting and needs not necessarily be specified. The parameter d does not constrain the width of the vector at this point, which allows a universal use of the function. The second part of the function is the declaration part, e.g., for local variables. Thus a variable, e.g., with the name 'result' could be provided for returning the result of the function with the RETURN assignment. Of course, the type of that variable must coincide with the type of the returned value in the function's declaration.

Differently from processes, in functions the variables are re-initialized at every function call. This is taken into account at synthesis.

```
VARIABLE result :  STD_LOGIC := '0';
```

The third and last part of a function is the part containing statements.

```
BEGIN
  FOR i IN d'RANGE LOOP
    result := result XOR d(i);
  END LOOP;
  RETURN result;
END;
```

4.9.2 Procedures

Procedures are sub-programs, too, very similar to functions when looking at declarations and structures. The declaration of procedures is legal in each declaration region; however, they are to be found most often in the declaration part of a package. The obvious difference from functions is that parameters in procedures can take the mode IN, OUT or INOUT. However, this direction of data flow must not be mixed up with the specifications in ports of an entity, because data flow direction has a different meaning here. Since parameters of a procedure are capable of transporting data into the procedure (mode IN) or to the world outside the procedure (mode OUT), a separate return value is not necessary. This is another essential difference from functions.

Going back to the examples mentioned before, a procedure is to test for even parity of the input vector d_in and assign a parameter error to '1' in the case of an error. At the same time, the

Table 4.23 Class and Mode of the parameters in sub-programs, and passing of parameters

Class	Mode			Passing the parameters:	
	IN	OUT	INOUT	VARIABLE	SIGNAL
VARIABLE	✓	✓ (Default)	✓ (Default)	✓	—
SIGNAL	✓	✓	✓	—	✓
CONSTANT	✓	—	—	✓	✓
Function					
Procedure					

checksum is to be re-calculated again and assigned to the output vector d_out. These two return values can only be accessed using a procedure.

```
PROCEDURE parity_generator(
  d_in:  IN STD_LOGIC_VECTOR,
  d_out:  OUT STD_LOGIC_VECTOR,
  error:  OUT STD_LOGIC) IS
BEGIN
  d_out := parity(
      d_in(d_in'LEFT-1 DOWNTO 0)) &
      d_in(d_in'LEFT-1 DOWNTO 0);
  IF (parity(
      d_in(d_in'LEFT-1 DOWNTO 0)) /=
      d_in(d_in'LEFT)) THEN
    error := '1';
  ELSE
    error := '0';
  END IF;
END;
```

All assignments in this example are assignments to variables. This is because OUT parameters of a procedure are variables, if not stated differently to be otherwise. Hence, having variables, this procedure can only be called in a sequential environment.

The third mode INOUT has a special meaning: it can pass parameters to be modified within the procedure and return the values using the same interface. INOUT does not model a tristate interface or bidirectional signals, but allows recursive algorithms. In functions it is simpler to read a signal or a variable and assign a new value to it:

```
y <= any_function(y);
```

In a procedure this parameter would have to take the mode INOUT in order to obtain the same behavior:

```
any_procedure(y);
```

Parameters of Functions and Procedures

Parameters of sub-programs can be a member of the classes SIGNAL, VARIABLE or CONSTANT. However, not all possible combinations of classes and modes are allowed (table 4.23). IN, OUT, and INOUT define the direction of data flow as seen by the sub-program, and the class defines the way the parameter interfaces to the outside world. Looking at the class membership, function parameters do not have restrictions, but have only the mode IN. Without a class specification a function parameter is a member of the class CONSTANT. Hence it can be an interface to a variable or a signal coming from the outside. On principle another class specification (e.g., signal or variable, instead of constant) would be possible, but it would limit a later use of the function.

These thoughts are also valid for the IN parameter at procedures. Procedure parameters having the mode OUT or INOUT can not be a member of the class CONSTANT, because their default class is VARIABLE. The declaration of a parameter to be a variable only permits a sequential use, e.g., in processes located in other procedures or functions.

4.9.3 Wired-Or Resolved with a Resolution Function

❑ *Example*: Or operation with a wired-or resolution function (fig. 4.60)

Resolution functions are used in cases in which a signal is driven by multiple sources. These functions determine which value the multiple driven signal has to take finally. During every cycle of simulation when the signal is active, which means a transaction is pending, the resolution function is called implicitly. As a parameter the function receives a vector which contains up to date driver values. Amongst those the function's algorithm calculates a return value to resolve the driver's conflict. This example declares a type MY_LOGIC and MY_LOGIC_VECTOR, plus a function wired-or to combine all drivers involved disjunctively. Finally, this is an application of the resolution function on MY_LOGIC. By then each signal of the type WIRED_LOGIC knows the values of the type MY_LOGIC and has an automatically built in resolution function (see definition of STD_LOGIC in appendix C.3).

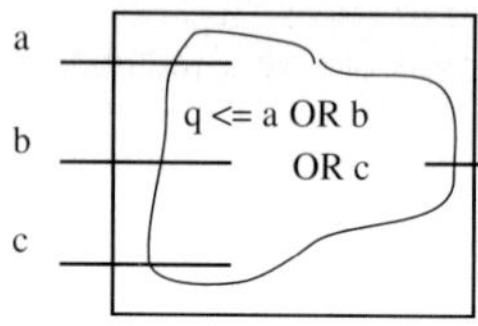

Fig. 4.60 Or operation using wired-or resolution function

```
PACKAGE wired_or_package IS

  TYPE MY_LOGIC IS ('0', '1');
  TYPE MY_LOGIC_VECTOR IS ARRAY
    (NATURAL RANGE <>) OF MY_LOGIC;
  -- function definition
  FUNCTION wired_or (involved_driver:
    MY_LOGIC_VECTOR)
RETURN MY_LOGIC;
  -- Every signal of Type WIRED_LOGIC
  -- takes implicit wired_or as
  -- resolution function
  SUBTYPE WIRED_LOGIC IS wired_or MY_LOGIC;

END wired_or_package;
```

```
PACKAGE BODY wired_or_package IS

  -- Function declaration (Implementation)
  FUNCTION wired_or(involved_driver:
    MY_LOGIC_VECTOR)
RETURN MY_LOGIC IS
  BEGIN

  IF(involved_driver'LENGTH = 0)
  THEN RETURN '0';
  ELSE
    -- For all involved drivers
    FOR i IN involved_driver'RANGE LOOP
      -- the result is final, when a driver
      -- takes the value '1'
      IF(involved_driver(i) = '1')
      THEN RETURN '1';
      END IF;
    END LOOP;
    -- No '1' found:  Result remains '0'
    RETURN '0';
  END IF;
END;

END wired_or_package;

LIBRARY IEEE;
USE IEEE.STD_LOGIC_1164.ALL;
USE WORK.wired_or_package.ALL;

ENTITY test IS
PORT(
  a,b, c:  IN WIRED_LOGIC;
  q:  OUT WIRED_LOGIC);
END test;

ARCHITECTURE test_arch OF test IS
BEGIN
  q <= a;
  q <= b;
  q <= c;
END test_arch;
```

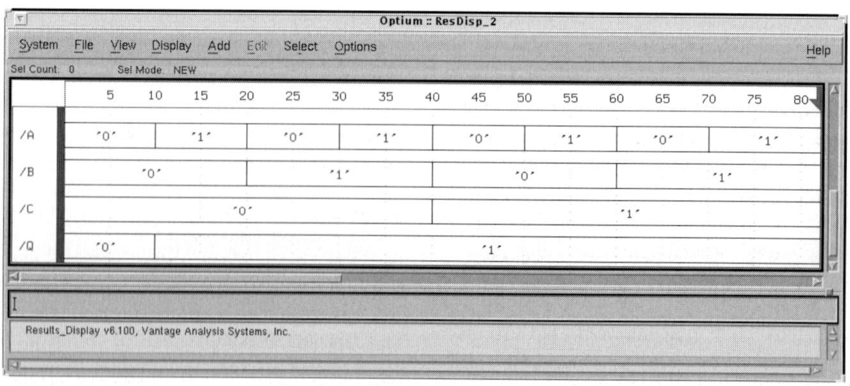

Fig. 4.61 Effect of the resolution function in simulation

4.10 Test Bench

The test of proper functionality of a VHDL model ideally starts during its development. To begin with, input signals like clock and data to stimulate the circuit and to produce response at the outputs are of interest. Later in the development cycle signal behavior at the output is of primary interest.

To set up a simulation environment a so called test bench, VHDL is highly suitable. The capabilities of VHDL can be used without any restrictions. From an external point of view the test bench is a closed box surrounding the object under test thus producing an autonomous system. The test bench should be a good and complete description of the environment the circuit is exposed to.

The most important characteristic of a test bench is an empty port list stated in a VHDL description one level of hierarchy above. Here the description under test is instantiated as a component. The purpose of a test bench is to produce suitable input signals and to check the outputs in a way that a time consuming visual control is not necessary. There are three parts which determine the structure of a test bench:

- A stimuli model or the definition of test vectors at the inputs;
- The VHDL description under test (device under test, DUT);
- A response model to check or store the results of the simulation.

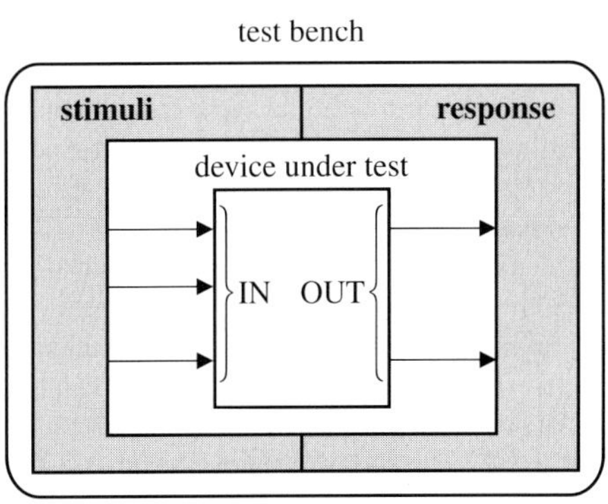

Fig. 4.62 Design of a test bench with a DUT

4.10.1 The Stimuli Model

The word 'Stimulus' designates a sequence of test vectors assigning signal values to the VHDL description under test (DUT) at well defined points in time.

Test vectors can be built in as prepared tables in the VHDL text or have an effect at run time as a result of a calculation. An interesting alternative is to store test vectors in a file, to be read line by line during simulation (section 4.10.3). This method has the advantage that, when modifiying the test vectors, the VHDL text does not have to be compiled again. In addition different test scenarios can be realized just by a change of the file.

❑ *Example*: Stimuli in a table:

```
-- Type of a line in the table
TYPE testvector IS RECORD
  in1, in2, in3:  STD_LOGIC;
END RECORD;

-- Table declaration of any numbers
-- of lines
TYPE testvectors IS ARRAY(NATURAL RANGE <>)
  OF testvector;

-- Create and initialize the table
CONSTANT stimuli:  testvectors :=
  (('0', '0', '0'),
   ('0', '0', '1'),
   ...
  );

CONSTANT timeincrement:  TIME := 50 ns;
PROCESS
BEGIN
  -- process table entries
  FOR i IN stimuli'RANGE LOOP
    in0 <= stimuli(i).in0;
    in1 <= stimuli(i).in1;
    in2 <= stimuli(i).in2;
    WAIT FOR timeincrement;
  END LOOP;
  WAIT;
END PROCESS;
```

Generating Pulse 'On Line':

```
-- Clock generator
PROCESS(clock_intern)
BEGIN
  IF(clock_intern = '0') THEN
     clock_intern <= '1' AFTER 20 ns;
  ELSE
```

```
      clock_intern <= '0' AFTER 20 ns;
    END IF;
  END PROCESS;

  clock <= clock_intern;
  -- Reset generator
  PROCESS
  BEGIN
    reset <= '0';
    WAIT FOR 100 ns;
    reset <= '1';
    WAIT;
  END PROCESS;
```

4.10.2 The Response Model

An automatic simulation run should include a response model supervising all signals at the output and comparing them with expected values.

To reach that goal is far more complicated than it appears to be at a first glance. It is not only necessary to know the expected values, but also it must be known at which point in time a comparison is to be made between the actual and the expected value.

In this context there must be an answer to the following questions: at which point in time is the simulation stable, thus providing meaningful results? How many clock cycles are necessary until the outputs respond to special input stimuli in the various states the circuit can enter?

A possibility to check the results is to implement the algorithms describing the VHDL model (DUT) once again in the response model. With this a comparison of results is possible during all phases of the design cycle (fig. 4.5). This procedure is not a test of the general functionality of the circuit in its later surrounding. Apart from that there is a real danger that both models used contain similar mistakes. Because of that reason it is a good idea to describe the reverse model in a completely different manner which does not have to be synthesized.

Both methods have in common that a comparison is made generating suitable messages with the help of assertion statements.

Another possibility to check simulation results is to store the values after a close and detailed visual inspection. The values would be used as a reference for comparisons to come. A close visual inspection is not feasible dealing with designs of a high complexity. Here, stored data could be checked using tools like C, C++, etc.

4.10.3 Package TEXTIO

For a flexible and universal design of a test bench file I/O functions are a must. Instead of coding stimuli or response data in the VHDL source it makes sense to read these from a file at run time. VHDL has a simple I/O system, providing line-oriented write and read operation which is the package TEXTIO located in the standard library std. Its most important sub-programs are:

```
PROCEDURE readline
    (FILE f:  TEXT; l:  INOUT LINE);
PROCEDURE read
    (l:  INOUT LINE; value:  OUT INTEGER);
PROCEDURE writeline
    (FILE f:  TEXT; l:  INOUT LINE);
PROCEDURE write
    (l:  INOUT LINE; value:  IN INTEGER);
FUNCTION endfile
    (FILE f:  TEXT) RETURN BOOLEAN;
```

Table 4.24 Predefined types of the line entries

Type	Example
BIT	value := '0';
BIT_VECTOR	value := "1010";
CHARACTER	value := 'A';
STRING	value := "abcdef"
INTEGER	value := 123;
TIME	value := NOW;

The procedure readline reads a complete line from textfile f and stores it in the variable 1. Looking at textline l read takes from left to right the next entry when being called and changes it to the corresponding VHDL type. After that the read pointer is set automatically to the next entry. Overloaded procedures support the most important types (see table fig. 4.24). A write step is the equivalent counterpart. The procedure WRITE passes a value to textline l and writeline writes the collected textline to the file f. The following process shows a first application of readline and writeline in a copy file function.

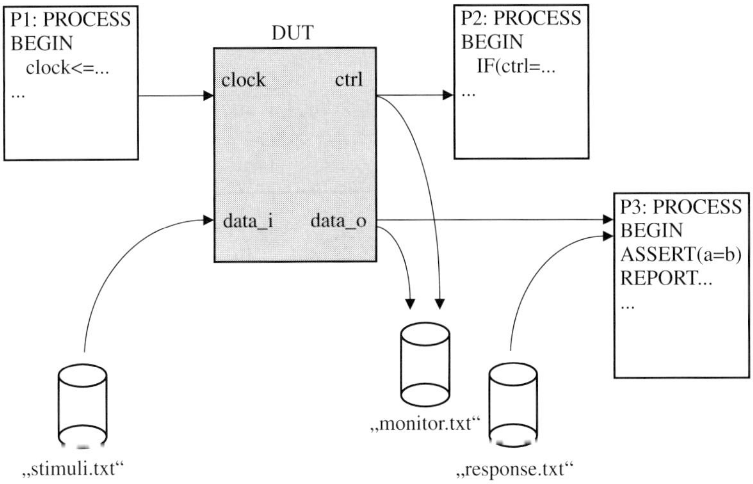

Fig. 4.63 TEXTIO functions in the test bench

```
USE STD.TEXTIO.ALL;

ENTITY file_copy IS
END file_copy;

ARCHITECTURE copy_arch OF file_copy IS
BEGIN

COPY: PROCESS
FILE source_file:  TEXT OPEN READ_MODE IS
  "file1.txt";
FILE target_file:  TEXT OPEN WRITE_MODE IS
  "file2.txt";
VARIABLE ln:  LINE;
BEGIN
  WHILE(NOT endfile(source_file)) LOOP
    readline(source_file, ln);
    writeline(target_file, ln);
  END LOOP;
  WAIT;
END PROCESS COPY;

END copy_arch;
```

4.10.4 Example of a test bench

The example shows a working test bench. A component 'multiplier' having the clock dependant function:

```
out1 = in1 x in2.
```

It is assumed that entity and architecture of that component is already visible in the working library WORK. To read the stimuli at port in1 the file entry is read as a string and subsequently changed to the type STD_LOGIC. This work around is necessary because in Standard VHDL there is no procedure readline, capable of reading STD_LOGIC directly.

```
LIBRARY IEEE;
USE IEEE.STD_LOGIC_1164.ALL;
USE STD.TEXTIO.ALL;

ENTITY test_bench IS
-- Empty port list, test bench
-- has no interface
END test_bench;

ARCHITECTURE test_bench_arch OF
test_bench IS
SIGNAL in1:  STD_LOGIC_VECTOR(7 DOWNTO 0)
  := (OTHERS => '0');
SIGNAL in2, out1:  NATURAL RANGE 0 TO 4;
SIGNAL clock:  STD_LOGIC := '0';

COMPONENT multiplier
PORT(clock:  IN STD_LOGIC;
  in1:  IN STD_LOGIC_VECTOR(7 DOWNTO 0);
  in2:  IN NATURAL RANGE 0 TO 4;
  out1:  OUT NATURAL RANGE 0 TO 1023);
END COMPONENT;

BEGIN

  -- clock generator
  clock <= NOT clock AFTER 50 ns;

  STIMULI: PROCESS(clock)
```

```
FILE testpattern:  TEXT OPEN READ_MODE IS
"stimuli.txt";
VARIABLE ln:  LINE;
VARIABLE spaces:  CHARACTER;
VARIABLE var1:  STRING(8 DOWNTO 1);
VARIABLE var2:  NATURAL RANGE 0 TO 4;
BEGIN

  IF(clock'EVENT AND clock = '1') THEN

    IF(NOT endfile(testpattern)) THEN

      readline(testpattern, ln);

      read(ln, var1);
      in1 <= string2std_logic(var1);
      -- ignore spaces
      read(ln, spaces);

      read(ln, var2);
      in2 <= var2;

    ELSE

      in1 <= (OTHERS => '0');
      in2 <= 0;

    END IF;
  END IF;
  END PROCESS STIMULI;

RESPONSE: PROCESS(clock)
FILE compare_pattern:  TEXT OPEN
READ_MODE IS "response.txt";
VARIABLE ln:  LINE;
VARIABLE var:  NATURAL RANGE 0 TO 1023;
BEGIN

  IF(clock'EVENT AND clock = '1') THEN
    IF(NOW > 100 ns) THEN
      IF(NOT endfile(compare_pattern))
      THEN
        readline(compare_pattern,
                 ln);
        read(ln, var);

        ASSERT var = out1
        REPORT "Comparison failed!"
        SEVERITY WARNING;
      END IF;
    END IF;
  END IF;
END PROCESS RESPONSE;
MONITOR: PROCESS(clock)
FILE protocol:  TEXT OPEN WRITE_MODE IS
"monitor.txt";
VARIABLE ln:  LINE;
VARIABLE spaces:  CHARACTER := ' ';
VARIABLE var1:  NATURAL RANGE 0 TO 1023;
VARIABLE simulationtime:  TIME;
BEGIN

  IF(clock'EVENT AND clock = '1') THEN
    var1 := out1;
    simulationtime := NOW;
    write(ln, var1);
    write(ln, spaces);
    write(ln, simulationtime);
    writeline(protocol, ln);
  END IF;
 END PROCESS MONITOR;

 DUT: multiplier PORT MAP
   (clock, in1, in2, out1);

END test_bench_arch;
```

„stimuli.txt“

```
00000000 0
00000001 1
00000011 2
00000100 2
00001000 3
00001111 4
11111111 4 ...
```

„response.txt“

```
0
1
6
8
24
60
1020 ...
```

„monitor.txt“

```
0 50 NS
0 150 NS
0 250 NS
1 350 NS
6 450 NS
8 550 NS
24 650 NS ...
```

Fig. 4.64 File formats of the test bench

```
FUNCTION char2std_logic (ch:  IN CHARACTER)
  RETURN STD_LOGIC IS
BEGIN
  CASE ch IS
    WHEN 'U' | 'u' => RETURN 'U';
    WHEN 'X' | 'x' => RETURN 'X';
    WHEN '0' => RETURN '0';
    WHEN '1'=> RETURN '1';
    WHEN 'Z' | 'z' => RETURN 'Z';
    WHEN 'W' | 'w' => RETURN 'W';
    WHEN 'L' | 'l' => RETURN 'L';
    WHEN 'H' | 'h' => RETURN 'H';
    WHEN '-' => RETURN '-';
    WHEN OTHERS =>
      ASSERT FALSE
        REPORT "Illegal Character found
        " & ch SEVERITY ERROR;
```

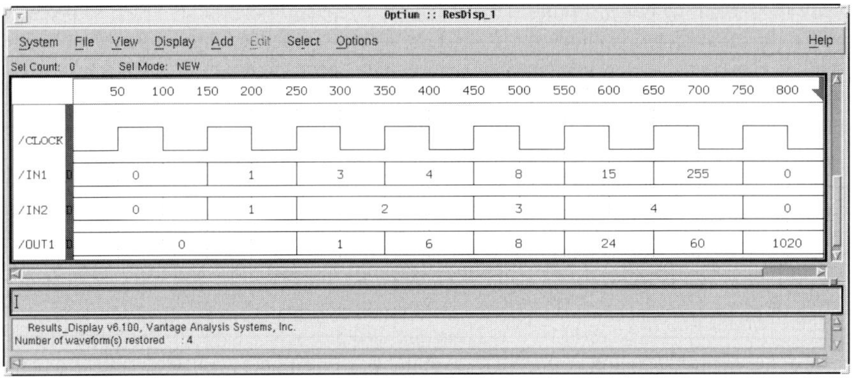

Fig. 4.65 Simulation output of the test bench

```
      RETURN 'U';
  END CASE;
END;
FUNCTION string2std_logic (s:  STRING)
RETURN STD_LOGIC_VECTOR IS
VARIABLE vector:
    STD_LOGIC_VECTOR(s'LEFT - 1 DOWNTO 0);
BEGIN
  FOR i IN s'RANGE LOOP
    vector(i-1) := char2std_logic(s(i));
  END LOOP;
  RETURN vector;
END;
```

4.11 Packages and Libraries

VHDL supports the encapsulation of data, components, procedures, and functions in a so called package. A package combines multiple used declarations and distributes them amongst several design units. Compared to a library, which is responsible for the re-use of complete designs, a package preferably contains the tools to manage the re-use. It can be compared with 'include files' of other program languages.

4.11.1 The Structure of Packages

A package has **two parts**:

- A package declaration;
- A package body.

The package declaration defines an interface to a package. Similarly to a table of contents, it lists what is available in the package. It deals with the visible aspects of the package and offers, using a type or function declaration (prototypes), a kind of operational manual for the user.

The package body contains the implementations of functions and procedures. The user only wants to know if the requested function is present and what type of input and output parameters are required. This part is optional, it can be omitted if there are only type or constant definitions in the package.

To access the contents of a package a USE statement must identify the library, and within the library the parts of the package to be made visible for the actual design. The most frequently used statement is the employment of packages STD_LOGIC_1164 from the IEEE library which offers the type STD_LOGIC.

```
LIBRARY IEEE;
USE IEEE.STD_LOGIC_1164.ALL;
```

Section 4.11.3 contains a complete example of writing and using a package (see Appendix C).

4.11.2 Libraries

As a rule, a complex VHDL design is distributed to many different text files. Each file contains one or more design units. There are **five classes** of design units:

- Entity declaration;
- Architecture;
- Configuration declaration;
- Package declaration;
- Package body.

In the first phase the VHDL compiler checks the syntax and semantics of each design unit prior to the compilation phase, which transforms the VHDL text into a form only a computer can read. Often the work flow of a VHDL text, that is, the separation of design units and the compiling to C++, leads to a linkable assembler code. When the simulation starts, the choice of a configuration or of an entity architecture pair decides what is to be simulated. Based on these instructions the simulator can choose the corresponding object files and link them to be simulated.

All files necessary for running a simulation form a library. Its file structure and management is a responsibility of the simulator or the computer system, they are not specified by the VHDL standard. The VHDL standard only covers attaching libraries and accessing design units contained there. A logical name will be assigned to the library. This name will be used in VHDL. Linking the logical name and the physical storage area is a job of the simulator. The library with the logical name WORK has a special meaning. This library represents the actual working library and defines where the compiled VHDL texts are going to be stored. So before the compiler can start, the symbolic name WORK must point to an existing library (fig. 4.66).

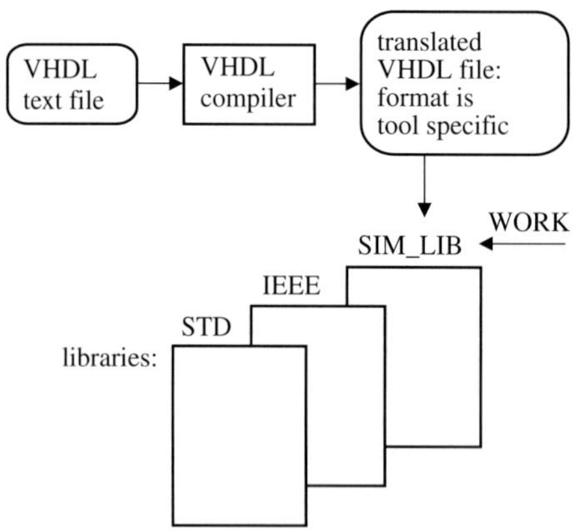

Fig. 4.66 Typical work flow when compiling a VHDL text

There is a pre-defined library STD. It contains the packages STANDARD and TEXTIO (Appendices C.1, C.2). Apart from that, the nine-valued logic system of the IEEE standard is widely spread. Its definition, providing overloaded operators, can be found in the library IEEE as package STD_LOGIC_1164. This library, however, is not a part of the language itself.

4.11.3 Overloaded Operator in a Package

❑ *Example*: Implementation of a function for adding two STD_LOGIC_VECTOR operands

Because this function is needed several times, it is part of a package.

The overloaded '+' operator is only seen from outside. The function vector_to_int is only locally available within the package body. The function presents an intermediate result, normally of no interest.

```
LIBRARY IEEE,
USE IEEE.STD_LOGIC_1164.ALL;

-- Package Declaration
PACKAGE math IS
  FUNCTION "+" (l, r:  STD_LOGIC_VECTOR)
  RETURN INTEGER;
  -- FUNCTION vector_to_int
    (S: STD_LOGIC_VECTOR) RETURN INTEGER;
END math;

PACKAGE BODY math IS

  FUNCTION vector_to_int
    (S: STD_LOGIC_VECTOR) RETURN
INTEGER IS
  VARIABLE result:  INTEGER := 0;
  VARIABLE product:  INTEGER := 1;
  BEGIN
    FOR i IN S'RANGE LOOP
      IF(S(i) = '1') THEN
        result := result + product;
      END IF;
      product := product * 2;
    END LOOP;
    RETURN result;
  END vector_to_int;

  FUNCTION "+" (l, r:  STD_LOGIC_VECTOR)
  RETURN INTEGER IS
  BEGIN
    RETURN( vector_to_int(l)
          + vector_to_int(r));
  END;
END math;

-- Start of next compilation unit
-- link libraries and packages again
```

```
LIBRARY IEEE;
USE IEEE.STD_LOGIC_1164.ALL;
USE WORK.math.ALL;

ENTITY adder IS
PORT(
  in1, in2:  STD_LOGIC_VECTOR(7 DOWNTO 0);
  out1:  INTEGER);
END adder;

ARCHITECTURE adder_arch OF adder IS
BEGIN
  out1 <= in1 + in2;
END adder_arch;
```

4.12 Advanced VHDL

4.12.1 Generics

The re-use of already developed, simulated, and working circuits is a more frequently requested primary goal. ASICs and FPGAs are capable of integrating highly complex systems on a small piece of silicon. But only by re-using available components, the capabilities of logic cells are fully utilized. Unfortunately the attempt to re-use components can go wrong, because of different operating conditions or an architecture which depends heavily on a technology available at an earlier point in time.

Generics help to parameterize a circuit. Generics are a special kind of constants for modifying the properties of components. E.g., when designing an 8 bit architecture, a future employment in a 16-bit environment can be taken into account. One can think of combining several components with similar functions into one design, and define special properties of the individual component at the time of instantiation by using suitable generics.

Generics are declared in the entity declaration prior to the port list. Hence generics are accessible in the interface description as well as in the associated architecture.

To explain the use of generics, the parity generator in section 4.5.3 will be modified in such a way that it can be used with signals having any bus width.

```
ENTITY parity_gen IS
GENERIC (width:  NATURAL);
PORT (
  data_in:
  IN STD_LOGIC_VECTOR(width-1 DOWNTO 0);
  data_out:
  OUT STD_LOGIC_VECTOR(width DOWNTO 0));
END parity_gen;

  ...
  gen:  PROCESS(data_in)
  VARIABLE parity:  STD_LOGIC;
  BEGIN
    par := '0';
    FOR i IN width-1 DOWNTO 0 LOOP
      parity := parity XOR data_in(i);
    END LOOP
    data_out <= parity & data_in;
  END PROCESS gen;
...
```

Within the architecture there is no difference between regular constants and generics. A well defined value is assigned to the generic map when the component is instantiated.

```
par1:  parity_gen
  GENERIC MAP(width => 8)
  PORT MAP (a, b);

par2:  parity_gen
  GENERIC MAP(width => 16)
  PORT MAP (c, d);
```

4.12.2 Attributes

Attribute classes are shown in table 4.25:

Table 4.25 Classes of predefined attributes

Attribute Class	Result is ...
Value	A constant value
Function	The return value of a function call
Signal	A new signal of the same type
Type	A type name
Range	A range

```
TYPE ascending IS 0 DOWNTO 255;
SUBTYPE down IS ascending
  RANGE 255 DOWNTO 0;
TYPE color IS (red, green, yellow, blue);
SIGNAL data:  STD_LOGIC_VECTOR(7 DOWNTO 0);
SIGNAL clk:  STD_LOGIC;
```

Attribute Class: Value

Table 4.26 Attributes of Types Integer or Enumerator (with sub-types)

Attribute	Result	Example
'LEFT	Left bound	ascending'LEFT=0 color'LFT=red
'RIGHT	Right bound	ascending'RIGHT=255 color'RIGHT=blue
'HIGH	Upper bound	ascending'HIGH=255 color'HIGH=blue
'LOW	Lower bound	ascending'LOW=0 color'LOW=red
'ASCENDING	TRUE if defined with an ascending range FALSE otherwise	ascending'ASCENDING=TRUE descending'ASCENDING=FALSE

Table 4.27 Value Attribute for Arrays (constrained)

Attribute	Result	Example
'LENGTH	Number of values in vector	data'LENGTH=8

Attribute Class: Function

Table 4.28 Function Attributes for the Types Physical or Enumerator

Attribute	Result	Example
'POS(v)	Position number of the value of the parameter	color'POS(yellow)=2
'VAL(v)	Value whose position number is corresponding to v	color'VAL(1)=green
'SUCC(v)	The value whose position number is one greater than that of the parameter v	color'SUCC(yellow)=blue
'PRED(v)	The value whose position number is one less than that of the parameter v	color'PRED(green)=red
'LEFTOF(v)	Left neighbor of v	color'LEFTOF(blue)=yellow
'RIGHTOF(v)	Right neighbor of v	color'RIGHTOF(red)=green

Table 4.29 Function Attributes for Arrays

Attribute	Result	Example
'LEFT	Index of element on the left bound of the array	data'LEFT=7
'RIGHT	Index of element on the right bound of the array	data'RIGHT=0
'LOW	Smallest index value of the Array (absolute)	data'LOW=0
'HIGH	Largest index value of the Array (absolute)	data'HIGH=7

4

Table 4.30 Function Attributes for Signal Objects

Attribute	Result	Example
'EVENT	A value that indicates whether an event has just occured	clk'EVENT AND clk = '1' Signal changes on '1'
'ACTIVE	A value that indicates whether signal S is active	
'LAST_EVENT	The amount of time that has elapsed since the last event occured	CONSTANT t_setup:TIME:=5 ns; data'LAST_EVENT < t_setup;
'LAST_VALUE	The amount of time that has elapsed since the last time at which signal S was active	clk'EVENT AND clk = '1' AND clk'LAST_VALUE = '0' Signal changes on '0' to '1'

Attribute Class: Signal

Table 4.31 Signal Attributes

Attribute	Result	Example
'DELAYED(t)	A signal equivalent to signal s delayed *t* units of time	SIGNAL clk_skew: STD_LOGIC; clk_skew = clk'DELAYED(1 ns);
'STABLE(t)	A signal that has the value TRUE when an event has not occurred on s for *t* units of time	IF(data'STABLE(t_setup) = TRUE)
'QUIET(t)	A signal that has the value TRUE when the signal has been quiet for s for *t* units of time	IF(data'QUIET(t_setup) = TRUE)
If *t* is not specified a simulation delta is assumed		

Attribute Class: Type

Table 4.32 Type Attribute

Attribute	Result	Example
'BASE	Base of type	descending'BASE=ascending descending'BASE'LEFT=0

Attribute Class: RANGE

Table 4.33 Range Attributes

Attribute	Result	Example
'RANGE	Returns the Index range of an Array	FOR i IN data'RANGE – in the example: 7 DOWNTO 0
'REVERSE_RANGE	Returns the reversed Index range of an Array	FOR i IN data'REVERSE_RANGE – in the example: 0 TO 7

4.13 References

[4.1] *Baker, L.*: 'VHDL Programming with Advanced Topics'. – New York: John Wiley & Sons, Inc., 1993

[4.2] *Bhasker, J.*: 'A VHDL Primer'. – Englewood Cliffs, NJ: Prentice Hall, 1994

[4.3] 'VHDL Language Reference Manual'; IEEE Std 1076-1993 (ISBN 1-55937-376-8)

[4.4] *Rushton, A.*: 'VHDL for Logic Synthesis'. – Maidenhead: McGraw Hill, 1995

5 Graphical Specification of System Behavior

FRIEDEMANN STOCKMAYER, HANS KREUTZER

5.1 Introduction

Since the beginning of the 80's schematic entry of digital circuits has been replaced by hardware description languages. This was driven by the spreading use of synthesis tools which needed a language for design entry. To avoid a variety of entry languages the hardware description languages VHDL and Verilog were standardized. The algorithms which can be coded using these languages had been presented graphically in the design of electronic circuits long before.

5.1.1 Clear Style of Representation

The graphical description of algorithms and behavior has the advantage that these algorithms can be presented in a way more transparent than code. The graphical display of sequences can be understood by humans more easily, whereas the coded display of sequences is more suitable for a computer. Although VHDL is a hardware description language it is becoming more and more difficult to obtain a general view when complexity increases. The following VHDL code and the

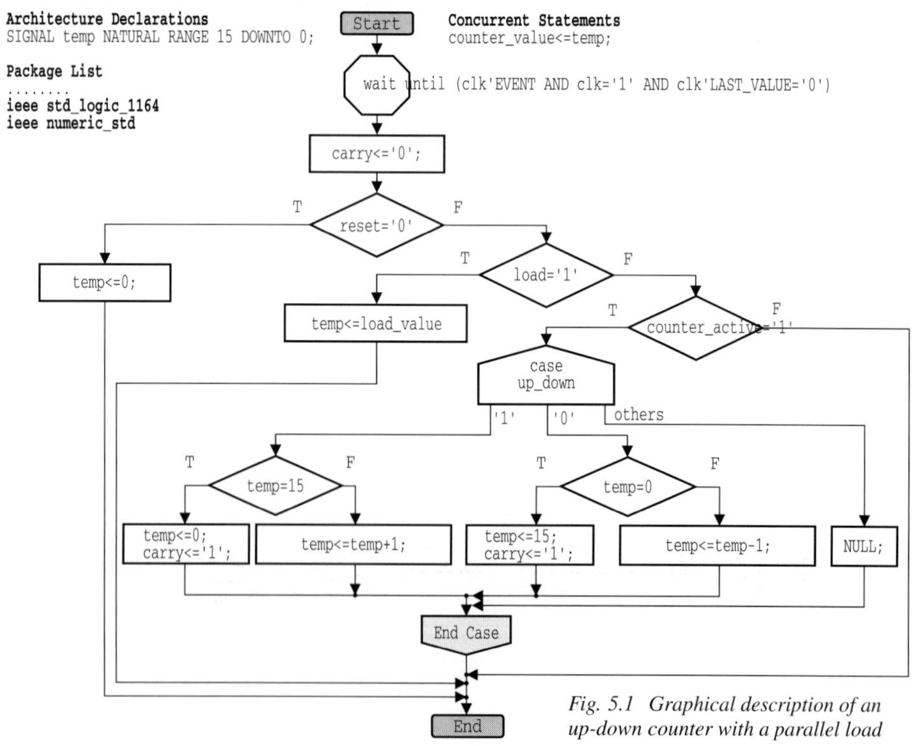

Fig. 5.1 Graphical description of an up-down counter with a parallel load

graphical representation in fig. 5.1 show a down-counter with a synchronous reset, a synchronous load, and an enable input.

```
ARCHITECTURE flow OF counter IS
SIGNAL temp NATURAL RANGE 15 DOWNTO 0;
BEGIN
process1 :  PROCESS
BEGIN
    WAIT UNTIL(clk'EVENT AND clk = '1' AND
               clk'LAST_VALUE = '0');
    carry <= '0';
    IF (reset = '0') THEN
        temp <= 0;
    ELSE
       IF (load = '1') THEN
          temp <= load_value;
        ELSE
          IF (counter_active = '1') THEN
             CASE up_down IS
             WHEN '0' =>
                 IF (temp = 15) THEN
                    temp <= 0;
                    carry <= '1';
             ELSE
                    temp <=
                    temp + 1;
             END IF;
          WHEN '1' =>
                 IF (temp = 0) THEN
                    temp <= 15;
                    carry <= '1';
                 ELSE
                    temp <=
                    temp - 1;
                 END IF;
             WHEN others =>
                 NULL;
             END CASE;
          END IF;
       END IF;
    END IF;
END PROCESS process1;
counter_value <= temp;
END flow;
```

5.1.2 Structure of the design

Today the so called top down specification is widely used. The graphical specification of behavior supports or enforces structuring the draft of the design. Especially when dealing with complex circuit designs structuring is necessary in order to avoid a high number of graphical elements on the same sheet.

Structuring the design can be carried out by a separation into parallel units of functionality on the same level of hierarchy, as well as hierarchically within given functional units.

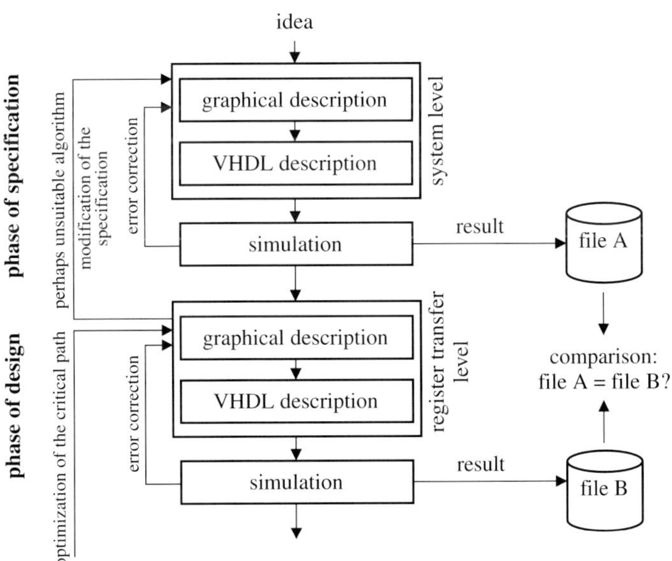

Fig. 5.2 Design cycle of a circuit with VHDL (see fig. 4.5)

5.1.3 The Design Cycle using a Graphical Specification

In principle the design cycle of a circuit presented in fig. 4.5 of chapter 4 is the same one. The only difference to fig. 5.2 is that the specification phase and the design phase can be done using a textual description (e.g., VHDL) or prior to that by a graphical description. Derived from the graphical description, a textual description (e.g., VHDL or Verilog) is generated automatically. It must be noted that the graphical description of the behavior is nothing more than the schematic representation of a VHDL or Verilog description. Therefore as a rule VHDL or Verilog syntax must be used when producing graphical input. Without knowing the hardware description language it is not possible to develop a circuit using graphical entry tools only.

5.2 Ways of Graphical Descriptions

The structuring of a design can be done by using block diagrams. The functional description of parts of the design can be given in a graphical manner using truth tables, flow charts, or state transition diagrams.

5.2.1 Block Diagrams

Normally block diagrams structure a design according to its functional units. In a complex design block diagrams define the structure of the circuit. Blocks are connected by signals. Block diagrams can show a hierarchy; each block can consist of several additional blocks (see fig. 5.3). If a block does not contain any other blocks the lowest level of hierarchy is reached. Therefore a functional description must be assigned to that block. This functional description can be a truth table, a flow chart, a state transition table, or HDL text.

5.2.2 Truth Tables

Truth tables (see the example of fig. 5.4), often called function tables, are used to describe combinatorial circuits. Thus decoders and multiplexers can be described clearly.

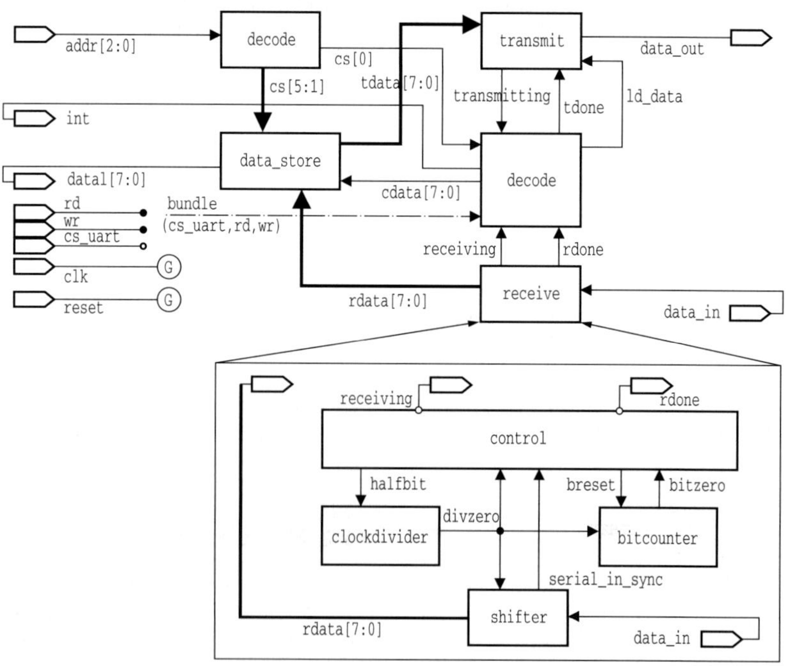

Fig. 5.3 Structure of a UART with a block diagram

	A	B	C	D	E	F
1	address	address	cs1	cs2	cs3	cs4
2	>=X0000"	<X2000"	'1'			
3	>=X2000"	<X3000"		'1'		
4	>=X3000"	<X7000"			'1'	
5	>=X7000"					'1'

Fig. 5.4 Example of a Table for a decoder

The VHDL code, generated automatically, is now presented. Its process declarations describing the truth table properties are entered in a pop up window not shown here.

```
LIBRARY ieee;
USE ieee.std_logic_1164.all;
USE ieee.numeric_std.all;

ARCHITECTURE table OF Memdecsto IS
BEGIN
    truth_process:  PROCESS(address)
    cs1 <= '0';
    cs2 <= '0';
    cs3 <= '0';
    cs4 <= '0';
    BEGIN
        IF (address >= X"0000")
           AND (address < X"2000") THEN
           cs1 <= '1';
        ELSIF (address >= X"2000")
           AND (address < X"3000") THEN
           cs2 <= '1';
        ELSIF (address >= X"3000")
           AND (address < X"7000") THEN
           cs3 <= '1';
        ELSIF (address >= X"7000") THEN
           cs4 <= '1';
        END IF;
    END PROCESS truth_process;
END table;
```

Example of a multiplexer (fig. 5.5) and generated VHDL code:

```
LIBRARY ieee;
USE ieee.std_logic_1164.all;
ARCHITECTURE table OF Mux IS
BEGIN
    truth_process:  PROCESS(s, d, c, b, a)
    BEGIN
        IF (s = "00") THEN
           y <= a;
        ELSIF (s = "01") THEN
           y <= b;
        ELSIF (s = "10") THEN
           y <= c;
        ELSE
           y <= d;
        END IF;
    END PROCESS truth_process;
END table;
```

	A	B
1	s	y
2	"00"	a
3	"01"	b
4	"10"	c
5		d

Fig. 5.5 Example of a table for a multiplexer

The examples show that a graphical entry of combinatorial circuits allows one to focus on the essentials of the function presented, whereas the behavioral description, manually written using VHDL, would be quite a burden. The automatic generation of VHDL code, derived from the graphical representation, takes a large workload away from the designer (writing code, thinking about syntax).

As a rule the generation of VHDL code can be controlled. For example, it is possible to employ **CASE statements** instead of IF constructs.

```
LIBRARY ieee;
USE ieee.std_logic_1164.all;

ARCHITECTURE table OF Mux IS
BEGIN
    truth_process:  PROCESS(s, d, c, b, a)
    BEGIN
        CASE s IS
        WHEN "00" =>       y <= a;
        WHEN "01" =>       y <= b;
        WHEN "10" =>       y <= c;
        WHEN OTHERS =>     y <= d;
        END CASE;
    END PROCESS truth_process;
END table;
```

Synthesis for those two VHDL versions can produce different circuits. A general prediction is difficult to make, because the result depends on the synthesis tool and the target technology.

5.2.3 Flow Charts

Flow charts can be employed, because of the flow structure, to describe any sequential flow. In VHDL, processes define a sequential behavior. Hence in VHDL, flow charts are mapped to processes.

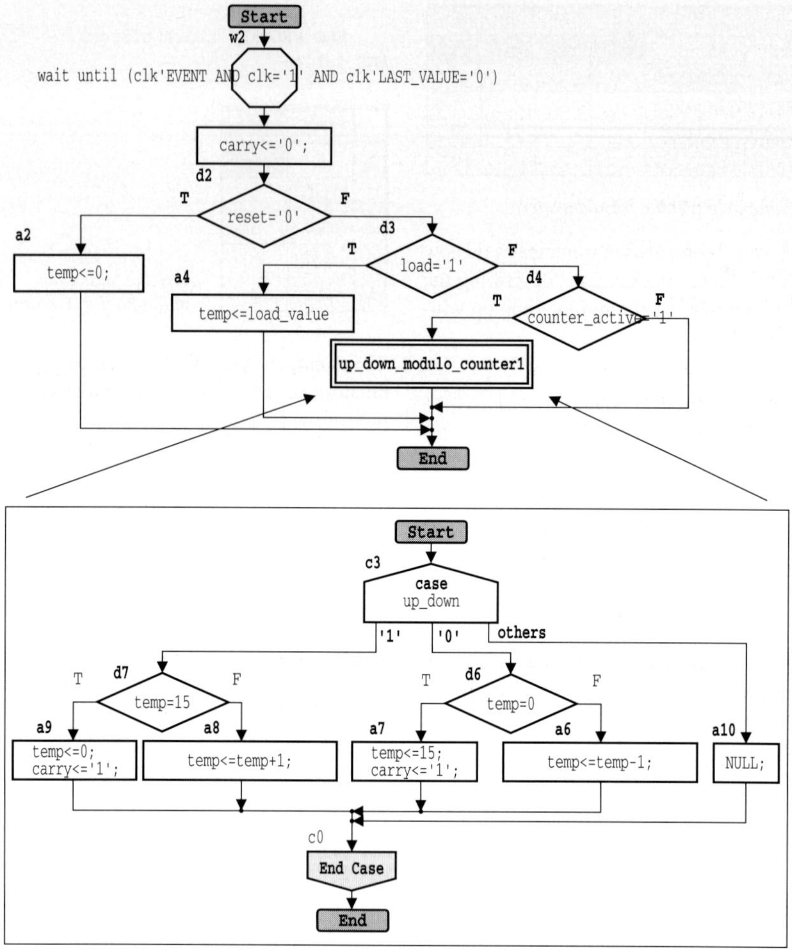

Fig. 5.6 Hierarchical flow chart of the counter in fig. 5.1

In a flow chart the following elements are typically used:

- assignment elements;
- decision elements;
- wait elements;
- loop elements.

All elements are connected with each other by arrows. Although flow charts can describe all kinds of functions, their prime use is to generate the so called test benches for testing a circuit during the specification phase and the design phase.

Similarly to block diagrams, the number of elements per sheet should be kept small in order to make the presentation readable. Therefore in large designs it makes sense to structure the flow charts hierarchically. The flow chart of the counter in fig. 5.1 is shown in fig. 5.6 using a hierarchical structure.

The automatically generated VHDL code of the counter is identical to that produced by the one derived from the flow chart fig. 5.1.

5.2.4 State Diagrams

State diagrams are a way of displaying a state machine graphically. In this case a state machine is used to model a sequential circuit. The sequential circuit is characterized by a finite number of states and state-transitions.

The state machine; being a well known and structured method of describing systems; allows one to describe many problems.

Principles of Sequential Circuits

There are two kinds of state machines used which can be combined:

- Mealy state machine; and
- Moore state machine.

These two kinds vary only in the way in which output signals are produced. Because of that it can be said that an output signal has a 'Moore' or a 'Mealy' behavior.

Differences of sequential circuits should be pointed out in brief, in order to explain the graphical representation shown in a flow chart.

State diagram of the Mealy state machine

Output signals produced by a Mealy state machine depend on its internal state and on its input signals.

Thus having a sequential circuit of Mealy type, output signals can change state, when an input signal changes state. This change of the output signal is independent of a change of the clock. Certainly if a clock event takes place the output signals can change too (see fig. 5.7).

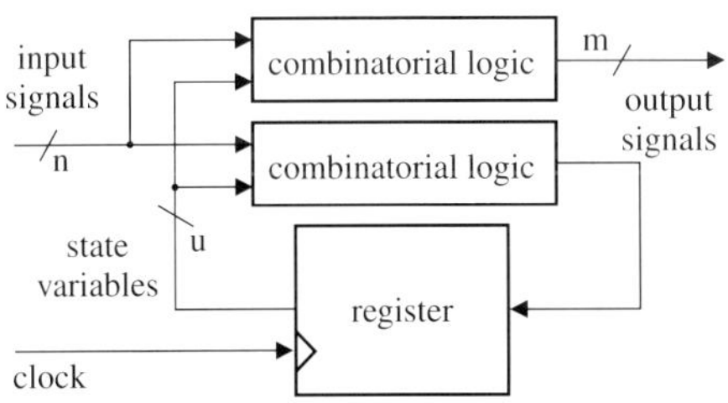

Fig. 5.7 Model of a Mealy state machine

A simple example of a counter is used to show the state diagram of a Mealy machine as can be seen in fig. 5.8. After three clock pulses the counter should produce an output pulse if the input signal carries a logical 1. Assuming the input signal shows a logical 0, the counter should remain in its present state.

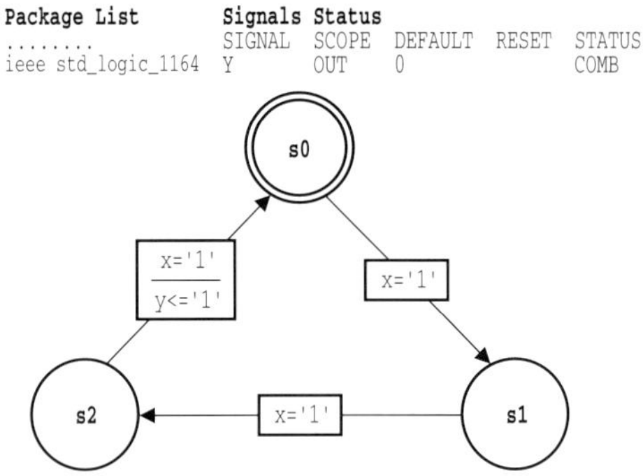

Fig. 5.8 State diagram of a simple Mealy state machine

The output signals are assigned to the state transitions. It is common practice to describe how output signals depend on the present state of the machine and the value of an input signal.

The following lines show the VHDL code derived automatically by the state diagram in fig. 5.8.

```
LIBRARY ieee;
USE ieee.std_logic_1164.all;
ARCHITECTURE fsm OF Mod_3_Counter IS
    TYPE state_type IS (s0,s1,s2);
    SIGNAL current_state,
    next_state :  state_type ;

BEGIN
    clocked :  PROCESS (clk,reset)
    BEGIN
        IF (reset = '0') THEN
           current_state <= s0;
        ELSIF (clk'EVENT AND clk = '1') THEN
           current_state <= next_state;
        END IF;
    END PROCESS clocked;
    nextstate :  PROCESS (current_state, x)
    BEGIN
        IF current_state = s0
           AND (x='1') THEN
           next_state <= s1;
        ELSIF current_state = s0 THEN
           next_state <= s0;
        ELSIF current_state = s1
           AND (x = '1') THEN
           next_state <= s2;
        ELSIF current_state = s1 THEN
           next_state <= s1;
        ELSIF current_state = s2
           AND (x = '1') THEN
           next_state <= s0;
        ELSIF current_state = s2 THEN
           next_state <= s2;
```

Hierarchy

Just the same as when drawing a flow chart, it is reasonable in a description of a state diagram not to put too many elements on one page. Because of that a hierarchy can be built up, in state diagrams.

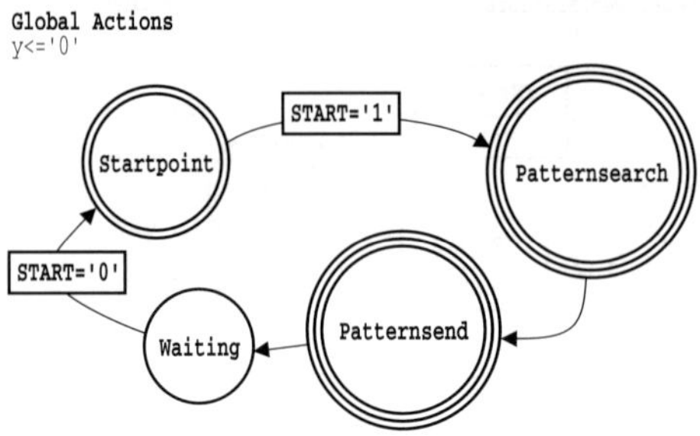

Fig. 5.16 Example of hierarchical state transitions (top level)

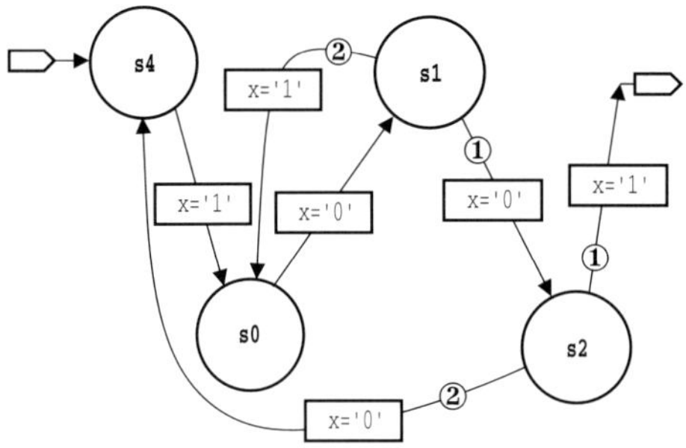

Fig. 5.17 Example of hierarchical state transitions (patternsearch)

Here the hierarchy is established by a state, that has been declared to be hierarchical, which holds additional states. It depends on the skill of the designer, to combine states in a reasonable way and to increase readability by that (fig. 5.16, 5.17 and 5.18).

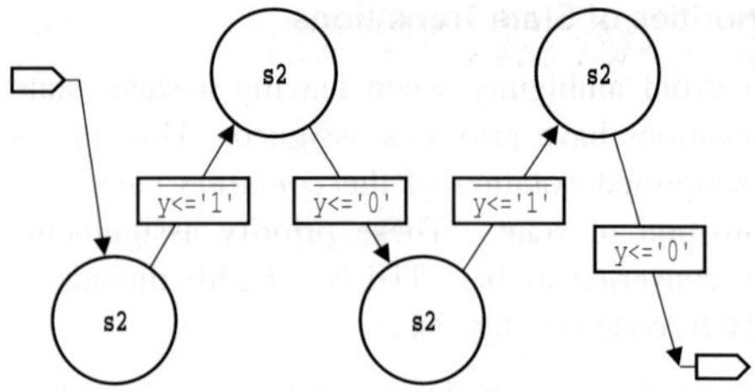

Fig. 5.18 Example of hierarchical state transition diagrams (patternsend)

Concurrent State Diagrams

If the behavior of a circuit can be described by not connected state diagrams, most of the graphical tools offer a possibility of realizing several state diagrams in one single block. These state diagrams share the same interfaces, the same local declarations, and the same references.

Setting of special properties to generate the HDL code

As shown in chapter 4, high level languages need signal declarations and conventions in order to code the states, as well as statements describing how to drive the clock and reset signals. Also the generator tool must have information dealing with the HDL style, such as for example the use of CASE or IF statements. As the number of possibilities is final, all these choices can be made in corresponding pop up windows using graphical input. The pop up windows allow a quick definition of the desired properties. Since the entry of these properties depends heavily on the tool employed and is, as a rule, self-explanatory, this will not be discussed in detail here.

Under the bottom line, a graphical specification of behavior having the properties described above can lead to a substantial increase in productivity. In addition, this method contributes significantly to the documentation aspect.

5.3 References

[5.1] *Hachtel, G. D.; Somenzi, F.*: 'Logic Synthesis and Verification Algorithms'. – Kluwer Academic Publishers, 1996

[5.2] Mentor-Graphics: 'Mentor/Renoir Tutorial' 98.4. Okt. 1998

[5.3] Summit Design Inc.: 'Visual HDL User's Guide'. Beaverton, OR, 1997

6 Synthesis

FRIEDEMANN STOCKMAYER, HANS KREUTZER

6.1 Introduction

Hardware description languages support the development cycle at a higher, more abstract level, as it was the case at the gate level. The interface to the silicon foundry or programmable logic is still a netlist at the gate level. Logic synthesis combines both levels, automatically compiling the source code from the high level language to an optimized circuit at the gate level. This process has two phases:

- *Compilation*: The RTL description is translated to a generic, which means a technology-independent netlist not yet optimized in this stage.
- *Optimization*: Mapping of the generic netlist to a target technology. The result must satisfy requirements in area and operating speed.

Not only is an efficient optimization algorithm a prerequisite for a successful synthesis, there must be an agreement on what type of hardware will be generated by special high level language constructs; last but not least, reproducible results must be ensured. Each synthesis tool uses a so called style guide, which is a collection of VHDL examples, as a reference. This manual does not explain in the first place which structure at the gate level is produced by an IF or CASE statement. Instead, it is a guide to telling what a process description should look like to describe a combinatorial circuit embedded within registers, or which statements of the high level language result in a pin which has tristate properties; just to name two examples.

The VHDL code fragments in fig. 6.1 also show the dependence on the result of the synthesis and the quality of the source code. Whereas the first suggestion compiles to an adder, subtracter, and a multiplexer, the second version only needs an adder and a multiplexer at input b. These structural differences can hardly be adjusted by the optimization tool.

```
...
c := a + b;
d := a - b;
IF(sub = '1')
THEN
   result <= d;
ELSE
   result <= c;
END IF;
```

```
...
IF(sub = '1')
THEN
   b := - b;
END IF;

result <= a + b;
```

Fig. 6.1 Two VHDL details with the same functionality but different results after synthesis

Section 6.2 shows a part of the guidelines for structuring code of the companies Altera and Synopsys. Some recommendations to support synthesis by a good partitioning of the design follow in section 6.3. The next chapters deal with optimization, explaining basic algorithms and strategies. Lastly, there is a table listing VHDL constructs which can be synthesized.

6.2 Examples of VHDL code which can be synthesized

D Flip Flop with an asynchronous Reset

(fig. 6.2)

```
DFFA: PROCESS(clk, reset)
BEGIN
    IF(reset = '0') THEN
        data_out <= (OTHERS => '0');
    ELSIF(clk'EVENT AND clk = '1') THEN
        data_out <= data_in;
    END IF;
END PROCESS DFFA;
```

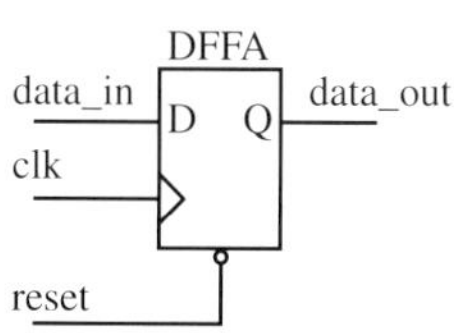

Fig. 6.2 D-FF with asynchronous reset

D Flip Flop with a synchronous Reset (fig. 6.3)

```
DFFS: PROCESS(clk, reset)
BEGIN
    IF(clk'EVENT AND clk = '1') THEN
        IF(reset = '1') THEN
            data_out <= (OTHERS => '0');
        ELSE
            data_out <= data_in;
        END IF;
    END IF;
END PROCESS DFFS;
```

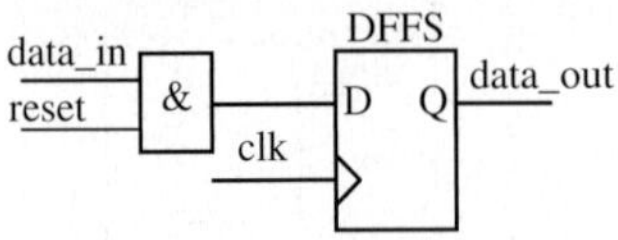

Fig. 6.3 D-FF with synchronous reset

Combinatorial logic with flip flop (fig. 6.4)

```
SYNC: PROCESS(clk, reset)
BEGIN
    IF(reset = '0') THEN
        data_out <= (OTHERS => '0');
    ELSIF(clk'EVENT AND clk = '1') THEN
        data_out <= in1 XOR in2;
    END IF;
END PROCESS SYNC;
```

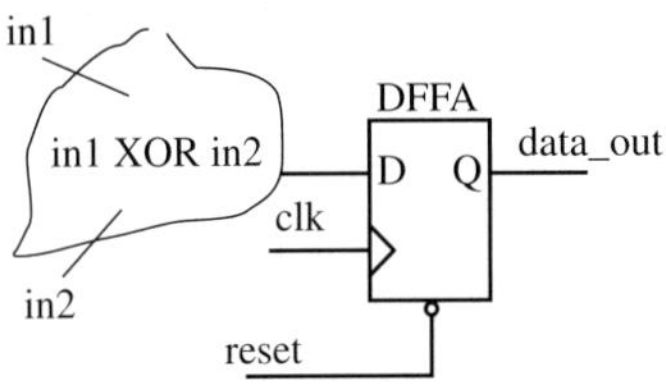

Fig. 6.4 Logic and Flip Flop

Flip Flop with feedback (fig. 6.5)

```
PROCESS(clk)
BEGIN
    IF(clk'EVENT AND clk = '1') THEN
        IF(ena = '1') THEN
            data_out <= a;
        END IF;
    END IF;
END PROCESS;
```

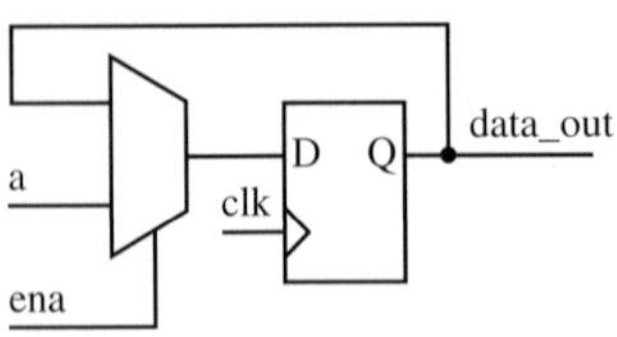

Fig. 6.5 Flip Flop with feedback

Multiplexer with inputs of the same priority (fig. 6.6)

```
MUXA: PROCESS(sel, a, b)
BEGIN
    CASE sel IS
        WHEN '0' => data_out <= a;
        WHEN '1' => data_out <= b;
        -- for all other states
        -- 'X', 'Z', 'U', ...
        WHEN OTHERS => NULL;
    END CASE;
END PROCESS MUXA;
```

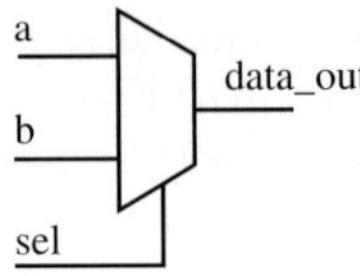

Fig. 6.6 Multiplexer

Multiplexer with priorized inputs (fig. 6.7)

```
MUXB: PROCESS(a,b,c,sel1,sel2)
BEGIN
    IF(sel1 = '1') THEN
        data_out <= a;
    ELSIF(sel2 = '1') THEN
        data_out <= b;
    ELSE
        data_out <= c;
    END IF;
END PROCESS MUXB;
```

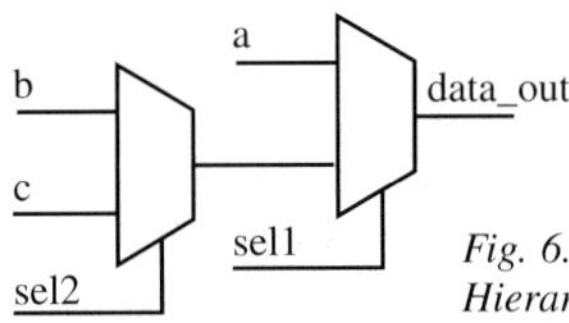

Fig. 6.7 Hierarchical multiplexer

Bidirectional data interface with a tristate driver (fig. 6.8)

```
ENTITY CPUIF IS
PORT( data_io:
    INOUT STD_LOGIC_VECTOR(7 DOWNTO 0);
    cs:  IN STD_LOGIC;
    rd:  IN STD_LOGIC);
END CPUIF;

ARCHITECTURE CPUIF_ARCH OF CPUIF IS
BEGIN
    -- reading of the interface
    -- is always possible
    data_in <= data_io;

    TRI: PROCESS(cs, rd, data_out)
```

```
    BEGIN
      -- write, when cs and rd are active
      IF(cs = '0' AND rd = '0') THEN
          data_io <= data_out;
      ELSE
          data_io <= (OTHERS => 'Z');
      END IF;
    END PROCESS TRI;

END CPUIF_ARCH;
```

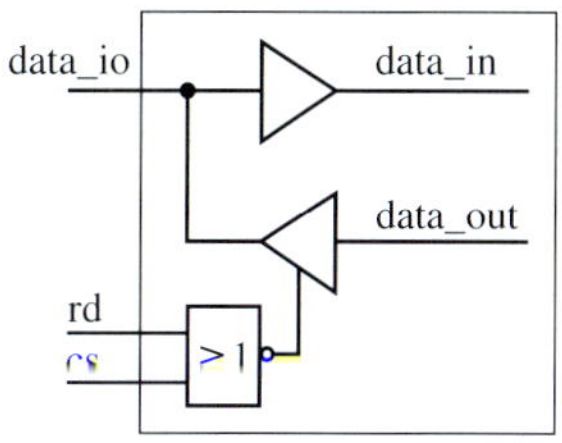

Fig. 6.8 Tristate

6.3 Partitioning

An important condition for a good result in the synthesis is a careful partitioning of the design. This means dividing the whole circuit into modules, which could show at the same time a hierarchy, or, in other words, a nesting of modules. The earlier partitioning is taken into account in the design cycle, the simpler synthesis will be. A change of the module boundaries often causes a re-definition of the interfaces which is time consuming and a source of error. So far there are no explicit rules for performing partitioning, but the following recommendations help the right decision to be made.

> Combinatorial parts connected with each other should be in the same module (fig. 6.9).

The example in fig. 6.9 contains three combinatorial parts which are; directly or indirectly; elements of the critical path (the signal path which has the maximum delay time). In this case, when optimizing the synthesizing compiler has more degrees of freedom to take the edge off a critical path by choosing an advantageous combination when combining logic. A starting point which is not as good can be seen in fig. 6.10. There the compiler can not combine or relocate logic across the hierarchical levels.

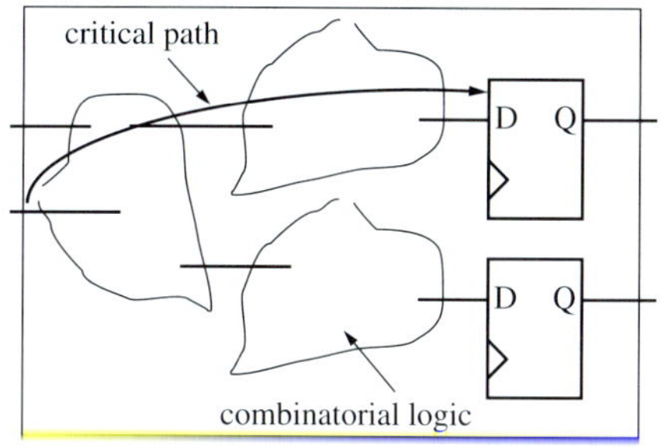

Fig. 6.9 Combinatorial logic connected together in one module

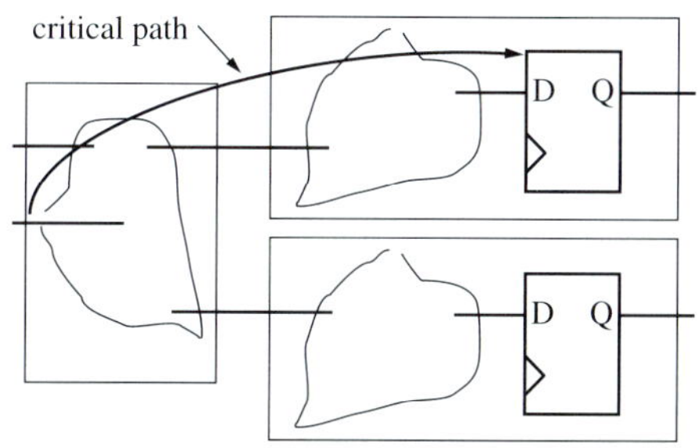

Fig. 6.10 Unfavorable distribution of modules

Still looking at the partitioning in fig. 6.10, another difficulty is the distribution of the delay amongst the individual combinatorial parts. Because of this the driver strength and the loads have to be taken into account. To optimize each individual module, precise specification of the maximum delay time (output- or input-delay), the expected load, and driving strength must be made at the interface q (fig. 6.11). Figure 6.11 uses some statements of the Synopsys compiler to do that.

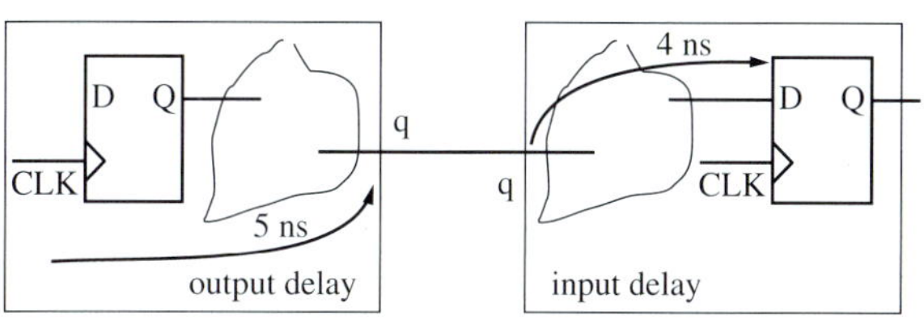

set_output_delay 5 -clock CLK q
set_load 0.5 q

set_input_delay 4 -clock CLK q
set_drive 0.2 q

Fig. 6.11 Consideration of the delay times and driver strength

Resource Sharing

The following VHDL example uses two operations of addition ($a + b$, or $c + d$). Depending on optimization settings, the compiler decides if there are two different adders to be used for these operations. If speed is of prior concern two independent adders will be realized. If a small chip area is desired, maybe one adder can be used for both operators. In order to leave this decision to the compiler, or; in other words; to keep as many degrees of freedom as possible, it makes sense to look at the same time at all components involved, in one module they share (fig. 6.12).

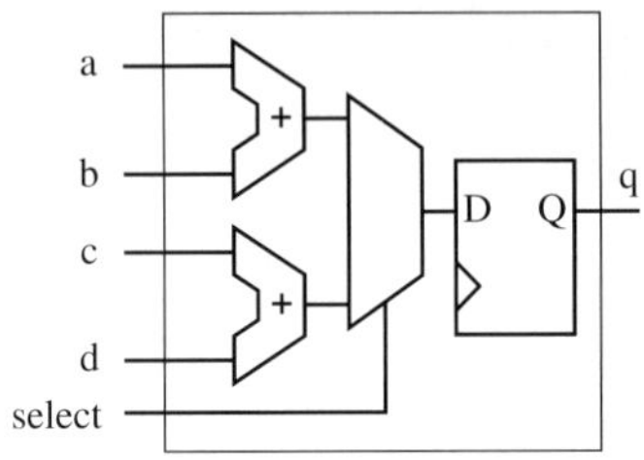

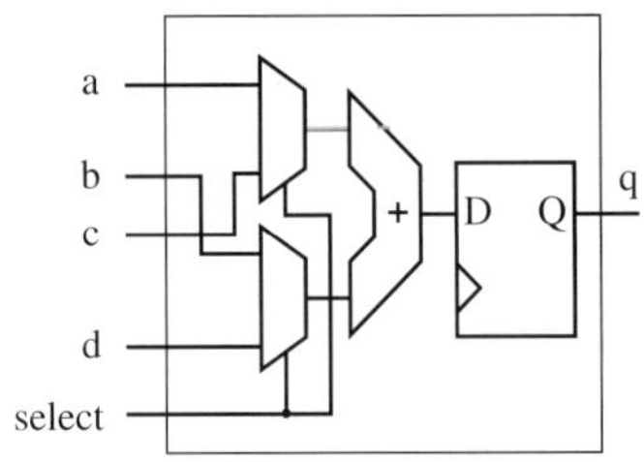

Fig. 6.12 Resource Sharing

```
IF(select = '1') THEN
     q <= a + b;
ELSE
     q <= c + d;
END IF;
```

Separate modules for different goals of optimization

The best result in synthesis can be achieved when circuit parts having different goals for optimization (e.g., small component area or high speed) reside in different levels of hierarchy. Generally speaking, both goals can not be realized at the same time, because normally the requirements are conflicting.

The goals of the optimization are described in the compiler using constraints, that is, requirements which must be fulfilled. The statement set_max_area = 0 causes the compiler, perhaps by structuring (section 6.5), to synthesize a circuit as small as possible.

Especially when looking at clock-synchronous circuits, the speed; in other words, the clock frequency; the circuit can operate with is of primary concern. Therefore the compiler must be informed of the name of the clock net and the intended clock period.

Separate modules for different optimization strategies

Not only different goals in optimization are one reason to accomodate the involved logic areas within separate modules. The right strategy to perform optimization can contribute to activating the last reserves in speed or area of a module. These are two strategies normally pursued; flattening or structuring (see section 6.5). Modules which show an explicite structure, e.g., several levels of XOR; should not be flattened out because the result would not show improved terms of speed or area. Modules with no regular structures (e.g., random logic, no adder, no large multiplexers) are more suitable for being flattened out.

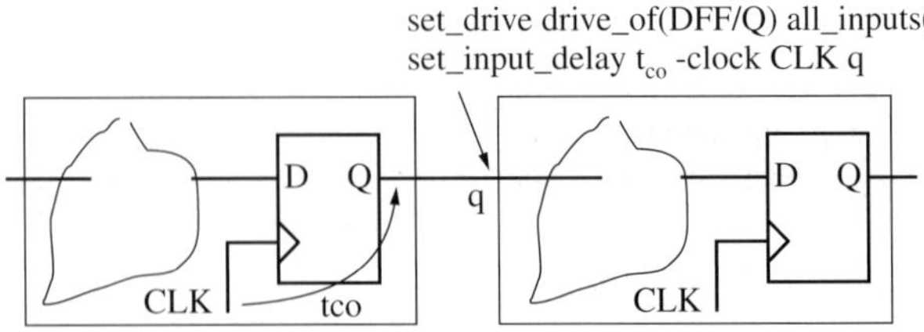

Fig. 6.13 Register at the module output

Registers at all outputs

Registers at all outputs of the modules reduce the number of constraints and simplify synthesis as a result. Which input load is to be taken into account at a module input can easily be decided using this recommendation, because as a rule it is the driving strength of a flip flop. The input delay of a module related to the clock depends only on the propagation delay of the register (clock to output, tco) of the preceding module and not to a lot of unknown and variable amount of combinatorial operations (fig. 6.13). Hence the critical path is only within the module. The optimization in time can be done more easily; for example by a new optimization run using changed, new constraints.

6.4 Modification of Hierarchy

It is not always possible to plan partitioning prior to the design in a way to obtain an ideal set of logical modules with a suitable hierarchy. For example, components originating from earlier designs do not fit properly into the actual concept, or a modification in the design changes partitioning. In such cases it can be necessary to change the hierarchy in parts or in the whole and to restructure anew. Restructuring can be done in two ways: either by changing the VHDL design, or by using suitable commands during synthesis. To perform that task the Synopsys compiler knows the commands group and ungroup, and allows an optimal partitioning to be found experimentally.

Ungroup Command

The ungroup command removes an existing level of hierarchy, e.g., when many modules having a small content obstruct optimization. The example in fig. 6.14 shows that case. After the command ungroup all cells are part of the same level of hierarchy.

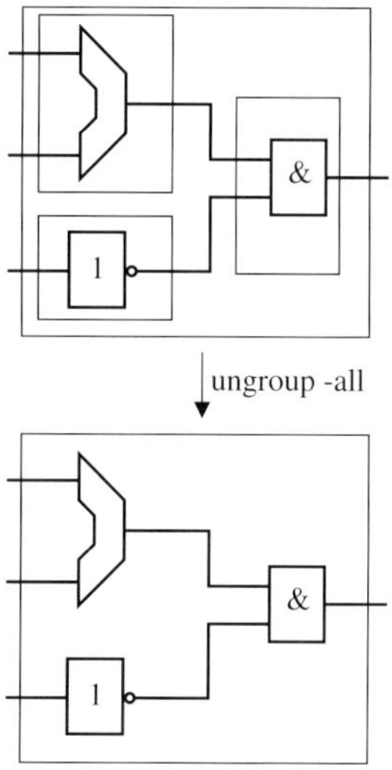

Fig. 6.14 Application of the ungroup command

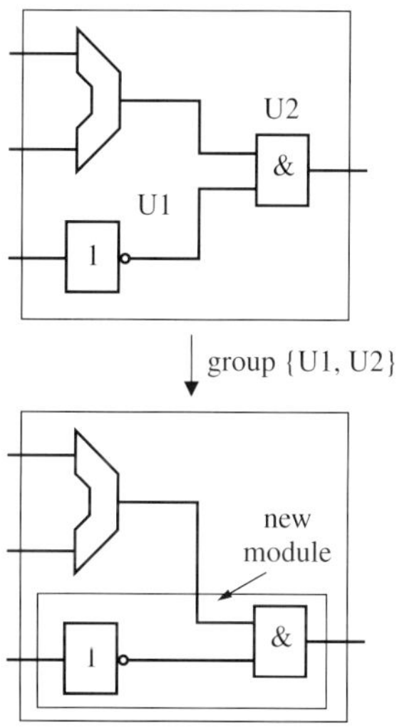

Fig. 6.15 Application of the group command

Group Command

Having a flat design, new levels of hierarchy can be established by using the group command. This statement allows one to collect objects of the present level of hierarchy in a new module. The command is very versatile because objects can be single gates as well as blocks, processes, or components, immediately after reading the source code.

6.5 Optimization

To yield the best results in optimization, it is important to give the compiler information about the real environment of the design. Constraints are declarations to define the special optimization goals for a specific circuit. Whether the design goal was met can easily be decided by checking characteristic quantities like timing, area, or capacity, or by comparing the result with previous results. During optimization there are two types of constraints in use:

- Design Rule Constraints (see table 6.1);
- Optimization Constraints (see table 6.2).

Design rule constraints are requirements depending on technology. They always exist implicitly because of the selected target library. They are general rules for ensuring a correct function. For the compiler these requirements have the highest priority (table 6.1).

Table 6.1 Design Rule Constraints in Synopsys

Constraint	Function
set_max_transition	Maximal tolerable time for a change of the logic level on a net (→ Buffering)
set_max_fanout	Sum of the loads connected to an output of a module (→ driving strength)
set_fanout_load	Expected load at the output of a module
set_max_capacitance	Maximal tolerable capacity on a net

Optimization constraints define explicitly individual goals of optimization. Realistic requirements regarding area and speed have better chances of being taken into account for the circuit (table 6.2).

Table 6.2 Optimization Constraints in Synopsys

Constraint	Function
set_max_delay	Maximal delay in the combinatorial part
set_max_area	Maximal Area (e.g., in gate equivalence)
set_input_delay	Extern delay at module input related to clock
create_clock	Identifies a module input to be the clock source and specifies the clock period
set_clock_skew	Defines clock skew at all flip flops

6.5.1 Consequence of optimization constraints

Synthesis is a complex process having many parameters which can be changed. Some of them are discrete, some are continuous. Discrete parameters are the RTL architecture, logic function, as well as many switches in the synthesis tool to control the algorithm of synthesis statically. Continuous parameters describe, e.g., input or output delay times, capacitive loads, or clock periods.

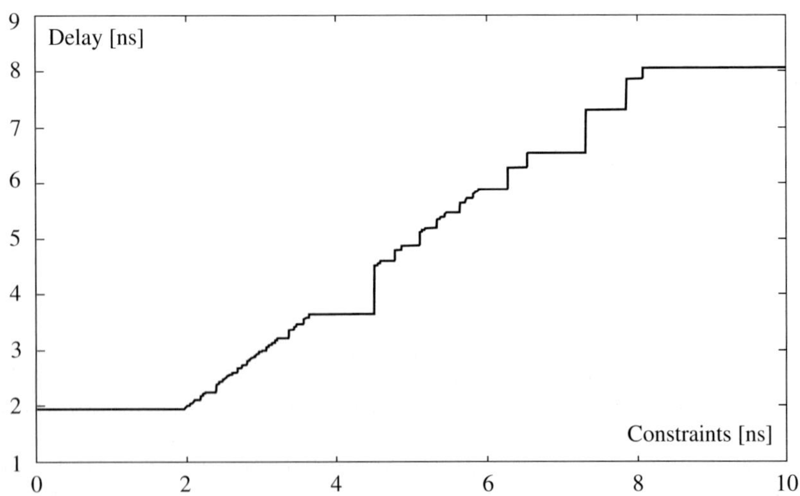

Fig. 6.16 Delay in the critical path as a function of the constraint

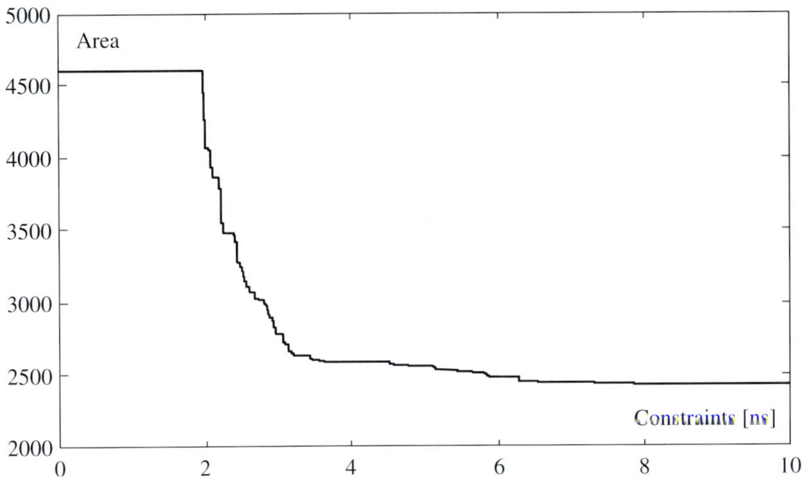

Fig. 6.17 Chip area as a function of a time constraint

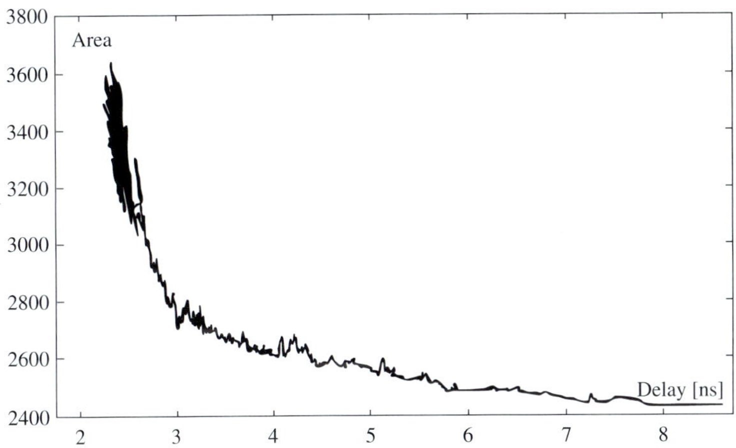

Fig. 6.18 Area as a function of the delay in the critical path [6.4]

To show the way a synthesis tool works in principle, it does not make sense to change all parameters at the same time. Looking at figures 6.16 ... 6.18, mainly the area and speed of a given design are of interest. As stated in [6.4], it is assumed that the clock period obtained (which is nearly equivalent to the maximal delay in the critical path) is a measure of the operating speed, and that all other parameters related to time depend on it.

Figure 6.17 supplements the picture in fig. 6.16, where for each synthesized netlist its area depending on the constraints is displayed. The smaller the delay in the critical path, the bigger the area, because of an increase in the degree of parallelism and the use of additional buffers.

Both results from fig. 6.16 and 6.17 combining yields the characteristic presentation in fig. 6.18. The graph connects individual points which correspond to a synthesis run (yielding a netlist) for a particular constraint. The behavior at the left side of the graph calls for some attention. In the attempt to realize the highest operating speed the results are chaotic. A minute change of a constraint causes a significant change in the area. This type of behavior can often be observed in algorithms having a randomly selected starting point.

6.5.2 Optimization strategies

Designs of digital circuits normally have hierarchical levels. The top level of this hierarchy then consists only of blocks being connected with each other. Some blocks process data, others are only responsible for control. The internal structure of the blocks is different, too. Both aspects require an individual treatment during synthesis. In clearly structured components (e.g., a huge adder) it is desirable to maintain the structure and to use it during opimization. Within non-structured blocks it is the job of the synthesis tool to remove redundant logic and thus improve the structure of the circuit.

The structure of a circuit description can be recognized at a logical level as well as at the gate level. The logical level is independent of technology and is described by a set of boolean equations. The gate level of a circuit is made by a real implementation of the logical level introducing the target technology (fig. 6.19).

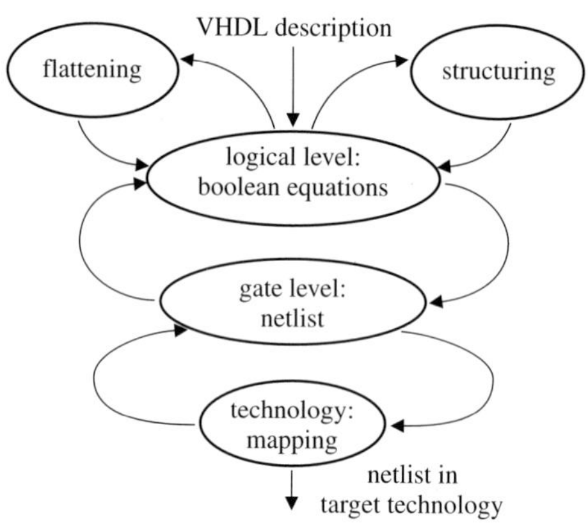

Fig. 6.19 The steps from VHDL description to the netlist

The purpose of an optimization on a logical level is to improve the structure of the digital design without changing the function. The first attempt leads to a reduction of the product terms as well as to smaller circuits and to smaller propagation delays. Because area and speed are the main criteria during a circuit's optimization the strategy used plays an important role. The optimization strategy consists of a meaningful application of the following steps or a combination of both:

- Flattening;
- Structuring [6.7].

Flattening

Flattening identifies an optimization step which solves all expressions in boolean equations enclosed in brackets by applying the distributive law and which removes all intermediate variables. An optimization of operating speed is the primary goal.

❑ *Example*:

Prior to flattening:

```
out = uv
u = a + b(c + f)
v = d + ē.
```

```
After flattening:
out = ad + bcd + bdf + aē + bcē + bēf.
```

The result of the flattening is, as is seen in the example, a sum of product terms. This does not always yield the circuit version having the highest operating speed, because some variables are present in more than one product term, which causes an increased load (fan out) by a large number of gate inputs connected to the driver involved. This aspect is shown in fig. 6.20a.

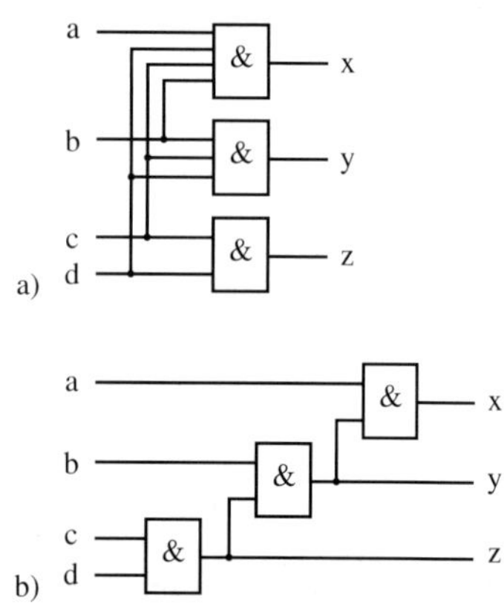

Fig. 6.20 Result of flattening a) with following structuring in b)

Flattening can be followed by an algorithm to minimize the boolean equations. Optimization tools can minimize each output separately, or deal with all outputs together at the same time during the process of minimization.

The first case produces an optimal result for the individual output. In the second case product terms

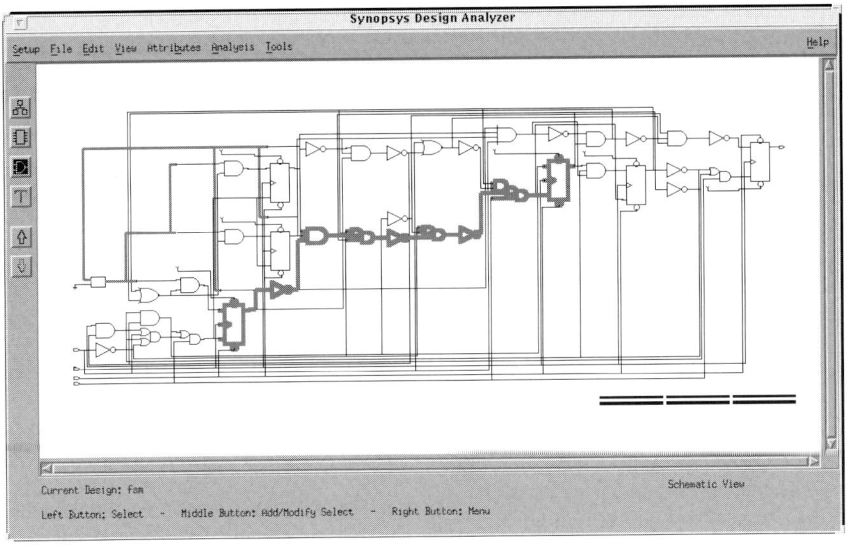

Fig. 6.21 Structuring without timing constraints

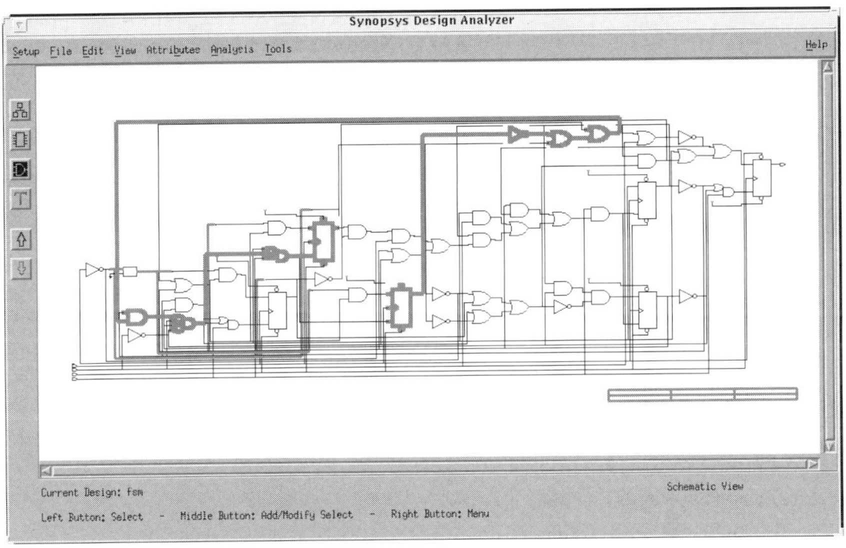

Fig. 6.22 Optimization of the critical path

needed more than once are shared. Identical product terms frequently appear when many outputs are depending on nearly all inputs. Minimization in this case is very efficient when looking at the chip area. In fig. 6.20b the total size in the example has not changed, but the loads are distributed more evenly.

Structuring

The primary goal of *structuring* is to find things which the circuit structure has in common. Therefore a flat representation will receive a structure by introducing intermediate variables.

The intermediate variable stands for a partial function and which can be used more than once to simplify the boolean equation. As a consequence of sharing partial functions less chip area is taken, but, on the other hand, the logical depth in the critical path can be increased in some cases.

❑ *Example*:

Before structuring:	After structuring:
$x = ab + ac$	$x = a\,v$
$y = b + c + d$	$y = d + v$
$z = \bar{b}\bar{c}\bar{e}$	$z = \overline{v + e}$
	$v = b + c$

Structuring is often recommended in the first optimization run. After that a detailed specification of delay times between input and output can be used to optimize the critical path without changing the uncritical regions. The attempt to satisfy additional constraints can lead to an increase in area, because run time optimization can cause an increase in parallelism. This parallelism means a duplication of one path or another (fig. 6.21 and 6.22).

6.5.3 Optimization of two-stage logic

The progress of logic synthesis and optimization visible in these days results from the investigation of combinatorial logic in two stages. The most popular method of this type of description is the sum product term, also known as the disjunctive function. Here, according to boolean algebra, AND terms form the product types in the first stage. These are added in a second stage using OR gates where, strictly speaking, inverters located at gate inputs form a third stage. Additional forms of combinatorial logic are OR/AND, AND/OR, NOR/NOR or NAND/NAND (fig. 6.23 and 6.24).

y = (a AND b) OR (NOT a AND c)

Product terms

Fig. 6.23 Two-stage combinatorial logic as the sum of products: $y = ab + \bar{a}c$

The practical realization of two-staged combinatorial logic can be found in components having a PAL structure (programmable array logic). Because of the direct mapping of the mathematical description to the silicon area, these structures are highly suitable for an automatic logic or layout synthesis. The relatively simple structure yields short delay times and a high operating speed of the realized circuit. In this case the operating speed can be estimated very precisely in an early design stage.

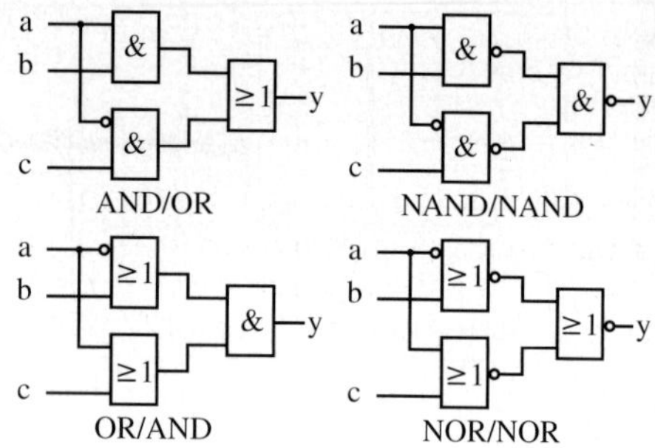

Fig. 6.24 Equivalent forms of two-stage logic

However, the two-stage logic of regular operators such as addition, subtraction, parity, is growing excessively in size when the number of inputs goes up. For example, when calculating parity the EXOR operation of N signals causes 2^{N-1} product terms. Nevertheless, optimized two-stage forms of representation are very often the starting point of multistage logic synthesis algorithms.

The goal of an optimization of a two-stage logic is to find a boolean function equivalent to the one present at the start of optimization, that has an optimum number of product terms and literals (direct or negated variables in the boolean function). The main attention is dedicated to the problem of finding a set of product terms which completely describes the function presented at the starting point, but shows the least redundancy possible. The solution has **two phases**:

- Decomposition of the boolean function to be optimized to a set of product terms, in such a way that none of these terms is present in another term (prime implicants);
- The choice of a minimal set of product terms to obtain a complete description of the original function.

Quine McCluskey Algorithm

An algorithm using both of these steps is the method of Quine McCluskey [6.2]. The approach depends on a suitable *combination of product terms*.

$$z = abc + \bar{a}bc = (a + \bar{a})bc = (1)bc = bc$$

Modern minimization programs are based on that method. Its mechanism will be explained best by an example: specific output values of a four bit counter abcd have to be decoded. Hence a boolean function y consisting of minterms, as in table 6.3, is examined. A minterm is a product term which contains all input values.

$$y = p_1 + p_2 + p_3 + p_4 + p_5 + p_6 + p_7 + p_8 + p_9$$

In the first step a comparison (pair by pair) is performed to find product terms different only in one position. This position is replaced by an 'X' (table 6.4).

Table 6.3 Minterms forming the function

count		Minterm
Decimal	Binary: *abcd*	
0	0000	$p_1 = \bar{a}\,\bar{b}\,\bar{c}\,\bar{d}$
5	0101	$p_2 = \bar{a}\,b\,\bar{c}\,d$
7	0111	$p_3 = \bar{a}\,b\,c\,d$
8	1000	$p_4 = a\,\bar{b}\,\bar{c}\,\bar{d}$
9	1001	$p_5 = a\,\bar{b}\,\bar{c}\,d$
10	1010	$p_6 = a\,\bar{b}\,c\,\bar{d}$
11	1011	$p_7 = a\,\bar{b}\,c\,d$
14	1110	$p_8 = a\,b\,c\,\bar{d}$
15	1111	$p_9 = a\,b\,c\,d$

Table 6.4 Summary of Minterms (first iteration)

0, 8	X000	A
5, 7	01X1	B
7, 15	X111	C
8, 9	100X	
8, 10	10X0	
9, 11	10X1	
10, 11	101X	
10, 14	1X10	
14, 15	111X	

The terms of the result of the first step are compared in a second iteration (table 6.5) in the same way. However, now the restrictions apply that if two terms should be combined 'X' must now be located at an identical position, and only one more position (which would be replaced by 'X') can be different. After that combination five terms showing no similarity would be left.

Table 6.5 Second iteration

8,9 10,11 8,10 9,11	10XX	D
10,11 14,15	1X1X	E

The five left terms are used to choose a minimal set to realize all minterms. Therefore in the last step it will be marked in table 6.6 which minterms are contained in the terms A–E (denoted by a mark ✓). Choosing from the columns A–E, columns can be removed in such a way that for each line at least one mark is still present. In the example this is column C. Its marks are part of column B and column E. If there are several possibilities for removing columns the columns having the highest number of literals (or the smallest number of 'X') should be removed. The remaining columns finally make up the solution (table 6.6).

The solution shows a significantly smaller number of product terms:

$$y = \mathrm{A} + \mathrm{B} + \mathrm{D} + \mathrm{E}$$
$$y = \bar{b}\,\bar{c}\,\bar{d} + \bar{a}\,b\,d + a\,\bar{b} + a\,c$$

Table 6.6 Remaining product terms A–E are 'prime implicants' (A, B, D, E are essential, C is redundant)

Minterm	A	B	C	D	E
	X000	01X1	X111	10XX	1X1X
0000	✓				
0101		✓			
0111		✓	✓		
1000	✓			✓	
1001				✓	
1010				✓	✓
1011				✓	✓
1110					✓
1111			✓		✓

6.5.4 Optimization of sequential logic

The focus of the preceding chapter has been on the optimization of combinatorial logic. Methods to optimize sequential circuits are now discussed. A model of synchronous sequential circuits is the state machine, because it contains a combinatorial part and registers typical of that type of description.

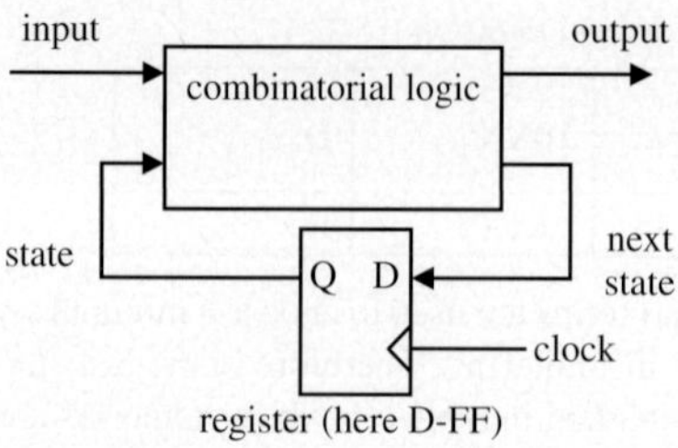

Fig. 6.25 Model of a state machine

State machines can be described by states and the present input signals. The combination of input signals and the present state (fig. 6.25) determines which state the machine is going to enter when the next clock event occurs. An approved form of presentation for state machines is the state transition diagram (fig. 6.26). Because of its level of abstraction, these diagrams describe behavior in the first place; they do not reflect the chip area needed or show information about possible clock speeds.

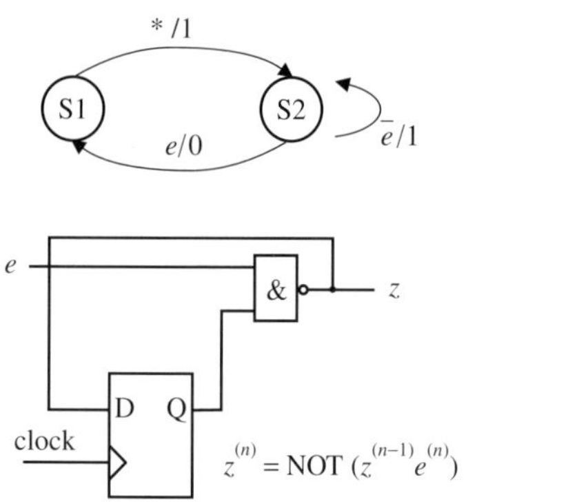

Fig. 6.26 Sequential model of state machine: state transition diagram und synchronous state machine

Minimization of states

When the overall complexity of a state machine is reduced, an economic use of resources is often a consequence. Fewer states means fewer storage elements. Fewer states means; generally speaking; fewer state transitions, which is a positive step when thinking about the optimization of chip area.

Algorithms for minimizing states are discussed in detail in [6.5]. A simple example explains the procedure for obtaining, starting with a given state transition diagram a state machine which has the same behavior but only a minimal number of states (fig. 6.27).

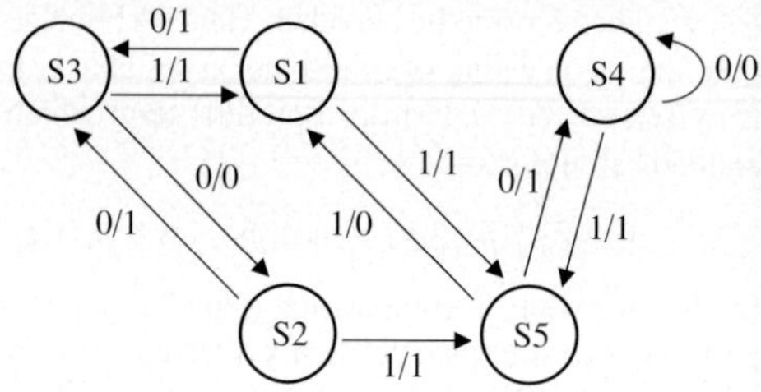

Fig. 6.27 State transition diagram before optimization (input signal/output signal)

The first step of a minimization of states is to search the table for states which have, for a specific input value, identical output values and identical 'next states'. In the example this is the case for the states S1 and S2. Because of that these states can be combined into one state.

Table 6.7 State transition table before optimization

Input	State	Next state	Output
0	S1	S3	1
1	S1	S5	1
0	S2	S3	1
1	S2	S5	1
0	S3	S2	0
1	S3	S1	1
0	S4	S4	0
1	S4	S5	1
0	S5	S4	1
1	S5	S1	0

Table 6.8 State transition table after optimization

Input	State	Next state	Output
0	S12	S3	1
1	S12	S5	1
0	S3	S12	0
1	S3	S12	1
0	S4	S4	0
1	S4	S5	1
0	S5	S4	1
1	S5	S12	0

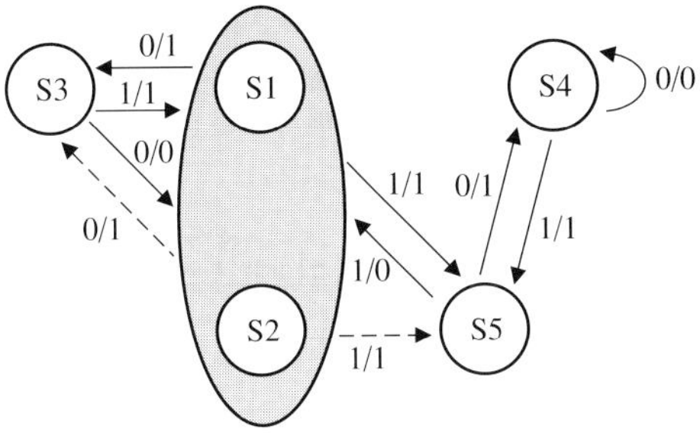

Fig. 6.28 State transition diagram after combining S1 and S2

Coding of states

To find a suitable coding of states is more difficult than it seems to be at first glance. The problem is to find a binary equivalent for every state in such a way as to optimize chip area and operating speed. The binary equivalent reflects the type of storage element (e.g., D, JK, or Toggle FF) and the type of coding of the states. There are **three possible types of coding states**:

- 1-Hot-Coding: 0001, 0010, 0100, 1000
  Each state is represented by exactly one flip flop.
  Advantage: High operating speed
- Binary Coding: 00, 01, 10, 11
  Using n state registers (flip flops) 2^n states can be coded.
  Advantage: small chip area
- Gray Coding: 00 01 11 10
  Only one state register changes its value on subsequent states.

❑ *Example*: State machine with a specification of coding

```
LIBRARY IEEE;
USE IEEE.STD_LOGIC_1164.ALL;

ENTITY counter IS
    PORT(
        reset:  IN STD_LOGIC;
        clk:    IN STD_LOGIC;
        cnt:    IN STD_LOGIC;
        co:     OUT STD_LOGIC);
END counter;

ARCHITECTURE counter_arch OF counter IS

TYPE STATES IS (S0, S1, S2, S3);
ATTRIBUTE enum_encoding: STRING;
ATTRIBUTE enum_encoding OF STATES : TYPE IS
-- -- 1-Hot:
-- "0001 0010 0100 1000";

-- user defined coding:
"1111 1110 1100 1000";

-- -- Gray code
-- "000 001 011 010";

SIGNAL current_state:  STATES;

BEGIN

    FSM: PROCESS(clk, reset)
    BEGIN

        IF(reset = '1') THEN
            current_state <= S0;
            co <= '0';
         ELSIF(clk'EVENT AND clk = '1')
        THEN
            -- co to '0', write only in S3
            co <= '0';
        CASE current_state IS
            WHEN S0 =>
                IF(cnt = '1')
                THEN current_state <= S1;
                END IF;
            WHEN S1 =>
                IF(cnt = '1')
                THEN current_state <= S2;
                END IF;
            WHEN S2 =>
                IF(cnt = '1')
                THEN current_state <= S3;
                END IF;
            WHEN S3 =>
                IF(cnt = '1')
                THEN current_state <= S0;
                co<='1'; END IF;
        END CASE;
        END IF;
    END PROCESS FSM;
END counter_arch;
```

6.6 Retiming

Retiming algorithms address the problem of optimizing clock frequency and the area of a synchronous circuit as well. The following basic possibilities can be chosen:

- Assigning a new location to the register while keeping the combinatorial logic (fig. 6.29);
- Distributing the combinatorial logic to available registers (fig. 6.30);
- Separating a large combinatorial logic into smaller parts (fig. 6.31).

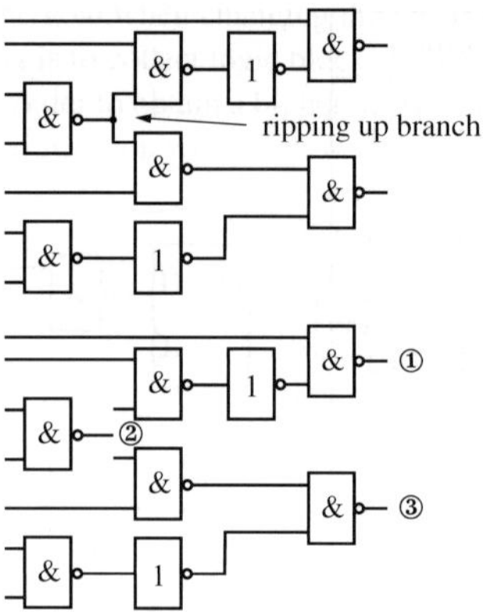

Fig. 6.34 Ripping up branches on the net

At such a starting point each node is compared with all patterns in the library. If there are several equivalent library elements the figures of merit of area or delay decides which one should be used to provide an optimal result under the bottom line.

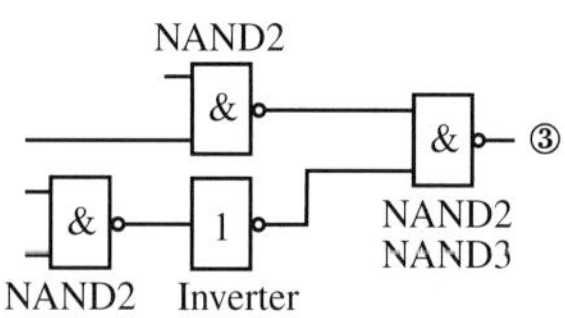

Fig. 6.35 Suitable library elements

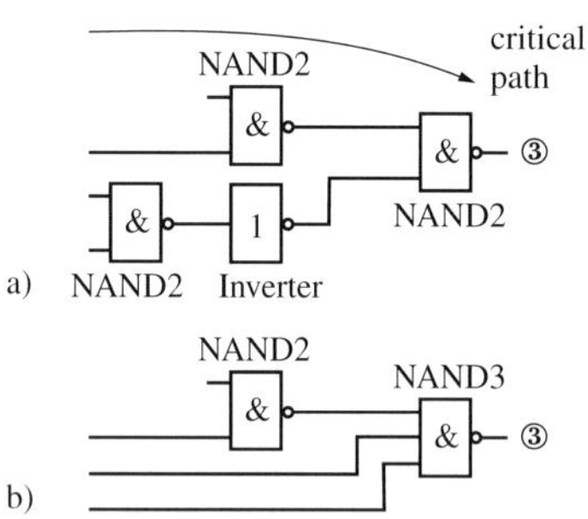

Fig. 6.36 Two different solutions of different targets a) optimizes critical path, b) optimizes area

There is at least one trivial solution because all basic functions are, at the gate level, part of the library. Suitable library elements for the branch having the starting point 3 (see fig. 6.34) are shown in fig. 6.35. Two solutions are possible. The solution according to fig. 6.36a is the one preferred if the assumed time-critical path has to be minimized, since the NAND2, compared to NAND3, offers a better propagation delay (fig. 6.32). The second alternative in fig. 6.36b, however, shows a smaller chip area.

6.8 Synthesizeable constructs, attributes, types and operators

Legend:
✓ Supported by synthesis tool;
— Not supported by synthesis tool;
✕ Not supported statement.

Table 6.9 VHDL constructs for synthesis

VHDL construct	
AFTER: statement of sequential or concurrent statements	—
Aggregates: Record and Array	✓
ALIAS	✓
Allocator	✕
Architecture Body	✓
ASSERT: sequential or concurrent	—
ATTRIBUTE: declaration and specification	✓
BLOCK	✓
CASE	✓
COMPONENT: declaration and instantiation	✓
CONFIGURATION: declaration	—
CONFIGURATION: specification	✓
CONSTANT: declaration	✓
ENTITY: declaration	✓
EXIT	✓
FILE: declaration	—
FUNCTION	✓
GENERATE	✓
GENERIC	✓
IF	✓
LIBRARY	✓
LOOP: without specification of iteration	✕
LOOP: (FOR)	✓
LOOP: (WHILE)	✕
NEXT	✓
PACKAGE: declaration and body	✓
PROCEDURE	✓
PROCESS	✓
Qualified Expression	✓
RETURN	✓
SIGNAL: simple, conditioned and selective assignment	✓
SIGNAL: sequential assignment	✓
SIGNAL: declaration (not in packages)	✓
SUBTYPE	✓
TYPE: declaration and conversion	✓
USE	✓
VARIABLE: declaration and assignment	✓
WAIT (only one wait in process)	✓

Table 6.10 VHDL attributes in synthesis

Object class	Attribute name	
Array	'HIGH	✓
	'LEFT	✓
	'LENGTH	✓
	'LOW	✓
	'RANGE	✓
	'REVERSE_RANGE	✓
	'RIGHT	✓
Signal	'ACTIVE	×
	'DELAYED	×
	'EVENT	✓
	'LAST_ACTIVE	×
	'LAST_EVENT	×
	'LAST_VALUE	✓
	'QUIET	×
	'STABLE	✓
	'TRANSACTION	✓
Type	'BASE	✓
	'HIGH	✓
	'LEFT	✓
	'LEFTOF	✓
	'LOW	✓
	'POS	✓
	'PRED	✓
	'RIGHT	✓
	'RIGHTOF	✓
	'SUCC	✓
	'VAL	✓

Table 6.11 VHDL types in synthesis

Type class	Type	
Enumerator	BIT	✓
	BOOLEAN	✓
	CHARACTER	✓
	STD_LOGIC STD_ULOGIC	✓
Integer	INTEGER	✓
	NATURAL	✓
Floating Point	REAL	×
Physical		×
Array	BIT_VECTOR	✓
	STD_LOGIC_VECTOR STD_ULOGIC_VECTOR	✓
Records		✓
File		×
Access		×

Table 6.12 In synthesis supported operators BIT and BIT_VECTOR, STD_LOGIC_VECTOR (with suitable package)

Operator class	
Logic	AND, OR, NAND, NOR, XOR, XNOR, NOT
Comparison	=, /=, <, <=, >, >=
Addition	+, −, &
Sign	+, -
Shift	SL, SRL, SLA, SRA, ROL, ROR
Multiplication	*, /, MOD, REM
Miscellaneous	**, ABS

6.9 References

[6.1] Altera: Altera/Synopsys User Guide. Application Note, July 1995

[6.2] *Devadas, S.; Ghosh, A.; Keutzer, K.*: 'Logic Synthesis'. – Maidenhead: McGraw-Hill, 1994

[6.3] *Hachtel, G. D.; Somenzi, F.*: 'Logic Synthesis and Verification Algorithms'. – Kluwer Academic Publishers, 1996

[6.4] *Landman, H. A.*: 'Visualizing the Behavior of Logic Synthesis Algorithms'. – San Jose: Toshiba America Electronic Components, Inc., California, 1998

[6.5] *De Micheli, G.*: 'Synthesis and Optimization of Digital Circuits'. – Maidenhead: McGraw-Hill, 1994

[6.6] *Nelson, K.*: 'High-Level Design Methodology Overview'. Synopsys Methodology Notes, March 1994

[6.7] Synopsys: 'Flattening and Structuring'. A Look at Optimization Strategies. Application Note 1994

[6.8] *Vahtra, K.*: 'ASIC Design Partitioning'. Synopsys Methodology Notes, January 1994

6

7 Hardware/Software Co-Design

DIRK JANSEN

As the complexity of IC design has increased, a trend to include ever larger portions of functionality on a single chip has been observed. Programs that once would have been considered 'software' are now encoded directly into memory within the chip structure. In the past a processor core was used which was controlled by discrete components elsewhere on the same circuit board. Extensive software support in form of compilers, assemblers, and simulators was also encoded into chips separate from the processor. Today these components, as well as peripheral modules (e.g., UART-serial interface, and PIO-parallel input/output interface) are used frequently in so called 'embedded processors'. These discrete components are inserted block by block directly into the circuit during the design phase.

7.1 The Concept of Design Re-Use

7.1.1 Why Re-use of Design Modules?

Today more functions can be readily integrated into an IC design than can be designed in a reasonable development time by a design team. Even a highly productive engineer does not have sufficient time to develop a circuit design from scratch. Ever shorter development periods demanded by today's market only exasperated the situation. This situation produces a so called '**productivity gap**'. This gap defines what is possible and what can be accomplished in a reasonable time by a designer. This gap is constantly expanding as improvements in what technology can allow and what can be reasonably done designer compete. A reduction in the gap can only be realized by new procedures and a change of paradigms, Fig. 7.1.

Currently ASICs contain 150k gates on average and 400kbits of memory. Designs of this complexity are typically described by 10,000 lines of code (VHDL). Five times as many lines of code are required for large, complex chips, but this task is still performed economically. In the future there will be 100,000 lines of code, delivering 500 k gates and with more than 1Mbit of memory. Even designs of this magnitude do not consume what is technologically possible in silicon using 0.18 μm or 0.13 μm technology, see table 7.1.

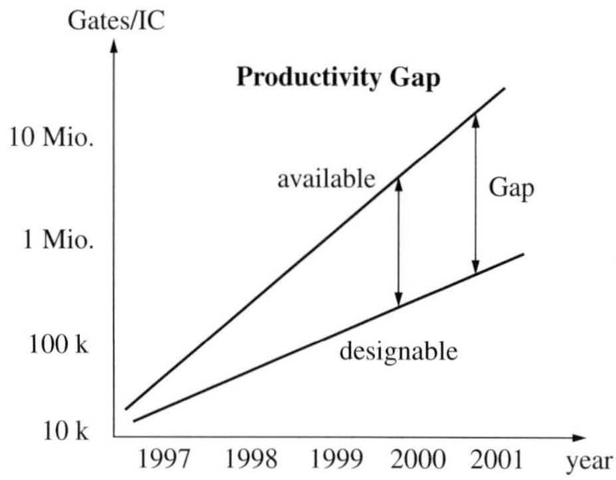

Fig. 7.1 The so-called 'Productivity Gap'

Table 7.1 Technology Trend of typical ASICs

Technology (μm)	Gate Number	SRAM in bit/cm^2
0.35	1 M	3.3 M
0.25	2 M	6.6 M
0.18	2.6 M	11 M
0.13	6.2 M	22 M

It has therefore only become possible to design with a high level of abstraction **by re-using** already developed and established modules in a block design. The growth in complexity must be tempered by partitioning tasks into building blocks. Concentration on interfaces and standardisation of the data exchange can be carried out to help reduce the overall risk when using this design framework.

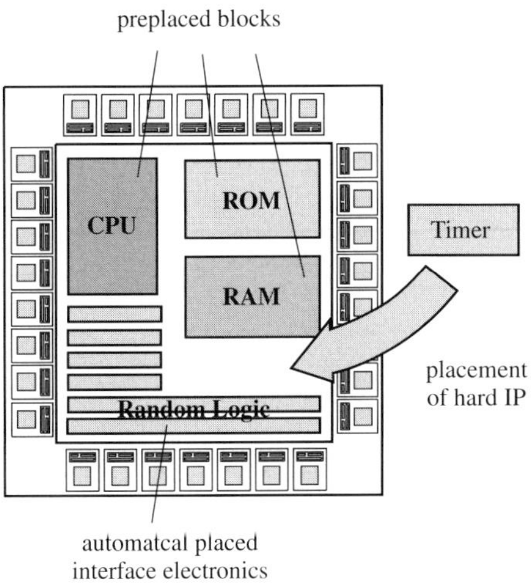

Fig. 7.2 Floor plan of a chip by placement of pre-wired blocks (IP) in an block design style as well as special adaptation electronics in a standard cell style

Programmable structures and processor cores, which have become common cell blocks, the principle of 're-use' is already applied. So called 'hard macros' are placed directly in the chip design; these blocks have been previously verified by use in a previous device and are completely pre-routed. Transferring these blocks into another technology with smaller device structures, other designing principles, etc. is difficult and can require substantial development costs. For example: the transition of a technology from 2 metal layers using 0.7 μm CMOS to a modern 0.35 μm technology or 0.18 μm technology with 3 or more metal layers forces a total re-organization of the netlist, a replacing of the cells, and a re-routing of the core. Synthesis tools can help with the new circuit layout, but timing verification must be redone.

What is left to Design? The main idea; the intellectual content of the design, or simply the *Intellectual Property* (IP).

As the use of VHDL and other hardware design languages has spread it has become possible to formulate an idea in an executable form. Simulators allow the function to be reconstructed and verified against a certain specification. Synthesis programs create netlists. Additional programs allow blocks to be placed and routed automatically. Functional cells can be generated in the target technology.

Descriptive formulation in high level languages is independent of technology, and easily modifiable. This methodology helps to contain development costs. A similar benefit can be seen in software programming where the use of high level languages (e.g., C or C++) help reduce development costs. IP could play the same role in IC design as libraries do in software programming. IP may be transferred between designers and become the subject of trade and licensing.

Fig. 7.3 Typical integrated circuit with placed functional modules (IP) (chip photo of Fhop_16 controller, designed by the University of Applied Sciences, Offenburg after [7.26])

Within the software area the transition to high level languages has led to losses in *performance* of the languages; programs are related to programs assembled in code beforehand. Block IC design also has a large area of reduced efficiency when compared to hand optimised designs. This loss, however, is compensated by a large increase in design productivity, and thus in development speed. Since chip area is plentifully available the new paradigm reads: **'Time to Market'**.

The application of IP modules strengthens this trend. The old paradigm of **'optimization to minimum chip area'** increasingly loses its meaning. It has already been affected by sub-optimal area consumption by using the *standard cell design style* compared to *full custom design style*.

7.1.2 Hard/Soft Macros, Virtual Components and Intellectual Property Devices (IP)

The term IP can best be translated as '*intellectual property subject*' and can be generalized by the concepts *soft macro* and *hard macro* which have existed in the design world for quite some time. Lewis in 1997 said that

> An IP is a design object, predominant value of which came out from the knowledge of the producer, and its revised design would take significant time.

The object is derived from IP, obtaining a piece of silicon by synthesis, placement and routing of standard cells. This is the *instances* of the IP. 'Small IP' in particular interface modules are sometimes called *virtual components (VC)*. Both are described as discrete components in technical data sheets and have guaranteed levels of performance. Today, IP can be differentiated as follows:

- **Hard IPs** are described at the mask level (e.g., by GDSII data). Typical examples are standard cells provided by the manufacturer, large processor cores (e.g., ARM 7TDI, Intel x86, and TI 320C25), and analogue or mixed signal cells; the latter have substantial dependence on the technology. Complete transceiver and RF front ends are available as hard IP. Common generators for RAM, ROM, MPLY, etc., belong to this class as well, although these are configurable in some way; even though they are configurable, the result of these generators are geometrically defined blocks which cannot be transformed or restructured.
- **Firm IPs** describe the function at netlist level with predefined placing, however without details of routings. Typical examples are nearly all cores used in FPGA's. Firm IPs are not directly mappable to other technologies and in this way they are technology specific.
- **Soft IP** describe the function of the module in a high level language (e.g., VHDL or Verilog). Netlists are produced by synthesis tools. An appropriate program maps to the target technology. Interface modules are typically for this style: basic digital blocks (e.g., counters, adders, shifters, *PCI interfaces)*, small controllers, like the *8051 & 6805-family*, digital filters, and more complex functional units like *MPEG decoder*, *viterby decoder*, etc. These blocks can be parameterized and are altered to meet the application requirements.

Sometimes there may be a floating transitions between types. With soft IP there are mechanisms which prevent copying, modification or even visualization of the core. This is done to protect copyrights of the IP provider. In many cases, to achieve the desired performance level (register transfer level, RTL) most of the code is written in a structural style, not a behavioural one. This looks like a netlist description. Together with scripts for synthesis, which control the mapping to the target technology and preplacement information, these IPs are more Firm IP than Soft IP.

Based on statistics of *design and re-use* [7.6] databases which keep their cores have been categorized with 49 % hard IP, 25 % firm IP and around 22 % as soft IP. The remaining 4 % are model descriptions at a purely abstract language level. It is predicted that as process standardization and protective mechanisms mature, the portion of Soft IP will strongly increase. The use of building blocks outside the founding organization places a high demand on documentation and quality of the design object. The more complex is the functional block, the greater the expenditure for documentation and interfacing. At the same time, a smaller probability of block re-use can be expected in future. Exceptions are *processor cores* and standardized interface building blocks (e.g., *PCI interface* blocks).

7.1.3 Virtual components (VC)

One of the most importent building blocks, besides the standard cells themselves, are the generators of *regular blocks*. Adders, multipliers, counters, shifters, FIFO Memories, and other well known digital blocks can be automatically generated through the use of VHDL statements or other similar mechanisms. Given the bit width of a data path and a general architecture selection, these blocks are derived from a prototype in a systematic fashion with a predefined performance.

Fig. 7.4 shows the input form of a generator as it is used in a typical design system for a FPGA (Actel). The generator produces a model in VHDL or Verilog, with timing data and anticipated performance. This information can be used by a digital simulator or by a synthesiser to generate a netlist for the block. Furthermore it creates a symbol for use in schematics. These automatically generated building blocks are standardized in the *Library of Parameterized Modules Standard (LPM)*. More information on LPM can be found in chapter 17 of this book.

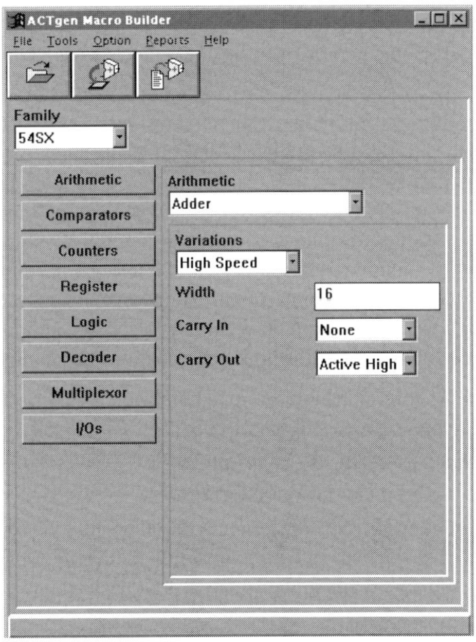

Fig. 7.4 Input form of a modul generator (Actel Corp.)

The generators enable detailed selection of an architecture (i.e., a *ripple carry* architecture, a *look ahead carry* architecture or a *carry select* architecture). This affects the timing, performance, and physical size of the block. Generators can be used for different levels. Some generators (e.g., MENTOR LOGIC BLOCKS) are primarily made for the simulation level, essentially they generate behavioural code. These generators are called *soft generators* because the synthesis process has to be performed at a later time on the results. Since these generators create graphic symbols automatically, they are usable during schematic design input, and they permit fast, clear representation on block schematics.

Using parameterized generators for mapping of arithmetic operators '+' or "*" is called synthesis. If this type of operator is found in the code the synthesis includes a *function call for a synthesis library component*. Attributes specify and select the appropriate architecture, whilst word size is taken directly from code declarations.

If these library generators are not used the arithmetic operators are analyzed during synthesis in accordance to the IEEE 1164 packages 'std_logic_arith' into their basic gates. This technique usually requires much more area and a time delay. Once disassembled, the architecture of a *carry select adder* or a *Booth multiplier* cannot be found again in a long synthesis run. Extensive IP libraries are offered for synthesis, see table 7.2 and 7.4.

Parameterizable IP for analogue components (e.g., A/D converters, switched capacitor filters, building blocks for signal processing components and operational amplifiers) are available to the modern designer. These are mostly hard IP generated and are provided in a predefined technology.

Table 7.2 gives an overview of available parameterized IP and generators.

Generators are preferentially found in the FPGA design systems of ALTERA, XILINX, ACTEL and others. In these cases it is easy to parameterize because the hardware is programmable. Generators are commonly provided by third party providers. These providers work closely with both the EDA companies and technology providers.

Table 7.2 Parameterizable Virtual Components (IP), short overview

Building Block	Type	Parameter	Provider
Single Port RAM	generator, hard IP	word size, volume	Dolphin Integration
Dual-Port-RAM	generator, hard IP	word size, volume	Dolphin Integration
ROM	generator, hard IP	word size, volume	Dolphin Integration
Multiply	generator, hard IP	word size, volume	Dolphin Integration
Multiply	generator, soft IP	word size, architecture	Synopsys, Mentor
Adder/Subtractor	generator, soft IP	word size, architecture	Synopsis, Mentor, Altera
Arithmetic Logic Unit	generator, soft IP	word size, architecture, control statements	Micro Tech Consulting, Mentor, Synopsys etc.
FIR Filter	fix and variable coefficients, soft IP	word size, grade, type, number of coefficients	Mentor, Cadence, Synopsys, Dolphin Int.
A/D Converter	sigma delta, succ. approx., direct conversion, hard IP	number of channels, resolution, sample rate, interface	LSI (CoreWare), Dolphin
D/A Converter	sigma delta, direct conversion D/A, hard IP	resolution, interface	CoreWare, Dolphin

7.1.4 Standardisation, the Virtual Socket Interface Alliance (VSIA)

Spreading the *re-use idea* is fundamental to the standardisation concept. Standardised should be

- the terms;
- the description forms;
- the models;
- the delivery items;
- the basic legal conditions;
- the EDA aspects in conjunction with the use of virtual components.

In this endeavour the *Virtual Socket Interface Alliance* (VSIA) was formed by the important, active companies in this area [7.27]. It is their task to establish standards and distribute information on IP and re-usable components from a central location, or their stated goal is: "*To specify or recommend a set of hardware and software interfaces, formats, and design practices for the creation of functional blocks that enable the efficient and accurate integration, verification and testing of multiple blocks on a single piece of silicon*" [7.27].

More details on this subject can be obtained from the VSIA web site at http://www.vsia.com. This website contains a wealth of information which is freely available for download, see table 7.3. Detailed specifications, however, are only accessible by members. VSIA represents the interests of IP providers and manufacturers, as well as IP users and customers. The VSIA organization tries to accommodate the interests of both sides. VSIA has demonstrated its influence on modelling concepts, deliverables, and on the manner in which IPs are implemented into EDA.

Table 7.3 Available documents from the Virtual Component Interface Alliance

Document	Status	Remarks
VSI Alliance Roadmap	Version 1.0 1997	Defines goals of VSIA
VSI System Level Design Model Taxonomy	Version 1.0 1998	Defines naming and requirements of models for system
VSIA Architecture Document	Version 1.0 1997	Defines terms, methods, model levels, verification- and test requirements for virtual components

Further information concerning design and use of IP can be found in '*The Re-Use Methodology Manual*', which is published by both Mentor Graphics and Synopsis [7.20].

7.1.5 Virtual Components as Commercial Objects

Every important EDA provider delivers virtual components and IP today. In most cases they operate as brokers who sell the technology of smaller design houses. A small group of 'spin-off' companies which gained knowledge working on design projects for the IP market now sell their know how of special cells. A partial list is shown in table 7.4. Everyday new companies are founded specializing in this area. A database of these speciality companies can be found at http://www.design-reuse.fr. This website is maintained by TIMA/Grenoble and at present contains more than 4 000 IPs in a huge database. For each a description, data-sheet, and address of the provider is given. Access to the data base is free of charge, but an on-line registration is required.

In IP marketing the business model of **licensing** is pursued, similar in structure to software licensing, license fees typically entitle the user to the use of the core in a single design. If additional designs are desired, the license must be renewed; additional licenses are typically provided at reduced costs. At present license fees are significant; they can begin at a few thousand dollars for an FPGA interface core to several $100,000 for a high quality RISC processors (ARM). Smaller controllers (e.g., the

Table 7.4 Some available virtual components

Provider	Design Objekts
ACTEL: CoreHDL	Firm IPs: I2C, CAN, USB, UART, ATM
ALTERA: OpenCore©, MegaCore© http://www.altera.com	Firm IPs: DSP, JPEG, Viterby-Decoder, FFT, NCO, PCI, CAN, USB, UART, Controller
Argonaut RISC Cores Inc.	RISC Processor
ARM Ltd: ARM Thumb Family http://www.arm.com	RISC Processor
Cadence Design Systems http://www.cadence.com	processor, controller, interfaces, DSP functions, protocol building blocks, logic blocks
Dolphin Integration: Circuitware© and FLIP© http://www.dolphin.fr	RAM/ROM generators, Logic IP, processors, PLL, analougue and RF Cores
Frontier Design BV http://www.frontierd.com	telecommunication cores, multimedia cores, DSP cores
Integrated Silicon Systems Ltd: ISS Cores http://www.iss-dsp.com	JPEG, FFT, DCT/IDCT, ADPCM, Reed Salomon, Viterbi, QPSK, FIR, DSP components
LSI-Logic: CoreWare© http://www.lsilogic.com	soft IP, OakDSP, ARM core, MIPS core, GigaBlaze©, HyperPHY©, interfaces, bus interfaces, AD converter, protocols, DSP functions
Mentor Graphics: INVENTRA© http://www.mentor.com	processors, controller, interfaces, DSP functions, protocols, logic blocks
MTC Micro Tech Consulting: CAST http://www.mtc.de	DSP IP, ASIC cores, processors, standard components
Phoenix Technologies Ltd: Virtual Chips© http://www.phoenix.com	PCI soft cores, IrDA cores, USB, models, BIOS software, application software
PIVOTAL: Fulcrum© http://www.pivotaltech.com	Hard IP: A/D converter, D/A converter, PLL, oscillator, band gap ref, charge pumps, temp. sensors, voltage regulators etc.
Stellar Semiconductor Inc: PixelSquirt© http://www.stellarsemi.com	3-D graphics cores, soft IP for graphics
Synchonicity Inc. http://www.syncinc.com	internet based IP supply and groupware
Synopsys GmbH: DesignWare© http://www.synopsys.com	processors, controller, DSP functions, protocol, logic blocks
XILINX http://www.xilinx.com	Firm IP: PCI, CAN, USB, UART, controller
FH-Offenburg http://www.asic.fh-offenburg.de	FHOP16 controller, UART, INTR, PIO, free download of VHDL IP under public license

8051 type) are available at around $20,000 and upwards. Owing to strong competition in the market and an increased availability of standardised basic blocks, a price reduction is anticipated in the future. As rule of thumb the price for a license is around 15 to 20% of the cost to design the block from the scratch. It should be noted that license fees cover more than just naked code; the fees may also cover additional services, employee training, documentation, and test time at a site or a workbench to verify the core performance. Adaptations for special customer needs are often made. FPGA cores frequently offer little or no support; for this reason they are often offered at much lower rates than real ASIC designs.

Significant importance in the market must be ascribed to the controller and processor cores, beginning with the small 8051 family through to the larger 32 bit RISC cores (e.g., the ARM family). These cores are used in millions of telecommunication devices (e.g., mobile phones, personal digital assistants (PDA) and other mass consumer applications). The cores, when combined with special electronics, form a SOC ASIC. Furthermore, analog cells (e.g., A/D converters, RF components, receiver frontends, and modem components for telecommunications) are important to the ASIC market volume. These blocks are often avoided by systems companies who are not able or willing to invest in analog skills or full custom design for their ASIC development team.

IP and processor cores are provided by universities at no charge under public license conditions and can be downloaded from the internet. A collection of links is maintained at the university of Hamburg, Germany at http://tech-www.informatik.uni-hamburg.de/vhdl/vhdl.html. Freely downloadable models and links to servers with similar information can be found at this website. The author, who has developed the FHOP16 microprocessor core, offers it and software tools for its implementation without charge at http://www.asic.fh-offenburg.de. A quick search on the internet yielded two processor cores of the 8051 family, one ERC32 core utilizing a SPARC architecture, a PIC clone, a 8085 clone and other less known cores, all available for free. Another valuable source is the website of the *Free Model Foundation FMF* at http://vhdl.org./vi/FMF. FMF distributes free VHDL-cores. A good source for VHDL models are obtainable at the *RASSP Project* http://rassp.srca.org. RASSP (*Rapid Prototyping of Application Specific Signal Processors*), a DAPRA project with a $150 million budget. European developers still dream of such governmental support. However, in Europe there is the *Open Systems Initiative* OMI; the OMI combines activities in this field and is a good source of software and VHDL models. A process similar to the software/operating system is anticipated in this field. LINUX, primarily used in universities, is now spreading into traditional markets. There is enough space for public domain development adjacent to commercial development.

Upon delivery, a virtual component should contain at least:

Soft IP

- VHDL code (or Verilog);
- synthesis script;
- test patterns for the simulation (stimuli), or workbench;
- description of signal behaviour and functionality;
- bus model with timing;
- documentation, data sheet with performance data and description.

Hard IPs

- description of geometry (Gds II);
- LVS-files (netlist or similar);
- used design rules;
- Simulation model in C or VHDL (description of behaviour);
- stimuli files or workbench;
- test pattern and associated results;
- documentation with data sheet and performance data.

These are minimum requirements. Typically there are numerous documents detailing the model. When comparing hard IPs and soft IPs, verification and implementation of hard IPs require additional metrics and supporting documents. In order for some metrics to be checked (e.g., a *layout versus schematics check* (LVS)) netlist information must be available. Since the provider does not want to give detailed connection diagrams away, complicated and pedantic mechanisms are commonly employed. Model information and cell layout are provided independently. The two may not even be derived from the same origin; as a consequence they may not be consistent. Even so, the cell is qualified by silicon prototyping and then verified by measurement. Performance data is reliable with a defined process. Hard IP corresponds mostly to the discrete components used on printed circuit boards.

Soft IPs are charmed, in that they can insert themselves smoothly into the design flow of the customer, thus the consistency problem is avoided. Performance data, however, depends strongly on the synthesis, and are not typically fully guaranteed. In order to protect the provider and limit the degrees of freedom in the synthesis, the VHDL code is generally delivered in a structural VHDL style. Mapping may be done on a generic library. With the delivered scripts the mapping is normally done without difficulty from the generic library to the target technology. A complete and exhaustive simulation, including timing, is required for each case. Because a simulation of the core cannot be done by the customer without detailed knowledge, complex cores such as microprocessors are not well suited for soft IP. A huge verification effort is required for both cases. A common solution of this verification issue is exhaustive workbench verification; unfortunately this hides detailed performance information about the core, but it does ensure the integrity of the functionality.

7.2 Design with Virtual Components and Processor Cores

7.2.1 Design to Target Technology FPGA

When looking at FPGA as a target technology a special situation exists: companies like ALTERA or XILINX are primary interested in selling chips. For this reason they do not offer design house activities. Their interest is to spread IP technology, so they provide and distribute cores at low cost. Libraries offered under these circumstances, and the *MegaCore*© project are concerned about the price; it is not a primary commercial objective for these companies. They work as brokers for distribution of cores developed by customers or public domain IP, without guarantee, and generally **promote the use of IP**.

Besides standard libraries, which contain logic series and well known building blocks (e.g., counters, registers, adders, etc.), more complex LPM modules are available by using the provided generators. Structural block diagram set-up with in co-operation with these components is easy and effective. In the case of a language driven design, where the target functions are described within the design language, operators are effectively replaced during synthesis by firm macros. The library from ALTERA [7.1] or a similar one from a competitor, offers a large selection of complex building blocks, from adders over multipliers to complete memory designs. Who really needs more? In addition, companies like MENTOR, SYNOPSYS, and others offer libraries specifically suited for FPGA design. Furthermore, target independent synthesis software for FPGA synthesis (i.e., LEONARDO Spectrum from Exemplar Logic Inc, now Mentor) is also available.

The *PCI interface* was one of the first commercial successful complex soft cores. This made the breakthrough for the use of IP technology. This particular core demonstrated the ease with which a design can be performed. Using IP in this way showed a benefit, the reduction of development time and risk, to the customer. This core permitted easy design of PC card interfaces utilizing FPGAs; this resulted in a huge market opening up for the FPGA companies.

FPGA cores are usually available as *firm IPs* in code form. They can be completely simulated on dedicated development systems of the manufacturer before a license is purchased and before the integration of the core into the design. In this situation the customer does not have to carry a design risk. The license is only necessary for programming the design into the target component (*Opencore*© program ALTERA).

A remarkable large market for huge FPGA with more than 100 k gates is the **emulation of soft cores and whole designs**, which are implemented for testing purposes on FPGA. Significant simulation times can be saved and design accuracy is increased. Emulation is still of advantage if the target frequencies in the FPGA technology cannot be realized. The gain in speed may, in relation to a classical digital simulation, be more than a factor of 100, sometimes 10,000 or more and with the

7

consumption. Substantial arithmetic performance is demanded in these applications, both in mobile phones, where signal processors are adopted, as well as in classical arithmetical tasks, such as in GPS receivers. Standard 8 bit controllers such as the 8051 family are not well suited and have poor performance. Computer game consoles place high demands on a processor, but to be competitive it must be provided at a low price. Often computer game consoles have more computing and graphics capabilities than many modern computers. Here we find RISC cores, such as the 32 bit ARM 7, although the licensing costs are still high. The ARM cores are available in a large family with large selection in performances, now adaptable to nearly all relevant CMOS technologies [7.2]. They are accompanied by high quality software tools, supported by many providers. The cores are now available on FPGA too, in the form of hard cores as well as in the form of soft cores in high level design language. The license costs are still high for small companies, so ARM is mostly found being used by big customers.

Smaller processor cores are mostly distributed as soft cores, but the effort for verification is significant and depends strongly on the support of the provider and the synthesis scripts used. So there are still hard cores available and used because of the lower design risk, if they are available in the target technology. Sometimes re-targeting is done by the supplier.

A substantial problem is still the integration of the building blocks belonging to the processor system such as interface, parallel interface, RAM, and ROM. Because these blocks originate from different suppliers and there is no real standardisation of the bus interconnection, numerous adjustments and critical simulations are needed to verify performance. ARM is here again in the forefront of the technology with its AMBA bus standard, which allows certain *plug in* performance. It is therefore very important to have an exact description of the bus, e.g., in the form of a *bus model*. There are many suggestions for such *bus standards* [7.27], but they may be adopted with new designs only.

Small and medium enterprises, for which commercial processor cores are too expensive, may fall back to the freely available public domain cores mentioned before, e.g., to the FHOP16 controller fig. 7.3 with 1.5 mm^2 area and a throughput of 5 MIPS, designed at the institute of the author, which is already used successfully in several applications. The use of hard IP requires that a semiconductor manufacturer be chosen to provide the chips. This leads to a long term binding agreement with the chosen manufacturer. The decision for a certain processor core is in this way a strategic decision at the management level and binds the company because of the high and long term investments in license, tools, and, not least, in the training of the engineers.

ASIPs require very high personnel and capital expenditure. In addition to the actual core the entire software development system, including compiler, must be developed. In order to simulate the core and, with the dedicated hardware, to validate its performance, special developments have to be made in the CAE area. Ordinarily only large enterprises operating world wide are able to do that.

7.2.3 Embedded Software

With the use of processor cores a new level of complexity is reached, since the functionality of the ASIC no longer depends on the gate logic, the architecture, or *the hardware* alone, rather than the software implemented on the chip, determines functionality. This is similar to a freely programmable computer system made of discrete devices. There are many advantages:

- Functionalities can be designed which would not be feasible with hardware alone because of their complexity;
- The software may have many different tasks which can be specified late in the development of the chip, and can be modified flexibly;
- The chip can be used for very different applications and recovers thereby a certain degree of universality, whereby potentially larger numbers of deliveries and with that lower costs may follow. At least this leads to substantial savings in development;
- If the software is downloaded on activation of the chip or is stored in a FLASH or EEPROM memory, then extreme flexibility is achievable. This level of flexibility is not possible with universal, discrete embedded systems.

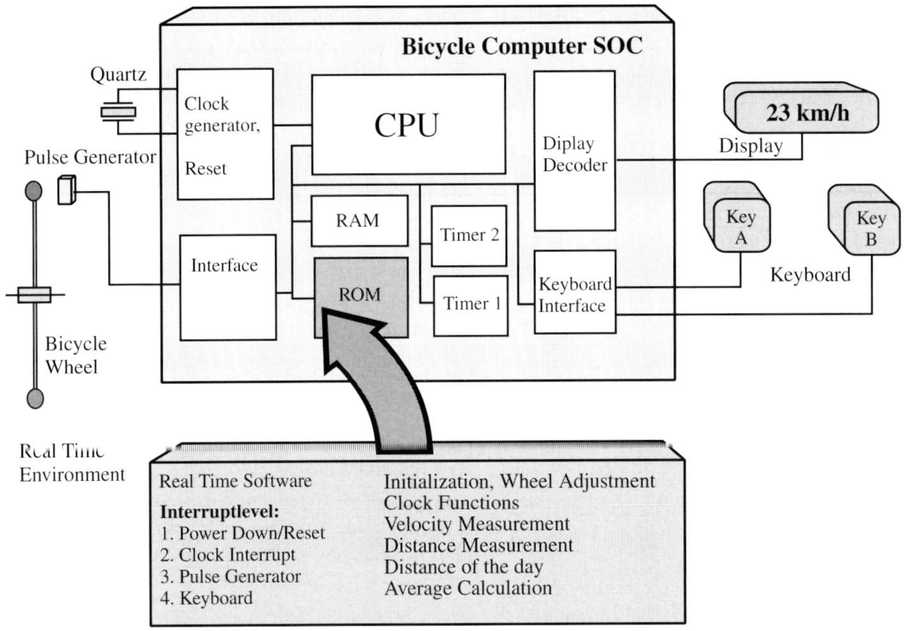

Fig. 7.5 Configuration of hardware and software in a typical application of a bicycle computer

Inclusion of software into the development process complicates the task of design and demands additional knowledge in microprocessor program design from the ASIC designer. A further task is uniting and verifying *hard and software together* in a common simulation system, a task which is still not yet solved satisfyingly.

Compared to discrete board designs, the available resources of RAM and ROM are a magnitude smaller on the chip, because they cost chip area and define by that the price of the product. The designer will concentrate to keep these modules, generally created with a RAM/ROM generator, as small as possible.

ASICs equipped with processor cores are represented on small **independent systems** which react to signals from the external world. They perform complicated time sequence and control functions. These systems are generally *real time applications* with several different interrupt sources and levels, priorities and repetitive tasks, which have to be performed in a non-deterministic way. A well known application is the bicycle computer, which integrates a clock, a distance meter, a velocity meter, and some more functions. It reacts on pulses from a sensor at the front wheel, takes commands from a small keyboard, and drives a LCD display. The required software is stored in a ROM. At a selling price of $10 such a chip must cost less than $1; this is a typical ASIC mass application fig. 7.5.

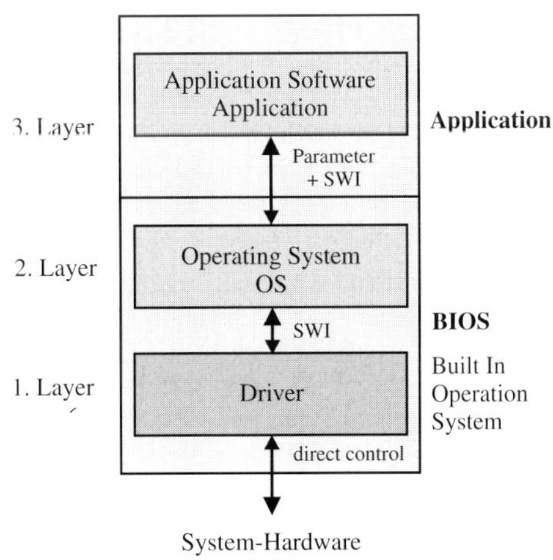

Fig. 7.6 Layer model for ASIC embedded software

Most of these application programs are still developed in **assembler** because of the small resources in RAM and ROM. But it is recommended to apply certain principles of *real time operation systems (RTOS)*. These principles are:

- Organize the software in layers with their own functionality and well defined interfaces between the layers;
- Separate the communication modules and the hardware related modules from the main application program and define the BIOS. Isolate the hardware from the application software with an abstract interface;
- Use a modular or object oriented style for the application software.

The layer model requires a minimum of three layers:

- The layer with the hardware driver;
- the operating system layer;
- the application layer.

In many cases a more detailed partitioning makes little sense. The **driver layer** contains all routines, which directly affects the hardware; it drives interfaces, obtains information from the real world (e.g., keyboard readout, display driver, sensor interface, real time clock, a/d converter readout, etc.) and drives special ASIC periphery modules. The software routines are normally small and specifically defined concerning their I/O addresses. The initialisation routine may also be connected to the drivers, and it sets up all the hardware after a reset or power down. These routines are generally combined in the **BIOS**[1], which is placed in the ROM of the chip. To be able to boot in the right way the BIOS instructions must be mapped to those addresses in the ROM where the processor starts execution after a reset. This may be at the bottom of the memory map at 0x0000 or at the top, depending on the processor architecture.

The procedures are called with arguments which may be transferred via the registers of the processor core. As far as being implemented in the core, a software interrupt call (SWI call) is preferred compared to a normal function call. With normal subroutine calls the addresses of the routines have to be known at compilation, which can be done by header files and an added BIOS library. A more independent solution is to use the index mechanism of software interrupts to isolate the BIOS routines which are provided as ROM ware, sometimes called *firm ware*, from the application program, with a common agreement on the interrupt number, where each vector stands for a certain routine. BIOS calls are now independent of version and no library or header file has to be provided for compilation (e.g., if the software interrupt 15 is activated this may call in each case a subroutine, copying memory from location A to location B, interrupt 18 may provide an output of data on pin number 3, etc.). The index table[2] is loaded during the initialisation phase by the associated BIOS routine and always contains the actual call addresses to the subroutines. Because this table is not fixed these vectors can be exchanged by the application program at run time, exchanging in this way complete routines and functionality. This adds to the flexibility, see fig. 7.7 for more details.

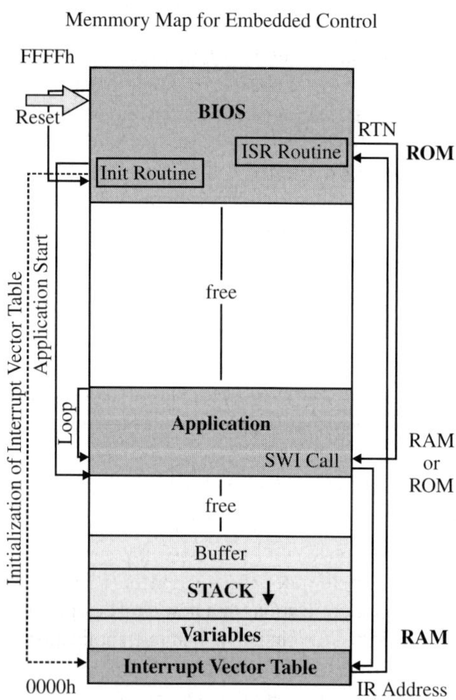

Fig. 7.7 Isolation of Bios by the software-interrupt-mechanism

[1] BIOS stands for **Built in Operating System** also **Basic Input/Output System**

[2] An interrupt vector table contains pointers to the interrupt service routines

In the **operating system layer** the more complex scheduler tasks are accommodated. This includes the administration of the interrupts and of the task priorities. The control of the sequence of tasks, called *scheduling*, and the administration of chip resources (e.g., the RAM) belongs further to this layer. One of the most important features is the ability to start tasks and to communicate through communication channels. This may be via simple keyboard or display operations, but more frequently these are complex, protocol driven communications via USB, CAN, or TCP/IP interface modules. The protocol machine states are organized here, as is the central *task scheduler*; in this case they are implemented as software routines. The operating system layer refers to the underlying driver layer concerning the driver routines and generates the calls for this subroutines. Further provided in this OS layer are useful routines which extend the features of the processor (e.g., routines for mathematical operations like division, sorting etc.). If there is a SWI mechanism, this part of the BIOS will be standardised, too, by using indirect calls as describe before. Otherwise there must be some kind of classical *subroutine call*, perhaps organized as a *jump table*.

The **application layer** finally contains the code for the tasks which form the actual application. With the concept of a structured programming style the application layer consists to a large extent of subroutine calls to the underlying layer. Very complex functionalities can be programmed by this hierarchical architecture. At the same time the code can be well maintained and easily modified, also with little memory. At this level cross compilers can be used effectively, which improves the productivity of the developers substantially, without large concessions in memory consumption.

By its isolation, the BIOS and the application program can be developed further independently. Once developed, the BIOS might be transferred with slight modifications to a new chip design, and in that way *re-used* in more than one application.

The **application program** can be arranged in the ROM, too. The chip can, however, be used only for the intended purpose. It is more elegant to download (*boot*) and execute the application program at the start-up of the chip, which may be the rising slope of the main supply voltage. Such a download may happen from a serial EEPROM as well as from a FLASH-ROM, if there is no reprogrammable memory on board the chip. Downloading is carried out by the initialising BIOS routine, which executes automatically after a reset. The required EEPROM is tiny and cheap, see fig. 7.8. The procedure of downloading is comparable to the initialisation of a RAM based FPGA, where the configuration information is stored in an EEPROM and loaded into the chip with the rising of the supply voltage.

If we include an *upload* feature into the BIOS, too, it is now possible to modify the application program on-line, to update dynamically, and so provide new features. The application program may be organized in *pages*, which are loaded by the operating system dynamically in a classical *overlay* style. The number of loadable pages and by this the size of the implemented program is thus nearly unlimited. As an alternative to an EEPROM also a FLASH ROM can be used for booting.

Programming of embedded software is done today using integrated development systems with **cross compilers** which usually run on PC platforms. As far as the core is compatible with discrete processors, the classical development systems can be used. In practice, however, because of the very limited resources of memory and the high performance demands, more than 70 % of the programs are written in assembly language. The use of high level language C-compilers still play a subordinated role, not at least because of the unacceptable *overheads* of code and the sub-optimal performance [7.19]. This is changing with the new 32 bit processors and larger (external) memory, because there is no alternative.

For widespread discrete processors **real time operating systems (RTOS)** are available as complete, configurable programs. For chip applications these are usually too large or have to be substantially modified. The advantage in using such predefined modules is then small. Recently there are first *RTOS kernel*, which can be used in SOC [7.25]. For the larger RISC processors there exist many *real time operating systems*, which may be adopted (e.g., for the ARM family), according to data given from the manufacturer [7.2], there

7

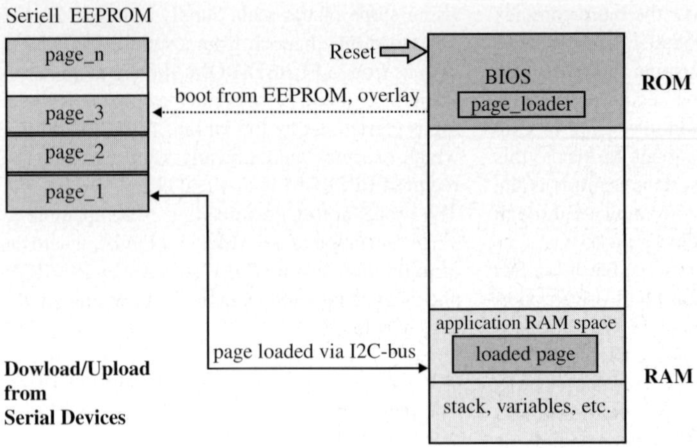

Fig. 7.8 Upload/Download of an application program from a serial EEPROM

are more than 20 different providers [1)]. For ASIP the manufacturers maintain their own OS. Huge ROM based operating systems like WINDOWS CE, PSION EPOC, NetBSD, or embedded LINUX are available, too, but are never integrated into the chip itself.

The programming and validating of the driver routines remain the task of the designer. In the meantime there is more support on the horizon. Now there are driver routines for reading and writing into FLASH ROM provided by the big manufacturers INTEL, AMD, and others, sometimes with all *file system* features, for the most important processor architectures. It has to be expected that other providers of IP or virtual components (VC) will deliver the driver routines in future for their products for dominating processor architectures too. This will ease and accelerate the development significantly.

7.2.4 Hardware/Software Co-Simulation

With the implementation of processor cores the classical *top down design flow* is considerably disturbed. At the beginning of the SOC design cycle, it must be decided on the partitioning of the task. What can be done in hardware and what has to be done in software? Which functionalities are made by programming and which have to be made in dedicated hardware blocks? Software is much cheaper, so there is a rule: '**what is possible to do in software shall be programmed**!'. Hardware blocks are only designed in cases in which requirements concerning speed, performance or functionality cannot be fulfilled by a solely software solution. In most cases it is easyer to decide. Enormous research efforts, are going into the field of software/hardware partitioning [7.18]. There are mixed solutions where a close interaction of the processor core with special hardware blocks is needed, sometimes these are termed *coprocessors*. These solutions require accurate cycle true hardware/software co simulation. Verification of these systems takes much more effort than pure hardware or pure software solutions.

One goal of partitioning is the **isolation of the two work areas** with the intention of being able to work on both tasks in parallel and as independently as possible. Such *concurrent engineering* makes sense because of the time needed for software development. A sequential processing of, first, hardware and, then, software development will blow up the overall development time. Frequently software development takes more than 50 % of the overall development effort [7.10]. The dependencies between hardware and software must be bridged by modelling constructs, so that most of the tasks may be designed independently of the counterpart.

For Software simulation, so called *virtual software components* may be used. Modern operating systems like WINDOWS or LINUX allow communi-

[1)] For the ARM family there are offered: Windriver VRTX, CHORUS, and JavaOS from Sun, Microtec VRTX, JMI, Embedded System Products Rtxc, and Integrated System pSOS.

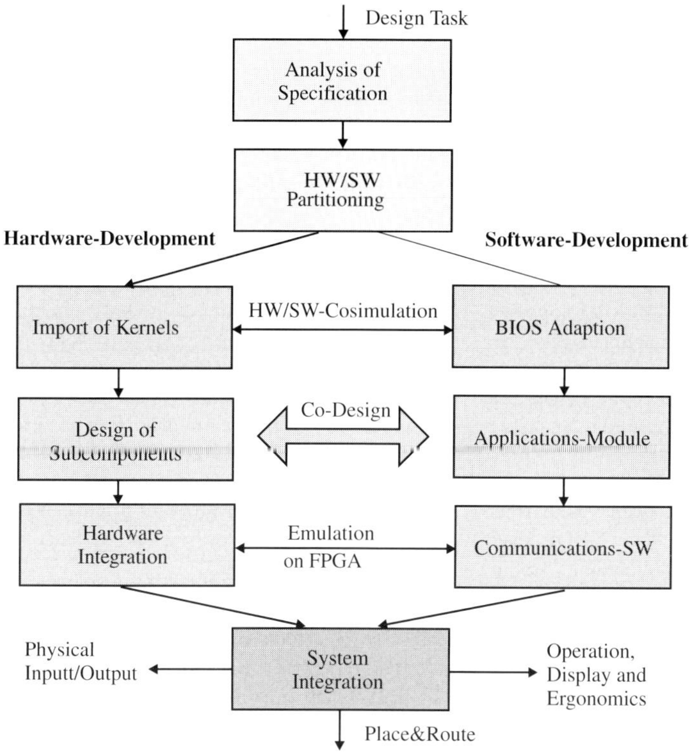

Fig. 7.9 Concurrent design of hardware and software

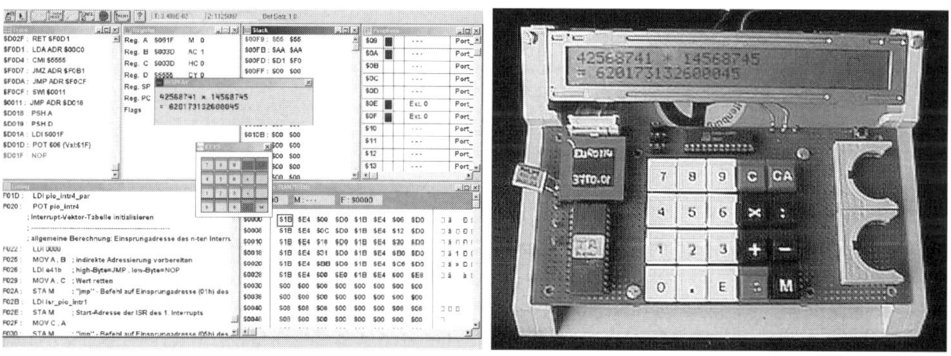

Fig. 7.10 Use of "virtual software components" in software simulation in the example of a pocket calculator program; a) screen representation, b) pocket calculator experimental setup

cation between independent tasks via an OLE or DDE mechanism. So hardware components like keyboard or display may be represented by a WINDOW program, which is then connected via OLE directly to a debugger/simulator. If an application is debugged in the simulator and keyboard entry is required, the appropriate data is typed in and channelled to the program as if there were a hardware connection of a mechanical keyboard. In the same way, data for the display is generated which may be channelled to a WINDOW screen. Figure 7.10 shows the virtual components 'keyboard' and 'display' of a pocket calculator system, which is simulated truly cycle by cycle on a debugger. In the simulation there may now be a detailed analysis and optimisation of the program done by breakpoints, register displays, stepwise execution and detailed reference to the source code. All blocks which are referenced by the BIOS driver layer may be chosen as virtual software components. Using the file features of the PC, complex scenarios may be recorded and replayed.

In the usual design flow the hardware development takes place to a large extent uncoupled from the software development, whereby the driver programs of BIOS must be verified in a classical functional digital simulation. Because these driver programs are usually quite short, the time effort is affordable. The content of the ROM must be loaded from the development system with suitable converter programs at the C or VHDL level into the ROM model before simulation. As an intermediate format the wide spread *INTEL Hex-format* is appropriate, which can be generated by many development systems and is also used in the programming of EPROM components.

If the design process is proceeding further, a total verification of all functions must be done in the next step. This must include all external circuitry and building blocks as well as the application programs. This can be done:

- by classical digital simulation at the *gate* or *register transfer* level with loaded BIOS and application program;
- on an integrated hardware/software design system;
- by *rapid prototyping* and emulation on FPGA.

With the assumption that the application program takes some kilobytes and the BIOS is of the same order of magnitude, a verification effort easily of some million processor cycles can be predicted. Each cycle may be divided again into some timing intervals, which would lead to computing times of hours, days, or weeks already on big computers. This is usually not acceptable. Some people call it the *verification trap*, the brute force method does not work. There is a lot of workaround to this problem. One can take behavioural models for the core instead of the RTL model, which speeds up simulation significantly. Timing simulation may be omitted, but the verification value is decreased. The periphery must always be present in RTL style. External interfaces are difficult to simulate, especially if human interaction is included. These inputs are non-deterministic and cannot be replaced by pre-programmed behaviour in the same way. One can state in summary that **a satisfying verification of SOC with classical digital simulation is not possible**.

Most important EDA software providers today delivers combined **hardware/software development systems** which allow a limited simulation of software controlled SOC. These systems use behaviour models, formulated at a high abstraction level in the C language. The problem of long simulation times and emulation of interaction scenarios is not solved, only improved. There is still the possible inconsistency between the used core and the model, derived from different origins. Real-time performance is far away. These systems will be discussed later in detail.

In practice, therefore, the emulation of designs in FPGA, designated as '**rapid prototyping**', is the preferred way of verification. The processor core may be included as a discrete component, as far as one exists, into the emulation. If the core exists as a soft IP it may be synthesized to FPGA and included into the emulation directly. Large designs may be spread over more than one FPGA, today circuits with up to 10 million gates are available for this purpose.

For pure digital circuits this **FPGA emulation** allows a realistic simulation, the FPGA can be connected to the real world and fed with real signals. Most of there emulations work in real time. Except that with very high clock frequencies real time is not possible, a slowdown in clock frequency is needed and the input stimuli have to be delayed. But it has to be remembered that the internal architecture of FPGA is quite different from an ASIC implementation, and a comparison concerning timing behaviour is not directly possible. In clock synchronous designs, which are state of the art and generally preferred, the behaviour is defined at the rising clock edge and this should be equal to FPGA or ASIC. So with timing problems the master clock has to be slowed down until the above conditions are fulfilled. Compared to digital simulation there is still a gigantic improvement and speed up.

Today FPGA emulation systems are offered on the market from several EDA providers. The standard development systems from ALTERA, XILINX and others may be taken for this purpose as well. Suitable electronic boards with prototype FPGA are available, where a design can be downloaded from the workstation, but in many cases they fit only for relatively smaller projects.

7.2.5 Placement and Routing of ASIC Cores

Only soft IP can be integrated into FPGA and *gate arrays*, perhaps firm IP with the associated placing rules. Using *standard cell design style* hard IP can be used too, as far as they are available in the selected technology. Hard IP providers port their cores into the target CMOS technology on request, so there are few restrictions in using a block design style for creating big designs.

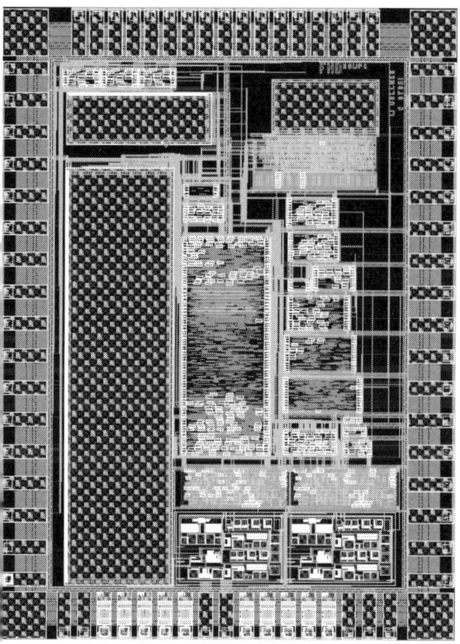

Fig. 7.11 Example of a chip design with pre-routed analogue and digital building blocks in a standard cell design style (DSWPC designed at the University of Applied Sciences, Offenburg)

Processor cores may be placed in one corner, the even larger RAM and ROM-blocks in the other and interconnected with a power supply ring. The other building blocks are then placed into the remaining gaps. If they do not fit, they are dissolved and flattened in hierarchy to use the silicon area as optimally as possible. With multiple metal layer routing, very dense designs can be created, although the block style is not optimal in respect of area. Analogue blocks need their own power supply system and are separated by guard rings with contacts to substrate and supply or ground, isolated as far as possible from the digital parts of the circuit. These blocks are preferred to be placed at the borders of the chip, in oder to obtain short interconnection wiring to the pad cells and to be able to organize their own separated, and isolated analogue supply. Analogue cells are not quite as densely routed when compared to digital cells and need more space on silicon. Fig. 7.11 shows a typical block cell design with analogue and digital building blocks.

7

7.3 EDA Systems for Hardware/Software Co-Design

The design of systems with IP cores, in particular with processor cores, requires a new dimension of performance from the used software tools:

- support in *design space exploration* after allocation of hardware and software;
- Support for *simultaneous, concurrent development* of hardware and software;
- Support for *verification* with different models on *different abstraction levels*.

7.3.1 Design Space Exploration Tools

The automatic partitioning of a given task in a software and a hardware part, controlled by a *formal executable specification* [7.9] is a fastidious task and still the subject of intensive research [7.4], [7.7]. These efforts are concentrated on complex applications of signal processors (ASIP) where algorithms are well defined by a C language model or in a special specification language like **SpecC** [7.9], [7.36], [7.37] or SpecCharts [7.39]. The mapping from such a functional description in a specification language to a given architecture may have several solutions and permits a number of variants, which must be evaluated concerning effort, silicon area consumption, and operational speed. These variants span the so called *design* *space*, which has to be explored for an optimal solution. Questions concerning the used cost functions and the given constrains are far from trivial. The results of these researches are programming systems such as CATHEDRAL [7.4] from CoWare, developed at the IMEC, which are available today in commercial versions. Furthermore PTOLOMY [7.5] and COSSAP, which are more related to the design, refinement, and optimisation of signal processor applications from formal algorithmic specifications. The output of these programs are synthesizable VHDL models and related control schedules (e.g., in the form of C codes or microcode). These software tools are able to process and verify very heterogeneous designs, but partitioning and resource allocation is mainly done by human interaction.

Designs in telecommunications are characterized frequently by the fact of the signals being processed at very **different data rates**. So the primary signal may be processed with a high sampling rate of above 100 k samples/sec, the control loops, tuning, amplitude regulation, and automatic frequency adjustment has low bandwidth and is comparably done with low rates. These control loops may be irregular and complex. The associated algorithms have various communication relations to the external world as well as to internal structures.

From the data rate it follows that mainly primary and close coupled receiver structures working at a fast rate are preferably implemented in hardware, the complex, but slower, control loops which need thousands of cycles, are better implemented as software on a processor core. These are the first and main criteria for partitioning and allocation.

The big differences in processing rate and time constants between the two areas, sometimes of more than a factor of 100,000 makes simulation and verification difficult. Whoever has tried to simulate a PLL with SPICE knows the problem of large cycle numbers, leading to long simulation times and huge data volumes.

Modern software systems, such as offered by CoWare, allow one to integrate good and verified blocks such as commercial processor cores taken from the library SYMPHONY, and support by this the demand for *re-use* of developed and qualified components [7.4].

7.3.2 Compilers for Irregular Target Architectures (Retargetable Compiler)

Today the variety of commercially available processor cores is very large and there are many associated software development systems. Central programs are the *assembler* for the selected core, able to generate object code for the target architecture, a *simulator/debugger* and a *C cross compiler*. The C compiler may have constraints concerning the ANSI standard (i.e., lack of programs for floating point processing). The systems offered from IAR Systems [7.33], KEIL [7.34], [7.35] Hewlett Packard, Mentor Graphics, and Cadence run on a PC under WINDOWS or LINUX and can be used with different processor architectures.

The main programming language is C, other languages such as PASCAL are rarely used. Only a low percentage of high level languages are used in *embedded applications*, for which assembly programming dominates, especially for driver development. Normally, compiler generated code is substantially larger than hand coded programs and usually has poor speed performance. Because of this handicap [7.10], an acceptable level should be below 20 % in memory requirement and a maximum of 30 % in speed performance. Other uses of a compiler would be inefficient. Good C compilers with small overhead and optimised libraries exists in the controller area (IAR, KEIL). In 32 bit systems (ARM) the compiler handicap is accepted because there is no alternative and adequate memory resources are provided, so most of the projects are programmed in C language.

Even if these compilers can be switched according to the target processor, they are not *re-targetable compilers*. A re-targetable compiler is a compiler which can be configured to new architectures and instruction sets with little effort. They permit development of the above mentioned ASIP processor cores with an application specific instruction set and architecture. This is an important research topic today [7.16], [7.15].

Retargetable compilers RTC may be generated with the classical compiler tools. Irregular architectures, non-symmetric and non-orthogonal register sets, unusual commands and addressing modes make it difficult to define a general solution. Lim

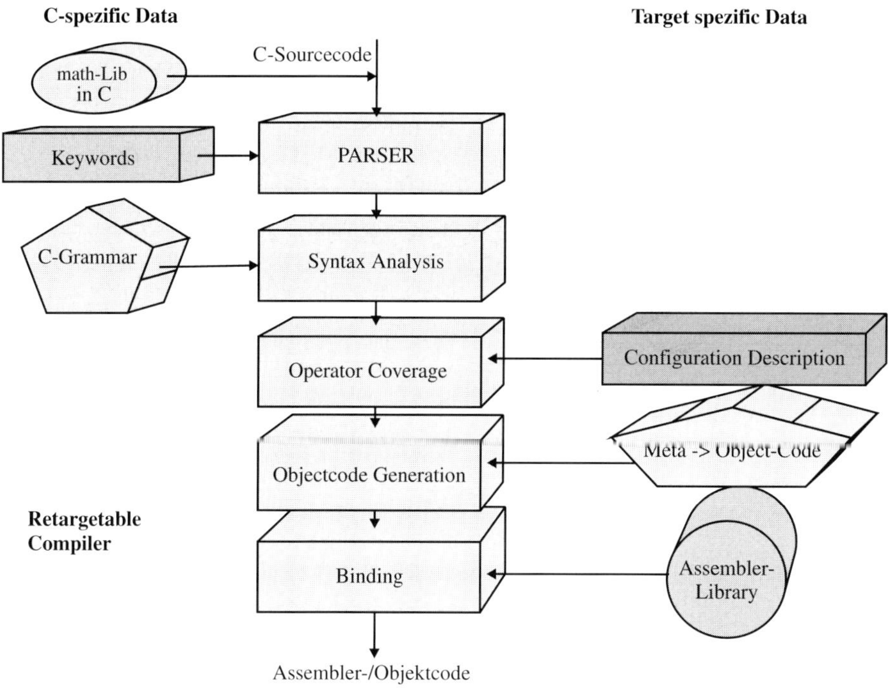

Fig. 7.12 Structure of a retargetable compiler

ited use of resources, different cycle and pipeline processing, and interdependencies between internal structures requires specialists to define the configuration files for RTC. Highly demanding are the SIMD[1)] architectures, and VLIW processor cores, which can mainly be found in the signal processing area, with more than one CPU, pixel processing cores and special instructions for multimedia applications. Companies which develop ASIP use nearly all home made software tools, which sometimes have their origin in university development (GNU, CATHEDRAL).

The typical *re-targetable compiler* consists of several layers (fig. 7.12):

- Parser;
- syntax analysis;
- operator covering;
- generation of object code;
- binding and linking.

The *Parser* reads the C source code and parses it for keywords, declarations, variables, etc. Following the syntax analysis the statements are checked for validity and meaning and then transformed into a kind of *meta-language*. In this meta-language, essentially only *identifiers* and *operators* exist, which are combined with control statements such as *if-then-else*. Up to here only the grammar of the C language is used. This part of the compiler is independent from the target architecture and is the same for all target systems.

In the next processing step the **identifiers** are related to memory space or registers, depending on their type, **operators** are 'covered' with functions, which can be processed by the hardware. There are:

- **unary operators** with one input value and one output value such as, e.g., assignment, negation, inverting;
- **binary operators** such as addition, subtraction, multiplication, division, etc.

All mathematical expressions, even with parentheses, can be resolved into a tree structure of unary

[1)] SIMD stands for 'single instruction multiple data' architecture, in which with one instruction several data words are processed at the same time.

and binary operators. Expressions with more than two inputs are mapped onto a sequence of binary operators, fig. 7.13. The resulting tree, or *data flow graph*, consists of operators at the nodes. These nodes can be processed by hardware functions, and identifiers at the edges, which are mapped to memory locations. The task of the compiler is to *cover the tree structure* in optimal form with the existing resources of the target processor. The compiler may also find a sequence of operations and memory assignments that achieve the result, and is more optimal with resource usage at the same time. As an example it may be better to keep intermediate data in registers than in memory. To do that the compiler must know the target architecture as well as the costs of usage and constraints in resources. This information is found in a configuration file, which must be provided.

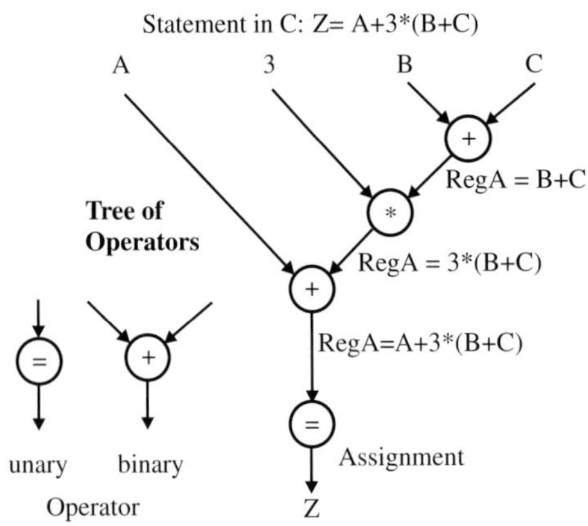

Fig. 7.13 Representation of expressions as a tree of operator relations

In the last step of the compilation run, object code in assembler mnemonics is generated for the statements found and sequenced. This part of the compiler, the **code generator**, is very specific to the target architecture and is adjusted to the following assembler and its syntax. An actual example for a retargetable compiler, where all machine dependent programs are concentrated on one configuration file only, is the **LCC** (*Little C Compiler*), created by FRASER and HANSON [7.29]. The associated book [7.30] describes in detail all that is needed to formulate the rules in the configuration file. Further work in this area is being done at the University of Dortmund, Germany, with the LANCE system [7.15], [7.16], [7.31], at the *Center for Embedded Computing Systems*, University of California, Irvine, with EXPRESSION [7.38], and the GNU compiler gcc, which has been ported successfully to different target architectures including ARM. Porting gcc requires some effort.

There is a strong interdependence between processor architecture, instruction set, and efficiency of code generation. The C language defines different data types like *char* (8 bit), *int* (16 bit), *long* (32 bit), *float* (32 bit), *double* (64 bit), etc. These data types can be held completely, or only partly, in one register. 32 bit processors are able to handle large word sizes better than 16 bit or even 8 bit controllers, in which the transportation of data needs more than one assembly statement. Thus a C compiler for 8 bit architectures is generally ineffective. On the other hand, 32 bit cores need significantly larger resources in memory that are not available in single chip SOC designs.

For one expression in *meta code* several assembler instructions may be inserted during the **coverage procedure**. This may also be the call of a subroutine from an attached assembler library, optimised for the core. Such libraries contain more, complex instructions such as division, floating point processing, and string processing. The use of a library is favourable, since these routines are used several times but only stored once. The *binder*, or *linker*, called part of the compiling software inserts (sometimes) only those parts of the library into the object code which are called in the application program. The generated assembler statements must be assembled and linked to the *executable*. The linker binds the routines from the library and allocates memory in the available RAM or ROM architecture. Depending on type and programming style, *globals* are allocated in RAM, *constants* are allocated in the ROM. Assembler and linker are sometimes closely integrated, so the different steps are hidden from the user. With *compilation* the whole process of generation of an executable from a high level language is designated. For integrated cores it is preferred to carry out these steps carefully in a definite sequence under the control of the designer; this is especially true if a RTOS or BIOS has to be included [7.3], [7.33], [7.34].

Large and complex functionalities are generally integrated at C language level. But binding of the *stdlib* of the ANSI C standard already requires a lot of memory space (about 5 ... 10 Kbytes); furthermore, the *stdio*, *string*, and *math* libraries also belong to the standard. C compilation without these libraries is always rudimentary. These libraries form the basis for every higher operating systems and must be supported in some way, perhaps with constraints.

For a *retargetable compiler* there must be at minimum three data sets specified:

- Configuration file with *description of the target architecture*, constraints, and cost functions, as well as a description of the data types and memory models,
- Mapping of the *metacode* to *object codes* of the target system,
- A library in object code of the target systems with elementary functions and definitions, preferably the *stdlib library* set from the ANSI-C-standard.

The usage of the C language is effective with architectures with word size of 16 bit and 32 bit and enough memory resources. With increasing density in actual chips, RISC cores with 32 bit word size, such as ARM 7, will be used in applications which are today still dominated by 8 bit controllers. It can be expected that high level languages will be used to program these systems without any compromise.

7.3.3 Integrated Development Environments (IDE)

Even if a software development system for the selected core exists, it is generally not coupled with a hardware simulator at the RTL or VHDL level. Because of the great importance of co-simulation of hardware and software, coupled systems are offered by the big EDA providers. One example is the **SEAMLESS Co-Verification Environment** from Mentor Graphics Inc.; similar development systems are offered from other providers, too. These are densely coupled programs for simulation of embedded code in a software simulator/debugger on the one hand, a gate level simulator (perhaps a VHDL-simulator like *modelsim*) on the other hand. Mentor Graphics reports speeds of about 5000 instructions/sec, which allows the complete simulation of smaller applications. The system can be delivered with compilers and simulators for all important processor cores and may replace dedicated and proprietary development systems of the core suppliers. The system may be coupled with hardware emulator boards on a FPGA basis, where part of the system may be represented in hardware directly.

For large application programs this kind of coupled simulation is still too slow, since the VHDL simulator run at the RT level is event controlled, whereby the number of the events is about proportional to the number of implemented gates. This makes RT simulation slow in comparison to pure behaviour simulation. One solution may be a **cycle based simulation**, in which behaviour models are generated from the true RT description for each clock cycle. Effects of timing are neglected. In a synchronous design, where the maximum time delay must be shorter than the delay given by the critical path through the network, this abstraction is allowed. With this mapping, a direct, cycle true executable code is generated, which shows speed similar to the application software code itself. Such a program is offered by Quickturn Design Systems Inc. under the name **SpeedSim**. An acceleration of the simulation by a factor of 10 ... 100 is not unusual, so hardware emulation or special hardware accelerators may be avoided. An advantage of this solution is that the model can be derived automatically from existing VHDL code, so that the method can be integrated into a consistent design flow. If one works with hand designed models the risk of inconsistencies and discrepancies is large.

7.4 System on Chip Designs (SOC)

In systems on a chip (SOC) one expects today not only a processor core and dedicated application-specific electronics, but increasingly also sensors, similar circuits, and possibly even micromechanical components. The chip may contain not only the functionality of the printed circuit board but an entire system. Examples are contactless RF ID tags, sensors with integrated evaluation electronics and several smart card applications. A high degree of integration together with simultaneous mass production leads to low prices, and the impact on the consumer market is accordingly large. There are many new products on the horizon, the market for ID tags alone is gigantic and already in start up.

7.4.1 Concept and Specification of SOC

The design of a SOC puts highest demands on the designer, not least because he has now to understand more technical disciplines than ever before and combine them in one model consistently. The specification of the analog blocks is one order of magnitude more difficult and does not follow a systematic methodology compared to the digital blocks. In the analog area, completely different terms are used, and the results are not as reproducible as in digital domain. So physical environmental conditions such as temperature, voltage supply, and process tolerance have an important influence on function and have to be considered or compensated. Circuits which work in the high frequency regime should fulfil further requirements, which may be difficult to test. It is therefore nearly impossible to write a formal specification for such a system, even more to process it in an EDA system automatically.

Usually analog cells are specified and qualified separately and later transferred as verified IP cores to the SOC design. Proceeding further in the development, they were described by behaviour models, perhaps in VHDL-AMS. Analogue IP, such as A/D converters, reference cells, or temperature cells are offered by specialized IP providers, e.g., see the **CircuitWare© Library** by Dolphin Integration.

7.4.2 Micro-Mechanical Systems: Combined Digital, Analogue and Mechanical World

With the inclusion of micro-mechanical functional blocks into a SOC, the highest complexity is reached. At present there are only few successful examples of *one chip products*. The same conditions apply to multi-chip modules in which the micro-mechanical chip is combined with a conventional CMOS IC on a ceramic substrate. The hybrid version is today more cost effective than the monolithic design, because the micro-mechanical process is in many cases incompatible with the standard IC process. Even if the processes are made compatible, many researchers are working on this topic, the questions of yield and silicon area consumption are still not answered satisfactorily. Mass applications such as the acceleration sensors for car airbags are nearly everywhere built as hybrid components.

For modelling these systems the mechanical components and their behaviour also have to be included into the simulation. For such applications, perhaps as a later follower of the famous SPICE simulator, **VHDL-AMS** (AMS stands for Analogue Mixed Systems) is used as a new standard modelling language, allowing a unified specification and simulation style for the analogue as well as for the digital world. VHDL-AMS will be described more, in detail, in another chapter. With this standard simulation language analogue and micro-mechanical components can be included into a functional system simulation in a consistent way.

7.4.3 SOC Simulation with VHDL AMS

The simulation of a complex SOC requires the cooperation of several engineering disciplines and puts high demands on the modelling quality. Everybody who uses SPICE knows that the accuracy of the results depends mainly on the extent reality is described by the models. This requirement is already difficult to accomplish with purely electrical circuits and becomes even more difficult with micro-mechanical components. A solution may be a separation of the tasks and moving of the modelling task completely to the cell-provider, who has worked out the cell in a detailed and expensive development and qualification process. The cell provider is the only one who can guarantee the specified characteristics.

In further development steps simulation has to be applied at every level. The existence of an analog cell will not disturb further, as far as the outputs of these cells are modelled consistently. Modern VHDL simulator tools such as *modelsim* have no problems with showing analogue signals in their trace windows, VHDL AMS support is already there.

7.4.4 The challenge: Testing of Systems on Chip

Testing of a SOC containing analog and micro-mechanical components together with digital electronics is a challenge. A functional test is in many cases only possible after complete assembly in a housing. In order to obtain a better yield, component tests at the IC level are very important.

As an entrance to the chip, the **JTAG concept** is generally accepted. By means of a few additional pins, used eventually as a secondary function for existing connections, all internal conditions of the chip are accessible and can be set or read. This philosophy is described in other chapters more in detail.

Analogue subsystems usually cannot be tested with JTAG; additional measures are required. Common is the integration of *stimuli generators* into the chip for the analogue components, whose *response* can be selected and read out via digital signals. These tests are limited usually to the demonstration of the general function of the cell, less to the compliance to the specified values. But they are useful, too, for self-testing of the chip in an application. Detailed testing can only be done on the complete, assembled system.

The **tolerances of analogue systems** are frequently compensated by adjustment and calibrating. Trimming (e.g., through laser ablation of structures on the chip) is common and established for analog components; for a SOC, however, it is a significant cost factor. To avoid trimming, the intelligence of the integrated processor can be used to compensate tolerances by calculation. The measured calibration values may be stored digitally and used with each measurement to improve the results. Such a system may be cheaper than a laser trimmed version. Although there is a digital processor added, the development costs are spent only once. Trimming has to be done with each unit. In this way, process tolerances, as well as temperature influence on sensors, can be compensated to high quality, it would be easy to use and have a calibrated output.

Such a concept needs consistent logistical and measurement techniques during manufacturing, but allows multi-dimensional compensation and service quality. The processor core may be used to add communication facilities to the SOC, allowing it to interconnect directly to networks (CAN) and providing standard digital interfaces to thumb analogue sensors. Via these interfaces self-tests may be initiated, using the abovementioned stimuli generators, providing signal integrity, which is needed for present day complex, highly networked systems. These so called *intelligent sensors* will be the first significant commercial market for SOC.

7.4.5 Example Design of a System on Chip

To demonstrate a system on a chip design, a development made in the institute of the author, called 'Thermologger', will be described. This is a smart card which is able to record or *log* temperature profiles [7.12] over shorter or longer time periods from hours up to years. All functions including storage are integrated on one chip. Applications are in the area of food transport, medical transport, temperature monitoring of rooms, as well as in agriculture and planting.

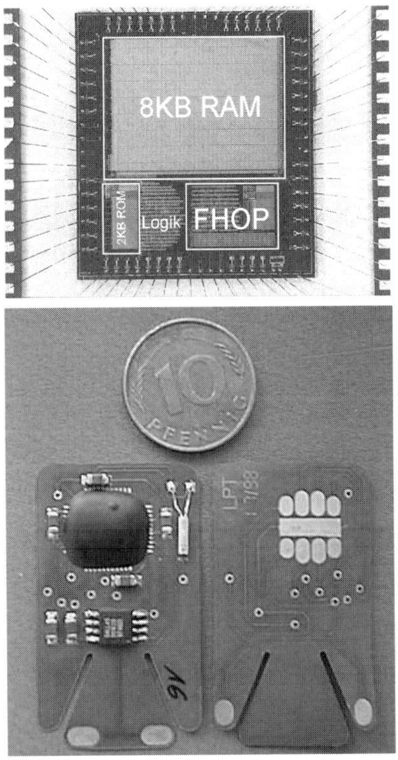

Fig. 7.14 Smart card SOC design 'Thermologger' after [7.12], chip photo and circuit board with mounted chip

Temperature is measured with an integrated temperature cell, converted to digital form and stored in a static CMOS RAM. For activation, as well for reading out the stored data the card is inserted into a reader. The microprocessor core in the SOC communicates with the reader which is connected to a personal computer PC via a serial interface. The PC controls the communication via passwords and stores configuration data such as activation date, naming of application, period of measurement, etc., on the card. After data collection, the card is read out by the PC and the data measured may be represented as a profile, a graph, or in textual form. The internal chip RAM is able to store about 7000 measurements, even more with data compression techniques, which is enough for most of the applications mentioned.

Control of data acquisition, of communication protocol, and the data compression is carried out by the integrated digital processor core. The core uses a built in BIOS stored in the ROM, which contains the driver for the periphery functions as well as the communications protocol engine. After connection to the PC the application routines are downloaded into the RAM, so the main program is not stored in the ROM but downloaded by activation and then executed. The main part of the software is in this way very flexible and allows one to adapt the system to very different applications.

After activation of the system it goes into a *power down mode* in which all functions are switched off, except for an interval timer which has been configured beforehand during activation. After the pre-programmed intervals of minutes or hours the system awakes and performs the next measurement. This is repeated until the memory is full or the smart card is inserted into the reader station. Because of the low power consumption during *power down mode* and the short active phases of measurement, a small battery is able to supply the SOC for more than a year.

A picture of the chip is shown in fig. 7.14. A mixed *block design style* was used in a CMOS 0.7 standard cells technology provided by ES2. The RAM memory cell is the largest part of the chip. The processor core FHOP_16 is to be seen as hardIP down right, the ROM completely on the left. The remaining electronics was dissolved in its hierarchy and flattened in a standard cell style.

The version shown does not yet contain the temperature sensor, which will be integrated with the next generation, together with an inductive interface, which allows a wireless communication with the chip and so a hermetical sealing. A commercial use of this real 'one chip design' is intended in future.

In this system, only the BIOS programs are developed in a real hardware/software codesign methodology and verified by functional simulation with the VHDL-simulator *modelsim*. The application program was developed, after the integrated circuit was already available. This was possible because of the download feature of the chip, allowing to implement programs into the RAM via the communication interface during runtime. The software uses several hardware-interrupt levels and software interrupts for the BIOS routines. The development took place in close interaction with the development of the reader and controlling PC-program.

7.5 References

The website of CECS (*Center Of Embedded Computer of Systems, University of California/Irvine* http://www.ics.uci.edu/~cecs/) provides a good idea of present academic activities in the area of embedded systems and **hardware/software co-design**. Furthermore, there is an exhaustive overview in the Proceedings of the IEEE, Vol. 85 No.3 March 1997. In addition the books [7.17], [7.9], [7.16], [7.22], [7.23] and [7.5] give a good summary and presentation of the research fields and processes. Information about the current commercial program systems of SYNOPSYS and MENTOR GRAPHICS may be obtained from company documentation or a summarizing book [7.20], or from the websites of these companies. An open forum for IP and other intellectual property devices is the international conference and exhibition 'The Intellectual Property in Electronics' [7.11]. There is an 'International Workshop on IP Based Synthesis and System design', organized by INPG/Grenoble. The best collections on IP and IP provider is maintained at **Design and Re-use**, a spin-off from TIMA/Grenoble [7.6], the so called *Yellow Pages*; not forgeting the website of the *Virtual Socket Interface Alliance* http://www.vsia.org/ with many documents, standards, and links to associated companies.

[7.1] ALTERA: Megacore Functions for High-Density Design. Firmenschrift ALTERA, 1998

[7.2] ARM Ltd.: Entwicklungsunterlagen der Firma ARM Ltd. – Cambridge, UK. http://www.arm.com/

[7.3] *Bischoff, O.; Jansen, D.*: Hochsprachen-Compiler für den FHOP. Berichtsheft des XVIII. MPC-Workshops in Albstadt-Sigmaringen, Hrsg. FH-Ulm, Januar 1998

[7.4] *Bolsens, I.; Hugo J. de Man et al.*: Hardware/Software Co-Design of Digital Telecommunication Systems. Proceedings of the IEEE, Vol. 85, No. 3, March 1997

[7.5] *Buchenrieder, K.; Sedelmeier, A.; Veith, V.*: Industrial HW/SW codesign, in Hardware/Software Co-Design. G. De Micheli and M. Sami, Eds. – Boston; Dordrecht; London: Kluwer, 1996

[7.6] Firmenschrift der Fa. Design And Reuse, Grenoble 1998, http://www.design-reuse.com/

[7.7] *Edwards, S.; Lavagno, L.; Edward A. Lee et. alt.*: Design of Embedded Systems: Formal Models, Validation, and Synthesis. Proceedings of the IEEE, Vol. 85, No. 3, March 1997

[7.8] *Fischer, M.; Vollmer, W.; Jansen, D.*: FHOP V2.0. Entwicklung der 2. Generation eines 16 Bit Mikroprozessor-Kerns auf der Basis des FHO-Prozessors FHOP. Berichtsheft des XX. MPC-Workshops in Furtwangen, Hrsg. FH-Ulm, Januar 1999

[7.9] *Gaiski, D.; Narayan, S.; Vahid, F.; Gong, J.*: Specification and Design of Embedded Systems. – Englewood Cliffs, NJ: Prentice Hall, 1994

[7.10] *Goossens, G. et. alt.*: Embedded Software in Real Time Signal Processing Systems: Design Technologies. Proceedings of the IEEE, Vol. 85, No. 3, March 1997

[7.11] The Intellectual Propperty in Electronics, Conference and Exhibition, Proceedings IP 98. – Frankfurt, 1998

[7.12] *Jansen, D.; Klump, T.*: Thermologger, eine Chipkarte zur Aufzeichnung von Temperaturverläufen. Proceedings der ES&S 1997 in Nürnberg, 1997

[7.13] *Jansen, D.; Gieringer, T.; Zimpfer, F.*: A microprocessor in four months. Proceeding of the IEEE International ASIC Conference and Exhibit. – Rochester: 1994

[7.14] *Jansen, D.; Vollmer, W.; Klöser, F.*: Application Specific System Engineering with the Embedded Microprocessor-Kernel FHOP. Proceedings der internationalen Konferenz EMAC 97 in Barcelona, 1997

[7.15] *Leupers, R.; Marwedel, P.*: Retargable Code-Generation for Digital Signal Processors. – Boston; Dordrecht; London: Kluwer, 1991

[7.16] *Marwedel, P.; Goosens, G. (Eds)*: Code Generators for Embedded Processors. – Boston; Dordrecht; London: Kluwer, 1996

[7.17] *De Micheli, G.; Sami, M.*: Hardware/Software Co-Design. – Amsterdam: Kluwer, 1996

[7.18] *De Micheli, G.; Gupta, Rajesh k.*: Hardware-Software-Co-Design. Proceedings of the IEEE, Vol. 85, No. 3, March 1997

[7.19] *Paulin, P. G. et al.*: Embedded Software in Real-Time Signal Processing Systems: Application and Architecture Trends. Proceedings of the IEEE, Vol. 85, No. 3, March 1997

[7.20] The Reuse Methodology Manual. Kluwer Academic Publishers, Boston, USA, 240 pp, May 1998, herausgegeben von Mentor Graphics und Synopsys

[7.21] *Ben-Romdane, M.*: Lego-Like Integration of IP-Cores. The Intellectual Property in Electronics, Conference and Exhibition, Proceedings IP 98. – Frankfurt, 1998

[7.22] *Rozenblit, J.; Buchenrieder, K.*: Codesign: Computer-Aided Software/Hardware-Engineering. – Piscataway, NJ: IEEE, 1995

[7.23] *Sanchez, E.; Tomassini, M.*: Toward Evolvable Hardware. – New York: Springer-Verlag, 1996

[7.24] *Saucier, G. et al.*: Hard IP Migration. The Intellectual Property in Electronics, Conference and Exhibition, Proceedings IP 98. – Frankfurt, 1998

[7.25] *Tindell, K.*: Echtzeit-Betriebssystem für Ein-Chip-Systeme. Elektronik 18/1999, WEKA-Fach"-zeit"-schriften-Verlag GmbH

[7.26] *Vollmer, W.*: Entwurf eines Mikrocontrollers mit dem embedded Prozessorkern FHOP. Berichtsheft des XVI. MPC-Workshops in Reutlingen, Hrsg. FH-Ulm, Januar 1997

[7.27] The Virtual Socket Interface Alliance, Verband der an der Entwicklung und Vermarktung von Intellectual Property und Virtuellen Komponenten interessierten Industrie, Internet Website: http://www.vsia.org/

[7.28] *Wolf, W.*: Hardware-software co-design of embedded systems. Proceedings of the IEEE, vol. 82, pp. 967–989, July 1994

[7.29] Website LCC-Compiler: http://www.cs.princeton.edu/lcc/

7

[7.30] *Fraser, C.; Hanson, D.*: A Retargetable C Compiler: Design and Implementation. Addison-Wesley Publishing Company, 1995

[7.31] Website LANCE: *Leupers, R*: http://ls12-www.informatik.uni-dortmund.de/~leupers/lanceV2

[7.32] Website GNU-Projekt: http://www.gnu.org

[7.33] Website IAR-Systems: http://www.iar.com

[7.34] Website Keil Software: http://www.keil.com

[7.35] *Baldischweiler, M.*: Der Keil C51-Compiler einschließlich V6.0 und uVision 2, Einführung und Praxis Teil 1 (Buch). – Grassbrunn, Keil Elektronik GmbH

[7.36] *Gajski, D. D.; Jianwen Zhu; Dömer, R.; Gerstlauer, A.; Shuqing Zhao*: SpecC: Specification Language and Methodology. – Kluwer Academic Publishers, 2000

[7.37] *Gajski, D. D.*: Prinziples of Digital Design. – Prentice Hall, 1997

[7.38] *Halambi, A.; Grun, P.; Ganesh, V.; Khare, A.; Dutt, N.; Nicolau, A.*: EXPRESSION: A language for Architecture Exploration through Compiler/Simulator Retargetability. International Conference DATE 99, www.ics.uci.edu/~cecs/

[7.39] *Gajski, D. D.; Dutt, N.; Wu, C. H.; Lin, Y. L.*: High Level Synthesis: Intoduction to Chip and System Design. – Kluwer Academic Publishers, 1994

8 Tabular Design Formats

GERT VOLAND

8.1 Netlist Formats

Netlists describe circuits in a textual list format. They are normally of structural type and have two main targets:

- Enumeration of all devices (or parts) including the input and output pins;
- Enumeration of all connections between the device pins.

Devices are uniquely identified by their part name and their instance name. Multiple devices with the same part name are distinguishable through their instance name (see chapter 3). Node names are either entered by the user or are generated automatically with running numbers by the schematic editor or the synthesis program. The path gets added to the part names and node names in hierarchical design structures.

A signal describes the electrical information on a node; in the following descriptions both expressions, signal and node, are used synonymously.

In principle, there are two different basic types of netlists: device oriented and node oriented netlists.

Device Oriented Netlists

In such a netlist a single device and all the nodes that are connected to it are listed in one line. In the following example, the instance name is placed first. The number of pins may vary from one device to another. In a positional pin list the exact sequence of pins is implicitly defined. Thus, the position in the list of nodes describes exactly to which pin it is connected.

In device oriented netlists the instance is named once only and the nodes as often as there are pins to which they get connected. An agreement has to exist regarding the position of pins because they are not named explicitly.

```
<instance name> <part name>
     <net name> <net name> ...
<instance name> <part name>
     <net name> <net name> ...
     /--- Pinpositionen ---/
...
```

Variations of the netlist formats exist where the pin names are explicitly assigned to the connected node. In such named pin lists the sequence of the pin assignment is irrelevant. In hierarchical designs the information of the hierarchy level is included in the instance names as well as in the node names. A common way of adding the level information is to list the sub-design names with a separator (for example a ‘ | ’):

```
<la>|<lb>|<lc>|<inst> <part>
    <la>|<lb>|<lc>|<net>
    <la>|<lb>|<lc>|<net> ...
<la>|<lb>|<lc>|<inst> <part>
    <la>|<lb>|<lc>|<net>
    <la>|<lb>|<lc>|<net> ...
    /--- Pinpositionen ---/
...
```

In the above list the following abbreviations have been used for a compact presentation:

```
la, lb, = levela, levelb, ...;
inst    = instance name;
part    = part name;
net     = net name.
```

Node Oriented Netlists

In such a netlist all the devices are listed in a first section together with their instance names (for example, section PARTS). Afterwards pins connected to one node are listed after the node name (for example, section NETS). Every single pin has to be identified with the individual device to which it belongs by the distinguishable instance name. In the following list, the pin name appears as a qualifier of the instance name. The instance

Based on those activities a first specification was published in 1985, EDIF 1 0 0. At the beginning all of the following design areas were supposed to be covered:

- Netlist;
- Mask layout;
- Symbolic Layout;
- Logic models;
- Test;
- Gate array;
- Parametric description;
- Algorithmic description;
- Documentation.

After the Electronic Industries Association (EIA) adopted the EDIF standards in 1986 the version EDIF 2 0 0 appeared in 1988, of which only the netlist version has reached a sufficient popularity. But the open definition of the netlist format proved to represent a certain weakness. EDIF defines the structure, the syntax and the semantics of the formats, but not the contents. As a result the suppliers of EDIF writers and those of EDIF readers have to explicitly define the content. This situation has led to a number of different non-compatible EDIF dialects.

In 1992 the EDIF Committee has been reorganized to become a division of the EIA. Today Prof. Hilary Kahn of the University of Manchester, UK, is the supervisor of the technical EDIF center. The activities concentrate on the following subcommittees:

- PCB Technical Subcommittee (Printed Circuit Boards);
- Test Technical Subcommittee;
- Schematic Technical Subcommittee;
- LPM Technical Subcommittee (Library of Parameterizable Modules).

Version EDIF 3 0 0 was published in 1993 supporting buses and correcting some weaknesses. Since 1996 EDIF 4 0 0 supports PCBs and MCMs (Multi-Chip Modules).

The idea behind the Library of Parameterizable Modules (LPM) was to enable the use of compact design descriptions for high level languages, such as VHDL, Verilog, or ALTERA AHDL. The set of 29 modules defined today are supposed to allow a design style independent of tool suppliers and silicon foundries (also see Chapter 17). More details about the latest and future activities can be found on the Web [8.3].

8.3.1 Structure and Elements of EDIF 2 0 0

Structure

The netlist format EDIF 2 0 0 has reached a wide distribution and will therefore serve for the following examples. The EDIF syntax is based on a LISP approach which can easily be scanned and interpreted by the reading program. One of the main criteria was the extendibility, a reason why the method of nested brackets has been used:

```
(Key Word{Integer Value | Text | Identifier
        | Nested Expression})
```

The opening parenthesis is always followed by an EDIF key word. The optional list of subsequent elements may contain integer values, text, EDIF identifiers, and further EDIF expressions. The structure of the format is in a sense object oriented because the assignment of a name or a value is only valid in the area of the last parenthetical key word. In two different libraries equally named cells may exist, but not in the same library.

This type of syntax is very verbose; a lot of text is used and it is difficult to write by hand. Although the number of pre-defined key words is relatively high [8.2], additional key words may be defined by the user with ‘keywordAlias’ or ‘keywordDefine’. The same is possible for identifiers, such as device names or node names, by the use of ‘rename’. Using that mechanism one can replace characters generated by a system that are not recognized on the target system (for example, leading numbers).

In principle such an EDIF netlist is node oriented. At first, all devices are listed including the pins and their direction (input, output, etc.). Then each node is listed with the pins connected to that node. Additional information is also placed in the list, such as the libraries used, the date, and the generating system.

Elements

The EDIF Reference Document [8.2] defines about 400 key words and identifiers. As an example, some of those are explained here:

edif Beginning of the EDIF file;

library Start of cell list there may be additional external references;

cell Definition of a cell;

viewType Type of description, such as NETLIST, MASKLAYOUT, etc.;

interface List of I/O pins;

port Pin names;

direction Signal direction of a port: INPUT, OUTPUT, or INOUT;

instance Placement name of a cell in a Design;

net Node name;

joined List of the connected nodes;

unit Type of physical unit, such as CAPACITANCE, CURRENT, DISTANCE, FREQUENCY, VOLTAGE, TIME, etc.;

property Physical properties (identifiers or values) that are passed to the reading program.

8.3.2 Example of an EDIF Netlist

List 8.2 gives an overview of the structure and of some details of an EDIF netlist. The EDIF key words are printed in bold.

The list has been generated after a synthesis and Place&Route run on an ALTERA system. The target system is of type 'Mentor Graphics' for Post-Place&Route simulation including back annotated delays (for example Delay_9). Using this EDIF list, the design can be transferred to the MENTOR system which must have the library information containing the functionality of all the cells for the logic simulation.

8

List 8.2 Example of a partial EDIF netlist containing primitive and hierarchical cells as well as delay elements. The comments after '<=' have been added by hand for explanation purposes.

```
(edif half_add                                         <= Start of EDIF file
    (edifVersion 2 0 0)
    (edifLevel 0)
    (keywordMap
        (keywordLevel 0))
    (status
        (written
            (timeStamp 1999 3 17 18 16 48)
            (comment "Target vendor:  Mentor Graphics")))
    (library ALTERA                                    <= Start of library definition
        (edifLevel 0)
        (technology
            (numberDefinition
                (scale 1                               <= Scale factor for delay
                    (e 100 -12)                           times (= 100*E-12 = 0.1 ns)
                    (unit TIME))))
        (cell AND2                                     <= List of all primitive cells
            (cellType GENERIC)
            (view view1
                (viewType NETLIST)
                (interface
                    (port IN1                          <= Input pin of cell AND2
                        (direction INPUT))
                    (port IN2
                        (direction INPUT))
                    (port Y                            <= Output pin of cell AND2
                        (direction OUTPUT)))))
```

```
(cell XOR2
    (cellType GENERIC)
    (view view1
        (viewType NETLIST)
        (interface
            (port IN1
                (direction INPUT))
            (port IN2
                (direction INPUT))
            (port Y
                (direction OUTPUT)))))
(...)                                         <= Further primitives

(cell half_add                                <= Hierarchical cells
    (cellType GENERIC)
    (view view1
        (viewType NETLIST)
        (interface
            (designator "EPM7128 J 84")       <= Device assignment (ALTERA)
            (port VCC
                (direction INPUT))
            (port A
                (direction INPUT)
                (designator "12"))            <= Pin assignment of EPM7128
            (...)
        (contents                             <= List of sub-designs
            (instance AND2_8                  <= Instance of AND2
                (viewRef view1                   (instance name is AND2_8)
                    (cellRef AND2))
                (property TPD
                    (integer 30)
                    (unit TIME)))
            (instance DELAY_9                 <= Delay cell for a Post-
                (viewRef view1                   Place&Route simulation with
                    (cellRef DELAY))             the propagation delay TPD
                (property TPD
                    (integer 15)
                    (unit TIME)))
            (...)                             <= Further instances in half_add
            (net N_17                         <= Start of the node list
                (joined
                    (portRef Y                <= Node N_17 connects to pin Y
                        (instanceRef DELAY_9))   of cell Delay_9
                    (portRef IN1                 and to pin IN1
                        (instanceRef AND2_8))))  of cell AND2_8
            (net A
                (joined
                    (portRef A)               <= Pin A is an external pin
                    (portRef IN1                 of the designs
                        (instanceRef DELAY_16))
                    (portRef IN1
                        (instanceRef DELAY_9))))
            (...)                             <= Further nodes

(design root
    (cellRef half_add                         <= Name of the design
        (libraryRef ALTERA))))                <= Library reference and
                                                 end of the EDIF file
```

8.4 The SDF Format

8.4.1 Objective of the SDF Format

SDF is an abbreviation for Standard Delay Format published by the non-commercial interest group, Accellera, a unification of Open Verilog International and VHDL International [8.4]. SDF is no language and no program but it describes how delay data should be formatted in a design process. The goal is to make that kind of data independent from the CAD programs of different suppliers.

The present version is 3.0. The format is kept relatively stable and only improvements are added so that data sets from older versions are still compatible. For that reason and because version 2.1 is widely used, the following descriptions deal mainly with this version 2.1.

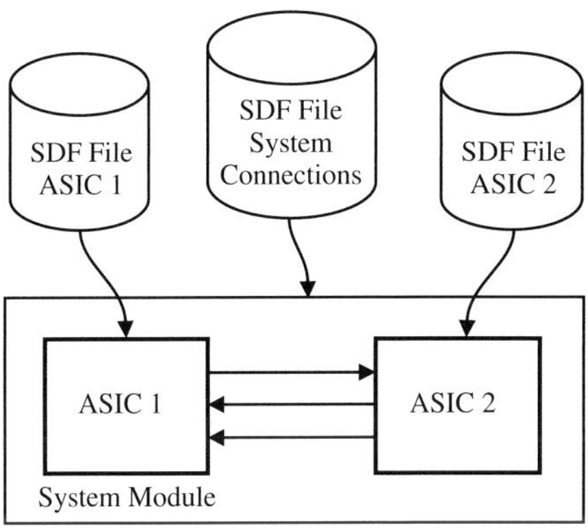

Fig. 8.1 The SDF format is possible for single devices, as well as for system modules

There are SDF formats for single devices (ASICs, such as FPGAs, gate arrays, and cell based standard products) and also for multiple system modules, as shown in fig. 8.1.

The SDF format must be independent of programs and still represent data of different types:

- Delay Times which could be associated with the following information:
  - module path;
  - device;
  - interconnect;
  - port.
- Clock Constraints:
  - setup and hold times;
  - recovery after asynchronous reset;
  - skew;
  - width;
  - period.
- Path and Skew Constraints;
- Incremental and Absolute Delays;
- Conditional Delays and Constraint Checks;
- Design Specific or Library Specific Data;
- Production Tolerances and Operational Limits:
  - data triple [best, nom and worst case];
  - temperature;
  - supply voltage.

SDF may be a standardized format, but the libraries, programs, and design flows which are involved have to be coordinated. In that sense, SDF eases a consistent implementation, but does not guarantee it.

8

8.4.2 SDF Files in a Design Flow

There are typically two types of timing specifications in design flows:

- Forward Annotation; and
- Back Annotation.

Forward Annotation

> Forward annotation describes the transfer of timing constraints generated during the specification process to the subsequent logic synthesis and layout generation steps. In the later steps the programs have to try and meet the constraints within their frame of optimization.

For example, a synthesis program is supposed to generate a structural design for a given maximum of operational clock frequency. Thereby the area is to be minimized without significantly exceeding the frequency.

Back Annotation

> Back annotation means the feed-back of delay information after successful synthesis or layout.

After a Place&Route run, the wire lengths and loads can be extracted and re-calculated in form of an SDF delay format for cells and interconnect. Finally, the data may be fed back into a timing analyzer or a digital simulator. In the case of back annotation, the SDF files always exist next to the netlist files, because the analysis tools need both data sets. Figure 8.2 demonstrates the principle of both types of annotation.

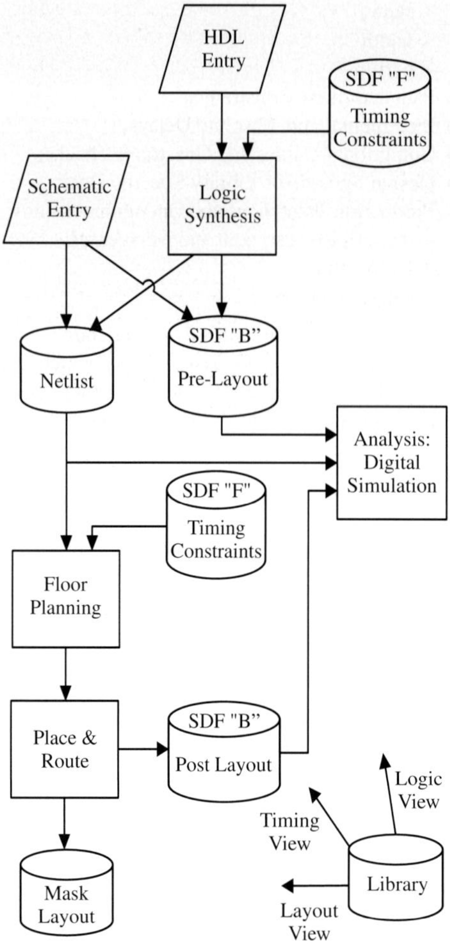

Fig. 8.2 SDF files in different design flows.
The letter 'F' means forward annotation and 'B' back annotation.

In the case of forward annotation into the synthesis program constraint data based on the target specification have to be transferred. The delay data after the synthesis run (Back Annotation Pre-Layout) are usually generated by formulas, taking into account the complete netlist. Depending on the estimated physical dimension, additional delay data are calculated for each instance of a cell. That type of calculation is often carried out within the simulator.

The timing constraints for the layout steps may consist of a few 'critical paths' based on the pre-layout digital simulations. A much more exact calculation can be done after the layout generation (Back Annotation Post-Layout). The extraction program for generating the post-layout data has to be able to identify the actual 'critical paths' in order to check the original target of maximum clock frequency.

8.4.3 Structure and Elements of SDF

SDF is a nested format of expressions in parentheses with specific SDF key words. The following values may be of type text string, value triple, or further parenthetical expressions. The exact definition of key words can be found in [8.5].

A Header Section at the beginning is always followed by Cell Entries.

Header Section

The key words are mostly self-explanatory:

```
SDFVERSION,    DESIGN,      DATE,
VENDOR,        PROGRAM,     VERSION,
DIVIDER,       VOLTAGE,     PROCESS,
TEMPERATURE,   TIMESCALE.
```

`VOLTAGE`, `PROCESS` and `TEMPERATURE` are followed be three values (triple) best:nom:worst. They describe the conditions for the following timing analysis. The header information has the character of documentation because only the values following the delay entries are used.

The value of `TIMESCALE` sets the unit for interpreting the timing data.

Cell Entries

For describing the cell and its instance name the following three key words can be used which correspond directly with the netlist:

```
CELL,  CELLTYPE,  INSTANCE.
```

`CELLTYPE` is used for the device name and `INSTANCE` describes the instance name.

The following key words are possibly attached to timing parameters:

```
DELAY,      TIMINGCHECK,    PATHPULSE,
ABSOLUTE,   INCREMENT,      IOPATH,
PORT,       INTERCONNECT,   NETDELAY,
DEVICE,     SETUP,          HOLD,
RECOVERY,   SKEW,           WIDTH,
PERIOD,     COND,           posedge,
negedge.
```

One of the most common elements of delay times is IOPATH, which describes the name of the input and output pins of the timing path. The format for an inverter could look like this:

```
(CELL (CELLTYPE "INV")
      (INSTANCE INV_27)
      (DELAY (ABSOLUTE
          (IOPATH IN Y (2:5:7))))
)
```

Propagation delays may be defined between the following logic levels:

```
0→1, 1→0, 0→Z, Z→1, 1→Z, Z→0, 0→X,
X→1, 1→X, X→0, X→Z, Z→X.
```

A complete characterization has to produce up to 12 value triples.

In addition to defining path delays, delay times may be attached to pins of cells or single nets.

The most important constraint checks, setup and hold, have to be applied between data and clock pins of flip flops. This is an example for a D flip flop:

```
(CELL (CELLTYPE "DFF")
      (INSTANCE top.mod1.dff_13)
      (TIMINGCHECK
          (SETUP din (posedge clk) (3:4:5))
          (HOLD din (posedge clk) (4:5:6)))
)
```

The instance name may consist of a hierarchical path.

8.4.4 An SDF Example

This example has been created on an ALTERA system. It serves the purpose of demonstrating SDF in a mixed design flow:

- The design has been created on a front-end design system with a high level language, such as VHDL or Verilog. It could have been a behavioral or RTL (Register Transfer Language) level design, which may have been verified by a high level simulation run before synthesis;
- After the synthesis run for an ALTERA target library the netlist has been transferred to the ALTERA system via EDIF;
- During the subsequent compilation the physical CPLD placement file and the timing data have been created;
- By using EDIF for the netlist and SDF for the set of timing data a final timing simulation can be done on the front-end design system;
- This final verification step has to ensure the conformity to the original timing constraints. An automatic comparison of the simulation runs of the behavioral (pre-placement) and the structural (post-placement) can also be conducted.

In this example the timing triple contains three times the same value, which prevents a true best-case/worst-case analysis. The conditions, which are valid for all three values, are explained in the header section:

```
VOLTAGE = 5V,
PROCESS = "worst",
TEMPERATURE = 25 .
```

According to the SDF definition there may be two triple timing sets. The first set is meant for rising signals ($0 \rightarrow 1$, $0 \rightarrow Z$ and $0 \rightarrow X$) and the second one for falling signals ($1 \rightarrow 0$, $1 \rightarrow Z$ and $1 \rightarrow X$).

Some functional cells, such as AND2, have values of 0.0 only. The ALTERA system has re-structured some logic to the specific CPLD. Some functions may exist in the netlist and the SDF file, but not in the optimized physical implementation, for example DELAY_24.

List 8.3 Example of a (partial) SDF file. It is based on the same design as the EDIF netlist in List 8.2.

```
(DELAYFILE
(SDFVERSION "2.1")
(DESIGN "half_add")
(DATE "07/22/99 12:13:04")
(VENDOR "Altera")
(PROGRAM "MAX+plus II")
(VERSION "Version 8.2 RC3 1/8/98")
(DIVIDER .)
(VOLTAGE :5:) (PROCESS "worst") (TEMPERATURE :25:)
```

```
(TIMESCALE 100ps)
(CELL
(CELLTYPE "TRIBUF")
(INSTANCE TRIBUF_2)
(DELAY (ABSOLUTE
(IOPATH IN1 Y (26:26:26) (26:26:26))
(IOPATH OE Y (0:0:0) (0:0:0) (45:45:45) (45:45:45) (45:45:45)
(45:45:45)) )
    ))
    (...)
(CELL
(CELLTYPE "DELAY")
(INSTANCE DELAY_24)
(DELAY (ABSOLUTE
(IOPATH IN1 Y (5:5:5) (5:5:5)) )
    ))
(CELL
(CELLTYPE "XOR2")
(INSTANCE XOR2_25)
(DELAY (ABSOLUTE
(IOPATH IN1 Y (0:0:0) (0:0:0))
(IOPATH IN2 Y (0:0:0) (0:0:0)) )
    ))
(CELL
(CELLTYPE "AND2")
(INSTANCE AND2_27)
(DELAY (ABSOLUTE
(IOPATH IN1 Y (0:0:0) (0:0:0))
(IOPATH IN2 Y (0:0:0) (0:0:0)) )
    ))
(CELL
(CELLTYPE "INV")
(INSTANCE INV_28)
(DELAY (ABSOLUTE
(IOPATH IN1 Y (48:48:48) (48:48:48)) )
    ))
(CELL
(CELLTYPE "DELAY")
(INSTANCE DELAY_29)
(DELAY (ABSOLUTE
(IOPATH IN1 Y (43:43:43) (43:43:43)) )
    ))
)
```

8.5 References

[8.1] *Tuinenga, P. W.*: 'SPICE, A Guide to Circuit Simulation & Analysis Using PSPICE'. – Englewood Cliffs, NJ: Prentice Hall, 1992

[8.2] EDIF, Electronic Design Interchange Format, Version 2 0 0'. EIA EDIF Steering Committee, 1989

[8.3] EDIF Technical Centre: www.edif.org

[8.4] Accellera Interest Group: www.ovi.org

[8.5] Standard Delay Format Specification, Version 2.1. Open Verilog International: www.eda.org/sdf

Overview EDA
1, 2

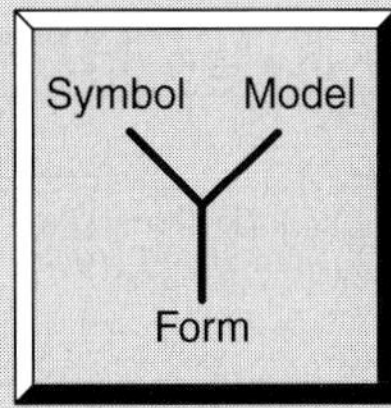

Symbolic Design
3

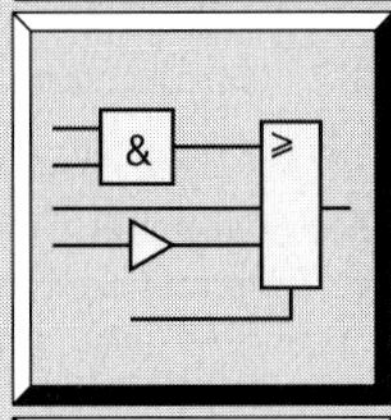

High Level Language Design
4, 5, 6, 7, 8

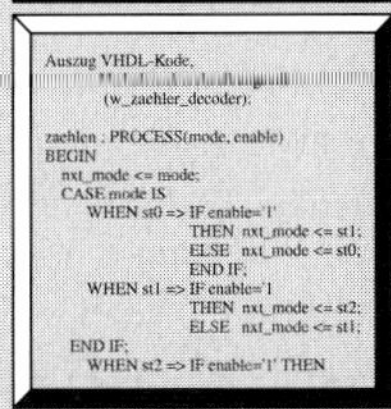

Modelling and Verifications
9, 10, 11, 12, 13, 14, 15

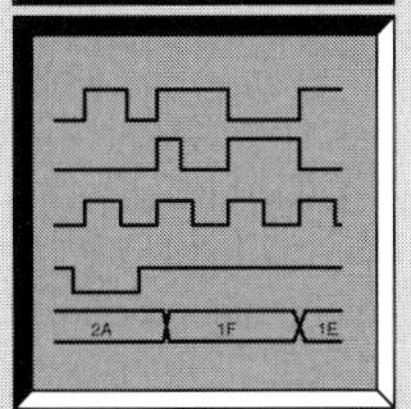

Implementation
16, 17, 18, 19, 20, 21, 22, 23, 24, 25

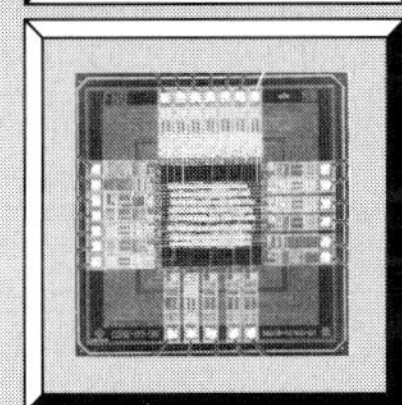

Tutorial
26

Appendix
A, B, C, D, E

9 Circuit Verification

WOLFGANG RÜLLING

There are several concepts of how to verify the correctness of a circuit. Since designing and manufacturing integrated circuits are very time-consuming and expensive tasks it is necessary to detect errors as soon as possible in order to avoid additional costs (e.g., see [9.2]).

We have to distinguish between different kinds of errors (fig. 9.1). The fundamental prerequisite for all following work is that the **specification of a design** is correct, such that it can be used for all following checks of correctness. Typically the specification is given at a very abstract description level. It specifies **what** has to be done, but it does not describe **how** it is done. In an extreme case the specification will not even determine whether a task has to be done in software or hardware.

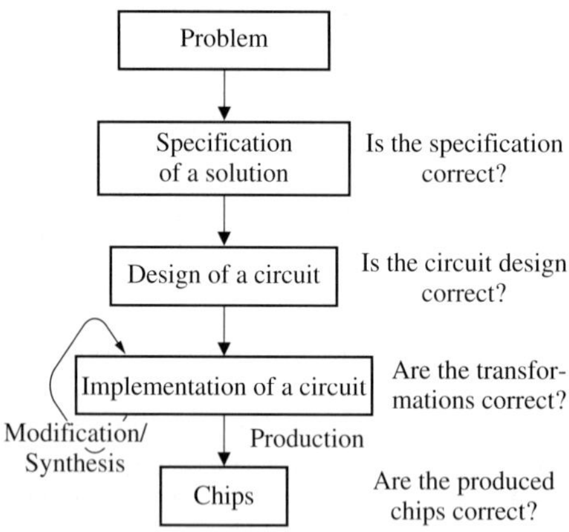

Fig. 9.1 Different tasks of verification on the route from a given problem to a produced chip

During the phase of **circuit design** the designer describes how to perform the specified task with a circuit. For sake of simplicity we assume the task is completely done in hardware and that no software/hardware co-design is necessary. The **design verification** (section 14.6) then has to compare the design produced against the specification to find out whether both are equivalent.

Typically the designer will perform several modifications and refinements to describe details of the circuit before he finally creates mask data for the manufacturing of chips. For every such manual or automated conversion step its correctness has to be verified in order that the chips produced will perform the specified task. In the following sections of this chapter we consider techniques available for such verifications. Please note that whenever an error occurs during the design process all chips produced will have the same faulty behavior. The job of **verification** is to localize such **design errors**.

Additional errors may occur during the **manufacturing of chips**. In this case only single chips of a series may be faulty, whilst others are working correctly. Obviously such faults can only be detected by a **chip test** checking the behavior of each individual manufactured chip. Since the test effort grows strongly with the complexity of the circuits, and since testing each individual chip is very time consuming, one uses special techniques to obtain reliable predicates about the correctness of chips in reasonable time. This topic is treated in detail in chapter 15. Generally the techniques for testing chips are not suitable for detecting design errors. Indeed, to detect production errors one has to assume the design to be correct.

9.1 Verification by Simulation

One method of checking the correctness of a circuit is to perform a **simulation** of the circuit. For this purpose one uses typical input patterns (**stimuli**) and checks whether the simulated outputs are identical to the intended outputs of the circuit. Any differences are owed to design errors or inaccurate simulations. In both cases one has to localize the error and to eliminate it. On the other hand, if the simulated results are identical to the desired values there is a certain confidence in the correctness of the design. But unfortunately this is no guarantee of correctness.

Since each simulation uses simplified models of reality a simulation cannot discover all possible errors. For example, the **digital simulation** considered in chapter 11 uses a simple description of gates and a highly simplified timing model. On the other hand the more precise **analog simulation** described in chapter 10 can discover more design errors but is extremely time consuming so it is not suitable for complex integrated circuits. Furthermore, using a very quick **high level simulation** supported by design languages such as VHDL (chapter 5), the simulated results and the behavior of the circuit may disagree because of awkward high level circuit descriptions. Such phenomena are discussed in section 9.1.1. For these reasons simulations should be treated with great caution, even if the simulated results are as expected.

A further problem with circuit simulation is the selection of suitable **stimuli**. In practice, even for fast simulators it is impossible to simulate a complex circuit considering all possible input patterns and all possible states of the circuit. For example, the exhaustive simulation of a multiplier for two binary numbers of length 32 bits would have to consider $2^{64} = 18{,}446{,}744{,}073{,}709{,}551{,}616$ input samples. Even with a simulation rate of 100 million multiplications per second the simulator would need 5,849 years. Thus in practice one has to restrict the simulation to the most relevant *stimuli*. Then in the case of a wrong selection of stimuli possible design errors may remain undetected.

A further problem is the evaluation of observed simulation results. For the multiplier example it is quite simple to check the outputs against mathematical software. But verification becomes more difficult if, for example, the circuit generates control signals for some other device. Then it may be harder to decide whether a complicated sequence of signals will have the desired effect. In this case one needs a detailed specification of a complete system, also including the environment of the chip to check the results. For example, using the language VHDL this can be done by a *test bench* (section 4.10). But because any misunderstanding of a circuit specification may result in a faulty test bench design errors may again remain undetected.

9.1.1 Verification by High Level Simulation

Using description languages such as VHDL one achieves a very fast high level simulation because of a high degree of abstraction. However, with high level simulation there is the risk that in the case of an awkward circuit modeling the simulated behavior may be quite different from the real behavior of a chip. This is not only owing to neglected details, such as, for example, signal delay times, but also because of different interpretations of VHDL statements by simulation and synthesis.

9

This is demonstrated by the following incorrect VHDL *examples* of behavioral descriptions. They lead to drastic mismatches between the results obtained by high level simulation and gate level simulation. Because of such examples the interpretation of simulation results should be carried out very carefully. As long as a simulation produces unexpected results giving hints of design errors it is very useful. However, when the simulation produces correct results one has to use more exact simulation methods to confirm the correctness.

For the VHDL description from listing 9.1 the incorrect results are owed to the treatment of undefined signals in conditional statements. In addition to the boolean values `'0'` and `'1'` the data types *std_logic* and *std_ulogic* from the *IEEE* library implementing standard 1164 also consist of signals `'U'`, `'X'` `'Z'`, `'W'` `'L'`, `'H'` and `'-'`. But for synthesis only the values `'0'`, `'1'` and `'Z'` are relevant. Thus the semantics of the language VHDL is different for simulation and synthesis [9.5].

The intended effect of process P is that signals a and b always have inverse values such that the signal z independently of s always becomes `'1'`. As a consequence the output y should always be identical to the input x. Indeed, synthesizing the circuit from the given VHDL description we obtain a circuit in accordance with `y <= x` consisting of only a single wire.

But during simulation the internal signal s obtains the value `'X'` because of a missing initialization. Then in both conditional statements of process P the `else` cases are executed assigning the value

'1' to signals a and b. Thus we obtain $z =$ '0' and therefore $y =$ '0', so that obviously the simulation results are in contradiction with the synthesized circuit.

List 9.1 Example of an awkward VHDL description in which the VHDL simulation will show a different behavior from that of the synthesized circuit

```
LIBRARY IEEE;
USE IEEE.std_logic_1164.all;
ENTITY experiment IS
    PORT (x:  in std_logic;
          y:  out std_logic);
END experiment;

ARCHITECTURE BEHAVIOR OF experiment IS
SIGNAL s,a,b,z:std_logic;
BEGIN
    P:PROCESS(s)
        BEGIN
        IF s = '1' THEN a <= '0';
                   ELSE a <= '1';
        END IF;
        IF s = '0' THEN b <= '0';
                   ELSE b <= '1';
        END IF;
        END PROCESS;
    z <= a xor b; - Simulation:  z <= '0',
                  - Hardware:    z <= '1'
    y <= x and z; - Simulation:  y <= '0',
                  - Hardware:    y <= x;
END BEHAVIOR;
```

The second *VHDL example* concerns the activation of processes, comparison of signals, and timing problems. For this the listing 9.2 presents an incorrect description of a modulo counter.

The designer intended to increment the state s of the counter with every rising clock edge, and when obtaining the value 6 to replace it by 0. Using the input *reset* the counter should be synchronously reset with the rising edge of *clk*. Figure 9.2 shows the expected behavior and the result of simulating the VHDL description.

It is noticeable that in the VHDL simulation with a rising edge of *clk* we achieve the state 6 and the correction to $s = 0$ occurs not before the next falling edge. This is because the assignment `s < = s+1` is performed concurrently, so that for the following test `if s = 6`... the signal s still holds the old value 5 instead of the new value 6. The next activation of process P with a successful test on $s = 6$ occurs with the next modification of *clk*, thus with the falling edge.

List 9.2 Incorrect VHDL description of a modulo 6 counter

```
ENTITY counter IS

    PORT (clk:  in Bit;
          reset:  in Bit;
          y:  out integer range 0 to 7
          );
END ENTITY;

ARCHITECTURE BEHAVIOR OF counter IS
        SIGNAL s:  integer range 0 to 7;
    BEGIN
        y <= s;
        P: PROCESS (clk)
        - s is missing in the sensitivity
        - list
            BEGIN
            IF clk'event AND clk='1'
                THEN IF reset='1'
                        THEN s <= 0;
                        ELSE s <= s+1;
                END IF;
            END IF;
            IF s=6
                THEN s <= 0;
            END IF;
    END PROCESS;
END BEHAVIOR;
```

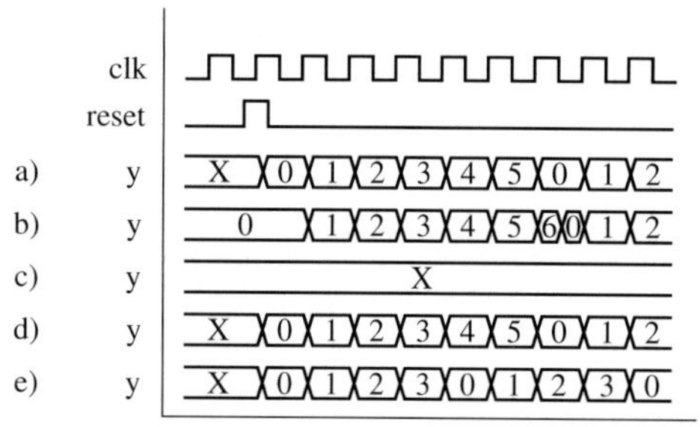

Fig. 9.2
a) Intended behavior of the circuit from listing 9.2;
b) Result of the VHDL simulation;
c) Gate level simulation of the synthesized circuit;
d) Observed behavior of an FPGA prototype realization;
e) Behavior of the chip with bad signal delays on wires

After synthesizing the VHDL description we obtain the circuit shown in figure 9.3. One easily recognizes the synchronous increment operation and the asynchronous modulo calculation. Obviously the correction from state 6 to state 5 is performed

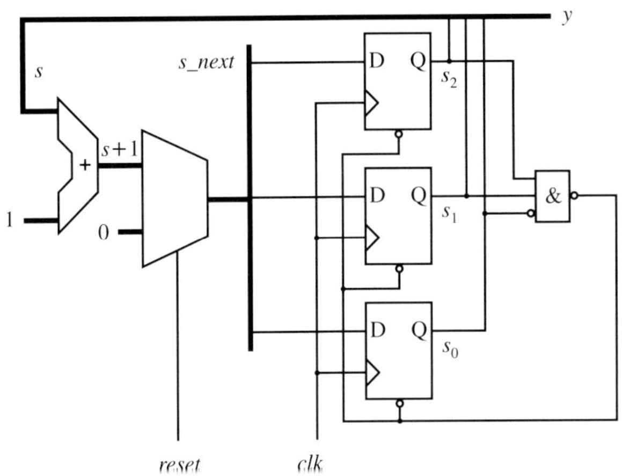

Fig. 9.3 Sketch of the synthesized modulo 6 counter based on the VHDL description from listing 9.2

immediately without waiting for the falling edge of the clock signal. The cause for this deviation between simulation and synthesis is the absence of signal *s* in the sensitivity list of process P. Including *s* in the list the modulo calculation will also be performed synchronously within the simulation. This modification has no effect on synthesis.

Another remarkable fact is that in the beginning of the simulation the counter is in state 0, even without use of the *reset* signal. In general this will not be true for the synthesized circuit. Indeed, the gate level simulation tells us that at the beginning the state is undefined. This deviation between both kinds of simulation is owed to the state modeling by a data type of *integer*. If we replace this data type by an *unsigned integer* the state will be undefined until it is initialized by signal *reset*.

All presented modifications of the behavioral description have no effect on synthesis, but are useful for obtaining a more realistic simulation. Although the VHDL simulation then shows exactly the desired behavior and the FPGA prototype implementation also works well, there is one more problem with gate level simulation. In that simulation the state remains undefined even if the *reset* signal is set. The reason for this unexpected phenomenon is a special feature of the synthesized circuit. Signal *s_next* of the counter is derived from the current state and from the signal *reset*. Figure 9.4 shows a small part of the combinatorial circuit computing the bit s_next_1. The labels are according to the applied input *reset* $= 1$. Correctly we obtain $s\_next_1 = \overline{s_1 + \overline{s_1}} = 0$ not depending on whether s_1 is 0 or 1. But since the gate level simulation uses $s_1 =$ '*X*' for the undefined state of the counter it derives $s\_next_1 =$ '*X*'.

Obviously in this case the gate simulation is too careful. When we manually initialize the counter to any defined value the gate level simulation also shows the intended behavior of the circuit. Nevertheless, the circuit described is poorly suitable in practice. Depending upon wiring and sizing of the gates we will obtain different capacities, and thus different signal delays, we may influence the calculation of s_next_2, s_next_1 and s_next_0. For example if signal s_next_1 is slower than the other two signals the transition from $s = 3 = (011)_{\text{bin}}$ to $s = 4 = (100)_{\text{bin}}$ will briefly produce the intermediate result $s\_next = (110)_{\text{bin}} = 6$ and thus start an asynchronous reset operation for the flip flops. This erroneous behavior is shown in figure 9.2 as case e).

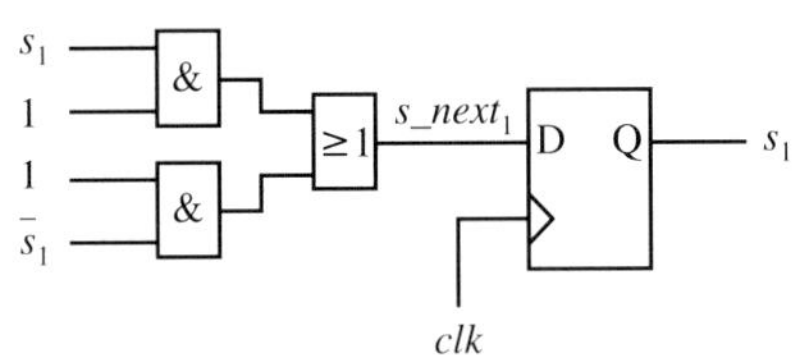

Fig. 9.4 Part of the circuit from figure 9.3 labeled with signals for the input reset = 1

Altogether the aforementioned examples show that simulation results should be very carefully interpreted and different kinds of simulation may produce contradictory results. The problems shown arise from an awkward or incorrect behavioral description of circuits given by the designer. Of course in practice the designer should try to avoid such problems by using well formulated descriptions. Nevertheless, there is no guarantee that a VHDL description will be error-free.

9.2 Concept of Formal Verification

As presented in section 9.1, it is not practicable to carry out an exhaustive simulation using all possible stimuli. **Formal verification** is a method of avoiding this problem. Here the correctness of the design is proved, for example, by showing the equivalence of two descriptions. This technique is explained in more detail in chapter 14. For this methodology there are convincing sample applications. For instance, if a circuit is manually modified to improve its time performance or to minimize the chip area then one can check by formal verification whether the circuit descriptions before and after the modification are equivalent [9.3], [9.1]. If, for example, the designer makes a mistake by cutting a wire, or by causing a short circuit between wires, or by using a wrong gate, then the modification will change the logical behavior of the circuit and the error will be detected by a comparison of both circuit descriptions.

Typically, the formal verification will generate a stimulus for each detected error so that the error can also be observed by simulating both circuits for the given stimulus. The great advantage of formal verification is that it is not necessary to try out a large number of stimuli to detect an error. Instead of this the relevant simulation inputs are calculated by the *verification tool*. Of course the verification task becomes simpler if both circuit descriptions are very similar.

The large problem of formal verification is that we need a precise specification of the desired chip behavior which can be formally compared to the actual circuit design. The creation of this specification may be just as difficult as designing the circuit. Furthermore, if specification and design both contain the same error it cannot be detected by formal verification.

With none of the verification methods do we obtain a guarantee for the correctness of a circuit's implementation. In practice all we can achieve is to avoid special types of design errors. For that reason we restrict the following presentation to the task of verifying special properties of a design. For example, we may verify the *timing* to ensure that for a specified clock rate there is enough time to achieve the next stable state from every current state of the circuit and for any input data.

9.3 Statical Analysis of Timing

Typically for logic synthesis the designer has to specify the desired clock rate, and after synthesis he has to check the generated **critical paths** in order to see whether the given timing constraints are actually satisfied. Those critical paths are calculated during synthesis using a **statical analysis of timing**. For this all paths within the circuit are checked for the worst case delay times of changing signals. The **critical path** found then defines the fastest possible clock rate for which the circuit is able to derive correct outputs.

Example: In figure 9.5 changes of signals are generated at primary inputs and at the outputs of flip flops. They are propagated through the circuit to the inputs of flip flops and to primary outputs. For example, there is a path from output Q of flip flop s_1 to input D of flip flop s_3. On this path there is only one *NOR* gate. In table 9.1 the relevant delay times and capacities of *NOR* gates are given as an example. In general the delay of output y owed to a rising edge at input x_i can be computed by

$$x_{i\,\mathrm{rising}} + c_y \cdot \Delta x_{i\,\mathrm{rising}}$$

Here c_y denotes the capacity to be driven by Signal y. As shown in the table, a rising edge at input x_2 causes a modification of the output signal y after a delay of typically 0.12 ns. On the assumption that the other input x_1 of the *NOR* gate is 0, a falling edge is generated at y. Furthermore, since the signal produced has to drive two inputs of gates (s_3 and g_2) each of 0.04 pF, we obtain a total signal delay of 0.12 ns + (0.5 + 0.04 + 0.04) · 1.02 = 0.73 ns from input x_2 of the *NOR* gate to input D of flip flop s_3, neglecting the capacities of the interconnections. In total we obtain a signal prop-

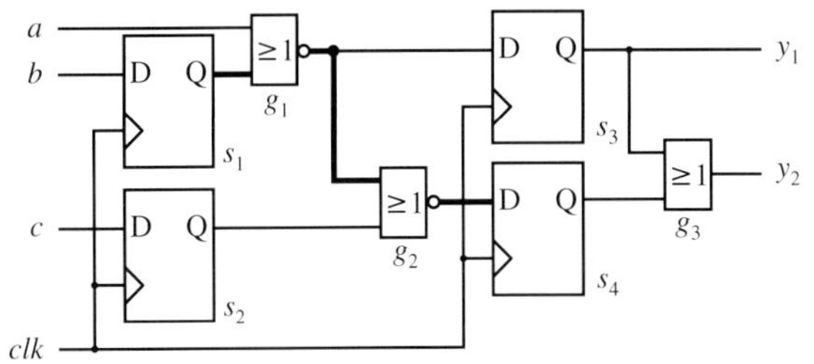

Fig. 9.5 Example for calculating the critical path of a circuit

agation time of 2.16 ns for the path from flip flop s_1 to flip flop s_3, as indicated in table 9.2.

Table 9.1 Delay times of a NOR gate $y = \overline{x_1 + x_2}$ for fast, typical, and slow implementations and the capacities of inputs and outputs

NOR	FAST	TYPICAL	SLOW
$x_{1\,\text{rising}}$	0.06 ns	0.23 ns	0.25 ns
$x_{1\,\text{falling}}$	0.08 ns	0.18 ns	0.35 ns
$x_{2\,\text{rising}}$	0.06 ns	0.12 ns	0.24 ns
$x_{2\,\text{falling}}$	0.06 ns	0.14 ns	0.27 ns
$\Delta x_{1\,\text{rising}}$	0.86 ns/pF	1.83 ns/pF	3.62 ns/pF
$\Delta x_{1\,\text{falling}}$	0.48 ns/pF	1.02 ns/pF	2.02 ns/pF
$\Delta x_{2\,\text{rising}}$	0.86 ns/pF	1.83 ns/pF	3.62 ns/pF
$\Delta x_{2\,\text{falling}}$	0.48 ns/pF	1.02 ns/pF	2.02 ns/pF

port	capacity
x_1	0.04 pF
x_2	0.05 pF
y	0.50 pF

Table 9.2 Example of the delays on the path from flip flop s_1 to flip flop s_3 for a rising or falling edge at flip flop s_1

signal	edge	delay in ns	total time in ns
s_1.*CLK*			0
$g_1.x_2$	rising	+1.43	1.43
s_3.*D*	falling	+0.73	2.16
s_1.*CLK*			0
$g_1.x_2$	falling	+1.35	1.35
s_3.*D*	rising	+1.18	2.53

With similar calculations we find a critical path running from Q_{s1} to D_{s4}, as emphasized in figure 9.5. It limits the maximal possible clock rate of the circuit. If necessary, the time performance of the circuit may be improved by **re-timing**. This means reallocating the combinatorial parts of the circuit to the paths between flip flops.

Hereby it is also possible to consider *timing constraints* given by the designer. For example, the designer might state that a primary input a will not be stable before 2 ns after the rising edge of the clock signal. Furthermore, it may be required that the primary output y must already be stable at 3 ns after the rising clock edge and should be able to drive some pre-defined capacityance. Any determined deviations from such requirements can be used to control synthesis.

The **timing verification problem** consists of checking whether the desired timing is still valid after generation of the layout, that is, after performing placement of cells and after routing wires. A possible approach of timing verification is to simulate the circuit considering also capacities and resistances of wires that are extracted from the layout. This would be called a *dynamical* timing analysis. As previously mentioned, this method is very time consuming because too many stimuli have to be considered. Also it does not help to restrict the simulation to critical signals found during logic synthesis, because there may exist different critical paths after routing.

Therefore, as a more efficient alternative one should perform a **static timing analysis** for the netlist extracted from the layout considering also real capacities and resistances of wires. Again, the more precise signal delays obtained on all paths have to be compared to the *timing constraints* specified by the designer. Whenever a path does not satisfy the required timing one can try to repeat the synthesis using more restrictive timing constraints for that critical path.

9.4 Detection of Critical Signals

Whenever possible, circuits should be designed in such a way that slight changes of delays of gates or wires do not affect the logical correctness of signals. A simple *example* demonstrating a critical timing is shown in figure 9.6.

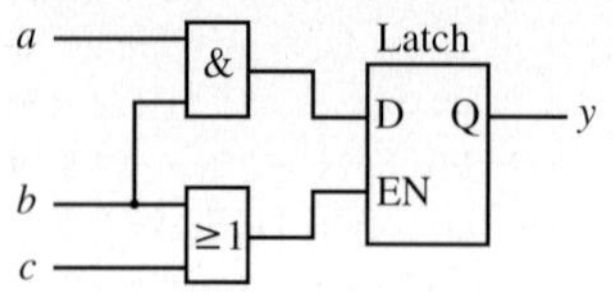

Fig. 9.6 Example of a circuit with critical timing

The *latch* transfers its input data D to output Q as long as input *enable* is 1. On changing *enable* from 1 to 0 the most recent input value remains stored in the latch. A problem arises if with the change of *enable* there is also a change of input D. In this case the state of the latch becomes undefined. In the example the output y strongly depends on the exact signal delays on the paths from b to D and from b to EN. For inputs $a = 1$ and $c = 0$ it is undefined whether a value 1 or 0 will be stored in the latch after changing the signal b from 1 to 0.

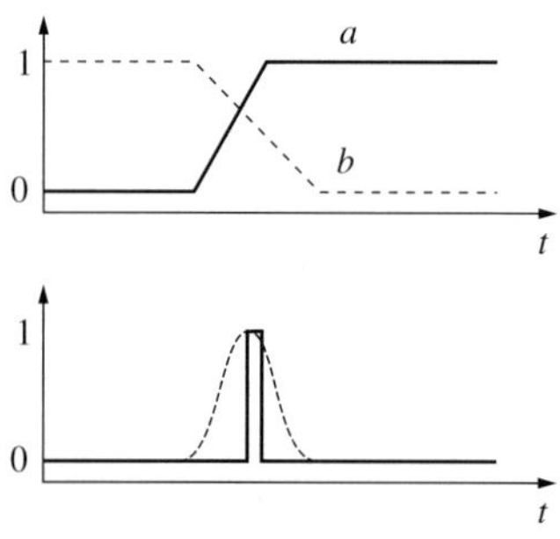

Fig. 9.7 Voltage and signals for an AND gate demonstrating the generation of a ***hazard*** *due to different delays of signals*

A similar effect can occur if several signals are changing simultaneously, such that for a very short time an incorrect signal (**hazard**) is generated. For example, if both inputs $a = 0$ and $b = 1$ of an *AND* gate simultaneously are changed to $a = 1$ and $b = 0$ then the output will be 0 before and after switching, but for a short time as a result of different delays of the input signals we might obtain $y = 1$. This timing is sketched in figure 9.7.

If the critical signal y is used to asynchronously set or reset a flip-flop this might produce a wrong stable state of other signals.

To detect such effects one can perform a **delay independent** simulation. That is an event driven simulation without considering delays in which a signal change from 0 to 1 or from 1 to 0 is simulated by interposing an intermediate state X. For example, a sequence 01 of signals is replaced by 0X1. This simultaneously describes a slow switching 001 and a fast switching 011. The delay independent simulator propagates X-values until a stable state of the circuit is reached. Only after this are the new values of signals used for a second simulation step. Whenever a delay-independent simulation reaches a well defined stable state of the circuit can we conclude that indeed in practice the same stable state will be reached by the circuit. In particular, that state does not depend on the exact delay of signals.

On the other hand, if there are undefined signals left after a complete simulation step we can conclude that the logical behavior of the circuit depends on the exact delay of signals. Thus the circuit might be sensitive to fluctuations of manufacturing parameters or to environmental influences. In the worst case this may result in incorrect outputs, and if there are feed back loops the circuit may even oscillate.

9.5 Verification with Programmable Devices

A quite different method of design verification is to implement a **prototype** using programmable devices. Such a **PLD** (*Programmable Logic Device*) consists of a certain number of configurable logic blocks which can be configured to behave like an elementary gate or to perform more complex logical functions. Furthermore, the logic blocks contain sequential elements (flip flops). The logic blocks are connected by a configurable interconnection network such that according to the specific configuration the logical behavior of any circuit can be modeled.

There are different types of programmable devices. In the case of an **FPGA** (*Field Programmable Gate Array*) the configuration is stored in a RAM and can easily be changed an arbitrary number of times. Therfore the FPGA may be used for an inexpensive circuit verification even in an early phase of the chip's design. For example, the behavior of a combinatorial logic block with four primary inputs and two primary outputs can be configured using a RAM (*Random Access Memory*) with $2^4 = 16$ data words of width 2 bits. Into this $2 \times 16 = 32$ bits one stores a complete *function table* representing the intended behavior of the logic block. During operation the input data x_1, x_2, x_3, x_4 are applied to the address lines of the memory and its data output drives the primary outputs y_1, y_2 of the logic block.

For other types of memory such as PROM (*Programmable Read Only Memory*) or EPROM (*Erasable PROM*) one obtains non-volatile programmable devices.

Although there are essential differences between PLDs and ASICs, programmable devices are used for design verification because there are fast and easy methods within ASIC design systems to convert an ASIC design into a PLD configuration. Using this approach a PLD can be used as a prototype implementation of a design.

Essentially, the verification method is to apply sample inputs to the programmable device and to compare its outputs with the expected values. This is very similar to the methodology of verification by simulation. Therefore, as described in section 9.1, there is the problem of selecting the most relevant input patterns. Although a PLD is much faster than a circuit simulation it is impossible to apply all possible sequences of input patterns.

One has to distinguish between two applications of programmable devices. First, they can be used as a so called **hardware accelerator** for simulation. In this case one needs a special simulation software to use the FPGA as a co-processor performing time consuming simulation tasks. In this way the designer can use the front end of a logic simulator, but results are computed much faster, therefore more extensive simulations become feasible. Depending upon the selected signals to be monitored and on the desired accuracy, the complete circuit, or only some sub-circuits, are performed on the FPGA. In this way the only advantage of the programmable device over the simulation software is the reduction of computation time.

A second more interesting approach is to use an FPGA prototype within the desired hardware environment of the ASIC. Then it is also possible to test the interaction of the circuit with other components. As a simple *example* we consider the control of several 7-segment LED displays using a *multiplexing* method. Normally a simulation will produce a sequence of output patterns, and the designer has to decide whether the patterns are correct. For example, he has to decide whether the data is applied to the correct display and whether the correct segments are selected. Furthermore, he has to decide whether the signals are consistent to the LED specification and whether the *timing* of *multiplexing* is well chosen, so that the display is bright and does not flicker. To check automatically all these effects by simulation a very fine modeling of 7-segment displays would be necessary.

9

More generally, the complete hardware environment of a circuit has to be modeled for simulation, which in most cases cannot be done in practice. However, the designer can obtain the same information by directly testing the prototype within the target hardware, or at least within a simplified environment on a *demonstration board.*

Thus the main advantage of a prototype is to check the interaction of a circuit with other components already available. This way one can also detect design errors that are the results misunderstandings of the specification and that cannot be found by simulation.

A problem of verification with PLDs is the different timing of ASICs and PLDs. This is because in a PLD every sub-circuit is implemented by reading results from a look up table. Thus the switching time of a logic block does not depend on the complexity of the modeled sub-circuit.

Since the interconnection between logic blocks consists of pre-defined wire segments connected or separated by transistors, the distribution of net capacities for a PLD is quite different from that of an ASIC, resulting in different signal delays. Only a small modification of the circuit can result in a quite different allocation of sub-circuits to logic blocks, producing large differences in the length of wires. For an ASIC the same modification may only locally affect few signals.

Thus a PLD implementation of a prototype is not suitable for checking the exact timing of a circuit design. In particular, one cannot determine the maximal possible clock rate of an ASIC by investigating the behavior of a PLD prototype.

9.6 References

[9.1] *Eveking, E.; Hinrichsen, H.; Ritter, G.*: 'Automatic Verification of Scheduling Results in High-Level Synthesis.' DATE'99, 1999

[9.2] *Fournier, L.; Arbetman, Y.; Levinger, M.*: 'Functional Verification Methodology for Microprocessors Using the Genesys Test-Program Generator.' DATE'99, 1999

[9.3] *Hendricx, S.; Claesen, L.*: 'Formally Verified Redundancy Removal.' DATE'99, 1999

[9.4] *Lipsett, R.; Schaefer, C.; Ussery, C.*: 'VHDL: Hardware Description and Design.' Kluwer Academic Publishers, 1989

[9.5] *Ten Hagen, K.*: 'Abstrakte Modellierung digitaler Schaltungen.' Berlin: Springer Verlag, 1995

10 Analog Simulation

HORST NIELINGER

The cover of the June 1998 edition of the *IEEE Spectrum* shows the 'father' of SPICE, Prof. Emer. DONALD O. PEDERSON of the University of California, Berkeley, on the occasion of his award of the IEEE medal of honor. His **Simulation Program with Integrated Circuit Emphasis** (abbreviated as SPICE) is without doubt the most important **analog simulator in the world of microelectronics**. The widespread usage of SPICE is owed on the one hand to the competent electrotechnical background and, on the other hand, to the admirable attitude of Prof. PEDERSON in giving the program practically for nothing to all who wanted to use it. It was his principle never to apply for a patent. His generosity is still effective today: the version of SPICE which runs on Personal Computers (PSPICE) is available practically cost free for interested users as the so called *evaluation version* (restricted to 10 transistors, 50 circuit nodes) [1)].

This cost effectiveness results in fascinating possibilities in engineering education because every student is practically able to install his own simulated laboratory on his PC. On the following pages the PSPICE evaluation version 8 with German symbols for the circuit elements [10.5] will be used to demonstrate some problems of the simulation of analog circuits in an exemplary way. An extensive discussion of SPICE simulation of analog circuits can be found in [10.17], which refers to the outstanding standard book on analog circuits [10.19]. But first an insight into the mathematical and electro-technical fundamentals of SPICE will be given which tend to stay in the background, with the nowadays usual over-emphasis of program handling problems. This will be more or less a takeover of the introduction to SPICE formulated by the author of this chapter in [10.6].

After that the core question of analog simulation will be dealt with, which is the **modeling of semiconductor circuit elements**, especially Bipolar Junction and MOS Field Effect Transistors. As in [10.17], the static behaviour will be emphasized because the dynamics of modern electronics are more or less determined by the connection delays. An extra section is dedicated to the operational amplifier, the most important component in the analog world. The very effective analog behavioral modeling (ABM) technique will be applied and the (erroneous!) macro model of the operational amplifier to be found in PSPICE will be discussed. A central problem when simulating analog circuits is the assessment of stability, and with respect to this problem the determination of the loop gain of a logarithmic amplifier will be discussed.

Originally it was intended to show the simulation of transmission line problems in connections of digital circuits on a printed circuit board, because these problems will occur in future at chip level when more transistors will be connected. Dealing with this kind of problem SPICE is used as a mixed mode simulator as it is not sufficient to describe a digital circuit only by its logical behavior when it co-operates with an analog element 'transmission line'. As there are better ways to handle mixed-mode problems nowadays, e.g., by VHDL-AMS [10.21] the discussion of mixed mode problems with SPICE has been omitted.

It should be noted that some circuit problems can only be simulated rather clumsily by SPICE. This is the case when widely spread time constants are found in a circuit, which will result either in convergence problems or extremely long computer time. Examples of such circuits are SC filters or PLL circuits in carrier frequency systems. For

1) Note that the power of the evaluation version has decreased in the course of time.

these types of circuits the use of ADS [10.7] is of advantage.

10.1 The SPICE Concept

SPICE is a universal simulation program for the analysis of electronic circuits. It is able to handle non-linear direct current analysis, dynamic transient analysis, and linear small signal alternating current analysis with different temperatures. In addition to simulation of passive components, independent and controlled sources, it allows the simulation of seven different semiconductor models.

10.1.1 Types of Analysis

The following types of analysis are possible in SPICE [10.19]:

- **DC Analysis**. DC Analysis stands for nonlinear direct current analysis, in which the operational point of an electronic circuit is determined. For this purpose inductors are replaced by short circuits and capacitors by open circuits. A DC analysis is automatically performed before an AC or TR analysis is started (see below) in order to determine defined starting conditions or small signal parameters for the operating point. In addition, the DC analysis can be used to calculate DC transfer curves by changing the values of an independent DC source and listing the resulting output variables of interest. It is also possible to determine the DC small signal sensitivities of certain output variables with respect to circuit parameter alteration.
- **AC Analysis**. AC Analysis stands for a linear small signal alternating current analysis. Here the AC output variable is determined in a frequency range wanted in which the nonlinear semiconductor models are replaced by linearised small signal models whose parameters are generated in the preceding DC analysis. A special case of AC analysis is that of noise analysis.
- **TR Analysis**. TR Analysis stands for large signal analysis of the transient behavior of a circuit. Examples are the calculation of the step or pulse response of a nonlinear circuit where the initial conditions are either determined by a preceding DC analysis or by given capacitor voltages or inductor currents, respectively.

10.1.2 Circuit Description by a Netlist

The input for the program SPICE is provided by a **netlist** formed according to fixed rules. Nowadays the netlist is usually generated by an auxiliary program, the schematic editor. But for error detection knowledge of the SPICE netlist structure is indispensable.

The netlist comprises, on the one hand, component lines which define the topology of the circuit, and the component values. The control lines, on the other hand, define the model parameter and the types of analysis. The first line in a netlist is always a title line in the form of a comment and will not be processed. The last line is always '.END'. In between, the order of lines is arbitrary.

Each component of a circuit is described by a **component line** that comprises its name, its connection node numbers and its value. The **first letter of the component name** signifies the **type** of component (e.g., R for a resistor), followed by up to seven characters which can be alphanumerical. Nodes are signified by non-negative numbers (letters are also allowed in newer SPICE editions), not necessarily consecutive. The **reference node (ground) always has the number 0**. Each node must have two connections (exceptions are MOSFET substrate and an open transmission line). Table 10.1 shows the structure of the most important component lines of a SPICE netlist.

Comment to the **netlist format**:

- The **component name** starts with the letter shown and can comprise up to eight alphanumerical characters.
- **N+ and N−** signify the connection nodes and define in their order a direction (when necessary, e.g., for a voltage source).
- **'Value'** is a numerical value that can be considered as Ohm, Farad, Henry, Volt, Ampere, Siemens, V/V, A/A respectively).
- **'Type'** = type of source: DC, AC, SIN, PULSE, PWL (piecewise linear).
- **NC+ and NC−** are the controlling nodes for the control voltage.
- **VNAM** is the voltage source through which the controlling current is flowing.

Table 10.1 The most important component lines of a SPICE netlist

Component	Name	Nodes	Specification
Resistor	Rxxxxxxx	N+ N−	Value
Capacitor	Cxxxxxxx	N+ N−	Value
Inductor	Lxxxxxxx	N+ N−	Value
Voltage Source	Vxxxxxxx	N+ N−	Type, Value(s)
Current Source	Ixxxxxxx	N+ N−	Type, Value(s)
VCIS	Gxxxxxxx	N+ N− NC+ NC−	Value
VCVS	Exxxxxxx	N+ N− NC+ NC−	Value
ICIS	Fxxxxxxx	N+ N− VNAM	Value
ICVS	Hxxxxxxx	N+ N− VNAM	Value
Diode	Dxxxxxxx	N+ N−	MNAME, AREA
BJT	Qxxxxxxx	NC NB NE NS	MNAME, AREA
JFET	Jxxxxxxx	ND NG NS	MNAME, AREA
MOSFET	Mxxxxxxx	ND NG NS NB	MNAME, L, W

10

- **MNAME** refers to a semiconductor model which is either user defined or taken from a library.
- **AREA** is an (optional) area scaling factor.
- **NC, NB, NE, NS** are nodes with which the collector, base, emitter and (optionally) substrate of a BJT are connected.
- **ND, NG, NS, NB** are nodes with which drain, gate, source and substrate of a JFET or MOSFET are connected.
- **L and W** are channel length and width in m.

10.1.3 History

The design of an electronic circuit normally needs an experimental verification because usually the first approach to the design is done with simplified models of the components so a check is necessary to determine whether the simplifications are suitable. As long as one had to deal with a few components only an experimental bread board test was appropriate. Dealing with integrated circuits, bread boarding is not possible any longer because soldering lumped components together leads to totally different electrical conditions compared to those of the integrated miniaturised circuit, owing to different stray capacitances and transmission line effects. Therefore, with the integrated circuit becoming more and more popular and economical, simulation of the circuit becomes more and more important. Components are described by mathematical models, numerical methods replacing typical measurements such as determination of frequency or time response. In addition a simulation program can provide information which is hard to achieve otherwise, e.g., how sensitively a circuit reacts to component changes (*Sensitivity Analysis*).

There are many different methods of how to analyze electronic networks, therefore there are many different simulation programs. A committee of the NTG (Nachrichtentechnische Gesellschaft = Communication Society), to which the author of this chapter belonged, tried to evaluate the different simulation programs for electronic circuits in 1972/73 [10.16]. Today practically only SPICE is quoted, and this because of three *reasons*:

- The user-friendliness of SPICE. The input language is very simple and adapted to the language of an electronic engineer. The program automatically inserts reasonable default parameters for semiconductors if the user does not or cannot specify them. A very clever error control generally avoids computer time-consuming erroneous circuit analyses.
- The competent scientific research which was done by the creators of the program. In addition to Prof. PEDERSON [10.11], one has to name

Prof. NAGEL of the University of California, Berkeley who investigated in his dissertation [10.12] the different methods of matrix construction, dealing with non-linearities and numerical integration, selecting the optimal methods respectively.

- The generous distribution of SPICE. Everybody who was interested could get the program practically free!

SPICE 1 was released by the *University of California, Berkeley, in May 1972* and was extensively used afterwards by students and the semiconductor industry. After 17 updates SPICE 2 was released (1975), a much more powerful program owing to dynamic memory organization and automatic time step control. In the meantime many new versions of SPICE were issued, signifying the success of the program and its adaptability to the different needs of the user. For several years there have been SPICE versions suitable for PCs 'PSPICE' [10.5]. With this option the range of users has increased tremendously and the program is a world wide success.

10.1.4 Mathematical Algorithms in SPICE

An extended discussion of the algorithms used in SPICE is given in [10.11]. It would be beyond the scope of this chapter to discuss all the algorithms in detail, but in order to convey a basic understanding of the mathematical background to the reader the following approach was chosen, namely that of fundamental mathematical principles discussed with reference to CALAHAN [10.2] in which SPICE specific modifications and specialities are mentioned.

Setup of Equation System

The equation system of SPICE is set up according to the node voltage analysis algorithm. The construction of the conductance matrix (with complex elements when using AC analysis) follows very simple rules.

- An element on the main diagonal a_{ii} is calculated as the sum of conductances connected to the node i.
- All other matrix elements a_{ij} are equal to the negative conductance connecting nodes i and j.

A simple example is to show the setup of the conductance matrix. Consider the electrical network shown in fig. 10.1. Wanted are the normalized node voltages x_1, x_2, x_3 which can easily be determined elementarily ($x_1 = 1/2, x_2 = 1/4, x_3 = 1/4$) with this simple problem. The rules given above lead to the following system of equations:

$$\begin{pmatrix} 3 & -1 & -1 \\ -1 & 2 & 0 \\ -1 & 0 & 2 \end{pmatrix} \cdot \begin{pmatrix} x_1 \\ x_2 \\ x_3 \end{pmatrix} = \begin{pmatrix} 1 \\ 0 \\ 0 \end{pmatrix}$$

$$\boldsymbol{A} \cdot \boldsymbol{x} = \boldsymbol{b}$$

(conductance matrix · voltage vector = current vector)

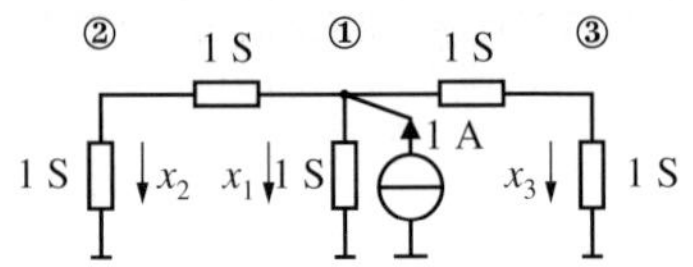

Fig. 10.1 Example: network for conductance matrix set-up and LU factorisation

This simple setup is only valid if all sources are **current sources**. In applications **voltage sources** are used much more and in SPICE a **modified node analysis** can be found. The right hand side of the equation system then consists of a vector which contains the given currents and voltages of the sources, the vector of the unknown quantities then contains the node voltages and the currents through the given voltage sources. The modified node analysis is considered the simplest and most effective setup method compared to many other approaches [10.11].

Solution of the equation system

The solution of the equation system is done in SPICE by means of the **LU factorization** algorithm with following forward and backward substitution. LU factorization means that the original matrix is factorized into two matrices where the L matrix only contains elements on and below the main diagonal which contains only one.

By this factorization the number of elements is the same as before so no additional memory is needed. Intermediate storage is also not necessary as the factorization algorithm uses successive elements of the original matrix to determine the new elements without having to use them afterward again. Thus the new elements can be stored in memory places where originally the old elements were to be found and is a very economical procedure especially when dealing with large networks.

In matrix notation the factorization results in:

$$\boldsymbol{A} \cdot \boldsymbol{x} = \boldsymbol{L} \cdot \boldsymbol{U} \cdot \boldsymbol{x} = \boldsymbol{b}$$

By definition of an intermediate solution vector

$$\boldsymbol{U} \cdot \boldsymbol{x} = \boldsymbol{y}$$

producing the following two equation systems

$$\boldsymbol{L} \cdot \boldsymbol{y} = \boldsymbol{b} \tag{10.1}$$

$$\boldsymbol{U} \cdot \boldsymbol{x} = \boldsymbol{y} \tag{10.2}$$

With the factorization algorithm described in [10.2] for the numerical example given above it is found.

$$\begin{pmatrix} 3 & -1 & -1 \\ -1 & 2 & 0 \\ -1 & 0 & 2 \end{pmatrix} =$$

$$\begin{pmatrix} 3 & 0 & 0 \\ -1 & 5/3 & 0 \\ -1 & -1/3 & 8/5 \end{pmatrix} \cdot \begin{pmatrix} 1 & -1/3 & -1/3 \\ 0 & 1 & -1/5 \\ 0 & 0 & 1 \end{pmatrix}$$

$$\rightarrow \begin{pmatrix} 3 & -1/3 & -1/3 \\ -1 & 5/3 & -1/5 \\ -1 & -1/3 & 8/5 \end{pmatrix}$$

$$\boldsymbol{A} = \boldsymbol{L} \cdot \boldsymbol{U} \rightarrow \text{LU matrix} \tag{10.3}$$

Using equation (10.1) and by forward substitution (zeroes above the main diagonal in the L matrix) gives the solution

$$y_1 = 1/3, \quad y_2 = 1/5, \quad y_3 = 1/4$$

This inserted into equation (10.2) and by backward substitution (zeroes below the main diagonal in the U matrix) gives the final and expected solution (see above)

$$x_3 = 1/4, \quad x_2 = 1/4, \quad x_1 = 1/2$$

As mentioned before, the element places of the A matrix store the respective elements of the L and U matrix, which is formally called the LU matrix in equation (10.3).

Special attention is drawn to the two zeroes in the A matrix. It is a typical property of electrical networks that they **lead to matrices which contain many zeroes** (sparse matrices). In our simple example this property is not so significant; however, larger networks lead to matrices which contain 90 % or more zeroes! Special so called **sparse matrix algorithms** were developed to handle effectively such problems in order to save computer time and memory. SPICE uses a pointer system which points to only non-zero filled memory places. There are even more savings possible when the property is used that matrices of electrical networks are symmetrical or nearly symmetrical, thus the pointer system can be constructed very effectively.

As seen in equation (10.3) the original two zeroes in the A matrix are lost during the LU factorization, one speaks of so called ‘fills’. In SPICE a re-ordering of rows and columns takes place after each step of the LU factorization in order to minimize the number of ‘fills’. In [10.2] an example is shown in which the number of fills in a 19 node problem was reduced by a re-ordering algorithm from 140 to 10.

Non-linearities

Dealing with non-linearities is carried out in SPICE by means of the **Newton Raphson algorithm**. Figure 10.2 shows an example of a nonlinear circuit in which the connection between the diode current and the diode voltage is given, on the one hand by the following equation

$$i = I_s \left(e^{v/V_T} - 1 \right)$$

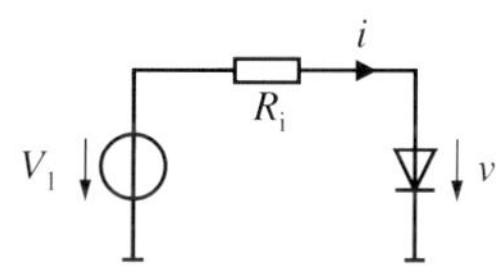

Fig. 10.2 Example: network for a nonlinear circuit

On the other hand, diode current and voltage must obey the linear boundary condition given by the generator voltage V_i and the inner resistance R_i. Figure 10.3 shows the diode and the generator characteristics. In order to determine the operating point (the intersection of the two characteristics) the Newton Raphson algorithm proceeds in the following iterative way: at a starting point v_m, i_m (in SPICE this is usually the point with the highest gradient of bending of the diode's characteristic) a tangent to the diode characteristic is constructed and the intersection with the generator characteristic determined. As shown in fig. 10.3, the voltage value of the intersection defines the starting value for the next iteration. This procedure generally converges to the desired operating point. From fig. 10.3

$$\frac{i_{m+1} - i_m}{v_{m+1} - v_m} = \left.\frac{\mathrm{d}i}{\mathrm{d}v}\right|_m = \frac{I_s\,\mathrm{e}^{\frac{v_m}{V_T}}}{V_T} = G_m$$

Solved for i_{m+1}:

$$i_{m+1} = G_m v_{m+1} + i_m - G_m v_m$$

This equation can be interpreted as a linearized equivalent circuit diagram of a diode, which is shown in fig. 10.4.

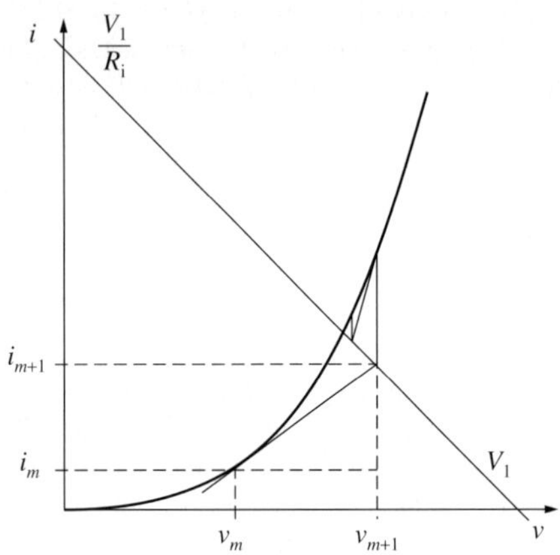

Fig. 10.3 Analysis of the circuit in fig. 10.2 by the Newton Raphson algorithm

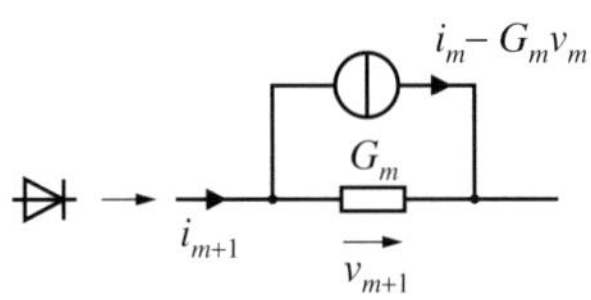

Fig. 10.4 Linearised equivalent circuit of a diode

The **analysis of a network containing diodes** is now carried out in the following way:

Each diode is replaced by its equivalent circuit diagram (fig. 10.4). For the point of highest grade of bending in the diode's characteristic the values i_0, v_0, G_0 are known, therefore an analysis of the linearized network can take place producing i_1, v_1, G_1. With these values a new analysis of the linear network is performed, and this procedure is repeated as long as the voltages across the diodes do not alter more than a given error voltage compared to the voltages of the previous iteration. Thus the analysis of nonlinear networks is reduced to an iterative analysis of linear equivalent networks. Sometimes an overflow of the allowed number range of the computer can occur owing to the exponential increase of the current with respect to the voltage in diode problems. Modifications of the Newton Raphson algorithm by so called 'simple limiting' in order to avoid overflow problems are described in [10.12].

Numerical Integration

The analysis of the transient behavior of electronic circuits is the most demanding and time consuming task for a simulation program. SPICE uses an **implicit integration method** for reasons of stability. The simplest example of an implicit method is the **backward Euler integration**. The procedure will be explained with a capacitor as an example. As is well known, the current and voltage of a capacitor are related in the way shown in fig. 10.5.

$$i = C\frac{\mathrm{d}v}{\mathrm{d}t}$$

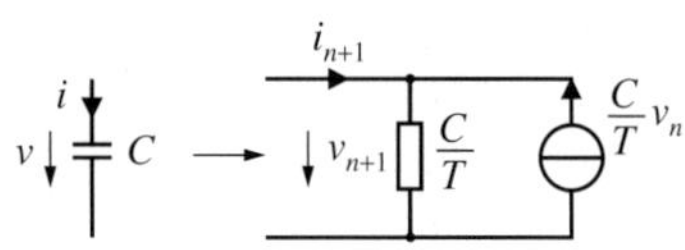

Fig. 10.5 Equivalent circuit of a capacitor

Figure 10.6 shows an assumed capacitor voltage function of time. The differential quotient for t_{n+1} is approximated by the difference quotient, i.e.,

$$\left.\frac{\mathrm{d}v}{\mathrm{d}t}\right|_{n+1} \approx \frac{v_{n+1} - v_n}{t_{n+1} - t_n} = \frac{v_{n+1} - v_n}{T}$$

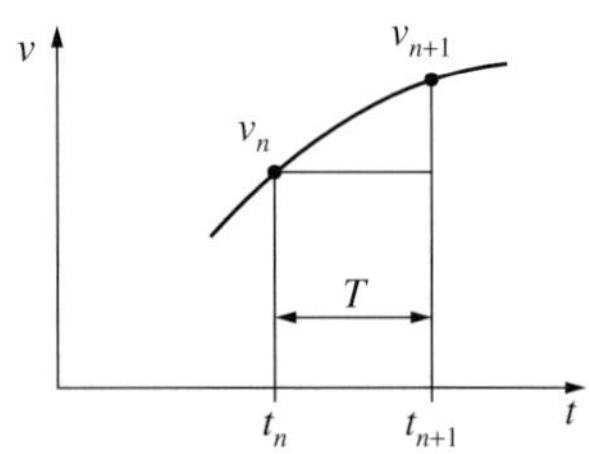

Fig. 10.6 Serves as explanation of implicit integration (see text)

Therefore the current through the capacitor at the timepoint t_{n+1} is

$$i_{n+1} = C\left.\frac{\mathrm{d}v}{\mathrm{d}t}\right|_{n+1} \approx \frac{C}{T}v_{n+1} - \frac{C}{T}v_n$$

This equation defines an equivalent circuit for the capacitor which is shown in fig. 10.5. By means of the equivalent network and the knowledge of the voltage at the time point t_n the voltage at the time point t_{n+1} can be determined, iterating this procedure until a given final time point is reached. Inductors can be handled in a similar way.

In [10.2] it is shown that the error connected with the backward Euler algorithm is dependent on the time step T in a linear way. SPICE uses **trapezoid integration**, an implicit algorithm in which the error is reduced quadratically when the time step decreases. A special problem is the **choice of the appropriate time step**, which is dependent on the time constants of the electrical circuit to be analyzed. These time constants can sometimes vary by several orders of magnitude. With a fixed time step one can either experience convergence problems or very high values of computer time if the time step is chosen according to the smallest time constant in the circuit. SPICE therefore has an automatic **time step control** which reduces the time step when steep slopes are involved and increases the time step during times of little activity. The intensive investigations of different integration methods and of different possibilities of time step control and the final optimal choice is certainly at least partly responsible for the success of SPICE.

10.1.5 The Program Structure of SPICE

Originally SPICE was written for the computer CDC 6400 and comprised 15,000 FORTRAN and assembler statements. The program consists of the root program, seven major segments, and a common memory region (Blank Common) which is shown in fig. 10.7. With reference to fig. 10.7 the necessary memory range is found by adding 16 K (octal) of the root to 6 K (octal) of the biggest segment (DCTRAN = 6 K (octal)) and for the Common. When the program starts, the memory assigned to the Common is 4 K (octal). Should it be registered during the course of analysis that this memory is not sufficient then the memory of the Common is extended up to the maximum possibilities of the computer. This so called dynamic memory extension is an additional advantage of SPICE.

The **tasks of the different segments** in fig. 10.7 are listed in the following:

- **READIN**. read in of the input and construction of element and node lists which serve for dynamic extension of memory;
- **ERRCHK** checks the input for plausibility, e.g., whether each node is connected with at least two elements. In addition, the model parameters are checked, the missing once replaced by default values;
- **SETUP** constructs the pointer system for processing the sparse matrix. Re-orders the matrix in order to minimize ‘fills’. Generates machine code for fast LU factorization and for forward and backward substitution;
- **DCTRAN** the biggest and most complicated segment. Controls Newton Raphson iterations when determining the operational point and the initial conditions for the transient analysis. Repeated DC analyses with altered inputs lead to DC transfer curves. Calculates TR analyses with automatic time step control;

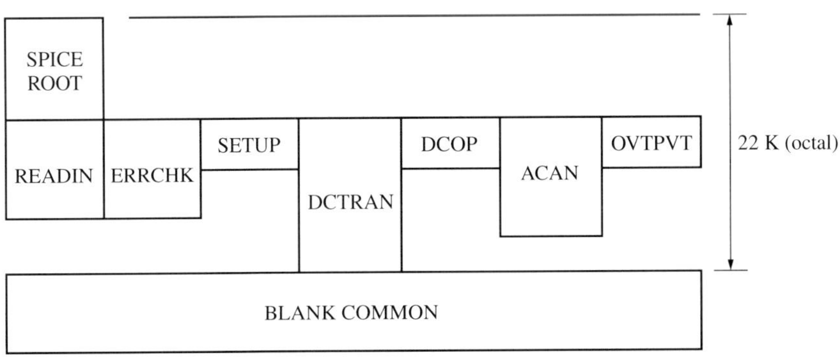

Fig. 10.7 Memory map of the program SPICE

- **DCOP** calculates linearized model parameters in the operating point. Controls sensitivity analysis;
- **ACAN** controls AC analysis of the (at the operational point) linearized network, also noise and distortion analysis;
- **OVTPVT**: prepares the output quantities, calls plot routines. Fourier analysis of a distorted periodical waveform is possible here.

In the meantime SPICE was re-written in 'C' at the University of California, Berkeley, and the program structure refined and redeveloped. For several years the company **Microsim** (now taken over by OrCAD) has been marketing a **PC version of SPICE** under the name **PSPICE or Design Center** [10.10], [10.18].

10.2 SPICE Transistor models

The main problem when simulating electronic circuits lies in the **modeling of semiconductor circuit elements**, especially transistors. One tries to map as precisely as possible the real circuit to the computer in order to avoid a very cost-intensive redesign of the integrated circuit. For this purpose it is not sufficient to signify a BJT by its current gain only or a MOSFET by its threshold voltage, but a lot of model parameters have to be used in order to catch the static, dynamic, and temperature behavior of transistors. The manufacturing companies of transistors or integrated circuits provide model parameters of the transistors produced or used, respectively, so by calling a transistor model by its component name much hidden information is activated. It is unsatisfactory for the users to apply these models without knowing the physical meaning of the different model parameters. There are, of course, many references in which the transistor model parameters are explained in detail (e.g., [10.8]). It does not make sense to repeat here all the details of transistor modeling, and in order to convey a fundamental idea of the complex topic the following effects will be omitted:

- Temperature effects;
- Bulk resistance effects;
- Dynamic effects (all transistor capacitances are assumed 0).

The last assumption is justified by technological progress. Transistor capacitances are nowadays of the order of fF, the capacitances of the connection lines on a chip are at least of the same order of magnitude. Therefore the dynamic behavior, i.e., mainly the delay, for digital circuits is nearly totally determined by the connections.

The following discussion is restricted to the Bipolar Junction Transistor (BJT) and the Metal oxide Field Effect Transistor (MOSFET) because these transistors are most important for modern integrated circuits. BJT and MOSFET will be introduced by using default model parameters in the evaluation version of PSPICE [10.5]. Even in the first version of SPICE sensible default parameters were inserted in case the user did not or could not specify parameters and thus the elementary properties of the transistor were correctly simulated. For illustration purposes the **large signal equivalent circuit** using the default parameters is shown. From that the **small signal equivalent circuit** is derived by differentiation at the operating point, because the latter is important for analog circuit design. The influence of a selected model parameter on the static behavior of transistors will be discussed. Dealing with MOSFETs the industrial standard **BSIM3v3-model** will be discussed by its characteristics and by its small signal parameters.

10.2.1 Bipolar Junction Transistor

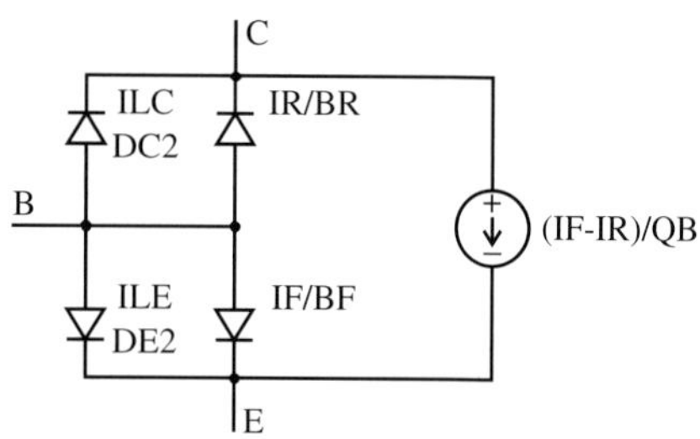

Fig. 10.8 Static large signal equivalent circuit of an npn transistor

Figure 10.8 shows the equivalent circuit for an **npn transistor** without capacitors and bulk resistances. Table 10.2 shows a list of respective static model parameters, SPICE names and SPICE default values [10.8]. The most important element in fig. 10.8 is the collector current source which is controlled by two diode currents:

$$I_F = I_S\left(e^{\frac{V_{BE}}{n_F V_T}} - 1\right) \tag{10.4}$$

Table 10.2 Static model parameters, SPICE names, and default values of the BJT

Model parameter	Symbol	SPICE name	Default
Ideal BJT			
Forward current gain	B_F	BF	100
Backward current gain	B_R	BR	1
Saturation current	I_S	IS	10^{-16} A
Forward emission coefficient	n_F	NF	1
Backward emission coefficient	n_R	NR	1
Early effect			
Forward Early voltage	V_{AF}	VAF	∞
Backward Early voltage	V_{AR}	VAR	∞
Recombination			
Saturation current of the non-ideal diode DE2	I_{SE}	ISE	0
Emission coefficient of DE2	n_E	NE	1,5
Saturation current of the non-ideal diode DC2	I_{SC}	ISC	0
Emission coefficient of DC2	n_C	NC	2
High current injection			
Forward knee current	I_{KF}	IKF	∞
Backward knee current	I_{KR}	IKR	∞
Exponent of high current injection	n_k	NK	0,5

$$I_R = I_S\left(e^{\frac{V_{BC}}{n_R V_T}} - 1\right) \quad (10.5)$$

The decrease of current gain at low currents caused by recombination in the junction is modeled by the two leakage diodes:

$$I_{LE} = I_{SE}\left(e^{\frac{V_{BE}}{n_E V_T}} - 1\right) \quad (10.6)$$

$$I_{LC} = I_{SC}\left(e^{\frac{V_{BC}}{n_C V_T}} - 1\right) \quad (10.7)$$

The normalized charge of majority carriers in the base zone is described approximately by the variable Q_B which is dependent of the two variables Q_1 and Q_2:

$$Q_B = \frac{Q_1}{2}\left[1 + (1 + 4Q_2)^{n_k}\right] \quad (10.8)$$

Q_1 takes into account the voltage dependence of the base width (Early effect):

$$Q_1 = \frac{1}{1 + \frac{V_{CB}}{V_{AF}} + \frac{V_{EB}}{V_{AR}}} \quad (10.9)$$

With increasing collector base voltage V_{CB} the base width, and therefore Q_1 or Q_B, respectively, will be smaller and the collector current will increase. The quantity

$$Q_2 = \frac{I_F}{I_{KF}} + \frac{I_R}{I_{KR}} \quad (10.10)$$

describes the increase of the charge of majority carriers in the base zone resulting from high current injection with which a decrease of current gain for high currents is associated. For small voltages and currents ($Q_2 \ll 1$) one finds $Q_1 \approx 1$ and $Q_B \approx 1$.

Default BJT

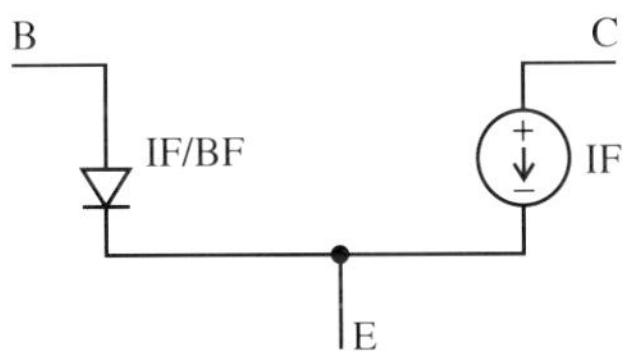

Fig. 10.9 Static large signal equivalent circuit of the default npn transistor in active forward operation

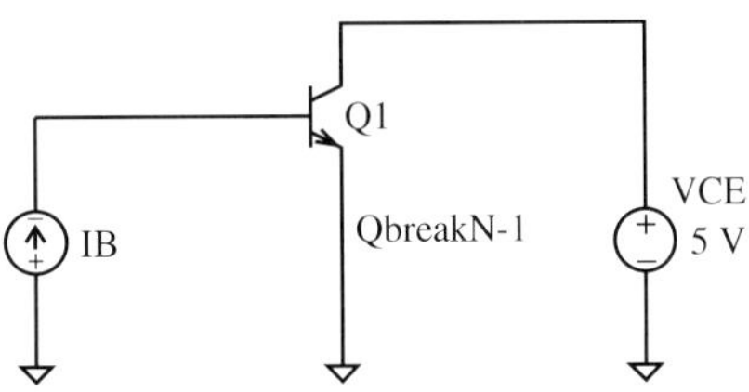

Fig. 10.10 Circuit for simulation of the family of output characteristics with parameters from Table 10.2

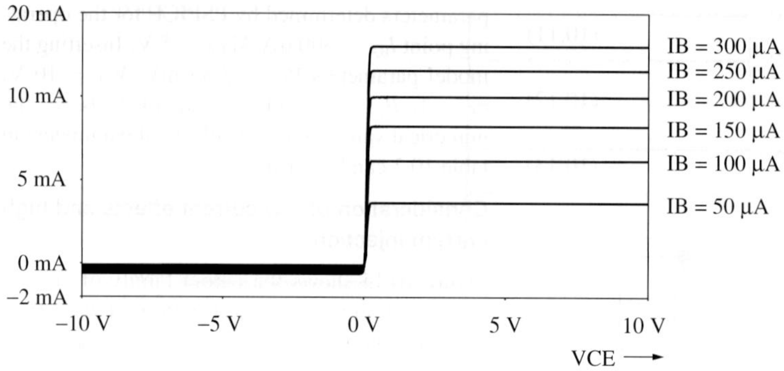

Fig. 10.15 Family of output characteristics of the npn transistor ($I_{KF} = 10$ mA)

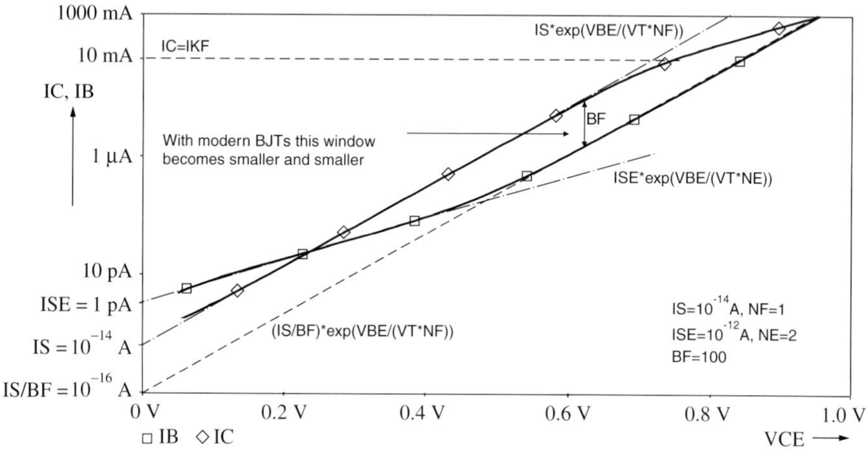

Fig. 10.16 Gummel plot

From fig. 10.16 one observes that the model parameter B_F (forward current gain) describes a constant relation between collector and base current only in a very restricted range which becomes even smaller for modern transistors. With higher collector currents the exponential character is lost owing to high current injection, with lower base currents the constant relation to the collector current is lost owing to recombination. Figure 10.17 shows the DC current gain $\beta_{DC} = i_C/i_B$ as a function of i_C.

This diagram was generated by means of PROBE, the graphical output program of PSPICE, using the same analysis values as in fig. 10.16. The asymptotic approximations were taken from [10.8].

Figure 10.17 shows how doubtful it is to signify a BJT by a single value for the current gain. Table 10.4 shows the parameters for the operating point $V_{BEA} = 0.7$ V, $V_{CEA} = 5$ V. Figure 10.18 shows β_{DC} and β_{AC} as a function for i_B of the transistor analyzed in fig. 10.16. The small signal current gain β_{AC} is calculated according to equation (10.14) by differentiation of i_C with respect to i_B. The cursor signifies the values of β_{DC} and β_{AC} for the operating point, which are consistent with the values shown in table 10.4.

Summarizing, it can be stated with respect to the **current gain**:

- For the default transistor:

  $\beta_{AC} = \beta_{DC} = B_F$;
- For the transistor with Early effect:

  $\beta_{AC} = \beta_{DC} \neq B_F$;
- For transistors in general:

  $\beta_{AC} \neq \beta_{DC} \neq B_F$.

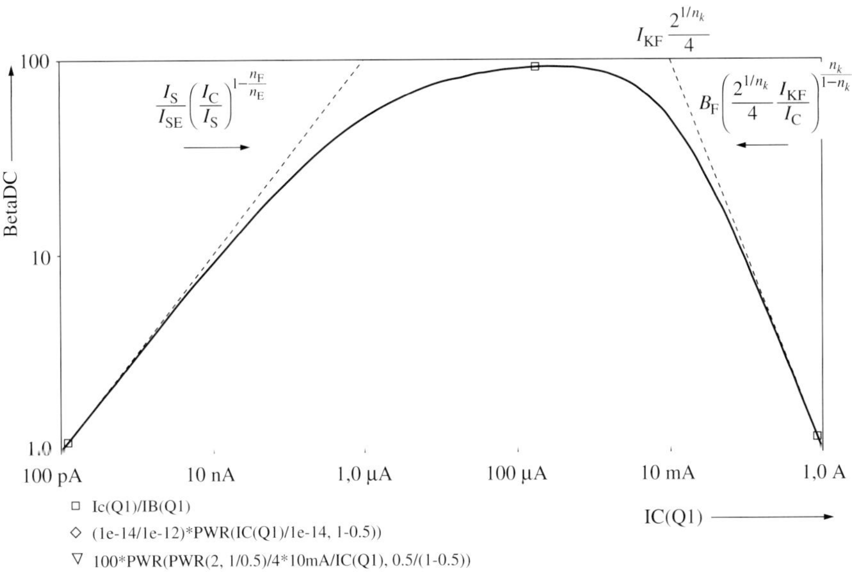

Fig. 10.17 Current gain β_{DC} as function of I_C

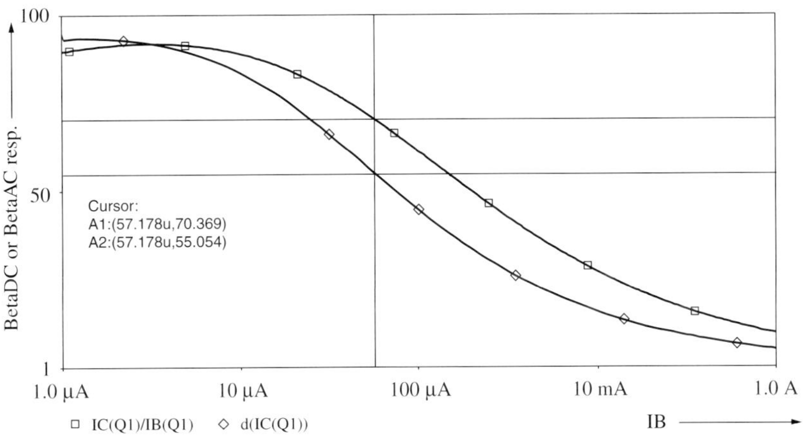

Fig. 10.18 Current gain β_{DC} and β_{AC} as function of I_C

10.2.2 MOSFET

There are several **MOSFET models** to be called up in SPICE. The parameter LEVEL signifies the kind of model to be applied [10.8]:

- **LEVEL = 1 (Default value)** signifies the physical 'first order model' (SHICHMAN, HODGES);
- **LEVEL = 2** signifies an extended **physical model** with additional equations and parameters mainly in order to take care of the effects caused by short channel length;
- **LEVEL = 3** signifies a **semi-empirical model** with parameters determined by adaptation to measured values. It is almost as powerful as LEVEL 2 but needs up to 40 % less computer time;
- **LEVEL = 4, 5, 6** signify today outdated intermediate versions of models for better simulation of **submicron technology** effects;

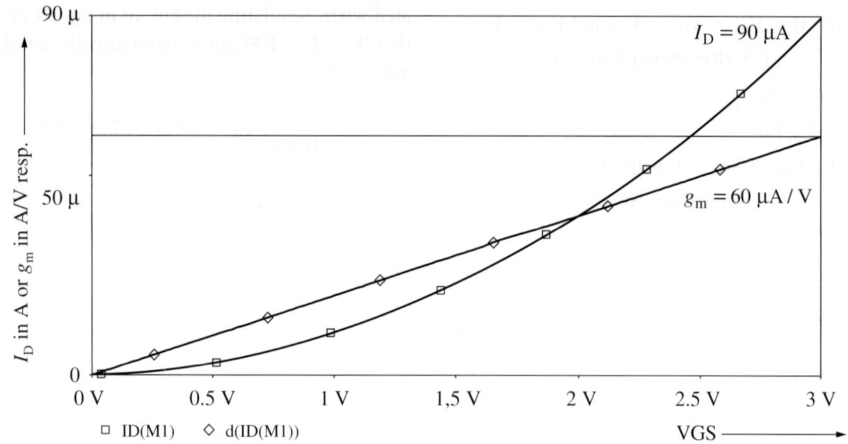

Fig. 10.23 Control characteristic of the default n-channel MOSFET and its derivative

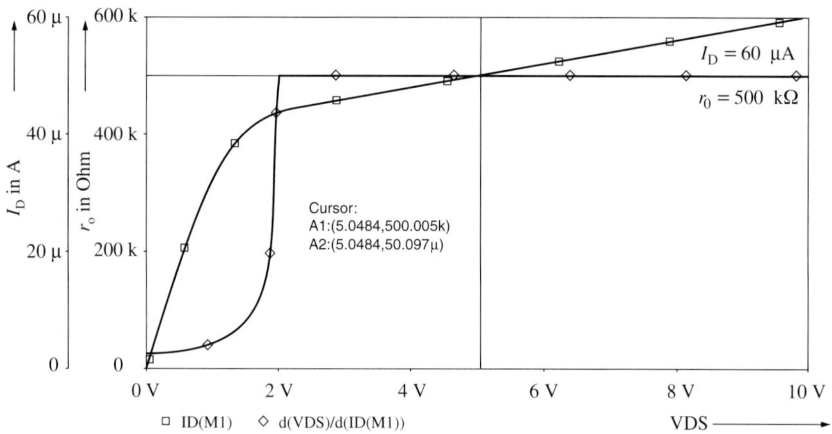

Fig. 10.24 Output characteristic and differential output resistance ($\lambda = 0.05$)

Consideration of channel length modulation parameter λ and bulk threshold voltage parameter γ

Table 10.8 lists the now valid model parameters of the MOSFET MbreakN-1 in fig. 10.21. Typical values according to [10.17] were chosen: $V_{T0} = 1$ V, $\gamma = 0.9$ V$^{1/2}$, $\lambda = 0.05$ V^{-1}. Figure 10.24 shows the output characteristic for $V_{GS} = 3$ V and the differential output resistance. Within the saturation range one finds $r_0 = 500$ kΩ. Figure 10.25 shows the control characteristic and its differential quotient whose linearity shows the quadratic character of the control characteristic. For the operating point ($V_{GS} = 3$ V, $I_D = 50\,\mu$A) $g_m = 50\,\mu$A/V.

Table 10.8 Model parameters for the circuit in fig. 10.21 ($VTO = 1$ V, $GAMMA = 0.9$ V$^{1/2}$, $LAMBDA = 0.05$ V^{-1})

MbreakN-X	NMOS
LEVEL	1
L	100.000000E-06
W	100.000000E-06
VTO	1
KP	20.000000E-06
GAMMA	0.9
PHI	0.6
LAMBDA	0.05

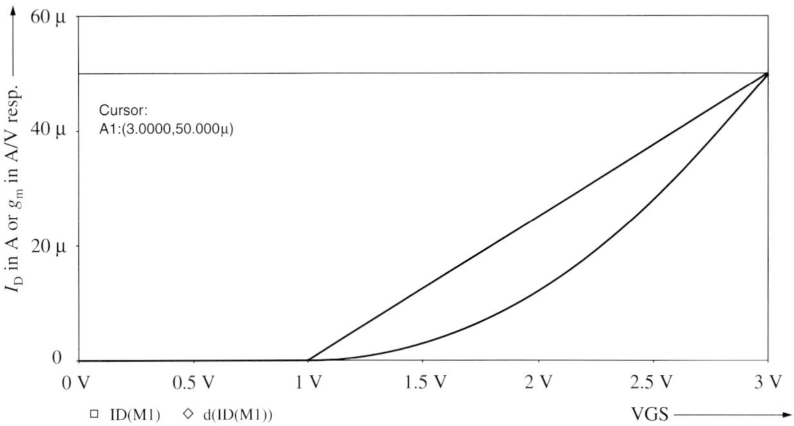

Fig. 10.25 Control characteristic and its derivative (VTO = 1 V)

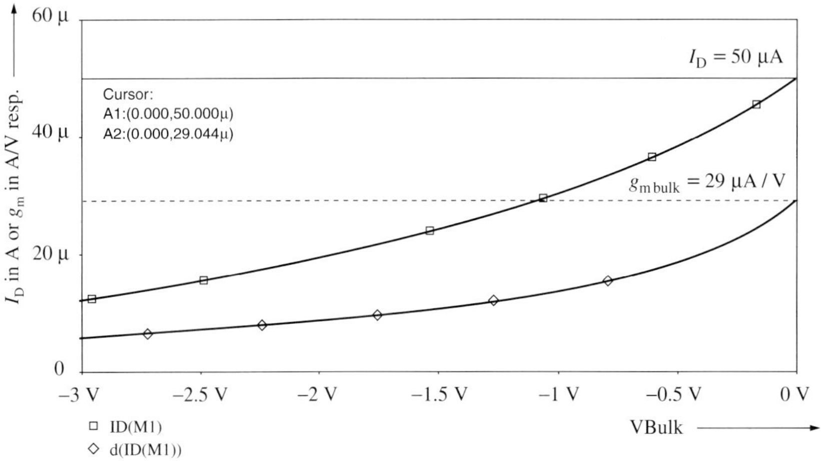

Fig. 10.26 Function of the drain current with respect to the bulk voltage and its derivative ($\gamma = 0.9$)

Figure 10.26 shows the influence of the bulk voltage on the drain current $I_D = f(V_{BS})$ and the corresponding differential quotient. The voltage V_{BS} has to be negative because the bulk diodes shown in the equivalent circuit (fig. 10.20) have to be reverse biased. In fig. 10.26 the cursor marks the value $g_{mb} = 29.044\ \mu A/V$. This value can be found in table 10.9 which lists the small signal parameters from the SPICE output file. The previously discussed values $g_m = 50\ \mu A/V$ and $g_{DS} = 1/r_0 = 1/(500\ k\Omega) = 2\ \mu S$ are also found. The corresponding small signal equivalent circuit is shown in fig. 10.27.

Table 10.9 Operating point and small signal parameters for the circuit in fig. 10.21 ($VTO = 1$ V, $GAMMA = 0.9\ V^{1/2}$, $LAMBDA = 0.05\ V^{-1}$)

NAME	M_M1
MODEL	MbreakN-X
ID	5.00E-05
VGS	3.00E+00
VDS	5.00E+00
VBS	0.00E+00
VTH	1.00E+00
VDSAT	2.00E+00
GM	5.00E-05
GDS	2.00E-06
GMB	2.90E-05

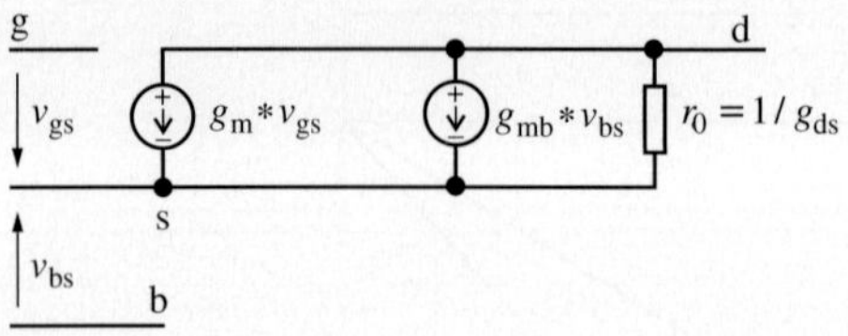

Fig. 10.27 Small signal equivalent circuit with reference to table 10.9

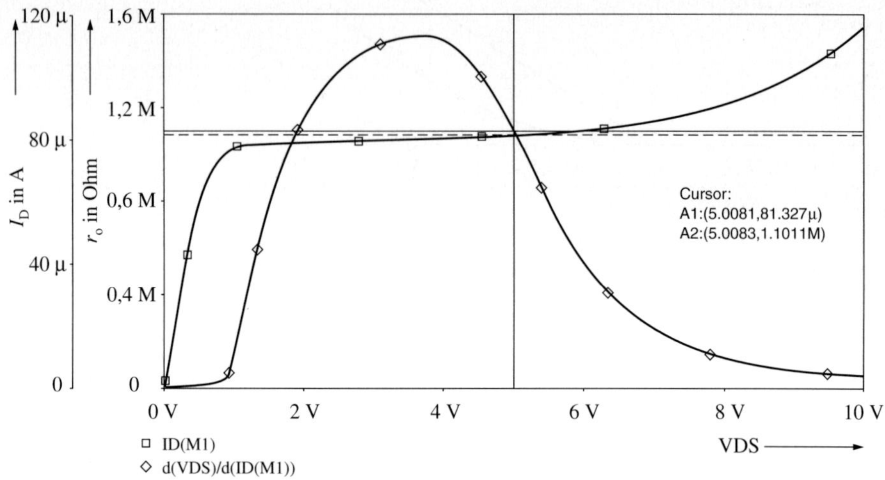

Fig. 10.28 Output characteristic and differential output resistance (BSIM3v3)

BSIM3v3 MOSFET Model

Table 10.10 lists the default model parameter to be found in the SPICE output file if the circuit shown in fig. 10.21 is simulated and the BSIM3v3 MOSFET model is used by defining the model parameter LEVEL = 7.

Table 10.10 Default model parameters of the BSIM3v3 MOSFET model (LEVEL = 7)

MbreakN-X	NMOS
LEVEL	7
L	100.000000E-06
W	100.000000E-06
VTO	0.7
KP	138.125800E-06
GAMMA	0
IS	1.000000E-15
JS	100.000000E-06
PB	1
PBSW	1
CJ	500.000000E-06
CJSW	500.000000E-12
CGSO	207.188600E-12
CGDO	207.188600E-12
TOX	15.000000E-09
XJ	150.000000E-09
DIOMOD	2
VFB	-1
U0	0.067
XPART	0
UC	-46.500000E-12
VSAT	80.000000E+03
VOFF	-0.08
A0	1
A1	0
A2	1
LDD	0
LITL	82.158390E-09
UC1	-56.000000E-12
DSUB	0.56
DLC	0
DWC	0
CF	72.989890E-12
NOIA	100.000000E+18
NOIB	50.000000E+03
NOIC	-1.400000E-12
VTM	0.025864

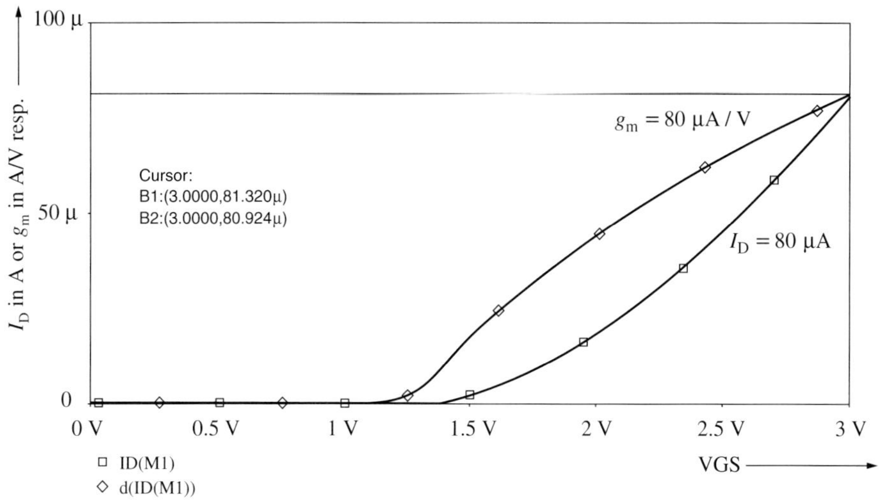

Fig. 10.29 Control characteristic and its derivative (BSIM3v3)

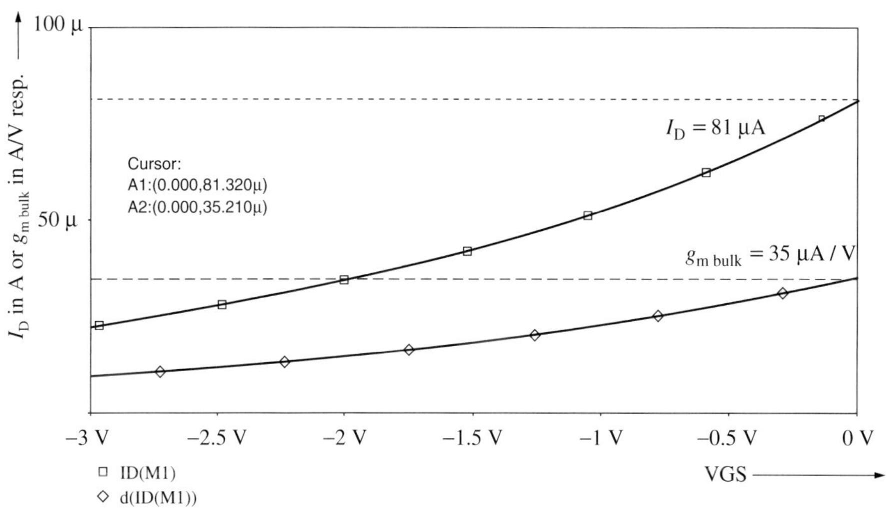

Fig. 10.30 Function of the drain current with respect to the bulk voltage and its derivative (BSIM3v3)

The same characteristics as before will be discussed in order to show the differences to the LEVEL 1 model. Figure 10.28 shows the output characteristics for $V_{GS} = 3$ V and the differential output resistance. The latter is (when dealing with the BSIM3v3 model, according to [10.8]) determined not only by channel length modulation but also by drain-induced lowering of barrier and bulk effect. The latter causes the exponential increase of the output characteristic with higher drain currents and therefore the lowering of the output resistance. Figure 10.29 shows the control characteristic and the corresponding differential quotient which is now not linear. Therefore one does not find a purely quadratic control characteristic anylonger, as with the LEVEL 1 model. Figure 10.30 shows the influence of the bulk voltage on the drain current and the corresponding differential quotient.

10

Table 10.11 Operating point and small signal parameter for the circuit in fig. 10.21 (BSIM3v3)

NAME	M_M1
MODEL	MbreakN-X
ID	8.13E-05
VGS	3.00E+00
VDS	5.00E+00
VBS	0.00E+00
VTH	1.27E+00
VDSAT	9.84E-01
GM	8.09E-05
GDS	9.04E-07
GMB	3.25E-05

Table 10.10 lists operating point and small signal parameter values which are marked by the cursor in the characteristics discussed above. One finds satisfactory correspondence. The small signal behavior is also described by the small signal equivalent circuit shown in fig. 10.27 whereas the small signal parameters of the BSIM3v3 model are determined in a much more complicated way, as before.

10.3 Models for Operational Amplifiers

10.3.1 Device Model

There are no principal difficulties to prepare a circuit diagram of an integrated operational amplifier for a SPICE input. Figure 10.31 shows for instance the circuit diagram of the classical μA741 operational amplifier [10.6] from which the SPICE input can be derived immediately.

Figure 10.31 is called a device model. The difficulty in applying such a model lies in the model parameters of the 20 to 30 transistors being not normally known and not being able to determined by measurements involving output nodes only. Even if the IC manufacturer provides all model parameters the number of transistors is the problem: if, for instance, filters consisting of several operational amplifiers are analysed one has to deal with a circuit of hundreds of transistors, which can result in convergence difficulties and excessive computing time.

10.3.2 ABM Models

Often it is not necessary to take all effects of an operational amplifier into account when interest is only in its analog behavior. The suitable models for this purpose are called 'Analog Behavioural Models' (ABM Models).

Ideal Operational Amplifier

To model an ideal operational amplifier a voltage controlled voltage source with a gain of, e.g., 200,000 can be used. Even with finite input and output resistances one speaks of an ideal operational amplifier which is shown in fig. 10.32.

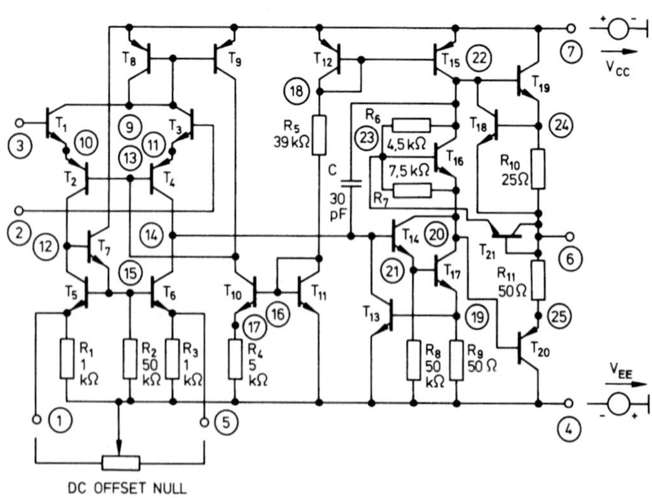

Fig. 10.31 Device model of the operational amplifier μA741

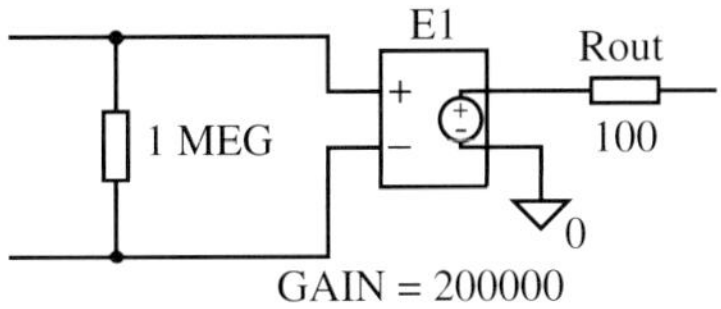

Fig. 10.32 Ideal operational amplifier (voltage controlled voltage source)

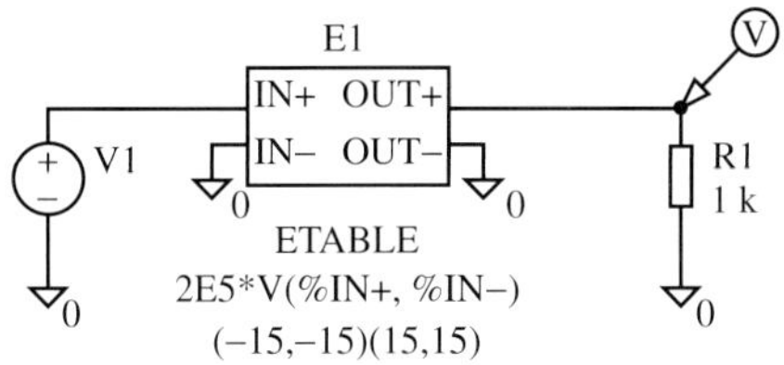

Fig. 10.33 Operational amplifier with saturation

Operational Amplifier with Saturation

If one wants to simulate oscillators using operational amplifiers one has to model saturation effects, otherwise the amplitude of the oscillating voltage would increase to infinity. Figure 10.33 shows the application of the ABM model 'ETABLE' for this purpose. By defining the values (−15, −15) (15, 15) saturation at the output to ±15 V is caused, as is shown in the simulated transfer curve fig. 10.34.

Operational Amplifier with Frequency Response

When simulating active filters the frequency response of the operational amplifiers used also has to be taken into account. For this purpose a convenient possibility is given when applying the ABM model 'ELAPLACE'. Figure 10.35 shows an example for the application where the Laplace transfer function is signified by two poles at 5 Hz and 2 MHz. Figure 10.36 shows the simulated frequency response with respect to magnitude and phase.

10

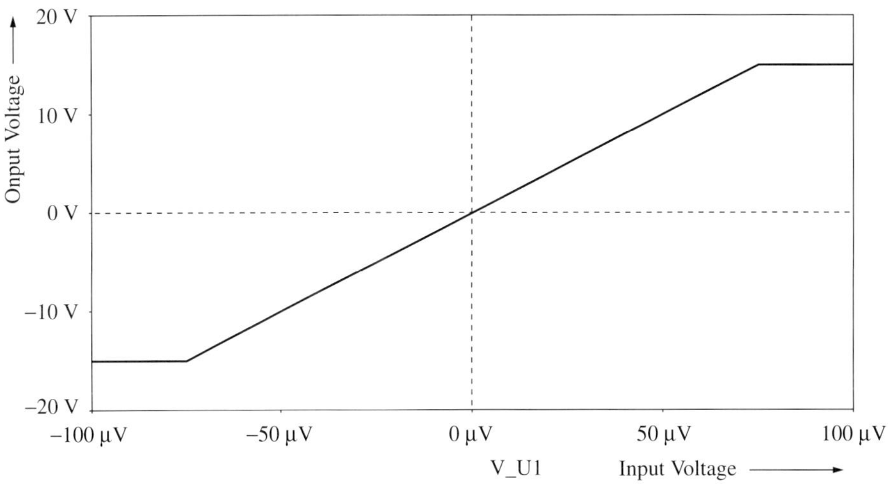

Fig. 10.34 DC transfer function for fig. 10.33

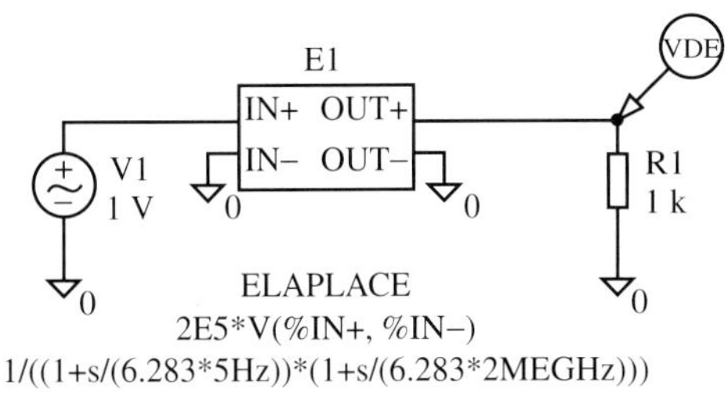

Fig. 10.35 ABM model of an operational amplifier with frequency response

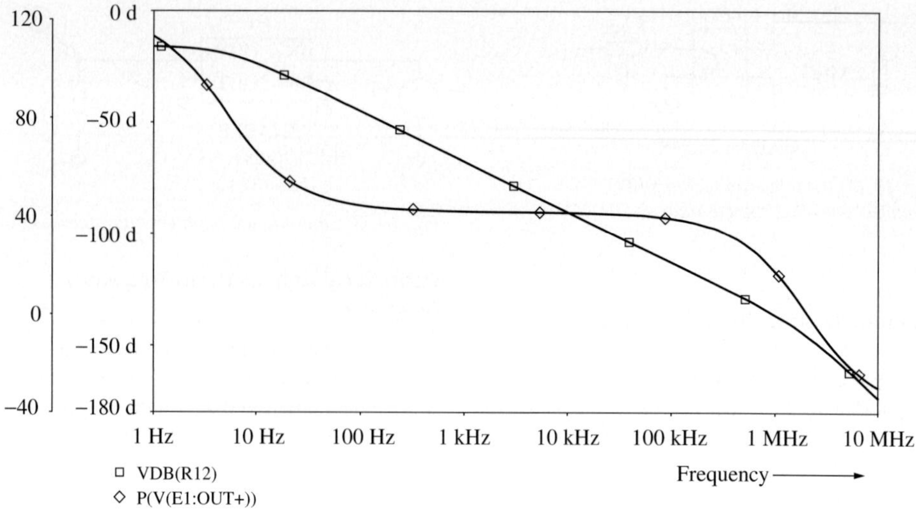

Fig. 10.36 Frequency response for fig. 10.35

10.3.3 Macro Model of the Operational Amplifier

Figure 10.37 shows two operational amplifier models with 100 % feedback paralleled at their inputs. The first is the ABM model 'ELAPLACE' discussed above, the second is the library model μA741 to be found in the PSPICE version 8 [10.5].

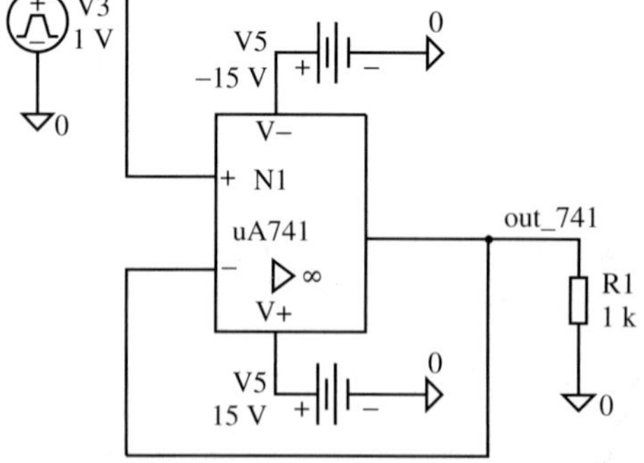

Fig. 10.37 Two different operating amplifier models paralleled

Figure 10.38 displays the frequency response with respect to magnitude of both models which show practically no difference.

If, on the other hand, the step response is of interest the differences are remarkable as is shown in fig. 10.39. The reason is to be found in the linear ABM model not being able to simulate the non-linear slew rate effect whereas this is possible with the library model of the μA741. This model is a so called macro model of the operational amplifier and represents a compromise between the device and the ABM model. The idea behind this is to simulate as many effects of the real operational amplifier as possible with a minimum of non-linear components. In addition, the model parameters may be determined from data sheet information of the manufacturer [10.6]. Listing 10.1 shows the SPICE netlist of the macro model μA741 to be found in the library EVAL.LIB of the PSPICE evaluation version 8.

Figure 10.40 shows the according circuit diagrams. Detailed information regarding the different components can be found in [10.13]. Summing up, the following may be said: the input behavior is simulated by the differential amplifier with the transistors Q1 and Q2. The other semiconductor elements are diodes which only serve to simulate saturation effects and do not affect the normal operation.

List 10.1 SPICE netlist for the macro model μA741

```
* connections:   non-inverting input
*                |  inverting input
*                |  |  positive power supply
*                |  |  |  negative power supply
*                |  |  |  |  output
*                |  |  |  |  |
.subckt uA741    1  2  3  4  5
*
  c1    11 12 8.661E-12
  c2     6  7 30.00E-12
  dc     5 53 dx
  de    54  5 dx
  dlp   90 91 dx
  dln   92 90 dx
  dp     4  3 dx
  egnd 99   0 poly(2) (3,0) (4,0) 0 .5 .5
  fb     7 99 poly(5) vb vc ve vlp vln 0 10.61E6  10E6 10E6 10E6  10E6
  ga     6  0 11 12 188.5E-6
  gcm    0  6 10 99 5.961E-9
  iee   10  4 dc 15.16E-6
  hlim 90   0 vlim 1K
  q1    11  2 13 qx
  q2    12  1 14 qx
  r2     6  9 100.0E3
  rc1    3 11 5.305E3
  rc2    3 12 5.305E3
  re1   13 10 1.836E3
  re2   14 10 1.836E3
  ree   10 99 13.19E6
  ro1    8  5 50
  ro2    7 99 100
  rp     3  4 18.16E3
  vb     9  0 dc 0
  vc     3 53 dc 1
  ve    54  4 dc 1
  vlim   7  8 dc 0
  vlp   91  0 dc 40
  vln    0 92 dc 40
.model dx D(Is=800.0E-18 Rs=1)
.model qx NPN(Is=800.0E-18 Bf=93.75)
.ends
```

The large gain of the operational amplifier is mainly represented by the current controlled current source FB, the controlled source GCM simulates the common mode gain. In [10.13] it was pointed out that there is an error in this kind of common mode simulation which is already to be found in the original paper [10.1] and in all PSPICE versions. It is wrong to take the common mode voltage from the resistor R_{EE} because this leads necessarily to an unwanted offset at the output as a result of a DC voltage drop in R_{EE} at the operating point. One has to derive the controlling voltage of the source GCM directly from the common mode voltage at the input, i.e., from the arithmetic mean value of the input voltages.

In order to illustrate the macro model the slew rate simulations will be discussed. A step function at the (+) input saturates the transistor Q2 and the current I_{EE} flows through $R_{C2} = R_{C1}$. Therefore $V_A = I_{EE} \cdot R_{C1}$ and the voltage controlled current source G_A charges the capacitor C_2 with I_{EE}. The slew rate is given by:

$$\frac{dV_{out}}{dt} = \frac{I_{EE}}{C_2} \approx \frac{15\,\mu\text{A}}{30\,\text{pF}} = 0.5\,\frac{\text{V}}{\mu\text{s}}$$

10

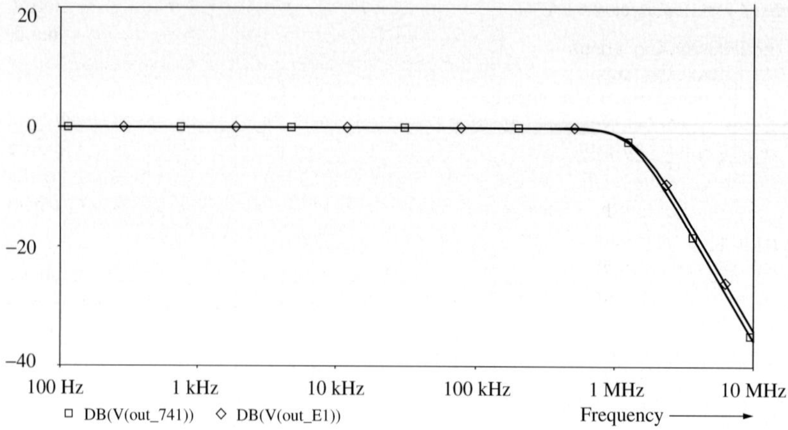

Fig. 10.38 Frequency responses for fig. 10.37

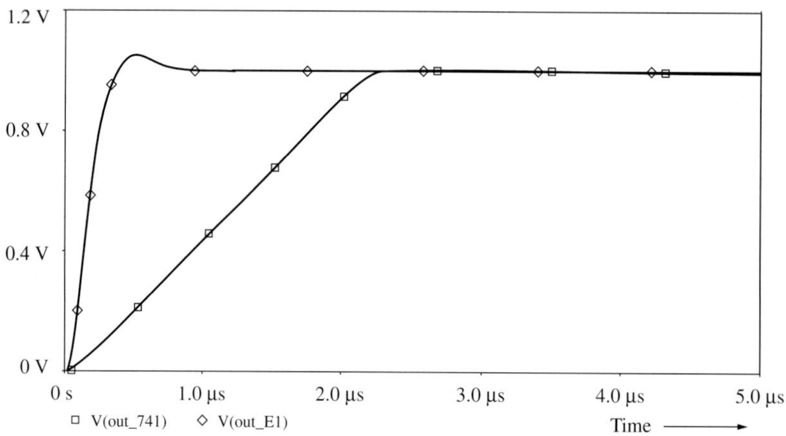

Fig. 10.39 Step responses for fig. 10.37

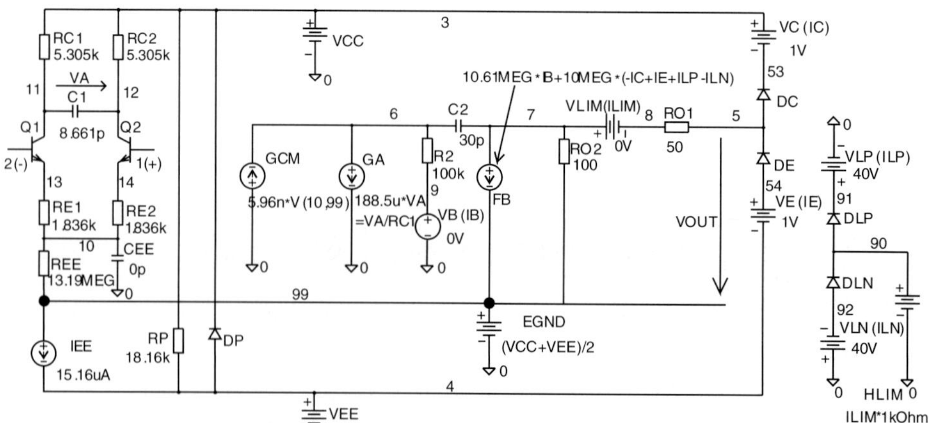

Fig. 10.40 Circuit diagram for the macro model µA741

10.4 Analysis of Loop Gain as Stability Criterion of Analog Circuits

In [10.17] the determination of the loop gain for stability analysis of analog circuits is discussed extensively. There it is necessary to open the circuit and to take care that by this opening the electrical conditions for the operating point and loading are maintained, which can be very difficult in some cases. Here an elegant method of determining loop gain will be discussed which does not need to open the circuit. It was first published in [10.20] and is not easily understood, repeated erroneously in a Microsim Application Note [10.9]. A didactically prepared presentation with the notation used here can be found in [10.14].

Figure 10.41 shows the circuit diagram of a logarithmic amplifier taken as an example for stability analysis. The additional AC voltage sources at the output are not disturbed because of their zero inner resistance as well as the AC current source to ground because of its infinite inner resistance. Assuming virtual ground at the inverting input of the operational amplifier with feedback one can consider at first sight the transistor Q_1 as a diode: the collector is virtually the base directly connected to ground. Therefore the output voltage can be taken as

$$V_a = -V_{BE} = -V_T \ln\left(\frac{I_1}{I_S}\right)$$

This result is verified by a *DC sweep analysis* $V_a = f(I_1)$ over many decades. If one defines a step function for I_1 ($0 \ldots 1\,\mu$A) and a TR analysis is performed, an immediately starting oscillation of $f = 200$ kHz will result. If, on the other hand, the transistor Q1 is, in fact, connected as a diode by disconnecting the base from ground and connecting it to the collector the step response analysis will lead to the expected result: after a rise time the output voltage is approximately equal to the expected base emitter voltage for $I_1 = 1\,\mu$A, where part of the current is the base current.

The explanation of the instability discussed above is given by the *loop gain analysis*, which is performed by means of the parametric AC analysis in PSPICE using the additional AC sources in fig. 10.41.

If the parameter 'VOLTS' = 0 the current source I_2 in parallel will send a current of 1 A into the system. It should be kept in mind that before an AC analysis is performed the circuit will be linearized so that saturation cannot occur. The resulting currents through the voltage sources V_1 and V_2 define a current transfer function

$$T_i = \frac{I(V_1)@1}{I(V_2)@1}$$

where the notation @1 signifies the first parameter value, i.e., VOLTS = 0. With VOLTS = 1 the current source is switched off and the voltage source V_1 in series will be effective with a voltage of 1 V. The thus resulting voltages at point A and B define a voltage transfer function

$$T_v = -\frac{V(A)@2}{V(B)@2}$$

where the notation @2 signifies the second parameter value, i.e., VOLTS = 1.

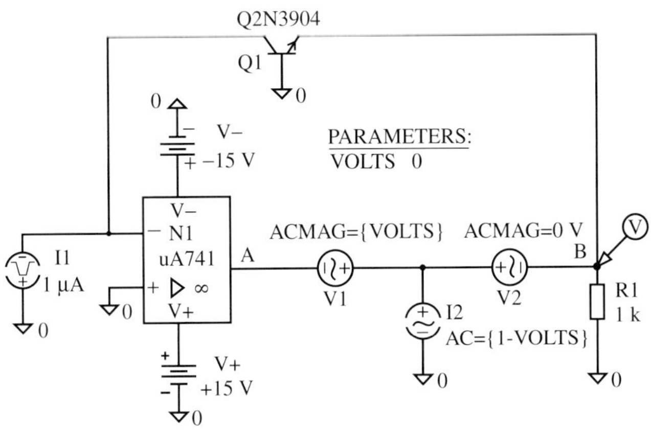

Fig. 10.41 Logarithmic amplifier

10

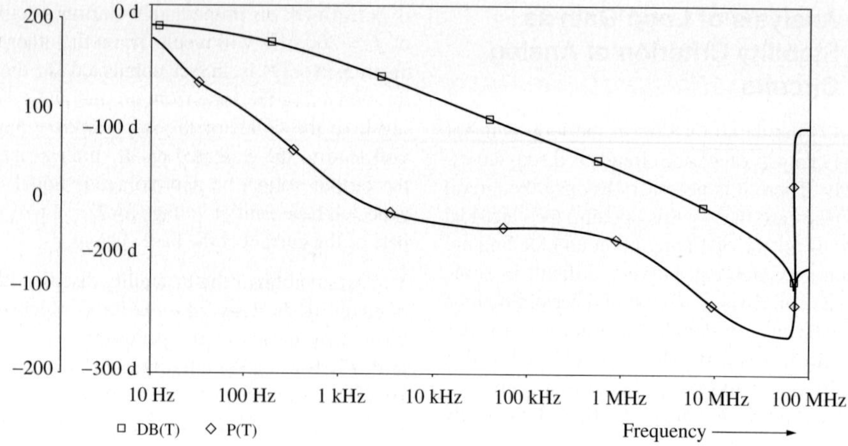

Fig. 10.42 Loop gain T with respect to magnitude and phase

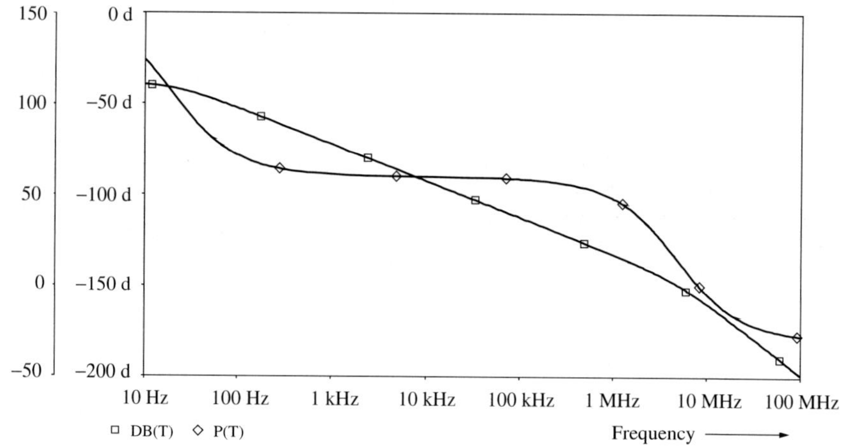

Fig. 10.43 Loop gain T with respect to magnitude and phase (transistor connected as diode)

In [10.14] it is shown that the *loop gain* T can be determined from the two transfer function discussed above by the formula

$$T = \frac{T_i T_v - 1}{2 + T_i + T_v}$$

which can be easily programmed as a so called *Macro* in the output program PROBE of PSPICE.

Figure 10.42 shows the thus determined *loop gain* T with respect to magnitude and phase. One realises that in a rather large frequency range the phase is approximately 180° and the magnitude (gain) much greater than 0 dB: the system must therefore be unstable. The physical explanation is given by the fact that the transistor as a common base circuit forms an additional amplifier and therefore alters the overall frequency response with respect to magnitude and phase. Table 10.12 lists values of the operating point ($I_1 = 1\ \mu\text{A}$) and the small signal parameters of transistor Q1.

From there the voltage gain v of the common base amplifier can be calculated:

$$v = g_m R_0 = 3.87 \cdot 10^{-5} \cdot 7.4 \cdot 10^7 = 2864 \rightarrow 69\ \text{dB}$$

and a low frequency pole at

$$f_p = \frac{1}{2\pi C_{BC} R_0} = \frac{1}{2\pi \cdot 3.64 \cdot 10^{-12} \cdot 7.4 \cdot 10^7}\ \text{Hz} = 591\ \text{Hz}$$

Table 10.12 Operating point and small signal parameters of the transistor Q1

NAME	Q_Q1
MODEL	Q2N3904
IB	2.32E-08
IC	1.00E-06
VBE	4.87E-01
VBC	1.99E-06
VCE	4.87E-01
BETADC	4.30E+01
GM	3.87E-05
RPI	1.36E+06
RX	1.00E+01
RO	7.40E+07
CBE	5.80E-12
CBC	3.64E-12
CBX	0.00E+00
CJS	0.00E+00
BETAAC	5.27E+01
FT	6.51E+05

With these values the frequency response shown in fig. 10.42 becomes plausible: the gain of 69 dB adds to the operational amplifier gain of 100 dB to approximately 170 dB. For frequencies > 1 kHz the additional pole causes an additional phase shift of 90° which adds to the phase shift of the operational amplifier of 90° in that frequency range. This example shows the possibilities of finding a deeper understanding of electronic circuits by means of simulation, for the concept of 'virtual ground' favoured by many over-practical orientated engineers has its shortcomings.

If the transistor Q1 in fig. 10.41 is connected as a diode, i.e., its base is disconnected from ground and connected to the collector, then the loop gain analysis will result in the frequency response shown in fig. 10.42. With this frequency response the circuit will be stable because the phase is well above 180° when the gain reaches 0 dB.

Figure 10.43 represents nothing else because the open loop gain of the operational amplifier itself, as one finds at the input a current source with infinite inner resistance, and the small signal equivalent series resistance of the diode is negligible compared to the high input resistance of the operational amplifier. Therefore the current transfer function T_i becomes very large and the loop gain $T \approx T_v$.

10

10.5 References

The following list of literature contains only sources which were used for this chapter. In addition there are an immense number of books and papers on SPICE or PSPICE, respectively.

As a first approach to PSPICE [10.5] is recommended. It leads in a didactically excellent way to first experiences of success and explains the program handling extensively.

For deeper understanding of semiconductor modelling [10.8] is recommended. The influence of static and dynamic model parameters on the performance of all technically interesting transistors is explained and discussed in detail.

[10.1] *Boyle, G. R. et al.*: Macromodeling of Integrated Circuit Operational Amplifiers. IEEE Journal of Solid-State Circuits, Vol. SC-9, No. 6, Dec. 1974, p. 353–363

[10.2] *Calahan, D. A.*: Rechnergestützter Schaltungsentwurf. – München: Oldenbourg Verlag, 1973

[10.3] *Cheng, Y. et al.*: BSIM3v3 Manual. Dept. of Electrical Engineering and Computer Science, University of California, Berkeley 1995, 96. http://www.nikola.com/bsim3v3.htm

[10.4] *Frank, W.; Thürmer, B.; Nielinger, H.*: Auf Polar(is)-Expedition. Simulation von gekoppelten Leitungen auf gedruckten Schaltungen. Elektronik 19, 1995

[10.5] *Heinemann, R.*: PSPICE. Elektroniksimulation. Lehrgang, Handbuch, Kochbuch. – München; Wien: Carl Hanser Verlag, 1998

[10.6] *Hoefer, E. E. E.; Nielinger, H.*: SPICE. Analyseprogramm für elektronische Schaltungen. – Berlin: Springer Verlag, 1985

[10.7] Website of Hewlett Packard: http://www.tmo.hp.com

[10.8] *Khakzar, H. et al.*: Entwurf und Simulation von Halbleiterschaltungen mit PSPICE. – 3. Auflage. – Ramingen-Mahmsheim: Expert Verlag, 1997

[10.9] o. Verf.: Simulating High-Q Circuits Using Open Loop Response. Tutorial, Application Notes and Design Ideas. Vers. 6.2, Microsim Corp., Irvine, CA, April 1995

[10.10] o.Verf.: Microsim PSPICE A/D Reference Manual. Microsim Corp., Irvine, CA, 1996

[10.11] *Nagel, L. W.; Pederson, D. O.*: SPICE. University of California, Berkeley, ERL-M382, 1973

[10.12] *Nagel, L. W.*: SPICE2: A Computer Program to Simulate Semiconductor Circuits. UC Berkeley, ERL-520, 1975

[10.13] *Nielinger, H.*: OP-Makromodell für korrekte Simulation des Gleichtaktverhaltens. Elektronik 11, 1989

[10.14] *Nielinger, H.*: Loop gain mit PSpice analysieren. Elektronik 10, 1999

[10.15] *Reiß, K.*: Integrierte Digitalbausteine, Kap. 8. – München: Siemens AG, 1970

[10.16] *Renk, K. D.; Steinkopf, U.*: Programme zur Analyse elektrischer Schaltungen – eine vergleichende Übersicht. NTZ 26, 1973

[10.17] *Roberts, G. W.; Sedra, A. S.*: SPICE for Microelectronic Circuits. – Orlando: Saunders College Publishing, Fort Worth, 1992

[10.18] *Santen, M.*: PSPICE Design Center. Arbeitsbuch. – Karlsruhe: Fächer Verlag, 1994

[10.19] *Sedra, A. S.; Smith, K. C.*: Microelectronic Circuits. – 3rd ed. – Orlando: Saunders College Publishing, Fort Worth, 1991

[10.20] *Tuinenga, P. W.*: SPICE – A Guide to Circuit Simulation and Analysis Using PSpice. (p. 59–65). – Englewood Cliffs: Prentice Hall, Int. Ed., 1988

[10.21] Website der IEEE 1076.1 Working Group: http://www.vhdl.org/analog/

11 Digital Simulation

Ermenfried Prochaska

ICs accommodate millions of transistors in a single device. Therefore it is a necessity to identify faults in the design process at the time when they are implemented. Faults already may have been introduced in the process of specification, in the implementation into digital logic, or even in the models used for simulation of the circuit.

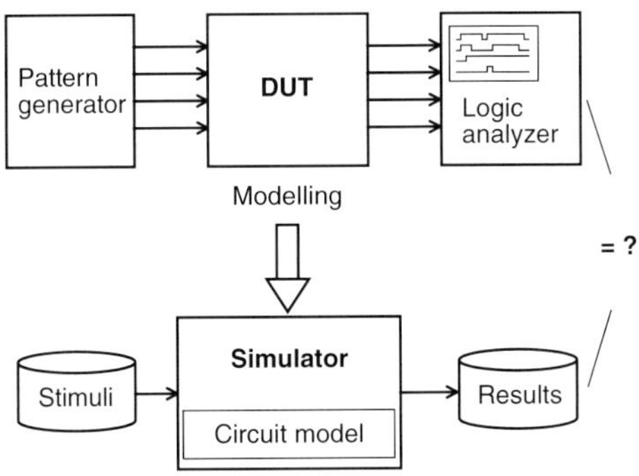

Fig. 11.1 Digital Simulation and ASIC Development

In designing digital circuits one of the most important tools for a designer is a simulator. A simulator allows the verification of a circuit (DUT, *device under test*) after different steps of development at gate level. However, one must be aware that every simulator is based on models. The result of a simulation is only as accurate as the model used.

Apart from verifying the intended function of a digital circuit, the designer must take into account the efficiency of his stimuli for the production test of his circuit. With special programs fault simulation will deliver the fault coverage of a given test program.

In the following we will use the terms *functional simulation, pre-layout simulation, post-layout simulation*, and *fault simulation*. Functional simulation applies for the simulation of the logic which the designer uses as source code. After optimizing the source code *pre-layout simulation* provides the reaction of the optimized logic. Consequently after mapping the logic to the target device, *post-layout simulation* enables the designer to simulate the actual logic of the IC used, which may differ from the original logic of the source code, owing to the logic modules available in the specific technology of the target device.

All logic circuits in this chapter have the prerequisite that the design engineer agrees with the goal

that his circuit is testable.

Expressed in figures, it would mean that the fault coverage *is about 98 %*. This in turn means that testability must be designed from the very beginning of the design phase. Additionally, because of the multiple chances of faults and the multitude of their effects in a complex circuit, the test pattern for production test must be considered. Only if the circuits produced react correctly to the stimuli will the customer accept the ICs.

11.1 Digital Simulation: Why?

One of the main benefits of digital simulation for the development engineer is the verification of his design right from the start of his project. In the following the power of digital simulation is demonstrated by considering the different stages of design process.

Digital simulation for design verification

Independently of the language of the source program, every circuit in the design project is translated into digital logic, which in turn is implemented in an IC. Digital simulation enables the design engineer to verify the function of his design at every stage. However, he has to provide stimuli to excite the circuit. Digital simulation documents the implemented function of the circuit with the related test pattern.

Digital simulators have to simulate the behavior of every circuit, e.g., a faulty circuit, as closely to reality as possible. In particular, the following faults should show up:

- Unwanted feedback;
- Oscillation;
- Asynchronous circuits;
- Spikes, hazards.

Considering costs, the number of stimuli (test vectors) should be kept as small as possible. The state of the art is to download 200 k test vectors in one process without reloading.

Finally, the designer verifies and documents the function of his circuit using digital simulation.

Fault simulation to proof the effectiveness of stimuli

The development engineer is responsible for the test of his circuit. Therefore, it is necessary to know about the effectiveness of test stimuli. Fault simulation initially was developed for the evaluation of production test patterns. Manufacturing faults were modeled as hard faults. The fault simulator simulates the faulty circuit once for every fault. The result is the number of faults detected by the test stimuli.

With fault simulation it is documented how many of the inserted faults are detected by the test vectors specified. In practice it was shown that the development engineer receives additional information by the fault simulation results, which lead to a deeper understanding of the behavior of the circuit. This in turn may lead to the recognition of development faults. Certainly, fault simulation detects simple faults such as open pins of a circuit. Fault simulation checks some aspects of consistency of the design. One important aspect which fault simulation discovers is which nodes of a circuit do not change their values during the complete testing process. And, as mentioned before, which faults are not detected with the given test vectors. This may be owing to an insufficient number of test vectors or may be a result of lacking testability of the circuit. It may, too, indirectly lead to the recognition of a specific function of the circuit being blocked under specific conditions. Oscillating parts are detected as well (section 11.6).

All development platforms provide specific programs for fault simulation.

Requirements which must be met by the design engineer

Testable digital design demands a systematic hierarchical structure of testable sub-modules. The fault coverage should be around 98 %. To produce an electronic product efficiently the production engineering unit will ask for testable products.

Modules consisting of combinatorial circuits must be constructed to have a fault coverage of 100 %. If this is not possible the modules must at least be reorganized to an appropriate size. With this approach test programs of modules are re-used for testing the complete circuit. Both development time and cost are thus reduced considerably.

The *Boundary Scan Test* is one approach to allowing easy access to internal nodes of a circuit via a serial link. Nodes may be set to specific values. Values of specific nodes can be accessed for further use. The 4 line port used for this purpose is TAP (*Test Access Port*). IEEE Standard 1149.1 *Test Port And Boundary Scan Architecture* specifies TAP. There are differences between 1149.1 and the JTAG (*Joint Test Action Group*) Standard which preceded 1149.1. In spite of additional costs for more complex hardware within an IC, the cost of testing (test vectors, test time) decrease which reduces over all costs considerably.

11.2 Integrated Circuits and Simulation Model

Digital simulation uses a model of the circuit which the designer has specified with his source code. This model uses the parameters of the target device. Examples for such parameters are gate delay, specific flip flops of the target technology, or the capacity of interconnections. If a development environment is used, all parameters are stored in a specific library. This parameter library for components is different for distinct brands and depends on specific technologies.

11.3 SDF Format for Standardized Digital Models

There are various activities for creating libraries which are not specific to a manufacturer. The goal of the initiative is to standardize and increase the quality of models for cell libraries concerning ASICS, independent of IC manufacturers.The objectives of the VITAL (VHDL Initiative Towards ASIC Libraries) initiative can be summed up in one sentence:

> Accelerate the development of sign off quality ASIC macrocell simulation libraries written in VHDL by leveraging existing methodologies of model development.

In 1992 there existed only a few ASIC cell libraries. The main problem was to specify the timing parameters in a standard way. VHDL was the language for source code chosen by VITAL. The *VITAL ASIC Modeling Specification* [11.8] was created. It allows the modelling of timing behavior of logic circuits in VHDL in a standardized and effective way. Thus standardized circuit models can be used in simulators of different vendors. They may be parameterized for diverse technologies of different vendors. VITAL uses constructs of VERILOG *Language* which is already an industry standard.

Details of VITAL files will be explained in chapter 19. Properties of models are interchanged in *Standard Delay Format* (SDF), which allows the timing behavior of logic circuits to be described.

The behavior of integrated circuits is determined by its logic and the varying parameters of production. However, the specifications of semiconductor manufacturers are considered to be conservative. In most cases this leads to the function verified via simulation being the function of the real hardware, provided that the parameters are used correctly.

The organization *Open Verilog International* (OVI) published its first Standard SDF in 1993. A newer version will become IEEE Standard 1497.

The definition of the SDF format is determined by the software tools used by the designer. Thus the information interchange between different development platforms becomes possible. The SDF format uses ASCII code which is widely generally accepted.

The SDF file is created by the *Timing Calculator* (fig. 11.2), which is part of the software tool used.

The SDF file contains only timing data. The timing behavior of the single logic elements is not described in the SDF file. It is in the library with cell timing models. In the simulator logic modules are analyzed and the specific delay times are calculated accordingly. Minimal, typical, and maximum delays are taken into account as well as the differences between rising and falling edges, provided that these parameters are specified in the timing models.

The *Timing Calculator* creates the SDF file. The *Timing Calculator* accesses models for simulation of the timing for interconnections. The parameters used for the timing model are evaluated. The result is on one hand the pin to pin delay and on the other hand the distributed delay times of every single path. The delay times may be calculated in the pre-layout phase or in the post-layout phase of the design. Either expected layout parameters or parameters of the actual layout (*Back Annotation*) can be used.

If there is a design change the SDF file must be generated again because it is directly based on data of the design.

The *Annotator* is the program which has to match data in the SDF file with the design. It may be part of the simulator or part of the development tool. Each region in the design identified in the SDF file must be located and its timing model be found. The values of the SDF file will be inserted into the models.

In addition to the timing information from back annotation forward annotation is also supported by SDF. Using forward annotation the design engineer may specify timing constraints, e.g., for the data path. The timing constraints will be observed in the development tools used later in the design process (fig. 11.3). Floor planning, layout, or routing may use this information. The result will be a circuit meeting its specification or a circuit with parts which are malfunctioning.

Specific behavior of logic elements, such as masking of a short pulse at the input which does not propagate to the output (*Output Pulse Propagation*) may be specified in a SDF file. In this case a limit can be specified, up to which a short impulse

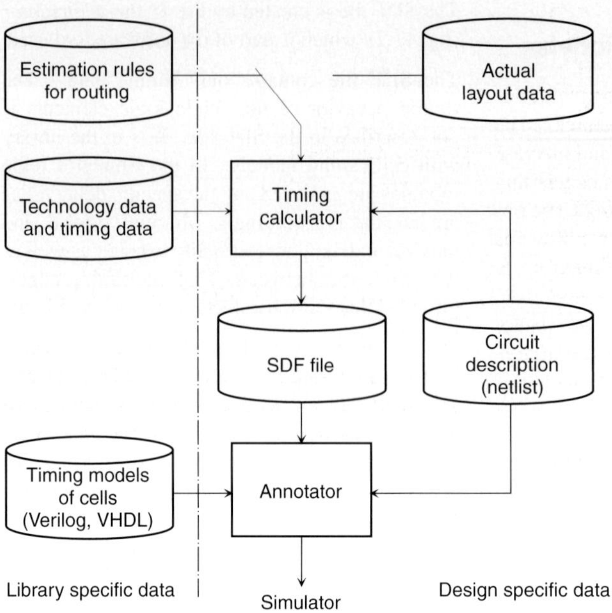

Fig. 11.2 SDF Files in Timing Back Annotation

will show up with its true logic value at the cell output.

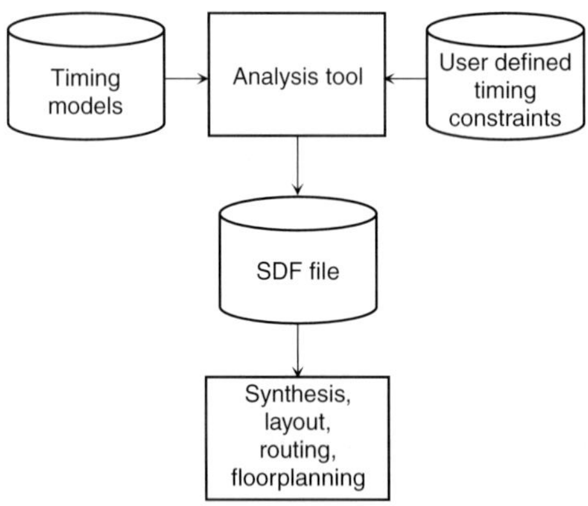

Fig. 11.3 SDF Files in Constraint Forward-Annotation

Modeling Timing Checks, such as setup, hold, and recovery timing checks are supported by SDF. The SDF file is therefore an important interface for circuit design. It will gradually gain importance if standardization of SDF continues.

11.4 Structure of a Digital Simulator

Every powerful development environment (*workbench*) for ASICs or FPGAs uses simulators. They are indispensable for circuit design at the complex level which integrated circuits of modern technologies. In this case the simulator already has the appropriate interfaces to the other development tools of the workbench. If the simulator needs data for timing simulation of the circuit, they are extracted from the layout and processed for the use of the simulator. If a simulator of a different development environment is used, the specific data must be adapted using pre or post processors.

11.4.1 Simulator for Logic Circuits for Verification of the Design

A simulator for logic circuits is an interactive program which uses a model of the circuit designed.

The circuit description, assumed to be in HDL (*hardware description language*), is the source for building the circuit specific simulation model (fig. 11.4). Logic elements are in the model library which is specific to the technology used

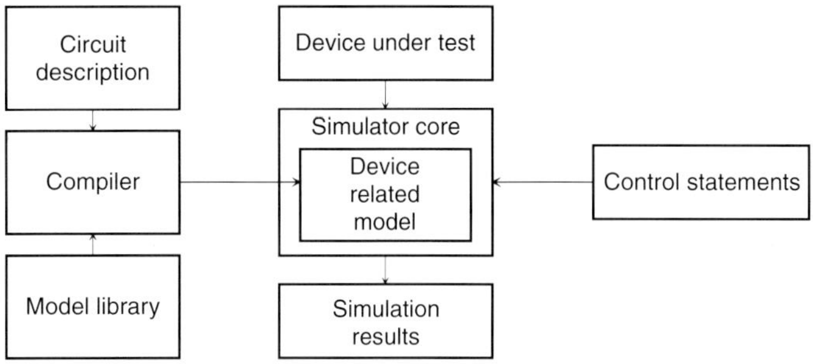

Fig. 11.4 Setup of a simulation system for digital simulation

to implement the ASIC. This task is performed by a program with similarities to a compiler for programming languages. It is necessary to build a specific model for simulation of a circuit. A universal simulator, which has to be tailored to simulate the behavior of a specific digital circuit, would use too much memory and its simulation speed would be too slow.

The kernel of the simulator calculates the reaction of the circuit according to the stimuli created by the designer. Depending on the data available for the logic model, different results will be obtained. Determined by control instructions, a simple functional simulation with unit time delay for each logic gate or flip flop can be executed. If timing data are available, timing simulation is able to produce results close to the real circuit.

The huge number of gates in an ASIC requires a decision by the designer about which part of his system he wants to simulate and at which level of accuracy in the development cycle. The simulation time depends greatly on the accuracy of the models and the complexity of the circuit. The cost for generating test stimuli and simulation time must be appropriate to the goal of verifying the intended function of the circuit. This trade off is difficult to evaluate in different design phases.

To achieve testability an ASIC must be designed in modules from the very beginning. The pretested modules are combined into a system whose function also has to be verified. The input pattern for testing, the stimuli, are defined as programs in a specific language or may be designed and modified using a graphic interface. There are tools for automatic generation stimuli.

The simulator calculates the reaction of the circuit to the test stimuli. The designer controls the type of simulation to be performed. He decides which part of stimuli is used and, in particular, which signals are displayed and documented as results. The user may specify any external or internal signal of the circuit to be displayed on the screen. For debugging it is advisable to select only relevant signals.

A difficult task is to evaluate the results of simulation and to decide if there are no faults within the design. The designer knows the function to be implemented, he also defines the test stimuli, and therefore he can check the simulated circuit reaction. If there are data available for direct comparison, from a description of the function in a different language like VHDL, both results can be compared automatically on the workbench.

11.4.2 Definition of Logic Levels (Logic Values)

Both terms logic level and logic values are considered equivalent.

The high number of gates in digital circuits demands in comparison with analog simulation a reduction of the accuracy for the logic values used in a simulator. The number of different logic values processed in a simulator must be adjusted to the necessities of different stages of the development cycle. The simulation time of circuits can be shortened considerably by reducing the number of logic values simulated. However, the designer has to assess that the simulation result still shows correct reactions of the circuit to be verified.

The simplest simulator for digital circuits uses only two different values, namely **0** and **1**. 0 represents the range of voltage recognized by a specific tech-

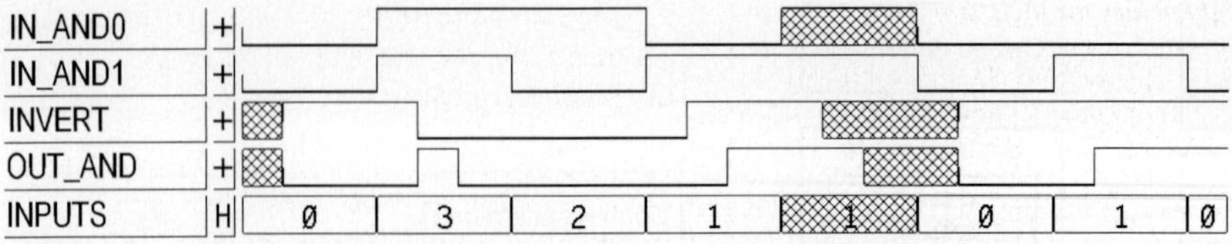

Fig. 11.5 Timing diagram using three-values logic system and unit time delay

nology as a logic 0. If a TTL-compatible circuit is used this range is from 0 V to 0.8 V.

A simulator which uses a two-value logic system simplifies every output of a logic gate and every state of a flip flop to a 1 or a 0. It is obvious that for a circuit with flip flops this is an insufficient description. After 'power on' the output of a flip flop may be 0 or 1 which can not be described appropriately.

A three-value logic system has the additional value X (0 or 1). A signal having a value X means that its logic value is unknown. It may be 0 or it may be 1. This value is shown graphically in simulators as a patterned area (fig. 11.5). Depending on the logic function an X may propagate to the output of a gate. In the literature different terms are used for this logic value. U (*unknown logic value*) is also used as a different term instead of X. The value U alternatively designates an uninitialized value in literature. For the user the logic system of a simulator is explained in detail with the expressions used in the relevant manuals.

X describes the behavior of hardware more detailed than only 0 and 1. But there are small differences. If a flip flop with two outputs q and /q is inspected, both outputs are X after power on. In a real circuit /q is the inverse q. The value X does not contain this information.

In simulation it is useful to assign a unit time delay for every gate or flip flop. The execution time is greatly reduced compared to simulators using individual time specifications. At an early stage of development simple faults such as missing variables or a wrong logic definition are recognized by using three-value logic and unit time delay simulation.

If more complex structures are analyzed, the value Z (high impedance) allows the simulation of the behavior of bus structures. The value Z may be either a logic 0 or a logic 1, or neither a 0 nor a 1. It will be processed in the simulator to evaluate the reaction of the circuit. It represents a bus which in reality is not driven by any circuit. It may represent any voltage in the range of a given system between 0 and V_{cc}. Naturally simulation time increases if four-level logic system is used in a simulator.

An example of the use of a six-value logic system is the Design Centre (PSPICE), an industry standard for simulation of analog circuits. The six logic values are given in table 11.1.

Table 11.1 Six-value logic system

Value in simulation	logic value
0	low
1	high
R	rising edge
F	falling edge
X	undefined
Z	high impedance

For example, by using R the rising edge of a signal is modelled.

A simulator must take into account how a logic level will be evaluated in regard to the respective driving circuit. If a signal is connected to V_{cc} with a pull up resistor and will be connected to ground by an active transistor, then it will change its value depending on the greater strength of the transistor. Another attribute is logic strength which is defined for a more accurate simulation of logic circuits.

In Verilog, eight different values are defined to describe logic strength (table 11.2).

It is remarkable that the logic strengths 4, 2, and 1 are assigned to capacitors. The capacity of interconnections in a chip can thus be taken into account in simulating a circuit.

Table 11.2 Eight values for logic strength (Verilog)

logic strength	strength number	designa-tion	abbrevia-tion
supply drive	7	supply	Su
strong drive	6	strong	St
pull drive	5	pull	Pu
large capacitor	4	large	La
weak drive	3	weak	We
medium capacitor	2	medium	Me
small capacitor	1	small	Sm
high impedance	0	high	Hi

There are more definitions for logic value systems. IEEE Std 1164 1993 defines a nine-value logic system (table 11.3).

Table 11.3 Nine-value logic system (IEEE Std 1164 1993)

Value in simulation	logic value
0	strong low
1	strong high
L	weak low
H	weak high
X	strong unknown
W	weak unknown
Z	high impedance
—	don't care
U	uninitialized

The result of digital values can thus be calculated with higher accuracy. This system is suitable for modeling CMOS circuits with transmission gates and tri-state circuits for busses.

If the logic AND gate is simulated the output value is resoved by a table (table 11.4).

But if these logic values are used who will provide the necessary models and their parameters for simulating a real circuit. The model parameters depend on the logic function and the technology used for implementation. With the growing standardization of model libraries this problem will be solved gradually. One important step in this direction is the *IEEE Std_Logic_1164 Package*, in which models for logic standard functions and their various kinds of parameters are defined. These definitions allow simulation in a nine-value logic system. Generally the designer depends on the model library attached to the system he uses for circuit development.

Table 11.4 Resolution table for resolved std_logic, taken from IEEE Std_Logic_1164 1993

	U	X	0	1	Z	W	L	H	—
U	U	U	0	U	U	U	0	U	U
X	U	X	0	X	X	X	0	X	X
0	0	0	0	0	0	0	0	U	0
1	U	X	0	1	X	X	0	1	X
Z	U	X	0	X	X	X	0	X	X
W	U	X	0	X	X	X	0	X	X
L	0	0	0	0	0	0	0	0	0
H	U	X	0	1	X	X	0	1	X
—	U	X	0	X	X	X	0	X	X

In section 11.6 the power of logic simulation is illustrated by design examples.

Performance of a logic simulation

The first simulators calculated signal values for every single node of a circuit in a time scale. This leads to simulation times which are not acceptable. Nowadays **event driven simulators** are used. Only if there is a change at the input of a logic component like an AND gate or a flip flop will a new logic value for the corresponding output be calculated by the simulator. Only those components are treated which are driven by the changing signal. In parallel, a timing wheel takes track of the elapsing time to represent the time scale chosen by the designer for his simulation process. All events are documented in an event list or event queue with the time related to the event.

11.5 Fault Simulation for Verification of Fault Coverage of Test Stimuli

The effectiveness of test stimuli can be evaluated with fault simulation.

The theory of fault simulation is described in detail in chapter 15.

For the fault simulation a circuit specific simulator will be compiled using the circuit description and the model library of the development environment. The designer decides which fault models of the

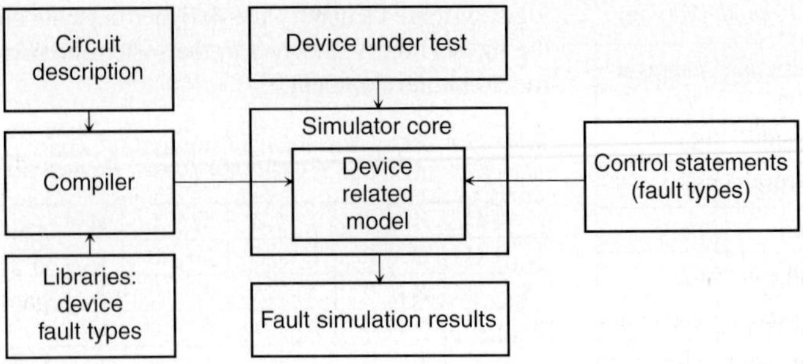

Fig. 11.6 Structure of a fault simulator

fault library or of interest should be used for fault simulation.

A simple logic fault model for verifying the effectiveness of test pattern is the 'stuck fault' model. There is the s.a.0 fault (stuck at zero) fault which assumes the logic value of a specific node permanently to be 0, which means it is connected to ground. If an s.a.1 fault (stuck at one) is inserted the node will have a permanent 1, that is, it is connected to V_{cc}.

Generally, single stuck faults are assumed for fault simulation. One fault is inserted and the fault simulator simulates the behavior of the complete circuit under this circumstances. The test stimuli are applied to the external inputs of the circuit now called *primary inputs*. The reaction of the circuit will be observed at the external outputs of the circuit, now called *primary outputs* (fig. 11.7).

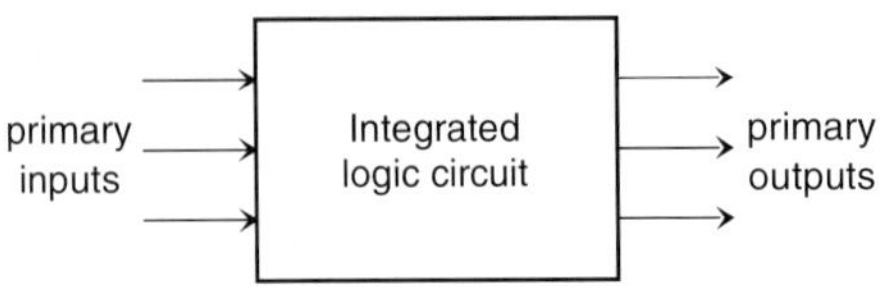

Fig. 11.7 Primary inputs and primary outputs

Naturally, before integrating a module into a complete circuit it can be treated likewise in order to evaluate the effectiveness of the test stimuli.

For reference the fault simulator will simulate the reaction of a fault-free circuit at every node. For each single fault inserted the simulator will simulate the respective faulty circuit again and monitor the circuit reaction. In order to save simulation time parts of the circuit which are not affected by the fault will not be simulated. If the simulated value at a primary output deviates from the respective value of the fault free circuit, simulation is stopped and the fault will be marked on the fault list as 'detected'.

To save time several faulty circuits are simulated parallely in a system.

More details about fault simulation are described in section 11.7 using example circuits.

11.6 Performance and Use of Logic Simulation

Two simple logic circuits, an AND with inverter and an R-S latch are used to demonstrate performance and use of logic simulation. The simple circuits are chosen to put the performance of a simulator into foreground and not the circuit by its complex function.

11.6.1 Inverter with AND

The logic circuit *inverter with AND* inverts the input signal EIN_UND0 to the variable INVERT. The AND gate has two inputs, INVERT and IN_AND1. The two outputs of the circuit are OUT_AND and INVERT (fig. 11.8, list 11.1). Output INVERT was added to show the value of the internal node explicitly.

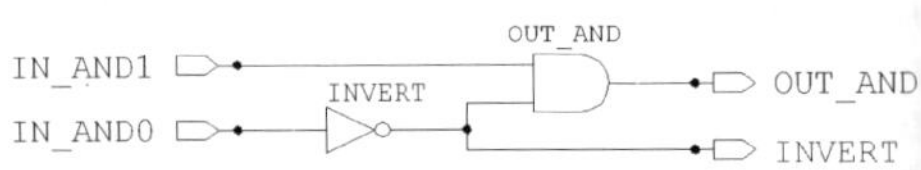

Fig. 11.8 Inverter with AND

List 11.1 Source code (HDL) of the circuit inv_and

```
; inputs
IN:            IN_AND0,IN_AND1
; outputs
OUT:           INVERT, OUT_AND
```

```
*BOOLEAN EQUATIONS
; Inversion
INVERT          = /IN_AND0
; logic AND
OUT_AND         = INVERT & IN_AND1
```

Functional simulation rapidly produces the reaction of the synthesized logic.

Functional simulation: circuit inv_and

Functional simulation will use a simple model with constant time delay (unit time delay) for every gate. Dynamic processes owed to different time delays of gates which may produce glitches can not be discovered at this stage. Additionally, all signal changes are assumed to be ideal step functions. The technology of the target device is not considered either.

Test stimuli are developed by the designer or a codesigner (list 11.2). The actual test program was entered using a graphic interface and altered manually afterwards. This speeds up the generation process because one can directly see the circuit reaction and thus develop test program accordingly. The ineffective test steps can be deleted manually.

Most work benches provide tools for inspecting the signals of every node of the hardware. This helps greatly identifying design faults during the design process.

Using test stimuli given in list 11.2 the functional simulation produces the timing diagram of fig. 11.9.

List 11.2 Test vectors for inv_and

```
module CommonStimulus(IN_AND0, IN_AND1);
    #0  IN_AND0 = 0;
    #0  IN_AND1 = 0;
    #10 IN_AND0 = 1;
    #10 IN_AND1 = 1;
    #20 IN_AND0 = 1;
    #20 IN_AND1 = 0;
    #30 IN_AND0 = 0;
    #30 IN_AND1 = 1;
    #40 IN_AND0 = 1'bx;
    #50 IN_AND0 = 0;
    #50 IN_AND1 = 0;
    #60 IN_AND1 = 1;
    #70 IN_AND1 = 0;
```

To recognize each combination of the two input variables easily, all inputs are bundled with as a additional variable INPUTS whose values are displayed in hexadecimal numbers. The numbers 0, 3, 2; and 1 show that all input combinations are applied in the first 4 steps of test stimuli.

Simulation begins with both inputs 0. After one unit time delay both outputs change from undefined to defined values, that is INVERT = 1 and OUT_AND = 0. The delay for output OUT_AND is only one time unit because IN_AND1 = 0 directly forces the output OUT_AND to 0, disregarding the unknown output value of INVERT.

Depending on the sequence of changing input variables a hazard may occur. This is the case when both circuit inputs change at time marked with ① from 00 to 11 (fig. 11.9); this is depicted as a change from 0 to 3 of the bundle variable INPUTS: After one time unit, the output OUT_AND changes to 1 and returns to 0 after one time unit. In particular, the changes of values for the first part of the input pattern are as follows:

Input UND_EIN0 changes its value at ① from 0 to 1. Therefore the output INVERT changes to 0 at ②, which happens 1 time unit after ①. During the time period between ① and ② IN_AND1 = 1 und INVERT = 1. Owing to this behaviour OUT_AND goes to 1 at ②. OUT_AND will return to 0 one time unit later, which happens at ③ because INVERT has changed to 0 at ②.

The pulse at output OUT_AND is shown clearly by the simulator. However, it depends on the value of the relevant parameter in the AND model used in the simulator. For the time of the parameter specified there the simulator will suppress this pulse.

This logic model need not necessarily show the function of the target device which will be shown in the following passage.

A short note concerning the input values in the timing diagram fig. 11.9: Input variables may also have the value X (unknown). In the design example the input IN_AND0 is set to X at ④. After one time unit INVERT changes to X which affects the output OUT_AND after another time unit to change to X as well. If IN_AND0 returns to 0, one time unit later both INVERT and OUT_AND will obtain 1 and 0 respectively, after one time unit. This is because a 0 – in this case IN_AND0 – at one input of an AND gate asserts a 0 at its output, regardless of the value of its other inputs.

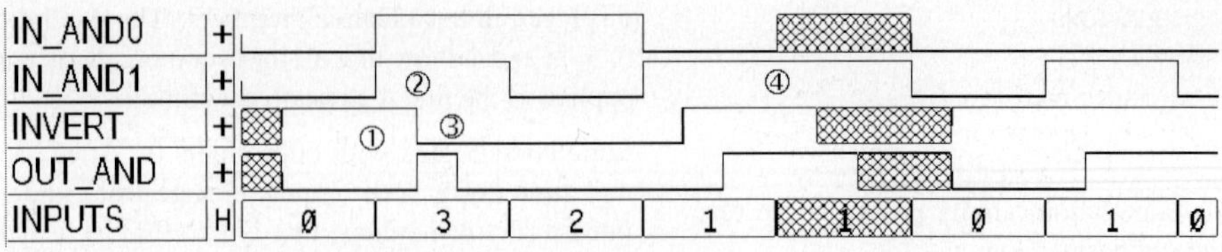

Fig. 11.9 Timing diagram functional simulation of inv_and

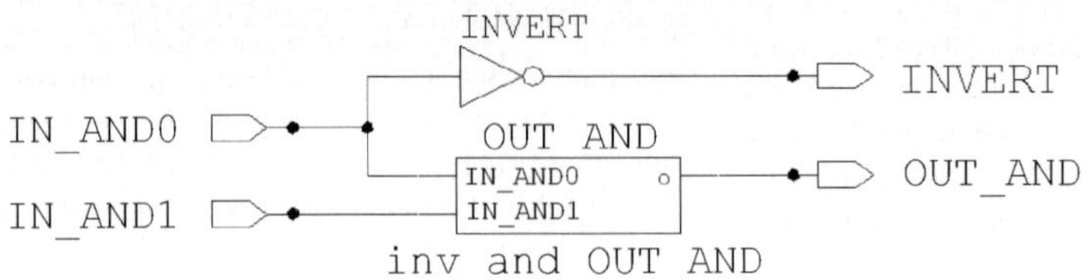

Fig. 11.10 Schematic of the generated logic for FPGA implementation

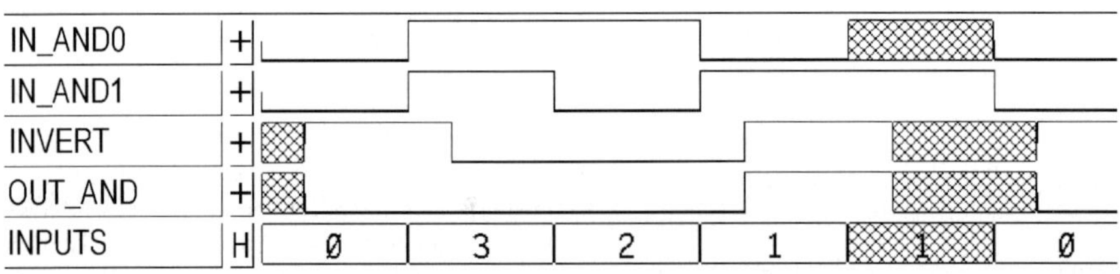

Fig. 11.11 Timing diagram of pre-layout simulation of the circuit inv_and

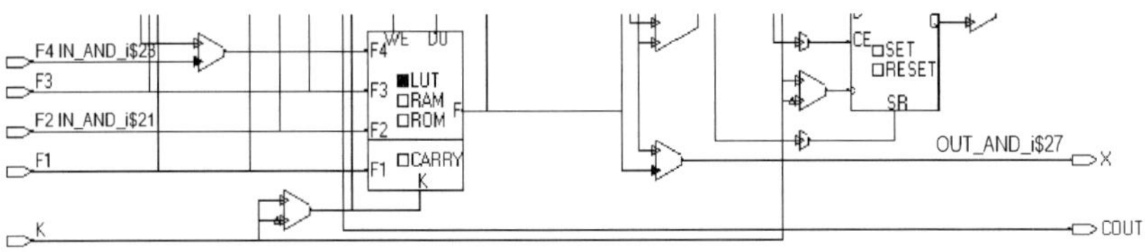

Fig. 11.12 Implementation of logic with a LUT within the CLB

Pre-layout simulation: circuit inv_and

After functional simulation the structure of the circuit must be tailored according to the target device. The logic of the circuit is converted in such a way that the specific logic modules of the target device are used exclusively. To demonstrate this effect the FPGA 4003 of XILINX is used as a target device. This process occurs similarly in the development of ASICs.

The FPGA uses *Configurable Logic Blocks* (CLB) which are the logic basis of this FPGA. In a CLB look up tables (LUT) multiplexers and flip flops can be configured to implement the intended logic function. In this case LOGIC2 does the synthesis for the circuit inv_and and generates the logic function (fig. 11.10) tailored for the FPGA. The designer has to verify the logic implementation with *pre-layout simulation*. Naturally test stimuli of the pre-layout simulation are used again to show the differences of the implementation.

In pre-layout simulation the load imposed on outputs is ignored. Load is imposed by the capacitance of interconnections and the input load of connected gates.

Most interesting is that the output INVERT is generated directly by an inverter. The reason for

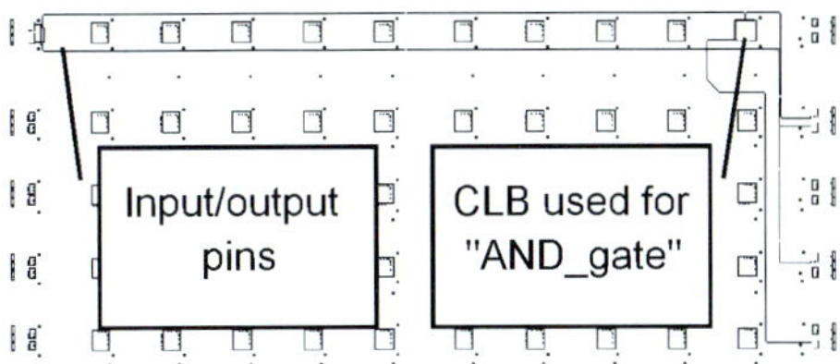

Fig. 11.13 Connections of inv_and within the FPGA

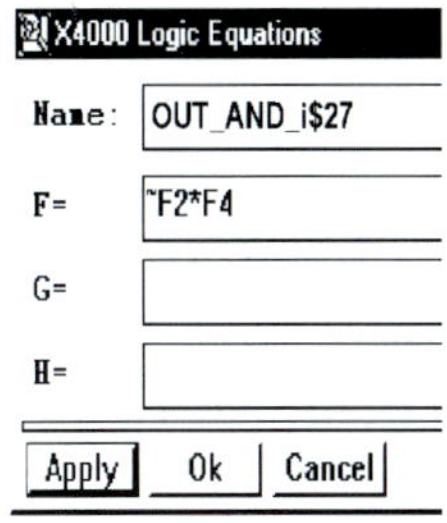

*Fig. 11.14 Function OUT_AND := /F2*F4*

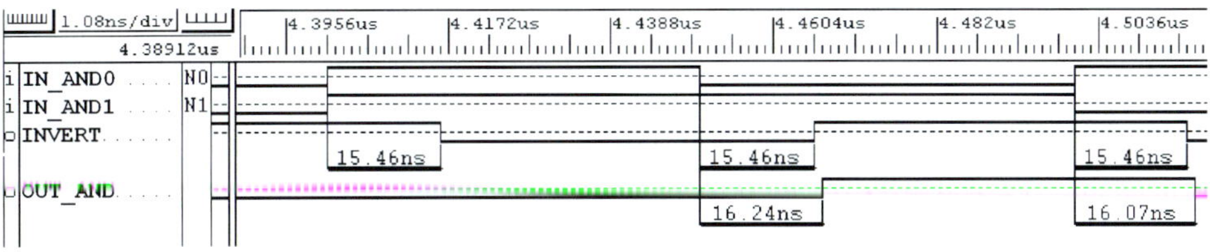

Fig. 11.15 Post-layout simulation for the circuit inv_and

his implementation is that a CLB has an inverter for each external output, which in this case is directly used and does not require any additional resources of the CLB. Outputs are inverted within the CLB without using additional resources. The logic function itself is implemented with an LUT. Logic functions with 4 inputs can be programmed in this LUT without any need for optimization.

Pre-layout simulation (fig. 11.11) shows that no pulse is generated after the change of INPUTS from 0 to 3. All changes of output values occur after one unit time delay.

As mentioned before, the logic function is implemented in a LUT within the CLB. Accordingly no inverters are used for the logic implementation. The logic is just programmed into the LUT. The output of a LUT will change after one unit time delay independent of the logic programmed in the LUT. In fig. 11.12 both inputs F2 (IN_AND0) and F4 (IN_AND1) feed the LUT and the output F of the LUT is connected to output X (OUT_AND).

The layout of the logic AND within the FPGA is shown in fig. 11.13. Logic functions of each signal in the FPGA can be evaluated by the workbench. The output F is equal to /F2*F4, using the internal variables of the FPGA. F2 refers to IN_AND0 and F4 to IN_AND1 respectively (fig. 11.14).

For debugging this feature is very useful.

Post-layout simulation: circuit 'inv_and'

For the *post-layout simulation* the design entry tool Foundation of Xilinx was used. Instead of the term post-layout simulation the term timing simulation is used (fig. 11.15). Before starting the simulation the circuit parameters are extracted from the actual layout.

The delays from input to output are significant. This is owed to the delay of input buffers and output buffers. These do not show up in functional and pre-layout simulation.

Output INVERT goes high prior to the output signal OUT_AND. The time delay for the output signal is slightly shorter than the path delay for the output OUT_AND. This shows clearly that the results of the simulation depend on the hardware of the implementation.

11.6.2 Design Example: RS Flip flop

The RS flip flop implemented with NOR gates (fig. 11.16) is another design example to demonstrate the capability of functional simulation with unit time delay.

It is special to this implementation that the output QSTAR is not the negated Q. Because in the literature there is no standard way of drawing the schematic of a RS flip flop, it should be noted that

11

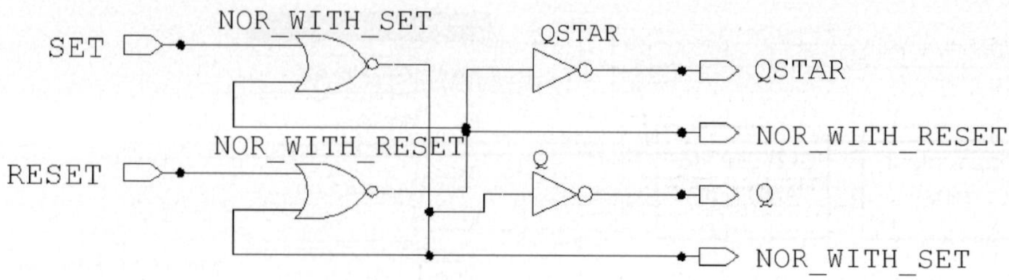

Fig. 11.16 RS flip flop

the input SET is in the upper left of the schematic and the output QSTAR is on the upper right.

Functional simulation of the RS flip flop circuit

In functional simulation the simultaneous change of both inputs SET and RESET from 11 to 00 shows that both outputs Q and QSTAR oscillate in phase (fig. 11.17). The flip flop is in an unstable condition. The input combination 11 for SET and RESET is not permissible, the simulator indicates the problem in this circuit. The explanation in detail is as follows. Starting at 30 ns, both inputs are 1. Taking the values of the inputs as they are means setting and resetting the flip flop at the same time. This is not a meaningful action. The reaction of the flip flop is that both outputs Q and QSTAR are 1. But considering this condition, what happens if at 40 ns both inputs go low? A logic 1 at the outputs is a logic 0 at the outputs NOR_WITH_SET and NOR_WITH_RESET after one unit time delay. Therefore all inputs of both NOR gates are 0, which in turn leads to a 1 for both NOR_WITH_SET and NOR_WITH_RESET after another unit time delay. This behavior of the circuit only shows up in simulation. The delay times of both gates are identical in the simulated model of the circuit, which is never true in a real logic circuit. But simulation indicates that an illegal input combination has occurred.

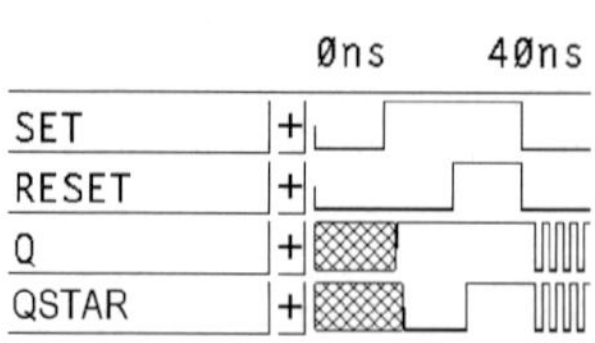

Fig. 11.17 Functional simulation: timing diagram of the RS flip flop

Post-layout simulation of circuit RS flip flop (timing simulation)

Pre-layout simulation is left out because no additional information is expected in this simple example. After layout all macro cells, output drivers, and the wiring are fixed. The actual parameters can be extracted from the circuit, which is an FPGA in this case. Using these parameters the circuit model is modified to match the performance of the circuit model with the real circuit. Mostly so called lumped models are used; for instance, a single capacitor is inserted instead of the distributed capacity of a wire. The process of inserting parameters extracted from the layout is called *back annotation*. Thus hardware oriented delay parameters of the target device can improve the accuracy of the simulation results.

The goal of post-layout simulation is to take dynamic behaviour into consideration

Varying delays of gates cause many events in the circuit at different times. The purpose of post-layout simulation is to analyse the dynamic behavior of the circuit. If post-layout simulation recognizes no unknown logic values and no oscillations in a circuit the designer is one step nearer to design a functioning circuit with reproducible results.

Post-layout simulation produces a result with stable states (fig. 11.18) for the given test stimuli.

Both outputs, called P_Q:PAD and P_QSTAR.PAD in the diagram have stable values. The reason is that in a real circuit two gates with their related delay times have different delay times. This is the difference compared to the results of the functional simulation. Therefore one should use both – functional simulation *and* post-layout simulation.

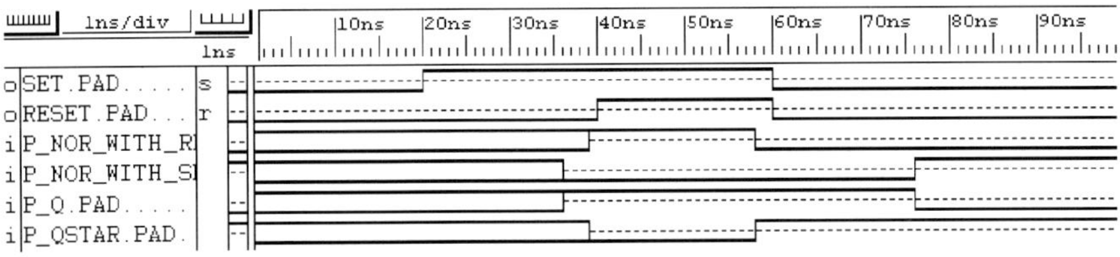

Fig. 11.18 Timing diagram of Xilinx post-layout simulation of the RS flip flop

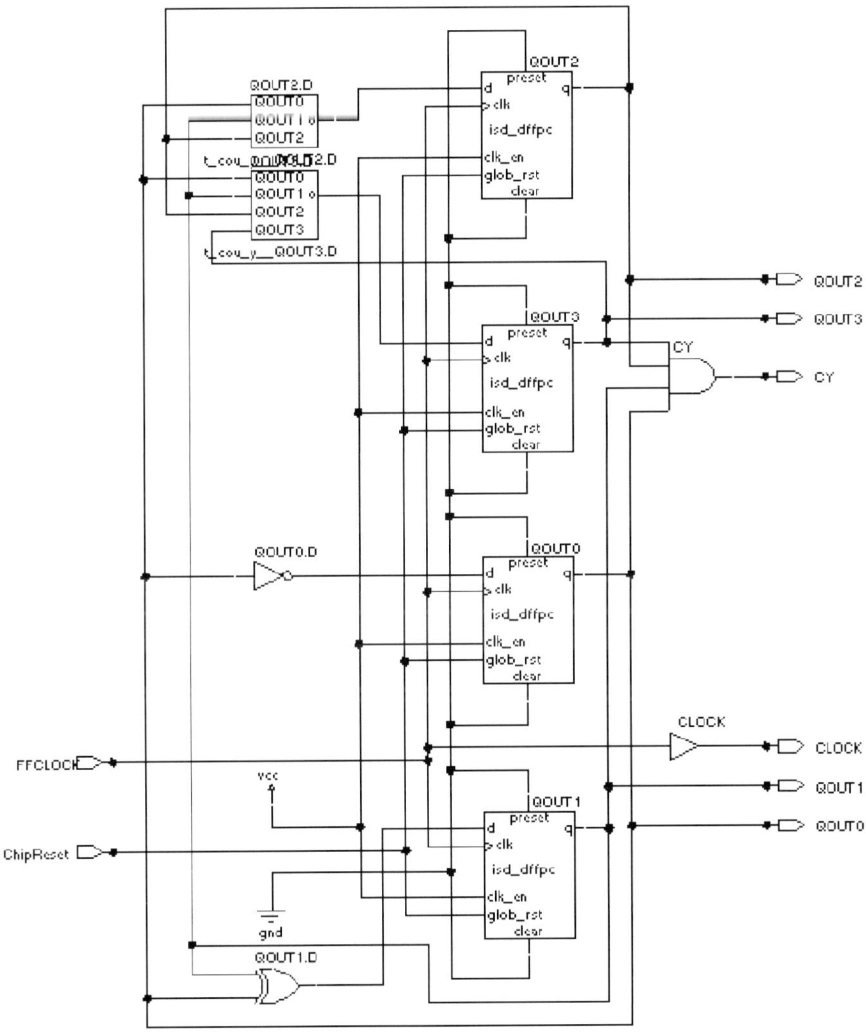

Fig. 11.19 Functional simulation of the Moore machine

Design example: 4-bit synchronous binary counter

The 4-bit binary synchronous counter with carry and reset is a design example which will be designed in two ways, as a *Medvedev machine* and as a *Moore machine*. Simulation is used to demonstrate the differences between their timing. In particular the path delays of both circuits are analyzed. The differences of internal delay times which are caused by all elements associated with the function are discussed.

The counter is implemented in a FPGA but all results are equally true in an ASIC design. Elements like input buffers, interconnections, and special clock lines exist in all implementations for logic circuits. The differences are in the absolute values which can be acquired in any technology by back annotation in the workbench used.

At first the 4-bit binary synchronous counter with carry and reset is implemented as a Moore machine. The carry signal is generated by decoding the state of the counter. The result will be compared with the second design (which is a Medvedev machine). Medvedev machines use flip flops for every output of the circuit. This means that an additional flip flop must be used for the carry output. It will be demonstrated that the speed of a circuit can be enhanced if additional hardware is used.

For the Moore machine it is expected that the delay time from clock edge up to the change of the value of the carry output is longer.

Functional simulation of both counters

It is shown in fig. 11.19 that the signal CY is decoded with an AND gate. The change of CY will occur after the outputs of the flip flops have changed. In contrast to this, in fig. 11.20 CY is the output of an additional flip flop. Therefore, all outputs will change synchronously.

In this case, unit time delay of 1 ns is used. Functional simulation of the Moore machine shows 2 ns time delay between clock edge and output change of the signal CY (fig. 11.21). The 2 ns add up from 1 ns delay for the output change of the flip flop and 1 ns for the gate delay of the AND gate. Opposed to this, only 1 ns time delay can be observed in the Moore machine (fig. 11.22).

Post-layout simulation: Moore machine

Post-layout simulation in XILINX foundation software is called timing simulation. The delay time 'minimum delay' was selected for this simulation. The output buffer has 8.5 ns time delay.

For the Moore machine in fig. 11.23 there is a delay of 20.2 ns from the positive clock edge to CY. The Medvedev machine shows 16.5 ns (fig. 11.24).

What is the reason for the different results in functional simulation and in post-layout simulation? In post-layout simulation extracted values for the capacitance and resistance of interconnections are observed as well as the fan out of the logic elements. Additionally delay times of buffers have to be taken into account for the real circuit in the IC because all signals of the counter are connected to the pins of the FPGA.

All timing values of delay paths are listed in detail for the simulation (list 11.3, list 11.4). For the Moore machine the delay time is 4.016 ns + 16.204 ns = 20.22 ns and for the Medvedev machine the delay time adds up to 4.023 ns + 12.486 ns = 16.509 ns.

Another model of the circuit is used to discuss delay times. In the block diagram of fig. 11.25 the inputs and outputs of this model are depicted in order to illustrate the path delays calculated in what follows. Every element that influences delay time is shown graphically in a diagram (fig. 11.26).

Delay time for the path FFCLOCK → CLOCK is subdivided in three parts

- the delay time from input pad to the output of the input buffer;
- the delay time of the connection between the output of the input buffer and the input of the output buffer;
- the delay time from input of the output buffer to the output pad.

In the simulation the delay time is calculated as 15 ns (list 11.5).

In detail, the delay time for the path FFCLOCK → Cy consists of:

- the delay time from input pad to the output of the input buffer;
- the delay time of the connection between the output of the input buffer and the clock input of the flip flop;

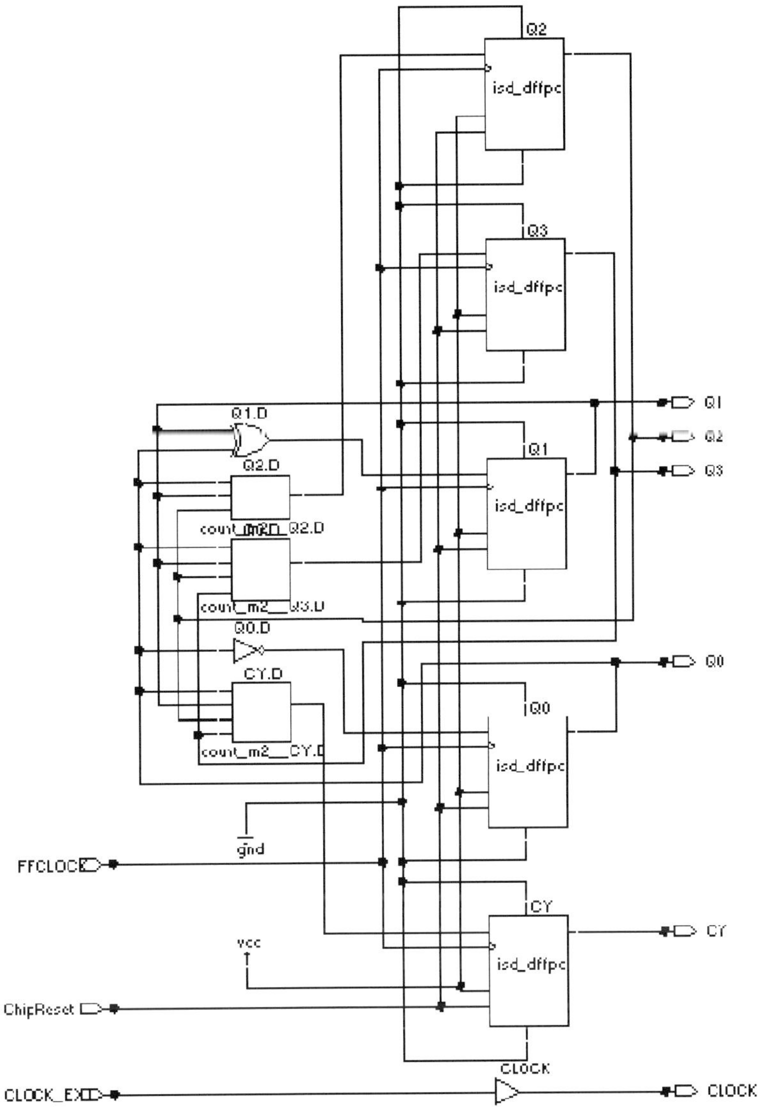

Fig. 11.20 Functional simulation of the Medvedev machine

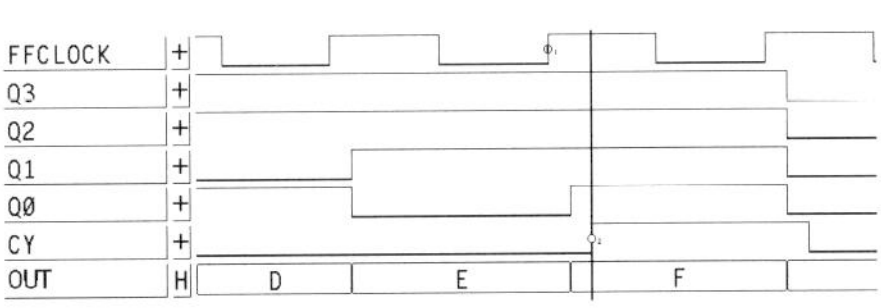

Fig. 11.21 Timing diagram Moore machine

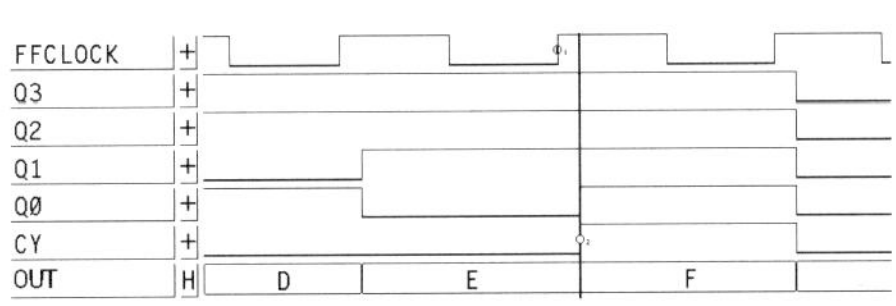

Fig. 11.22 Timing diagram Moore machine

- the delay time from clock edge to the output of the flip flop;
- the delay time of the connection between the output of the flip-flop and the input of the LUT;
- the delay time of the LUT;
- the delay time of the connection between the output of the LUT and the input of the output buffer;
- the delay time from the input of the output buffer to the output pad.

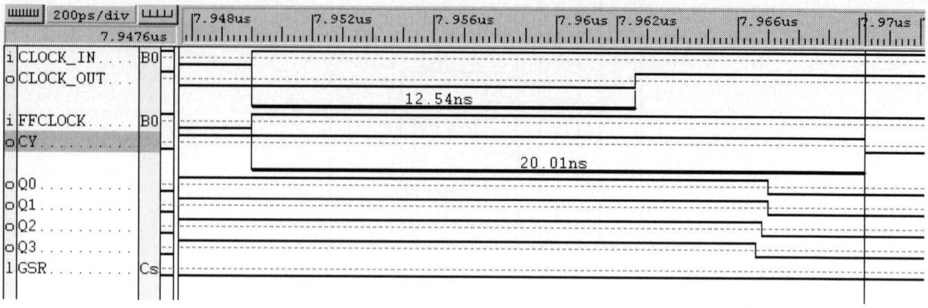

Fig. 11.23 Positive edge of CY signal Moore machine

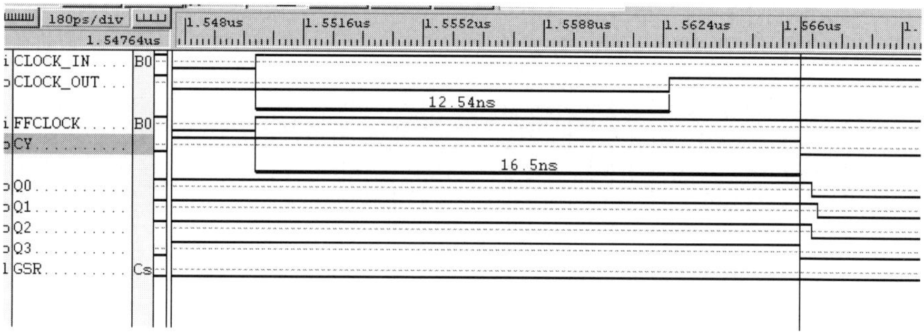

Fig. 11.24 Positive edge of CY signal Medvedev machine

List 11.3 Delay time Moore machine FFCLOCK → CY

```
Delay:      4.016ns FFCLOCK to Q0_i$69

Path FFCLOCK to Q0_i$69 contains 2 levels of logic:
Path starting from Comp: P78.PAD
To                    Delay type          Delay(ns)  Physical Resource
                                                     Logical Resource(s)
-------------------------------------------------  --------
P78.CLKIN             Tclkin                 0.000R  FFCLOCK
                                                     FFCLOCK.PAD
BUFGP_TR.I            net (fanout=1)         2.780R  FFCLOCK
BUFGP_TR.O            Tclk                   0.000R  c$59
CLB_R9C9.K            net (fanout=4)         1.236R
FFCLOCK_i$60
-------------------------------------------------
Total (0.000ns logic, 4.016ns route)         4.016ns
      (0.0% logic, 100.0% route)

--------------------------------------------------------------
```

```
Delay:    16.204ns Q0_i$69 to CY

Path Q0_i$69 to CY contains 3 levels of logic:
Path starting from Comp: CLB_R9C9.K (from FFCLOCK_i$60)
To                   Delay type         Delay(ns)  Physical Resource
                                                   Logical Resource(s)
-------------------------------------------------  --------
CLB_R9C9.XQ          Tcko                   2.820R  Q0_i$69
                                                   Q0_i$69/I$1
CLB_R9C10.F2         net (fanout=6)         1.369R  Q0_i$69
CLB_R9C10.X          Tilo                   2.000R  CY_i$71
                                                   CY_i$71
P66.O                net (fanout=1)         1.515R  CY_i$71
P66.PAD              Tops                   8.500R  CY
                                                   CY
                                                   CY.PAD
-------------------------------------------------
Total (13.320ns logic, 2.884ns route)     16.204ns
      (82.2% logic, 17.8% route)
```

List 11.4 Delay time Medvedev machine FFCLOCK → CY

```
Delay:     4.023ns FFCLOCK to CY_i$75

Path FFCLOCK to CY_i$75 contains 2 levels of logic:
Path starting from Comp: P78.PAD
To                   Delay type         Delay(ns)  Physical Resource
                                                   Logical Resource(s)
-------------------------------------------------  --------
P78.CLKIN            Tclkin                 0.000R  FFCLOCK
                                                   FFCLOCK.PAD
BUFGP_TR.I           net (fanout=1)         2.780R  FFCLOCK
BUFGP_TR.O           Tclk                   0.000R  c$63
CLB_R9C10.K          net (fanout=4)         1.243R
FFCLOCK_i$64
-------------------------------------------------
Total (0.000ns logic, 4.023ns route)        4.023ns
      (0.0% logic, 100.0% route)

-----------------------------------------------------------------

Delay:    12.486ns CY_i$75 to CY

Path CY_i$75 to CY contains 2 levels of logic:
Path starting from Comp: CLB_R9C10.K (from FFCLOCK_i$64)
To                   Delay type         Delay(ns)  Physical Resource
                                                   Logical Resource(s)
-------------------------------------------------  --------
CLB_R9C10.XQ         Tcko                   2.820R  CY_i$75
                                                   CY_i$75
P66.O                net (fanout=1)         1.166R  CY_i$75
P66.PAD              Tops                   8.500R  CY
                                                   CY
                                                   CY.PAD


-------------------------------------------------
Total (11.320ns logic, 1.166ns route)     12.486ns
      (90.7% logic, 9.3% route)
```

List 11.5 Delay time for the path FFCLOCK → clock Moore machine

```
--------------------------------------------------------------
Delay:     14.986ns FFCLOCK to CLOCK

Path FFCLOCK to CLOCK contains 3 levels of logic:
Path starting from Comp: P78.PAD
To                   Delay type          Delay(ns)  Physical Resource
                                                    Logical Resource(s)
-------------------------------------------------  --------
P78.CLKIN            Tclkin                 0.000R  FFCLOCK
                                                    FFCLOCK.PAD
BUFGP_TR.I           net (fanout=1)         2.780R  FFCLOCK
BUFGP_TR.O           Tclk                   0.000R  c$59
P61.O                net (fanout=4)         3.706R
FFCLOCK_i$60
P61.PAD              Tops                   8.500R  CLOCK
                                                    CLOCK
                                                    CLOCK.PAD
-------------------------------------------------
Total (8.500ns logic, 6.486ns route)       14.986ns
      (56.7% logic, 43.3% route)
--------------------------------------------------------------
```

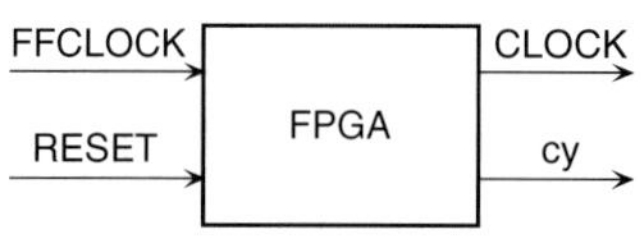

Fig. 11.25 Block diagram as an FPGA

In simulation a delay time of 20.2 ns is calculated.

It has been shown in detail how delay times are simulated and how they are affected by the various elements of a circuit. The comparison of both counters in table 11.5 shows that the gate count using gate equivalent is 60 for the Medvedev machine and 54 for the Moore machine.

Table 11.5 Comparison of Medvedev and Moore machine for the 4-bit counter

	c_medved	c_Moore
Target Device	XC4003E	XC4003E
Target Package	PC84	PC84
Target Speed	−3	−3
Flip Flops	5	4
4-Input LUT	5	5
Equivalent Gate Count	60	54
Simulation Clock	10 MHz	10 MHz
Timing Model	Average	Average
FF-Clock ↑ ⇒ CY ↑	16.5 ns	20.2 ns

4-bit counter as ASIC

Both counters will be implemented in an ASIC in order to illustrate differences in speed. Additionally some properties of the technology used are demonstrated. Furthermore some features of the workbench are shown:

- Technology: Alcatel;
- Mietec 0.5 µm;
- Library: MTC35000;
- Workbench: Mentor Design_Architect Version C1;
- Model Technology Modelsim Simulator Version 5.1g.

Logically the schematics of the ASIC implementation remain unchanged (fig. 11.27, fig. 11.28). In principle, simulation shows very similar results if the differences of the absolute values of delay times are not taken into account (fig. 11.29).

Signals of internal nodes are designated by Modelsim with special names. They are listed together with the signals in table 11.6 for the Medvedev machine to simplify the interpretation of the timing diagram.

There is no tool for listing the details of the delays for individual elements which is equivalent to the one used in the FPGA. Individual delay times can be extracted from the timing diagram and the list must be created manually.

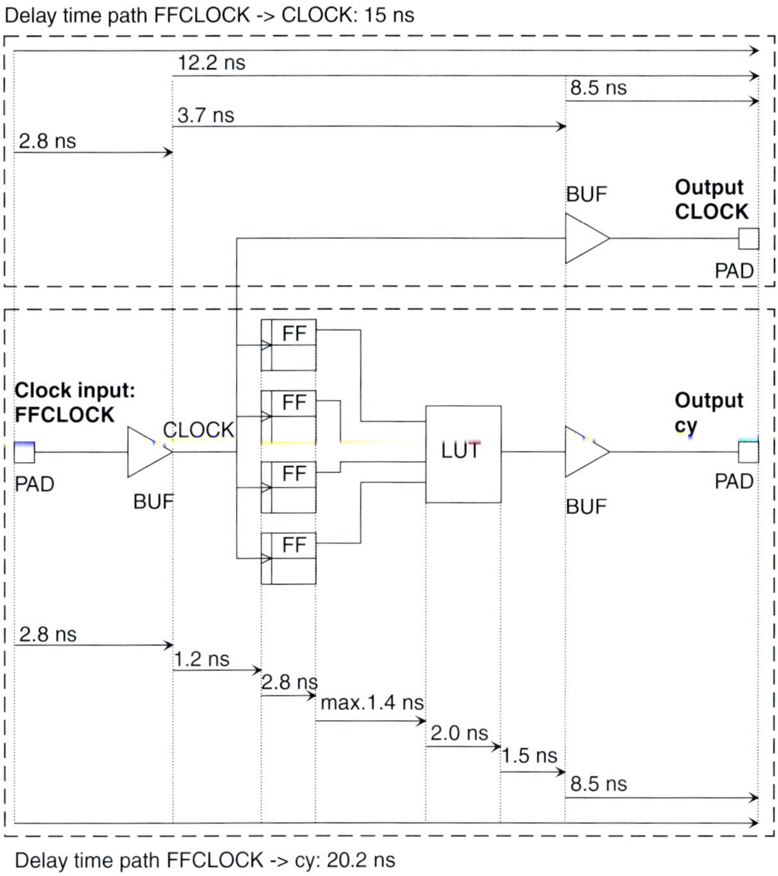

Fig. 11.26 Graphic illustration for the path FFCLOCK → clock and FFCLOCK → Cy Medvedev machine

Table 11.6 Special names in Modelsim for signals of Medvedev machine

Internel Designator	Signal Name
i_03624/a	input of the input buffer
i_03624/z	output of the input buffer
i_0363/cp	CLOCK input CY Flip Flop
i_0363/d	D input CY Flip Flop
i_0363/q	Q output CY Flip Flop
i_03631/a	input of the output buffer
i_03631/z	output of the output buffer

Note that no input and output buffers are included in this simulation. The delay time from clock edge of FFCLOCK to output Q0 is 1500 ps (fig. 11.29) in the Medvedev machine. There are no delay times included in the simulation, the delays of the input and output buffers which are 425 ns and 420 ns, respectively, are not included either. The path delay from the input pin of the clock to the output is 425 ps + 1,500 ps + 420 ps = 2,345 ps.

This demonstrates the high speed of an ASIC.

If the simulation results of the Moore machine implemented in an ASIC is analyzed the results are similar. In this case path delay from input pin of the clock to the output is 420 ps + 1,496 ps + 658 ps + 420 ps = 2,994 ps (fig. 11.30).

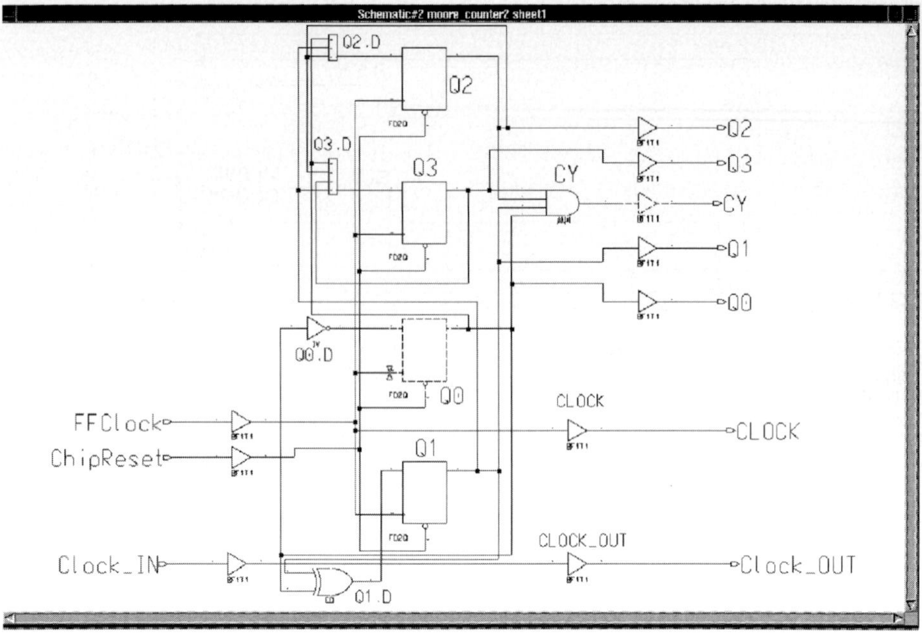

Fig. 11.27 4-bit counter Moore machine as ASIC

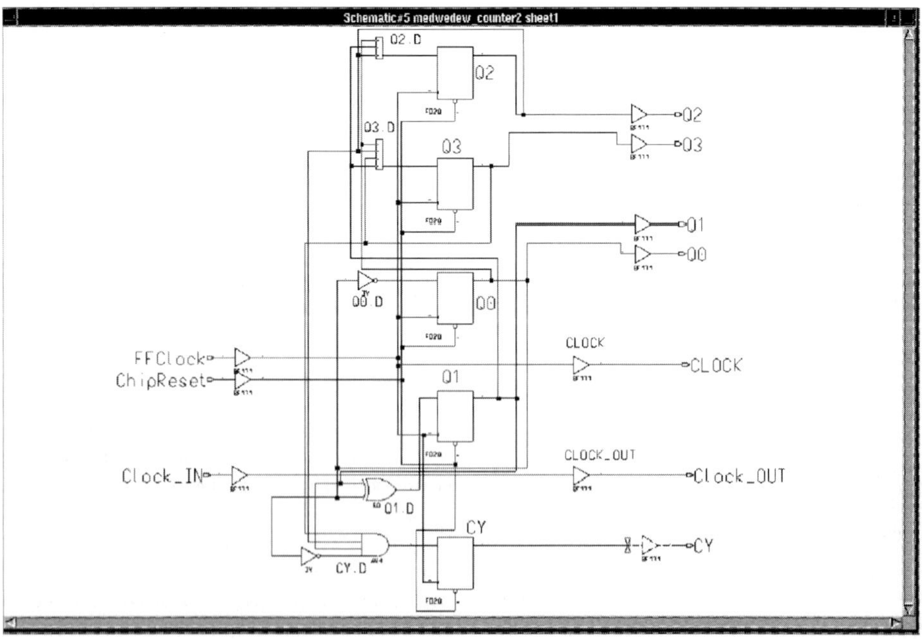

Fig. 11.28 4-bit counter Medvedev machine as ASIC

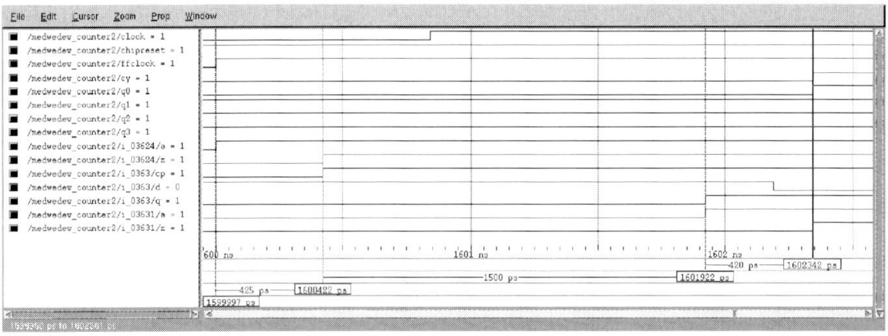

Fig. 11.29 Timing diagram (Modelsim) for post-layout simulation of the 4-bit counter Medvedev machine as ASIC

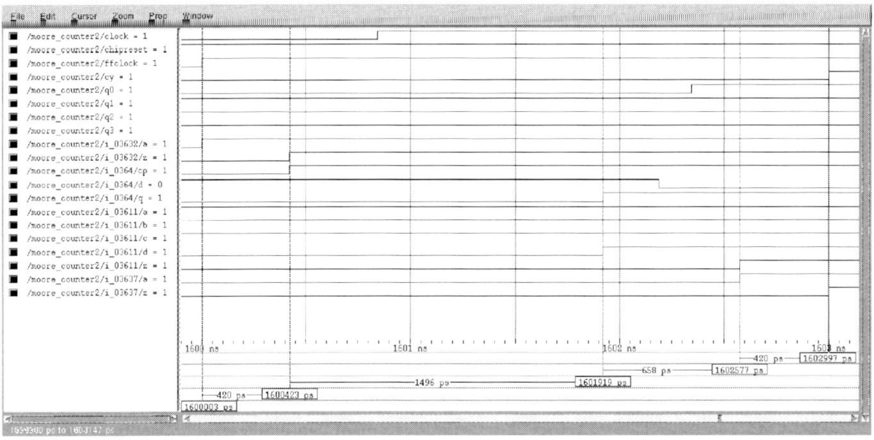

Fig. 11.30 Timing diagram (Modelsim) for post-layout simulation of the 4-bit counter Moore machine as ASIC

11.7 Verification of Testability with Simulation

Theory of fault simulation is explained in detail in chapter 15. In the following the performance of simulation tools is demonstrated with design examples.

Results of fault simulation will give answers to the following questions:

- How effective are the test stimuli (test vectors) used?
- What about the testability of the design?

For fault simulation the software *Quickfault* of Mentor Graphics was used. The design examples are parts of an IC which was implemented at IMS (Institut für Mikroelektronik Stuttgart) in an 0.8 μ process.

For economic reasons in fault simulation mainly simple *stuck at 1* and *stuck at 0* faults are used in simulation. But what about the ability of detecting faults for a given design in a given technology? In practice it is a good approach to use these two models. Up to now there is still no conclusive explanation for this effect.

Fault coverage is a term for giving the percentage of the number of faults defined by the designer which are detected with the applied test stimuli. This is an example of fault coverage which shows that one must read the specifications thoroughly to rate a statement concerning fault coverage: For fault simulation the designer decides to insert only s.a.1 faults in 100 out of 120 nodes in his circuit. After fault simulation 95 faults are detected. The fault coverage is 95 % because only the inserted

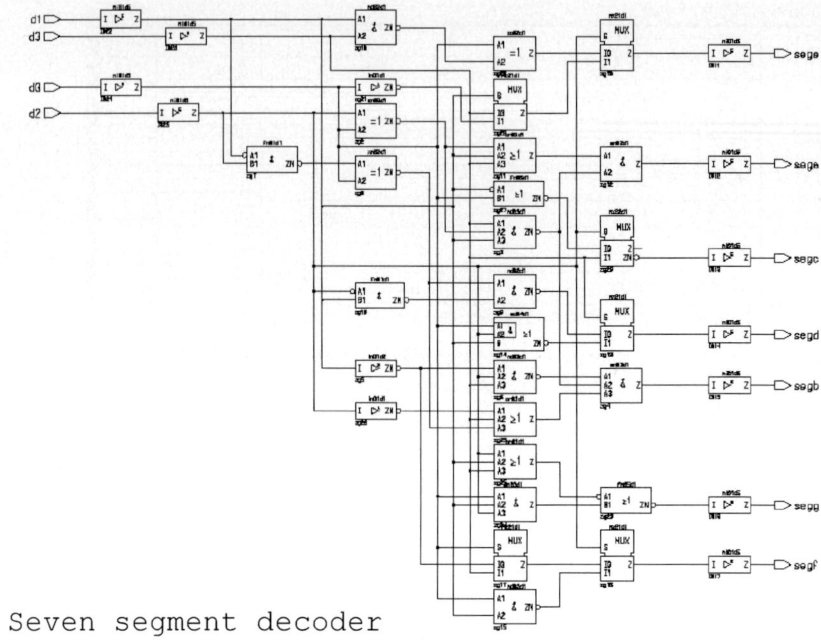

Fig. 11.31 Schematic for seven segment decoder in fault simulation

faults are considered. The number 95 does obviously not refer to s.a.0 faults which have not been simulated!

To estimate the fault coverage of a design it must be checked at how many nodes and which faults have been inserted.

11.8 Design Example: Decoder for a Seven Segment Display

Combinatorial logic parts in a system should be arranged in a way that all possible combinations of inputs may be applied. For such parts 100 % fault coverage is desirable. A seven segment decoder is a good example of such a circuit module.

Logic function of a seven segment decoder

A seven segment decoder is a combinational logic circuit with 4 inputs and 7 outputs to drive 7 segments of a display (fig. 11.31).

In order to detect faults test stimuli must be designed in such a way that at least each single node of the circuit will change once from 0 to 1 during simulation. Additionally the faulty response of the circuit must be recognized at least at one primary output.

In order to show different considerations of fault simulation and fault detection, at first test stimuli are developed which are not yet optimal. They will be improved in a second step.

Fault simulation with incomplete test stimuli

The circuit has 7 outputs sega, segb, segc, segd, sege, segf, segg and 4 inputs $d_3, \ldots, d_0$. For each step of simulation one test vector is defined. In list 11.6 the test vector is called *db*. One combination of a binary word with 4 bit $(d_3, \ldots, d_0)$ will be assigned to a test vector. In this case, input d_3 was deliberately not activated in order to show the effect in fault simulation.

List 11.6 Incomplete test program for seven segment decoder

```
OUTPUT sega,segb,segc,segd,sege,segf,segg;
INPUT d3, d2, d1, d0;
VECTOR db = d3,d2,d1,d0;
db = '0000'B;$
db = '0001'B;$
db = '0010'B;$
db = '0011'B;$
```

Table 11.7 Explanations of faults

Total Faults:	236	Sum of all s.a.0 and s.a.1 faults inserted by the fault simulator. The designer may exclude faults prior to the simulation process which do not appear in the given number.
Untestable Faults:	2	In most cases outputs which are not connected at all. Output Z of element zg26 MUX typ mx22d1, (fig. 11.34)
Testable Faults:	234	Fault which can be tested in theory.
Undetected Faults: (16.67 %)	39	Faults not found by the test stimuli.
Detected Faults: (83.33 %)	195	Faults detected by the test stimuli.
Possible Faults: (0.00 %)	0	Faults which may be recognized in a faulty hardware circuit depending on the initial values of storage elements of the circuit.
Oscillatory Faults: (0.00 %)	0	Oscillations in the circuit.
HM Dropped Faults: (0.00 %)	0	Faults which are in the fault list but are not simulated by the fault simulator.


```
db = '0100'B;$
db = '0101'B;$
db = '0110'B;$
db = '0111'B;$
/* does not detect all faults */
```

Input d_3 is not changed by test stimuli. 8 test vectors for 4 inputs are defined. Test stimuli and simulation result for the outputs are shown in fig. 11.32. The timing diagram of the fault-free circuit shows that because of the simultaneous change of d_0 and d_1 spikes at the outputs are created (at 204 ns spice at segf, fig. 11.33).

The first step of fault simulation shows the reaction to test stimuli. If outputs do not change it indicates that test vectors are not yet complete. Ths is not the case in the actual example.

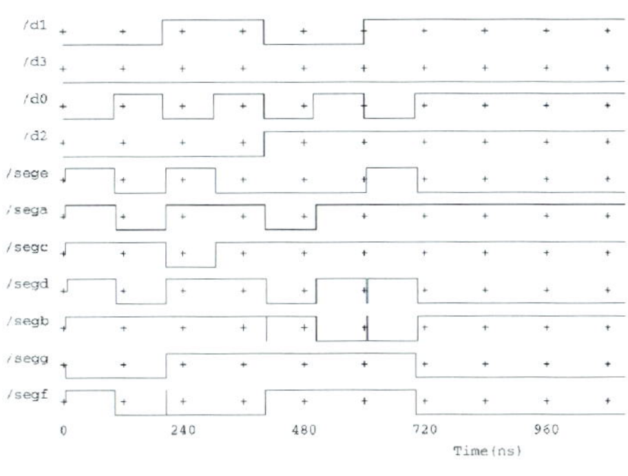

Fig. 11.32 Timing diagram of seven segment decoder

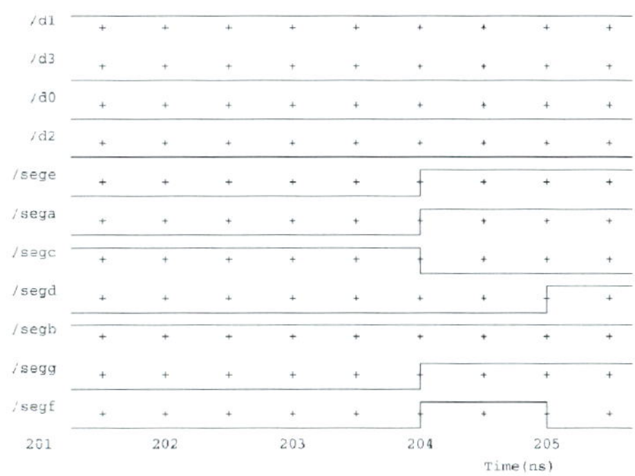

Fig. 11.33 Spike at 204 ns

Two kinds of faults were inserted as usual for fault simulation, s.a.0 and s.a.1. The delay time for fault simulation was adjusted to 100 ps. One result of fault simulation is the fault protocol in list 11.7.

List 11.7 Log of fault simulation with 8 test vectors

```
Total Faults              :  236
  Untestable Faults       :  2
  Testable Faults         :  234
    Undetected Faults     :  39    (16.67%)
    Detected Faults       :  195   (83.33%)
    Possible Faults       :  0     (0.00%)
    Hyperactive Faults    :  0     (0.00%)
    Hypertrophic Faults   :  0     (0.00%)
    Oscillatory Faults    :  0     (0.00%)
    HM Dropped Faults     :  0     (0.00%)
```

A fault coverage of 83 % was achieved with the test vectors defined.

Unfortunately software vendors have no standard definitions which describe different kinds of faults. Therefore the terms used in Quickfault will be explained briefly in table 11.7.

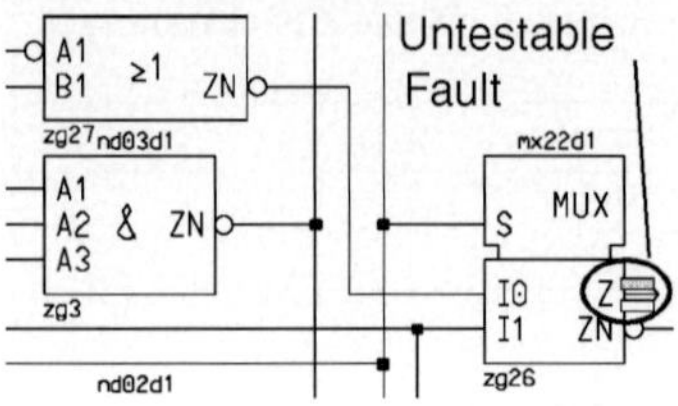

Fig. 11.34 Untestable fault at output Z of mx22d1

After fault simulation all undetected faults can be displayed graphically in the schematic (fig. 11.35) directly at the elements concerned. The designer may think about the reasons why a fault was not detected and change test vectors accordingly. It is shown in fig. 11.35 that undetected faults are at elements which are connected to input d_3.

Fault simulation with augmented test stimuli

As a second step input d_3 will be included in the test stimuli (list 11.8). All 16 combinations of 4 inputs are applied to the circuit.

The log of the fault simulator (list 11.9) shows fault coverage of 100 %.

List 11.8 Test vectors with 100 % fault coverage

```
CIRCUIT seg7;
TIMEDEF 1 PERIOD = 100NS ;
OUTPUT sega,segb,segc,segd,sege,segf,segg;
INPUT d0,d1,d2,d3;
VECTOR db = d3,d2,d1,d0;
db = '0000'B;$
db = '0001'B;$
db = '0010'B;$
db = '0011'B;$
db = '0100'B;$
db = '0101'B;$
db = '0110'B;$
db = '0111'B;$
/* undetected faults left */
db = '1000'B;$
db = '1001'B;$
db = '1010'B;$
db = '1011'B;$
db = '1100'B;$
db = '1101'B;$
db = '1110'B;$
db = '1111'B;$
db = '0000'B;$
/* all faults detected */
```

List 11.9 Part of the log of fault simulation with 16 test vectors: Fault 100 %

```
Total Faults              : 236
  Untestable Faults       : 2
  Testable Faults         : 234
    Undetected Faults     : 0        (0.00%)
    Detected Faults       : 234      (100.00%)
```

Another interesting question is: how many faults are detected at which test vector? This information is displayed in a histogram (fig. 11.36). It shows the effectiveness of each test vector. In fig. 11.36 two faults are detected in test cycle 14 and 15 which is indicated by a small line. At test cycle 16 nothing is detected. Test vector 16 is obsolete and it could be removed.

11.9 Design Example: 16-bit Counter with Carry

Sequential logic is a challenge for testing because the state of the storage elements and the input combination determine the next state of the circuit. The sequence of test stimuli greatly affects its effectiveness. A 16-bit counter is used because its function is well known.

If one wants to set a sequential circuit to a specific state sometimes many test vectors have to be applied in a definite sequence. A 16-bit counter is a good demonstration of the benefit of a scan path. Moreover, it shows clearly that dividing the 16-bit counter into modules of 4 bits simplifies the design of the test pattern.

11.9.1 Module 4-bit Counter

The 16-bit counter will be designed by using 4 identical modules of 4-bit counters. The correct function of the 4-bit counter has to be verified once and henceforth it may be used multiply without additional effort.

Functions of the 4-bit counter

The 4-bit counter module (fig. 11.37) was synthesized by *Autologic* (Mentor). With signal rb = 0 (r bar, *active low*) the counter is reset to 0000b. Clock input is cp (*clock pulse*). If the input en = 1 (*enable*) the counter will count up. For en = 0 the actual state does not change with the clock. Output

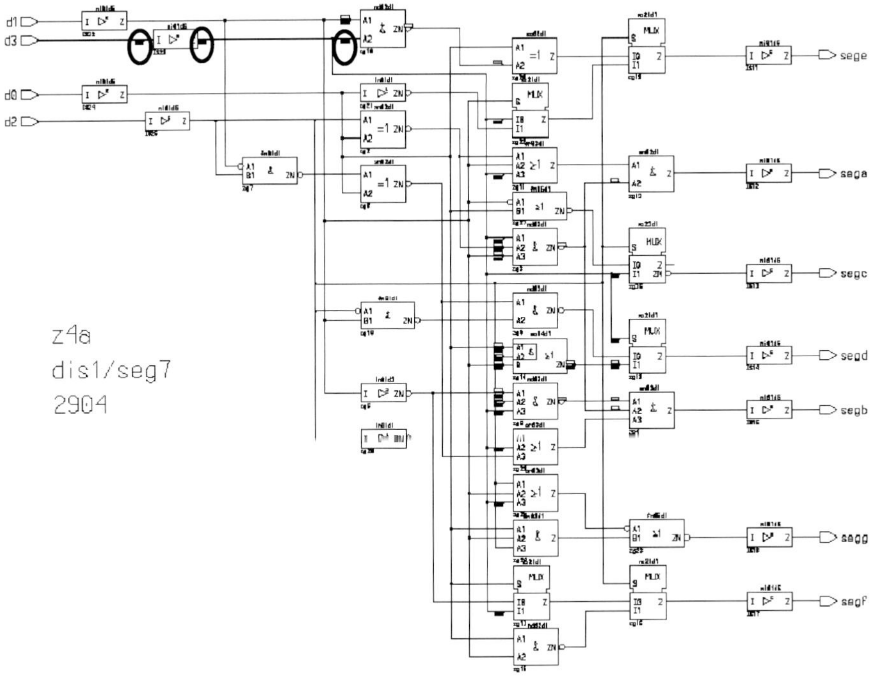

Fig. 11.35 Undetected faults because of incomplete test stimuli

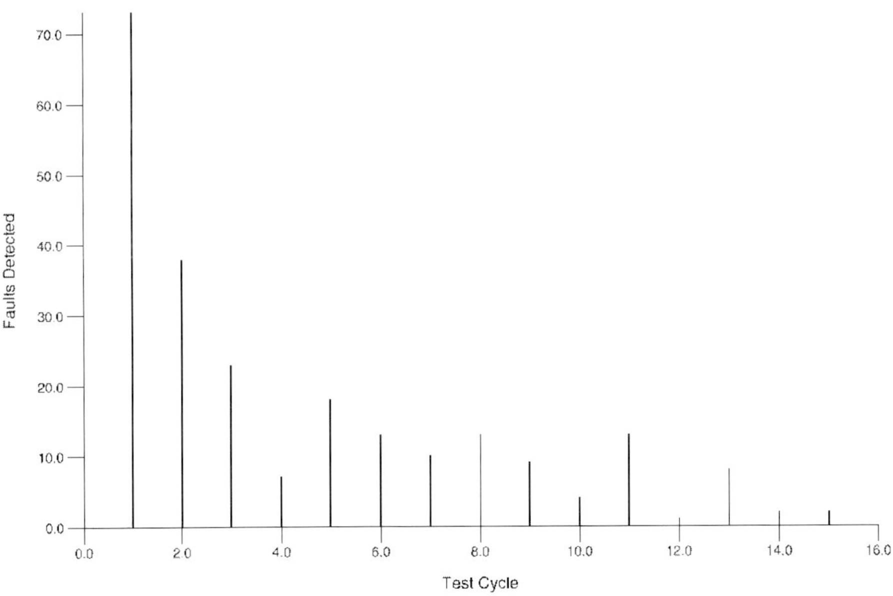

Fig. 11.36 Histogram for fault coverage of 100 %

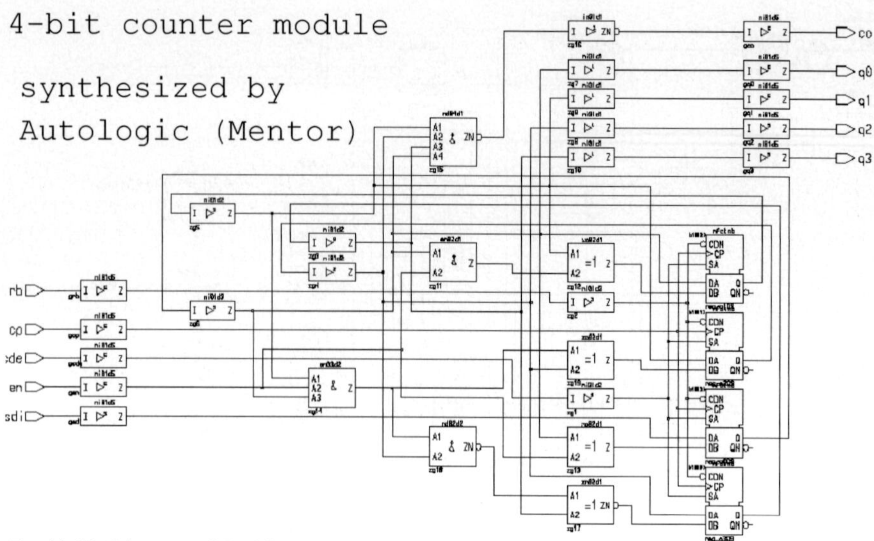

Fig. 11.37 Schematic of the 4-bit counter module

co (carry out) is set to 1 after the change of $q_3 \dots q_0$ from 1110b to 1111b. The counter has a scan path. To scan data in, sde = 1 (*scan data enable*) is used. Data of the input sdi (*scan data input*) are clocked in serially with clock cp into the flop-flops of the module.

Design of the first test stimuli for the 4-bit counter

The following considerations influence the design of test stimuli. With rb = 0 the 4 flip flops are reset to 0000b. For every state of the counter it is necessary to find out if it does not change while en = 0. The carry bit must be activated. To achieve the necessary states the designer draws up the necessary test vectors (list 11.10). Comments explain the vectors described. There effectiveness of test stimuli is checked with fault simulation.

List 11.10 Test vectors first version, 111 test vectors

```
CIRCUIT cnt04;
TIMEDEF 1 PERIOD = 100NS ;

OUTPUT co,q0,q1,q2,q3;
INPUT rb,cp,sde,sdi,en;

/*-----------------------------------*/
/* Definiton of inputs               */
/*-----------------------------------*/
rb=1; cp=0; sde=0; sdi=0; en=0$

/*-----------------------------------*/
/* RESET aktivated                   */
/*-----------------------------------*/
rb= 0 1$

/*-----------------------------------*/
/* ...Counter counts and             */
/* stops for...18 states             */
/*-----------------------------------*/
DO 18 TIMES {en =0$ cp = 1 0$
             en =1$ cp = 1 0$}
```

Discussion of the results of the fault simulation with the first version of test stimuli

With the first outline of the test stimuli a fault coverage of 85 % is achieved (list 11.11).

List 11.11 Log of fault simulation with the first version of test vectors

```
Total Faults        :  194
Unsimulated Faults  :  0

RESULTS FROM LAST RUN
--------------

Run Time                 :  14000.0ns
Total Faults             :  194
  Untestable Faults      :  8
  Testable Faults        :  186
    Undetected Faults    :  20       (10.75%)
    Detected Faults      :  158      (84.95%)
    Possible Faults      :  8        (4.30%)
```

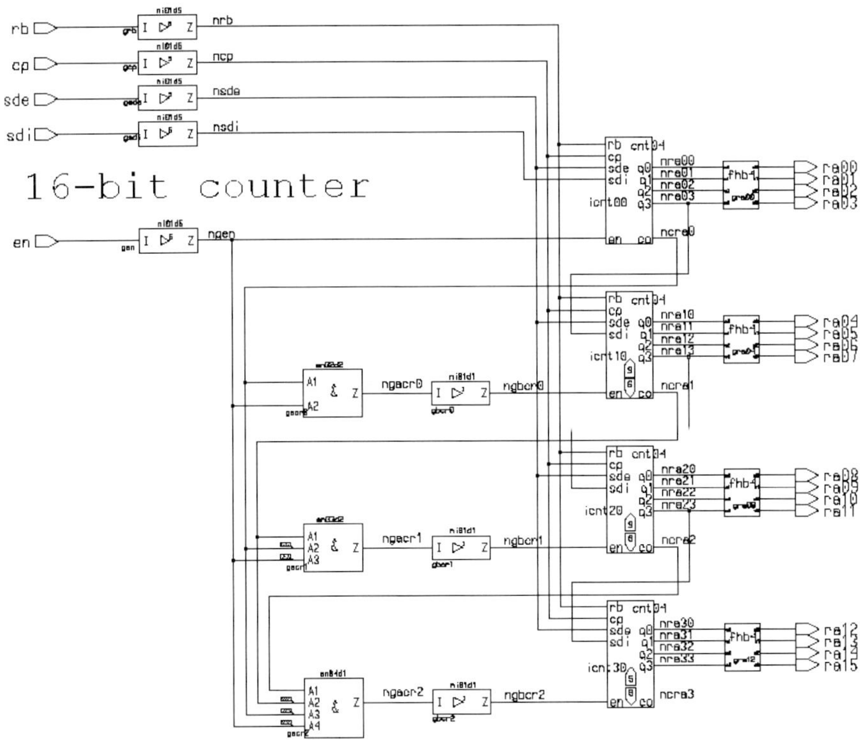

Fig. 11.38 4-bit counter with undetected faults; first version of test stimuli

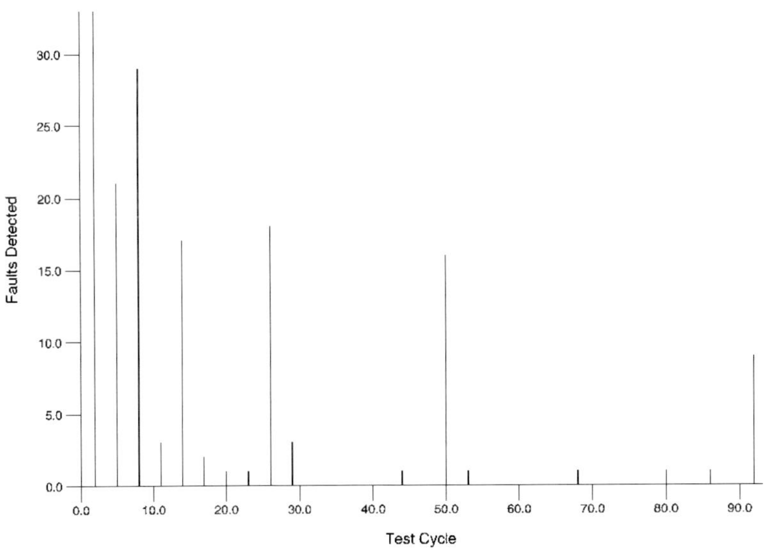

Fig. 11.39 4-bit counter histogram; first fault simulation

11

As explained previously, in fig. 11.34 untestable faults are outputs with no connections to elements. Obviously their behavior can not be observed at primary outputs. Possible faults will be explained later in this chapter. To find out undetected faults the schematic is examined

Most of the undetected faults are connected to the scan path which was not activated until now. The histogram (fig. 11.39) shows that after test vector 92 no additional faults have been detected. Activating test vector 92 the counter reaches 1111b. The test vectors after test cycle 92 can be deleted without affecting the number of detected faults.

Explanation of test vectors

Every fault simulator produces a timing diagram of the fault free circuit with the given test stimuli. This timing diagram is used to discuss the test vectors and the reaction of the circuit.

Applying the first test vector with rb = 1 and the rest of the input signals are 0, the output signals with undefined (depicted with an arrow between 0 and 1) will go to 0. With rb = 0 the circuit is reset and all outputs remain at 0. If rb goes to 1 again, it becomes inactive. The next vector is the first of a loop which is executed 18 times, en = 0, and the counter will not count with the next positive clock edge. Thus there is no change created by the next test vector in which cp goes to 1. Neither does the next test vector change any value when cp returns to 0 again. But the following test vector sets en to 1 and therefore q_0 goes to 1 with the next clock pulse. Clock cp goes low again and nothing changes. This was the first pass of the loop which will be executed 18 times. There are 6 test vectors for each pass in which the counter is incremented.

As mentioned before, the schematic shows that faults which belong to the scan path are not detected. Which raises the question of why the scan path has not yet been used in test. Because each state of the counter and its transition to the following state must be tested, naturally this test is performed by incrementing the counter. Therefore it is not necessary to reach a distinct state of the counter using the scan path. But the situation is different if at state 1111b the function of the reset will be tested. In this case, to reach 1111b from 0000b, 15 clock pulses are necessary which must be compared with 4 clock pulses to scan in 1111b directly. Obviously the scan path reduces the number of test vectors greatly. This will be explained in what follows.

In complex circuits scan path is state of the art.

Some details to possible faults

It is interesting that 4 % of the faults are called '*possible faults*'.

Possible faults will be explained according to the definition in the fault simulator *Quickfault*:

> Possible faults are faults which are related to the hardware test of the finished circuit.

They are related to the reset path of this counter, that is to the paths which deal with signal *rb* (fig. 11.41).

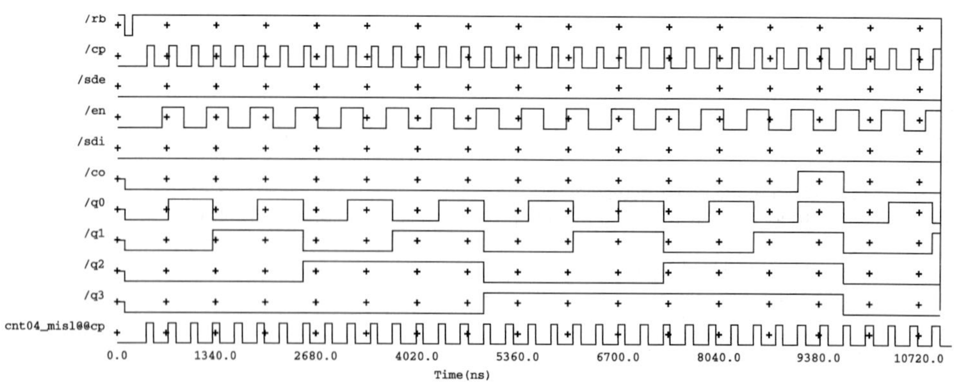

Fig. 11.40 Timing diagram of 4-bit counter, first fault simulation

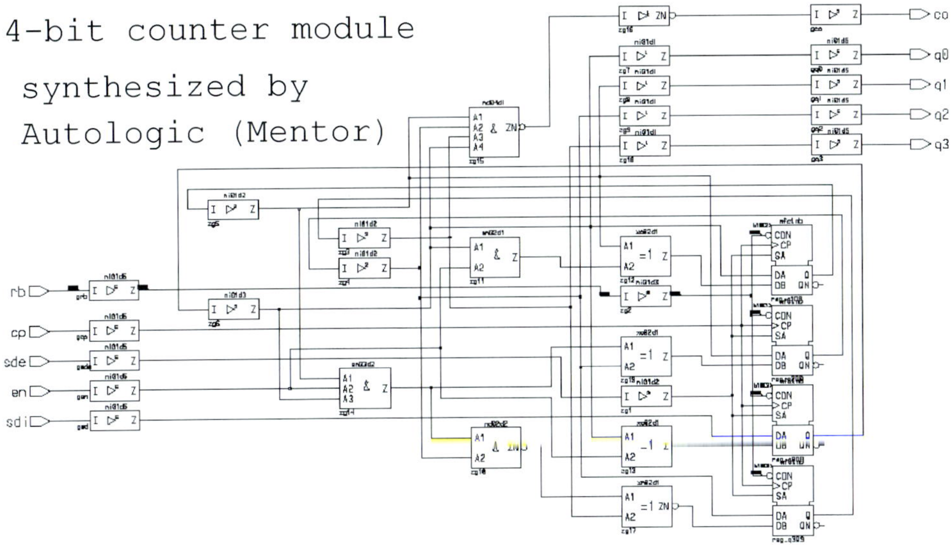

Fig. 11.41 4-bit counter with possible faults

What is the special quality of a possible fault? If the power supply is switched on the output value of a flip flop is undefined, it is either 0 or 1. After reset the output q of every flip flop is set to 0. If accidentally the output value of the flip flop is already 0 and at the same time there is a fault in the reset path, a test system would recognize the expected correct value 0. In this case a test system would not detect the fault with this test vector. That is why this sort of fault is called a *possible fault*.

If the test program is improved the number of possible faults will drop to 0 and the number of detected faults will increase.

Improvement of test stimuli

With the help of the scan path the test stimuli are enhanced. If all flip flops with the help of the scan path are set to 1, then after reset all q outputs of the flip flops will go to 0. If there is a fault in the reset path this fault will be detected and the possible fault will be removed from the fault list.

In addition the scan path is activated for the first time. All flip flops are set using *sdi*, which is the serial input of the scan path. With this action faults of the scan path will be detected likewise.

Using scan path the counter is set to 14, 1110b (/* data shift */). This state could be reached similarly by counting but this comparable procedure needs more clock pulses.

Subsequently the counter counts to 15 (/* next state */) and is reset afterwards. Test vectors of the enhanced test stimuli are in list 11.12.

List 11.12 4-bit counter enhanced test stimuli: Fault coverage 100 %

```
CIRCUIT cnt04;
TIMEDEF 1 PERIOD = 100NS ;

OUTPUT co,q0,q1,q2,q3;
INPUT  rb,cp,sde,sdi,en;

/*-----------------------------------*/
/* Definition of inputs              */
/*-----------------------------------*/
rb=1; cp=0; sde=0; sdi=0; en=0$

/*-----------------------------------*/
/* RESET activated                   */
/*-----------------------------------*/
rb= 0 1$

/*-----------------------------------*/
/* ...,counts, does not count...     */
/*-----------------------------------*/
DO 18 TIMES {en =0$ cp = 1 0$
en =1$ cp = 1 0$}

/*-----------------------------------*/
/* SCAN mode, shift from q0.>.q3     */
/*-----------------------------------*/
sde=1; en=1$
sdi=1$ cp = 1 0$         /* data shift */
```

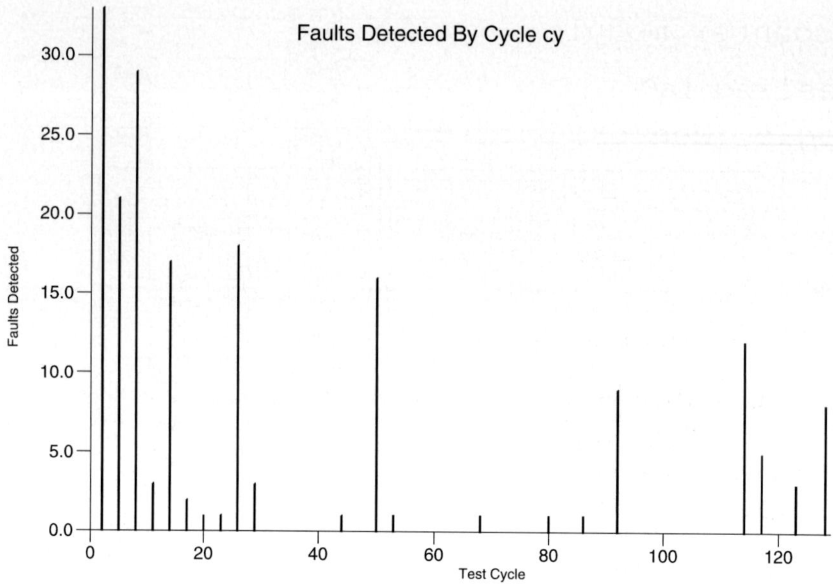

Fig. 11.42 Histogram of fault detection for 4-bit counter with 100 % fault coverage

```
sdi=1$ cp = 1 0$
sdi=1$ cp = 1 0$
sdi=0$ cp = 1 0$
sde=0$ cp = 1 0$          /* next state */
/*-----------------------------------*/
/* counter is 1111                   */
/* for reset test                    */
/*-----------------------------------*/
rb = 0 1$
```

4-bit counter with enhanced test stimuli: fault coverage of 100 %

With the additional test vectors all s.a.0 faults and s.a.1 faults are detected. The number of detected faults increased to 186 because all possible faults could be detected. Consequently their number dropped to 0 (list 11.13). Note that the number of untestable faults is still 8, because the outputs have not been connected to primary outputs.

List 11.13 4-bit counter final result

```
CURRENT STATUS
---------------------
Total Faults         :  194
Unsimulated Faults :  0
RESULTS FROM LAST RUN
---------------------
Run Time                :  13000.0ns
Total Faults                   :  194
  Untestable Faults            :  8
  Testable Faults              :  186
    Undetected Faults          :  0      (0.00%)
    Detected Faults            :  186    (100.00%)
    Possible Faults            :  0      (0.00%)
```

In fig. 11.42 a histogram shows the distribution of number of detected faults related with individual test pattern

An additional look at the timing diagram shows the behavior of the fault-free circuit. Taking the originally designed test pattern, the final state of the counter is 2, that is 0010b (fig. 11.43). To clock in test data *sde* must be 1. At the same time the signal *en* must be 1 during the process of loading flip flops as well. To load a 1 the signal *sdi* must be set to 1. With the next clock pulse a 1 is shifted in the first flip flop of the scan path. The result is the state 5 which is 0101b because the 1 of q_1 was shifted with this clock pulse into q_2. After reaching state the 1110b the scan in process is finished with sde = 0. The counter reaches 1111b with the next clock pulse and the signal carry changes to 1 as expected. In this state the function of reset is tested with rb = 0. With this test vector all *s.a.1* and *s.a.0* faults in the reset path are detected.

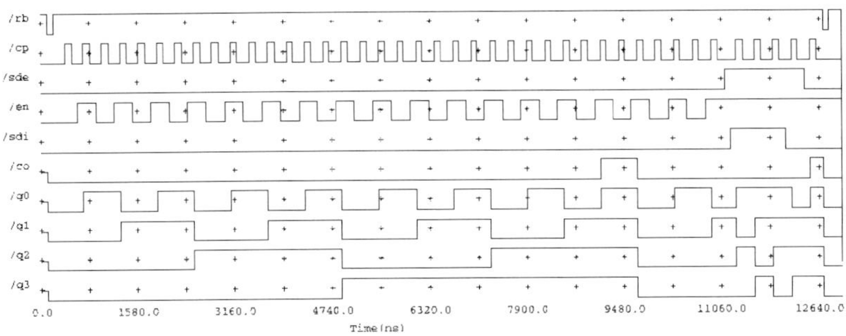

Fig. 11.43 Timing diagram of 4-bit counter: fault coverage 100 %

11.9.2 Design Example 16-bit Counter

The advantages of a modular and structured design using pre tested modules are demonstrated in the following. The 16-bit counter is designed with cascaded modules of the 4-bit counter.

Specification of the 16-bit counter

The new module 16-bit counter (fig. 11.44) uses the same inputs as the module 4-bit counter: rb, en, cp, sde, sdi.

The counter is reset with input signal rb = 0 (*reset bar*). The clock is connected to the input signal cp. Only if en = 1 is the counter counting up. To use the scan path the input signal *sde* must be set to 1. The 4 modules of 4-bit counters are concatenated, forming a serial path. The input signal *sdi* provides the serial data for the scan path. Output q_3 of 3 modules are connected to input *sdi* of the subsequent module. Counter outputs are $q_{15} \ldots q_0$. Serial data out of the scan path is q_{15} of the counter.

Advantages to use scan path for fault simulation of the 16-bit counter

The counter could also be tested just by incrementing the counter using 65536 clock pulses. Additionally control signals like input en would have to be changed equally, which adds up to a huge number of test vectors. This number of test vectors could be reduced greatly by utilizing scan path.

In order to test the function of the 16-bit counter all carry out signals *co* must function and be processed likewise correctly by the subsequent module. As an example the output signal *co* of the third counter module, which is called *icnt20*, is considered. The output signal *co* is called *ncra2* and is connected to logic which controls the input *en* of the fourth module called *icnt30*. If this carry would be activated by incrementing the counter, 32768 clocks would be needed.

If the scan path is used only 16 clock pulses are applied to set the state for testing *ncra2*.

In the following all flip flops are loaded by using a macro in order to keep the loading procedure standardized.

Macro for loading the counter

The scan path is used many times in testing. It is useful to write a macro for loading the counter.

The macro is called *SCANin* (list 11.14). The state of the counter is the parameter $id_{15}, \ldots, id_{00}$ and it is loaded serially through input *sdi*. To enable loading, *sde* must be 1. Here is an example for loading the counter with 0000000011111111: the macro call is SCANin(0.0.0.0.0.0.0.0.1.1.1.1.1.1.1.1). The most significant bit that corresponds to output pin *ra15* is the first bit to be loaded. It is the first 0 in this pattern. Clock pulses are included in the macro.

List 11.14 Macro SCANin

```
/*-----------------------------------*/
MACRO SCANin(id15,id14,id13,id12,
             id11,id10,id09,id08,
             id07,id06,id05,id04,
             id03,id02,id01,id00)
{
sde=1$
sdi=id15 $ cp = 1 0$    /* data shift */
sdi=id14 $ cp = 1 0$
sdi=id13 $ cp = 1 0$
sdi=id12 $ cp = 1 0$
```

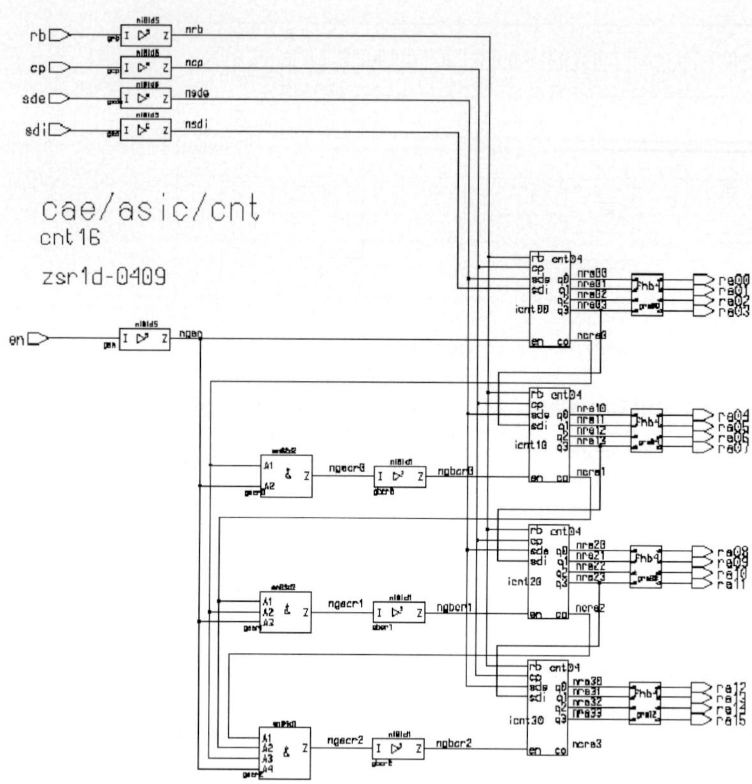

Fig. 11.44 Schematic of 16-bit counter module

```
sdi=id11 $ cp = 1 0$
sdi=id10 $ cp = 1 0$
sdi=id09 $ cp = 1 0$
sdi=id08 $ cp = 1 0$
sdi=id07 $ cp = 1 0$
sdi=id06 $ cp = 1 0$
sdi=id05 $ cp = 1 0$
sdi=id04 $ cp = 1 0$
sdi=id03 $ cp = 1 0$
sdi=id02 $ cp = 1 0$
sdi=id01 $ cp = 1 0$
sdi=id00 $ cp = 1 0$
}
```

Testing least significant 4-bit module icnt00

At the beginning of the test the least significant 4-bit module icnt00 is activated. The first part of the test is the same as the one for the original 4-bit counter. That shows that using modules is an additional advantage in test pattern design. Scan path is not yet used.

Testing carries between 4-bit modules

The functions of three carry bits ncra2 ... ncra0 have to be tested. The procedure is explained for *icnt10*. The initial state fort his test is 00ffh. Scan in is switched off by sde = 0. With en = 0 and cp = 1 the counter must keep its state unchanged for this test vector. The next test vector is cp = 0 and en = 1. With the following cp = 1 the counter counts to 0100h. The same procedure applies for the other modules.

Testing individual modules icnt30, ..., icnt10

Module icnt00 was tested already with the test vectors of the original 4-bit counter.

The other tree modules are tested in a similar way. For this purpose a structure for testing is set up in (table 11.8).

Details of the test vectors are explained on the basis of the module *icnt20*. By means of the scan

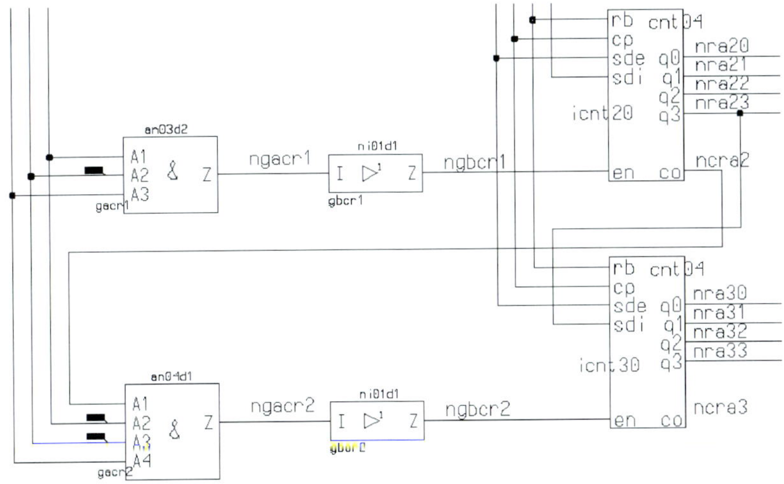

Fig. 11.45 Three undetected faults

path the pattern 00ffh is loaded. With en = 0 and cp = 1 the counter must keep its state. The next test vector with cp = 0 and en = 1 prepares counting. With cp = 1 the counter will count to 0100h. After that test the counter is loaded by the scan path with 01ffh. Serial loading is switched of with sde = 0. Again, the next test vector with cp = 0 and en = 1 prepares counting. With cp = 1 the counter will count to 0200h. At the end of this part of the test cp is reset to 0.

The similar procedure is performed for icnt10 and icnt30.

Testing the remaining undetected faults

With the test vectors there are three faults left undetected in the circuit (list 11.15).

List 11.15 Result of fault simulation with 96 % fault coverage and possible faults

```
Total Faults              :  896
  Untestable Faults       :  50
  Testable Faults         :  846
    Undetected Faults     :  3       (0.35%)
    Detected Faults       :  809     (95.63%)
    Possible Faults       :  34      (4.02%)
    Hyperactive Faults    :  0       (0.00%)
    Hypertrophic Faults   :  0       (0.00%)
    Oscillatory Faults    :  0       (0.00%)
    HM Dropped Faults     :  0       (0.00%)
```

In order to find test vectors to detect these faults, the marked faults of the schematic are examined.

The three faults are at the inputs of the AND gates (an04d1, an03d2) controlling the input en of module *icnt20* and *icnt30* (fig. 11.45). In order to detect these faults the outputs ncra0 and ncra1 must be forced to 0. The signals refer to the inputs A2 of the AND gate *an03d2* and the inputs A3 und A2 of the AND gate *an04d1* (fig. 11.45). This can be accomplished by the three test vectors in list 11.16. Because of the explanations of the macro SCANin the commands need not be discussed in detail.

List 11.16 Test vectors to detect the last three faults

```
SCANin(1,1,1,1, 1,1,1,1, 1,1,1,1, 0,0,0,0);
sde=0$ en=0$ cp=1 0$ en=1$ cp=1 0$

SCANin(1,1,1,1, 1,1,1,1, 0,0,0,0, 1,1,1,1);
sde=0$ en=0$ cp=1 0$ en=1$ cp=1 0$

SCANin(1,1,1,1, 0,0,0,0, 1,1,1,1, 0,0,0,0);
sde=0$ en=0$ cp=1 0$ en=1$ cp=1 0$
```

The result of the expanded test shows fault coverage of 96 % (list 11.17). Untestable faults refer to the outputs which are not connected to primary outputs. As explained previously the possible faults are the s.a.0 faults of the flip flops. If additional test vectors similar to the test of the original 4-bit counter would be added, a fault coverage of 100 % is achieved.

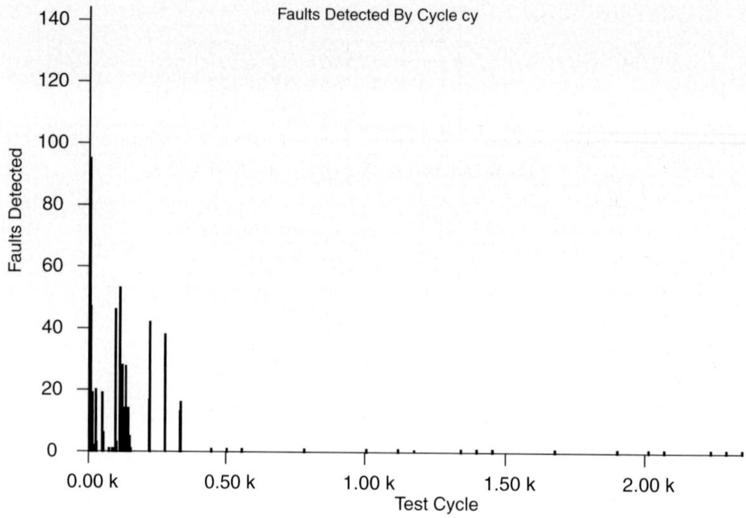

Fig. 11.46 Histogram of fault detection for 16-bit counter with 96 % fault coverage

List 11.17 Result of fault simulation all faults detected, possible faults excluded

```
CURRENT STATUS
----------------------
Total Faults                  :  896
Unsimulated Faults            :  0

RESULTS FROM LAST RUN
----------------------
Run Time                      :  400000.0ns
Total Faults                  :  896
  Untestable Faults           :  50
  Testable Faults             :  846
    Undetected Faults         :  0     (0.00%)
    Detected Faults           :  812   (95.98%)
    Possible Faults           :  34    (4.02%)
```

11.9.2.1 Final Remarks to the Fault Simulation of the 16-bit Counter

All faults are detected. Untestable faults and possible faults had not been removed. The structured approach listed in table 11.8 shows the structure of the test program for fault simulation. After reset of the counter the first counter module *icnt00* is examined. Before testing the other counter modules, the carry bit between two counter modules is activated. Now the counting of the modules *icnt10*, *icnt20* and *icnt30* is examined. Finally, the three remaining faults are checked by forcing specific carry bits to 0. This procedure can be modified for testing similar structures.

In fig. 11.46 a histogram shows the distribution of numbers of detected faults related to individual test patterns.

11.10 Necessity of Various Design Tools in the Design Process

The way in which a designer designs his circuit greatly depends on the workbench and the tools he uses. Logic simulation and fault simulation are essential as powerful tools in the development process of a circuit.

There are several approaches to describing the function of a circuit. In many areas source code for describing a circuit usually is written in VHDL. It is still necessary to analyze the synthesized circuit at the schematic level during the verification process. A designer who has to write test vectors for his circuit depends in many cases on the hardware which was being synthesized.

The display of a histogram shows clearly the effectiveness of the test vector and helps to optimize the test procedure. Faults printed into a schematic help to identify the problems associated with them. Therefore graphical representation of complex results is an important feature which eases the use of design tools.

Table 11.8 Structure of the test of the 16-bit counter

Circuit part or signal to be tested	Expected result	Remarks
all flip flops	Reset of the counter, state 0000h	rb = 0
module icnt00 ($q_3 \ldots q_0$)	Counting of module icnt00	Original test vectors of 4-bit counter
carry ncra0	Counter counts from 000fh to 0010h	Scan in 000fh, cp und en
carry ncra1	Counter counts from 00ff to 0100h	Scan in 00ffh, cp und en
carry ncra2	Counter counts from 0fffto 1000h	Scan in 0fffh, cp und en
al modules	Counter counts from ffffh auf 0000	Scan in ffffh, cp und en
icnt10	Counter increments its state by one	Scan in 0000b to 1111b for icnt10
icnt20	Counter increments its state by one	Scan in 0000b to 1111b for icnt20
icnt30	Counter increments its state by one	Scan in 0000b to 1111b for icnt30
not ncra0	ncra0 = 0	Scan in mit fff0h
not ncra1	ncra1 = 0	Scan in mit ff0fh
not ncra2	ncra0 = 0 and ncra2 = 0	Scan in mit f0f0h

11.11 Limits of Digital Simulation and Treatment of Complex Circuits

To develop ASICs the use of powerful tools is mandatory. So a development engineer may design his circuit step by step and integrate the intermediate results in his software environment. He verifies the design at various levels to identify faults as soon as possible if they occur during the design process. Thus the cost of eliminating faults is minimized.

If digital simulation guarantees that the design is fault-free, it depends on various factors:

- Accuracy of simulation models and the related parameters;
- Quality of test vectors;
- Definition of the function of the circuit.

It is obvious that simulation results depend heavily on the models used.

In the same way the result of a simulation represent the reaction of the circuit to the test vectors defined by the designer. With them he is able to create critical situations for finding out whether the reaction of the circuit is correct.

A difficult task is to define the function of the circuit. For real circuits it is quite impossible to describe the complete function. There is always the danger of overlooking parts of the function which may malfunction because of insufficient definition.

Complex circuits implemented in ASICs, such as controllers with various interfaces are difficult to simulate as a complete circuit. The simulation time can be extremely long. In this case it is advisable to generate code for an FPGA. If the source code is in VHDL the additional effort can be justified. In this case the emulation of the hardware substitutes for the simulation of the complete circuit. Finding design faults may be more difficult in emulation. One advantage of the FPGA is that the circuit's speed is close to that of an ASIC. For testing, stimuli are applied by a pattern generator. A logic analyzer and oscilloscope help to evaluate the reaction of the circuit. Additionally, the real environment can be used as a test bed for the hardware (fig. 11.47).

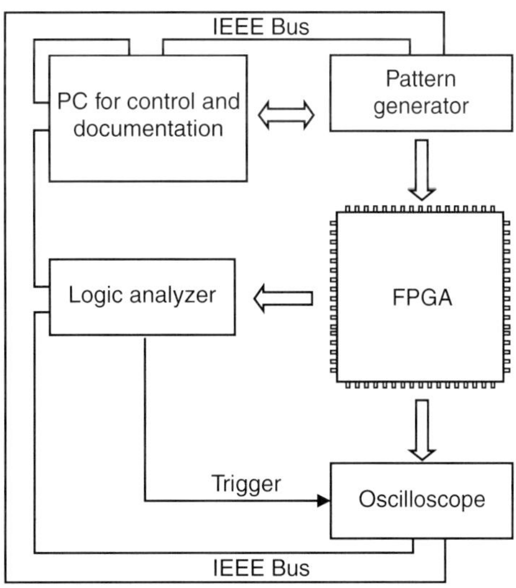

Fig. 11.47 Emulation with FPGA

11.12 References

[11.1] *Abramovici, M.; Breuer, M. A.; Friedmann A. D.*: 'Digital System Testing And Testable Design.' – New York: Computer Science Press, 1990

[11.2] *Eveking, H.*: 'Verifikation digitaler Systeme.' – Stuttgart: B. G. Teubner, 1991

[11.3] *Prosser, F. P.; Winkel, D. E.*: 'The Art of Digital Design.' – Englewood Cliffs: Prentice Hall, 1987

[11.4] Industry Brochure: 'LOGIC2, Einführung.' Industrieschrift 1995.

[11.5] *Erhardt, D.; Schulte, J.*: 'Simulieren mit PSPICE.' – Wiesbaden: Vieweg, 1995

[11.6] *Smith, M. J. S.*: 'Application-Specific Integrated Circuits.' – Reading, MA: Addison-Wesley, 1998

[11.7] Design Automation Standards Committee of the IEEE Computer Society, P1497 'DRAFT Standard for Standard Delay Format (SDF)'

[11.8] IEEE 1076.4 TAG (Technical Action Group): 'Standard VITAL ASIC Modeling Specification.' IEEE P1076.4

[11.9] IEEE Standard 1149.1: 'Test Port And Boundary-Scan Architecture.' IEEE 1149.1

12 Mixed Signal Simulation

MARTIN RIEGER

12.1 Overview

The term **mixed signal simulation** signifies a simulation of a circuit where parts of which are described at different levels of abstraction and are simulated simultaneously [12.6]. In this chapter particularly the simultaneous simulation of analog and digital parts of a circuit is discussed.

The *necessities and motivations* for applying mixed signal simulation to a system are listed below.

- Many important circuit blocks are intertwined analog and digital circuits, e.g., Analog Digital Converters (ADC), Digital Analog Converters (DAC) or Phase Lock Loops (PLL).
- More and more digital circuit blocks are described at an abstract level, e.g., in VHDL. If other digital and/or analog parts of the system have to be simulated, the result is again mixed signal simulation.
- Sometimes the different parts of a system have a different development maturity. The more elaborate parts are described at a logical level, for instance, whereas other parts are described only at a more abstract level.
- Another reason for using different abstract levels is given by complex components (e.g., a CPU) being available only as models at a high level of abstraction whereas other circuit parts have to be simulated at a more detailed level.
- One can reach the optimal simulation speed and accuracy only if the critical parts which determine the accuracy are described at a level of higher abstraction, and therefore can be simulated much faster.

12.2 Simulation on different levels of abstraction

The levels of abstraction discussed in this chapter are

- transistor (or circuit) level;
- logic level;
- register transfer level;
- system level (see fig. 12.1) [12.5].

The models in the different levels of abstractions represent different views of the parts of the system. The quantities used for description differ according to the level of abstraction (e.g., voltages or logic levels, respectively), also to the algorithms applied. Therefore one needs connecting **interfaces** when simulating at different levels of abstraction. When climbing up the hierarchy the complexity increases, the details are more or less lost, and the simulation effort decreases.

At the **transistor (or circuit) level** the models of the components are described by algebraic equations often in the form of differential equations. Time and value range are continuous, the quantities used at the analog nodes are voltage and current.

The circuit used as an example for transistor level description in fig. 12.1 represents a two input NAND gate. The performance of circuits of this kind can be checked by analog (or circuit) simulation. For that purpose the simulator has to solve the non-linear, coupled differential equation which describes the circuit. This process is very time consuming. The most popular analog simulator is SPICE.

At the **logic level** the models of the components are described by Boolean equations, truth tables, rise, fall, delay times, etc. Time and value range are discrete. The quantities used at the digital nodes are named 'Bit' and are logic levels, e.g., with the values 0, 1, X, R, F ... These values can be associated with different driving forces.

The circuit used as an example for logic level in fig. 12.1 represents a D flip flop consisting of NAND gates. It could be possible to simulate one of the NAND gates (e.g., the one leading to the output Q) at transistor level. Simulators for logic circuits mostly use event driven algorithms, which leads to faster simulations compared to analog simulators.

- one group of simulators uses a **separate simulation** program for each level of abstraction and the communication is carried out by means of an interface program (example: Continuum of Mentor Graphics).
- another group of simulators covers all levels of abstraction in one **comprehensive simulator** which comprises all necessary models, signal descriptions, and algorithms (examples: PSPICE, VHDL-AMS).

In a recent publication a new comprehensive simulator is presented which supports the separation of a circuit with respect to different levels of abstraction [12.7].

12.3.2 Separate Simulators

In this approach a separate simulator is used for each level of abstraction. The single simulators are coupled by means of interface programs and a co-ordination program. Generally this program has a graphical user interface which enables the user to input the circuit description with the simulation control and to organise the output of the results. The co-ordination program also performs the overall control of the separate simulators. This approach is suitable for systems with clearly defined circuit blocks which can be assigned to the different levels of abstraction like in fig. 12.4. Typically these circuit blocks are rather large. The connection between the blocks is formed by many different signals. The approach with separate simulators has several advantages, e.g., the simulation within one level of abstraction can be performed with simulators which are existent and reliable.

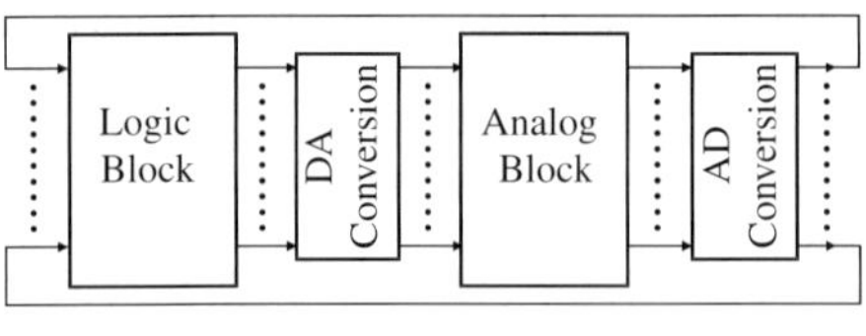

Fig. 12.4 Partition of circuit blocks when using separate simulators

In this way it is also possible to simulate systems which comprise more than two levels of abstraction, provided that the necessary simulators are available. For example, an analog simulator could be used for the transistor level, a logic simulator for the logic level, and the VHDL simulator for RTL. As mentioned above, an example of this approach is the program Continuum of Mentor Graphics.

12.3.3 Comprehensive Simulator (Example: PSPICE)

In this approach all levels of abstraction which exist in a certain system are covered by one simulator only. The simulator also comprises interfaces in order to map the different signal types of the different levels of abstraction to each other. This approach is especially successful if the levels of abstraction are strongly intertwined, as is the case in fig. 12.5. Here the single blocks are relatively small and there are multiple connections between the blocks. The approach with a comprehensive simulator is efficient when dealing with small systems with intertwined levels of abstraction. The overhead for changing simulators is not necessary, which is a decisive advantage when dealing with small systems.

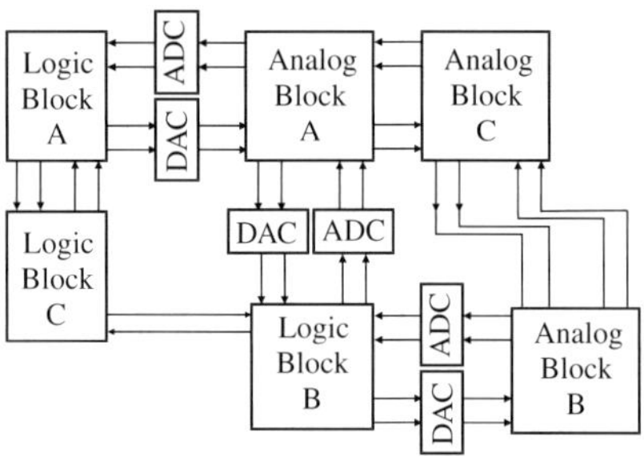

Fig. 12.5 Partition of circuit blocks when using a comprehensive simulator

The following paragraph discusses **PSPICE** as an example of a comprehensive simulator. The components of the logic level are described as time models (for delay times) and as I/O models (for input/output properties). If the digital component has connection to analog components automatically AD or DA interfaces, respectively, are inserted, the models of which are included in the model description of the digital component.

A digital node can assume the states listed in Table 12.1.

Table 12.1 CMOS logic states with PSPICE

State	Meaning
0	false
1	true
R	rising slope
F	falling slope
X	unknown
Z	high impedance

The simplified **model of a NAND gate** with interfaces is shown hierarchically in fig. 12.7.

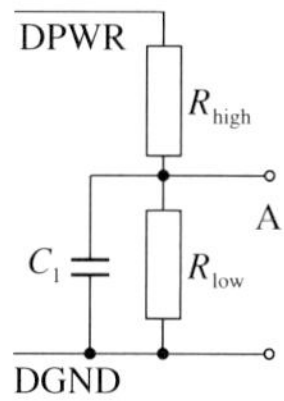

Fig. 12.6 Equivalent circuit for a DA interface

```
.subckt MyNand A B Y
U1 nand(2) DPWR DGND A B Y
+D_MyGate IO_My .ends

.model D_MyGate ugate (
+   tplhmn=20      pstplhty=25ps
+   tplhmx=30      pstphlmn=25ps
+   tphlty=45      pstphlmx=40ps)

.model IO_My uio (
+drvh=25.5          drvl=18.7
+inld=0.01pF
+AtoD1="AtoD_My"    AtoD2="AtoD_My_NX"
+DtoA1="DtoA_My"    DtoA2="DtoA_My")

.subckt AtoD_My  A D  DPWR DGND
+    params: CAPACITANCE
X1  DGND A     My_CLAMP
X2  A     DPWR My_CLAMP
C1  A DGND {CAPACITANCE+0.01pF}
O0  A DGND DOMy DGTLNET=D IO_My
.ends

.subckt My_Clamp A C
G_Clamp  A C TABLE {V(A,C) }
+ 0.0, 1nA
+ 0.5, 1uA
+ 5.5, 1A
.ends

.model DOMy doutput (
+  s1name="0"   s1vlo=-1.5   s1vhi=1.5
+  s2name="R"   s2vlo=1.5    s2vhi=2.55
+  s3name="R"   s3vlo=2.45   s3vhi=3.50
+  s4name="X"   s4vlo=1.50   s4vhi=3.50
+  s5name="1"   s5vlo=3.50   s5vhi=7.0
+  s6name="F"   s6vlo=2.45   s6vhi=3.50
+  s7name="F"   s7vlo=1.50   s7vhi=2.55 )

.subckt DtoA_My  D A  DPWR DGND
N1  A DGND_OL DPWR_OH DINMy DGTLNET=D IO_AC_DTOA
C1  A DGND {CAPACITANCE+.01pF}
R1  A DGND 1G
.ends

.model DINMy dinput (
+  s0name="0"  s0tsw=10.7ps  s0rlo=1.0   s0rhi=100K
+  s1name="1"  s1tsw=10.7ps  s1rlo=100K  s1rhi=1
+  s2name="X"  s2tsw=10.7ps  s2rlo=104   s2rhi=100
+  s3name="R"  s3tsw=10.7ps  s3rlo=104   s3rhi=100
+  s4name="F"  s4tsw=10.7ps  s4rlo=104   s4rhi=100
+  s5name="Z"  s5tsw=10.7ps  s5rlo=200K  s5rhi=200K )
```

Fig. 12.7 Hierarchical representation of a NAND gate

The logic NAND function is realized in subcircuit ‘MyNand’ with the Boolean function ‘nand(2)’ where A and B are inputs, Y is an output and DPWR, DGND are connections to power and ground, respectively. The timing model ‘D_MyGate’ describes delay times of the NAND gate on the logic level. A special timing model can be assigned to every logical model. The time values signify the minimal, typical, and maximal delay times, respectively, for the rising and falling slope. The given timing model refers to fast 0.5 μm CMOS technology.

The I/O model ‘IO_My’ describes the driving strength at the output (for the 1 state ‘drvh’, for the 0 state ‘drvl’) and the capacitive load represented by an input (‘inld’). In addition, the connection to two AD and DA interfaces is performed, the selection of one of the interfaces is made by a parameter. One of the AD and of the DA interfaces is described in the following paragraph.

The AD interface ‘AtoD_My’ connects the analog node ‘A’ to the automatically inserted node ‘D’ at logical level. The parameter CAPACITANCE is determined by the value of the capacitive load ‘indl’ from the I/O model and applies this load to the analog node. Between the analog node and the power connection a clamping circuit ‘My_Clamp’ is inserted, which is realised by a voltage controlled current source ‘G_Clamp’. The core of the AD interface is formed by the digital output ‘DOMy’, which the different analog voltage ranges are assigned to an appropriate logical value, e.g., ‘0’ for the range $-1.5\ \text{V} \ldots + 1.5\ \text{V}$.

The DA interface ‘DtoA_My’ is inserted between a node ‘D’ of a digital output and an analog node ‘A’. The equivalent circuit of an analog output is shown in fig. 12.6.

The actual DA conversion is carried out in the digital input model ‘DINMy’. For each logical value of the digital output the appropriate resistance values are assigned to the resistors R_{high} and R_{low}. These form a voltage divider with respect to the power supply voltage, and thus the appropriate analog output voltages according to the logical values are generated, e.g., for logic ‘0’ $R_{\text{high}} = 100\ \text{k}\Omega$ and $R_{\text{low}} = 1\ \Omega$, therefore the full power supply voltage is effective at the output (see fig. 12.6).

If the input of the DA interface changes from the old logical value to a new one ‘i’ (with ‘i’ = ‘0’, ‘1’, ‘X’, ‘R’, ‘F’, ‘Z’) the resistors change to their appropriate new values within the time ‘sitsw’. The value of C_1 is taken from the parameter ‘indl’ of the I/O model.

The NAND gates previously discussed are chained in a circuit shown in fig. 12.8, embedded in an analog environment. In the lower half the circuit is modeled at transistor level, in the upper half at logic level. Each connection from the logic level to the transistor level is made via an interface. The differences in modeling become clear if the reaction of the different circuit parts to a pulse pattern applied at the input ‘in’ is appreciated.

The technological background for the circuit discussed is 0.5 μm CMOS technology. On the transistor level the transistors are modeled with the SPICE MOS model (LEVEL = 3) [12.1]. At the logic level the models are taken from fig. 12.7, in which the timing model in D_MyGate is adjusted to the SPICE model.

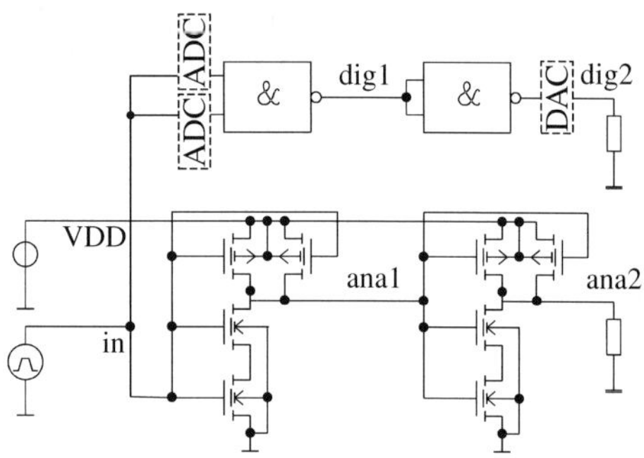

Fig. 12.8 Chain circuit of two NAND gates as logic gates (above) and analog circuit (below)

Considering the graph shown in fig. 12.9, the analog stimulus ‘V_{in}’ is converted into the logical value ‘in’ directly at the input of the first logical gate according to the table given in ‘DOMy’ (fig. 12.7). Because of the relatively long rise and fall times of ‘V_{in}’ not only ‘0’ and ‘1’ are found but also ‘R’ and ‘F’. The digital node ‘dig2’ follows the stimulus with the delay defined in the model. The value of the resistors at the node ‘dig2’ map the logical value to the corresponding analog voltage ‘V_{dig2}’ (see fig. 12.6 and ‘DINMy’).

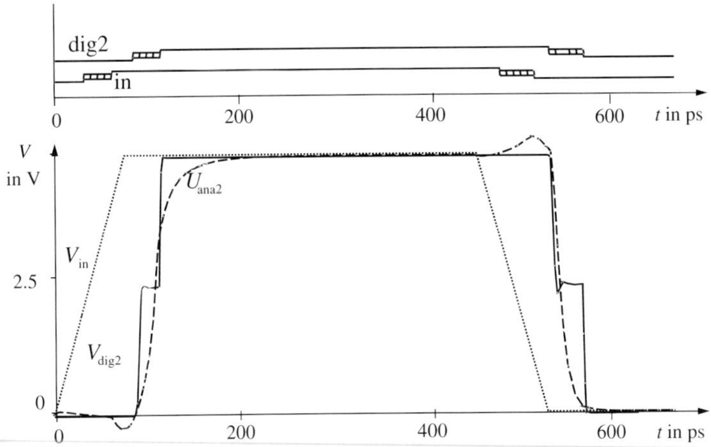

Fig. 12.9 Simulation result for fig. 12.8

The time functions of the voltages 'V_{dig2}' and 'V_{ana2}' correspond to each other quite well especially when considering the delay time at half the maximum voltage value, but not so convincingly with respect to the shape of the rising and falling slope.

The advantage of the digital model is given by the shorter computer time needed for simulation. The circuit with logical models (fig. 12.8 upper part) needs less than 30 % of the computer time compared to the circuit with discrete transistors (fig. 12.8 lower part). If one can do without the analog embedding and works only at logic level, then the computer time needed reduces to 10 %. The savings in computer time will be even more significant, the bigger the circuits are to be modeled at the logic level.

12.4 Application

12.4.1 First example: CMOS Ring Oscillator

In this paragraph a 'ring oscillator' simulation at the transistor and logic levels will be discussed.

Ring oscillators often are used as oscillators with electronically controlled frequency. In addition they are used for the experimental test of gate delay times.

Figure 12.10 shows a five-stage ring oscillator, the simple stages of which are formed by NAND gates used as inverters.

The output of one stage is connected to all inputs of the next stage. The capacitor C_1 takes care of a reliable start of the oscillation after Vdd is applied. The **oscillation frequency** is determined by

$$f = \frac{1}{n \cdot (t_r + t_f)} \tag{12.1}$$

where n is the number of stages and t_r, t_f are the rise and fall time, respectively. These times, and with them the frequency can be altered by variation of V_{DD}. The capacitive load formed by the wiring in not shown in fig. 12.10. The oscillation frequency can be lowered considerably by this load. The technological background is again 0.5 µm CMOS technology. The transistors are defined by a SPICE MOS model (LEVEL = 3) [12.1]. The geometry relation W/L was chosen as 10 for the NMOS and as 30 for the PMOS transistors.

The simulated voltages at the nodes 'in' and 'a1' are shown in fig. 12.11. The switching behaviour is determined by threshold voltages of 1 V and 4 V. With $V_{DD} = 5$ V the rise time is $t_r = 50$ ps and the fall time is $t_f = 100$ ps measured between the threshold voltages mentioned above. Equation (12.1) gives a frequency $f = 1.33$ GHz which corresponds well with the simulated frequency of 1.4 GHz.

If lower oscillation frequencies are required either the gate outputs can be loaded capacitively or the number of stages can be increased.

The power supply voltage V_{DD} has a great influence on the oscillation frequency as can be seen from

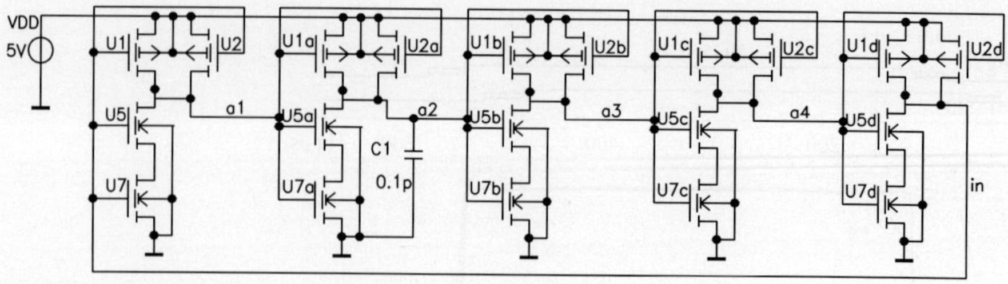

Fig. 12.10 Ring oscillator with CMOS NAND gates represented on the transistor level

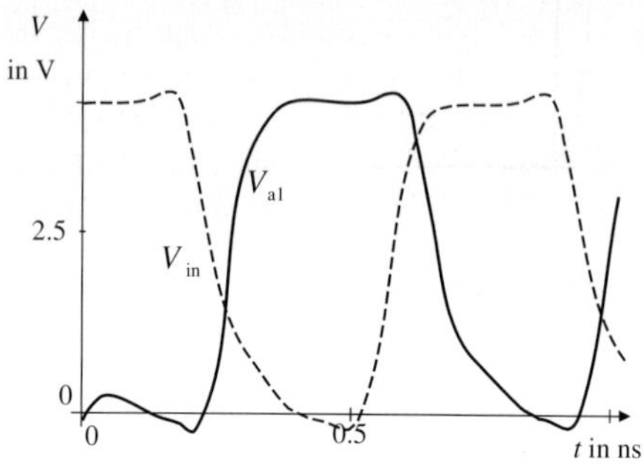

Fig. 12.11 Simulation result for fig. 12.10

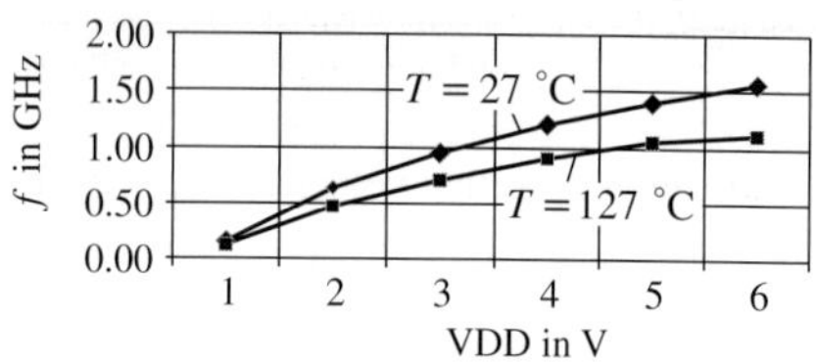

Fig. 12.12 Tuning characteristic for the circuit in fig. 12.10

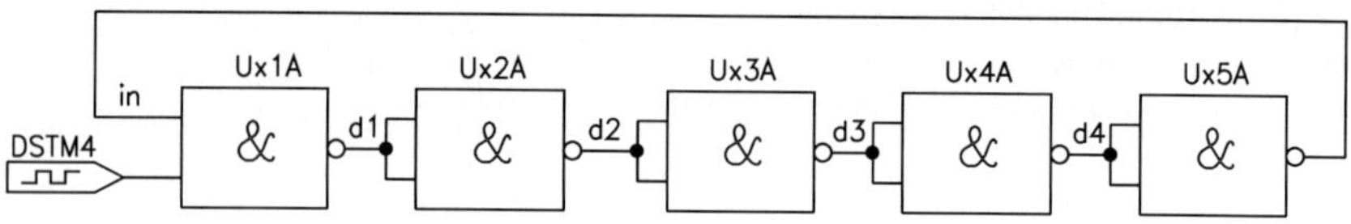

Fig. 12.13 Ring oscillator with CMOS NAND gates represented on the logic level

fig. 12.12. The tuning characteristic of the frequency as a function of V_{DD} is nearly linear, a fact which enables the ring oscillator to be used as a VCO (voltage controlled oscillator). The oscillator frequency is highly dependent on the temperature, therefore the oscillator is hardly usable without being embedded in a controlling PLL circuit.

The modeling of the ring oscillator at transistor level has the advantage that several influences can be studied, e.g., those of V_{DD}, temperature, or capacitive loading.

The effort with respect to simulation time is more than ten times as great as with a simulation of a logic level ring oscillator model, which is shown in fig. 12.13. The fundamental structure of coupled NAND gates is identical to the one shown in fig. 12.10. The stimulus at the input of gate 'Ux1A' is only necessary for a guaranteed start of the oscillation. As expected, the logical circuit also generates an oscillator signal, but the frequency is different from that found with the circuit in fig. 12.10. In order to produce results which are sufficiently precise at logic level the NAND model from fig. 12.7 has to be improved considerably. The timing model 'D_MyGate' has to be extended in order to cover influences of V_{DD} and temperature. In the case of a ring oscillator a small but critical circuit it is advisable to do the modeling at the transistor level.

12.4.2 Second Example: Phase Lock Loop (PLL)

The **simulation of a PLL** discussed in this paragraph is an example for a mixed signal simulation with components defined at different levels of abstraction. The reason for this lies, on one the hand,

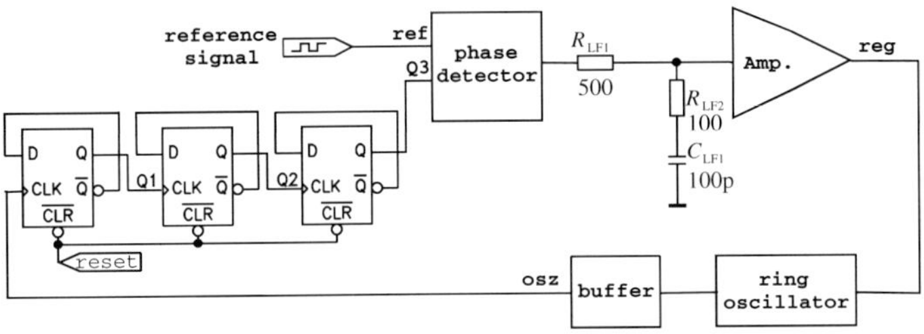

Fig. 12.14 PLL with ring oscillator (fig. 12.10), frequency divider, phase detector, and analog loop filter

in the different nature of the components and, on the other hand, in the degree of going into details of the different circuit blocks. By means of a PLL a VCO is locked to a reference oscillator with respect to frequency and phase. The most important applications for a PLL are to be found in the fields of frequency synthesis and synchronization [12.8].

The PLL circuit discussed here is shown in fig. 12.14. The VCO which is realized by the ring oscillator discussed in the previous paragraph (fig. 12.10) is controlled with respect to its frequency via its power supply voltage. This component is described on the transistor level. The oscillator signal is directed via a buffer to a frequency divider. The purpose of the buffer is, on the one hand, to decouple the ring oscillator from the frequency divider. On the other hand, the buffer has to adapt the output voltage amplitude of the ring oscillator to the logic values of the frequency divider because a change in the control voltage of the VCO not only changes the output frequency but also the output voltage amplitude. Without this adaptation the frequency divider which is modeled at the logic level could produce undefined states. The buffer is modeled partly at transistor, partly at logic level. The frequency divider consists of three slope controlled D flip flops and divides by a factor of eight. At the beginning of the simulation the flipflops are reset.

The digital phase detector modeled at the logic level compares frequency and phase of the divided VCO signal with those of a reference signal which is provided in the form of a digital stimulus.

The output of the phase detector is smoothed by an analog passive loop filter. The following amplifier is described as an analog behavioral model, i.e., a simple voltage controlled voltage source. The actual design of the amplifier has to be done in a later stage of development. The **dynamic properties** of the PLL, important for the transient and locking behavior, are described by a few characteristic constants only [12.9].

The **natural frequency** ω_n of the PLL is determined by:

$$\omega_n = \sqrt{\frac{K_{\text{VCO}} K_{\text{D}} K_{\text{Amp}}}{n(\tau_1 + \tau_2)}} \tag{12.2}$$

where:

K_{VCO}	control slope of the VCO, $K_{\text{VCO}} = \dfrac{2\pi \cdot \Delta f}{\Delta V_{\text{reg}}}$;
K_{D}	gain of the phase detector;
K_{Amp}	gain of the amplifier;
n	dividing factor of the frequency divider;
τ_1	time constant: $\tau_1 = R_{\text{LF1}} \cdot C_{\text{LF1}}$;
τ_2	time constant: $\tau_1 = R_{\text{LF2}} \cdot C_{\text{LF2}}$.

The **damping factor** ζ is determined by

$$\zeta = \frac{1}{2}\omega_n \left(\tau_2 + \frac{1}{K_{\text{VCO}} K_{\text{D}} K_{\text{Amp}}} \right) \tag{12.3}$$

The following parameter values were used for the simulation of the PLL shown in fig. 12.14: $K_{\text{VCO}} = 2\pi \cdot 300\ \text{MHz/V}$, $K_{\text{D}} = 2/\pi\ \text{V}$ and $K_{\text{Amp}} = 1$, $\tau_1 = 50\ \text{ns}$, $\tau_2 = 10\ \text{ns}$.

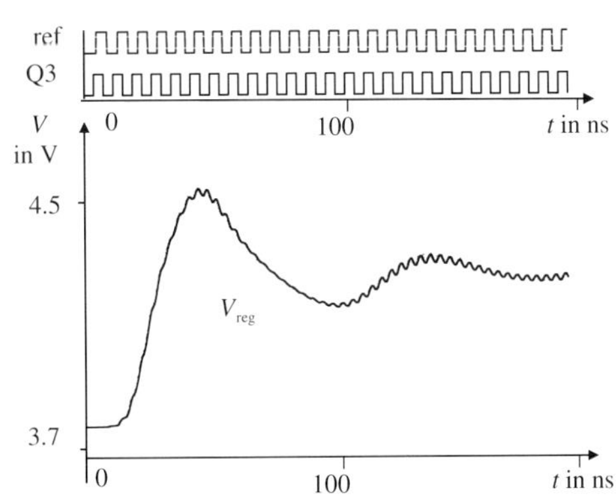

Fig. 12.15 Simulation result for fig. 12.14

With these values the natural frequency $f_n = 8$ MHz and the damping factor $\zeta = 0.3$.

The simulation result for a reference signal with a period of 8 ns is shown in fig. 12.15 with an end time of 200 ns. The derivable values for f_n and ζ correspond well with the calculated values above.

The PLL is an example of a so called '**stiff system**' in which the time constants involved lie widely apart. The frequency of the VCO is usually at least two orders of magnitude greater than the natural frequency of the PLL. The time step of the simulation is determined by the smallest time constant, the duration of the simulation by the largest time constant. This leads to simulation with many steps, therefore here it is of special importance that the computer time needed to perform a single step is minimized. This is accomplished in the example just discussed because decisive parts of the system are modeled at the logic level.

12.5 References

[12.1] *Smith, M.*: 'Application-specific integrated circuits'. – Reading, Massachusetts/Harlow, England u. a.: Addison-Wesley, 1997

[12.2] *Reifschneider, N.*: 'CAE-gestützte IC-Entwurfsmethoden'. – München [i. e.]; Haar; London u. a.: Prentice Hall, 1998

[12.3] *Hoppe, B.*: 'ASIC-Design: Realisierung von VLSI-Systemen mit Mentor V8'. – Berlin; Heidelberg u. a.: Springer-Verlag, 1999

[12.4] *Spiro, H.*: 'CAD der Mikroelektronik'. – München; Wien: Oldenbourg Verlag, 1997

[12.5] *Khakzar, H.*: 'Simulation und Synthese logischer Schaltungen'. – Ehningen: expert-Verlag, 1991

[12.6] *Horneber, E.-H.*: 'Simulation elektrischer Schaltungen auf dem Rechner'. – Berlin; Heidelberg u. a.: Springer-Verlag, 1985

[12.7] *Buch, P.; Kuh, E.*: 'SYMPHONY: A Fast Mixed Signal Simulator for BiMOS Analog/Digital Circuits'. 10th Intern. Conference on VLSI Design, 1997

[12.8] *Grebene, A.*: 'Bipolar and MOS integrated Circuit design'. – New York; Chichester u. a.: John Wiley & Sons, 1984

[12.9] *Best, R.*: 'Theorie und Anwendungen des Phase-locked Loops'. – Aarau: AT-Verlag, 1987

13 System Simulation

PETER SCHWARZ

13.1 Overview

> *System simulation* means the simulation of very complex and often also heterogeneous systems.

To emphasize these aspects, sometimes it is called *overall system simulation*. In system simulation models at the higher abstraction levels are generally used. Of course, there are many modeling and simulation aspects in common with those described in the foregoing chapters 10 (Analog Simulation), 11 (Digital Simulation) and, most of all, 12 (Mixed-Signal Simulation). But there are also typical specialties, which led to the development of specialized system simulators.

13.1.1 Reasons for System Simulation

There is more than one reason for an overall system simulation. They are determined by the typical problems of the *top-down* and *bottom-up* oriented design style:

- In the *top-down design*, a verification or validation of the system specification is necessary. If it is possible to refine the specification in the first design step so that it can be simulated at an abstract level, we speak about an *executable specification*. Signal-processing algorithms described by formulas or block diagrams are typical examples. Only after a few more design steps will there be enough detailed information about the implementation so that circuit simulators can be used at *gate* or *transistor level*, see also fig. 12.1.
- In the *bottom-up design* transistor and gate counts often become so large that a circuit simulation at this level would take huge computational times (days or weeks). In the digital domain simulation accelerators (hardware accelerators) could be used, but usually they are very expensive. A widely used method is the development of more abstract models with reduced accuracy which is still sufficient for a system simulation.
- Independently of the applied *bottom-up* or *top-down design* methodology, the interactions of the system, which is to be designed, with its environment have to be analyzed together. For example, in the design of telecommunication modem circuits the data transfer between the subscribers over wires has to be considered, and in the design of micro-electro-mechanical systems **(MEMS)** the influence of external forces and temperatures has to be included in the system simulation.

It will not always be possible to divide a system description so clearly into different abstraction levels, as is described in fig. 12.1. But in such cases system simulation can also be understood as a *multi-level simulation*: models at lower and higher levels of abstraction are included into one simulation model and the simulator has to simulate them simultaneously.

13.1.2 Modeling for System Simulation

The most important requirement for carrying out a system simulation is a powerful *multi-level, mixed-signal simulator* [13.5], [13.66]. However, there exist some demands which should be satisfied:

- A large number of signal and data types: *continuous-time*, *discrete-time*, *continuous-value*, *discrete-value*; bit and character strings, abstract data types (see chap. 12) such as *words* and *records*.
- The possibility of describing the behavior of new system components (their models are not available in a library, yet) by the user.
- This *behavioral modeling* should be supported by a modeling language (Hardware Description Language, HDL), but, additionally, an interface to include C and C++ programs should be available (*Foreign Language Interface*).

- The *graphical front-end* should accept block diagrams and circuit schematics and generate automatically a textual netlist. Buses and different data types (e.g., as element parameters) have to be supported. Ideally all the element types shown in fig. 12.1 should be available as models in a library.

Unfortunately the widely used simulators of the SPICE family (PSpice, HSpice, iSpice, ...) do not satisfy most of these demands. They can be used for system simulation only in very special cases. Much better are other simulators like Saber, ELDO, Spectre, and Smash, which are not so strongly focused on purely electrical circuit simulation. These simulators have some mixed-mode simulation capabilities, their libraries contain analog and digital models, and they have (proprietary) modeling languages for behavioral modeling. But these simulators have their roots in analog circuit simulation, and therefore the digital simulation capabilities are not very powerful compared with logic simulators (see chapter 11).

If a very efficient *mixed-signal simulation* is required, the coupling of different simulators may be recommended. Powerful analog simulators such as **Saber** or **ELDO** may be coupled with powerful logic simulators such as **ModelSim**, **VSS** or **Verilog** via special interface software. In this way the most powerful simulation algorithms of both domains may be combined. Another advantage is the direct usability of the analog and digital circuit descriptions (e.g., as simulator-specific netlists) which have not to be converted into the netlist formats of other simulators. There are also no problems with the compatibility of the model libraries of different simulators: each simulator uses its own input language and model library. This data consistency is extremely important in an industrial design flow.

In system simulation different modeling approaches may be used.

- **Behavioral models:** the behavior of the elements is described by mathematical formulas or tables. The notation may be carried out in the form of a program written in a universal programming language (FORTRAN, C, C++) or in a simulator-specific description language (HDL) such as **Mast** for Saber, **HDL-A** for ELDO, or **SpectreHDL** for Spectre. Behavioral models may be used at all levels of abstraction, from the specification level downwards to the transistor circuit level [13.3], [13.13], [13.21], [13.29], [13.48], [13.56].
- **Structural models:** a complex model results from the interconnection of simpler components. For these basic components simulator-internal models (the so-called *primitives*) exist. Such basic components are, e.g., resistors, capacitors, inductors, or different types of controlled sources. In the SPICE world this type of modeling is often called *macromodeling* [13.9], [13.20]. While older simulators only support this type of modeling, modern simulators permit additionally a behavioral description formulated by the user. A *mixed structural and behavioral description* is particularly efficient: the system is modeled by the interconnection of subsystems which are modeled by behavioral descriptions or, hierarchically, by the interconnection of other basic elements [13.24].

Particularly for analog systems a further distinction between **basic properties** of physical effects and **their transformation** into models is important. Physical systems may be classified [13.40], [13.58], [13.76] into:

- **Conservative systems**: their system quantities are *flow quantities* and *difference quantities* (e.g., in electronics: currents and voltages);
- **Non-conservative systems:** consisting of interconnected elements with inputs and outputs, only one type of quantity (*signals*), and mostly a directed signal flow.

A **flow quantity** (a *through quantity*) is measured at one point of the physical system (e.g., an electrical current or a hydraulic flow). A **difference quantity** (an *across quantity*) is measured between two points, e.g., a mechanical displacement or an electrical voltage. In table 13.1 some of the most important flow and difference quantities are shown. These concepts will be discussed more detailed in section 13.3.

The term **conservative physical systems** is motivated by compatibility constraints for *flow* as well as *difference quantities* being valid. In electrical systems these constraints are the well-known Kirchhoff's current law and voltage law. Similar conservation laws for flow and difference quantities also hold in many other physical domains.

This fact can be reflected in different classes of system models: *conservative system models* with *conservative signals* (flow and difference signals) and *non-conservative systems* (with *non-conservative signals*). We use the term *signal* as a model of all kinds of physical quantities which are able to exchange energy or information between the subsystems. We will use the term *network* to describe a system model consisting of elements, connections between these elements, and conservative signals. To stress that these networks are models of physical systems in different domains, sometimes the term *generalized Kirchhoffian network* is used [13.73].

Table 13.1 Flow and difference quantities in different physical domains

Physical domain	Flow quantity	Difference quantity
electrical	current	voltage
mechanical-rotational	force	velocity
mechanical-translational	torque	angular velocity
pneumatic	volume flow	pressure
thermal	heat flow	temperature

Non-conservative systems only have one type of signal and there are no conservation laws for these signals. Block diagram descriptions of control or communication systems as well as signal flow graphs are the well-known examples. If the equations describing the system behavior are known the construction of an *equivalent block diagram* or a *signal flow graph* is very simple. Vice versa, setting up the system equations from a block diagram description is a relatively easy task because of the directed signal flow in all elements and the absence of any reverse interactions between the interconnected elements. Equation-oriented simulators such as **Matlab/Simulink** or **MatrixX** are very popular for the analysis of non-conservative systems.

For **conservative systems** the formulation of mathematical equations is substantially more difficult, especially if one is interested in as small as possible a number of system variables (*state equations*, *state-space description*). Therefore the manual setup of system equations for an equation-oriented simulator is extremely cumbersome and can not be recommended. Instead, specialized simulators, oriented towards conservative systems, should be used. The system description usually consists of a description of interconnected subsystems and elements: textually as a netlist, graphically as a circuit diagram. It is the task of the input processor of such a simulator to transform this structural system description into a system of nonlinear differential-algebraic equations (DAEs). During the simulation the DAE system is solved numerically in the simulator.

These different modeling possibilities will be illustrated by means of the simple example of a vibrating system (a damped oscillator). If the final implementation of this system is not yet defined we start with a mathematical description, a differential equation of second order:

$$M \cdot \frac{\mathrm{d}^2 x}{\mathrm{d}t^2} + D \cdot \frac{\mathrm{d}x}{\mathrm{d}t} + K \cdot x = y$$

x output signal;
y input signal;
M, D, K are constants (parameters).

A behavioral model written in the language MAST (applied in the simulator **Saber** [13.2], [13.48]) could be formulated as

```
template oscillator x y = M, D, K
input nu y
output nu x
# default values of the parameters:
number M = 1, D = 1, K = 1
{
# auxiliary variables xp = dx/dt
equations
  {
       xp = d_by_dt(x)
    x:  y = d_by_dt(M*xp) + D*xp + K*x
  }
}
```

Without dealing with further details, it is briefly mentioned that the model description – starting with the keyword *template* – contains an interface description and an internal part (which is not necessarily visible from the outside). The behavioral description is formulated in a way which is very similar to mathematical expressions.

But also with *structural* models the system can be described. In fig. 13.1 are shown:

- a block diagram for Matlab/Simulink;
- an electrical circuit for the simulation, e.g., with Spice.

In the electrical circuit the input signal y is realized as a current source i_{inp}, whereas the output signal x corresponds to the voltage at the node 1.

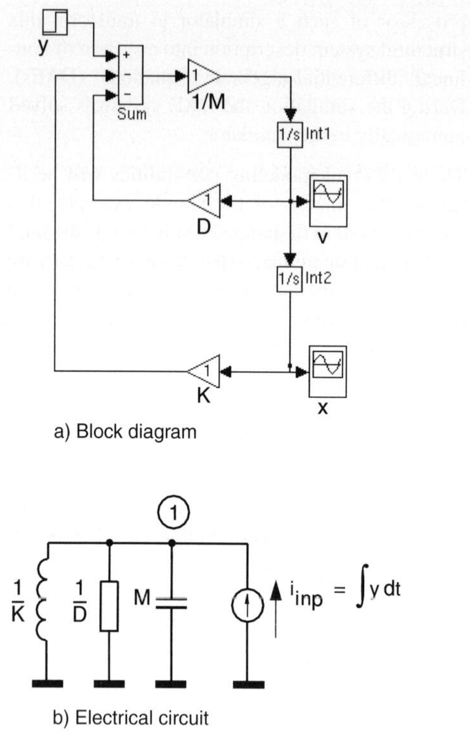

Fig. 13.1 Structural models of the vibrating system

From these examples it is also evident that one can select a conservative *or* a non-conservative model for the description of an abstract system. However, for the modeling of the system realized one should select this type of model, which corresponds best to the nature of the most essential physical quantities, e.g., choose an *electrical circuit*, or a *block diagram*, or a *mechanical network* as the most appropriate model type.

From these few remarks on mixed-signal simulation at system level it is evident that the present possibilities for system modeling and simulation, the number of available simulators and proprietary modeling languages lead to a confusing variety. Also the practical handling of the simulator coupling is sometimes difficult. These are the reasons for the on-going standardization efforts for mixed-signal modeling languages, which include also the system level (see section 13.1.4). These standardized modeling languages are increasingly supported by powerful mixed-signal, multi-domain simulators.

13.1.3 Overview of System Simulators

Just the various applications of the system simulation do not allow the application of 'the one system simulator'. Rather the nature of the technical system which has to be simulated and the possibilities of its modeling does influence the selection of the system simulator crucially. The following list gives a small, far from complete outline of efficient simulators.

General continuous systems

which are described mathematically by ordinary differential equations [13.16], [13.6]:

Matlab	with graphical input preprocessor **Simulink** [13.50];
MatrixX	with graphical input preprocessor **SystemBuild** [13.49];
ASCL, CSMP	input: equations in textual form [13.10].

The graphical input description capabilities mainly refer to block diagrams with a directed signal flow, as they are used, e.g., in control engineering and with data processing algorithms, see also fig. 13.1a). These simulators are particularly well suitable for non-conservative systems, for which the differential equations can be set up comparatively simple [13.10], [13.41]. Implicit differential equation systems and algebraic dependencies which are usual within conservative systems are difficult to be handled with these simulators.

General discrete systems

The signals are usually defined as discrete-value and discrete-time.

- **GPSS, Simscript, SIMPLE++**: Discrete Event simulators with queues and other models commonly used in queuing theory [13.8];
- **Statechart oriented simulators**: they use generalized automaton models; the states and the signal conditions under which transitions be-

tween the states occur are graphically described (e.g., with the simulator **Renoir** [13.51]);
- **Petri net simulators**: for simple, colored, timed and other Petri net classes [13.44];
- **Ptolemy**: an object-oriented simulation framework, in the research stage [13.12].

These simulators are not only used in electronics, but mainly in the analysis of communication, logistics, and production systems. An important application field is the performance evaluation (performance analysis) on the basis of simulations [13.8]. Basically also logic simulators, which are widely used in digital electronics, are quite suitable for the simulation of general discrete systems.

Analog circuit simulators

They were first developed for the analysis of electrical circuits. Newer simulators have modeling capabilities with behavioral description languages (**MAST**, ...) and are very well suitable for other, also non-electrical, systems:

Simulator	Provider	Behavioral Description Language	References
Saber	Analogy	MAST	[13.2]
ELDO	Mentor Graphics	HDL-A	[13.51]
Spectre	Cadence	SpectreHDL	[13.15]
Smash	Dolphin	ABCD	[13.67]

The widely used simulators of the **SPICE** family unfortunately have no behavioral modeling languages and new models can be only introduced by the construction of *equivalent circuits* (*macromodeling*). For some SPICE versions one can formulate at least simple equations as the behavioral description of new elements.

Digital circuit simulators

Primarily developed for digital circuits, described at gate level, these simulators have been very powerful also at the *register transfer* and *system level* by their extension with efficient hardware modeling languages such as VHDL and VerilogHDL:

Simulator	Provider	References
Leapfrog	Cadence	[13.15]
VSS	Synopsys	[13.69]
Verilog	Cadence	[13.15]
ModelSim	Mentor Graphics	[13.51]

Special operation modes, e.g., *cycle-based simulation*, or restriction to the simulation above the gate level (e.g., *instruction set simulation*) lead to substantial increases of the simulation speed and allow an efficient system simulation.

Some of the above-mentioned simulators for continuous or analog systems were extended to mixed-signal simulators:
- **Saber and ELDO**: analog networks and small digital systems modeled by digital basic elements and behavioral descriptions which are not too complicated;
- **Matlab**: addition of a Statechart simulator **Stateflow**.

All three simulators, however, do not use the standardized languages VHDL or VerilogHDL for the digital subsystems. Therefore the coupling of analog circuit simulators with VHDL or Verilog simulators is more efficient.

The modeling and simulation approaches described, on the one hand, are basically suitable for very different systems of electronics, computer and control engineering, as well as communication technology. On the other hand, this general applicability leads to
- not all relevant questions being able to be effectively be processed (e.g., in communications technology);
- a substantial modeling effort has to be spent (e.g., in microelectronics, mechatronics and again in communications technology).

These two aspects are discussed more exactly in the following sections 13.2 and 13.3.

13.1.4 A Perspective: VHDL-AMS

VHDL has been established world-wide (just like VerilogHDL) as a digital specification and modeling language – not least because of the standardization by IEEE, the development of efficient simulators, and the definition of a synthesizeable language subset. Many of the above-mentioned demands from the view of system simulation are already satisfied by this language. The insufficient handling of the analog, continuous-time and continuous-value signals and elements must be seen, however, as a substantial lack. It was consequent to extend the digital VHDL to **a**nalog and **m**ixed-**s**ignal systems: **VHDL-AMS** (similar extensions were defined also for VerilogHDL). The IEEE standardization of VHDL-AMS was finished in 1999 [13.72], the first VHDL-AMS simulators were launched in the last three years. They already support a large extent of the new language [13.2], [13.51], [13.64], [13.67]. Thus the variety of modeling languages for system simulation will be drastically reduced, as soon as more and more simulators support this language in the future. For a systematic design, however, not only standardized modeling languages and simulation possibilities are important, but also their embedding in the design flow. Present design systems support VHDL and SPICE transistor netlists perfectly, but the variety of proprietary mixed-signal description languages could not be included in the same way. Only standardized languages like VHDL-AMS give the chance to integrate the system simulation seamless into a top-down design flow.

Without dealing with VHDL-AMS in detail [13.19], [13.33], [13.72], [13.77], [13.79], [13.80], some substantial concepts are to be presented here, nevertheless, briefly. VHDL-AMS is a strict superset of VHDL, i.e., all constructs of the digital VHDL are available and are also needed in mixed-signal simulation. For the description of analog components some semantic extensions of the existing concepts were defined:

- ENTITY: Interface description: the port list can contain also conservative and non-conservative terminals carrying continuous signals.
- ARCHITECTURE: *Quantities, terminals, nature* declaration, and *simultaneous statements* are new parts of the body of an architecture description. Some attributes for discrete-time and continuous-time signals are defined.
- CONFIGURATION: This basic concept of VHDL can also be used in VHDL-AMS for the selection of one of the various architecture bodies corresponding to an entity.
- LIBRARY: Collection of frequently used declarations of types; pre-defined functions (contained in *Packages*) and models.

Additionally some new constructs were introduced to describe further port types, objects, and data types, as well as new types of statements. To the most important ones there belong:

- TERMINAL: Ports carrying *conservative quantities* (*through and across quantities*, e.g., currents and voltages in electronics). Terminal ports can be defined as being of an *electrical, thermal, kinematic*, ... nature. Internal nodes of an element can be declared with a terminal statement.
- NATURE: *Natures* represent *physical domains* (comp. table 13.1). Terminals and conservative quantity types are defined by the NATURE statement. Types of *through quantity* and *across quantity* at conservative clamps are determined with ACROSS and THROUGH. Additionally, the reference node (REFERENCE), belonging to a terminal type, is defined.
- QUANTITY: *Non-conservative* ports (to describe a directed signal flow in block diagrams), distinguished between the signal directions IN and OUT.
- free QUANTITY: Additional internal quantities (states);
- branch QUANTITY: Branch definition by the specification of the start and the end node as well as the through and across quantities of the branch.
- ==: With this symbol an *equality relationship* is introduced:

  left side == *right side*

  of the equation. It is not an assignment but an implicit equation which has to be solved numerically. The equation is shifted to the DAE solver of the simulator. The unknown quantities in the equation (the simultaneous statement) are calculated independently of whether they occur on the left or on the right side or on both.
- PROCEDURAL: All statements in a procedural block are executed sequentially;
- DOT: Attribute for the temporal derivative $\mathrm{d}/\mathrm{d}t$ (identifier): identifier 'DOT;
- LIMIT: Step size limitation in the analog simulation by numerical integration;
- BREAK: Handling of discontinuities in the analog simulator.

A simple example (a linear resistor with the terminals $p1$ and $p2$ and the resistance value *value_r*) will give a first impression of a behavioral description in VHDL-AMS:

```
LIBRARY DISCIPLINES;
USE DISCIPLINES.ELECTROMAGNETIC_SYSTEM.all;
-- contains, e.g., the declaration
-- of the terminal type ELECTRICAL
ENTITY r IS
  GENERIC (value_r :         REAL := 1.0 );
  PORT (TERMINAL p1, p2 :  ELECTRICAL   );
END ENTITY r;

ARCHITECTURE variant_1 OF r IS
  QUANTITY v ACROSS i TROUGH p1 TO p2;
BEGIN
  v == value_r * i;
END ARCHITECTURE variant_1;
```

For a linear capacitor essentially only the voltage-current relationship is to be formulated newly, whereby the temporal derivative d/dt is expressed with the DOT attribute. Additionally, a given initial value *vc0* of the capacitor voltage can be used or this initial value is calculated as the DC bias point by the simulator. If no parameter value *vc0* is given, the simulator uses the smallest real number (REAL'LOW) and detects by it that *vc0* has to be calculated statically (the current through the capacitor has to be zero).

```
ENTITY c IS
  GENERIC (value_c :  REAL := 1.0
           vc0 :  REAL := REAL'LOW);
  PORT     (TERMINAL p1, p2 :  ELECTRICAL );
END ENTITY c;

ARCHITECTURE var1 OF c IS
  QUANTITY u ACROSS i TROUGH p1 TO p2;
  BEGIN
    IF DOMAIN = QUIESCENT_DOMAIN USE
      IF vc0 /= REAL'LOW USE
        u == vc0;
      ELSE
        i == 0;
      END USE;
    ELSE
        i == value_c * u'DOT;
    END USE;
END ARCHITECTURE var1;
```

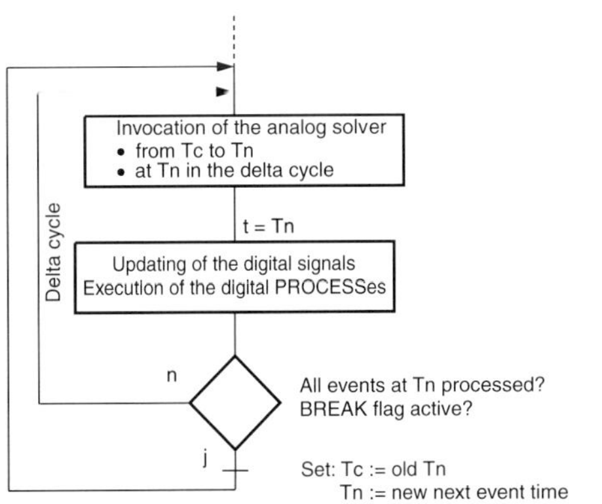

Tc = current time (actual simulation time), Tn = next event time

Fig. 13.2 Mixed-signal simulation cycle

Additionally, in the IEEE standard the principal time-domain, mixed-signal simulation cycle is defined (fig. 13.2). The knowledge of this simulation cycle is important in the construction of mixed-signal models, particularly in the modeling of the interfaces between analog and digital subsystems and during the initial value calculation.

Figure 13.3 shows a system frequently used in electronics: a sigma-delta converter. The digital signal at the output of the comparator (an A/D converter) is fed back into the analog circuit section over a D/A converter, which generates an analog electrical signal.

The VHDL-AMS models of the two converters are:

```
ENTITY d2a_verz IS
  -- parameters (delays)
  GENERIC    ( tv_01 :  TIME := 2 ns;
               tv_10 :  TIME := 3 ns;
               tf_01 :  REAL := 5.0E-9;
               tf_10 :  REAL := 6.0E-9);
```

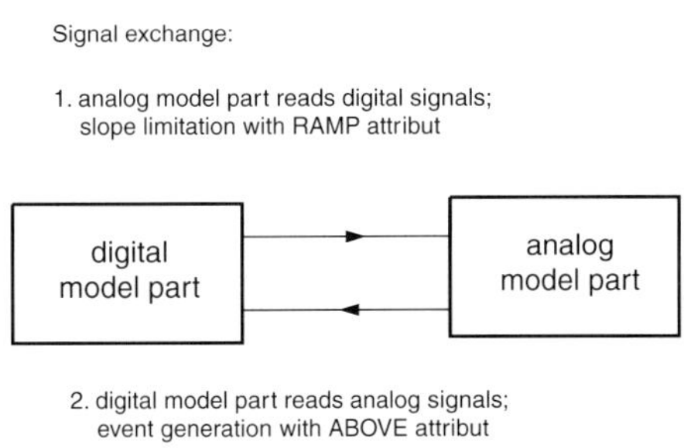

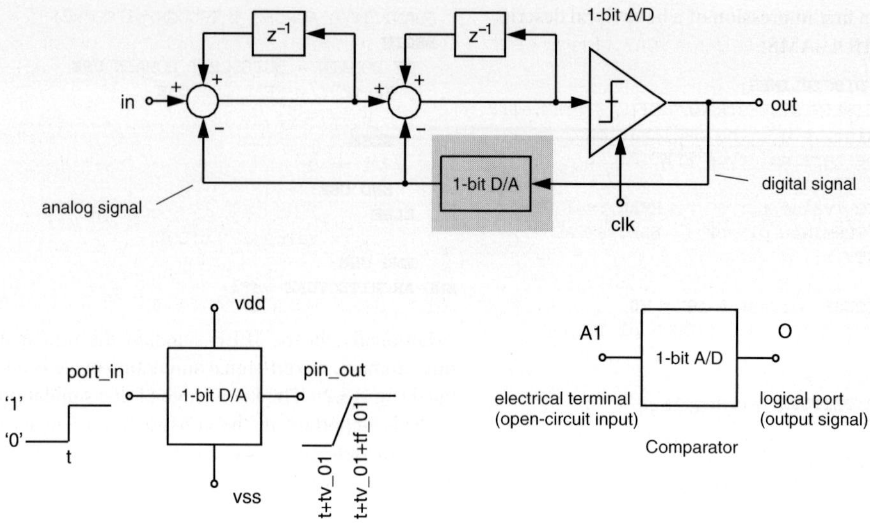

Fig. 13.3 Sigma-delta converter

```
  PORT (TERMINAL vdd, vss,
        pin_out :  ELECTRICAL;
        SIGNAL port_in :  IN BIT);
END ENTITY d2a_verz;
ARCHITECTURE v1 OF d2a_verz IS
  QUANTITY v_pin_out ACROSS i_pin_out
           THROUGH pin_out;
  -- internal variable:
  SIGNAL xout :  REAL := 0.0;
  BEGIN
    -- statements in the PROCESS section
    -- are elaborated sequentially
    PROCESS BEGIN
      WAIT ON port_in;
      IF port_in = '1' THEN
        xout <= 2.7 AFTER tv_01;
      ELSE
        xout <= 0.0 AFTER tv_10;
      END IF;
    END PROCESS;
    v_pin_out == xout'RAMP (tf_01, tf_10);
END ARCHITECTURE v1;

ENTITY comparator IS
  GENERIC (threshold :  REAL := 2.5 );
  PORT    (TERMINAL A1 :  ELECTRICAL;
           SIGNAL O : OUT BIT:= '0');
END comparator;
ARCHITECTURE ideal OF comparator IS
  QUANTITY  v1 ACROSS A1;
  SIGNAL    S : BOOLEAN;
  BEGIN
    S <= v1'ABOVE(threshold);
    PROCESS (S) BEGIN
      IF S = TRUE THEN
        O <= '1';
      ELSE
        O <= '0';
      END IF;
    END PROCESS;
END ideal;
```

For the consideration of delays and finite signal slopes in the D/A converter, the language element AFTER is used in the digital sub-model and the attribute RAMP in the analog sub-model. In the model of the A/D converter the attribute ABOVE(threshold) is applied to the analog signal v1 to produce an event for the digital simulator, which influences the digital simulation.

Even if these examples were only small, nevertheless they give an impression of the power of the new modeling language, e.g.:

- clear model structure (separation of the interface from the core of the model): ENTITY and ARCHITECTURE;
- notation of the behavioral description in equation-like manner;
- mixture of structural and behavioral descriptions at different abstraction and modeling levels;
- use of the simulator's equation solver also for the handling of equations in the model;

- clearly defined interactions between analog and digital simulation algorithms (mixed-signal simulation cycle).

Further supplements with respect to the digital VHDL concern the analysis in the frequency domain (small-signal AC analysis), the bias point calculation (DC analysis), and the noise analysis. VHDL-AMS is thus not limited to the modeling for transient simulation. All of these advantages make VHDL-AMS not only interesting for the mixed-signal *circuit* simulation but also for the *system* simulation.

13.2 System Simulation in Communication System Design

13.2.1 Introduction

Telecommunication systems are characterized frequently by a large complexity and variety of their subsystems and components. The different analysis and simulation tasks are just as varied [13.38], [13.45]. Some examples are shown in table 13.2, arranged in accordance with the degree of abstraction of their description. The design tasks reach from the specification engineering of global communication networks down to microelectronic implementations of analog and digital circuits.

For simulation purposes at first it must be distinguished between *system* and *circuit* design. At the circuit level we have to differentiate additionally between logic and electrical level, or, corresponding to the type of signals, between digital and analog. Very efficient simulation tools are available for these individual design tasks. Without any completeness or a valuation, some of them are mentioned:

- COMNET III, OpNet and BONES for the simulation of communication nets;
- SDT (with the modeling language SDL) for very abstract descriptions of hardware/software systems;
- Matlab/Simulink, MathCad for signal processing algorithms (described as formulas);

13

Table 13.2 Simulation of telecommunication systems

Design level	Level of abstraction, modeling level	Subsystems and components to be described	Simulation and analysis tasks
System	Queuing theory, information theory, Control (Petri Nets), *digital (+ analog)*	World-wide operating switching and communication systems (SDH, ATM, ISDN, DAB, HDTV, ...); Net nodes, modems	Information flow, capacity calculation, net planning, distortion analysis, effectiveness estimation of error correcting algorithms, protocol verification, deadlock detection
Logic	Control and data flow automata, register transfer gates switching gates, *digital*	Microprocessors, digital signal processors (DSP), controller, digital filters, integrated circuits: standard cell and gate array ICs, full custom ICs (ASICs)	Arithmetic-logical data processing, instruction set verification, digital signal processing, filter characteristics, static and dynamic behavior
Electrical circuit	Linear networks, filters, transmission lines, non-linear networks, transistor circuits *analog*	Analog and mixed-signal ICs, full custom ICs, realized in CMOS or bipolar technology, printed circuit boards (PCB) building blocks: filters (LC, SC), modulators and demodulators PLL, VCO, mixers, ...	Frequency and time domain behavior of voltages and currents calculation of transients on transmission lines, impulse shaping, cross-talk, thermal-electrical effects

- COSSAP and CoCentric System Studio [13.69], SPW [13.15], Matlab/Simulink, DSPStation [13.51], TESLA and Ptolemy [13.12] for a block-oriented system simulation (the formulas are transferred partly into block diagrams and partly into textual equations);
- SpectreRF [13.15], [13.43] for analog systems and circuits (particularly in the high frequency domain);
- ADS [13.1] for circuit and system simulation, for analog as well as digital signal processing;
- SWITCAP, SCAP, and TUSCA for switched capacitor circuits.

The traditional simulation at gate and transistor circuit level is executed with the simulators mentioned in the previous sections, thus, e.g.:

- VERILOG, Leapfrog, LSIM and VSS for digital system and gate simulation;
- SPICE, ELDO, SABER, SPECTRE and MDS for analog circuits.

Additionally, mixed-signal simulators are increasingly used, above all if they have HDLs for behavioral modeling.

Even in the design of complex communication systems, a global view over more than one abstraction level is necessary and all design phases (from concept to implementation) have to be included in the design methodology. But there are different deficiencies:

- During the transition from the system to the circuit design, the description methods and languages, models and tools (e.g., the simulators) must usually be changed, therefore, a common system and circuit design is hardly possible.
- The simultaneous development of hardware and software (HW/SW co-design) as well as the inclusion of already existing real hardware (e.g., DSPs or FPGAs) into the simulation of the whole system or its emulation is not sufficiently supported.

Improved possibilities for HW/SW co-simulation will be offered by the new description languages SystemC and SystemC-AMS [13.84], [13.85], [13.86].

It is an aim of the software providers to reduce the gap between *system* and *circuit design*. Examples are particularly the simulation systems SPECTRE-RF and ADS, which already support circuit *and* system simulation of communication systems. This is realized by multi-level simulation and compatible model libraries at both levels. If there is, e.g., a subsystem 'mixer', then there exists an abstract system-level model 'mixer' as well as a corresponding low-level model in form of a transistor circuit. Both models realize the behavior of a mixer but with different accuracy and simulation time. The parameters of the system model 'mixer' can be gained by simulation of the transistor circuit almost automatically. In the simulation of a system both models can be exchanged with each other.

In the following two typical simulation tasks in communication engineering are to be regarded more in detail: the **bit-true simulation** of digital signal processing algorithms and the **mixed-signal simulation** based on simulator coupling.

13.2.2 Simulation of Signal Processing Algorithms

A simple signal processing system is a *digital filter*. The transfer function of a FIR filter of 3rd order is:

$$y(k) = C_0 \cdot x(k) + C_1 \cdot x(k-1) + C_2 \cdot x(k-2) + C_3 \cdot x(k-3)$$

$x(k)$ is the input signal of the filter at the time point k; $x(k-1)$, $x(k-2)$ and $x(k-3)$ are the earlier input signal values delayed by 1, 2, or 3 clock cycles, respectively, $y(k)$ is the output signal, and the filter coefficients are C_0 to C_3. Such an algorithm can be simulated easily with any tool (e.g., Matlab or a self-written C program). Signal values and coefficients are usually defined as floating-point numbers of high accuracy (for the sake of comfort or to examine the filter function only basically). But the final technical implementation will usually come with a fixed-point or an integer arithmetic to reduce the hardware costs or the computing time, or both. These different number representations may have large influences on the filter characteristic and should be examined by simulation carefully.

Bit-true simulation means that the variables and operators in the algorithms are regarded exactly in the same way, as they will be used later in the circuit implementation. The bit-true simulation is prepared and realized by:

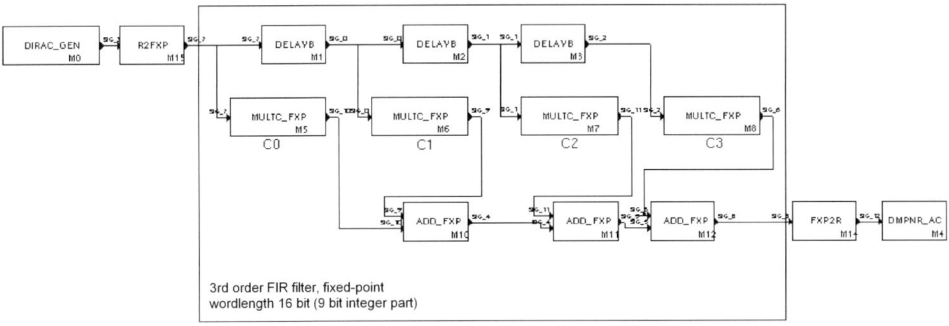

Fig. 13.4 COSSAP model of a 3rd order FIR filter

- the definition of the fixed-point, integer or floating-point *number format*;
- the *number of bits* to be used for each signal;
- the *sign* handling (e.g., a one's or two's complement representation);
- in connection with the operators: the number format of intermediate results and perhaps, followed by a *rounding procedure*;
- the overflow handling and other error handling mechanisms in case of arithmetic problems.

It is not of principal importance whether the implementation of the algorithms takes place later

- as software on a digital signal processor (DSP) or
- as hardware, consisting of arithmetic-logical units (ALUs), registers, and gate logic.

The simulation models for both kinds of signal processing mechanisms must be adaptable to it.

For this task simulators like COSSAP, SPW, and also Matlab (especially in connection with the Communication Toolbox) can be used. In fig. 13.4 a COSSAP representation is shown, where the formula was interpreted as a block diagram. This diagram was sketched with the schematic entry tool of the simulator. The three applied block types correspond to the operators in the filter function and may be found in the COSSAP model library:

- DELAYB: Signal delay with one clock cycle;
- MULTC_FXP: Multiplication of a signal value by a constant;
- ADD_FXP: Addition of two signals.

These models realize a fixed-point arithmetic in which the word length, the number of the integer digits, as well as rounding procedures and overflow handling can be determined. In this way numerical effects can be examined. For the simulation of the whole filter additional models are necessary which deliver or convert the test signals and output the results. In the example the elements DIRAC_GEN (pulse generator), R2FXP and FXP2R (converter between real and fixed-point signals) and DMPNR_AC (graphical output) are used. They also belong to the library of the simulator.

The efficiency of a system simulator is particularly determined by the efficiency of the simulation method and by the content and quality of its model libraries, too. Frequently the design task consists of the realization of a certain subsystem or module. To verify its design the interfaces to the entire system also have to be considered. In the ideal case these interfaces and the rest of the system can be described completely with sufficient accuracy by simulator models, so that a complete test environment for the module can be created with reasonable effort.

SPW [13.15] and COSSAP [13.69] are system simulators with very large model libraries specialized in communication technology. SPW contains, e.g., models for:

- basic mathematical operations (addition, multiplication, division) for fixed- or floating-point numbers;
- digital filters (high-pass, low-pass and band-pass filters with Butterworth, Bessel, or other characteristics);
- adaptive filters (LMS and RLS algorithm);
- modulators and demodulators (e.g., PSK, QAM, GMSK, FM);

- coder and decoder (e.g., BCH, convolutional, Reed-Solomon, Viterbi);
- signal generators (e.g., white noise, sine waves, modulated carrier, unit impulse);
- result output and post-processing (FFT, eye and scatter diagram display in the signal calculator).

Similar models are also offered by COSSAP and the *Communication Toolbox* of Matlab/Simulink [13.50]. In each case there are very extensive libraries with some hundreds of models. Beyond that the simulator providers offer more and more model libraries which are adapted exactly to a certain communication standard. They must usually be bought optionally. With SPW are, e.g., libraries for GSM, ADSL and Wideband CDMA.

The simulation efficiency is also influenced by the simulation algorithms. SPW and Matlab operate like logic simulators for clocked systems: in each clock cycle the input signals of the blocks are used for the calculation of their output signals which may be used in the next clock cycle. In COSSAP another simulation principle is implemented, which will be presented briefly. COSSAP operates as a **data driven simulator**. The system model consists of interconnected blocks (fig. 13.4) for which behavioral models exist in the library. At each block the interconnection terminals are uniquely partitioned into inputs and outputs. The individual blocks operate autonomously in the sense that they always calculate the *maximally possible number of output data* according to the number of input signal values actually available. The output signal values are then passed to the successor blocks. There is no global view on the system and no supervisor algorithm, no explicit clock pulses and not even a *time*. Instead, each block operates if as many data as necessary are available at its input. The user must interpret the data sequences in a time frame. This sounds somewhat complicated, but it leads to a very fast simulation (particularly in systems without feedback loops or with large delays in feedback loops). The relations between the input and output signals are described as C procedures in a COSSAP typical way. For the bit-true simulation, many operators are pre-defined as blocks and do not have to be programmed by the user.

Occasionally the designers of digital circuits do not have access to one of the specialized communication engineering simulators mentioned above. One reason could be the prices which are quite high, corresponding to the large power of these simulators. Then the use of logic simulators remains as an alternative, but these logic simulators must be extended by libraries for signal processing components. The mathematical signal processing algorithms must be formulated in the modeling language of the simulator (or in its graphical input mode as block diagrams). In [13.25] such a library for VHDL simulators is described. It contains the following model classes (comparable to a subset of the larger libraries in COSSAP and SPW):

- amplifiers;
- generators: function tables for deterministic time functions, random bit sequences;
- converters: A/D, D/A, sample-and-hold elements; pulse-shaping elements: GAUSS, cos**2;
- coder/decoder: binary, ternary, trellis; scrambler/de-scrambler;
- modulators/demodulators: FSK, ASK/PSK, QAM; AM, PM, FM;
- filters: digital (FIR, IIR) and analog; adaptive filters;
- communication channels with and without random disturbances.

Blocks for the mathematical basic functions:

- ADD, SUB, MUL, DIV;
- ABS, INV, SQR, SQRT, POW;
- EXP, EXP2, EXP10; LN, LD, LG;
- SIN, COS, TAN, ASIN, ACOS, ATAN; SINH, COSH, TANH;
- FALT, SUM, DIFF (Basic algorithms of signal processing);
- D_TO_FIX, D_TO_FL, Conversion functions (conversion of digital signals into fixed-point and floating-point numbers).

The models of the basic mathematical functions are supplied both as blocks, which can be inserted into a circuit, and as C functions, which the user can include into own behavioral models. Block models are delivered for operations with floating-point numbers and parameterizeable fixed-point numbers, C functions only for parameterizeable fixed-point numbers. Similar libraries are also offered as part of commercial design systems (e.g., in the Communication Toolbox of Matlab/Simulink, in the Telecom Libraries of Mentor Graphics for

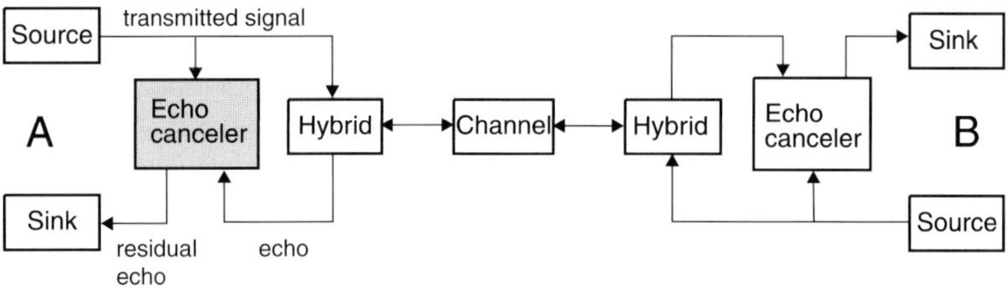

Fig. 13.5 Echo compensation in full-duplex transmission

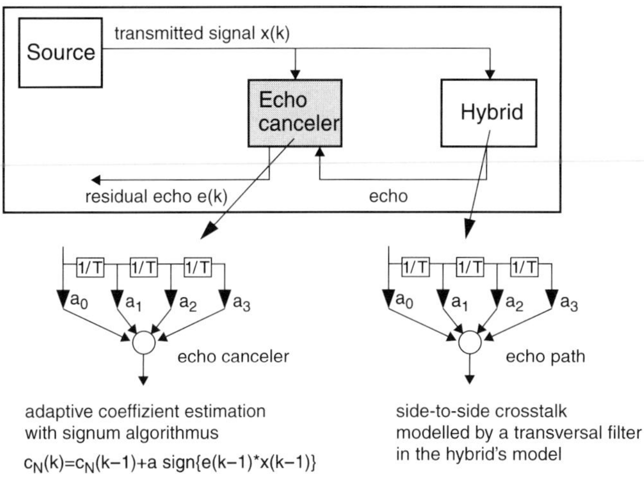

Fig. 13.6 System simulation of the echo canceller

the mixed-signal simulator ELDO, and of Analogy for the mixed-signal simulator *Saber*), but they are not programmed in VHDL. The use of such models for communication systems in connection with general-purpose VHDL simulators is not only a 'less than ideal solution' if no better simulators are available. It also offers the advantage that the signal processing algorithms can be simulated together with the digital system parts – a very important aspect within an overall system simulation!

The bit-true simulation was used in the design of an echo canceller for a telephone system [13.42]. The main problem of a full-duplex transfer over a two-wire line is the cross modulation of the transmitted signal on the received signal in the hybrid four-wire terminating circuit. In fig. 13.5 the communication system with two transceivers (A and B) and a channel is shown.

The non-ideal impedance matching between the hybrid four-wire terminating circuit and the transmission line is the reason that a part (the 'echo') of the transmitted signal of station A is fed back into its own receiver. The echo canceller estimates the echo to be expected and then it may be subtracted from the received signal of the station B. Because the echo canceller in each device must be adapted to the concrete circuit and transmission line, an adaptive filter is used. In the adaptation phase of the echo canceller the transmitted signal of the station B is equal to zero, therefore it is sufficient in the following to consider only one transceiver (subsystem A in the fig. 13.5). For estimating the echo different methods of adaptive digital signal processing are available, which differ according to the implementation effort and their numerical accuracy. The resulting accuracy depends both on the applied algorithm and on the number type and word length chosen for the arithmetic processing unit in the target hardware (DSP, ALU and gate logic). These hardware resources define the limit frequency of the communication system. All these effects can be determined by the bit-true system simulation.

For the verification of the echo canceller a simplified system model (fig. 13.6) is used. The echo arising in the hybrid four-wire terminating circuit is modeled by a transversal filter. A transversal filter structure was also chosen for the echo canceller and the filter coefficients c_N are determined according to the *signum algorithm*. The stimulus signal values $x(k)$ were assumed to be white Gaussian noise. The results of the system simulation (fig. 13.7) show the influence of the fixed-point arithmetic on the behavior of the echo canceller. With a favorably selected number format (a) one can obtain a well-compensated echo. After a transient time of approximately 15 ms the filter has been adapted sufficiently well to the interface transceiver channel. With improper parameter selection (b) overflow effects impair the function of the echo canceller.

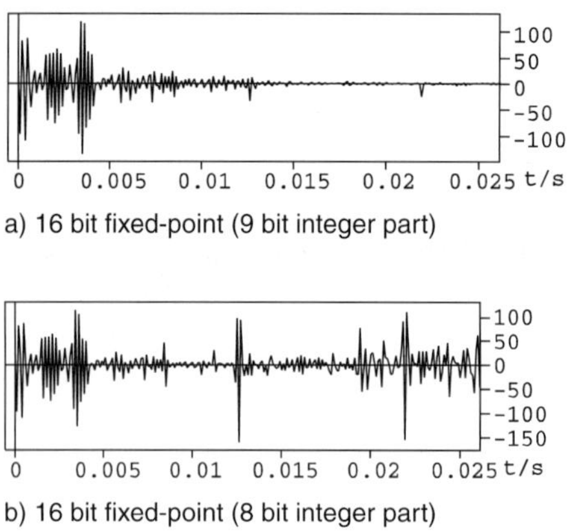

Fig. 13.7 Residual echo dependent on the number format

Frequently the signal processing algorithms which are to be implemented are formulated in such a way that they can be simulated completely by system simulators like COSSAP and SPW. That is the case whenever

- the hardware operates clocked;
- the Z-transform is used to transform signals and filter characteristics between time and frequency domain; or
- there is an approximate transformation of ordinary differential equations into difference equations (to be solved by the system simulator).

That is, however, not the case if the nonlinear behavior of transistors and the pulse distortions on transmission lines have to be analyzed very exactly since these effects influence the system performance substantially. Figure 13.8 shows symbolically a communication system for which multi-level, mixed-signal modeling and simulation are necessary to verify the correctness of the system.

It should be possible to carry out a **mixed-signal simulation** with PSpice (or comparable simulators) as described in chapter 12. But these simulators are focused on transistor circuits with analog and digital functionality. Most communication systems contain very large circuits to realize the signal processing and it is not possible to simulate these circuits at transistor level. Higher level models for system simulation are necessary. On the other hand, it is not possible to simulate transistor circuits with Spice-like accuracy in COSSAP, CoCentric System Studio, or SPW. Different methods are under investigation to realize a true multi-level mixed-signal simulation of communication systems.

A first approach is the development of more accurate component and subsystem models qualified for system simulators. This is increasingly supported by software implemented in widely used CAD systems. Cadence offered sophisticated models with many parameters (so-called 'k-models') for the simulator SPW. These models describe the dynamic linear and the static nonlinear behavior of analog subsystems (usually transistor circuits) very precisely [13.31]. But the parameter adjustment of the models is very complicated and a complex task which has to be supported by special software ('model generators', see fig. 13.9). This generator software needs as input those signals which are calculated for the nonlinear transistor circuits with an analog circuit simulator (here SpectreRF). With the system simulator SPW the whole system performance can be calculated at higher abstraction level with reduced accuracy but with a very high simulation speed.

An alternative approach is the use of analog simulators for transistor circuits, which were extended by more abstract system models for digital signal processing tasks, as specified above. This joint simulation at circuit and system level (*multi-level simulation*) was applied in the design of an ADR satellite broadcasting receiver (**A**STRA **D**igital **R**adio) [13.11]. With this method the application

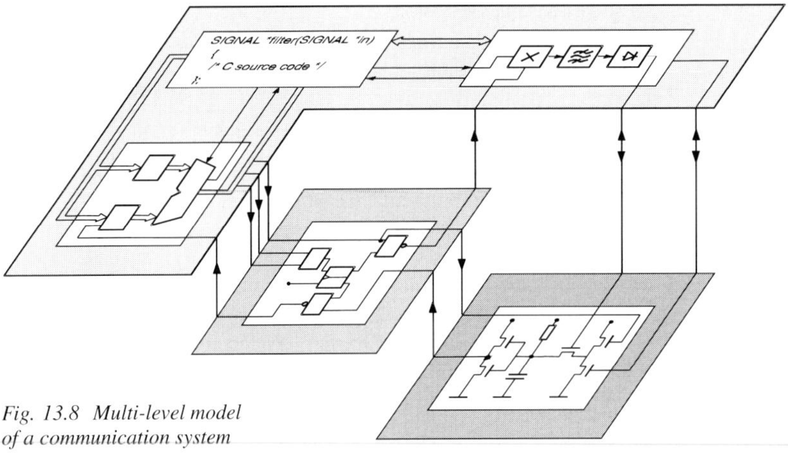

Fig. 13.8 Multi-level model of a communication system

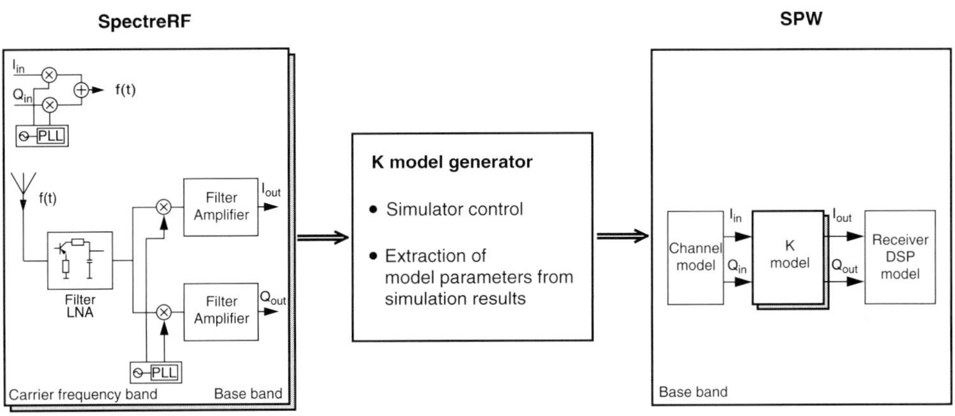

Fig. 13.9 Automatic generation of SPW models

of COSSAP or SPW could be avoided which cannot simulate transistor circuits with such a SPICE-like accuracy as was needed in the design tasks. Figure 13.10 shows the block diagram of the entire receiver.

The influence of the properties of the FM demodulator (which was, therefore, modeled as a transistor circuit) on the bit error rate (BER) of the disturbed data transmission has to be investigated. For the simplification of the modeling and for saving computing time, MUSICAM coding and the error protection in the digital channels were not considered in the model. A binary pseudo-random sequence was used to determine the bit error rate of the ADR transmission. Figure 13.11 shows the model of the entire ADR system and the different sampling rates.

For a pure system simulation the simulation of the high frequency circuit parts (from the FM modulator to the FM demodulator) can be carried out in the baseband. Instead of the frequency-modulated carrier of 489.5 MHz, the magnitude and phase of the FM signal are transmitted as a complex signal (with 'carrier frequency equal to 0 Hz'). In this way the very short sampling time (0.5 ns) of this signal can be avoided in the simulation and the simulation could be accelerated drastically.

A baseband modeling of the RF module is not possible in the favored multi-level simulation, since the FM demodulator processes directly the

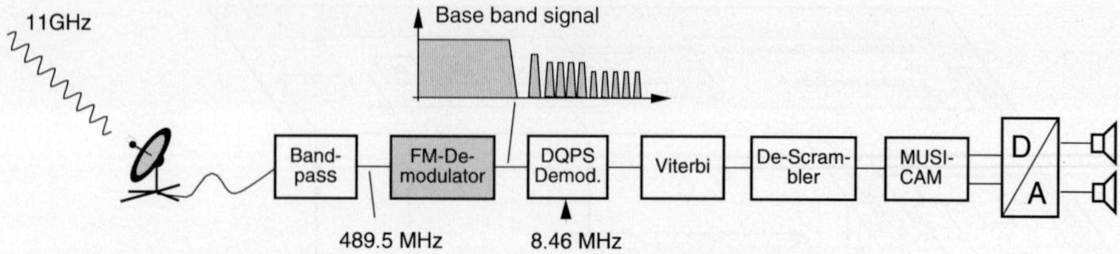

Fig. 13.10 Satellite receiver of an ASTRA transponder

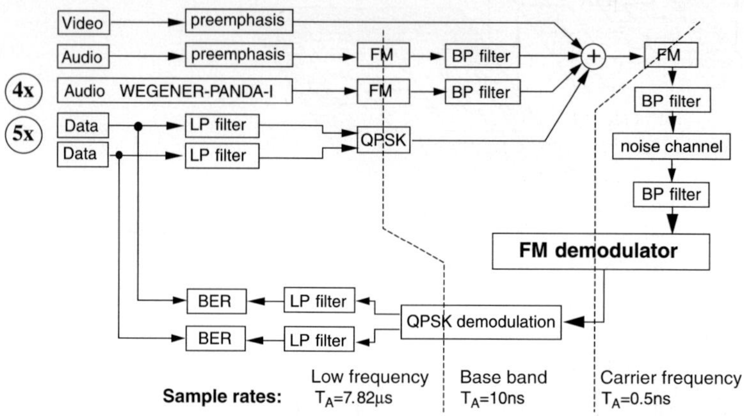

Fig. 13.11 Model of the entire ADR system

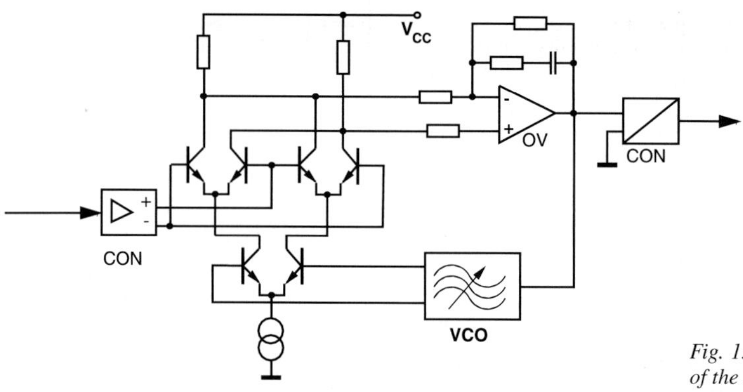

Fig. 13.12 Circuit model of the FM demodulator

frequency-modulated signal, i.e., the modulated carrier in the passband. In order to reduce the computing time, nevertheless the system was modeled as a *multi-rate system*. The signals in the system are divided into three classes according to their typical frequencies. In the same way the system is partitioned into subsystems with appropriate sampling rates. Intersection points are determined by the modulators and demodulators which realize the transition to other frequency ranges and thus to other sampling rates. Figure 13.11 shows the model of the entire ADR system modeled according to these considerations.

To determine the influence of the FM demodulator on the system performance, a transistor circuit model has been inserted into the system model. Figure 13.12 shows an implementation version of the FM demodulator. By the joint simulation of the transistor circuit with the other, more abstract, modeled subsystems the influence of circuit parameters on the bit error rate can be determined. The evaluation of the transmission performance results, e.g., from the inspection of 'scatter diagrams' or 'constellation diagrams' (fig. 13.13). For the QPSK transmission method which had to be examined, the received signal values are represented as points in quadrant fields. In an ideal data communication without distortion, exactly one point is located in each quadrant, so that a unique decision on the received signal value is possible. In the case of a disturbed signal transmission, 'clouds' are displayed in the scatter diagram. A doubtfree detection of the signal values is possible if these clouds are clearly separated from each other. In fig. 13.13 it is shown that the characteristics of the disturbed signal transmission are very similar with all three selected modeling accuracies. So it is allowed to choose the most simple kind of modeling related to the smallest simulation time and still sufficient accuracy.

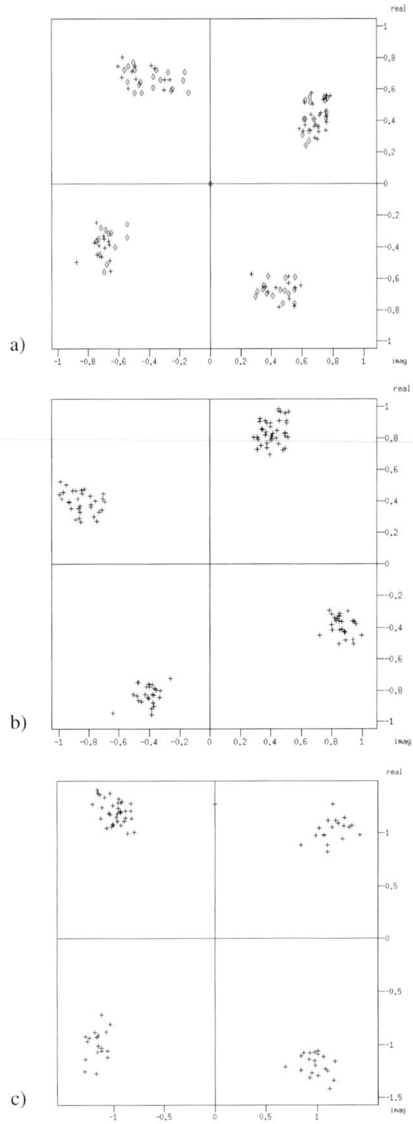

Fig. 13.13 Scatter diagrams of various model variants

The large computing times may not be kept secret, however. A realistic determination of the bit error rates requires the transmission of at least 10^5 bits for each ADR channel and thus a simulation of approximately 40 ms real time. Thus about $8 \cdot 10^{10}$ simulation steps would have to be executed because of the high clock rate of the FM transmission. This is also not possible with the multi-rate modeling in reasonable simulation times. After further model modifications finally the following computing times were measured:

- pure system simulation with the analog simulator: 1 hour 15 minutes;
- multi-level simulation: 45 hours.

This illustrates why the application of a multi-level mixed-signal simulation for such complex functionalities can be recommended only if the required accuracy cannot be gained by other kinds of system verification.

If both ways

- new and highly accurate models for a system simulator;
- extension of a circuit simulator by models at system level

are not possible (simulation and modeling software may not be available, transistor circuits may not be simulated sufficiently accurately at system level, there is a large part of digital control logic, ...), *simulator coupling* has to be considered.

13.2.3 Simulator Coupling

For the realization of the multi-level simulation (see fig. 13.8) *simulator couplings* were developed (fig. 13.14). Hereby the most appropriate simulator can be used for each subsystem (at algorithmic, logic, or transistor level). For the correct interaction of the simulators a special coupling software is necessary. It

- synchronizes the cooperation of the simulators (e.g., it must be awaited on the slowest simulator);
- executes the necessary data conversion (e.g., between real variables at the electrical level, Boolean variables for modeling the digital gate circuits, integer or fixed-point variables at the algorithmic level); and
- organizes the data transfer between the simulators, which can also reside on different computers in a local or global network.

Different operating system services can be used for the software coupling: files, sockets, pipes, TCP/IP, ... up to pvm (parallel virtual machine) or Internet based communication tools.

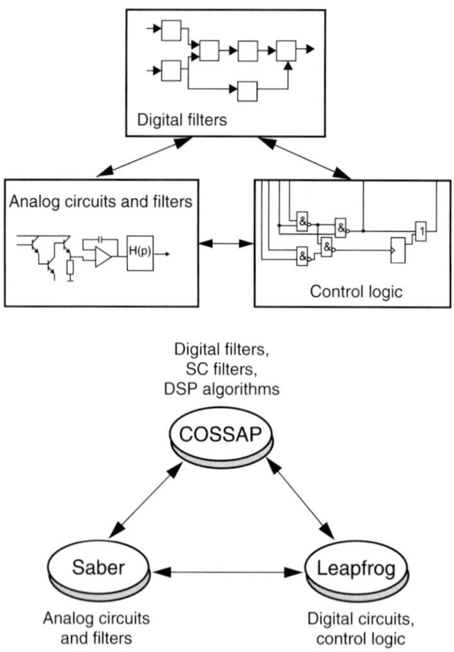

Fig. 13.14 Simulator coupling for simulation of communication systems

Only for some simulators is such a coupling software offered commercially (e.g., for Saber-Verilog, Spectre-Verilog, COSSAP-VSS). An experienced programmer often can develop this coupling software, but the principal suitability of the involved simulators must be presupposed. Mostly this is given if a C interface for modeling new elements exists for the simulator. This interface can be used frequently for simulator coupling too. In this way also was the coupling shown in fig. 13.14 implemented [13.26]. It is a border case of simulator coupling if additional autonomous simulation algorithms are included into a communication system simulator (e.g., COSSAP) as subroutines [13.61].

An example will demonstrate the application of simulator coupling. In fig. 13.15 the block diagram of a SLICOFI circuit (**S**ubscriber **L**ine **I**nterface **C**ircuit with C**O**DEC **F**ilter – an Infineon product) is shown. It is the link between analog telephone signals on a two-wire line and the PCM transmission in digital nets. Special modeling problems are:

- analog modeling of the high voltage section of the SLIC (voltage peaks up to 160 V are to be processed);
- high clock rate in the digital filter part SICOFI (over-sampling);
- multi-rate circuits (decimation and interpolation filters);
- word length effects in the digital signal processing (some DSPs realize filter functionality);
- transient simulation, AC and DC analysis have to be carried out.

Figure 13.16 shows a part of the simulation model for transient simulation and the behavior in the frequency domain (AC simulation). A separate simulation of the individual subsystems does not enable a complete verification, since thereby, e.g., the permanent adaptation of some digital filters to the analog voltage-current relations on the transmission lines can not be simulated. By the application of the three coupled simulators COSSAP, Saber, and (alternatively) Leapfrog or Verilog (see fig. 13.14) even those effects can be simulated which can be explained *only* by the interaction of the subsystems. The computing time was in the range of some hours and resulted particularly from the analog simulation at transistor level, necessary for handling non-linear transistors and linear transmission line effects.

Similarly large computing times result frequently from the coupling of analog and digital simulators to realize a mixed-signal system simulation. It is hoped that the new VHDL-AMS and Verilog-

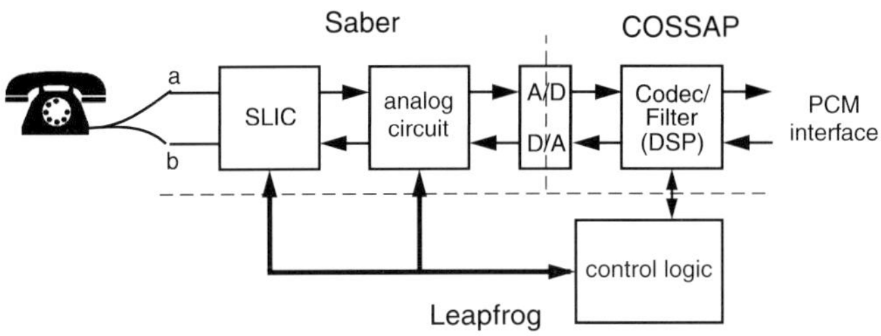

Fig. 13.15
SLICOFI block diagram

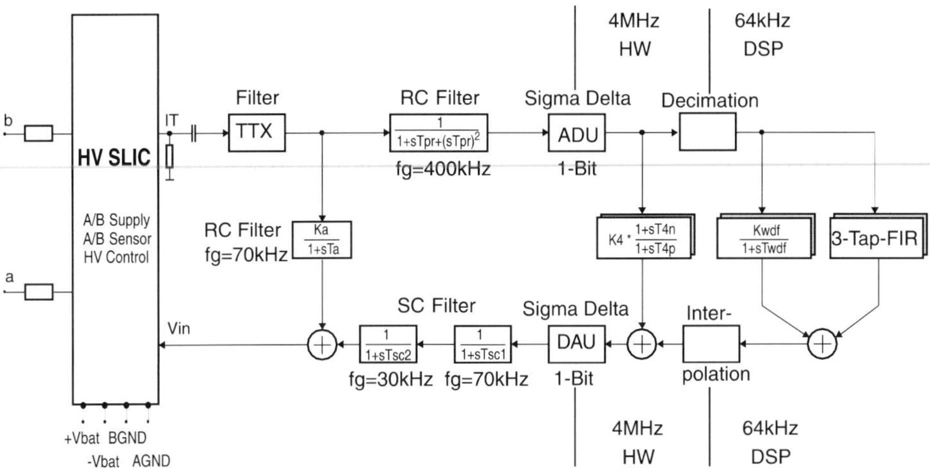

Fig. 13.16 SLICOFI model for transient and AC simulation

AMS simulators can reduce these computing time problems, since they will probably consist of a closer union of highly efficient analog and digital simulators. Their performance could be better than a self-programmed coupling via C interfaces and a resulting large communication overhead.

13.3 System Simulation in Microsystems Technology

13.3.1 Requirements to the Simulation

Microsystems technology and the design of micro-electro-mechanical systems (MEMS) are emerging engineering disciplines which increasingly win importance for electronic engineers also [13.27], [13.28], [13.57]. Airbag sensors in automobiles, which are connected with other sensors and control units by field buses, represent just one (but typical) example. Sensors and actuators in automatic control systems and in medicine (example: the implanted insulin pump) are mostly combined with electronic signal processing, partly also with wireless data transmission. The functional principles of micro-mechanical sensors and actuators are closely connected with electrical phenomena, e.g., for determining the mechanical deflection by capacitance measurement and for controlling the position of a mechanical component by applied voltages. Because of

- the participation of electrical phenomena;
- the production of many classes of MEMS with micro-electronic technologies (e.g., the standard CMOS fabrication process);
- the efficiency of EDA tools in computer-aided design of complex and also heterogeneous systems;

it is worthwhile examining the applicability of system simulation approaches, established in electronics, also in microsystems engineering.

At a first glance the modeling strategy adapted best to the MEMS design problems is a description as spatially distributed systems. Bending beams, membranes and plates, pipes, or electrical fields between moving electrodes are characterized by their geometrical dimensions. They can be described mathematically by partial differential equations (PDEs) which are to be solved numerically by discretization in space and time. Simulation with the methods of the FEM (Finite Element Method) [13.14], [Z-T94], the FDM (Finite Difference Method), the BEM (Boundary Element Method) and others is very accurate but needs also extremely long computing time. After discretization tens or hundreds of thousands of equations or ordinary differential equations (ODE) must be solved.

Widely used FEM simulators with large application areas are

ANSYS [13.4], **COSMOS/M, MSC/NASTRAN, CAPA, ABAQUS, ALGOR, INERTIA** and **RASNA**.

Other simulators are specialized in magnetic field calculations (e.g., MAFIA) or fluidic simulation (FLOTRAN). In the microsystems technology the *interaction* of effects in different physical domains (mechanics, electronics, optics, fluidics, ...) very often has to be investigated. Many FEM simulators cannot handle such complex systems or need unacceptable large computing time. For the simulation of such complex systems basically two ways exist, which are sketched in fig. 13.17.

An important method is the simplification of the descriptions at the 'lower' levels by **abstraction** and thus the transition to 'higher' levels by suitable modeling. Also the **transformation** of models at the same abstraction level from one physical domain into another belongs to this method. Thus, e.g., *mechanical networks* can be transformed into *electrical networks* and then analyzed with a circuit simulator. The model of the entire system can be simulated with a network simulator if all subsystems can be treated in this way.

Continuous-time and continuous-value processes dominate in micro-system engineering. Therefore,

- analog simulators such as *Matlab/Simulink* and *MatrixX* (for systems, which can be described by block diagrams); and
- circuit simulators such as *Spice, Saber* or *ELDO* (using analogy relations between the different physical domains)

are frequently used as system simulators. Also simulators from the mechatronics and mechanics domain:

ADAMS, DYMOLA-MBSDYM, MEDYNA, NEWEUL, ITI-SIM [13.36], **SIMPLORER** [13.64], ...

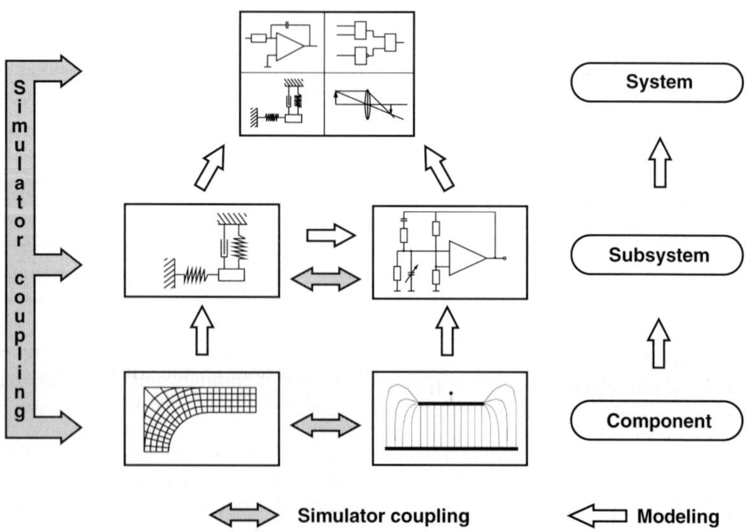

Fig. 13.17 Modeling and simulator coupling for MEMS simulation

can be applied, but they are not very common in the EDA world and therefore not considered any further here.

The other method is **simulator coupling**, once again. It can be recommended whenever the modeling task is too complex or leads to inaccurate results. Sometimes it is just the completely different physical domains of the subsystems and the experiences of the designers in handling special simulators which let the simulator coupling appear as the most obvious form of an interdisciplinary mode of operation.

13.3.2 Modeling for MEMS Design

The goal is the description of the phenomena in such a way that the multi-domain problems can be analyzed with only *one* simulator (the simulator coupling as an alternative is regarded in the next paragraph). Figure 13.18 gives an overview of modeling methods used at present or are investigated in research projects in the microsystem engineering [13.59], [13.60], [13.62], [13.63], [13.64], [13.78], [13.81].

Two of these methods are to be presented here in more detail:

- modeling with generalized Kirchhoffian networks;
- model export from FEM simulators and order reduction.

The assumption that the ordinary differential equations describing a microsystem are easy to formulate, and then simulators for general continuous systems (*Matlab/Simulink*, *MatrixX*...) can be used, is unfortunately often not fulfilled. The reason for this is that these systems are usually **conservative systems**. Therefore the use of network models (section 13.1.2) emphasizes itself.

Modeling with Generalized Networks

The key to the understanding of network applications as models of non-electrical systems (not only for microsystems technology, but, e.g., also for mechatronics), may be found in the concepts of '*through quantity*', '*across quantity*', and '*generalized Kirchhoff laws*'.

By **networks** we understand such systems in electronics which consist of **elements** (one-ports, multi-poles) and **connections**. Voltages and currents are assigned to the terminals of the elements and to the connections. Conservation laws apply to the voltages and currents, i.e., Kirchhoff's *mesh and node law* (or *current and voltage law*):

- **Kirchhoff's voltage law, KVL**: The sum of all voltage drops over the elements in a closed mesh is zero;
- **Kirchhoff's current law, KCL**: The sum of all currents flowing into a node is zero.

This concept can be extended to conservative non-electrical systems. The following generalizations are necessary:

Instead of currents, *flow quantities* are considered (e.g., *strength* in mechanics, *flow rate* in fluidics). A *flow quantity* is measured at one place, therefore it is also called a *1-point quantity* or a **through quantity**.

Instead of voltages, *difference quantities* are considered (e.g., *displacements* in mechanics, *pressure differences* in fluidic systems). A *difference quantity* is always measured between two points, therefore it is also called a *2-point quantity* or an **across quantity**.

For the introduced flow and difference quantities the generalized mesh and node laws apply in the non-electrical systems too. These conservation laws are well known from physics, e.g., from Newton's mechanics, and can be brought into the form of Kirchhoff's law.

The relations between *flow quantities* and *difference quantities* at an element are determined by the model equations of the element (network-element relations).

Flow and difference quantities at the terminals of the elements and on the connections between the elements can be multi-dimensional (e.g., forces and moments in three-dimensional space).

In fig. 13.19 these concepts are summarized. Since electrical networks belong to generalized networks too, simulators for electrical circuits can be used very effectively for the analysis of non-electrical systems. In table 13.3 the most important one-port elements in generalized networks are represented. In the treatment of coupled phenomena, additionally the interactions between different physical domains are to be modeled. Transformers and gyrators are used for the transformation between

13

Table 13.3 Analogy relations in generalized networks

	Symbol	General network (NW)	Mechanical NW translation	Mechanical NW rotation	Electrical NW	Fluidic NW (incompressible)	Thermal NW
Signals	f	flow quantity (FQ)	force	torque, momentum	current	volume flow rate	heat flow rate
	d	difference quantity (DQ)	velocity	angular velocity	voltage	pressure	temperature
	q	integrated flow quantity	translational momentum	angular momentum	charge	volume	heat energy
	ψ	integrated difference quantity	displacement, position	angle	magnetic flux	pressure momentum	—
Network elements	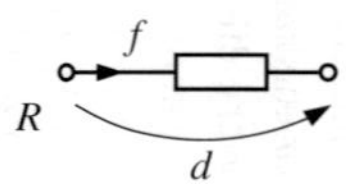	resistor, resistance e.g., $R = d/f$	frictional resistance (damping)	frictional resistance (damping)	resistance	fluid resistance	thermal resistance
	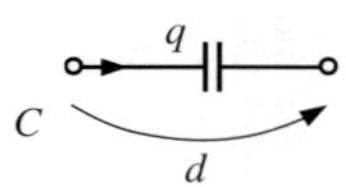	capacitor, capacitance e.g., $C = q/u$ energy $E_q = q^2/2C$	mass kinetic energy	rotational mass kinetic energy	capacitance electrical energy	fluid mass kinetic energy	thermal capacitance
	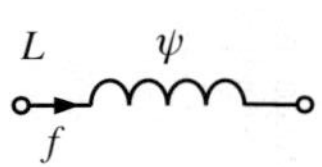	inductor, inductance e.g., $L = \psi/i$ energy $E_\psi = \psi^2/2L$	translational stiffness potential energy	torsional stiffness potential energy	inductance magnetic energy	fluid inertance kinetic energy	—
	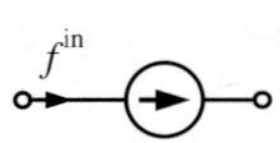	FQ source f^{in} = given function	force source	momentum source	current source	volume flow source	heat flow source
	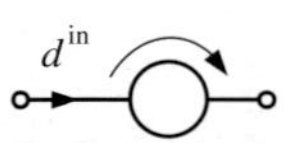	DQ source d^{in} = given function	velocity source	angular velocity source	voltage source	pressure source	temperature source

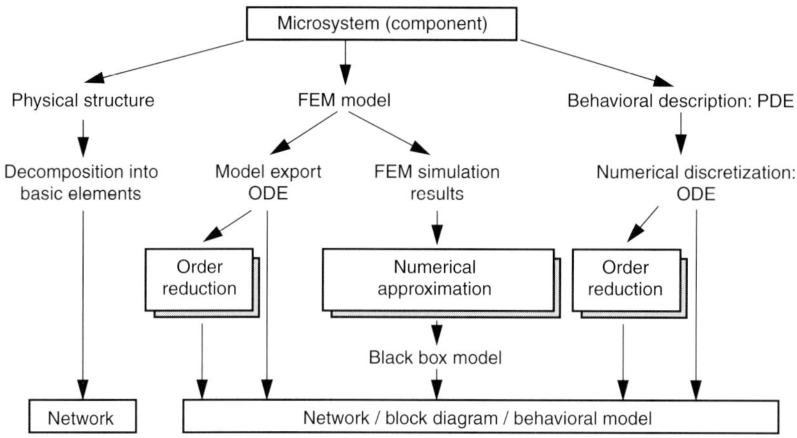

Fig. 13.18 Modeling methods for microsystems technology

Physical domain	Flow quantity f	Difference quantity d
electrical	current	voltage
hydraulic	volume flow	pressure
mechanical	force	velocity
thermal	heat flow	temperature

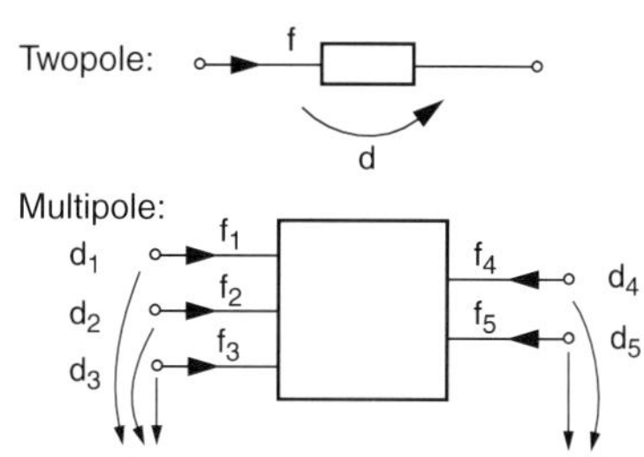

KIRCHHOFF's mesh and node law:

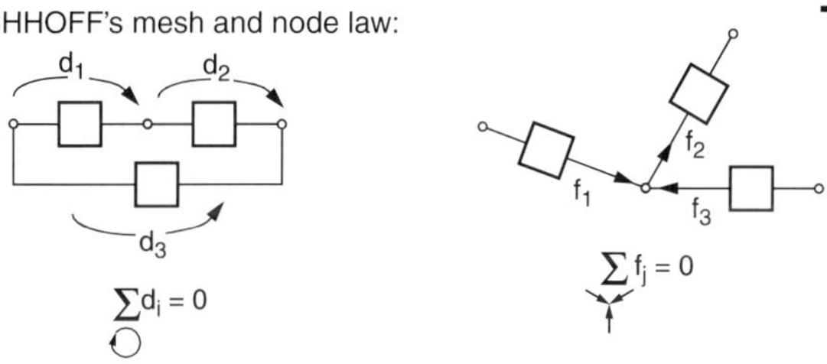

Coupling between different physical domains by transducers (transformers, gyrators)

Fig. 13.19 Generalized Kirchhoffian networks

the domains. The relationships between the networks in different physical domains are also called *analogy relations* (table 13.3). The application of the concept of generalized networks for the simulation of non-electrical systems has a long and successful tradition and is therefore promising success for micro-system engineering [13.28], [13.40], [13.46], [13.58], [13.73], [13.83].

The applicability of the network concept is not only based on the use of formal analogy relations, because, e.g., the elements in the different physical domains are described by ordinary differential equations of similar structure. Above all it bases on the fundamental physical principles of

- the hierarchical decomposability of a system into components (or, more general, into subsystems);
- the description of the 'terminal behavior' of the components (in a special form as in electrical four-pole theory, well-known to electrical engineers); and

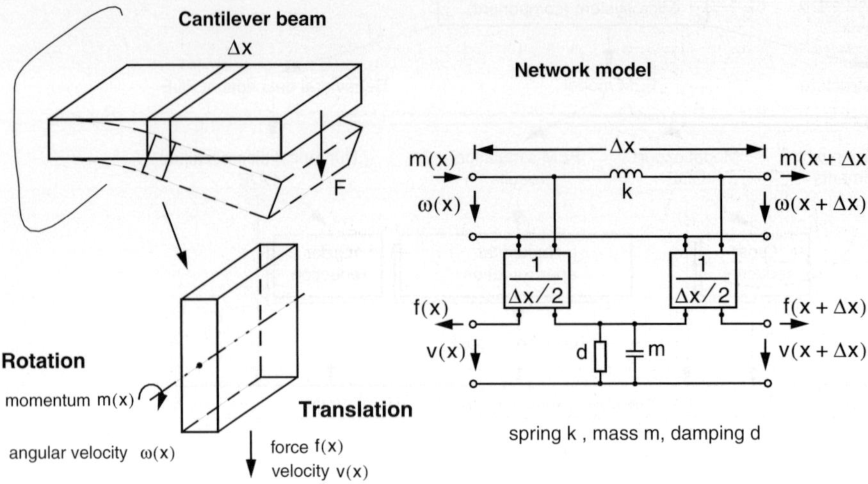

Fig. 13.20 Network model of a beam segment

- the validity of conservation laws for the *flow quantities* and the *across quantities* at the interfaces between the components.

Here also a relation is to be seen with object-oriented modeling [13.16].

Particularly important in microsystems engineering is another modeling step which leads from spatially *distributed* systems to networks with spatially *concentrated* elements. Mathematically this spatial discretization corresponds to the transition from partial differential equations (PDE) to ordinary differential equations (ODE). A further discretization in the time domain by the designer is not necessary, since the generated ordinary differential equations can be solved in the time domain most effectively by analog simulators. The steps for the construction of a network model are:

1. **Partitioning of the whole system into subsystems** (or components) which can be simpler modeled;
2. **Field quantities** are 'concentrated' on quantities which are assigned to the connections between the subsystems and their terminals: replacement of the electrical field strength by an electrical voltage, . . . ;
3. **Definition of the terminals** of the subsystems or components (interface definition);
4. **Modeling of the terminal behavior** of the subsystems (by behavioral or structural models);
5. **Use of the subsystem models** as components in a **superior network** which corresponds to the whole system.

In fig. 13.20 a typical mechanical microsystem component – a bending beam – is shown together with a network model for one segment of the bending beam [13.46]. Actually this segment should be infinitesimally small, but by the spatial discretization it has a finite length Δx. Experience shows that a rough partition into 4 or 8 of these segments is most often completely sufficient for the accuracy desired in system simulation [13.53].

Fig. 13.21 Acceleration sensor

But in the design of more complicated MEMS it is often very difficult to model them by such simple networks. Then it is more favorable to describe

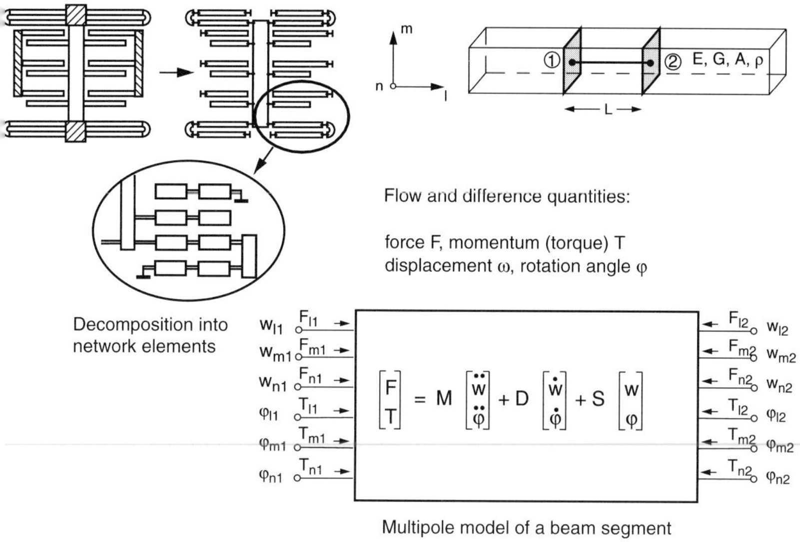

Fig. 13.22 Network model of the acceleration sensor

he mathematical relations for a subsystem with a ehavioral model [13.54], [13.62]. With the aid f the above-mentioned modeling languages it is hen possible to use these models in appropriate imulators. This procedure is described for a mirosystem with a relatively complicated structure, translational acceleration sensor, in fig. 13.21 13.53], [13.70].

or the whole system a network model is contructed which consists of models for the homogeeous bending beams and the connections between em (fig. 13.22). Since both translational and otational forces and moments, respectively, in the ree-dimensional space are to be examined, multioles with the according number of terminals have be defined. Their terminal quantities are used the behavioral models. These equations are o large to be discussed here in detail. They orrespond, however, exactly to the mathematical quations used in the text books of mechanics for e description of bending beams and other basic omponents. Therefore only the interface and a imple mathematical model of a bending beam f length L in the form of a matrix equation are hown. From such network elements the main etwork is assembled as the model of the whole ystem. In contrast to the simple structural model fig. 13.20, in which only the translation in x-direction and the rotation around the z-axis are considered, translation and rotation in all three directions in space are included here.

Model Export

The way just described is directly plausible, but it has also its difficulties. The formulation of the mathematical relations for the modeling of the basic components is everything but simple, even if these equations are to be found in the literature. The formulas are usually only given for simple geometrically homogeneous structures. More complicated structures may be simulated numerically by solving the partial differential equations. FEM simulators have complicated algorithms (meshing algorithms) in order to set up the describing equations also for inhomogeneous structures . Therefore another modeling approach consists of using these simulator-internal equations and exporting them to set up externally a behavioral model [13.32]. But the number of equations used in these models are too large to be used directly in system simulation where many of these components have to be considered simultaneously. A substantial simplification of the equations by **order reduction** (i.e., the reduction of the number of equations and variables) is necessary to achieve short computing times. The price is the effort for model reduction and a reduced simulation accuracy which is,

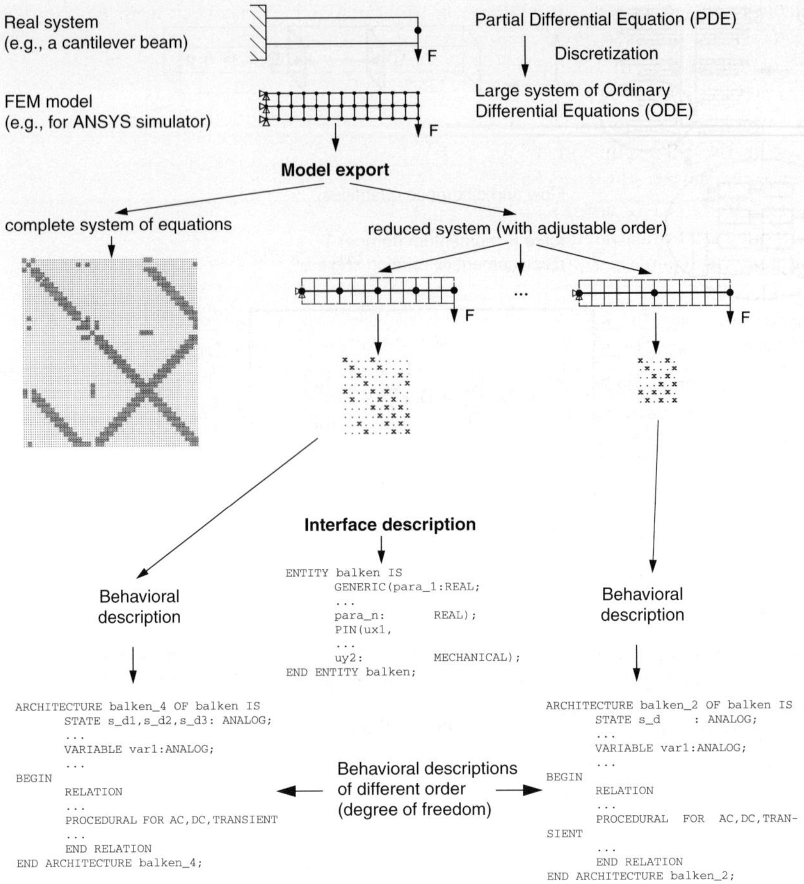

Fig. 13.23 Model generation by order reduction

however, sufficient for system simulation [13.81], [13.82]. In fig. 13.23 this way is sketched.

Order reduction is a sophisticated mathematical problem, therefore algorithms and programs for it are still in the research stage. So it is very important to note that at least for special, but particularly important, classes of problems some of today's FEM simulators support the automatic generation of behavioral models by the export of a reduced set of internal equations. E.g., the FEM simulator ANSYS has a 'sub-structuring mode' to export equations from ANSYS, already reduced with the Guyan algorithm [13.4]. Figure 13.24 shows a detail of a very inhomogeneous mechanical microsystem. A model for the static deformation in the x, y and z directions and the dynamic behavior under the effect of an external force F_e has to be formulated.

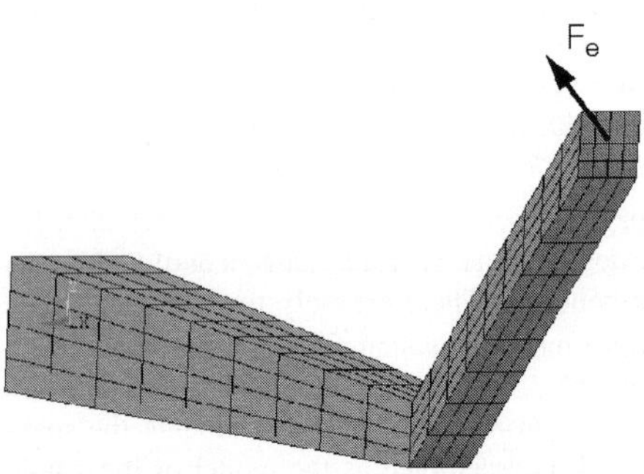

Fig. 13.24 Inhomogeneous part of a micro-mechanical system

It is hardly possible to set up an analytic model by hand. It would remain – as described above – a cumbersome partition into many small homogeneous segments and joining them to an overall network model, compare fig. 13.22. With the aid of the order reduction in ANSYS a simple model for system simulation can be generated automatically. The matrices noticed in fig. 13.22 are calculated in ANSYS and exported after order reduction. For example, a simplified set of equations with 69 lines and columns is produced:

```
Stiffness matrix K[i,j] (i, j, value):
1 1 5.648093e-02     1 2 2.340571e-02
                  ... 69 69 1.263220e-01
Mass matrix M[i,j] (i, j, value):
1 1 4.395064e-21 1 2  5.371975e-22
                  ... 69 69 3.884850e-20
Damping matrix D[i,j] (i, j, value):  ...
```

A post-processor can transform these data into a VHDL-AMS model:

```
library DISCIPLINES;
  use DISCIPLINES.KINEMATIC_SYSTEM.all;
  use DISCIPLINES.ROTATIONAL_SYSTEM.all;

ENTITY angle IS
# Declaration of mechanical nodes:
  PORT (
  # Knoten 1 x direction
  TERMINAL t1_1 :  kinematic;
# Knoten 1 y direction
  TERMINAL t2_1 :  kinematic;
  ...
  TERMINAL t45_343 :  kinematic );
END angle;

ARCHITECTURE arch1 OF angle IS
# v = displacement
  QUANTITY v_t1_1 ACROSS i_t1_1 THROUGH t1_1;
  QUANTITY v_t2_1 ACROSS i_t2_1 THROUGH t2_1;
  ...
# auxiliary quantities pv = dv/dt velocity
  QUANTITY pv_t1_1 :  REAL;
  QUANTITY pv_t2_1 :  REAL;
  ...
BEGIN
# Equilibrium of forces at node 1,
# x direction:
  -i_t1_1 == 4.3950638e-21*pv_t1_1'DOT
  + (-5.3719751e-22)*pv_t2_1'DOT + ...
  + 7.9716188e-26*pv_t69_471'DOT
  + 5.6480929e-02*v_t1_1
  + 2.3405710e-02*v_t2_1 + ...
  + (-5.5181300e-08)*v_t69_471;
  ...
# Equilibrium of forces at node 5,
# y direction::
  0.0 == 5.0286069e-24*pv_t1_1'DOT +
  (-9.9876969e-25)*pv_t2_1'DOT +
  ...   + 5.0616559e-18*v_t69_471;
  ...
# auxiliary quantities pv = dv/dt
  pv_t1_1 == v_t1_1'DOT;
  pv_t2_1 == v_t2_1'DOT;
  ...
END ARCHITECTURE arch1;
```

This model of a mechanical component can be used now, e.g., together with the model of an electrical subsystem for the simulation of the electromechanical microsystem (MEMS) with a VHDL-AMS simulator. It contains, however, only numerical parameters which may be calculated within ANSYS. In the case of modifications of the mechanical dimensions or of the material parameters the generation of this VHDL-AMS model has to be repeated.

13.3.3 Simulator Coupling for Microsystems Engineering

From the examples of the previous sections it is recognizable that the modeling for *one* simulator is possible, but often not quite easy. Therefore the second possibility for simulation of heterogeneous systems, the coupling of simulators, will be demonstrated by typical examples also for microsystems engineering.

In fig. 13.25 an electro-mechanical acceleration sensor is shown, whose seismic mass ('proof mass') suspended by both sides moves between the two plates under the influence of acceleration forces [13.17], [13.23]. The capacitances between the mass moved and the two electrodes are measured with a special circuit. Their changes are a measure for the displacements and thus for the acceleration forces. Additionally, the sensor is controlled, i.e., depending on the displacement of the movable mass a voltage is applied between the mass and the electrodes in order to bring the mass back into the central position. It is also shown in the picture what effect the control has on the behavior of the sensor after the appearance of a stepwise rising and then remaining constant acceleration force:

- large overshooting without control;

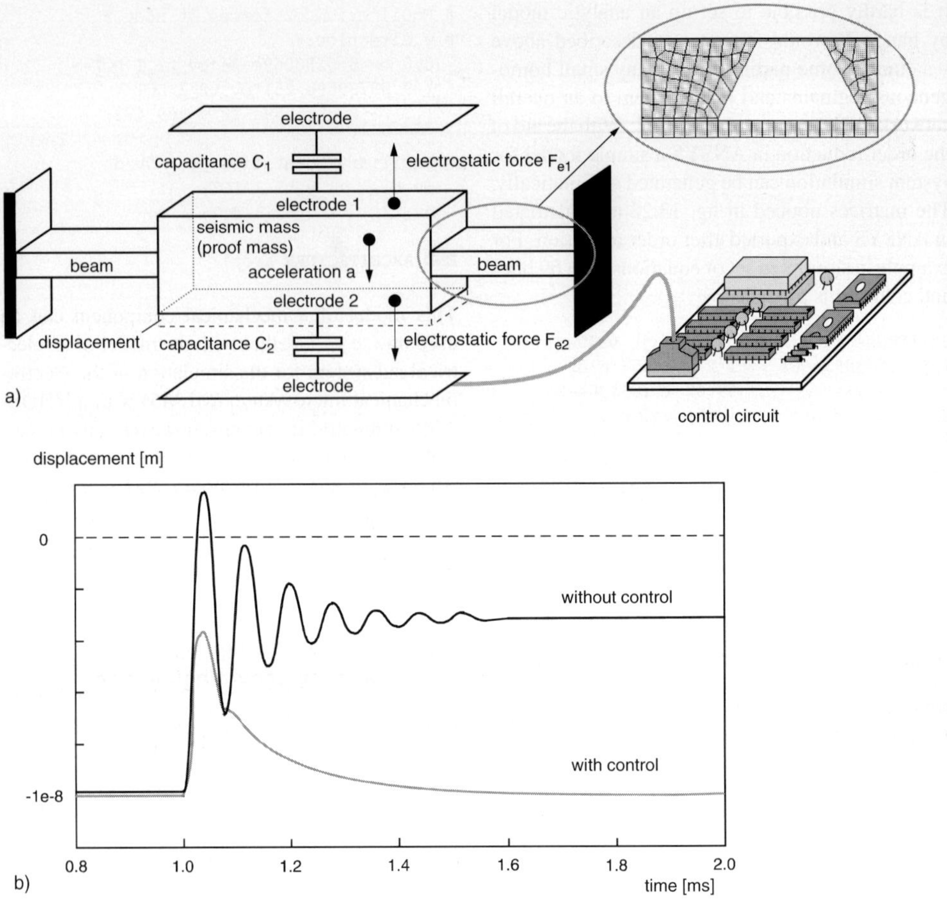

Fig. 13.25 Controlled electro-mechanical acceleration sensor

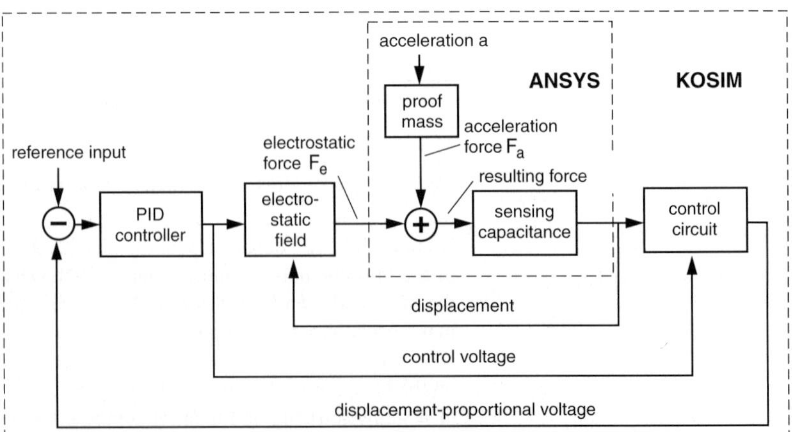

Fig. 13.26 Block diagram of the electro-mechanical acceleration sensor

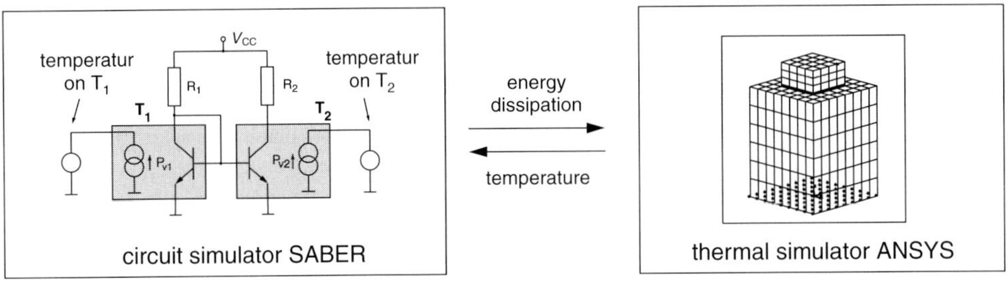

Fig. 13.27 Thermal-electrical interactions in an integrated circuit

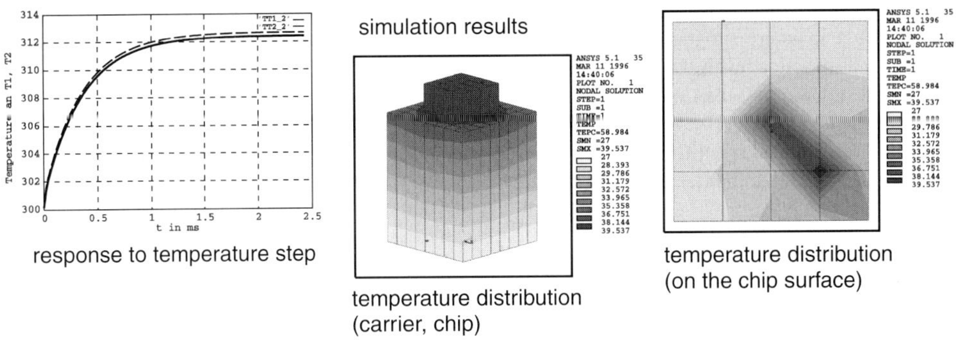

Fig. 13.28 Results of thermal-electrical simulation

- substantially reduced overshooting and achieving the initial position with control.

The calculations of these heterogeneous microsystems (in which particularly the mechanical-electrostatic interactions are to be analyzed and considered especially during the automatic controller design) can be realized very well by coupling a FEM simulator (e.g., ANSYS) to a circuit simulator (e.g., Saber). Figure 13.26 shows the block diagram with the data interfaces between the two subsystems where the data exchange between the two simulators takes place. Both simulators calculate the new output signals as a reaction of their input signals. The simulators operate in parallel and have to be synchronized for data exchange. Particularly for mechatronics systems and microsystems, couplings of commercially available simulators have been developed increasingly in the last years [13.36], [13.64].

As a result of the continually decreasing dimensions of integrated circuits the **electrical-thermal interactions** have to be considered. On the one hand, the energy dissipation and temperature of electronic elements depend on their currents and voltages; on the other hand, elements' currents and voltages are influenced vice versa by their individual temperatures (particularly in the junctions of bipolar transistors). Additionally, very often it is not enough just to calculate the thermal-electrical behavior only at one time point but rather also the dependences on the real input signals or several typical operating points must be simulated. From this point of view integrated circuits appear as typical heterogeneous dynamical systems with coupled effects in different physical domains.

In fig. 13.27 a part of an electronic circuit is shown which can be simulated, e.g., with *Saber*. There is also shown the thermal chip model, which can be calculated, e.g., with the FEM simulator ANSYS. The common simulation is realized by a simulator coupling [13.74]. The simulators exchange at each time point the energy dissipation and the temperature of those elements which are particularly power consuming and temperature-sensitive, respectively. The temperature distribution on the chip at one time point as well as the temporal changing of the temperature at two transistors are shown in fig. 13.28.

The **advantages** and disadvantages may be summarized (similarly as already stated in section 13.2 for communication systems):

For each subsystem *the* simulator can be selected which fits optimally to the physical domain (mechanical, thermal, magnetic, electronic) and the chosen abstraction level (PDE, ODE, discrete models, ...). Model libraries, input language, simulation algorithms, coupling capabilities with other simulators, and visualization of the results are to be considered in choosing the 'best' simulator. Thereby the modeling effort is reduced considerably in many situations. The modeling for higher abstraction levels, based on experiences and FEM simulations, may take days, weeks or yet longer if no model generator is available. This effort can be dropped if a FEM simulator can be used for the subsystem modeled at PDE level.

Also **disadvantages** stand against these advantages. Up to today only a few simulators could be coupled without any additional implementation effort of the user. The simulation times are mostly **considerably higher** than in simulations with only one simulator, and coupling software needs more computing time. On the other hand, the frequently used FEM simulators need **substantially larger computing times,** anyway, because of their more detailed models.

Apart from these technical advantages **further advantages** can arise which are related to license conditions and the very high purchase costs of simulators. Coupling of simulators may be realized not just on one processor or in a local processor grid. It is today's state of the art to connect computers by the use of world-wide networks. Thus, e.g., in research projects the coupling of circuit simulators in Dresden with FEM simulators located in Chemnitz in Germany and in Linz in Austria via the *Internet* was implemented. Also the cooperation of simulators and optimization programs via the Internet has been already realized [13.68].

13.4 User Specific Representation of Simulation Results

The above-mentioned simulators provided by EDA industry have extensive possibilities for the graphic representation of simulation results. The most important representation forms in electronics and control theory are:

- timing functions of analog and digital signals (transient simulation);
- attenuation and phase characteristics of linear circuits (AC analysis);
- frequency spectra (obtained, e.g., from the conversion between time and frequency domain by means of Fourier transforms);
- root locus and pole-zero representations (e.g., for stability analysis of control systems).

The display tools of such simulators offer a variety of possibilities for trace handling: zooming; measuring of curve characteristics by markers and rulers; as well as 'calculator functions' for adding, multiplying, calculating the logarithm, ... of traces.

All of these possibilities are needed for system simulation, but they are not sufficient. *Eye diagrams* and *scatter diagrams*, e.g., are important in the design of telecommunication systems (fig. 13.13). Eye diagrams supply – similarly to scatter diagrams – information about how well disturbed signals can be detected. The result of a transient simulation for 0-1-sequences of the input signal will be periodically written in the same diagram, one above the other. The transients in the system, distortions, and random disturbances lead to characteristics of the output signal, as is represented in fig. 13.29. A separation of the 0 and the 1 level by a *decider circuit* will be possible if the 'eye' inside the diagram, i.e., the central white area between the curves, is large enough. With some simulators an eye diagram can be generated by clever use of the waveform tools for the transient simulation. Similarly, special display tools for scatter diagrams have to be used or developed.

For simulation in the microsystems technology domain (MEMS) it is important to have two- and three-dimensional representations of mechanical deformations, pressures, temperatures, distribution of electrical currents, electric potential fields, and others. These visualization capabilities are part of the standard scope of FEM simulators.

But for other complex systems very specialized or problem-dependent display tools have to be programmed. There is a good software support for writing one's own visualization tools [13.18]:

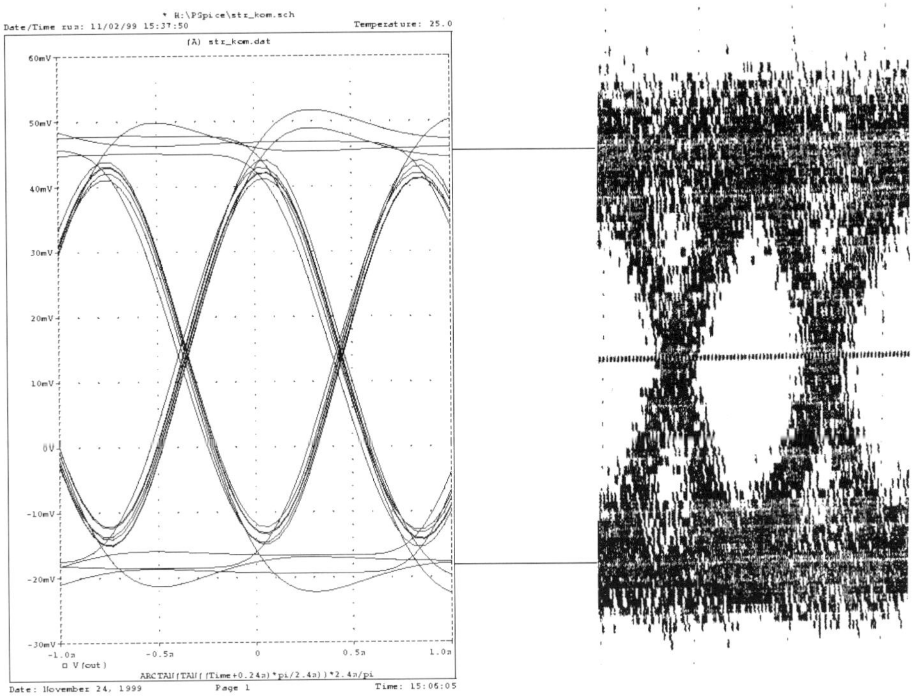

a) b)

Fig. 13.29 Eye diagrams: a) simulated, without noise; b) measured, with noise

- public domain software: gnuplot [13.30];
- Java libraries [13.37];
- for larger visualization tasks: commercially available libraries such as IMSL [13.35] and IDL (Interactive Data Language) [13.34];
- the use of manifold visualization routines of Matlab/Simulink and MatrixX, which has open interfaces to the user and other simulators.

Just for system simulation the desired visualization of simulation results transcends one- or two-dimensional curve representations considerably. E.g., it can be very helpful for a designer of a complex control system for automobiles or aircrafts to see simulation results not only as number series or traces, but also to visualize the results as pictures like a speedometer, an altimeter, or a compass. An input of a signal value can also be achieved by copying the position of a real knob or a pedal or by a moveable glider on the display. After the simulation of motion sequences (e.g., of robots or vehicles) a 'cine-like' visualization is useful, which is often called **animation**. Figure 13.30 shows a representation of a robot arm whose position will change as a result of the simulation (for the simulation of mechatronic systems the simulator DYMOLA [13.22] has been used).

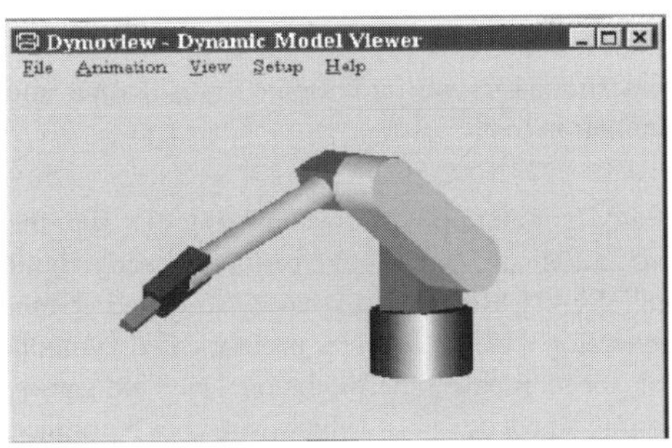

Fig. 13.30 Motion animation of a robot arm

Because its independence of a software platform and its wide use in the Internet, Java will become widespread. Java includes class libraries of 2D and

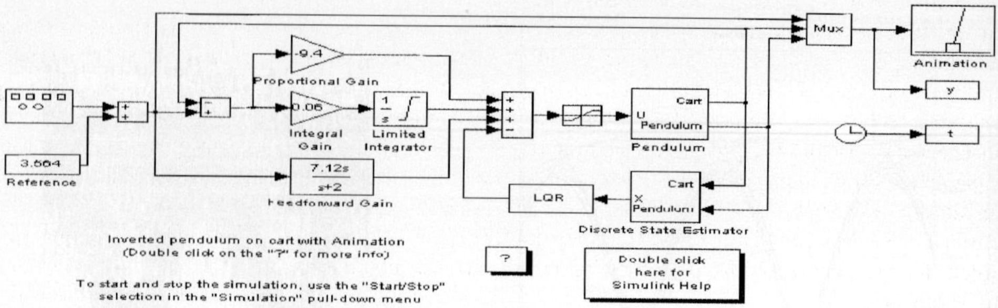

Fig. 13.31 Control of the inverse pendulum

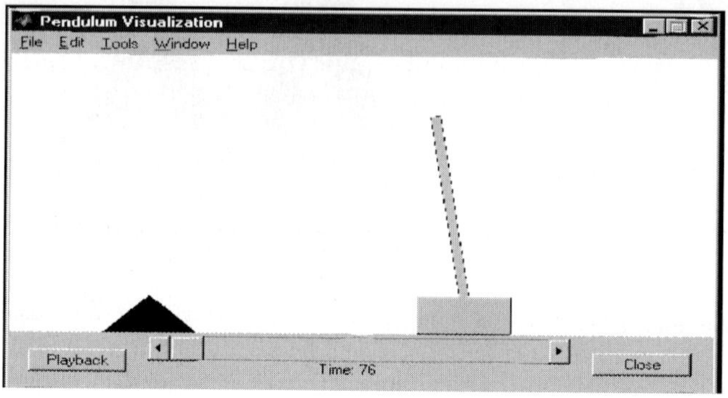

Fig. 13.32 Animation of the inverse pendulum

3D graphics, so the user can program his own very complex visualization tools easily.

VRML is a standardized file format for 3D scenarios and the description of projection areas, that corresponds to the movement of an observer going through the space. Only by the use of such standards it is possible to couple general-purpose system simulators with additional visualization and animation tools.

Many simulation systems have tools for the animation of simulation results already built in. STATEMATE, e.g., has a so-called 'panel generator' [13.7]. It offers prefabricated symbols for speedometers, thermometer, level indicators, setting devices, …, which can be combined together into a graphic representation of complex systems (e.g., an airplane cockpit or a measurement arrangement). A calculated speed will then be shown as a speedometer deflection. VAPS [13.71], a visualization and animation software of the company *Virtual Prototyping*, can be used for such problem-specific tasks.

Matlab/Simulink offers similar animation possibilities. In fig. 13.31 a control system for the control of an 'inverse pendulum' is shown.

A simply supported rod, mounted on a wagon, is held in balance by moving the wagon in such a way that a falling rod would be set back standing up. Therefore the wagon has to be moved to the falling direction of the rod and the inertial forces will set the rod upright. The design and dimensioning of the control system are challenging tasks and it is important to check the design by simulation. In fig. 13.31 a system description for Matlab/Simulink is shown. The movement of the wagon and the position of the rod is visualized by the block 'animation'. Figure 13.32 shows a snapshot of the falling rod.

More detailed and quantitative information about the simulated processes in the system can surely better be shown by traces and value tables of simulation results. But a first glance animation of complex *sequences* can also be an essential addendum, e.g., in the use of simulation for educational purposes. In this respect flight simulators and state-of-the-art play stations show a top performance, but the EDA tools also achieve respectable capabilities. But the effort for the elaboration of an informative and also aesthetic visualization and animation environment is quite high: days and weeks for the concept and for programming have to be scheduled.

Of course, in this short chapter only some aspects of system simulation and of problem-specific visualization of simulation results could be demonstrated. But the reader will probably be faced with this problem more often in future. Besides the use of EDA tools for the design of new circuits and systems, lifelong learning activities will play a big role in the future. Computer aided education will extensively use the Internet, will include interactive simulation, and its efficiency will be essentially determined by graphical communication.

13.5 References

Instead of publications, Internet site addresses have been included in the bibliography as references to simulators or other tools. Owing to the rapid development of such software (and the occasionally rapidly changing tool names and trade marks), the best way to actual information is to check the homepage of the providers.

[13.1] http://www.tm.agilent.com/tmo/hpeesof/products/ads/

[13.2] http://www.analogy.com

[13.3] *Antao, B.; Brodersen, A.*: 'Behavioral simulation for analog system design verification'. IEEE Trans. VLSI 3(1995)3, 417–429

[13.4] http://www.ansys.com

[13.5] *Antao, B. (Ed.)*: 'Modeling and Simulation of Mixed Analog-Digital Systems'. Dordrecht: Kluwer 1996

[13.6] *Atherton, D. P.; Borne, P.*: 'Concise Encyclopedia of Modeling and Simulation'. Pergamon Press, Oxford 1992. A global overview for continuous-value simulation (solution of differential equations, Z-transform, identification)

[13.7] http://www.ilogix.com

[13.8] *Banks, J. (Ed.)*: 'Handbook of Simulation'. Wiley, New York 1998. A comprehensive overview of discrete simulators (simulation methods and a short description of simulators such as GPSS/H, SIMSCRIPT and SIMPLE++)

[13.9] *Boyle, G. R.; Cohn, B. M.; Pederson, D. O.; Solomon, J. E.*: 'Macromodeling of integrated circuit operational amplifiers'. IEEE J. Solid-State Circuits SC-9(1974)6, 353–363

[13.10] *Breitenecker, F.; Ecker, H.; Bausch-Gall, I.*: 'Simulieren mit ACSL'. Vieweg, Braunschweig 1993. See also: http://acslsim.com

[13.11] *Bacher, T.; Engelmann, F.; Knoechel, U.; Schwarz, P.*: 'Multi-Level-Simulation beim Entwurf eines ASTRA Digital Empfaengers'. 4. GMM/ITG-Fachtagung 'Analog '96', Berlin 1996, 181-188

[13.12] *Buck, J.; Ha, S.; Lee, E. A.; Messerschmitt, D. G.*: 'Ptolemy: A framework for simulating and prototyping heterogeneous systems'. Intern. J. Computer Simulation. 4(1994) 155–182

[13.13] *Berge, J.-M.; Levia, Oz; Rouillard, J. (Hrg.)*: 'Current Issues in Electronic Modeling'. Kluwer, Dordrecht, since 1995. Many volumes comprising current developments in the area of the modeling at all abstraction levels (10 volumes so far):
1. Model Generation in Electronic Modeling. 1995
2. Modeling in Analog Design. 1995 (also the roots of VHDL-AMS are presented)
3. High-Level System Modeling: Specification Languages. 1995
4. High-Level System Modeling: Specification and Design Methodologies. 1996
5. Hardware Component Modeling. 1996
6. Meta-Modeling: Performance and Information Modeling. 1996
7. Object-Oriented Modeling. 1996
8. HW/SW Co-Design and Co-Verification. 1997
9. Models in System Design. 1997
10. Analog and Mixed-Signal Hardware Description Languages. 1997

[13.14] *Bathe, K. J.*: 'Finite Element Procedures'. Prentice-Hall, New Jersey 1996

[13.15] http://www.cadence.com

[13.16] *Cellier, F. E.*: 'Continuous System Modeling'. Springer, New York/Berlin 1991.

[13.17] *Clauss, C.; Gruschwitz, R.; Schwarz, P.; Wuensche, S.*: Simulation mikrosystemtechnischer Aufgaben mit gekoppelten Simulatoren. 2. Chemnitzer Fachtagung 'Mikrosystemtechnik – Mikromechanik & Mikroelektronik', TU Chemnitz-Zwickau 1995, 92–101

[13.18] *Chen, J. X.; Frieder, O.*: 'Applications of computer graphics software tools'. IEEE Trans. Computing in Science & Engineering 1(1999)6, 82–87

[13.19] *Christen, E.; Bakalar, K.*: 'VHDL-AMS – A Hardware Description Language for analog and mixed-signal applications'. IEEE Trans. CAS-II, 46(1999)10, 1263–1272

[13.20] Connelly, J. A.; Choi, P.: Macromodeling with SPICE. Prentice Hall, New Jersey, 1992

[13.21] *Duran, P. A.*: 'A Practical Guide to Analog Behavioral Modeling for IC System Design'. Kluwer, Dordrecht 1998

[13.22] http://www.Dynasim.se

[13.23] *Eccardt, P. C. et al.*: 'Coupled finite element and network simulation for microsystem components'. Proc. MICROSYSTEM Technologies '96, Potsdam, Sept. 1996, 145-150

[13.24] *Schwarz, P.; Einwich, K.; Haase, J.; Prescher, R.*: 'Mixed-mode design: experiences with multi-level macromodeling'. Published in: Huijsing, J.H. et al.(Eds.): Analog Circuit Design. Proc. Workshop Advances in Analog Circuit Design, Villach 1995. Kluwer, Boston 1995, 181–203

[13.25] *Engelmann, F.; Jentschel, H.-J.; Schwarz, P. (Hrg.)*: Proc. Workshop 'Modellierung und Simulation in der Nachrichtentechnik'. Dresden, November 1995

[13.26] *Einwich, K.; Schwarz, P.; Trappe, P.; Zojer, H.*: 'Simulatorkopplung fuer den Entwurf komplexer Schaltkreise der Nachrichtentechnik'. 7. ITG-Fachtagung "Mikroelektronik fuer die Informationstechnik", Chemnitz, 18./19. Maerz 1996, 139-144

[13.27] *Fischer, W.-J. (Hrg.)*: 'Mikrosystemtechnik'. Vogel, Wuerzburg 2000

[13.28] *Gerlach, G.; Doetzel, W.*: 'Grundlagen der Mikrosystemtechnik'. Hanser, Muenchen 1997

[13.29] *Goedecke, M.; Hamad, H.; Huss, S. A.*: 'A methodology for the development of system-level simulation models for analog functional blocks'. AEÜ 49(1995)2, 72–80

[13.30] http://www.cs.dartmouth.edu/gnuplot_info.html

[13.31] *Hartung, J.; Knoechel, U.*: 'Approaches to consider analog RF components in system level simulation of mobile communications'. Proc. ANALOG '02, Bremen 2002, 219-224. See also: http://www.cadence.com/rf_notes.html

[13.32] *Haase, J.; Reitz, S.; Schwarz, P.*: 'Behavioral modeling for heterogeneous systems based on FEM descriptions'. Proc. IEEE Intern. Workshop Behavioral Modeling and Simulation BMAS99, Orlando, FL, October 1999

[13.33] *Huss, S. A.*: Model Engineering in Mixed-Signal Circuit Design. Kluwer, Boston 2001

[13.34] http://www.rsinc.com

[13.35] http://vni/products/imsl

[13.36] http://www.iti.de

[13.37] http://java.sun.com/products/java-media/2D

[13.38] *Jeruchim, M. C.; Balaban, P.; Shanmugan, K. S.*: 'Simulation of Communication Systems'. Plenum Press, New York 1992

[13.39] *Karnopp, D. C.; Margolis, D. L.; Rosenberg, R. C.*: 'System Dynamics: A Unified Approach'. Wiley, New York 1990

[13.40] *Koenig, H. E.; Blackwell, W. A.*: 'Electromechanical System Theory'. McGraw-Hill, New York 1961

[13.41] *Korn, G. A.*: 'Interactive Dynamic System Simulation'. McGraw-Hill, New York 1989

[13.42] *Knoechel, U.; Tannert, U.; Haufe, J.; Schwarz, P.*: 'Verifikation nachrichtentechnischer Systeme mit Systemsimulation und HW/SW-Cosimulation'. GI/ITG/GME Workshop, Paderborn 1998, 175–184.

[13.43] *Kundert, K. S.; White, J. K.; Sangiovanni-Vincentelli, A.*: 'Steady-State Methods for Simulating Analog and Microwave Circuits'. Kluwer, Dordrecht 1990

[13.44] *Leszak, M.; Eggert, H.*: 'Petri-Netz-Methoden und Werkzeuge'. Informatik-Fachberichte 197, Springer-Verlag, Berlin, 1989

[13.45] *Lee, E. A.; Messerschmitt, D. G.*: 'Digital Communication'. Kluwer, Dordrecht 1994

[13.46] *Lenk, A.*: 'Elektromechanische Systeme' (3 vol.). Verlag Technik, Berlin 1971–1973

[13.47] *Lorenz, G.; Neul, R.*: 'Network-type modeling of micromachined sensor systems'. Proc. MSM98

[13.48] *Mantooth, H. A.; Fiegenbaum, M. F.*: 'Modeling with an Analog Hardware Description Language'. Kluwer, Dordrecht 1994

[13.49] http://www.isi.com/products/matrixx/

[13.50] http://www.mathworks.com

[13.51] http://www.mentor.org

[13.52] Modelica: see http://modelica.org ; many links to Modelica-related publications

[13.53] *Neul, R. et al.*: 'A modeling approach to include mechanical microsystem components into system simulation'. Proc. Design, Automation & Test Conf. (DATE'98), Paris, 1998, 510–517

[13.54] *Pelz, G. et al.*: 'MEXEL: Simulation of microsystems in a circuit simulator using automatic electromechanical modeling'. Proc. Microsystem Technologies, VDE-Verlag, Berlin 1994, 651-657.

[13.55] *Rammig, F. J.*: 'Systematischer Entwurf digitaler Systeme'. Teubner, Stuttgart 1989

[13.56] *Rammig, F. J.*: 'System Level Design'. In: Mermet, J. (ed.): Fundamentals and Standards in Hardware Description Languages. Kluwer, Dordrecht 1993, 109–151

[13.57] *Reichl, H.; Obermeier, E. (Eds.)*: 'MICROSYSTEM Technology 98'. Proc. 6. Intern. Conference Potsdam VDE-Verlag, Berlin 1998

[13.58] *Reinschke, K.; Schwarz, P.*: 'Verfahren zur rechnergestuetzten Analyse linearer Netzwerke'. Akademie-Verlag, Berlin 1976.

[13.59] *Romanowicz, B. F.*: 'Methodology for the Modeling and Simulation of Microsystems'. Kluwer, Dordrecht 1998

[13.60] *Senturia, S.; Aluru, N. R.; White, J.*: 'Simulating the behavior of MEMS devices: computational methods and needs'. IEEE Trans. Computational Science & Engineering, January 1997, 30–54

[13.61] *Schwarz, P.; Clauß, C.; Einwich, K.;Knoechel, U.; Matz, K.*: 'Hybride Simulation nachrichtentechnischer Systeme'. 12. Symposium Simulationstechnik Zuerich, 15.-18.9.1998, 67–74

[13.62] *Schwarz, P.; Haase, J.*: 'Behavioral modeling of complex heterogeneous microsystems'. Proc. 1st Intern. Forum on Design Languages (FDL'98), Lausanne, Sept. 1998, 53–62

[13.63] *Senturia, S. D.*: 'CAD challenges for microsensors, microactuators, and microsystems'. Proc. IEEE 86(1998)8, 1611–1626

[13.64] http://www.ansoft.com/about/academics/simplorer_sv/

[13.65] *Senturia, S. D.*: 'Microsystem Design'. Kluwer, Boston 2001

[13.66] *Saleh, R.; Jou, S.-J.; Newton, A. R.*: 'Mixed-Mode Simulation and Analog Multilevel Simulation'. Kluwer, Dordrecht 1994

[13.67] http.//www.dolphin.fr

[13.68] *Schneider, P.; Parodat, S.; Schneider, A., Schwarz, P.*: 'A modular approach for simulation-based optimization of MEMS'. Proc. SPIE Conf. Design, Modeling, and Simulation in Microelectronics, Singapore 2000, pp. 71–82.

[13.69] http//:www.synopsys.com

[13.70] *Teegarden, D.; Lorenz, G.; Neul, R.*: 'How to model and simulate microgyroscopic systems'. IEEE Spectrum 35(1998)7, 67–75

[13.71] http://vaps.org

[13.72] Information about VHDL-AMS (VHDL – Analog and Mixed Signal Extensions): http://www.vhdl.org/analog/ und http://www. vhdl-ams.com/

[13.73] *Wachutka, G.*: 'Tailored modeling: a way to the 'virtual microtransducer fab'?' Sensor and Actuators A 46-47 (1995), 603–612

[13.74] *Wuensche, S.; Clauss, C.; Schwarz, P.; Winkler, F.*: 'Electro-thermal simulation using simulator coupling'. IEEE Trans. VLSI 5(1997)3, 277–282

[13.75] *Zienkiewicz, O. C.; Taylor, R. L.*: 'The Finite Element Method' (2 vol.). McGraw-Hill, New York 1994

[13.76] *Schwarz, P.*: 'Physically oriented modeling of heterogeneous systems'. Mathematics and Computers in Simulation 53 (2000), 333–344

[13.77] *Haase, J.*: 'Rules for analog and mixed-signal VHDL-AMS modeling'. Proc. FDL'03, Frankfurt 2003

[13.78] *Haase, J.; Bastian, Reitz, S.*: 'VHDL-AMS in MEMS design flow'. Proc. FDL '02, Marseille, France, September 24-27, 2002

[13.79] *Cooper, R. S.*: 'The Designer's Guide to Analog & Mixed-Signal Modeling'. Avant! Corp., Beaverton 2001

13

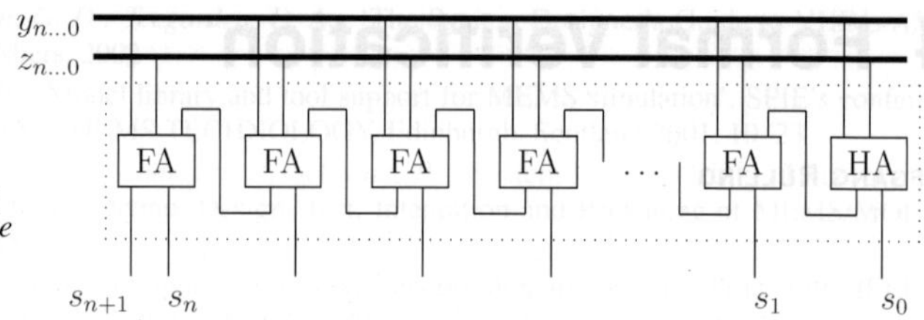

Fig. 14.3 Implementation of the final addition using a carry ripple adder for binary numbers of length n bits

14.1 Model Checking

In a proof of correctness using the method of **model checking** one has to prove that the property

$$a+b+c+d = y+z \quad (14.1)$$

is satisfied for the 4 to 2 reduction and

$$y+z = s \quad (14.2)$$

holds for the final addition. Combining both properties we obtain the logical correctness of the complete circuit.

Depending upon the complexity of a circuit such a proof can be performed fully automatically or may require interactive assistance from the user. In the given example the shown decomposition of the problem into two simple conditions (14.1) and (14.2) is such a possible assistance.

Model checking generally requires that the designer knows precisely how the circuit works. He should be able to do the proof of correctness by himself, but can delegate the details of the proof to a *verification tool*.

For the given *example* the 4 to 2 reduction could alternatively be implemented by using a sub-circuit from the circuit shown in figure 14.4 instead of the circuit from figure 14.2. Both circuits produce quite different intermediate values y and z because different addition methods are used. Nevertheless, both circuits are correct, because the sum $y+z$ is the same for both circuits. Thus the 4 to 2 reduction is an example of a design task in which the result should be specified in terms of abstract properties to maintain enough freedom in looking for a good design solution.

14.1.1 Example: 'Pentium Bug'

In [14.8] the verification method of **model checking** is impressively demonstrated by the example of a well known bug within the division unit of the first Intel Pentium processor. The division unit performs the sequential *SRT algorithm* producing two bits of the result within every clock cycle using a look up table. It is very difficult to check the device by simulation because the input patterns must be skillfully selected in order to check all entries of the look-up-table.

On the other hand, it is well known that during the SRT procedure all intermediate results have to satisfy an easily formulated arithmetical condition. Thus the formal verification of the division unit by model checking has to check whether, for the implemented hardware, the arithmetical condition of the *SRT algorithm* really holds. In this way a few minutes of computational time are enough to detect the bug. Furthermore, using the simulation inputs generated by the *verification tool* it is also easy to localize the fault within the design.

In this *example* a well known proof of correctness for an algorithm is used to verify essential properties of the designed hardware.

14.2 Equivalence Checking

Using the method of **equivalence checking** to prove the correctness of a design one compares the **implemented circuit** against another circuit which is known to be correct (the **specification**). For the example from figure 14.1 describing a circuit to add four binary numbers the specification might be the circuit from figure 14.4 using a sequence of three *carry ripple* adders. Then equivalence-checking has to prove that both circuits really implement the same function. There are different approaches to carrying out that proof. For example, both circuits can be represented by a normalized notation to simplify the comparison. This method will be described in this section.

A quite different comparison method results from test techniques for manufactured chips presented in section 15.4.1. For example, such a *chip test*

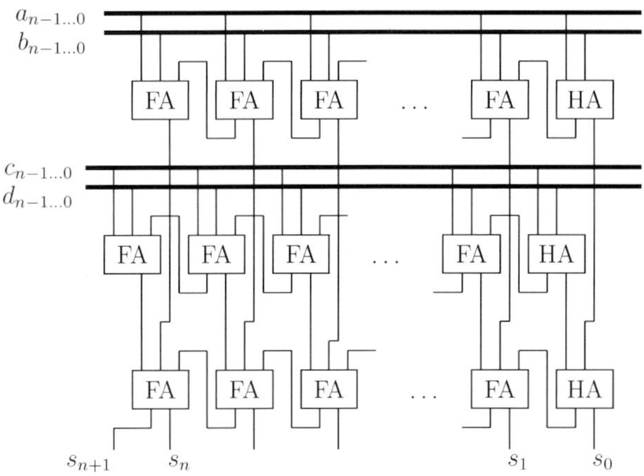

Fig. 14.4 Correct circuit to add 4 binary numbers of length n bits

has to check whether there is a stuck at 0 fault at the primary output y. For that test an *automatic test pattern generator* (ATPG: Automatic Test Pattern Generation) has to generate input patterns to derive the result $y = 1$.

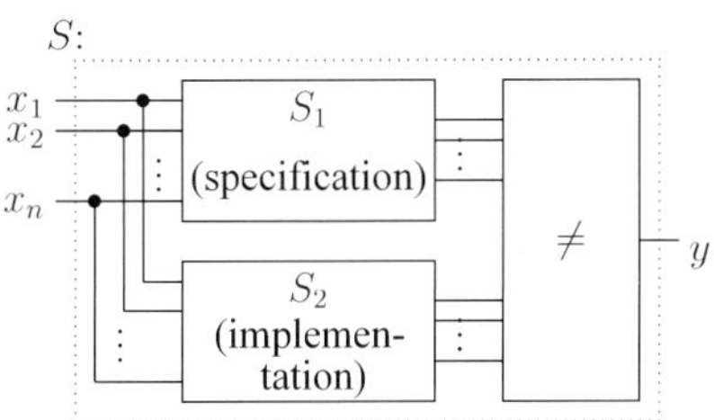

Fig. 14.5 Exactly when both sub-circuits S_1 and S_2 produce different outputs for some input $(x_1, x_2, \ldots, x_n)$ circuit S generates $y = 1$

Then the **ATPG method** can be used for equivalence checking in the following way. As shown in figure 14.5, one defines a virtual circuit S combining the circuits to be compared so that output y becomes 1 exactly when the results of *specification* and *implementation* differ for an input pattern. Then one starts a test pattern generation to test circuit S for the error y stuck at 0. Whenever a test pattern is found it demonstrates that *specification* and *implementation* are not equivalent. On the other hand, if the generator determines that there is no such test pattern then the equivalence between both circuits is proven.

Unfortunately, in the case of $S_1 \equiv S_2$ this type of the equivalence proof is very time consuming. Therefore the ATPG method is mostly used to demonstrate the differences $S_1 \not\equiv S_2$ for two circuit descriptions.

14.3 Fundamental Techniques

In this section fundamental techniques are presented which are used by *verification tools*. To simplify the description we only consider combinatorial circuits. Special problems arising from sequential circuits are presented in section 14.4.

14.3.1 Decision Diagrams

A typical problem arising in designing circuits is to verify the correctness of a circuit modification or of a design refinement. As an *example* we consider the circuits shown in figure 14.6. S_2 is a refinement of sub-circuit S_1 used within the 4 to 2 reduction from figure 14.2.

A simple way of comparing both circuits S_1 and S_2 is to derive boolean formulas for the three output signals and to compare the normalized representations. For the *example* using the *full adder* function $(sum, carry) = FA(a, b, c) = (a \oplus b \oplus c, ab + ac + bc)$ and the *half adder* function $(sum, carry) = HA(a, b) = (a \oplus b, ab)$ we obtain the following **expressions for S_1:**

$$y_1 = ab + ac + bc = abc + ab\bar{c} + a\bar{b}c + \bar{a}bc$$

$$\begin{aligned} y_2 &= (a \oplus b \oplus c) \cdot d \\ &= a\bar{b}\bar{c}d + \bar{a}b\bar{c}d + \bar{a}\bar{b}cd + abcd \end{aligned}$$

$$\begin{aligned} y_3 &= (a \oplus b \oplus c) \oplus d \\ &= a\bar{b}\bar{c}\bar{d} + \bar{a}b\bar{c}\bar{d} + \bar{a}\bar{b}c\bar{d} + \bar{a}\bar{b}\bar{c}d + abc\bar{d} \\ &\quad + ab\bar{c}d + a\bar{b}cd + \bar{a}bcd \end{aligned}$$

Similary **for S_2 we obtain**:

$$
\begin{aligned}
\widehat{y}_1 &= ab + (a \oplus b)c = ab + (a\overline{b} + \overline{a}b) \cdot c \\
&= abc + ab\overline{c} + a\overline{b}c + \overline{a}bc \\
\widehat{y}_2 &= ((a \oplus b) \oplus c) \cdot d \\
&= a\overline{b}\overline{c}d + \overline{a}b\overline{c}d + \overline{a}\overline{b}cd + abcd \\
\widehat{y}_3 &= ((a \oplus b) \oplus c) \oplus d \\
&= a\overline{b}\overline{c}\overline{d} + \overline{a}b\overline{c}\overline{d} + \overline{a}\overline{b}c\overline{d} + \overline{a}\overline{b}\overline{c}d + abc\overline{d} \\
&\quad + ab\overline{c}d + a\overline{b}cd + \overline{a}bcd
\end{aligned}
$$

Since the normalized formulas of S_1 and S_2 are identical both circuits obviously perform the same logic function. Thus indeed S_2 is a correct refinement of S_1. Please note that this kind of equivalence checking is done by performing boolean transformations. Therefore *theorem provers* known from computer science, are well suited for that task. They are often used within *verification tools.*

Unfortunately symbolical transformation of boolean expressions is very time consuming because normalized expressions may become very long. As an alternative representation one can use **binary decision trees (BDT)** or the more compact **binary decision diagrams (BDD)** [14.1]. As an example figure 14.7 shows the decision tree and the decision diagram of the function $f = a \oplus b \oplus c \oplus d$. Starting at the root node value $f(a, b, c, d)$ can be calculated by traversing edges labeled with corresponding input values. In figure 14.7 the path corresponding to $f(1, 0, 1, 1) = 1$ is emphasized.

By the example it is noticeable that the size of the decision tree for the operator $\oplus$ grows exponentially with the number n of arguments and has 2^n leafs. On the other hand, the size of the decision diagram only grows linearly with n. Here the diagram only consists of $2n + 1$ nodes. For this reason the BDD representation is well suited to comparing two circuits efficiently. A normalized compact representation is the so called **reduced ordered binary decision diagram (ROBDD)**. A BDD is called *reduced* if equivalent subtrees of the BDT are represented by the same node and nodes without branches are omitted. In figure 14.8 an example is shown using the function $y = a \cdot b + c$.

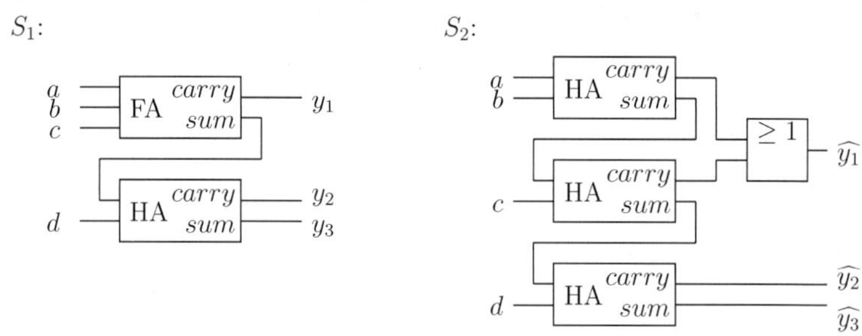

Fig. 14.6 S_2 is a refinement of cell S_1 from the 4 to 2 reducer from figure 14.2

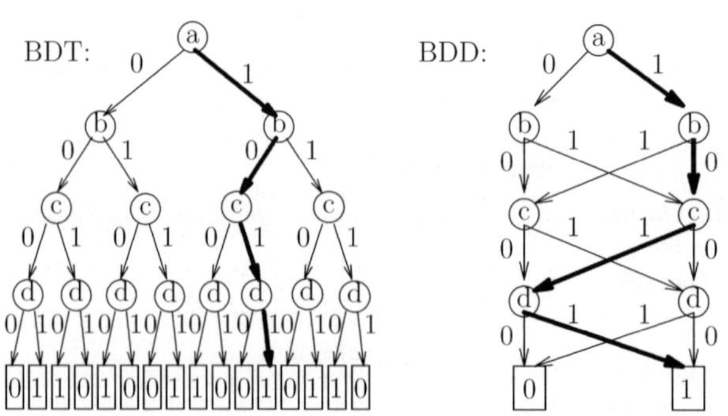

Fig. 14.7 BDT and BDD for the example $f = a \oplus b \oplus c \oplus d$

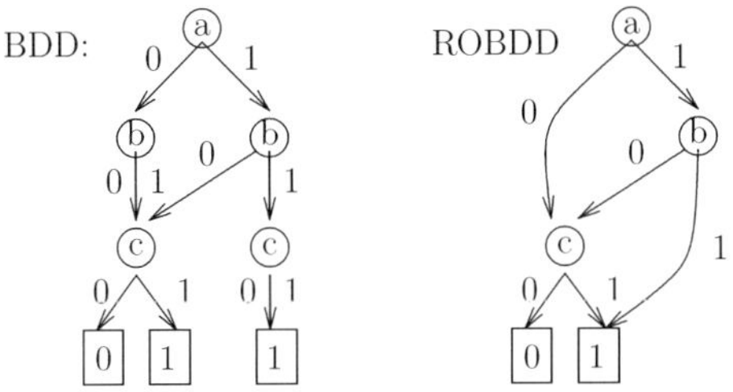

Fig. 14.8 Representation of function $y = a \cdot b + c$ by a BDD and a ROBDD

A BDD is *ordered* if the order of input variables is well defined. In this case the representation by a ROBDD is uniquely determined. Thus for equivalent functions the ROBDD representations are identical. Since the size of diagrams depends strongly on the chosen order of variables one uses heuristics to determine a favorable order [14.10], [14.5].

14.3.2 Signatures

Using decision diagrams one can easily derive characteristic numbers, called signatures, representing properties of the circuit structure. For example, in the BDD of figure 14.8 there are three different paths from root a to output value 0 and five paths from a to 1. This means that for the 8 possible inputs the result 0 is obtained for 3 times and the result becomes 1 for 5 times.

Similar characteristic numbers can be used to identify circuit inputs. In particular, if one has to compare sub-circuits there is no explicit allocation of input variables such that the order needed for a ROBDD is unknown. In this case one tries to identify the inputs on the base of **signatures** representing the circuit structure. For example the **signature function** $|f_{x_i}|$ describes how often the circuit produces the result 1 assuming that $x_i = 1$ holds. For this counting we assign constant 1 to x_i and consider the BDD for the function with the remaining free inputs. Accordingly the signature $|f_{\overline{x}_i}|$ describes how often the circuit result becomes 1 for $x_i = 0$.

For the *example* from figure 14.8 using the function $y = a \cdot b + c$ the signatures of inputs a, b and c are as follows:

$$\begin{aligned} |f_a| &= |f_b| = 3, \quad & |f_{\overline{a}}| &= |f_{\overline{b}}| = 2, \\ |f_c| &= 4, & |f_{\overline{c}}| &= 1 \end{aligned} \tag{14.3}$$

For two equivalent implementations of the function $y = a \cdot b + c$ we can localize c in both circuits because of its signatures. But for inputs a and b such an identification is not possible because these inputs are exchangeable without changing the behavior of the circuit.

In addition to signature functions for input variables there are other signatures to distinguish between arbitrary internal signals. A systematic investigation of the usability of signatures is given in [14.7].

14.4 Sequential Circuits

A simple approach for equivalence checking of sequential circuits is to find a mapping between the flip flops of both circuits. For example, this works well if one circuit is obtained by small modifications from the other. In this case the mapping may already be given because of the known names of nets. Alternatively one can decompose the circuit at the flip flops into combinatorial sub-circuits and then identify the flip flops by comparing structures of adjacent sub-circuits or signatures of signals. As a result the problem of equivalence checking is reduced to check the equivalence of combinatorial sub-circuits.

For completely different structured circuits comparison becomes more difficult. As an *example* the next section considers the special case of finite state machines.

14.4.1 Equivalence of Finite State Machines

The sequential circuits M in figure 14.9 and M' in figure 14.10 have the same inputs and outputs, and because all flip flops are driven by the rising edge of a clock signal *clk* the circuits can be interpreted as finite state machines (FSM).

To compare both circuits we can consider the state diagrams (figures 14.11 and 14.12) of the finite state machines describing the circuit behavior. Then the problem of equivalence checking is reduced to the question of whether both automata are equivalent. In order to answer this question we need a formal description of the problem. Firstly, the following definition describes an **FSM without output**.

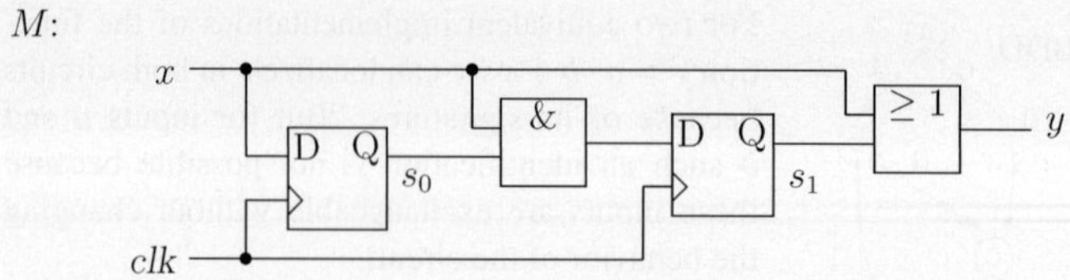

Fig. 14.9 The sequential circuit M can be interpreted as a finite state machine with input x and output y

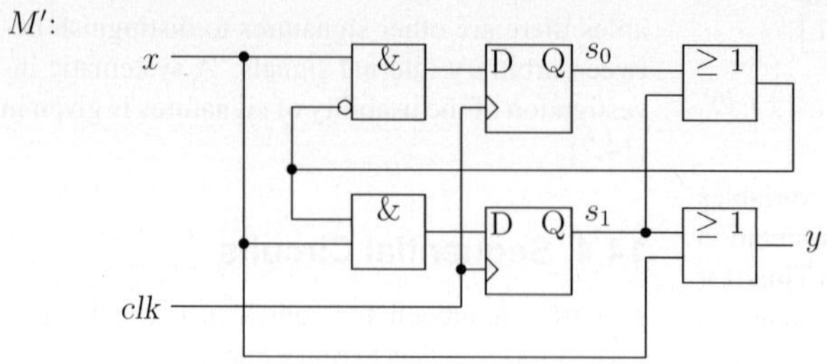

Fig. 14.10 Like M of figure 14.9 the sequential circuit M′ can be interpreted as a finite state machine with input x and output y

An **automaton** $M = (\Sigma, X, S, \delta)$ consists of a set Σ of states, a set X of input symbols, a start state $S \in \Sigma$, and a transition function $\delta : (\Sigma \times X) \to \Sigma$.

If the current state of the automaton is $z \in \Sigma$ and the current input is $x \in X$, the automaton changes to the next state $\delta(z, x)$. Starting at state $z_0 = S$ we obtain for a sequence of n input signals $x_1, x_2, \ldots, x_n$ a sequence $z_0, z_1, z_2, \ldots, z_n$ of states with $z_i = \delta(z_{i-1}, x_i)$ and $i = 1, 2, \ldots, n$.

For an *automaton with output* we have to distinguish between a **Moore automaton**, with an output function only depending on the current state and the more general **Mealy automaton** with an output function depending on the current state and on the current input.

A **Mealy automaton** $M = (\Sigma, X, Y, S, \delta, out)$ consists of an automaton (Σ, X, S, δ), a set Y of output symbols and an output function $out : (\Sigma \times X) \to Y$.

For an automaton in state $z \in \Sigma$ the function $out_z : X \to Y$ is called the *output function in state* z.

For the circuit M from figure 14.9 we obtain a state diagram with four states shown in figure 14.11. Obviously, if we use $(0,0)$ as starting state only three of the four states can be reached in practice. Since the output signal y is defined by $y = s_1 + x$ we obtain the output function for a given state as is shown in table 14.1.

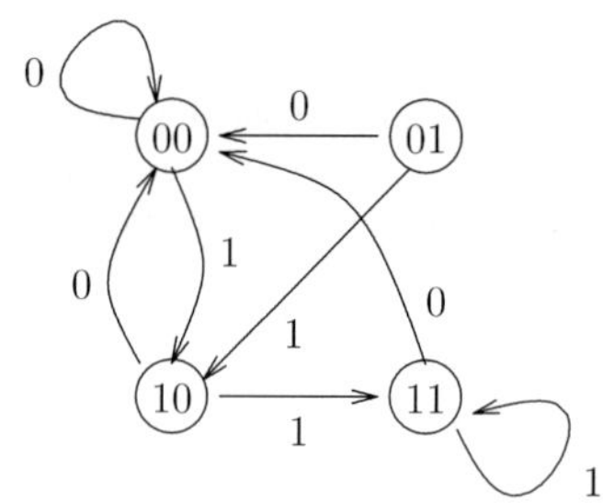

Fig. 14.11 State diagram of the automaton corresponding to the circuit M of figure 14.9 (neglecting the output)

Table 14.1 Output function $y = out_z(x)$ for the Mealy automaton M

state z	$y = out_z(x)$
$(s_0, s_1) = (0,0)$	$y = x$
$(s_0, s_1) = (1,0)$	$y = x$
$(s_0, s_1) = (0,1)$	$y = 1$
$(s_0, s_1) = (1,1)$	$y = 1$

Because of the following definition two automata are equivalent if they both have the same output behavior. However, the automata may use quite different codings of states, and even the number of states may differ.

Two **Mealy automata** $M = (\Sigma, X, Y, S, \delta, out)$ and $M' = (\Sigma', X, Y, S', \delta', out')$ with the same set X of input symbols and the same set Y of output symbols are called **equivalent** if after applying any input sequence both automata perform the same output function.

Starting at state S, respectively S', and using an arbitrary input sequence $(x_1, x_2, \ldots, x_k)$ the automaton M will reach some final state z and M' will reach z'. Then the equivalence of the two machines requires that the output functions of the states reached $out_z: X \to Y$ and $out'_{z'}: X \to Y$ are identical.

In accordance with this definition the above circuits M of figure 14.9 and M' of figure 14.10 are equivalent automata. To demonstrate this fact we construct the state diagram of M' also including the output function (figure 14.12).

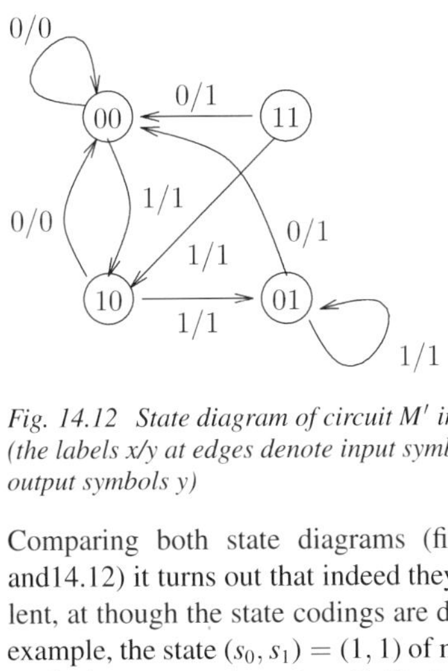

Fig. 14.12 State diagram of circuit M' in figure 14.10 (the labels x/y at edges denote input symbols x and output symbols y)

Comparing both state diagrams (figures 14.11 and14.12) it turns out that indeed they are equivalent, at though the state codings are different. For example, the state $(s_0, s_1) = (1, 1)$ of machine M is equivalent to state $(s_0, s_1) = (0, 1)$ of machine M'. Considering the different state codings the output functions are always the same for both machines. This is shown in table 14.2.

Table 14.2 Equivalent states of machine M in figure 14.9 and M′ in figure 14.10

automaton M (s_0, s_1)	automaton M' (s_0, s_1)	common output function
$(0,0)$	$(0,0)$	$y = x$
$(1,0)$	$(1,0)$	$y = x$
$(0,1)$	$(1,1)$	$y = 1$
$(1,1)$	$(0,1)$	$y = 1$

Using start states $S = (0, 0)$ and $S' = (0, 0)$ both machines are equivalent according to the definition. Please note again that in general it is not necessary that both machines have the same number of states. Furthermore, the output functions only have to be identical for reachable states. Therefore it is important that the **start states** of the circuits are well defined. Also, in practice it should be possible to reach the start state from any other state. Without that reset ability the behavior of the circuits may be undefined in practice.

Altogether, the equivalence proof of the sequential circuits M and M' of figures 14.9 and 14.10 is performed by constructing the state diagrams and proving the equivalence of both finite state machines using predefined start states.

14.4.2 Modeling of Concurrent Processes

Important properties of sequential circuits that have to be verified are, for example, whether a given state is **reachable** and whether the circuit is **dead lock free**. A dead lock is a state that cannot be left for any input.

14

As an *example* we consider an asynchronous communication between two hardware units such that each receipt of a message has to be acknowledged. In such a system a *dead lock* occurs if both components simultaneously become inactive, because both are waiting for an acknowledge. Of course a good implementation of a communication system should be *dead lock free*. Such synchronization properties can be modeled by Petri nets.

A **Petri net** consists of **places**, **transitions**, and oriented **edges** from places to transitions and from transitions to places. In figure 14.13 an *example* of a Petri net is given. The places are denoted by circles and the beams denote the transitions. A state may contain tokens which are individually drawn as dots or which may be indicated by an integer denoting the total number of tokens at a place. In the example the place S_1 contains two tokens and place S_6 contains one token.

A *transition* can **fire** by removing one token from every incoming edge and producing a new token to every outgoing edge. In the example only transition T_1 is able to fire. Doing this, a token is removed from S_1 and places S_2 and S_3 will get an additional token. Afterwards only transitions T_1, T_2 and T_3 are able to fire (figure 14.14).

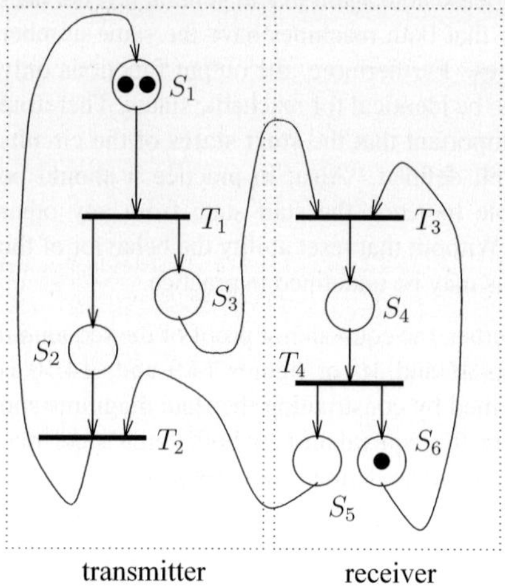

Fig. 14.13 Example of a Petri net with 5 places and 4 transitions

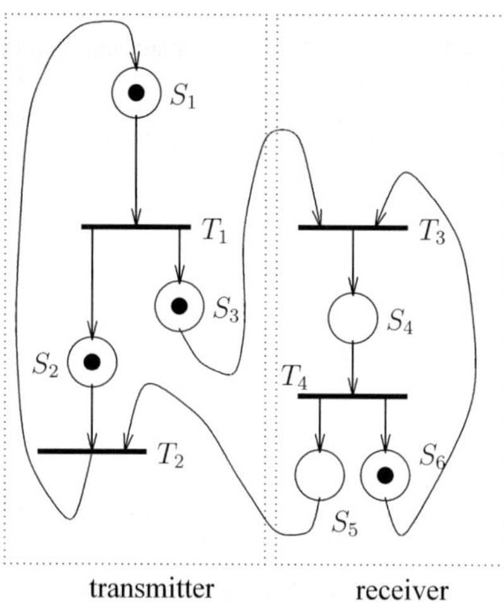

Fig. 14.14 Subsequent state of the Petri net from figure 14.13 after transition T_1 has fired

If several transitions are able to fire then they may fire in any order. This way one models the unknown duration of concurrent processes and arbitrary signal changes at the primary inputs of a circuit.

Interpreting the given example as a communication process between two components, T_1 denotes a transmission of data and T_2 denotes the receipt of an acknowledge. In the given model the transmitter has to wait for an acknowledge after at most two transmissions.

For example, it is of interest whether the protocol is correct or whether under certain circumstances the transmission process may stop because of a *dead lock* such that none of the transitions is able to fire.

In the given *example* we have the special case of a **finite Petri net** such that during its operation the total number of tokens always remains limited. Since finite Petri nets can be also described as finite state machines, one can use methods from automata theory to decide whether a Petri net is *dead lock free*. Thereby the possible allocations of tokens correspond to the states of a FSM.

14.5 Correctness of Synthesis Steps

With automated design tools there are many steps of converting and modifying a design to optimize or refine it or to add some additional information. During the sequence of steps it should be proven that the output produced is never in contradiction with the input. Then we have a **correctness by construction**. But since in practice the design systems are very complex the correct interaction of all synthesis steps cannot always be proven in advance.

Therfore an explicit verification for each application of a synthesis step becomes necessary. For this one may compare the design before and after a modification. In this special case one can use the fact that both descriptions are very similar and that the type of differences is well known.

For example, the logical behavior of a circuit should not be changed because of the automatic insertion of a *scan path* and after *layout generation* the transistor netlist obtained should be equivalent to the former gate level netlist. For these two applications of synthesis tools we now sketch how the verification can be done in practice.

14.5.1 Generating a Scan Path

After automatically **generating a scan path** the functionality of the circuit is enlarged, improving its testability. Furthermore, there are additional inputs and outputs. Thus the generated circuit is intentionally *not* equivalent to the circuit without *scan path*. Therefore we have to introduce additional specifications about the usage of the new primary inputs in such a way that the new circuit has to behave like the former one. Then using equivalence checking we can prove that the former functionality is also provided by the modified circuit. It is more difficult to verify the correctness of extended functionality because there is no equivalent circuit given by the designer. Thus we need an automatic generation of a *high level* specification for the *scan path* to prove the equivalence. Of course that *scan path* specification itself has to be verified by the designer (section 14.6).

14.5.2 Layout Synthesis

The **layout synthesis** generates a geometrical description from a given netlist. In a verification step called '**layout versus schematic**' the circuit information is extracted from the *layout* and then both circuits are compared. Here circuit extraction needs special attention (chapter 22). Considering the design rules the extractor generates a transistor netlist which may also include resistances and capacities of wires.

Then the verification problem is to compare a transistor netlist (neglecting resistances and capacities) to a gate level netlist. This can be done by replacing subnets of the transistor netlist by gates or flip flops to convert the transistor netlist to a gate level netlist. Also the other way is possible to convert a gate level netlist into a transistor netlist. After that both netlists have the same format and the equivalence check can be performed as shown above.

However, the needed computation time will grow exponentially in the size of the netlist. This effort can be reduced if the mapping of components and nets from one netlist to the other is known. In this case the structure of both netlists can be compared in linear time.

Indeed the mapping needed can be logged during the automatic placement and routing in such a way that it can be used for verification. This example demonstrates that the verification of circuit modification is heavily simplified if the modification steps are well known. In this sense a **cooperation between synthesis and verification** is useful.

14.6 Design Verification

The aim of **design verification** is to establish the correctness of the first design at a high level of abstraction. This proof of correctness is the basis of the verification of all following design modifications and refinements. It presupposes that a **specification of the problem** is given in a precise well defined form.

The special difficulty of design verification is that typically the problem specification does not contain hints on how to solve the problem. On the one hand this is because at the time of formulating the problem perhaps no idea of the solution is known. On the other hand, it is a good idea to specify the problem in a quite different way from describing a solution so that it becomes unlikely that the same errors are made within both descriptions.

As an *example* we consider a **specification for dividing numbers**. It may specify that for given numbers a and b a value y has to be computed in such a way that the property $a = y \cdot b$ holds. Of course, further information about range and type of numbers and about the precision of the result are necessary. In this way division is specified as the inverse operation of multiplication. The essential point with this specification is that nothing is said about the algorithm for dividing numbers. The specification only describes *what* has to be computed but not *how* it can be done.

For example, if we use the *SRT algorithm* mentioned in section 14.1.1 to solve the given problem then design verification exactly corresponds to proving the correctness of that algorithm. The crucial steps of the proof have to be given by the designer or must be derived by a **theorem prover**.

For a precise and well defined *problem specification* one may use a special logic language like **HOL** (HOL: High Order Logic) [14.3], [14.6] which is suitable for automatic theorem provers. Because the semantics of all statements has to be well defined in terms of a precise mathematical definition such specification languages typically consist of only a few types of statements. Therefore the formulation of a precise problem specification may be as complex as a circuit design.

More powerful languages like VHDL are easier to use because, for example, behavioral descriptions can be written in an easily read notation. Indeed, there are extensions of VHDL which enable the designer also to formulate verification tasks [14.9]. For example the well known *test bench* used to check a design by simulation is then replaced by a *verification bench* to describe all relevant properties of a circuit which have to be verified. Furthermore, there are compilers for converting VHDL descriptions into precise logic languages. To accelerate the speed of verification it is even possible to generate automatically more abstract description levels [14.4].

But for all these approaches it has to be noted that in practice the meaning (semantic) of design languages is often not precise enough for a formal verification. For example, the same VHDL description may be interpreted differently by a simulator and a synthesis tool, as demonstrated in section 9.1.1. As a consequence a VHDL specification, and also a design verification, may be faulty because of a wrong interpretation of the specification.

14.7 References

Much information about formal verification can be found on the internet. For example *homepage* http://lal.cs.byu/ of the 'Laboratory for Applied Logic' at Brigham Young University contains many links to workgroups, literature, and conferences concerning the topic.

[14.1] *Bryant, R. E.*: Graph-Based Algorithms for Boolean Function Manipulation. IEEE Transactions on Computers, Vol. C-35, No. 8, 1986

[14.2] *Eveking, H.*: Applied Formal Hardware Verification Methods. Tutorial, DATE'99, 1999

[14.3] *Gordon, M. J. C.; Melham, T. F.*: Introduction to HOL. Cambridge University Press, 1993

[14.4] *Hsieh, Y.-W.; Levitan, S. P.*: Model Abstraction for Formal Verification. DATE'98, 1998

[14.5] *Meinel, C.; Stangier, C.*: Increasing Efficiency of Symbolic Model Checking by Accelerating Dynamic Variable Reordering. DATE'99, 1999

[14.6] *Melham, T. F.*: Abstraction Mechanisms for Hardware Verification. Technical Report No. 106, University of Cambridge Computer Laboratory, 1987

[14.7] *Mohnke, J.*: A Signature-Based Approach to Formal Logic Verification. Dissertation, University of Halle, 1999

[14.8] *Payer, M.; Voges, J.*: How We Cracked the PENTIUM Bug within 5 Minutes with CVE. Technical Report, Siemens AG, HL CAD SV, Mch B, München, 1997

[14.9] *Reetz, R.; Schneider, K.; Kropf, T.*: Formal Specification in VHDL for Hardware Verification. DATE'98, 1998

[14.10] *Thornton, M. A.; Williams, J. P.; Drechsler, R.; Drechsler, N.*: Variable Reordering for Shared Binary Decision Diagrams Using Output Probabilities. DATE'99, 1999

15 Design for Testability

WOLFGANG RÜLLING

This section gives an overview of various aspects of testing chips. It considers the modeling of errors, automatic test pattern generation, and how to design circuits that can easily been tested. Usually in ASIC design systems there are special tools for supporting the testing of chips. Examples are the programs *Quick-Fault* and *Fast-Scan* from *Mentor Graphics*, *Verifault* from *Cadence* and *TestCompiler* from *Synopsys*. Additionally, synthesis tools typically support **Scan Path** and **Boundary Scan** techniques (sections 15.5 and 15.7). Section 15.6 gives some more specialized test structures which may be implemented by the chip designer, or are already part of parameterized cell generators.

15.1 Importance of Chip Testing

In spite of all efforts to increase the quality of chip manufacturing, **production faults** can not be avoided, and therefore also **defect chips** will be produced. For example, pollution can cause shorts between wires or interruptions of individual lines. The probability of such defects drastically grows with the size of a chip. In figure 15.1 the damaged chips on a wafer are marked by a cross.

Let p_{defect} be the probability of a square unit of chip area being damaged by pollution. Then the probability for a correctly produced chip with area N becomes $p = (1 - p_{\text{defect}})^N$ assuming that failures occur independently for all square units.

With increasing complexity of fabricated circuits the chip test gains ever more significance, since the rejection rate rises drastically (figure 15.2). Of course one has to detect faulty chips as soon as possible. Therefore a first production test is performed on the wafer in order to avoid *packaging* costs for defective chips.

Since each individual fabricated chip must be tested, testing is very time consuming and thus expensive. Therefore one tries to reduce the total number of test patterns per chip as far as possible. Furthermore, testing complex VLSI chips is difficult because the behavior of an extremely large number of devices (> 1 million) has to be tested using only a few pads (< 100).

15

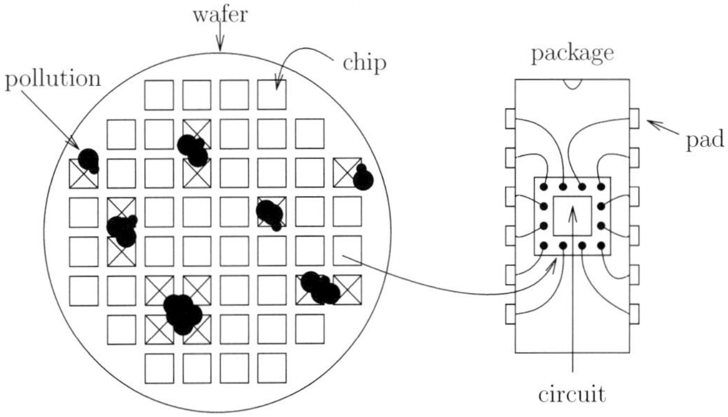

Fig. 15.1 Pollution on a wafer

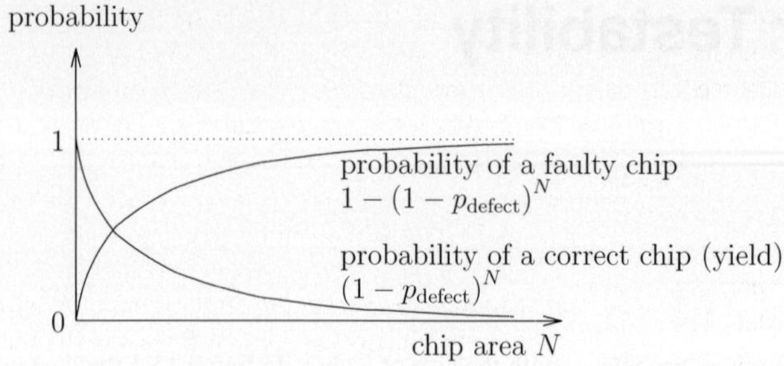

Fig. 15.2 Probability of faulty chips depending on the chip area

15.2 Black Box Test

The aim of testing is to check whether a chip is in accordance with its specification. A **black box test** checks the chip behavior without using knowledge about the structure of the implemented circuit. This will be demonstrated by a simple *example*.

A *full adder* can be specified by its function table given in table 15.1. The black box test of a *full adder* circuit will use all $2^3 = 8$ possible input patterns to check whether the chip really derives correct output signals *carry* and *sum* for all inputs.

Table 15.1 Function table of a full adder

a	b	c	*carry*	*sum*
0	0	0	0	0
0	0	1	0	1
0	1	0	0	1
0	1	1	1	0
1	0	0	0	1
1	0	1	1	0
1	1	0	1	0
1	1	1	1	1

For combinatorial circuits the number of test patterns needed for a *black box test* grows exponentially with the number of primary inputs. For large circuits this is a much too high expenditure. Thus, for instance, testing an adder with a data width of 16 bits already needs $2^{32} = 4\,294\,967\,296$ **test patterns**. Therefore the question arises of how to test the circuit with fewer *test patterns*. Unfortunately, without knowledge about the circuit implementation this is not possible. For example the combinatorial circuit could be implemented in terms of a ROM containing the data of the function table. The primary inputs then correspond to address lines of the ROM and the primary outputs are the data lines.

In this way the implementation of a full adder only needs a ROM with 8 words of length 2 bits. Using the input word (a,b,c) as a 3 bit address the ROM will output the data word (*carry*, *sum*) in accordance with table 15.1. Since an error could be situated in each individual ROM cell, one cannot proceed with fewer test patterns. In order to reduce the number of test patterns one needs additional assumptions about the possible types of faults. For this purpose we consider **fault models**.

15.3 Fault Models

Fault models specify which kinds of faults have to be detected by a test. Often one uses **structural models** where faults are equivalent to a structural modification of the circuit. The most well known *example* is the **model of 'stuck at' faults**. In that model a fault causes the signal of a wire to be fixed at the value 0 or at the value 1.

Such structural fault models are very abstract representations of physical **defects** occurring during chip manufacture. For example, it does not matter whether a stuck at fault is caused by a short between a wire and a supply line or by a line interruption. Instead of directly detecting a physical defect, one tries to observe its effect on the circuit behavior. Whenever a defect has an effect other than described by the fault model it will not be detected using the model.

The **quality of a fault model** can be defined by the number of faults covered. Since more thorough fault models need higher test effort, because more possible faults have to be considered, a trade off between quality and cost of a test must be found in practice. For that reason one often tests only for some limited percentage of the faults (98 %) described by the fault model. This percentage is called **fault coverage** of the test.

Since fault models represent simplifications and not all possible defects of a chip are covered, a high fault coverage does not necessarily correspond to a high **defect coverage**. Therefore even after successfully applying a **complete test** with fault coverage of 100 % one can not be sure that the tested chips actually operate correctly for all possible inputs.

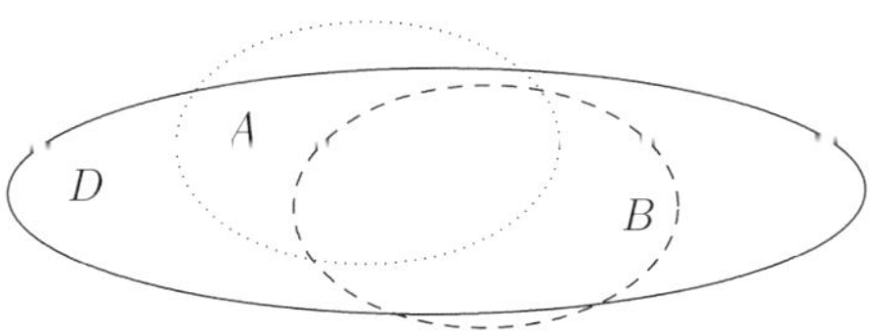

Fig. 15.3 Fault models A and B and set D of defects

Depending upon the fault model used different sets of test patterns are needed. In order to achieve a good defect coverage one can, for example, generate test samples for a fault model *A* until a good fault coverage is also achieved for another model *B*. In this case one can hope that the produced test patterns are generally well suited and do not cover only a special type of faults.

15.3.1 Stuck at Faults

The **single stuck at fault model** defines as only possible a fault which at the gate level of a circuit fixes exactly one wire α to value 0 or to value 1. This is called an α stuck at 0 or an α stuck at 1 fault.

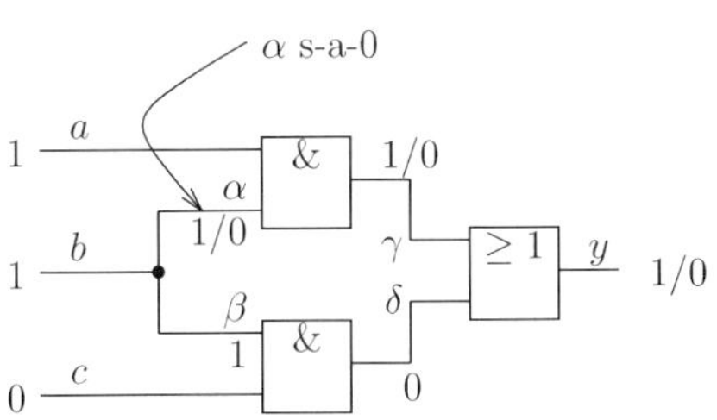

Fig. 15.4 Example of an α stuck at 0 fault

Figure 15.4 shows an example of a circuit such that wire α is permanently fixed to 0. This fault can only be observed if one tries to set wire α to 1. This is called the **stimulation** of the fault. In the *example* the fault can be stimulated by $b = 1$. Furthermore, the primary inputs of the circuit have to be set so that the faulty behavior can also be observed at a primary output. Thus the fault has to be **propagated** from the fault location α to a primary output. In the *example* it is only by $a = 1$ that we can propagate the fault through the upper *AND* gate. Otherwise for $a = 0$ we would obtain $\gamma = 0$ for the correct circuit and also for the faulty circuit. Accordingly we also need $c = 0$ to observe the fault at output y. Thus for the input pattern $(a, b, c) = (1, 1, 0)$ the fault α s a 0 is stimulated and propagated to output y. In this case the input is a suitable **test pattern** for detecting the fault.

Since for the *single stuck at fault model* exactly one wire is stuck to a specific logic value, the model describes exactly $2 \times m$ possible faults for a circuit with m wire segments. A **complete test** has to detect every such fault.

In counting wire segments special attention has to be paid to **multi-terminal nets**. For example, in the circuit shown in figure 15.4 there is a branching node transmitting signal b over wire segments α and β to two gates. A fault at location α does not affect wire segments b and β. Thus for a better understanding branching nodes should be considered as special gates transmitting an input signal to several outputs, as shown in figure 15.5. Then obviously in the example circuit there are eight wire segments (a, b, c, α, β, γ, δ and y) and the *single stuck at fault model* consists of $2{\times}8 = 16$ possible faults.

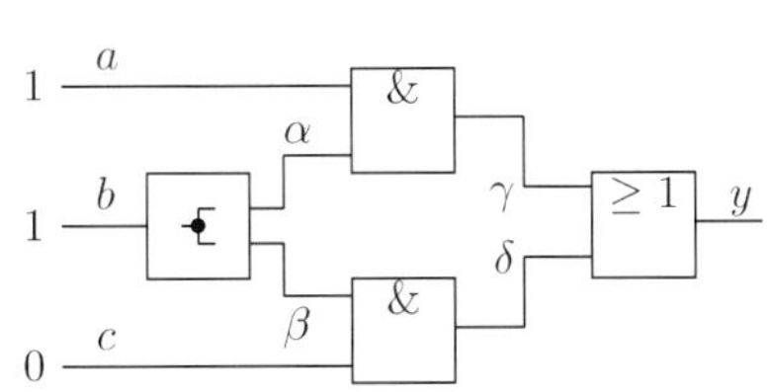

Fig. 15.5 Representation of branching nodes as special gates

With the special treatment of multi-terminal nets frequently occurring interruptions of wire segments can be handled. An interruption of the wire segment b has effects on signals α and β, but an interruption at α has no effect on b and β.

An extension of the fault model considered is the **multi stuck at fault model** also considering stuck at faults on several wires at the same time. For example the combination (a s-a-0, b s-a-1) is a possible fault within this model. With m wires and two possible faults per wire there are $4\times[m\times(m-1)]/2!$ possible **double faults**, $8\times[m\times(m-1)\times(m-2)]/3!$ possible **triple faults**, and, finally, $3^m - 1$ possible **multiple faults**. Of course the model of multiple faults is more thorough because it contains many more faults than the single fault model. However, it is not clear whether the increased test effort substantially improves defect coverage.

15.3.2 Cellular Fault Model

For the **cellular fault model** one assumes that a gate has a wrong combinatorial behavior. As an *example* table 15.2 describes the logical behavior of a *NAND* gate. A possible cellular fault might be a faulty entry within the table. Then in the example indicated the *NAND* gate would behave like an *XOR* gate. Of course, this modification does not correspond to a realistic defect. As an interesting property of the cellular fault model please note that every single stuck at fault is also a cellular fault. Thus the cellular fault model is a real extension of the stuck at fault model.

Table 15.2 Example of a cellular fault of a NAND gate

		correct	faulty
a	b	y	y
0	0	1	0
0	1	1	1
1	0	1	1
1	1	0	0

In order to stimulate all possible cellular faults one obviously has to drive every gate with all possible input combinations. As a consequence substantially more test patterns are needed than for the stuck at fault model. In figure 15.6 this is demonstrated by a very simple *example*. Though a complete test for stuck at faults consists of only four test patterns, a complete test for cellular faults needs six test patterns. All test patterns are given in table 15.3.

Table 15.3 Set of test patterns for stuck at and cellular fault model

a	b	c	complete test for single stuck at fault model
1	1	1	detects a s-a-0, b s-a-0, c s-a-0, d s-a-0, y s-a-0
0	1	1	detects a s-a-1, d s-a-1, y s-a-1
1	0	1	detects b s-a-1
1	1	0	detects c s-a-1
a	b	c	complete test for cellular fault model
0	0	1	test pattern (0, 0) for cell A and (0, 1) for cell B
0	1	1	test pattern (0, 1) for cell A
1	0	1	test pattern (1, 0) for cell A
1	1	1	test pattern (1, 1) for cell A and (1, 1) for cell B
0	0	0	test pattern (0, 0) for cell B
1	1	0	test pattern (1, 0) for cell B

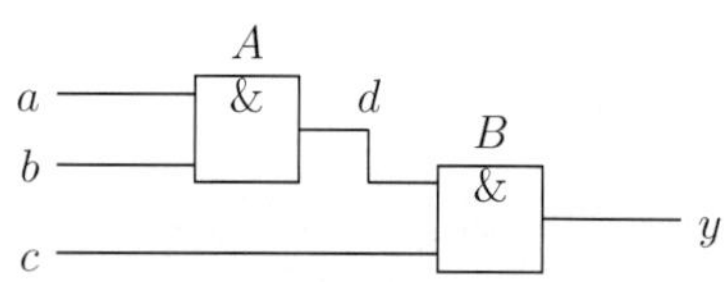

Fig. 15.6 Circuit to demonstrate the cellular fault model

15.3.3 Hard Bridging Faults

The **fault model of bridging faults** considers short circuits between two wires α and β of a circuit (figure 15.7). As long as both wires represent the same logic value the fault has no effect. However, a short circuit between signal 0 and signal 1 may change both signals to 0. This is called an *AND* bridging fault because the resulting faulty signal can be obtained by an *AND* operation. If on the other hand the resulting signal is 1 we have an *OR* bridging fault. Figure 15.8 shows the modeling of an *OR* bridging fault.

To stimulate the fault one assigns different logic values to wires α and β obtaining a change of α' or β' in the event of a fault. To observe the fault the signal modification has to be propagated to a primary output. However, observing the fault becomes more difficult if it causes a feedback within the circuit. Then for an extreme case the circuit might oscillate.

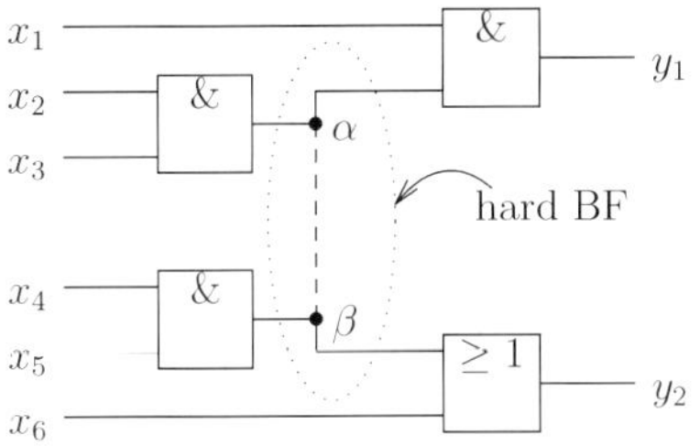

Fig. 15.7 Example of a hard bridging fault (BF)

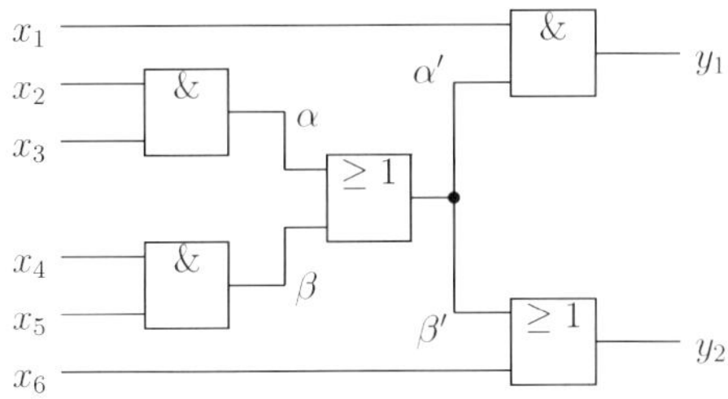

Fig. 15.8 Modeling the fault from figure 15.7 as an OR bridging fault

For a circuit consisting of m wires there are $m \times (m-1)$ possible pairs of wires for short circuits. Thus the fault model of *bridging* faults describes a quadratic number of possible faults. In practice one often restricts the fault model to short circuits between adjacent wires of a layout. However, this presupposes that in addition to a netlist geometrical layout data is also available.

Also for *bridging* faults there are several extensions of the fault model. For example, shorts between three or more wires may be considered. Another extension is the *weak bridging fault* considered in section 15.3.4. A further aspect of *bridging* faults will be discussed in more detail in section 15.8 in connection with the method of **IDDQ tests**. Since short circuits produce a faulty current, such faults can also be detected because of an increased power consumption of the chip instead of observing output signals.

15.3.4 Parametric Faults

In addition to the *hard bridging fault* described in section 15.3.3 as a short circuit between wires, there are also *weak bridging faults* coupling two wires by a resistance, as shown in figure 15.9.

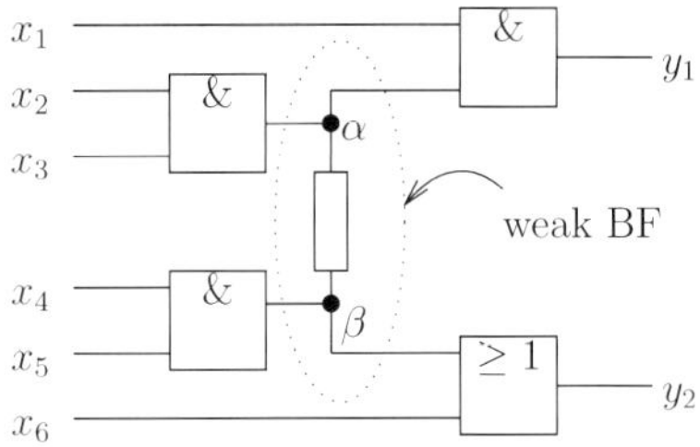

Fig. 15.9 Weak bridging fault

Here the circuit behavior depends strongly on the strength of the faulty resistance and may vary with the kind of physical defect. Opposite to structural faults, such faults are called **parametric faults**.

For the extreme case of $R = 0$ Ohm we have a *hard bridging fault*, and in the other extreme of $R \to \infty$ Ohm the circuit is correct. For other resistances the signal values obtained may not be well defined, so that a signal may be interpreted as logic 0 by one gate and as logic 1 by another. Such faults are hard to detect. It may even happen that the logical behavior of the circuit is correct but that there is a slight increase of power consumption. In this case the fault is classified as an **IDDQ fault**. Such faults are described in more detail in section 15.8.

Another effect of weak bridging faults may be an increased gate delay. This time the fault is called **gate delay fault**.

Accordingly one speaks of **path delay faults** if over a long path through the circuit we obtain an increased signal delay. Eventually individual gate delays are still acceptable; however, altogether they add up to a large *path delay* such that for high clock rates the circuit may produce incorrect results. Obviously in this case the fault can be observed but cannot clearly be localized within the circuit.

15.3.5 Transistor Faults

Depending on the level of abstraction of a circuit description different kinds of faults can be formulated. Refining a circuit from gate level to transistor level additional gate internal wires become visible and can be included into the fault model. Because of *stuck at* faults at the gate connection of a transistor the transistor may permanently be on or permanently be open. In these cases the faults

For the *example* of determining all test patterns for the fault 'α s-a-0' of the circuit given in figure 15.14 we first have to represent output y as a function depending on α. In accordance to equation (15.1) we then obtain

$$\begin{aligned}\alpha \cdot \frac{dy}{d\alpha} &= (x_1x_2) \cdot [(0 + x_3x_4) \oplus (1 + x_3x_4)] \\ &= (x_1x_2) \cdot (x_3x_4 \oplus 1) \\ &= (x_1x_2) \cdot \overline{x_3x_4} \\ &= x_1x_2\overline{x}_3 + x_1x_2\overline{x}_4 = 1\,.\end{aligned}$$

This equation supplies all input vectors such that the fault 'α s-a-0' can be detected by an incorrect output. Thus there are exactly three suitable test patterns $x = (1,1,0,0)$, $x = (1,1,0,1)$, and $x = (1,1,1,0)$.

Accordingly, using the approach $\overline{\alpha} \cdot dy/d\alpha = 1$ equation (15.2) supplies the test patterns for the fault 'α s-a-1':

$$\begin{aligned}\overline{\alpha} \cdot \frac{dy}{d\alpha} &= \overline{x_1x_2} \cdot \overline{x_3x_4} \\ &= (\overline{x}_1 + \overline{x}_2) \cdot (\overline{x}_3 + \overline{x}_4) \\ &= \overline{x}_1\overline{x}_3 + \overline{x}_1\overline{x}_4 + \overline{x}_2\overline{x}_3 + \overline{x}_2\overline{x}_4 = 1\,.\end{aligned}$$

Thus the fault 'α s-a-1' can be detected with all test patterns of the following kind: $(0, *, 0, *)$, $(0, *, *, 0)$, $(*, 0, 0, *)$, and $(*, 0, *, 0)$.

It may also happen that the approaches shown above fail to derive test patterns. As an *example* we consider the **redundant circuit** given in figure 15.15. Here we have $y = x_1x_2 + (x_2 + x_3)$, and for the fault location $\alpha = x_2$ we obtain the Boolean difference $dy/d\alpha = x_1\overline{x}_2\overline{x}_3$.

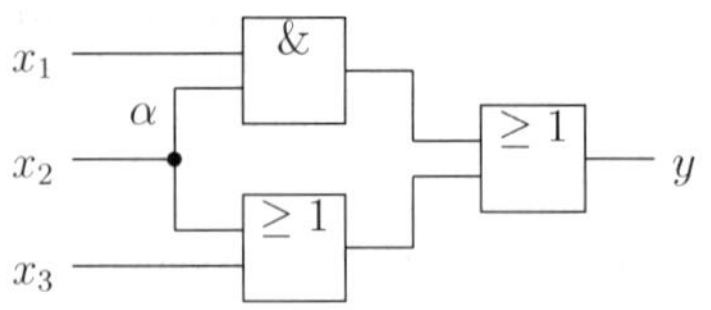

Fig. 15.15 Example of a redundant circuit

Using the characteristic equation of test patterns for fault 'α s-a-0' we obtain

$$\begin{aligned}\alpha \cdot \frac{dy}{d\alpha} &= 1\,, \\ x_2 \cdot x_1\overline{x}_2\overline{x}_3 &= 1\,, \\ 0 &= 1 \quad \text{contradiction!}\end{aligned}$$

Obviously the equation can not be satisfied for any input of the circuit. As a consequence the fault 'α s-a-0' is undetectable. The reason for this phenomenon is the redundancy of the circuit. Indeed, with some simple transformations we obtain the following representation of y

$$\begin{aligned}y &= x_1x_2 + (x_2 + x_3) = (x_1x_2 + x_2) + x_3 \\ &= x_2(x_1 + 1) + x_3 = x_2 + x_3\,,\end{aligned}$$

demonstrating that the output y does not depend on x_1.

Of course, for this *example stuck at* faults at wire x_1 are not detectable. More formally this can be seen from the fact that $dy/dx_1 \equiv 0$ holds, meaning that output y does not depend on x_1 for any input.

If a circuit has several primary outputs then of course it is sufficient to observe a fault at any one of the outputs. This simple fact leads to the generalized characteristic equations for *stuck at* faults.

For a **combinatorial circuit** with m primary outputs $y_1, \ldots, y_m$ test patterns for **single stuck at faults** at a wire α can be computed by solving the following characteristic equations:

- Characteristic equation for test pattern concerning fault 'α s-a-0':

$$\alpha \cdot \left(\frac{dy_1}{d\alpha} + \frac{dy_2}{d\alpha} + \cdots + \frac{dy_m}{d\alpha}\right) = 1\,; \qquad (15.3)$$

- Characteristic equation for test pattern concerning fault 'α s-a-1':

$$\overline{\alpha} \cdot \left(\frac{dy_1}{d\alpha} + \frac{dy_2}{d\alpha} + \cdots + \frac{dy_m}{d\alpha}\right) = 1\,. \qquad (15.4)$$

To compute Boolean differences the transformation rules shown in table 15.6 concerning the 'exclusive OR' operation ($\oplus$) may be useful. Furthermore, the following **concatenation rule** holds for Boolean differences:

$$\frac{dz}{dx} = \frac{dz}{dy} \cdot \frac{dy}{dx}\,. \qquad (15.5)$$

It can be used if in a circuit there is exactly one path from x to z and y is an intermediate signal on that path (figure 15.16).

As an *example* we consider the circuit given in figure 15.17. To compute the Boolean difference dz/dx using the concatenation rule we only have to compute the simple terms dz/dy and dy/dx. Since output y of the *AND* gate depends on its inpu

x exactly when its other input c is 1, we obtain $\mathrm{d}y/\mathrm{d}x = c$. Accordingly output z of the *OR* gate depends on its input y if its other input is 0. Thus we have $\mathrm{d}z/\mathrm{d}y = \overline{de}$, and finally $\mathrm{d}z/\mathrm{d}x = c \cdot \overline{de}$.

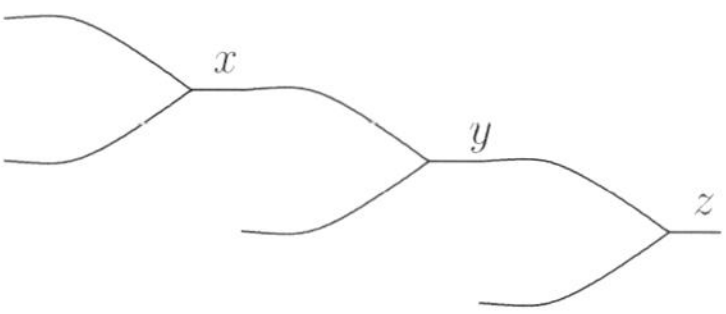

Fig. 15.16 Illustration of the rule $\frac{\mathrm{d}z}{\mathrm{d}x} = \frac{\mathrm{d}z}{\mathrm{d}y} \cdot \frac{\mathrm{d}y}{\mathrm{d}x}$ *for Boolean differences*

Table 15.6 Transformation rules for the XOR operation

$a \oplus 0 = a$
$a \oplus 1 = \overline{a}$
$a \oplus b = b \oplus a$
$(a \oplus b) \oplus c = a \oplus (b \oplus c)$
$a \oplus (a+b) = \overline{a}b$
$a \oplus (ab) = a\overline{b}$

Table 15.7 Boolean differences for elementary gates

AND gate	$\frac{\mathrm{d}(a \cdot b)}{\mathrm{d}a} = b$
OR gate	$\frac{\mathrm{d}(a+b)}{\mathrm{d}a} = \overline{b}$
EXOR gate	$\frac{\mathrm{d}(a \oplus b)}{\mathrm{d}a} = 1$
Inverter	$\frac{\mathrm{d}\overline{a}}{\mathrm{d}a} = 1$

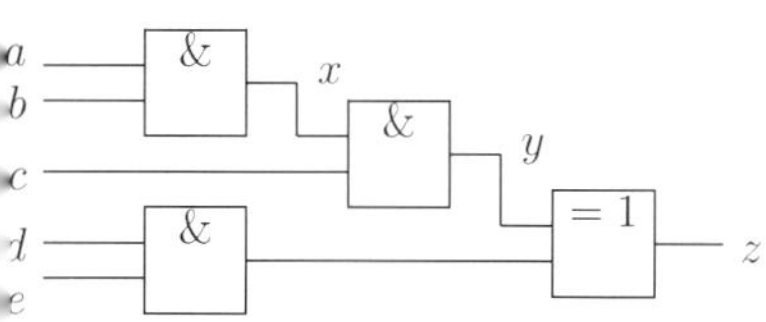

Fig. 15.17 Circuit to demonstrate the application of the chain rule

Also for other error models there are characteristic equations using Boolean differences to deduce test patterns. This will be demonstrated for the *example* of **hard AND bridging faults** (figure 15.7). For simplicity we at first assume the fault does not cause a feedback and thus call it a **combinatorial bridging fault**. To detect a short circuit between two wires α and β both wires must have different logic values and then the falsified signal has to be propagated to a primary output.

For an '*AND* bridging fault' the falsified signal becomes $0 \cdot 1 = 0$. In the case of $\alpha = 1$ and $\beta = 0$ the fault would falsify signal α, and in the case of $\alpha = 0$, $\beta = 1$ the signal β would be falsified. From this fact we immediately obtain characteristic equation (15.6) to compute test patterns. Accordingly characteristic equation (15.7) can be derived for an '*OR* bridging fault'.

For a combinatorial circuit with m primary outputs $y_1, \ldots, y_m$ the following characteristic equations hold for hard combinatorial bridging faults (α, β):

- Characteristic equation for test patterns concerning *AND* bridging faults:

$$\alpha\overline{\beta} \cdot \left(\frac{\mathrm{d}y_1}{\mathrm{d}\alpha} + \frac{\mathrm{d}y_2}{\mathrm{d}\alpha} + \cdots + \frac{\mathrm{d}y_m}{\mathrm{d}\alpha} \right) \qquad (15.6)$$
$$+ \overline{\alpha}\beta \cdot \left(\frac{\mathrm{d}y_1}{\mathrm{d}\beta} + \frac{\mathrm{d}y_2}{\mathrm{d}\beta} + \cdots + \frac{\mathrm{d}y_m}{\mathrm{d}\beta} \right) = 1\,;$$

- Characteristic equation for test patterns concerning *OR* bridging faults:

$$\alpha\overline{\beta} \cdot \left(\frac{\mathrm{d}y_1}{\mathrm{d}\beta} + \frac{\mathrm{d}y_2}{\mathrm{d}\beta} + \cdots + \frac{\mathrm{d}y_m}{\mathrm{d}\beta} \right) \qquad (15.7)$$
$$+ \overline{\alpha}\beta \cdot \left(\frac{\mathrm{d}y_1}{\mathrm{d}\alpha} + \frac{\mathrm{d}y_2}{\mathrm{d}\alpha} + \cdots + \frac{\mathrm{d}y_m}{\mathrm{d}\alpha} \right) = 1\,.$$

It is more difficult to derive test patterns for *bridging* faults causing a feedback within the circuit, because in this case the circuit behavior becomes sequential. If a test pattern generates oscillating signals because of the feedback the output signals are hard to predict, and thus the fault can only be detected by an *IDDQ measurement* (section 15.8) Therefore we here only consider test patterns producing stable states. For this situation the characteristic equations are (15.8), (15.9) and (15.10).

For a **combinatorial circuit** with primary output y and two wires α and β with $\beta = f(\alpha)$ the following three characteristic equations supply **test patterns for the *AND* bridging fault** (α, β).

$$\frac{\mathrm{d}y}{\mathrm{d}\beta}\overline{\alpha}\beta = 1\,, \qquad (15.8)$$

$$\frac{\mathrm{d}y}{\mathrm{d}\alpha}\overline{\left(\frac{\mathrm{d}\beta}{\mathrm{d}\alpha}\right)}\alpha\overline{\beta} = 1\,, \qquad (15.9)$$

$$\frac{\mathrm{d}y}{\mathrm{d}\alpha}\frac{\mathrm{d}\beta}{\mathrm{d}\alpha}\alpha\beta = 1\,. \qquad (15.10)$$

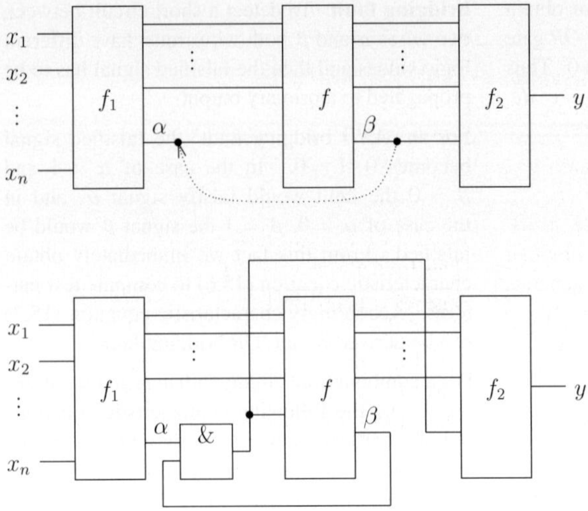

Fig. 15.18 Sketch of a circuit with an AND bridging fault causing a feedback

However, for equation (15.10) we need the additional assumption that there is an input 'reset' to force $\alpha = \beta = 0$ as an initialization of the circuit.

Figure 15.18 illustrates a feedback within a circuit owed to a bridging fault.

15.4.2 Undetectable Faults

In section 15.4.1 it was mentioned that for redundant circuits there may be undetectable *stuck at* faults. Obviously such faults have no effect on the circuit's behavior. However, undetectable faults may have an influence on the testability of a circuit. This will be demonstrated using the *example* of an undetectable *bridging* fault [15.19]. For example, an *AND* bridging fault between the two inputs of an *AND* gate cannot be detected because its effect is just the same as that of the gate. From equations (15.6) and (15.7) for combinatorial bridging faults one can directly derive the conditions for undetectable combinatorial bridging faults.

For example, combinatorial *AND* bridging faults (α, β) are undetectable exactly in the case of the following condition:

$$\alpha\overline{\beta}\frac{\mathrm{d}y}{\mathrm{d}\alpha} = \beta\overline{\alpha}\frac{\mathrm{d}y}{\mathrm{d}\beta} = 0 . \tag{15.11}$$

Accordingly, combinatorial *OR* bridging faults (α, β) are undetectable exactly in the case of the following condition:

$$\overline{\alpha}\beta\frac{\mathrm{d}y}{\mathrm{d}\alpha} = \overline{\beta}\alpha\frac{\mathrm{d}y}{\mathrm{d}\beta} = 0 . \tag{15.12}$$

Figure 15.19 gives an example of a non-redundant circuit implementing function $y = x_1x_2 + \overline{x}_1x_3$. This circuit can be completely tested on *stuck at* faults using the test patterns $\{010, 001, 101, 111\}$.

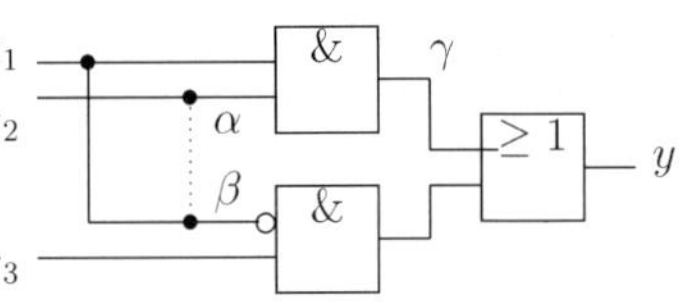

Fig. 15.19 Example of a circuit such that the OR bridging fault (α, β) causes a complete test for stuck at faults to fail

In the case of the sketched *OR* bridging fault the incorrect circuit will compute the function $\hat{y} = x_1(\overline{x}_1 + x_2) + (\overline{x}_1 + x_2)x_3$. With a simple transformation we obtain $\hat{y} = x_1x_2 + \overline{x}_1x_3 + x_2x_3 = y$ demonstrating that the faulty circuit exactly behaves like the correct one. Thus the *bridging* fault can not be observed. Of course, this can also be proved by checking equation (15.12).

As a consequence of this undetectable *bridging* fault the complete test given above does not work any longer. For example, originally the fault 'γ s-a-0' has been detected using test pattern (111) and observing the faulty result 0 at output y. However if the *stuck at* fault and the *bridging* fault are present at the same time the correct result $y = 1$ is produced for the test pattern.

This does not mean that it has become impossible to detect the fault 'γ s-a-0', but that the complete test used originally is not well chosen. Therefore in constructing a test for *stuck at* faults one should pay attention that the test cannot become ineffective because of other undetectable faults. In practice it is often helpful not to use minimal sets of test patterns (**minimal tests**) but to generate several test patterns for each *stuck at* fault.

15.4.3 Test Pattern Generation by Path Sensitization

With the method of Boolean differences described in section 15.4.1 it is possible to derive test patterns for every detectable fault of the respective fault model. However, this method is quite complex. More favorable and much easier to implement is a method called **path sensitization** that is also applied for the **D algorithm** of test pattern generation. This method again uses the idea of Boolean differences but avoids transformations of Boolean formulas.

A test pattern for the fault 'α s-a-0' causes the wire α to change from logic value 1 in the correct circuit to the faulty value 0. This situation of a signal changing from a correct '1' to a faulty '0' will be denoted by the new signal value D. Accordingly $\overline{D}$ denotes the case that a signal '0' in the correct circuit is replaced by '1' in the faulty circuit.

Furthermore, in the new signal set $\{0, 1, D, \overline{D}\}$ consisting of four values, symbols '0' and '1' denote signals that are not affected by the fault. In table 15.8 the meaning of all four symbols are summarized. To propagate such signal values through a circuit we use extensions of the common Boolean operators *AND*, *OR*, and *NOT* given in tables 15.9, 15.10, and 15.11.

Table 15.8 Signal values used for the D algorithm

signal of correct	signal of faulty circuit	symbol
0	0	0
1	1	1
1	0	D
0	1	$\overline{D}$

Table 15.9 Function table of the AND operator $y = a \cdot b$

AND	0	1	D	$\overline{D}$
0	0	0	0	0
1	0	1	D	$\overline{D}$
D	0	D	D	0
$\overline{D}$	0	$\overline{D}$	0	$\overline{D}$

Table 15.10 Function table of the OR operator $y = a + b$

OR	0	1	D	$\overline{D}$
0	0	1	D	$\overline{D}$
1	1	1	1	1
D	D	1	D	1
$\overline{D}$	$\overline{D}$	1	1	$\overline{D}$

Table 15.11 Function table of the inverter $y = \overline{x}$

x	y
0	1
1	0
D	$\overline{D}$
$\overline{D}$	D

The method of path sensitization tries to construct a **D chain** from the fault location α to a primary output. α is labeled by D or $\overline{D}$ depending on the fault type and then one tries to propagate symbol D or $\overline{D}$ to a primary output by applying appropriate input patterns. The resulting path from α to a primary output is called a **D chain**. In figure 15.20 such a D chain is shown for the fault 'α s-a-0'.

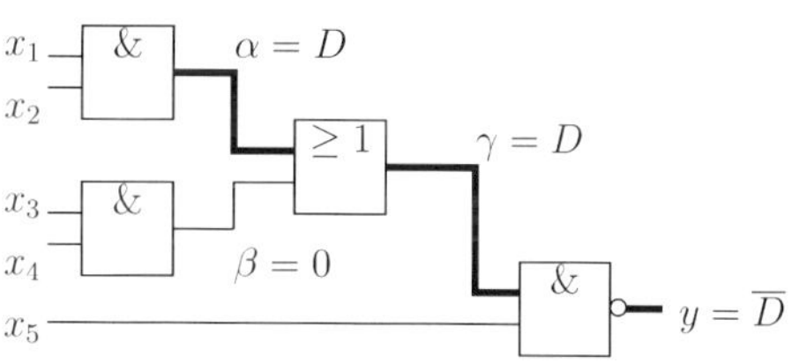

Fig. 15.20 D chain for the fault 'α s-a-0'

According to table 15.10 only with $\beta = 0$ or $\beta = D$ can the *OR* gate supply a suitable output signal. Since to observe the fault at wire α one cannot presuppose that there is also a faulty change of signal *beta*, only $\beta = 0$ can be used to propagate the fault. Accordingly $x_5 = 1$ is needed to complete the D chain through the *NAND* gate to output y.

We then conclude that the D chain found only exists for the case of $\overline{\beta} \cdot x_5 = 1$. Thus the condition

$\overline{x_3x_4} \cdot x_5 = 1$ has to be satisfied. Since this condition also describes when the output y depends on α, the existence of a D chain from α to y is equivalent to $\mathrm{d}y/\mathrm{d}\alpha = 1$.

Therefore the construction of D chains is a simple method of computing Boolean differences. In particular, for the circuit in figure 15.20 we obviously obtain $\mathrm{d}y/\mathrm{d}\alpha = \overline{x_3x_4} \cdot x_5$.

As a consequence the characteristic equations for computing test patterns for *stuck at* 0 faults (Eq. (15.3)) and for *stuck at* 1 faults (Eq. (15.4)) can also be formulated in terms of D chains. For the *example*, the input pattern $x = (1, 1, 0, 0, 1)$ is a suitable test pattern for 'α s-a-0' because it stimulates $\alpha = 1$ and generates the sketched D chain.

The following theorem holds:

> A fault 'α s-a-0' (resp., 'α s-a-1') can be detected exactly in the case when:
> - the fault can be stimulated, at fault location α signal D (resp. $\overline{D}$) can be generated; and
> - a D chain can be generated from fault location α to a primary output.
>
> An input pattern stimulating the fault and generating a D chain is a test pattern for the fault.

In practice, however, constructing a D chain is not always easy. In particular, for branching nodes it is not clear which branch to use for the D chain. If one decides arbitrarily for a path and does not succeed in constructing the D chain then one has to revise the decision and has to try alternative paths. Therefore if one needs n decisions between two possible branches, then in the worst case one has to consider 2^n different approaches for D chains. This method is used by the **D algorithm for test pattern generation**:

- **Input**: A combinatorial circuit S and a *stuck at* fault at wire α connecting a gate output a to a gate input e
- **Output**: One test pattern to detect the fault if it is detectable. Otherwise a message is generated to indicate that the fault is undetectable.
- **Method**: Remove wire α.

  Search (using *back tracking*) a stable state of the modified circuit with signals from $\{0, 1, D, \overline{D}\}$, such that following conditions hold:

  1. $a = \begin{cases} 0 & \text{if } \alpha \text{ s-a-1}, \\ 1 & \text{if } \alpha \text{ s-a-0}; \end{cases}$
  2. $e = \begin{cases} \overline{D} & \text{if } \alpha \text{ s-a-1}, \\ D & \text{if } \alpha \text{ s-a-0}; \end{cases}$
  3. For all primary inputs: $x_i \in \{0, 1\}$;
  4. There is a primary output y_i with $y_i \in \{D, \overline{D}\}$.

  If a solution is found report the labels of primary inputs $x_1, \ldots, x_n$ as test pattern. Otherwise, report that the fault is undetectable.

Of course, it is also possible that all attempts to construct D chains fail because no such chain exists for the given fault. Then the fault can not be observed at primary outputs because in spite of the fault the circuit always generates correct results. In this case there is an unintentional redundancy within the circuit such that the circuit can be simplified, or the circuit was designed to become fault redundant in such a way that it is insensitive to certain kinds of faults.

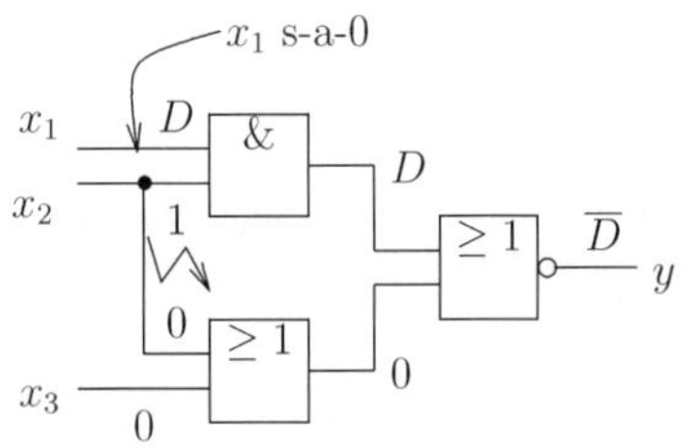

Fig. 15.21 Redundant circuit with no D chain for fault 'x_1 s-a-0'

A simple *example* of a redundant circuit is given in figure 15.21. On trying to construct a D chain from x_1 to y it turns out that $x_2 = 1$ is necessary to propagate D through the *AND* gate, but on the other hand $x_2 = x_3 = 0$ has to hold for propagation through the *NOR* gate. Therefore there is no D chain and the output y does not depend on x_1. As a check of that result one can transform the circuit function $y = f(x_1, x_2, x_3)$ into $y = \overline{x_1x_2 + (x_2 + x_3)} = \overline{x_2(x_1 + 1) + x_3} = \overline{x_2 + x_3}$ in order to recognize that indeed y does not depend on x_1, and thus $\mathrm{d}y/\mathrm{d}x_1 \equiv 0$ holds.

> The D algorithm determines for a given combinatorial circuit with n primary inputs, and for a given *stuck at* fault, whether the fault is detectable at primary outputs. For detectable faults it generates a test pattern for the fault. In the worst case the algorithm needs a computational time which depends exponentially on n.

The description of the D algorithm stated above does not contain details of how to find a stable state of the circuit, a consistent signal assignment to all nodes using signals from $\{0, 1, D, \overline{D}\}$. Instead, the '*back tracking* method' known from computer science is mentioned. The actual implementation of the search is crucial for the attainable rate of pattern generation. Indeed, one applies heuristics to favor more promising approaches whenever decisions have to be made, and one uses a *branch and bound* method to detect wrong decisions as soon as possible.

One such optimized implementation of the D algorithm is called the **PODEM algorithm** (PODEM: Path Oriented Decision Making). For that implementation signal assignments are restricted to primary inputs in order to limit the solution space for *back tracking*. However in spite of all the tricky implementations the worst case computational time of the D algorithm is exponential in the number of primary inputs of the circuit[1]. Thus the D algorithm is a very time consuming method.

In the following an efficient search for a test pattern is demonstrated using the *example circuit* from figure 15.22. At first the primary inputs are labeled with the signal set $\{0, 1\}$ and the terminals a and e of the faulty wire are labeled with $a = \{1\}$ and $e = \{D\}$ ($a = \{0\}$ and $e = \{\overline{D}\}$ for the fault 'α s-a-1'). Then these sets are propagated forward through the circuit obtaining a label for each node denoting a set of possibly assigned signals. This way one recognizes at which primary output the fault eventually might be observed. This is because D chains can only reach outputs having the symbol D or $\overline{D}$ in its signal set. Figure 15.23 gives the signal sets obtained after forward propagation. It can be seen that the fault 'α s-a-0' can at most be observed at output y_2 but cannot be observed at output y_1.

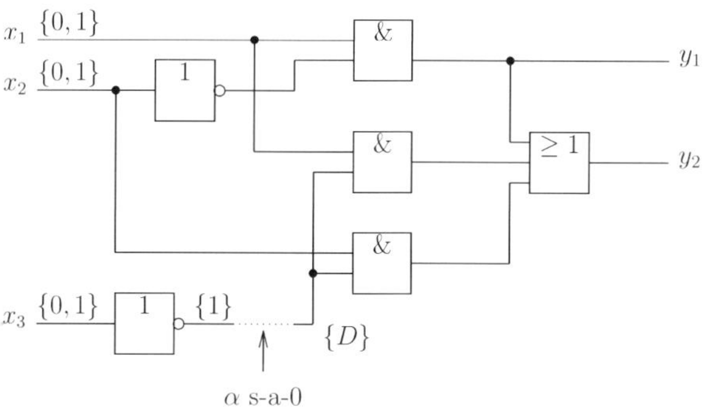

Fig. 15.22 Example for the D algorithm

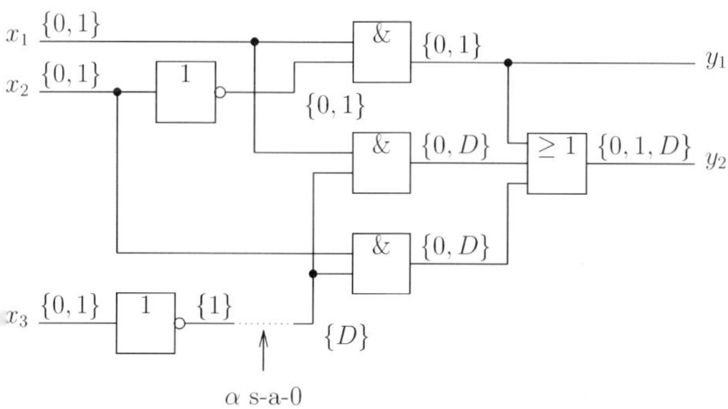

Fig. 15.23 Because of the forward propagating sets of signals it can be recognized that fault 'α s-a-0' can not be observed at output y_1

[1] In computer science one distinguishes between complexity classes *P* and *NP*. The decision problem of whether, for a given circuit and a given *stuck at* fault, there is a D chain belongs to the class *NP*. For problems of this class, so far only quite slow algorithms with exponential computational time are known. On the other hand, *P* denotes the class of problems with faster polynomial time algorithms. The question of whether also for *NP hard* problems there are polynomial algorithms has not been able to be answered up to now.

The *forward propagation* of signal sets used through a gate is performed by applying the gate function $y = f(x_1, x_2, \ldots, x_n)$ to all combinations of possible input signals. Thus for signal sets $M_{x_1}, \ldots, M_{x_n}$ at inputs we obtain the signal set $m_y = \{f(s_1, \ldots, s_n) \| s_1 \in M_{x_1}, s_2 \in M_{x_2}, \ldots, s_n \in M_{x_n}\}$ at the gate output y (figure 15.24).

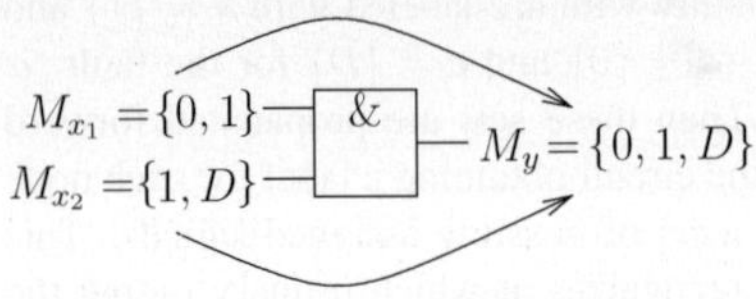

Fig. 15.24 Example of forward propagation through an AND gate

For example, if the two inputs of an *AND* gate are labeled by $\{0,1\}$ and $\{1,D\}$ then the output has to be labeled with $\{0\cdot 1, 0\cdot D, 1\cdot 1, 1\cdot D\} = \{0,1,D\}$. Such calculated signal sets may be too large because for the sake of simplicity one does not consider whether each of the input combinations can actually occur in practice.

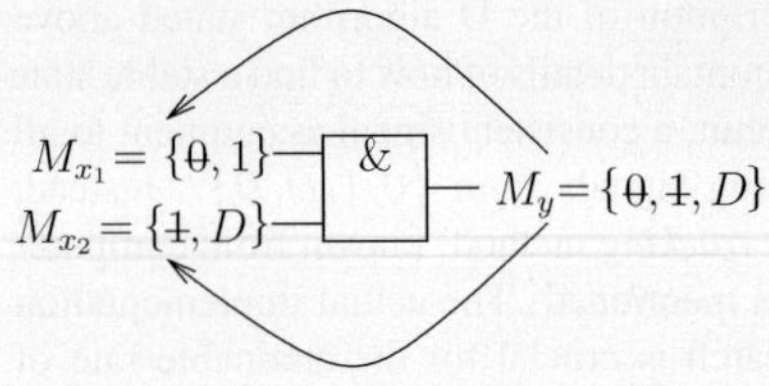

Fig. 15.25 Example of backward propagation of signal sets through an AND gate

After forward propagation one can delete signals which obviously can not be used to construct a D chain. For example, in figure 15.25 values 0 and 1 can be deleted from the signal set at output y_2 because a D chain needs a value D or $\overline{D}$. Then using forward and *backward propagation* of signal sets one determines the effects of these cancellations on other nodes. By backward propagation one deletes such input values of a gate that can only be used to create deleted output values. In this way for the *example* we obtain the signal sets shown in figure 15.26.

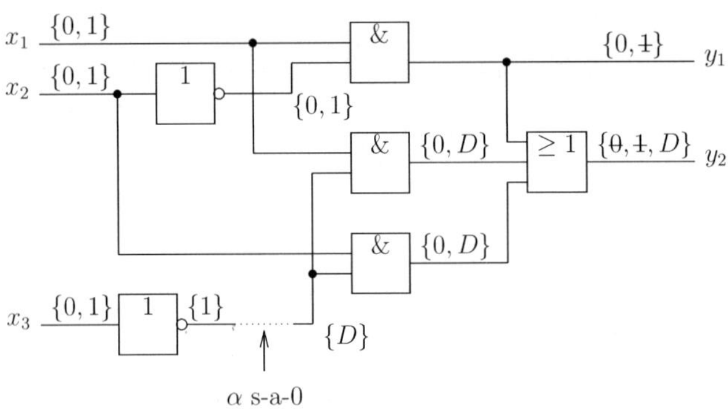

Fig. 15.26 Signal sets after removing irrelevant signals

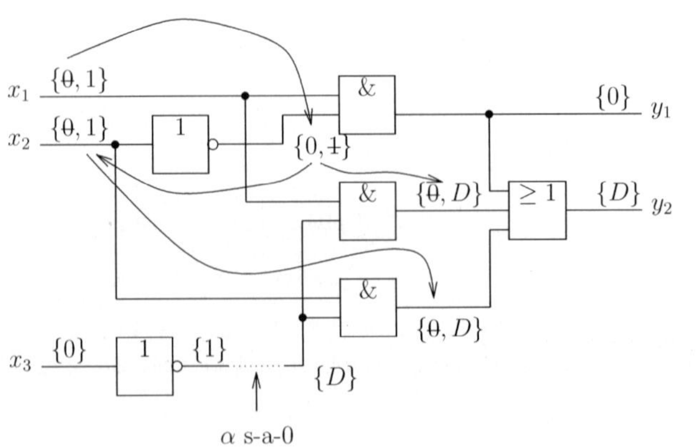

Fig. 15.27 Signal sets after guessing $x_1 = \{1\}$

Now one has to try out several additional assumptions. For example, one may delete the signal 0 from $x_1 = \{0, 1\}$, thus defining $x_1 = \{1\}$ and obtaining effects on other signal sets as shown in figure 15.27. Since the signal set of the output y_2 furthermore contains the signal D, the arbitrary definition of $x = 1$ can be maintained for the time. Coincidentally, all signal sets are reduced to containing only one signal. Therefore a test pattern $x = (1, 1, 0)$ has been found and the D algorithm terminates.

On the other hand, if instead of $x_1 = 1$ one would have decided on $x_1 = 0$ for the given example, the same procedure would have derived another test pattern for the fault 'α s-a-0'. Thus in the *example* it never becomes necessary to revise a wrong decision.

With slight modifications the D algorithms can also be used for other fault models such as hard *bridging faults* or *cellular faults*.

15.4.4 Fault Simulation

Fault simulation is another method of test pattern generation. Whilest the D algorithm constructs one test pattern for a given fault, the method of *fault simulation* determines all faults which can be detected using a given input pattern. At first, according to the respective fault model a list of all possible faults is created. For example in the circuit presented in figure 15.28 containing five wires there are ten possible *single stuck at* faults. The five *stuck at 0* faults will be denotes as a_0, b_0, c_0, d_0, y_0 and for the five *stuck at 1* faults we introduce the notation a_1, b_1, c_1, d_1, y_1.

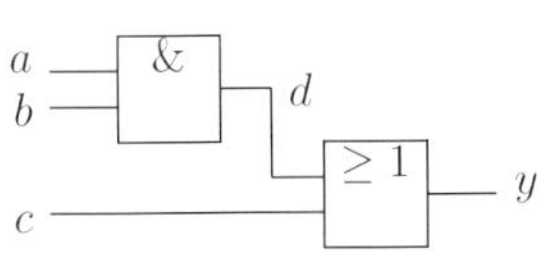

Fig. 15.28 Example of a circuit for fault simulation

As shown below, the algorithm of **fault simulation** sequentially considers all possible input patterns (for the *example* there are 8 input patterns) and computes the output for the correct circuit and the output for all faulty circuits:

- **Input**: A combinatorial circuit and a fault model;
- **Output**: Test patterns for all detectable faults;
- **Method**:

fault_set ← all possible faults in
accordance to the fault model
Until *fault_set* $\neq \emptyset$ do
{
Choose an input x
If x detects a fault from *fault_set*
{report x to be a test pattern
and delete all faults from *fault_set*
that can be detected by x
}
}

For the example of figure 15.28 the protocol of a fault simulation is presented in table 15.12. In the table erroneous output values are marked by underlining. Whenever a fault is detected, a test pattern for the fault has been found, and this case of a faulty circuit is no longer simulated.

In the first column it is indicated whether an input detects a fault not discovered before. Finally, the set of indicated patterns defines the test of the circuit. For the example we obtain a **complete test** $\{(000), (001), (010), (100), (110)\}$, a test with *fault coverage* of 100 %, for the *single stuck at* fault model.

However, in practice the input patterns used for fault simulation are not arranged in numerical order because in this way fault simulation might start with a lot of irrelevant patterns. For example, it may be necessary to set an *enable* input to 1 and *reset* to 0 in order to obtain relevant patterns.

Thus one often starts fault simulation using *stimuli* defined by the circuit designer for *verification by simulation*. Such simulation inputs should represent typical stimuli that are also relevant for practical applications of the circuit. Afterwards fault simulation continues to work with random patterns at primary inputs.

During fault simulation one first observes a large increase of fault coverage, because typically many faults can be detected by many different patterns. Such faults are **easy to detect**. After that it needs more time to obtain a new test pattern. That is because **hard to detect** faults can only be observed by a few patterns and there is only a low prob-

15

Table 15.12 Protocol of fault simulation for the circuit given in figure 15.28

	correct I/O				output of faulty circuits										detected faults	fault coverage
	a	b	c	y	a_0	b_0	c_0	d_0	y_0	a_1	b_1	c_1	d_1	y_1		
*	0	0	0	0	0	0	0	0	0	0	0	$\underline{1}$	$\underline{1}$	$\underline{1}$	c_1, d_1, e_1	30 %
*	0	0	1	1	1	1	$\underline{0}$	1	$\underline{0}$	1	1				c_0 ,e_0	50 %
*	0	1	0	0	0	0		0		$\underline{1}$	0				a_1	60 %
	0	1	1	1	1	1		1			1					
*	1	0	0	0	0	0		0			$\underline{1}$				b_1	70 %
	1	0	1	1	1	1		1								
*	1	1	0	1	$\underline{0}$	$\underline{0}$		$\underline{0}$							a_0, b_0, d_0	100 %
	1	1	1	1												

ability of finding such a pattern by chance. For that reason fault simulation is often aborted when the fault coverage obtained is good enough or a predefined computational time is exceeded.

Please note that because fault simulation permanently administrates a list of undetected faults on every occasion the fault coverage actually achieved is known. Thus the algorithm can also be used to determine the fault coverage for a given set of test patterns, for this one differentiates between **(surely) detected faults** and **potentially detected faults**. For example, fault simulation may achieve undefined signals X owing to a transistor fault, such that it is not known whether in practice a value 0 or 1 will be observed. Then a fault is detected with high probability if a sequence of X values corresponds to a lot of well defined signal changes in the correct circuit. In this case it is very unlikely that during chip test a faulty circuit coincidentally produces the correct output sequence.

15.4.5 Optimization of Test Pattern Generation

It is possible to combine the advantages of fault simulation and the D algorithm. In practice one first performs a fault simulation using *stimuli* from circuit simulation and using arbitrary inputs until too much computational time is needed to find additional test patterns. Afterwards random pattern generation is replaced by the D algorithm to derive suitable test patterns for remaining faults. This way one limits the number of calls of the D algorithm to faults which are hard to detect, and, furthermore, all faults which are detected by the same test pattern are deleted from the fault set.

Another strategy of optimization is to use some preprocessing to identify sets of *equivalent faults* or *dominating faults*.

Equivalent faults: Two faults of a combinatorial circuit are equivalent if they can be observed by exactly the same set of test patterns.

For test pattern generation one reduces computational time and storage requirement if only one fault of a set of equivalent fault is taken into the error set. It is still better only to consider *dominating faults*.

Dominating fault: A fault f **dominates** another fault g if any test pattern suitable for observing f can also be used to observe g.

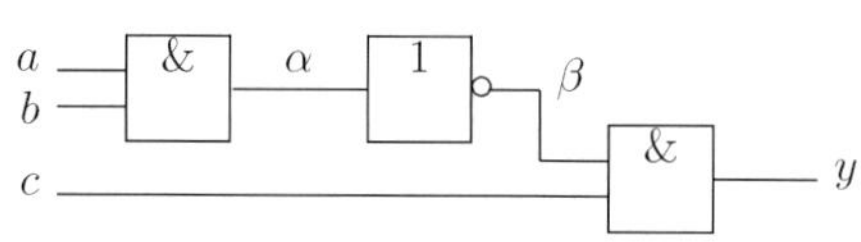

Fig. 15.29 Example of a circuit with 12 possible stuck at faults; the faults 'α s-a-0' and 'β s-a-1' are equivalent

This will be demonstrated by a simple *example*. Since the circuit given in figure 15.29 consists of six wires, there are twelve possible *stuck at* faults. Amongst them the faults 'α s-a-0' and 'β s-a-1' are equivalent.

This directly results from the characteristic equations of test patterns. For 'α s-a-0' we have $\alpha \cdot \mathrm{d}y/\,\mathrm{d}\alpha = 1$. With $\beta = \overline{\alpha}$ and applying the concatenation rule for Boolean differences we obtain exactly the characteristic equation $\overline{\beta} \cdot \mathrm{d}y/\,\mathrm{d}\beta = 1$ of the fault 'β s-a-1', proving that both faults are equivalent. Accordingly α s-a-1 and β s-a-0 are also equivalent.

More general *stuck at* faults at the input of an inverter are equivalent to faults at the output. For an *AND* gate all *stuck at* 0 faults at an input or the output are equivalent. Furthermore, all *stuck at* 1 faults at inputs of an *AND* gate are dominating the *stuck at* 1 fault at the output.

As a result of such rules for the example it is sufficient only to derive test patterns for the following four faults: 'α s-a-0'; 'a s-a-1';'b s-a-1'; 'c s-a-1'. Thus a fault simulation only has to consider four faults instead of twelve. Accordingly the D algorithm only needs four instead of twelve procedure calls.

Generally one can prove that for a *fan out free* combinatorial circuit it is sufficient to derive test patterns for all faults at primary inputs. The **following theorem** extends this result to arbitrary combinatorial circuits:

> In order to detect all *single stuck at faults* of a combinatorial circuit it is sufficient to apply test patterns for all faults at primary inputs and for faults at outputs of branching nodes.

A further optimization method for test pattern generators concerns the propagation of signals through combinatorial circuits. Considering gates in a favorable order, one has only to perform each gate function for one time. Such an optimal sequence of gates can be obtained by **topological sorting** of the gates.

Some implementations of fault simulation also use the respective word length of computers to simultaneously perform several simulations. Typically, bit wise Boolean operations can be applied to complete data words. For example, with a word length of 32 bits *one* operation of a computer can simultaneously simulate an *OR* gate for 32 different pairs of arguments.

In spite of all the optimizations mentioned, test pattern generation remains an extremely hard problem and requires high computational time. Therefore the techniques described in section 15.4.6 and section 15.6 are of special interest to design circuits in way such that they become easy to test. In particular, for extreme cases one can avoid any test pattern generation.

15.4.6 Controllability and Observability of Signals

For a *complete test* concerning *stuck at* faults it is necessary to change every signal of a circuit at least one time. The effort needed to control a signal using only primary inputs is measured in terms of **controllability**.

For example, one may compute the **entropy** of signals. If p_0 and p_1 are the probabilities for signal x to become 0, resp., 1, then the *entropy* $H(x)$ of signal x is defined as a value in the range between 0 and 1.

$$H(x) = -(p_0 \log_2 p_0 + p_1 \log_2 p_1).$$

$H(x) = 0$ means that the signal value never changes and $H(x) = 1$ indicates that the signal is perfectly random. Accordingly, the **observability** of signal x describes the influence of random changes, owed to a fault, on primary outputs.

Using such measures based on entropy, one can estimate whether a circuit is testable with random patterns, how difficult it is to derive suitable test patterns by fault simulation. Amongst other applications such measures can be used in ASIC design tools to control automatic logic synthesis to generate easy to test circuits.

As *examples* some **measures of controllability** (CC: Cost of Controllability) **and observability** (CO: Cost of Observability) are presented. $CC_0(s)$ and $CC_1(s)$ describe the effort of assigning logical 0, resp., logical 1 to signal s considering the number of relevant primary inputs and the number of gates which have to be passed. For example, to set output y of an *AND* gate to logic 1, both inputs must be set to '1' and signals have to be propagated through one gate. Thus for the *AND* gate one defines $CC_1(y) = CC_1(x_1) + CC_1(x_2) + 1$. In order to set y to '0' only one input of the *OR* gate has to be set to '0'. Therefore one defines $CC_0(y) = \min[CC_0(x_1), CC_0(x_2)] + 1$. The controllability of primary inputs is defined to be '1'.

Defining the observability the costs of primary outputs are set to 1. Then the observability of signal s results from the effort of propagating the signal to a primary output, to construct a D chain. This measure does not depend on the logic value of s.

Tables 15.13 and 15.14 summarize the definitions of controllability and observability for elementary gates [15.17], [15.48].

Table 15.13 Definition of signal controllability

gate	controllability
$y = x_1 \cdot x_2$	$CC_0(y) = \min[CC_0(x_1), CC_0(x_2)] + 1$ $CC_1(y) = CC_1(x_1) + CC_1(x_2) + 1$
$y = x_1 + x_2$	$CC_0(y) = CC_0(x_1) + CC_0(x_2) + 1$ $CC_1(y) = \min[CC_1(x_1), CC_1(x_2)] + 1$
$y = \overline{x}$	$CC_0(y) = CC_1(x)$ $CC_1(y) = CC_0(x)$

Table 15.14 Definition of signal observability

gate	observability
$y = x_1 \cdot x_2$	$CO(x_1) = CO(y) + CC_1(x_2) + 1$ $CO(x_2) = CO(y) + CC_1(x_1) + 1$
$y = x_1 + x_2$	$CO(x_1) = CO(y) + CC_0(x_2) + 1$ $CO(x_2) = CO(y) + CC_0(x_1) + 1$
$y = \overline{x}$	$CO(x) = CO(y)$

Alternative definitions of measures are normalized so that all values are in the range between 0 and 1. In addition to combinatorial measures one also defines similar measures to describe the sequential reachability and observability of states also considering the number of needed clock cycles [15.48].

For circuit designers there are simple methods to improve the controllability and observability of signals. For example, the combinatorial depth of a circuit can be limited by introducing a shift register to partition the circuit into smaller subcircuits. Introducing a *test mode* and additional ports for sequential write and read operations, internal signals become controllable and observable, thus increasing the testability of the circuit. For sequential circuits this technique is known as the **Scan Path method** and is described in more detail in section 15.5.1.

Another approach to improving the controllability of signals uses special **signal codings**. To explain that method we will construct combinatorial circuits such that only two test patterns are sufficient for setting all signals both to logic 0 and to logic 1 [15.22]. However, this may need a large amount of additional hardware. Therefore in practice one will not use this extreme case of a complete observability with only two input patterns but will use more refined codings.

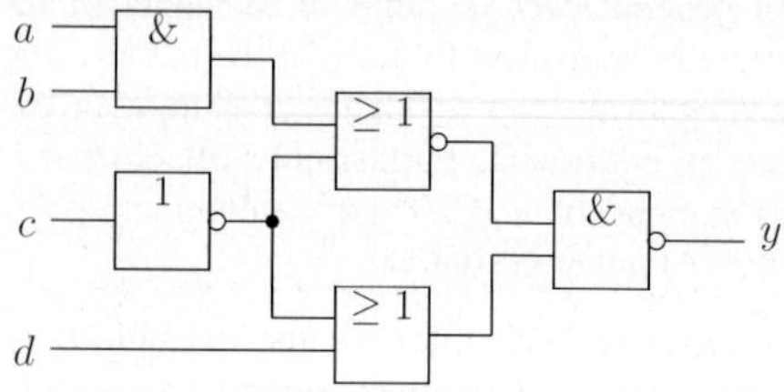

Fig. 15.30 Example of a circuit for the red/black coloring method

As an application example we consider the combinatorial circuit shown in figure 15.30 consisting of primary inputs a, b, c, d, primary output y, and five gates. Using the method of **red/black coloring** its controllability will be improved. At first we double all signals and gates obtaining 'red' inputs and outputs a_r, b_r, c_r, d_r, y_r and the 'black' inputs and outputs a_s, b_s, c_s, d_s, y_s. Accordingly we introduce red and black gates. As shown in figure 15.31, wiring is arranged so that all ports of gates are connected to signals of the same color. Only for inverting outputs is the color changed.

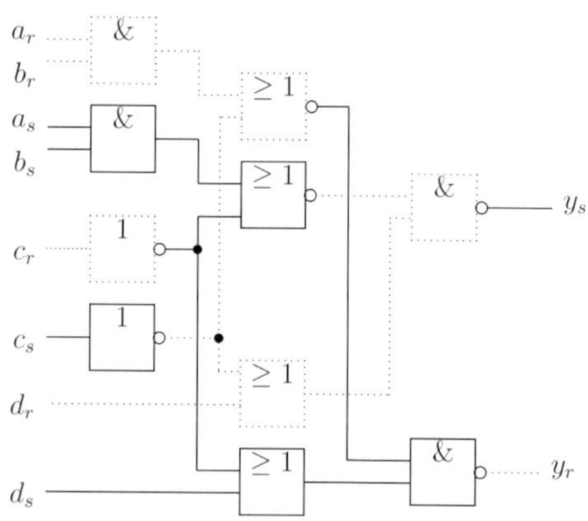

Fig. 15.31 Intermediate state for the red/black coloring method, with doubled signals and gates

Using the assignment $a_r = a_s = a$, $b_r = b_s = b$, $c_r = c_s = c$, and $d_r = d_s = d$ at primary inputs we obtain $y_r = y_s = y$. Thus the new circuit twice calculates the same result as the original circuit. Now we remove all gates and signals which are not needed for computing y_r and obtain the smaller circuit shown in figure 15.32. Again that circuit is equivalent to the original circuit but there is an additional interesting property. If all 'red' inputs are set to '0' (resp., '1') and all 'black' inputs are set to '1' (resp., '0') then all signals of the same color also have the same value. Thus using the two input patterns $a_s = b_s = c_s = d_s = 0$, $c_r = 1$, and $a_s = b_s = c_s = d_s = 1$, $c_r = 0$ every signal within the circuit can be set to '0' and '1'. Indeed, only two input patterns are sufficient for full controllability of all wires.

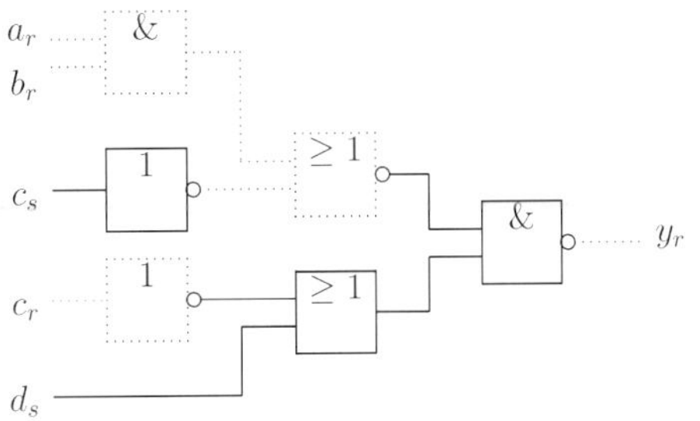

Fig. 15.32 Circuit modified by the method of red/black coloring

For the *example*, the signal coding mentioned before replaces the signal c by the pair of signals (c_r, c_s). For **normal operation** of the circuit one uses $c_r = c_s$; however, for **test mode** one uses $c_r \neq c_s$. Because of the coding there are additional input patterns available for testing. Please also note that using the presented method only controllability has been improved but not observability. In particular, for the generated circuit all possible single *stuck at* faults are stimulated by the two test patterns but it is not clear whether the faults can be observed at primary outputs.

15.4.7 Test Pattern Sequences

For *transistor faults* (section 15.3.5) and *bridging faults* (section 15.3.3) the behavior of a combinatorial circuit may become sequential. In such a case it is not always possible to detect a fault by applying only one test pattern.

For example, a transistor *stuck open* fault in a combinatorial CMOS gate may isolate the output node in such a way that for some input patterns the output will not represent the desired result, but the previously computed value.

To detect such effects the output node has to be *pre-loaded*. If the intended output value is '1' (resp., '0') the output node has to be set to '0' (resp., '1') before applying the test pattern. Then because of the changing output signal one can be sure of observing the actual output and not the previous one.

Using this method two inputs are needed for any test pattern. To reduce the number of necessary inputs one can arrange the test patterns in such a way that whenever possible one test pattern performs the pre-loading for the next one.

For example the test pattern set $\{11, 10, 01\}$ describes a complete *stuck at* test of a *NAND* gate. Table 15.15 gives a suitable sequence consisting of five inputs such that the output signal is correctly pre-loaded for every test pattern. This way *stuck open faults* are also detected.

Table 15.15 Test pattern sequence for a complete test of a CMOS NAND gate concerning stuck at and stuck open faults

a	b	z	
0	0	1	first input to pre-load y
1	1	0	test pattern (1,1) and pre-loading of y with 0
0	1	1	test pattern (0,1)
1	1	0	again pre-loading of y with 0
1	0	1	test pattern (1,0)

Because of the next **theorem** the method of using pairs of test patterns also works well for large circuits as long as only one gate is faulty. In general, if more gates are defective two input patterns may not be enough because pre-loading of an internal signal may also fail as a result of a second fault.

> In a non-redundant combinatorial circuit with gates in complementary CMOS logic every single *stuck open* fault can be detected using a pair of test patterns. This also holds for multiple *stuck open* faults as long as only one gate is affected.

15.5 Sequential Circuits

Since sequential circuits can store data there are many internal states that are often hard to control and to observe from outside. For that reason

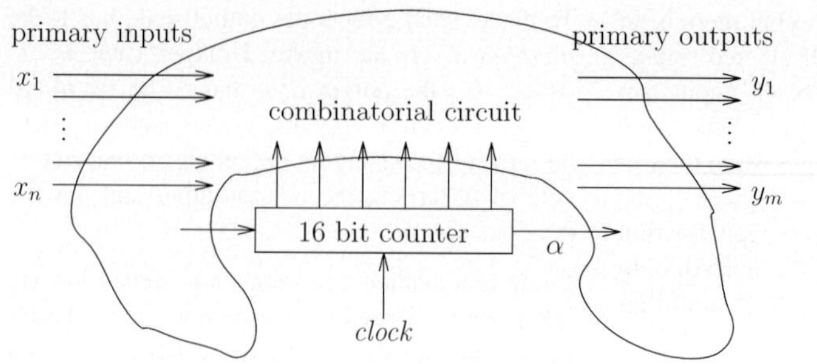

Fig. 15.33 Sketch of a sequential circuit with a counter as the only sequential component

testing sequential circuits is, in general, a difficult task. As an *example* figure 15.33 sketches a circuit which is partitioned into a 16 bit counter and a purely combinatorial sub-circuit. The only input of the counter is an *enable* signal that is used to increment the counter at the rising edge of a clock signal.

To test the *overflow* output α of the counter for *stuck at* faults one needs in each case 2^{16} clock periods to invert signal α. In the worst case the input patterns needed to enable the increment operation may even depend on the state of the counter. Then one needs a sequence of 2^{16} non-trivial test patterns to control signal α and obviously the sketched circuit is hard to test.

In practice one has to avoid circuits which are so hard to test and there are two different approaches to how this can be done. On the one hand, circuit extensions can be used to improve controllability and observability of signals. For example, this can be done using the **Scan Path** method described in section 15.5.1. This method is often part of design tools for improving testability in a way such that test pattern generators for combinatorial circuits (section 15.4) can also be used for sequential circuits.

An alternative approach is to directly design circuits in such a way that they become easy to test. Such design techniques are called **design for testability** and are presented in section 15.6.

15.5.1 Scan Path

To simplify the test problem many ASIC design systems only allow the designer to use special flip flops as sequential elements and they forbid other kinds of feedbacks. If other sequential elements are needed, for instance such as RAM blocks, they must be separable in such a way that distinct tests can be applied to circuit and storage blocks. Thus we only have to consider sequential circuits of the type sketched in figure 15.34.

Separating such a circuit at flip flops (broken lines in the sketch) we obtain several combinatorial sub-circuits S_i. Therefore the circuit can be represented by a purely combinatorial part and a storage block consisting of flip flops as shown in figure 15.35.

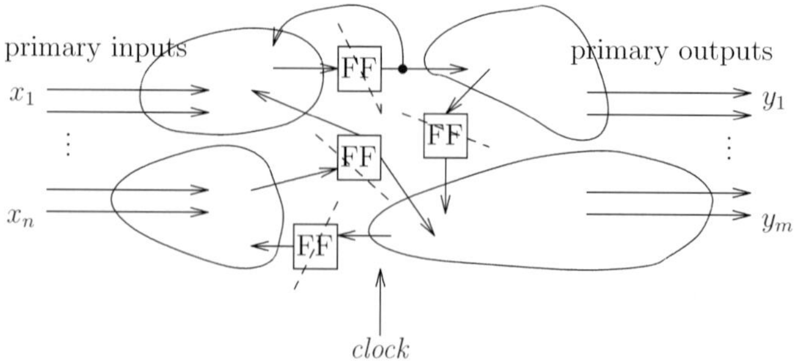

Fig. 15.34 Decomposition of a sequential circuit at flip flops

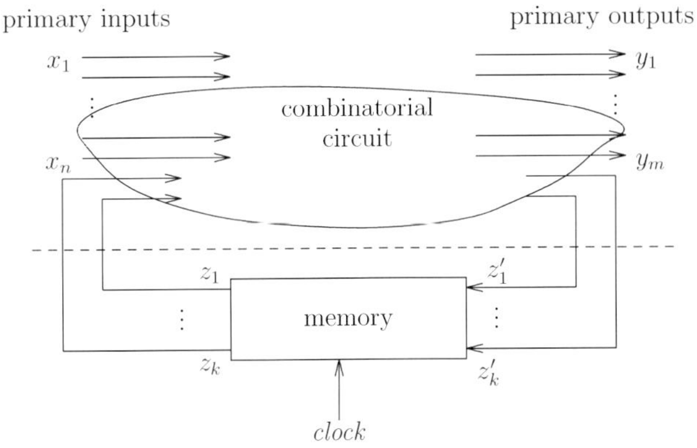

Fig. 15.35 Building a storage block consisting of all flip flops

The essential idea of the **Scan Path** technique is to introduce a **test mode** in order to separate the storage block consisting of k flip flops from the combinatorial part of the circuit. Then the remaining combinatorial sub-circuit has $n+k$ primary inputs $(x_1, \ldots, x_n, z_1, \ldots, z_k)$ and $m+k$ primary outputs $(y_1, \ldots, y_m, z'_1, \ldots, z'_k)$. Of course in practice one cannot insert additional *pads* for each flip flop. Therefore for test mode all flip flops are connected to build a *shift register* which is also called a **Scan Path**. With only one additional *input pad* '*scan_in*' all flip flops can be sequentially loaded with data and simultaneously the former data can be sequentially observed at the *output pad* '*scan_out*'. Thus using the *Scan Path* all flip flops are easy to control and easy to observe.

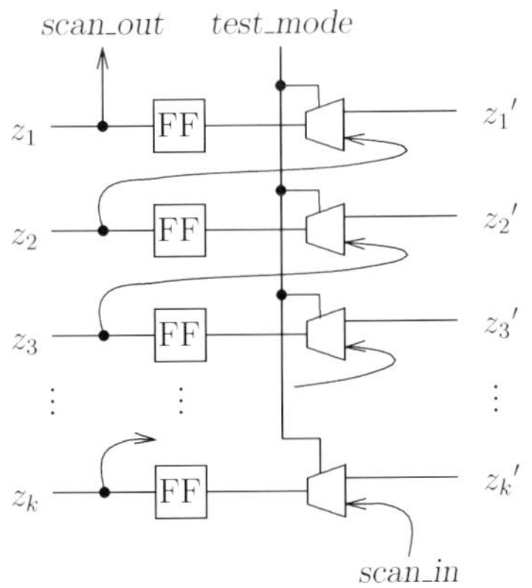

Fig. 15.36 Implementation of a Scan Path

In this way a *Scan Path* only needs three additional *pads*: '*scan_in*'; '*scan_out*'; and '*test_mode*'. However, for this implementation we have to assume that all flip flops are using the same clock signal which can be externally controlled and which is also available during *test mode*. Thus when using this type of *Scan Path* it is not allowed to derive clock signals from data signals. Other requirements for the circuit will be considered in more detail in section 15.5.2.

Using a *Scan Path* it is very simple to apply an automatic test pattern generation. At first one ignores all flip flops and applies the automatic test pattern generation for combinatorial circuits (section 15.4) considering $n+k$ primary inputs and $m+k$ primary outputs. Then n signals of a derived test pattern are directly applied to the circuit and the other k input signals have to be sequentially written into the *Scan Path* during test mode. Afterwards for one clock period *test_mode* is set to 0 to enable normal operation of the circuit and to transfer the results $z'_1, \ldots, z'_k$ into flip flops. Then the output signals $y_1, \ldots, y_m$ can be observed at primary outputs and $z'_1, \ldots, z'_k$ are read sequentially using *test_mode* $= 1$. Simultaneously with sequentially reading, the next test pattern can be written to the shift register. In this way, to apply l test patterns one needs a total of $l \cdot (k+1) + k$ clock cycles. At the same time the *Scan Path* is also tested.

A disadvantage of that test method is the slow operation rate of the circuit during test because k preparing clock cycles are necessary to apply one test pattern. In this way timing problems within the circuit may not be detected because critical signals are applied much earlier than in normal operation. This effect can be avoided by using *Master/Slave*

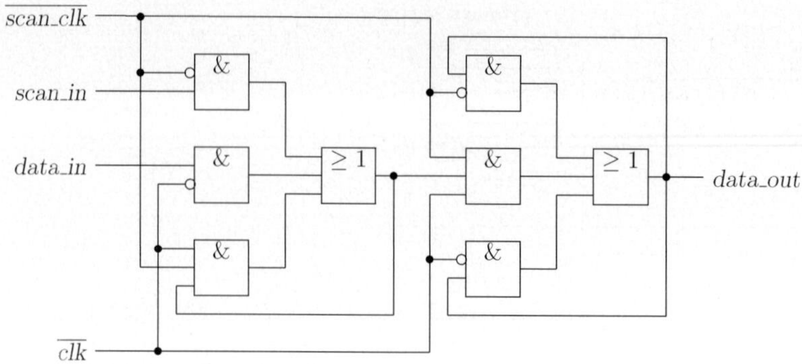

Fig. 15.37 Master/Slave flip flops with two separate clocks (signal $\overline{scan\_clk}$ is used by the Scan Path during test mode and $\overline{clk}$ is used during normal operation.)

flip flops, so that data from the *Scan Path* is applied to the circuit only after test mode is terminated.

Another reason for using more complex flip flops for a *Scan Path* is to introduce a separate clock signal for test mode. In this way one avoids the restriction that all flip flops of the sequential circuit have to be triggered by the same clock signal. In figure 15.37 a *Master/Slave* flip flop with two clock signals is shown.

With **Level Sensitive Scan Design** (LSSD) introduced by IBM even three clock pulses are used, one for the normal operation of the circuit, one for the *master*, and one for the *slave* of the flip flop.

Automatic **synthesis of a Scan Path** by a design system also has to define the arrangement of flip flops within the *Scan Path*. This information is also needed in order to correctly apply test patterns to the combinatorial part of the circuit. In principle an arbitrary arrangement of flip flops can be used. But then automatic placement and routing of layout synthesis may become more difficult. Therefore synthesis of a *Scan Path* should at least consider the rough placement of flip flops within a floor plan to create a *Scan Path* with short interconnections (figure 15.38). That method of considering geometrical information of a preliminary layout for *logic synthesis* is not unusual. For example, synthesis of *clock trees* is often done also using a floor plan.

Frequently used modifications of the *Scan Path* technique are to include only a subset of flip flops (**partial Scan Path**) or to distribute flip flops to **several Scan Paths**. This way using additional hardware (additional *pads* or *multiplexing* to existing *pads*) sub-circuits can be tested separately and the total time for testing can be reduced.

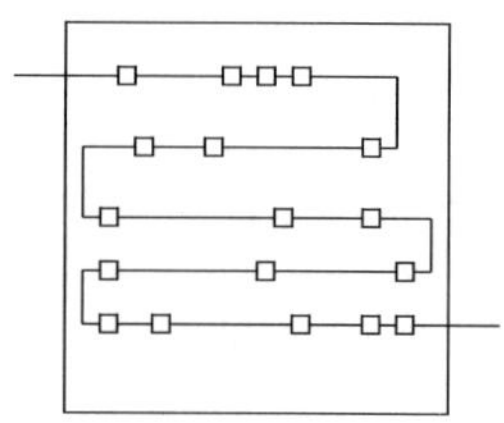

Fig. 15.38 Sketch of a Scan Path with an arrangement of cells derived from a floor plan

15.5.2 Requirements for a Scan Path

To improve the testability of a circuit by an automatic insertion of a *Scan Path* the circuit has to satisfy some special requirements. Since the circuit state will be controlled by the *Scan Path* the circuit should only contain such sequential components as can be included in a shift register. For example, it is not allowed to use *latches* that are set asynchronously.

Of course all edge triggered flip flops within a *Scan Path* have to be sensitive to the same kind of edge (rising or falling) and if the same clock signal is used in normal mode and test mode then all flip flops have to be triggered by the same external clock signal.

If, as an exception, there are different clock signals *clk1* and *clk2* in a circuit, as shown in part a) of figure 15.39, one has to add a multiplexer and an external selection signal *clk_select* to create a unique clock signal to be used for the Scan Path during test mode. This modification is shown in part b) of figure 15.39.

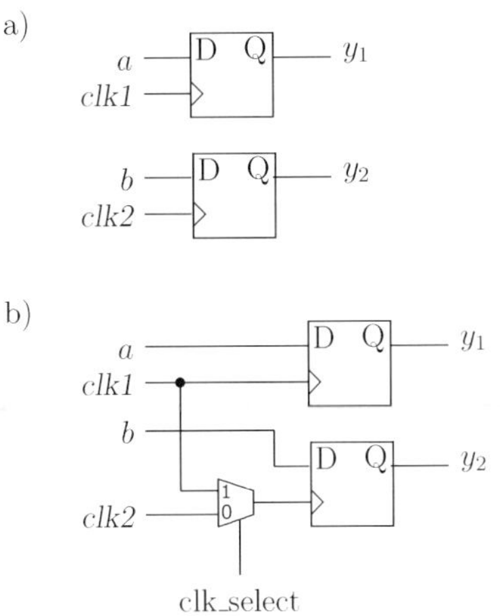

Fig. 15.39 a) Circuit with a clock signal derived from data signals, b) circuit modification to introduce a Scan Path

To automatically generate test patterns for such a circuit the generator needs some additional information about how to activate the *Scan Path*. In particular, for the example one has to specify that $clk\_select = 1$ is necessary for using the Scan Path.

Something similar applies when using asynchronous *reset* inputs of flip flops. Such *reset* signals also have to be controlled by primary inputs and should be inactive during test.

Also flip flops with an *enable* input need special attention. Since the *enable* signal should not disturb *Scan Path* operations the *enable* input of flip flops has to be replaced by an additional multiplexer at the data input, as shown in figure 15.40.

Since RAM blocks also can not be included into a *Scan Path* they have to be un-coupled during test mode so that they can be tested separately. Thus, all data, address, and control lines of the memory have to be made controllable and observable. For example, registers built from flip flops are a suitable interface between circuit and memory, because they can be included in the *Scan Path*. In this way during test mode the circuit will read data from the Scan Path instead of reading from memory. Accordingly, output data will not be written into RAM but will be written to the *Scan Path*.

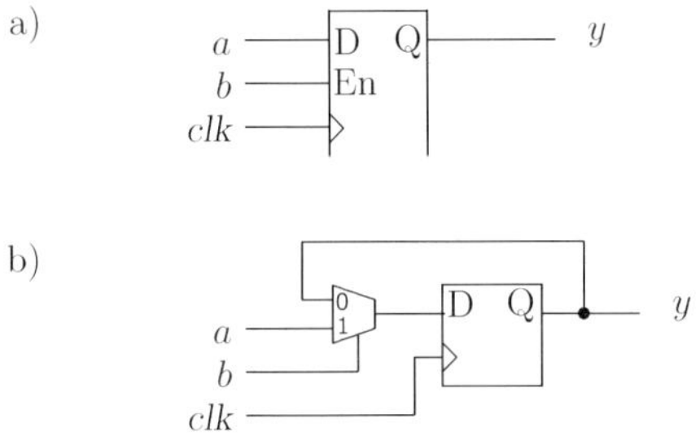

Fig. 15.40 Flip flop with ***enable*** *input and an alternative circuit suitable for a Scan Path*

15.5.3 Testing Sequential Circuits without Scan Path

The Scan Path presented in section 15.5.1 enlarges chip area and delay times and results in longer test sequences. Therefore one also tries to include only those flip flops in the *Scan Path* that are actually hard to control or hard to observe. For example, it is demonstrated in [15.24] that often the fault coverage obtained for a complete *Scan Path* can also be achieved by a suitably chosen relatively short **partial Scan Path**.

As an alternative to a *Scan Path* one can try to achieve a good controllability and observability using only available data lines. However, this means a substantially larger execution time for test pattern generation because the generation problem no longer can be reduced to the special case of combinatorial circuits. Exemplarily, the difficulties occurring will be demonstrated for flip flops used in data registers and finite state machines.

Logic synthesis of *high level* synthesis tools has to allocate data of an algorithm to hardware registers. For this allocation task there are several optimization criteria such as chip area needed, attainable clock rate, number of clock cycles needed for a computation and rate of data flow. A further criterion of synthesis is the testability of the generated circuit. For this criterion synthesis tools have to estimate the controllability and observability of internal data (section 15.4.6). Then with **data path synthesis** data has to be distributed on available registers, in such a way that easy and hard to

15

control (resp., to observe) variables share the same registers. In this way all registers become controllable and observable well without additional hardware [15.10]. Only if such an allocation cannot be found must additional control signals be introduced to improve testability [15.15].

For **finite state machines** there are further difficulties. While using a *Scan Path* it is easy to set all state machines of a circuit to well defined states, but without a *Scan Path* it becomes harder to reach the initial state of an automaton from any current state. To test for a fault the test pattern generator not only has to determine suitable states for stimulating and propagating the fault, but it also has to create sequences of input patterns to reach those states.

When several automata are serially connected within a circuit the problem arises of how to derive an input sequence to create some output sequence needed for the test. To simplify such problems one should introduce to all automata **reset states** which are easy to attain, thus improving the testability of the circuit [15.36].

A further difficulty arises for finite state machines if the specification only defines attainable states of normal operations. For example, in figure 15.41 a *Mealy* automaton with three states is given. For a binary coding of states one needs two flip flops, and thus the obtained circuit actually has four states. If in any way (because of random events at 'power on') the circuit does indeed reach the fourth state then its behavior is not defined.

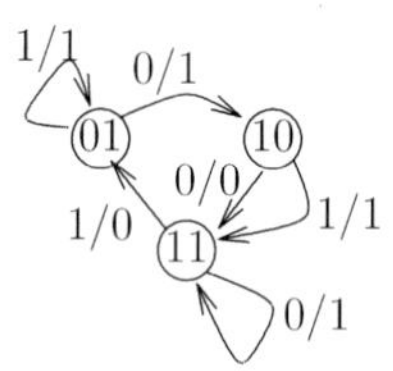

Fig. 15.41 Mealy automaton with three states (01), (10), and (11); edges are labeled by 'input/output'

The state transmission function and the output function are both implemented by combinatorial circuits. To observe a *stuck at* fault within the combinatorial part of the automaton it then may become necessary to enter an illegal state. Since illegal states do not occur during normal operation such a fault is not relevant for the correct operation of the automaton. Therefore it is called a **sequentially redundant fault** [15.12]. But unfortunately such irrelevant faults may lead to an extremely large execution time for pattern generation using only a gate level circuit description. This is because it is very hard to determine the **non-observability** of faults, resp., the **un-reachability** of states.

Similarly to a *logic minimization* eliminating redundant signals of combinatorial circuits that cause non-detectable faults, a synthesis tool should also perform a circuit optimization to eliminate *sequentially redundant faults*.

As an *example* we consider the automaton specified by table 15.16 which has three states, an input x, and an output y. As a special feature we assume that in practice input $x = 1$ does not occur for state (10). Therefore the appropriate entry is missing in the specification. For logic synthesis this obviously means an additional degree of freedom.

Table 15.16 Specification of an automaton with three states (01), (10), (11) (for state (10) the input $x = 1$ does not occur)

state	next state/output	
	$x = 0$	$x = 1$
01	10/1	01/1
10	11/0	- - /-
11	11/1	01/0

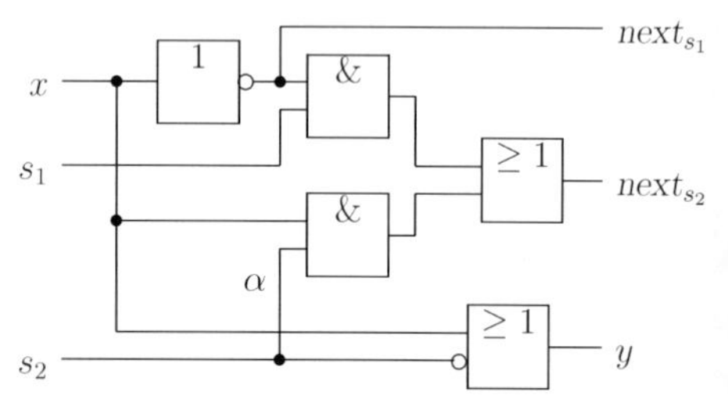

Fig. 15.42 Implementation of the combinatorial part for the automaton specified by table 15.16 containing the functionally redundant fault 'α s-a-1'

Figure 15.42 shows an implementation of the state transmission function and of the output function. For this combinatorial circuit fault 'α s-a-1' can only be detected using the test patterns $(x, s_0, s_1) = (1, 1, 0)$ or $(x, s_0, s_1) = (1, 0, 0)$. But according to the specification in table 15.16 for both inputs there is no admissible state transition such that the fault can not be observed if the automaton is part of a more complex sequential circuit. Such a fault, which can only be detected by illegal state transmissions, is called a **functionally redundant fault**.

In figure 15.43 a modified implementation of the automaton is given. It also satisfies the specification of table 15.16, but does not contain functionally redundant faults. For this implementation the specification was completed in such a way that for the state (10) and the illegal input $x = 1$ the next state becomes (11) and the output is $y = 1$. Then the automaton corresponds to the state diagram shown in figure 15.41. The *example* demonstrates that this kind of logic synthesis has large influence upon the execution time of test pattern generation. In particular, it is shown that a favorable implementation for test pattern generation may even need fewer gates.

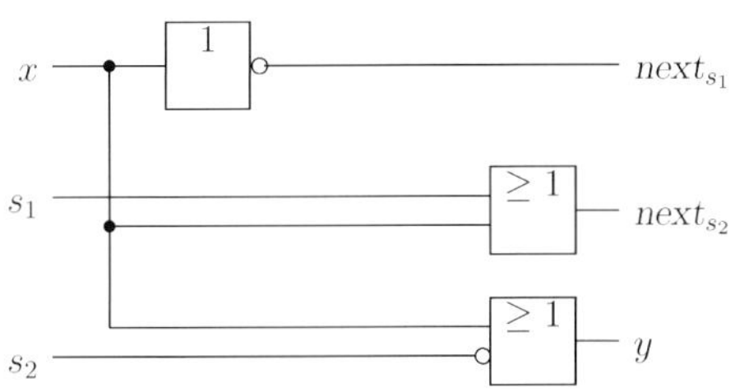

Fig. 15.43 Testable transmission function for the specification of table 15.16

In [15.27] the effect of logic synthesis on the performance of test pattern generation is investigated for the *example* of *re-timing* operations. In order to shorten critical paths and to achieve faster clock rates combinatorial parts of a circuit are equally distributed between flip flops. This operation of logic synthesis is called **re-timing**. In particular, *re-timing* can insert additional flip flops. For example, the logical behavior of both circuits presented in figure 15.44 is identical, but signal paths are differently partitioned amongst clock periods.

Although *re-timing* does not modify the logical behavior of a circuit and thus has no effect on testability of *stuck at* faults, it turns out that often the performance of test pattern generation is drastically reduced by *re-timing*. This is because for every additional flip flop the size of the search space for test patterns doubles and the percentage of relevant states becomes smaller.

To solve this problem one can use adaptive ATPG tools that use heuristics to avoid irrelevant states (resp., eliminate irrelevant flip flops).

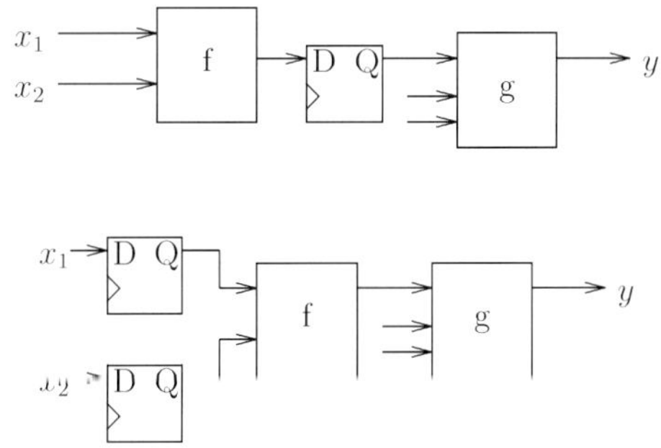

Fig. 15.44 Example of additional flip flops inserted by re-timing

15.6 Design for Testability

All methods presented so far for improving testability are generally applicable circuit modifications which may be performed after completing a circuit design. In this section we now present techniques of how to consider testability in an early phase of circuit design. 15

In section 15.6.1 circuits are designed in a way such that simple well known tests can be applied. Then sections 15.6.2, 15.6.3, and 15.6.4 consider techniques to derive **self checking circuits**. Often such circuits have a special mode of operation for generating test inputs and to check the computed outputs for correctness. For example, in this way chips may test themselves at **power on reset**. Also a self-test may be performed during the normal operation of a circuit. For this purpose one uses codes to simplify the detection of faulty results.

15.6.1 Universal Test

A **universal test** is a test which can be applied to a large class of circuits. A typical *example* is testing RAM memory. Here the principal type of test sequences is always the same and there are only slight differences depending on the actual size of a RAM block. Thus the test pattern sequence is defined in a parameterized way such that for a given data length and storage size it is easy to generate suitable test patterns. In a similar way it is also possible to derive test patterns for the class of *carry ripple* adders or *carry look ahead* adders that are parameterized by the data length.

It is more interesting that such *universal tests* also exist for a class of **easy testable PLAs** (PLA: Programmable Logic Array) [15.2], [15.6], [15.11], [15.46]. Since PLAs can be used to implement arbitrary Boolean functions they represent a very powerful class of circuits. In particular, PLAs are also used to implement state transition functions of finite state machines (FSM: Finite State Machine). When an arbitrary combinatorial circuit is implemented by an easily testable PLA then no explicit test pattern generation is necessary, because, depending on the actual size of the PLA, a pre-computed *universal test* can be used.

First we describe the structure of a conventional **PLA** (not designed for testability) for a Boolean function $(y_1, \ldots, y_m) = f(x_1, \ldots, x_n)$. The normalized representation for each signal y_i is defined as a sum of products over input variables and inverted input variables. Thus the function can be implemented using two levels of gates. As shown in figure 15.45 a PLA consists of an AND block and an OR block. In the AND block there are k columns, each computing one term of the normalized representation. Then the OR block combines such terms obtaining signals y_i.

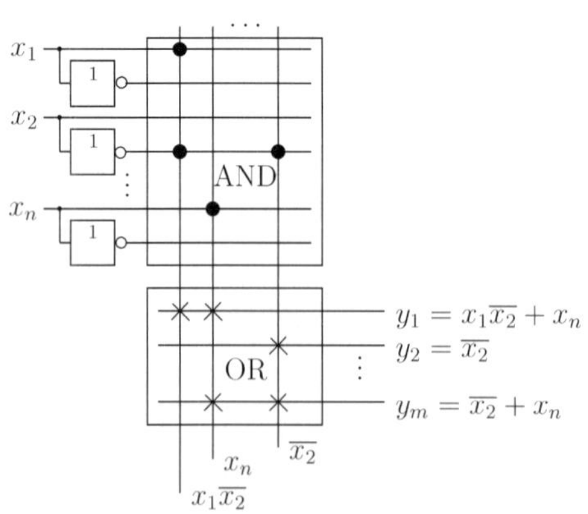

Fig. 15.45 Sketch of a PLA

To consider the testability of transistor networks we use a modified single *stuck at* fault model. Possible faults are a stuck at fault at a line or that a transistor is permanently conducting or permanently locking.

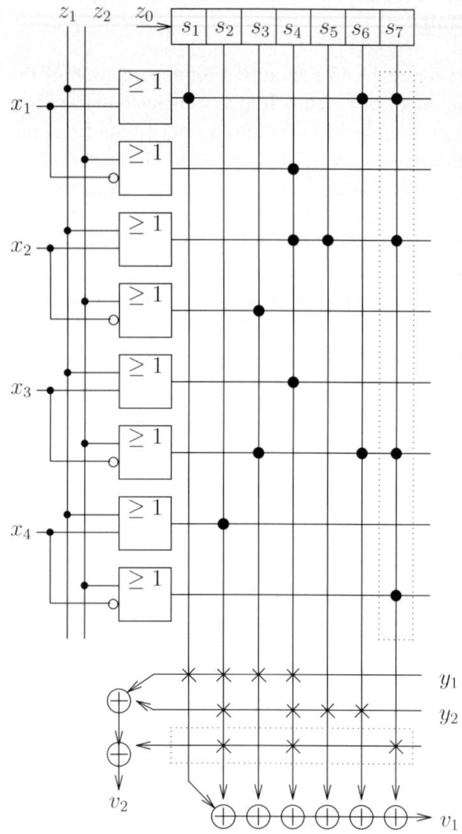

Fig. 15.46 Sketch of an easy to test PLA (⊕ symbols denote XOR gates.)

To improve the testability of a PLA one extends the circuit such that each column of the AND block becomes separately addressable and faults can be observed at additional test outputs v_1 and v_2 (figure 15.46). For that purpose a shift register with input z_0 is inserted above the AND block. It will be serially initialized to $(s_1, s_2, \ldots, s_k)$ such that for $s_i = 1$ the *AND* operation is performed for column i and for $s_i = 0$ column i always computes signal '0'.[1)] Furthermore, for the horizontal lines of the AND block the input signals are combined with test inputs z_1, resp., z_2, such that for *test mode* horizontal lines can easily be set to '1'. Finally, there

[1)] Indeed the signal s_i is used to pre-load a line which will be unloaded depending on the input variables.

Table 15.17 Universal test for single faults of an easy to test PLA (k denotes the number of columns in the AND block. The functions even(k) and odd(j) determine whether the argument k is an even or odd number and return Boolean results 0 (false) or 1 (true).)

	$x_1 \ldots x_{i-1}$	x_i	$x_{i+1} \ldots x_n$	z_1	z_2	$s_1 \ldots s_{j-1}$	s_j	$s_{j+1} \ldots s_k$	v_1	v_2
1 pattern	$* \ldots *$	$*$	$* \ldots *$	$*$	$*$	$0 \ldots 0$	0	$0 \ldots 0$	0	0
$2k$ patterns	$0 \ldots 0$	0	$0 \ldots 0$	1	0	$0 \ldots 0$	1	$0 \ldots 0$	1	1
	$1 \ldots 1$	1	$1 \ldots 1$	0	1	$0 \ldots 0$	1	$0 \ldots 0$	1	1
$2n$ patterns	$0 \ldots 0$	1	$0 \ldots 0$	1	0	$1 \ldots 1$	1	$1 \ldots 1$	$even(k)$	$*$
	$1 \ldots 1$	0	$1 \ldots 1$	0	1	$1 \ldots 1$	1	$1 \ldots 1$	$even(k)$	$*$
$k-1$ patterns	$* \ldots *$	$*$	$* \ldots *$	1	1	$1 \ldots 1$	1	$0 \ldots 0$	$odd(j)$	$*$

is an additional column in the AND block with transistors placed in such a way that for every row there is an odd number of transistors. Accordingly, there is an additional row in the OR block such that in every column of the OR block there is an odd number of transistors. Figure 15.46 shows such an easy to test PLA for the *example* of functions

$$y_1 = x_1 + x_4 + \bar{x}_2 \cdot \bar{x}_3 + \bar{x}_1 \cdot x_2 \cdot x_3 \,,$$
$$y_2 = x_2 + x_4 + x_1 \cdot \bar{x}_3 + \bar{x}_1 \cdot x_2 \cdot x_3 \,.$$

The circuit needs $n = 4$ data inputs (x_1, x_2, x_3, x_4) and computes $m = 2$ output bits y_1 and y_2 using six intermediate product terms. Thus the AND block consists of $k = 6 + 1 = 7$ columns. For the normal mode of the circuit one uses $z_1 = z_2 = 0$ and $s_1 = s_2 = \ldots = s_k = 1$. The good testability of the circuit is because each transistor can be selected separately and the additional column and row is used to calculate **parity bits** v_1 and v_2 such that several transistors can be tested simultaneously.

This way a complete test for single faults only needs the $3k + 2n$ test patterns given in table 15.17. Because of the computed parity bits the test patterns do not depend on the Boolean functions implemented. Thus the given test is a universal one and can be applied to any PLA of the type described.

15.6.2 Signature Analysis

Signature Analysis is a method derived for self-testing circuits. One collects, for example, intermediate data at certain lines of the circuit to compute a **signature**. For external evaluation the signature may be written to primary outputs or its correctness can also be checked immediately on the chip.

A simple computation of signatures can be achieved using **linear feedback shift registers** (LFSR). Figure 15.47 presents an *example* of a LFSR of length $l = 4$ bits. It is used to convert a long sequence of observed signals into a comparatively short signature.

Test sequence:
1. reset signature analysis;
2. input of a long sequence of test patterns;
3. output of a computed short signature;
4. compare signature to the desired value.

Table 15.18 Behavior of the signature analysis from figure 15.47 (An input sequence of length 8 bits is converted to signature (0110) of length 4 bits.)

clock	input	state			
0		0	0	0	0
1	1	1	0	0	0
2	0	1	1	0	0
3	1	0	1	1	0
4	1	0	0	1	1
5	0	0	0	0	1
6	0	1	0	0	0
7	0	1	1	0	0
8	1	**0**	**1**	**1**	**0**

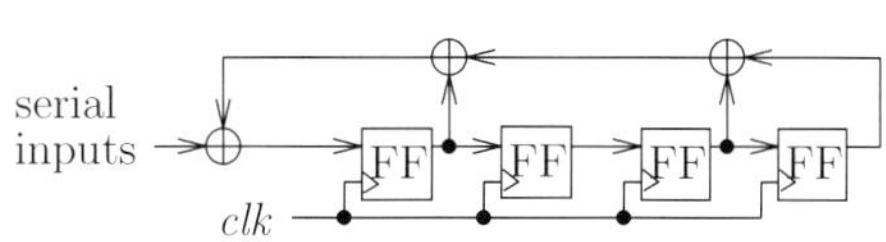

Fig. 15.47 Signature register of length 4 bits with sequential input

In table 15.18 this is demonstrated by a small example of converting an input sequence of length eight bits into a signature of length four bits.

The correct signature for a given sequence of test patterns can be determined by a circuit simulation and may be coded in hardware within the circuit in such a way that finally the chip itself can perform a signature check. If a wrong signature is observed then obviously the chip is defective. For a correct signature the chip may be correct or the correct signature is derived because of a compensation of several incorrect bits. To estimate the probability of such a hidden fault we assume for the sake of simplicity that after the first faulty bit we obtain a random sequence of states such that all possible signatures are generated with equal probability. Then the probability of obtaining a correct signature of length l, despite faulty input bits, becomes $1/2^l$. Thus for a sufficiently long signature the probability of an undetected fault becomes extremely small.

However, for a complete self-test, instead of applying external test patterns the patterns should be created on the chip. Since this is very difficult for complicated sequences of test patterns one often uses a **pseudo-random sequence** of test patterns which can be generated by a linear feedback shift register. In section 15.6.3 this technique will be considered in more detail.

In addition to the signature analysis described with sequential input there are other versions with parallel input (MISR: Multiple Input Signature Register). Furthermore, there are versions like BILBO (Built In Logic Block Observation) with additional multiplexers at shift register cells such that signature analysis can operate in different modes. This way, for example, a serial output of the computed signature can also be obtained.

15.6.3 On-Chip Generation of Test Patterns

To generate test patterns locally for a self-test on the chip one uses linear feedback shift registers as applied for signature analysis. If, for example, the input of the LFSR register shown in figure 15.47 is fixed to logical 1 and the register is initialized to '0000' then the LFSR generates the state sequence given in table 15.19. Of course a sequence generated in the manner described has to become periodic. For the *example* the period is of length 12.

Such arbitrary looking sequences of numbers are called **pseudo-random sequences**. In the case of a self-test they are used to replace random test patterns. The quality of pseudo-random sequences depends on the length of period. For a feedback shift register of length l the maximal attainable period length is $2^l - 1$. It can be obtained by selecting special feedbacks, such that the operation of the LFSR is equivalent to a polynomial division with some **prime polynomial**. For example, in [15.44] **minimal primitive polynomials** are listed such that only few feedback loops with *XOR* gates are needed.

Table 15.19 Application of the LFSR from figure 15.47 to generate a pseudo-random sequence of 4 bit data words with a period of length 12

clock	input	state			
0		0	0	0	0
1	1	1	0	0	0
2	1	0	1	0	0
3	1	1	0	1	0
4	1	1	1	0	1
5	1	1	1	1	0
6	1	1	1	1	1
7	1	0	1	1	1
8	1	1	0	1	1
9	1	0	1	0	1
10	1	0	0	1	0
11	1	0	0	0	1
12	1	0	0	0	0

In section 15.4.4 it was mentioned that using a random test pattern only 'easily detectable' faults can be observed. Therefore the *fault coverage* of self-testing circuits can be improved by **combining deterministic sequences and random sequences**. For this purpose one uses a test pattern generator to derive additional deterministic test patterns for those faults that are not detected by a random pattern sequence. Such deterministic patterns then may be stored in a ROM on the chip to be used as additional patterns for a self-test. In [15.8] a modified method is presented for using an OR matrix for re-coding the state of a shift register in such a way that the generated pseudo-random sequence also includes deterministic patterns. Figure 15.48 gives a sketch of the hardware used. At first a logic 1 is shifted through the register without using feedback loops, thus creating test patterns which correspond to single columns of the OR matrix. Afterwards feedback is used to create pseudo-random patterns using the OR operation on columns.

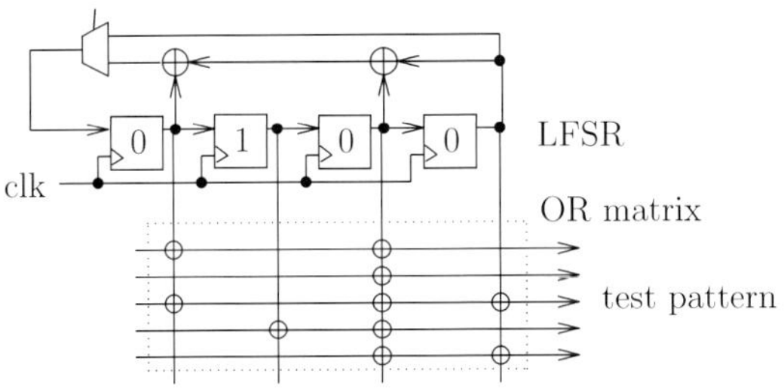

Fig. 15.48 Re-coding of pseudo-random patterns to generate also deterministic patterns defined by an OR matrix

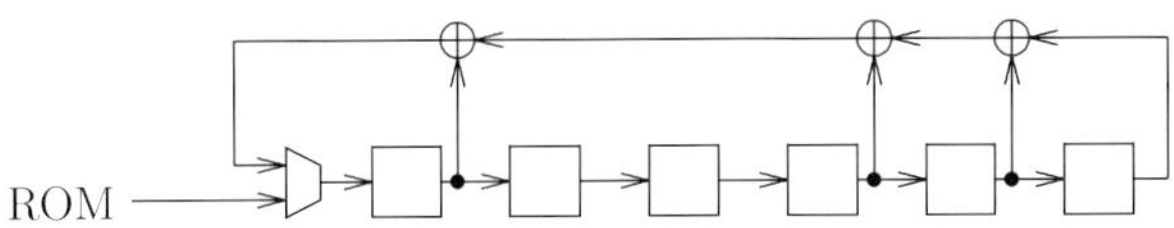

Fig. 15.49 LFSR with serial initialization to create pseudo-random patterns

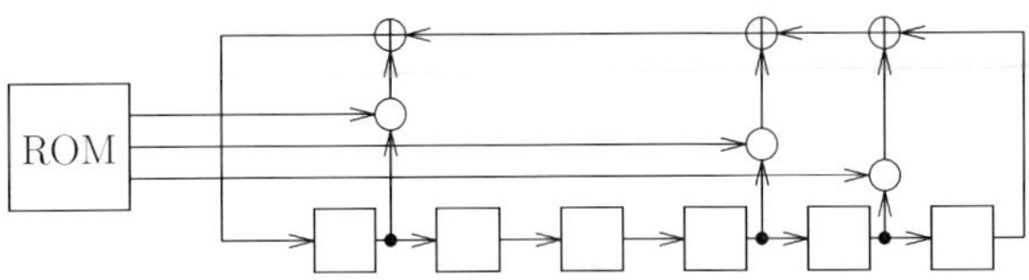

Fig. 15.50 LFSR with programmable feedback

Other approaches try to create all desired deterministic test patterns using the **LFSR method**. In this case the fault coverage obtained is defined by a deterministic test. For that purpose one can re-initialize the shift register, such that successively several sequences with different initial values are created. Alternatively, the type of feedback can be dynamically modified to obtain sequences with special properties [15.16]. Such implementations are sketched in figure 15.49 and 15.50. The ROM data needed can be derived from a deterministic test by solving a linear system of equations.

To test a circuit for *delay* faults or *stuck open* faults one needs pairs of test patterns. In this case also the arrangement of test patterns within a sequence is important. Indeed the LFSR method can also be used to create test pattern sequences such that pre-defined **pairs of test patterns** are included [15.45], [15.9].

An alternative method of generating pseudo-random sequences of test patterns by LFSR is to use a **field of cellular automata** [15.20], [15.14]. In such a field communication between adjacent automata ensures that input patterns are modified in a complicated way. For a suitable initialization and definition of interaction between adjacent cells one also obtains test pattern sequences containing predefined patterns or even predefined pairs of test patterns.

All methods mentioned are gaining significance by applying the *Boundary Scan* method considered in section 15.7. For chips which are equipped with that test interface there is already additional hardware at *primary inputs* which can also be used for test pattern generation. In this way only little additional hardware is sufficient for achieving an *on-chip* generation of test patterns.

15.6.4 Application of Codes

Special **codes** are used, for example, to transmit data such that a receiver is able to detect transmission errors. A simple *example* is the introduction of a *parity bit*. With other codes it is even possible not only to detect errors, but also to perform an automatic error correction. As an *example* we consider code $C = \{(000), (111)\}$ which only consists of two code words selected from eight possible binary words of length 3 bits. Figure 15.51 gives a graphical representation of this code. It is easy to see that any path between both code words needs three edges. Thus the code is of **Hamming distance** 3. If, owing to an error, one or two bits of a word are faulty this can be detected because the resulting word is not a code word. Assuming that only one bit is faulty the error can be corrected by replacing the wrong word by the adjacent code word.

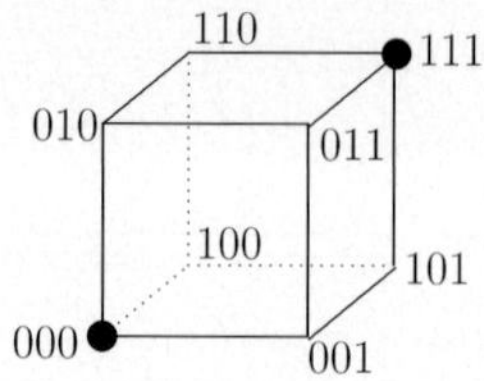

Fig. 15.51 Graphical representation of code $C \subset \{0,1\}^3$ with Hamming distance 3

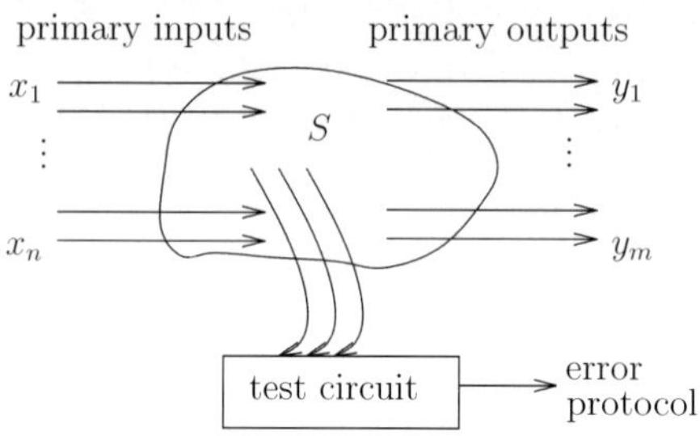

Fig. 15.52 Extension of circuit S by a test circuit to check whether the data words obtained belong to a code

The **redundancy** of a code can also be used to detect errors or even to correct errors for signals on a chip. For the following application we assume that a circuit S is implemented in such a way that intermediate data words or results have to be code words. Then a self-test can be performed as shown in figure 15.52 by using an additional test circuit to check whether indeed all data words belong to the code.

For this approach the problem arises that a fault may also occur within the test circuit, in such a way that the self-test will not work correctly. Thus the test circuit should be implemented to be **totally self-checking** in order also to detect its own faults.

As an *example* the circuit S may use an 'm of n code'. Such a code is defined by the set of all binary words of length n with exactly m bits set to '1'. It consists of $C^n_m = \binom{n}{m} = \dfrac{n!}{m!(n-m)!}$ code words. As an example the 3 of 6 code consisting of 30 code words is given in table 15.20. With such a code one can detect single errors (one bit of the code word is wrong) and all multiple errors when all faulty bits of a code word have the same logic value. A fault causing erroneous data words of this type is called an **uni-directional fault**. For example it may be owed to either *stuck at* 0 faults or *stuck at* 1 faults at several lines.

Table 15.20 'm of n code' for $n = 6$ and $m = 3$ (The code consists of $C^6_3 = \binom{6}{3} = \frac{6!}{3!(6-3)!} = \frac{6\times5\times4}{3\times2\times1} = 20$ code words.)

0	0	0	1	1	1	1	0	0	0	1	1
0	0	1	0	1	1	1	0	0	1	0	1
0	0	1	1	0	1	1	0	0	1	1	0
0	0	1	1	1	0	1	0	1	0	0	1
0	1	0	0	1	1	1	0	1	0	1	0
0	1	0	1	0	1	1	0	1	1	0	0
0	1	0	1	1	0	1	1	0	0	0	1
0	1	1	0	0	1	1	1	0	0	1	0
0	1	1	0	1	0	1	1	0	1	0	0
0	1	1	1	0	0	1	1	1	0	0	0

In order to design a self-checking test circuit the test circuit has also to produce code words.

Definition of a test circuit:

A **test circuit** computes a function $g\colon \{0,1\}^n \rightarrow \{0,1\}^k$ with the property that input data of a code $A \subset \{0,1\}^n$ is mapped to output data of a code $B \subset \{0,1\}^k$. Furthermore, for inputs $x \notin A$, illegal inputs, the output satisfies $g(x) \notin B$.

The property

$$g(x) \in B \qquad \text{for } x \in A\,,$$
$$g(x) \notin B \qquad \text{for } x \notin A\,,$$

is called the **separability property** of function g.

As an *example* A may be the *3 of 6* code and for B we use the *1 of 2* code ($B = \{(0,1),(1,0)\}$). Then the test circuit computes a function g accepting six input bits and producing two output bits y_1 and y_2. If in the input x there are exactly three ones then we obtain an output with $y_1 \neq y_2$. Otherwise, the output satisfies $y_1 = y_2$. The circuit implementation of g shall be done in a way such that following definition is satisfied.

Definition of a self-checking test circuit:

A test circuit is called **self-checking** for a fault set F if for each fault $f \in F$ there is at least one input $x \in A$ such that the circuit produces an output $y \notin B$.

Because of this definition every fault within a test circuit can be detected using only the available inputs $x \in A$. This is an important property, because the inputs of the test circuit can only be controlled by another circuit S producing only outputs of code A. Furthermore, the faults within a test circuit can be detected by observing 'obvious wrong outputs' $y \notin B$.

As a consequence of this definition the output y of a self-checking test circuit consists of at least two bits. Otherwise a *stuck at* fault at the only output line could cause a constant output $y \in B$ and thus would remain undetected.

Definition of a 'fault secure' test circuit:

A test circuit is called **fault secure** for a fault set F, if for each fault $f \in F$ the circuit will never produce for an input $x \in A$ an output $y \in B$ with $y \neq g(x)$.

Because of this definition, for any correct input $x \in A$ any faulty test circuit will either produce the correct output $y = g(x) \in B$ or an 'obvious wrong output' $y \notin B$.

Definition of a totally self-checking test circuit:

A test circuit is called **totally self-checking (TSC)** for a fault set F if the circuit is self-checking and fault secure for F.

In figure 15.53 a totally self-checking test circuit for the 3 of 6 code is given. The circuit corresponds to the following equations.

$$g_1(x) = (x_1 + x_2 + x_3) \cdot (x_4x_5 + x_4x_6 + x_5x_6) + x_1x_2x_3\,,$$

$$g_2(x) = x_4x_5x_6 + (x_1x_2 + x_1x_3 + x_2x_3) \cdot (x_4 + x_5 + x_6)\,.$$

More generally the circuit for any m of $2m$ code can be derived according to equation (15.13).

$$g_1 = \sum_{\substack{j=0 \\ j\,\text{odd}}}^{m} T(k_l \geqq j) \cdot T(k_r \geqq m-j)\,,$$
$$g_2 = \sum_{\substack{j=0 \\ j\,\text{even}}}^{m} T(k_l \geqq j) \cdot T(k_r \geqq m-j)\,. \qquad (15.13)$$

Here the input $(x_1, \ldots, x_n)$ is partitioned into a left and a right part and function $T(k_l \geqq i)$, resp., $T(k_r \geqq i)$, checks whether in the left, resp., right, part there are at least i bits set to '1'.

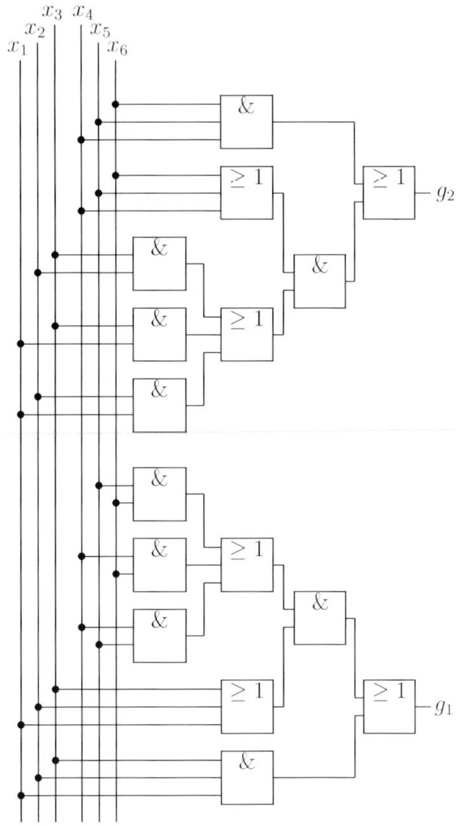

Fig. 15.53 TSC circuit for the 3 of 6 code

Function g has the property that for inputs with m ones the result $(0,1)$ or $(1,0)$ is derived depending on whether there is an even or odd number of ones in the left part of the input. In this case exactly one term

$$T(k_l \geqq j) \cdot T(k_r \geqq m-j)$$

in equation (15.13) becomes true. If the input contains not enough ones the result becomes $g(x) = (0,0)$ and if there are too many ones it becomes $g(x) = (1,1)$.

For the given implementation it is remarkable that the output bits $g_1(x)$ and $g_2(x)$ are computed by two disjoint sub-circuits. For this reason a *stuck at* fault within the test circuit can only influence one bit of the output and thus will never invert both output bits. Furthermore, a faulty input bit that might effect both output bits by definition will lead to the output $g = (0,0)$ or $g = (1,1)$. Thus the

15

code words (0, 1) and (1, 0) are never interchanged because of a fault.

Test circuits for an m of n code with $n \neq 2m$ can be designed by re-coding the inputs to reduce the problem to the case of a m of $2m$ code considered above [15.38], [15.26], [15.33].

Using other codes arbitrary combinatorial circuits can be made totally self-checking. For example, in [15.4] a circuit modification is presented such that any fault within a circuit can be observed at an odd number of output bits. In this case faults are detected by *parity* checking at primary outputs.

Another variant of self-checking circuits uses **fault indicators** registering whether at any time a given test property has been violated. In this way such faults can also be detected that occur for only short time periods and do not produce wrong results. For example, this technique can be used for **path delay faults**. In [15.35] such a fault indicator is presented for the 1 of 3 code.

A generalization of totally self-checking circuits are **strong fault secure circuits** [15.41], [15.33] that are defined to also handle multiple faults.

A further improvement leads to **fault tolerant circuits**. Such circuits will produce correct results even if there are 'small' faults within the circuit. To derive such circuits one can use a code with a large *Hamming distance* such that single erroneous bits can be automatically corrected. For example, fault tolerant circuits are of interest for safety-relevant applications where a fault may not lead to a total failure of a complete system. Also, if hardware overhead for *fault tolerance* is not too large this technique can be used to improve the yield of chip production. This is because in spite of a fault such chips may further be used as correct chips without fault tolerance.

15.6.5 Principle of Multiple Computation

A very simple method of using redundancy to detect faults is to perform the desired function several times using separate hardware and to compare the results obtained. This is a special case of a code such that each bit of the correct result is represented several times. However, this simple approach causes a large *overhead of hardware*. Sometimes there are more skilful implementations of multiple computations.

For example, when using **RNS arithmetic** (RNS: Residue Number System) for fast addition and multiplication of large numbers all numbers are represented as vectors with k relatively small components. Then the actual computation can be performed componentwise using k relatively small arithmetic units, and afterwards the final result is assembled from k independent intermediate results $y_1, \ldots, y_k$. In order to detect faults within arithmetic units one can extend the number representation by an additional redundant component. Then using $k+1$ instead of k arithmetic units one obtains a redundant representation $y_1, \ldots, y_k, y_{k+1}$ of the intermediate result.

The final result y can be determined from every k components of the intermediate result. Thus using the same hardware, but different data the result y, resp. $\hat{y}$ can be determined twice ($(y_1, \ldots, y_k) \rightarrow y$ and $(y_2, \ldots, y_{k+1}) \rightarrow \hat{y}$). Whenever one of the components y_i is erroneous, owing to a fault in an arithmetic unit, both results will differ and the fault is detected.

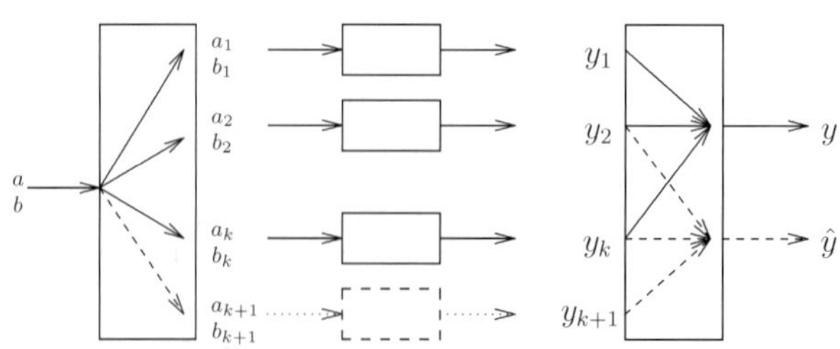

Fig. 15.54 Sketch of a circuit for redundant RNS arithmetic (Inputs a and b are represented by vectors of length k resp., k + 1. The actual computation is performed componentwise. The result y is uniquely determined by each subset of k intermediate results. Only little additional hardware is necessary for computing a second version ŷ of the result.)

15.7 Boundary Scan

For the **Boundary Scan** method [15.31] additional cells are inserted between a circuit and the *pads* on a chip, thus separating the circuit from its environment. In this way a chip can still be tested when it is already assembled within a larger system. Furthermore, the **Boundary Scan** can be used to check the wiring on a *Board* or to test a complete system. Indeed, testing complete systems was a particularly important aim of **JTAG** (Joint Test Action Group) for the development of *Boundary Scan* in the period from 1985 till 1988. The result obtained is IEEE standard 1149.1 defining a *standard test interface*.

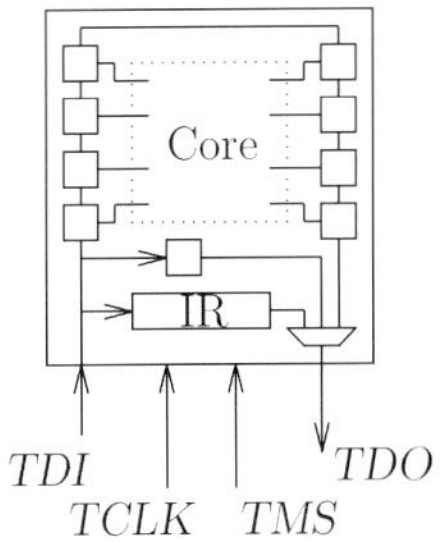

Fig. 15.55 Sketch of a Boundary Scan

As sketched in figure 15.55, the *Boundary Scan* (*BS*) of a chip is essentially a shift register which can be used to apply serially a test pattern to primary inputs and to observe serially primary outputs. Thus there is a large similarity to a **Scan Path**. However, instead, to improve controllability and observability of internal flip flops a *BS* is used to control and observe the *pads* of a chip. For a *BS* there are four additional *pads* which are denoted as **Test Access Port** (**TAP**). *TDI* (*testdata_in*) and *TDO* (*testdata_out*) are used for serial input and output of data and input *TCLK* (*test_clock*) serves for the clock supply of *BS*. Finally, the input *TMS* (*test_mode_select*) is used to select between several modes of operations.

Figure 15.56 demonstrates how to connect the BS pads of several chips on a board so that the *Boundary Scans* of all chips are combined to create a *Boundary Scan* of the complete *board*.

Basically the following **applications** are possible.

- **Test a chip** within a larger system (**in system verification**). Thereby the *Boundary Scan* is used to apply serially test patterns to the chip using input *TDI* and to output serially the obtained results to *TDO*;
- **Test the wiring on the board**. For this purpose *BS* data is written to a *pad* to stimulate a wire and then the same data again is read at other *pads* into the *Boundary Scan*;
- **Test assembly** of a board. In this mode chips can be forced to output their identification number to the *Boundary Scan*. This way the arrangement of chips within the *Boundary Scan* can be checked, which kind of chip is inserted at a special location;
- **Self-test of chips using signature analysis**. For *signature analysis* (section 15.6.2) one needs feedback shift registers at primary inputs and primary outputs of a circuit. Those registers are used to generate pseudo-random test patterns at primary inputs and to compute signatures at outputs. If, anyway, a *chip* is equipped with a *Boundary Scan* the *BS* cells can also be used for self-test, such that less additional hardware is required;
- **Combination with Scan Path**. If on a *chip* a *Boundary Scan* is available then an internal *Scan Path* may be integrated into the *Boundary Scan* such that for *test mode* internal flip flops also become accessible via *TDI* and *TDO*. This way additional *pads* for the *Scan Path* can be avoided.

15

To accelerate sequential input and output over the *Boundary Scan* it is desirable to include only currently relevant chips into the *Boundary Scan* of a system. Therefore there is a **BYPASS mode** to replace the **Boundary Scan register** of a chip by only one shift register cell. In the *example* given in figure 15.57 five of six chips on a board are in **BYPASS mode**, such that only the pads of one chip are included in the *Boundary Scan*.

In addition to the obligatory **Boundary Scan register** and the **bypass register** of length 1 there may be other optional registers for special applications. By a multiplexer one of these registers is selected to become the **data register** (DR) of the *Boundary Scan Path*. This selection depends on the content of a special **instruction register** described in section 15.7.2.

From the applications sketched it can be seen that *Boundary Scans* are mainly introduced to detect faults at *board* level or even *system* level. But there

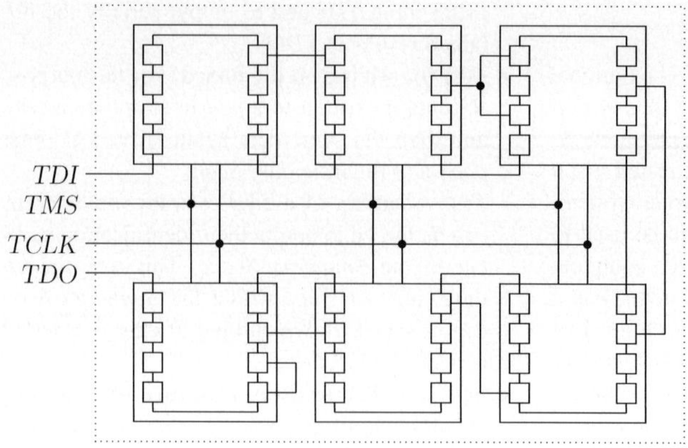

Fig. 15.56 Boundary Scan on a board with several chips which are suitable for BS

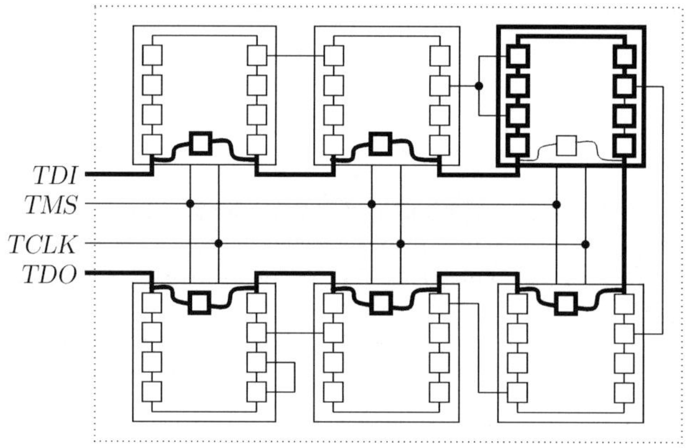

Fig. 15.57 Configuration of a Boundary Scan to test a single chip on a board (The other chips are in BYPASS mode.)

are also advantages for chip tests when a *Scan Path* or *signature analysis* is combined with a *Boundary Scan*. Furthermore, the *BS* technique can also be used to improve the testability of a circuit by using an internal *Boundary Scan* to partition the circuit into several sub-circuits that can be tested separately.

15.7.1 Boundary Scan Cells

There are several kinds of cells used for *input pads*, *output pads*, *tristate pads*, and *clock pads* within a *Boundary Scan*. As an example figure 15.58 gives a cell for an *output pad*. During normal operation mode the circuit output *y* is written to the *pad* '*y_Pad*'.

Alternatively, signal *y* of the circuit can be transmitted into the data register (DR) of the *Boundary Scan* to be written serially to *TDO*. Furthermore, it is possible to transfer data serially from *TDI* into the data register and to write that data to the output pad instead of *y*.

The signal *shift_DR* is used to determine whether the data register (DR) is loaded serially or in parallel. Anyway, loading is performed at the rising edge of *clock_DR*. In the case of $shift\_DR = 0$ the data comes from signal *y*, otherwise it comes from the previous BS cell via *serial_in*.

The signal *update_DR* is used to transfer data from the data register into the second flip flop such that it can be applied to the pad. Because of this

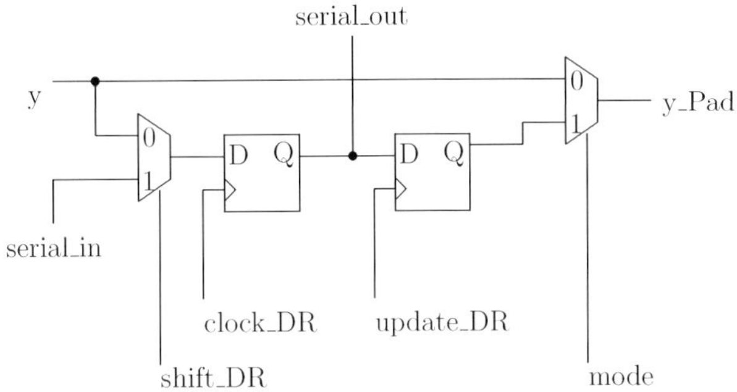

Fig. 15.58 Boundary Scan cell for a primary output to support EXTEST mode and INTEST mode

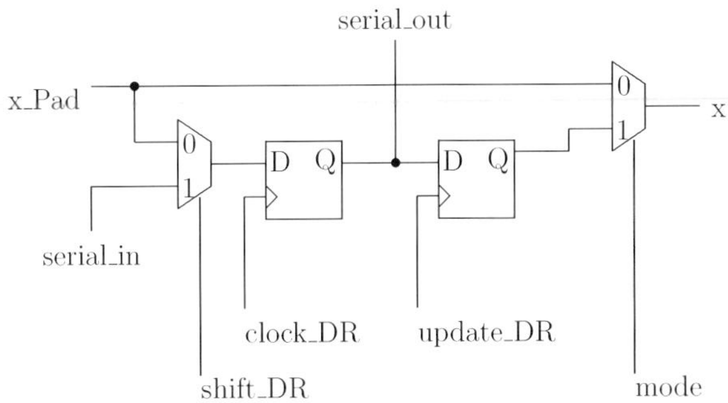

Fig. 15.59 Boundary Scan cell for a primary input to support EXTEST mode and INTEST mode

Master/Slave mechanism the *pad* signal is changed only after the data register has completely been updated by its serial input. Thus no undefined intermediate values are applied to *pads*.

The *Boundary Scan* cell for an *input pad* looks exactly the same (figure 15.59). Here, alternatively a pad signal or the content of the data register is applied to the primary input x or to the circuit.

15.7.2 TAP Controller

For configuring the *Boundary Scan* of a *chip* there is a **TAP controller**. That is a finite state machine deriving control signals for *Boundary Scan* cells and which can be used to select the mode of operation. Figure 15.60 gives the state diagram of a *TAP controller*. Its clock signal is *TCLK* and signal *TMS* (*test_mode_select*) is used as input of the automaton.

The state *Test Logic Reset* is a *reset* state of the automaton used to initialize the test hardware. This state can always be reached by forcing the input signal *TMS* to the value '1' for five clock cycles. The function of state *Run Test/Idle* depends on the current command in the *instruction register* and may be used for example to start a self-test of the chip (**RUNBIST** mode) or to put any register data like an identification number on the BS.

In the following the other states shall be explained using the *example* of a chip test (**INTEST** mode) sequentially reading the current data from the data register and applying the next test pattern.

Starting at *reset* state one reaches state *Capture DR* using input $TMS = 010$. This state is used for a parallel loading of the data register. This way the output signals of the circuit and the current external signals at input pads are written into the data register with the rising edge of the signal *TCLK*.

Then for a data register of length k the input *TMS* is set to '0' for k clock periods such that in the state *Shift DR* the current content is written sequentially to *TDO* and simultaneously the next test pattern is read from *TDI*.

After this, using input $TMS = 11$ one reaches the state *Update DR* applying the new content of the data register to the primary inputs of the circuit,

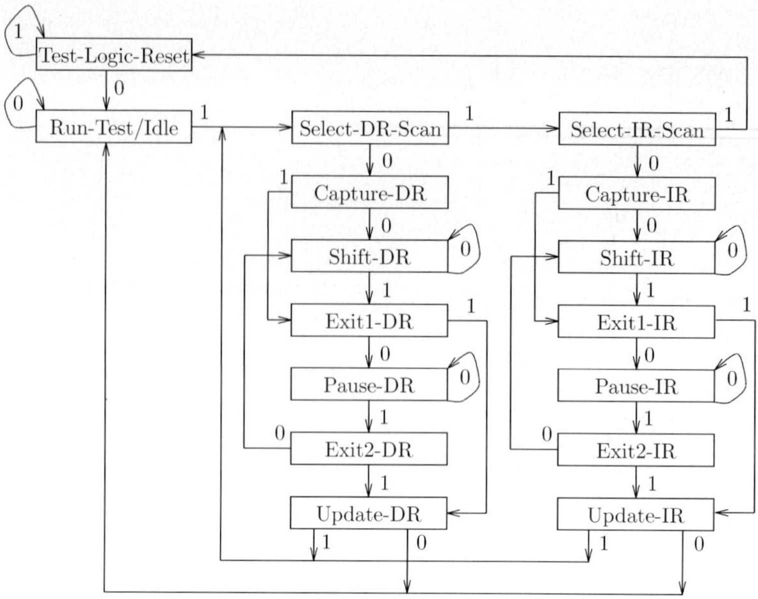

Fig. 15.60 State diagram of a TAP controller

resp., to the output pads, with the falling edge of *TCLK*. Finally, with $TMS = 10$ one again reaches the state *Capture DR* for the next observation of data and to insert another test pattern.

In addition to the *data register* there is also an *instruction register* for determining the operation mode of the *Boundary Scan*. According to the state diagram of the *TAP controller* the states *capture IR*, *shift IR*, and *update IR* are used to reload sequentially the instruction register by *TDI*. The length of the *instruction register* is at least 2 bits and, depending on the number of implemented operation modes, may be larger. According to the standard, the intended instructions are **BYPASS**, **EXTEST**, and **SAMPLE/PRELOAD**. The already explained **INTEST** mode for chip test is optional.

In **BYPASS** mode the *Boundary Scan* register is replaced by one single shift register cell. The *EXTEST* mode, for example, can be used to test the wiring of a board. For that purpose data from the *Boundary Scan* is written to output pads and again is read into the *Boundary Scan* at other input pads. With the optional *INTEST* mode for testing chips, data from the *Boundary Scan* is applied to primary inputs and data from the primary outputs are transfered to the *Boundary Scan*. Thus the only difference between *EXTEST* mode and *INTEST* mode is to test external or internal hardware of a chip.

SAMPLE/PRELOAD mode has a special meaning. It can be used to take a snapshot of current pad signals by copying all pad data into the BS. Later on that data may be sequentially written to TDO without disturbing the normal operation of the chip. On the other hand, this mode can also be used to reload sequentially the BS with data to prepare a later usage by *EXTEST* or *INTEST* mode. Additional modes can be implemented, for example, to temporary include an internal *Scan Path* in the *Boundary Scan* such that the internal flip flops of the chip can be observed and controlled by the *Boundary Scan*.

A quite different application of a *Boundary Scan* is presented in [15.14]. Here the BS cells are extended in such a way that for a special mode they represent a linear field of automata. Then at every clock cycle a machine state depends on the states of adjacent cells, thus creating a pseudo-random sequence of test patterns, which in addition to the primary inputs can also be applied to the *Scan Path*.

15.8 IDDQ Test

With the **IDDQ** test method one determines the power consumption of a chip at a stable state (**quiescent current**). Then a chip is called faulty if there is a significant increase of current for some test pattern. Such an increase of current might be owed to a physical defect of the chip. Nevertheless, it is conceivable that despite the defect the functional behavior of the chip is correct. Thus the method of *IDDQ testing* is rather a **defect oriented** method than an **error oriented** method. It may also be used to improve the *reliability* of chips (section 15.10). Within the model of IDDQ faults all conceivable faults are considered which may increase power consumption. For example, the fault model includes **bridging faults**, **gate oxide shorts**, **transistor stuck on faults**, and some **stuck at faults**.

15.8.1 Functional Undetectable Defects

With *functional tests* one tries to stimulate a fault and to propagate resulting erroneous signals to a primary output. But for an **IDDQ test** one uses the fact that, when stimulated, many faults cause an increased power consumption, such that it is not necessary to propagate the fault, but only to measure the flow of current for the chip.

This method will be demonstrated for the *example* of a CMOS inverter shown in figure 15.61. In any stable state exactly one of the two transistors is conducting and therefore the output y is either connected to *VDD* or to *VSS*. Apart from parasitic effects there is only a flow of current within the transient state of switching signals.

In the case of a *stuck on* fault at the transistor T_1 for the input $x = 1$ both transistors are conducting and there is an increased quiescent current between *VDD* and *VSS*. Depending on the resistance of transistor channels, the value of the output signal y results from the voltage divider built by T_1 and T_2.

It is also possible that despite the fault the voltage at the output y may be interpreted as the correct logic value. Thus the logical behavior of the circuit may be correct. However, because of the effect of *electron migration* the fault may later cause failures after a longer period of operation. Therefore on using the *IDDQ test* it is possible to detect defects that can not yet be detected by functional tests.

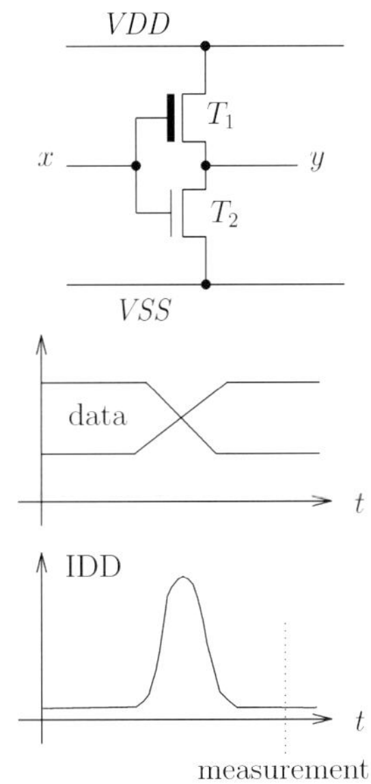

Fig. 15.61 Transistor netlist of an inverter in complementary CMOS logic

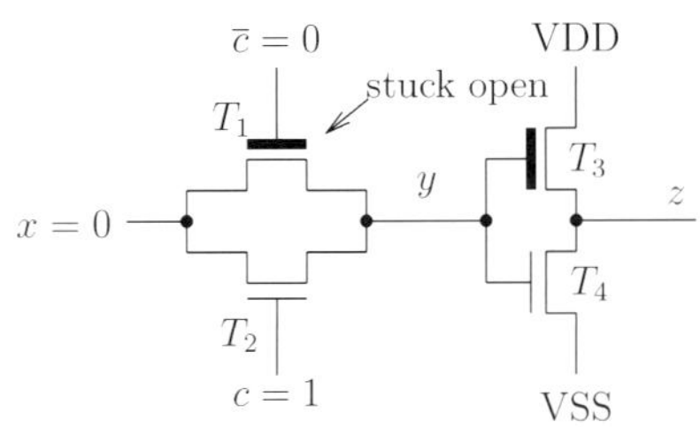

Fig. 15.62 Example of a circuit such that the fault 'T_1 stuck open' causes an erroneous current through T_3 and T_4

An increased current can even be caused by a transistor *stuck open* fault. This is shown by the *circuit example* given in figure 15.62. Here the p-transistor T_1 and the n-transistor T_2 form a **transmission gate**, transmitting the input value x to the internal node y for of $c = 1$. Here the n-transistor is well suited to transmit the value 0 and the p-transistor is well suited to transmit 1. With the fault 'T_1 stuck open' and the input $x = 1$ there is a bad transmission of '1', only generating a 'weak 1' at node y because the voltage at y is reduced by the threshold voltage of transistor T_2. As a consequence it may happen that the transistor T_3

of the succeeding inverter is not perfectly locked, and therefore there is an erroneous current between *VDD* and *VSS*.

In a similar way also interruptions of wires may cause an increased power consumption because of undefined signal levels. Often such faults are also detected by functional tests as *stuck at* faults. Further faults that cause an increase of quiescent current are **bridging faults**, and **gate oxide shorts**.

15.8.2 IDDQ Test Patterns

An important advantage of the IDDQ test is that generating test patterns is much simpler than for functional tests. This shall be demonstrated for the *example* of a hard combinatorial *bridging* fault (section 15.4.1) at two wires α and β. To detect the fault by an IDDQ test it is sufficient to drive both signals α and β by different values. This will cause a high current because of the short circuit. Thus the characteristic equation for test patterns simply becomes $\alpha \oplus \beta = 1$ and is correct for all kinds of *bridging* faults.

For an automatic IDDQ test pattern generation with common test pattern generators it is very easy to model the *bridging* fault. One only has to extend the circuit by an *XOR* gate performing $\alpha \oplus \beta$ and to introduce the output of the *XOR* gate as an additional primary output to be tested for 'stuck at-0'. Each pattern producing the signal 1 at the new output can be used as a test pattern.

Since for computing IDDQ test patterns fault propagation can be omitted, there are more possible test patterns for a fault than for functional tests. Thus an IDDQ test needs fewer test patterns. Unfortunately, in practice a sufficiently precise current measurement is relatively time consuming, and therefore to limit test effort one has to keep the number of IDDQ measurements as small as possible. Thus for a given number of measurements one determines a set of test patterns obtaining a maximal fault coverage.

Since the model of *stuck at* faults does not determine a unique kind of physical defect, some *stuck at* faults might increase quiescent current, whilest others do not. If, for example, an 'α s-a-0' fault is owed to a short between wire α and the supply line *VSS* then stimulating the fault by $\alpha = 1$ will, of course, produce high current. But this may not be true for an interruption of a wire. Thus the IDDQ method cannot replace functional tests but can extend such tests to improve defect coverage. One should never use IDDQ measurements to reduce the number of functional test patterns.

As an extreme example we consider the **red/black coloring** method presented in section 15.4.6. If all *stuck at* faults could be detected by IDDQ measurements then the circuits obtained would be completely testable for *stuck at* faults with only two test patterns.

Multiple faults do not cause additional problems for IDDQ testing. For example it can be shown that when simple design rules are respected [15.21] every test pattern detecting a single *bridging fault* *F* will also detect any multiple *bridging fault* with *F* is involved.

IDDQ test pattern generation also has to calculate the intensity of quiescent current. For this one may use an extended *switch level* simulation also considering realistic resistances of transistors. With a given accuracy of measurement such a simulation can be used to determine whether the erroneous current is sufficiently high to be recognized. On the other hand, such simulations can also be used to determine the accuracy needed for an IDDQ measurement.

In order to receive meaningful results IDDQ tests should be restricted to such test patterns producing a low power consumption for correct chips. Section 15.8.5 considers in more detail which properties have to be satisfied so that a circuit is well suited for IDDQ tests.

15.8.3 IDDQ Threshold Value

One problem with IDDQ measurements is that of deciding for which amount of quiescent current a chip should be classified as faulty. Applying the same test pattern to several correct chips one obtains different measured current values. Figure 15.63 shows the typically obtained standard distribution (Gauss curve) for current measurement.

The average value of that distribution denotes the typical quiescent current of a correct chip. But because of deviations during manufacture actual values will differ from the expected value. Repeating the measurement for several chips all containing the same IDDQ fault, the curve of the distribution would be shifted to the right by the amount of the erroneous current.

The threshold value for an IDDQ measurement should be determined according to the expected erroneous current. In figure 15.63 this is the value I_0. The area below thc distribution curve to the right of I_0 then corresponds to the probability that a correct chip is classified as defective. This probability can be used to define the threshold value I_0. Then one has to compare the costs of both kinds of erroneous decisions: What are the expected costs if a defect chip remains undetected and what does is cost to classify a correct chip as faulty?

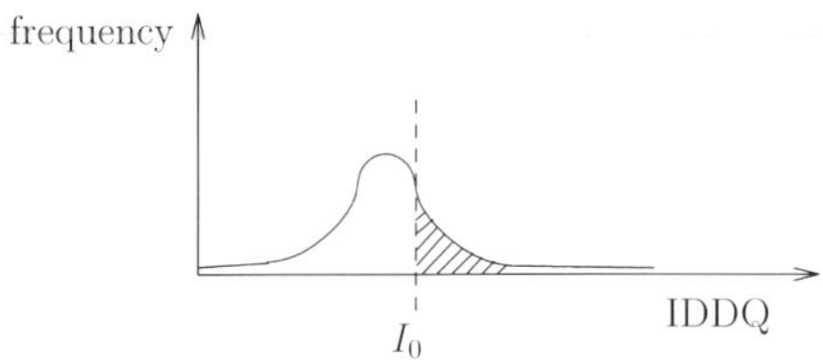

Fig. 15.63 Number of measured IDDQ values applying the same test pattern too many correct chips (The shaded area corresponds to the probability of classifying a correct chip as faulty because its quiescent current is higher than I_0.)

15.8.4 Principle of IDDQ Measurement

In figure 15.64 several techniques are shown to measure a quiescent current [15.37]. When inserting a resistor into the power supply of a chip the voltage drop at the resistor is a measure for the flow of current. But since such a resistor within a supply line will reduce the applied voltage it has to be shorted by a transistor for normal operation of the chip.

As an alternative approach the resistor can be replaced by a capacitor. Again, for normal operation it is shorted and unloaded. For testing, the transistor is opened and the capacitor is loaded by the quiescent current. Then, after a fixed period of waiting, the voltage at the capacitor is a measure for the quiescent current. If it extends a certain threshold value the chip fails the IDDQ test.

It is common to all approaches that IDDQ measurement with sufficient accuracy is relatively time consuming. Thus in practice one has to restrict the number of measurements to a small subset of IDDQ test patterns, thereby obtaining a significant improvement of defect coverage.

In [15.23] a measurement technique is proposed using the fast *pin electronics* of functional test equipment to observe the power supply also. This way it is possible to perform an IDDQ test without hardware overhead. As shown in figure 15.65, one uses several driver channels of the tester for the power supply of the chip. One of them is configured to become a monitor channel. For IDDQ measurement the *VSS* drivers are turned into the high impedance state. Then after a fixed time of waiting the increase of voltage observed at the pin *VSS* (figure 15.66) is again a measure for the quiescent current.

As an alternative to external IDDQ tests the current can also be measured on the chip using special sensors (**BICS**: Built In Current Sensor) [15.28], [15.47]. Since the current measurement should not obtain too much influence over the performance of normal operation, the circuit is partitioned into several sub-circuits which can be observed separately. With this technique self-tests are also possible.

15

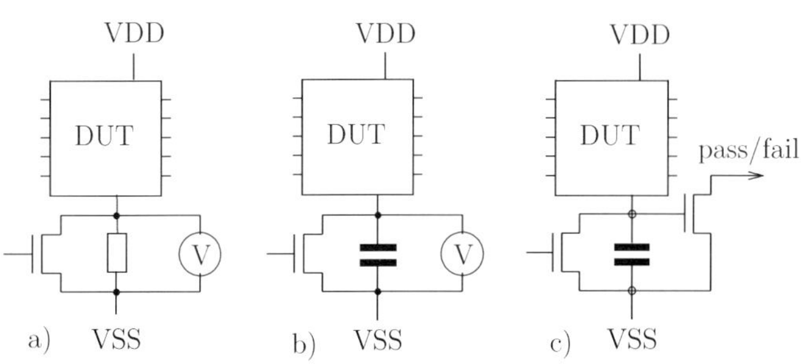

Fig. 15.64 Techniques for IDDQ measurement:
a) current measurement by voltage drop at a resistor;
b) loading a capacitor by the quiescent current;
c) same procedure producing a digital test result.

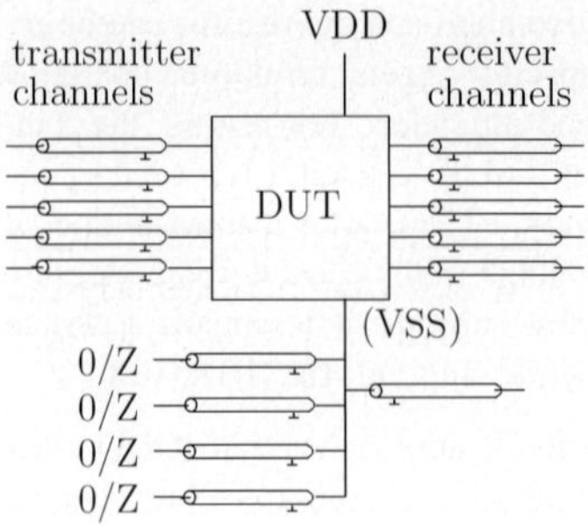

Fig. 15.65 IDDQ measurement using a functional tester

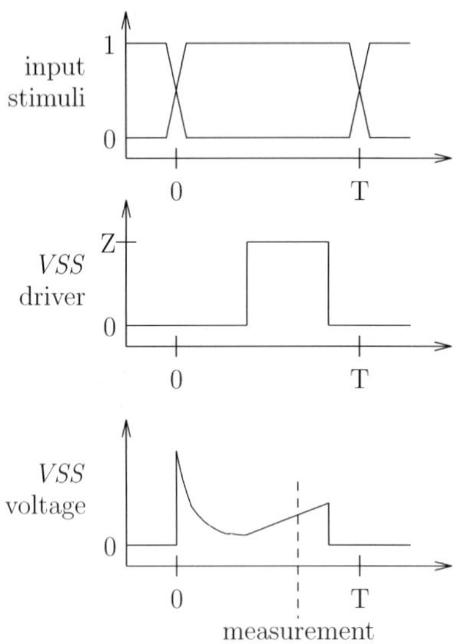

Fig. 15.66 Timing diagram of an IDDQ measurement

15.8.5 IDDQ Testability

In order to apply an IDDQ test the circuit has to satisfy special properties. For example, as mentioned above, the correct circuit should have a very low quiescent current such that the erroneous current is easily detectable.

Therefore the circuit may not use oscillators, and whenever there are dynamic storage blocks they have to be separated for the test. Also *pull up* resistors have to be disabled for the test mode, and for *pad* drivers, analog cells, and bipolar subcircuits a separate power supply is needed because they typically have a high power consumption.

Buses always should carry well defined signals. Otherwise additional drivers have to be provided to force buses to default values whenever there is no actual write operation. Thereby, of course, conflicting write operations must be avoided.

Because of the necessary time for exact current measurement the circuit must be able to work at a slow clock rate. Therefore if the chip itself is monitoring the system clock this must be deactivated for the test.

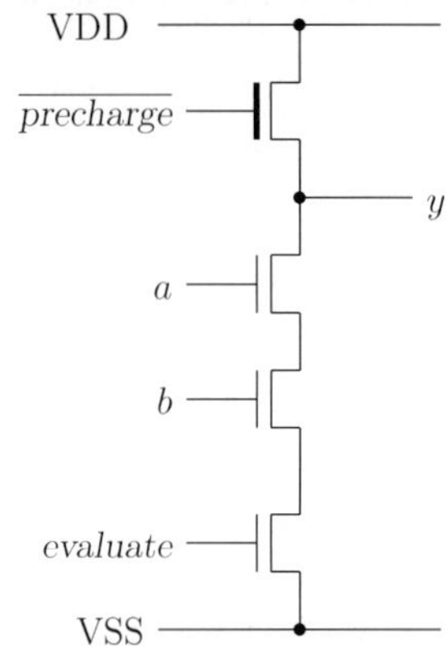

Fig. 15.67 Example of a NAND gate in dynamic logic (The signal $\overline{precharge}$ is used to preload output node y to logic 1 and with evaluate = 1 output y can be discharged depending on the input values a and b. Thereby signals $\overline{precharge}$ and evaluate should not overlap.)

Also circuits in **dynamic logic** are not suited to IDDQ testing. Apart from problems concerning the clock period this is because for faults of the evaluation logic neither in the preload phase ($\overline{precharge}$ = 1) nor in the evaluation phase (*evaluate* = 1) is there a path between *VDD* and *VSS*.

Furthermore, for **regular structured circuits** such as storage blocks, IDDQ tests are not of interest because there are already specialized tests available with high defect coverage.

15.9 Further Parameter Tests

Since one reason for an increased quiescent current is that of illegal signal levels, the observation of voltage levels at critical signals is also an alternative to IDDQ tests. For example, a simple sensor for voltage levels consisting of only three inverters is presented in [15.43]. Figure 15.68 shows that this **BIVS** (Built in Intermediate Voltage Sensor) only produces an output of '1' if the input signal can not clearly be interpreted as logic 0 or logic 1.

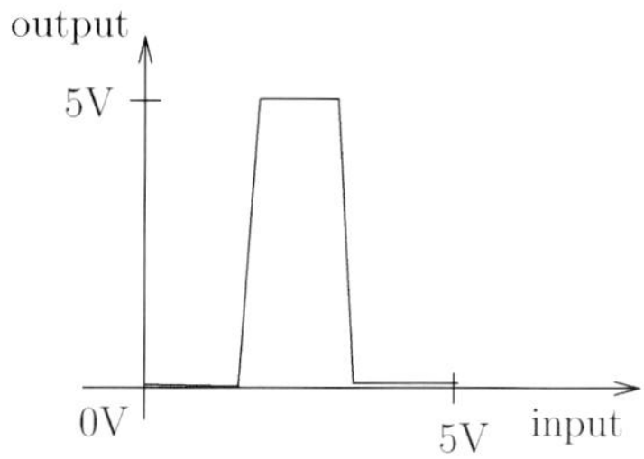

Fig. 15.68 Behavior of a BIVS to detect illegal voltage levels

In section 15.8 it has been discussed which defects may increase the quiescent current of a circuit. Of course faults can also cause an increased current during the phase transient states. To discover such effects one uses **IDDT** tests, observing *transient current*.

For example, in [15.42] an *on chip* sensor is presented to measure current peaks during switching operations. In particular, it is suitable for chips with low power supply.

A combination of IDDQ and IDDT tests consists of measuring the total energy needed by a chip to perform a set of test patterns. For this task a method is described in [15.39] such that the chip under test obtains its power from a capacitor which is reloaded whenever its voltage drops under a critical limit. Then the number of test patterns which can be executed between two reload operations defines an **energy signature** which can be used to decide whether the chip is correct. Advantages of this method are a fast test procedure and a simple measurement process, because IDDQ and IDDT currents are accumulated.

15.10 Reliability of Chips

Of course a chip should not only behave correctly immediately after production but should also work correctly for a long period of use. Thus one requires a high *reliability* of chips. In figure 15.69 the typical time of failure during operation is sketched. Since for any kind of technical products one can determine the average span of life, the failure rate rises when at the end of the expected span of life. But more interesting is that relatively many chips fail during the first period of operation. This is because of physical **defects** which do not immediately cause an erroneous behavior but cause increasing deviations from the optimal behavior during operation.

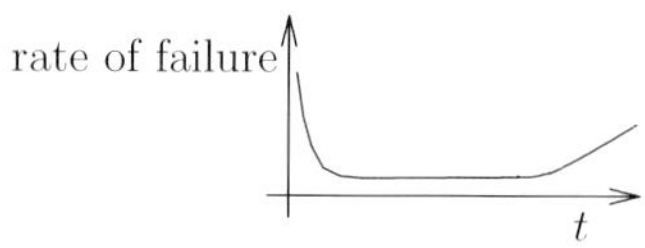

Fig. 15.69 Typical time of failure for chips during operation

Therefore **burn in** tests are used to test chips under extreme conditions. For example, one uses high temperature and an increased voltage of the power supply. Other possible stress factors are, for example, mechanical vibrations or extreme variations in temperature. Using such extreme environmental conditions one can simulate a longer period of operation such that early failures will already occur during the *burn in* test and not during the period of use. After the *burn in* test the usual chip tests are repeated to eliminate chips which have failed in the meantime. In particular, an *IDDQ test* is useful after *burn in*, because during *burn in* IDDQ faults may become more obvious.

Since *burn in* tests are relatively time consuming and expensive one is interested in detecting unreliable chips by a simpler test. One such possibility is the IDDQ test. Indeed, experimentally it can be shown that chips with increased quiescent current will fail earlier [15.29]. In particular, many of the chips which do not pass an IDDQ test will not even pass a functional test after applying a *burn in* phase. As a consequence applying an IDDQ test will improve the reliability of chips, but unfortunately the *burn in* test can not be completely replaced by an IDDQ test.

When using an IDDQ test to improve reliability, then in accordance with the burn in test one should use a supply voltage as high as possible. In this way defects may become more obvious. On the other hand, one should not increase temperature because this causes an increase of quiescent current that does not correspond to a shorter span of life.

15.11 References

An introduction to testability of digital circuits can be found in [15.48]. [15.1] gives a more detailed description to the topic. Both books also contain exercises with solutions. [15.52] also considers algorithmic concepts and complexity questions. In [15.44] many different techniques are discussed and there are many references to test literature up to 1986.

For the method of IDDQ testing a practical presentation is given in [15.37]. Furthermore, [15.29] contains a composition of publications from technical journals concerning this topic. A detailed presentation of the *Boundary Scan* architecture is given in [15.31]. This book also contains a collection of technical articles about related topics.

Further information about the current development in the area of design for testability can be found, for example, in papers presented at the following conferences: *Design Automation Conference* (*DAC*), *International Conference on Computer Aided Design* (*ICCAD*), *European Design and Test Conference* (*ED&TC*), *Design, Automation & Test in Europe* (*DATE*) and *International Conference on Computer Design* (*ICCD*).

[15.1] *Abramovici, M.; Breuer, M. A.; Friedman, A. D.*: Digital Systems Testing and Testable Design. IEEE Press, 1990

[15.2] *Agarwal, V. K.*: Easily Testable PLA Design. In VLSI Testing, T. W. Williams (Editor), North-Holland, 1986

[15.3] *Ballew, W. D.; Streb, L. M.*: Board-Level Boundary Scan: Regaining Observability with an Additional IC. In IEEE Transactions on Computer-Aided Design, Vol. 11, No. 1, S. 68–75, 1992

[15.4] *Bolchini, C.; Salice, F.; Sciuto, D.*: A Novel Methodology for Designing TSC Networks based on the Parity Bit Code. ED&TC'97, 1997

[15.5] *Cirit, M. A.*: Switch Level Random Pattern Testability Analysis. 25th ACM/IEEE Design Automation Conference, S. 587–590, 1988

[15.6] *Cockburn, B. F.; Brzozowski, J. A.*: Switch-level testability of the dynamic CMOS PLA. IEEE Transactions on Computers, Vol. C-30, No.1, 1981

[15.7] *David, R.*: A Totally Self-Checking 1-Out-of-3 Checker. IEEE Transactions on Computers, Vol. C-27, No. 6, S. 570–572, 1978

[15.8] *Dufza, C.; Viallon, H.; Chevalier, C.*: BIST Hardware Generator for Mixed Test Scheme. ED&TC'95,1995

[15.9] *Dufza, C.; Zorian, Y.*: On the Generation of Pseudo-Deterministic Two-Patterns Test Sequence with LFSRs. ED&TC'97,1997

[15.10] *Flottes, M. L.; Hammad, D.; Rouzeyre, B.*: High-Level Synthesis for Easy Testability. ED&TC'95, 1995

[15.11] *Fujiwara, H.; Kinoshita, K.*: A Design of programmable Logic Arrays with Universal Tests. IEEE Transactions on Computers, Vol. C-30, No.1, 1981

[15.12] *Fummi, F.; Dciuto, D.; Serra, M.*: Sequential Logic Minimization Based on Functional Testability. ED&TC'95, 1995

[15.13] *Fujiwara, H.*: Design and Analysis of Fault-Tolerant Digital Systems. The MIT Press, 1985

[15.14] *Gloster, C. S.; Brglez, F.*: Boundary Scan with Built-in Self-Test. IEEE Design & Test of Computers, Februar 1989, S. 36–44, 1989 (also in [15.31])

[15.15] *Gu, X.; Larsson, E.; Kuchinski, K.; Peng, Z.*: A Controller Testability Analysis and Enhancement Technique. ED&TC'97, 1997

[15.16] *Huang, L.-R.; Kuo, S.-Y.; Chen, I.-Y.*: A Gauss-Elimination Based PRPG for Combinational Circuits. ED&TC'95, 1995

[15.17] *Johnson, B. W.*: Design and Analysis of Fault-Tolerant Digital Systems. Addison-Wesley, 1989

[15.18] *Jha, N. K.*: A Totally Self-Checking Checker for Borden's Code. In IEEE Transactions on Computer-Aided Design, Vol. 8, No. 7, S. 731–736, 1989

[15.19] *Kodandapani, K. L.; Pradhan, D. K.*: Undetectability of Bridging Faults and Validity of Stuck-At Fault Test Sets. IEEE Transactions on Computers, January 1980, S. 55–59, 1980 (also in [15.29])

[15.20] *Kagaris, D.; Tragoudas, S.*: Cellular Automata for Generating Deterministic Test Sequences ED&TC'97,1997

[15.21] *Lee, K.-J.; Breuer, M. A.*: Constraints for Using IDDQ Testing to Detect CMOS Bridging Faults. IEEE VLSI Test Symposium, S. 303–308, 1991, (also in [15.29])

[15.22] *LaPaugh, A. S.; Lipton, R. J.*: Total Stuck-at-Fault Testing by Circuit Transformation. 20th Design Automation Conference, 1983

[15.23] *Laquai, B.; Richter, H.; Werkmann, H.*: A Production-oriented Measurement Method for Fast and Exhaustive Iddq Tests. ED&TC'97, 1997

[15.24] *Lin, X.; Pomeranz, I.; Reddy, S. M.*: Full Scan Fault Coverage With Partial Scan. DATE'99, 1999

[15.25] *Mao, W. H.; Ciletti, M. D.*: DYTEST: A Self-learning Algorithm Using Dynamic Testability Measures to Accelerate Test Generation. 25th ACM/IEEE Design Automation Conference, 1988

[15.26] *Marouf, M. A.; Friedman, A. D.*: Efficient Design of Self-Checking Checker for any m-Out-of-n Code. IEEE Transactions on Computers, Vol. C-27, No. 6, S. 482–490, 1978

[15.27] *Marchok, T. E.; El-Maleh, A.; Maly, W.; Rajski, J.*: Complexity of Sequential ATPG. ED&TC'95, 1995

[15.28] *Maly, W.; Patyra, M.*: Built-in Current Testing. IEEE Journal of Solid-State Circuits, S. 425–428, 1992 (also in [15.29])

[15.29] *Malaiya, Y. K.; Rajsuman, R.*: Bridging Faults and Iddq Testing. IEEE Computer Science Press, 1992

[15.30] *Muehdorf, E. I.; Savkar, A. D.*: LSI Logic Testing – An Overview. IEEE Transactions on Computers, Vol. C-30, No. 1, 1981

[15.31] *Mander, C. M.; Tulloss, R. E.*: The Test Access Port and Boundary-Scan Architecture. IEEE Computer Society Press Tutorial, 1989

[15.32] *Mucha, J.*: Testprobleme bei höchstintegrierten Schaltungen. Springer Informatik-Fachberichte, IFB 89, S. 37–46, 1984

[15.33] *Rao, T. R. N.; Fujiwara, E.*: Error-Control Coding for Computer Systems. Prentice Hall, 1989

[15.34] *Rajsuman, R.; Jayasumana, A. P.; Malaiya, T. K.*: CMOS stuck-open fault detection using single test patterns. 26th ACM/IEEE Design Automation Conference, S. 714–717, 1989

[15.35] *Paschalis, A.; Gaitanis, N.; Giszopoulos, D.; Kostarakis, P.*: A Totally Self-Checking 1-out-of-3 Code Error Indicator. ED&TC'97, 1997

[15.36] *Pomeranz, I.; Reedy, S. M.*: On the use of Reset to Increase the Testability of Interconnected Finite-State Machines. ED&TC'97, 1997

[15.37] *Rajsuman, R.*: Iddq Testing for CMOS VLSI. Artech House, 1995

[15.38] *Reddy, S. M.*: A Note on Self-Checking Checkers. IEEE Transactions on Computers, October 74, S. 110–1102, 1974

[15.39] *Rius, J.; Figueras, J.*: Exploring the Combination of IDDQ and IDDT-Testing: Energy-Testing. DATE'99, 1999

[15.40] *Savir, J.*: Good Controllability and Observability Do Not Guarantee Good Testability. IEEE Transactions on Computers, Vol. C-32, No. 12, S. 1198–1200, 1983

[15.41] *Smith, J. E.; Metze, G.*: Strongly Fault Secure Logic Networks. IEEE Transactions on Computers, Vol. C-27, No. 6, S. 491–499, 1978

[15.42] *Stopjaková, V.; Manhaeve, H.; Sidiropulos, M.*: On-chip Transient Current Monitor for Testing of Low-Voltage CMOS IC. DATE'99, 1999

[15.43] *Tang, J.-J.; Lee, K. J.; Liu, B.-D.*: Built-in Intermediate Voltage Testing for CMOS Circuits. ED&TC'95, 1995

[15.44] *Tsui, F. F.*: LSI/VLSI Testability Design. McGraw-Hill Book Company, 1986

[15.45] *Wurth, B.; Fuchs, K.*: A BIST Approach to Delay fault Testing with Reduced Test Length. ED&TC'95, 1995

[15.46] *Wehn, N.; Glesner, M.; Caesar, K.; Mann, P.; Roth, A.*: A Defect-Tolerant and Fully Testable PLA. 25th ACM/IEEE Design Automation Conference, 1988

[15.47] *Wunderlich, H.-J.; Herzog, M.; Figueras, J.; Carrasco, J. A.; Calderón, A.*: Synthesis of Iddq Testable Circuits Integrating Built-in Current Sensors. IEEE Journal of Solid-State Circuits, S. 425–428, 1992

[15.48] *Wojtkowiak, H.*: Test und Testbarkeit digitaler Schaltungen. B. G. Teub"-ner Verlag, 1988

[15.49] *Williams, T. W.*: Design for Testability. VLSI Testing, T. W. Williams (Editor), North-Holland, 1986

[15.50] *Williams, T. W.; Parker, K. P.*: Design for Testability – A Survey. IEEE Transactions on Computers, Vol. C-31, No. 1, 1982

[15.51] *Wunderlich, H.-J.*: Probabilistische Verfahren für den Test hochintegrierter Schaltungen. Informatik-Fachberichte 140, Springer-Verlag, 1987

[15.52] *Wunderlich, H.-J.*: Hochintegrierte Schaltungen: Prüfgerechter Entwurf und Test. Springer-Verlag, 1991

Overview EDA
1, 2

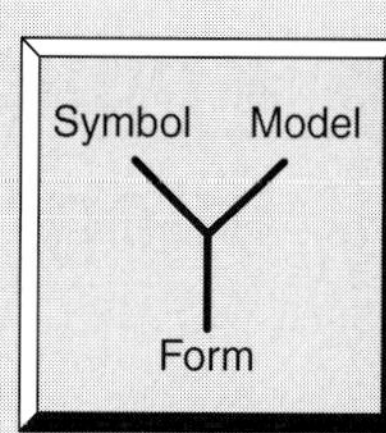

Symbolic Design
3

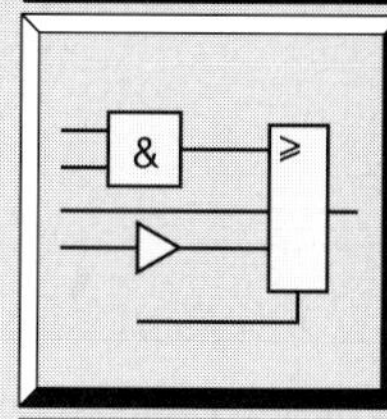

High Level Language Design
4, 5, 6, 7, 8

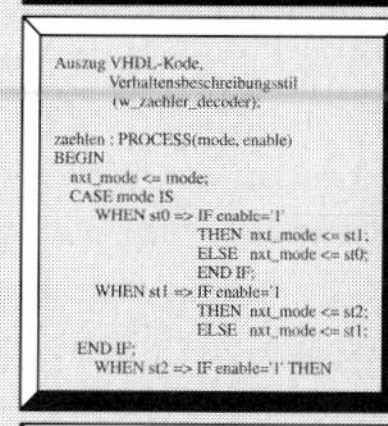

Modelling and Verifications
9, 10, 11, 12, 13, 14, 15

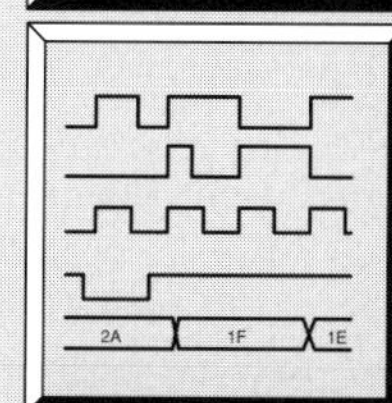

Implementation
16, 17, 18, 19, 20, 21, 22, 23, 24, 25

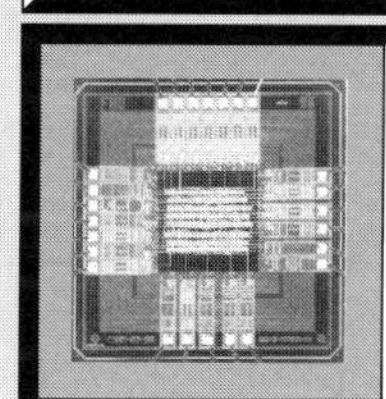

Tutorial
26

Appendix
A, B, C, D, E

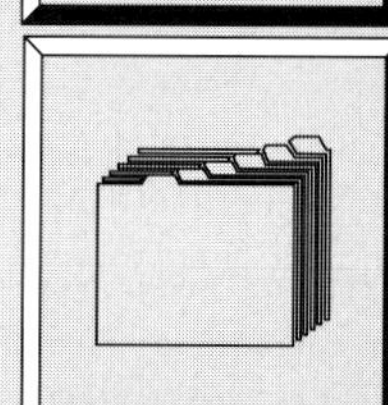

16 Application Specific Integrated Circuits (ASICs)

MARTIN RIEGER

16.1 Introduction

The abbreviation ASIC stands for Application Specific Integrated Circuits. Compared with standard circuits an ASIC is designed and manufactured according to specifications resulting from its application. The ASIC user is able to specify the circuit function at design level and/or by means of configuration options. If the function of an IC is only controlled by software then it is not called an ASIC [16.7].

Figure 16.1 shows a possible **classification of monolithically integrated circuits**. Examples of standard ICs are operational amplifiers (standard analog), TTL and CMOS Logic ICs (standard logic), microprocessors, RAM and ROM. Examples of ASICs are Field Programmable (FPGAs), Gate Arrays, Macro Cell, Standard Cell ICs and Full Custom ICs [16.5].

In some cases the classification standard IC/ASIC is difficult. If an ASIC design is so successful that it sells in large numbers it finally becomes a standard IC. Also the classification ROM/FPGA is not always clear.

16.1.1 Technological Characteristics of ASICs

Some technological characteristics show the dynamic progress in the field of ASICs.

The so called **feature size** of a certain technology is a measure of the length of the smallest MOS transistor, sometimes also for the minimal allowable distance between two connection lines on the chips. The feature size λ is in the order of the length of a transistor. With decreasing feature size the area needed is reduced and the transistor performance improves, at the same time the manufacturing will be more demanding.

The silicon chip of the IC, the so called **die** varies in its area from some mm^2 to several cm^2. The bigger the die and the smaller λ the more transistors can be integrated into one IC.

As a measure of the complexity of an IC one can consider the number of **gate equivalents**. A gate equivalent is equal to a NAND gate with two inputs.

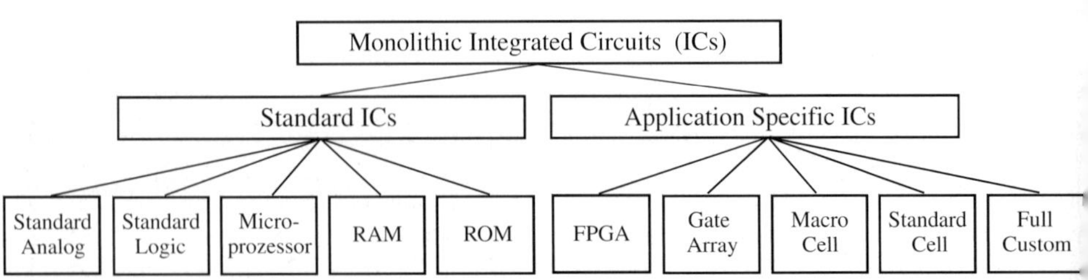

Fig. 16.1 Classification of monolithically integrated circuits

The **Semiconductor Industry Association** (SIA), an organization supported by all major semiconductor manufacturers, forecasts the progress of the semiconductor technology by means of so called **road maps** and specifies the needed processes and installations [16.9]. Table 16.1 shows the road map for microprocessors from which it can be derived that the integration density and the speed will allow one to build ASICs with more complex functionality and lower prices. The SIA considers the ASICs as one of the motors of technology with respect to integration density, performance, and package. For this compare chapter 1 and 2.

Table 16.1 Road map of semiconductor technology

Year of first production	2001	2005	2010	2016
feature size [nm]	150	80	50	25
max. clock rate [GHz]	1.7	5.1	11.5	28.7

16.1.2 Design Goals for ASICs

Compared to system realizations with standard ICs ASICs show several advantages. The most important design goals for a system are [16.7]:

- cost effective solution;
- low design effort with respect to time and cost;
- high performance, i.e., functionality, speed;
- low power dissipation;
- small volume and weight;
- high reliability;
- low cost for mounting and test.

In order to reach the goals with ASICs certain conditions have to be fulfilled. The methods of design and realization of ASICs the so called **design styles** have to be chosen optimally with respect to the system environment. The preferred design style is dependent on the expected number of parts, on the design time provided, and on the performance desired.

With respect to the most important goal of cost effectiveness the cost of design and manufacturing have to be considered. A related analysis will be carried out in section 16.3.

There are trade offs between different goals, e.g., higher performance generally demands a higher design effort and also results in a higher power dissipation.

16.1.3 Case Study: CD Player, ASIC as Key Component

Figure 16.2 shows a block diagram of a CD player without showing the operating and display unit [16.12]. The CD player consists of a 'mechadeck' comprising mechanical and optical components, of an ASIC for signal processing and control tasks, and of several standard ICs. The player is able to communicate via a bus interface with the processor of the operating and display unit. A further interface to the outer world is formed by the stereo audio signal.

The CD rotates within the 'mechadeck'. The 'pickup' to be found there also comprises the laser and six photo-diodes for reading the CD. The position of the pickup has to be controlled with respect to track and z-coordinate (focus).

The core of the CD player is formed by the ASIC 'CD one Chip', which takes over all important functions with respect to signal processing and position control. Embedded into the ASIC are:

- a standard microprocessor;
- a signal processor (DSP);
- memory (SRAM);
- a digital analog converter (Audio-DAC);
- a preamplifier for the read signals of the photo diodes.

The rather complicated decoding and error-correction of the read signal by means of the Reed-Solomon algorithm is realized in a hard wired form. The DSP is partly needed for the complicated calculation of the control signals for track and focus from the read signals. The ASIC contains mechanisms for automatic adjustment which enables the construction of a nearly adjustment-free CD player.

The CD player comprises as standard ICs an operational amplifier and some driver ICs. This poses the question of, why those functions realized with standard ICs are not integrated into the ASIC 'CD one Chip'. On one hand, the technology used in the ASIC is not suitable for some functions, on the other hand, the used standard ICs are so cheap that a further integration does not make sense.

The ASIC 'CD one Chip' meets many of the design goals discussed in the previous paragraph. The IC was designed in full custom style, which

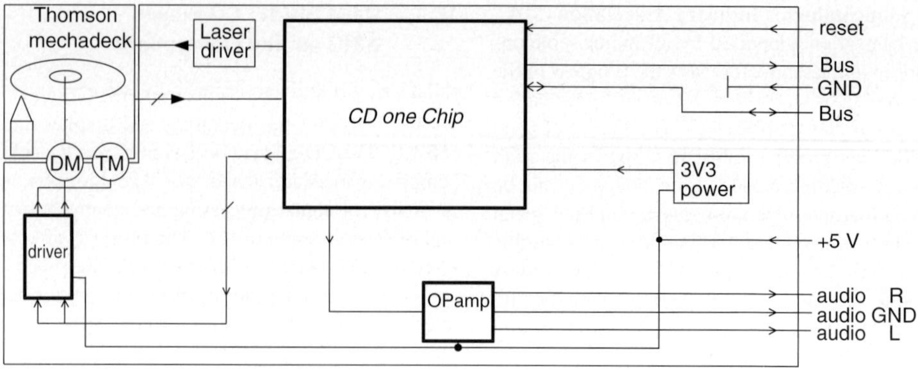

Fig. 16.2 Block diagram of a CD player

meant a high design effort, but led to a very compact layout. In a 0.5 µm technology only 30 mm^2 chip area was needed to realize the complex functions. Taking a sales volume of the 'CD one Chip' in the order of several millions into account this approach led to a cost effective solution.

Compared with a solution with standard ICs the advantages are:

- lower number of components;
- lower power dissipation;
- smaller volume and weight;
- higher reliability;
- lower cost for mounting and test.

In addition, by using an ASIC a certain protection against copying the design by a competitor is given.

Compared with a predecessor model with standard ICs the discussed CD player needs 30 % less PCB area, the overall cost savings are 20 %.

16.2 Design Styles

In this section the different design styles (fig. 16.1) will be presented and discussed with respect to their advantages and disadvantages.

The term semi custom ASICs covers the following different design styles:

- standard cells;
- macro cells;
- gate array;
- PLD;
- FPGA.

Compared with full custom ASICs the design of semi custom ASICs predominantly uses predesigned cells from libraries.

Often several design styles are used simultaneously when designing an ASIC, e.g., part of the IC can consist of macro cells, the rest of standard cells.

An integrated circuit is manufactured by means of 12 to 25 **masks**, dependent on the complexity of the technology used. More than half of those are needed to structure transistors and other components. The rest define the connections via metal lines and contacts. The choice of the design style depends on whether all masks for a certain ASIC have to be custom specific or only a part of them. In the latter case wafers can be produced with non-custom specific masks and kept in stock, which saves design and manufacturing time.

16.2.1 Full Custom Design Style

For a full custom ASIC, the design and layout are specific. In the extreme case every component is inserted individually in design and layout. This maximum design freedom enables an optimal design with respect to performance, size and power consumption. The cell layout is possible in whatever form is required. For a full custom ASIC all masks are specific.

A full custom design shows the **advantages** of optimal performance, optimal component density, and flexible forms. With a very high number of parts this design style is most economical.

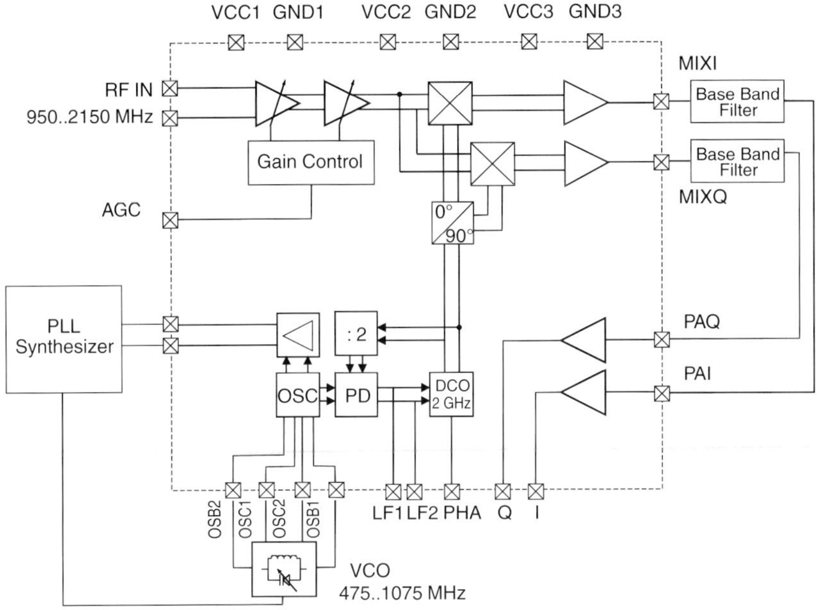

Fig. 16.3 Block diagram of a full custom ASIC for direct conversion of digital satellite TV signals

The **disadvantages** of the full custom design style are:

- high design effort with respect to time and cost;
- high design risk using untested cells.

Design and layout of a full custom ASIC demands a lot of experience. The full custom design style is most common with mixed signal and analog applications.

As an example of a full custom ASIC an IC for direct conversion of digital satellite TV signals will be considered, where the block diagram is shown in fig. 16.3 [16.1]. The incoming signal at RF IN is first amplified and then converted into two orthogonal base band signals by means of two mixers, which are fed with two orthogonal oscillator signals. These signals are generated by an internal delay cell oscillator (DCO), which is synchronized by means of an internal phase detector (PD) with an oscillator connected to an external tank circuit. After external base band filtering the signals are fed back into the IC and amplified in order to reach a sufficiently high level for digitization. The input signals and the oscillator signals of the DCO are in the frequency range from 950 to 2,150 MHz. The specifications with respect to linearity, to tuning range of the oscillators, and to deviation from quadrature were so tight that only a full custom design was possible. The design realized by means of a 20 GHz process demanded a very high design effort with several iteration steps.

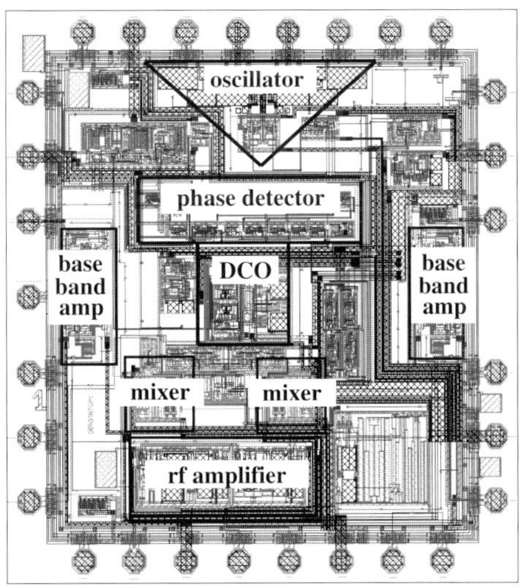

Fig. 16.4 Layout of the direct conversion IC of fig. 16.3

Figure 16.4 shows the layout of the direct conversion IC. The individual cell outlines typical for full custom design can be recognized. In order to design such a layout a lot of experience is needed because it is so difficult to meet the tight specifications regarding cross talk and quadrature deviation. Even in this pronounced full custom layout repeating cells can be recognized, which is caused partly by symmetry.

The design was accomplished without using pre-designed cells from a library, but, of course, experience gained from previous designs was of great value.

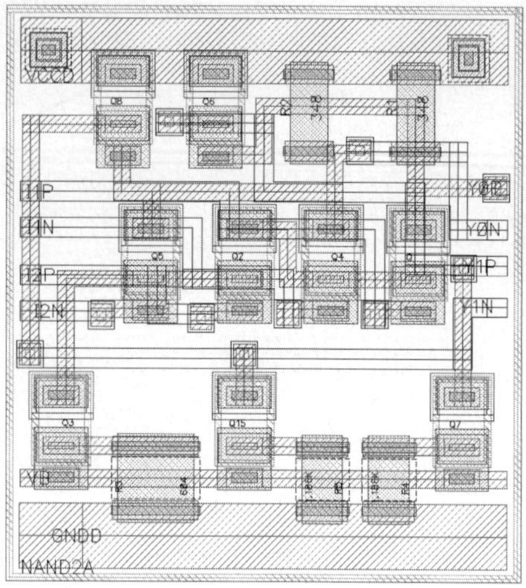

Fig. 16.5 NOR standard cell in ECL technology

16.2.2 Standard Cells

The term standard cell signifies a pre-defined block, which in general represents a logic function of low complexity (e.g., NAND, NOR or FlipFlop). Figure 16.5 shows a NOR standard cell in ECL technology [16.8] which is used in a standard cell layout (fig. 16.6). The power supply lines are the lines above and below the logic circuit. Input and output are to be found on the left hand and right hand side, respectively.

Standard cells are designed as full custom designs and stored in libraries after having been tested and optimized. The libraries can either be prepared individually or be provided by the manufacturer of the technology, or can be bought from a suitable company. The placement of standard cells is principally free but normally they are placed in **rows of cells** with 'running through' supply lines. Therefore the cells have to have the same height (see fig. 16.6). Figure 16.6 shows in the upper left part some full custom cells which can be integrated without complying with the standard cell pattern. All masks are specific for a certain ASIC.

The standard cell design is supported by one or several libraries of optimized base cells for which simulation and layout data are available. Some cells can be parameterized. Design and layout in standard cell design are well supported by EDA tools. For details of libraries see chapter 19.

Figure 16.7 shows the top layout of a circuit which consists of standard cells and macro cells discussed in paragraph 16.2.3. Several cell rows can be distinguished. Between the rows wide wiring channels are provided. In order to allow wiring in vertical direction gaps in the cell rows are provided, the so called feed throughs. The standard cell design has some **advantages** compared with the full custom design:

- the design time is much shorter;
- because of the lower design effort this approach is economical with a medium number of parts;
- the design risk is lower because of the use of a tested library.

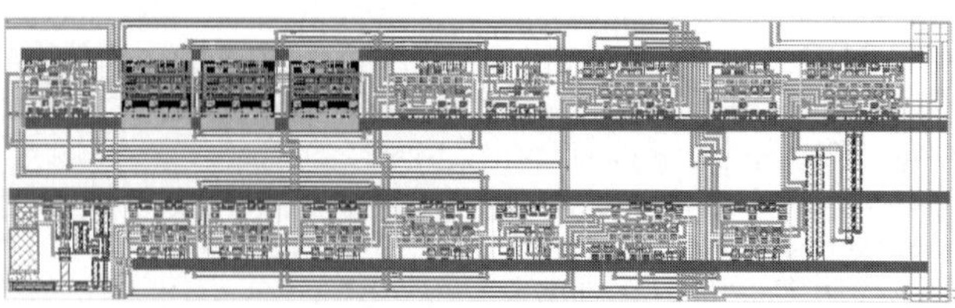

Fig. 16.6 Rows of standard cells using NOR standard cell form fig. 16.5

Though the standard cell design appears attractive there are still some **disadvantages**:

- transistors cannot be designed independently and optimally, thus the performance is sub-optimal;
- with a small number of parts this approach can be uneconomical as all masks have to be produced.

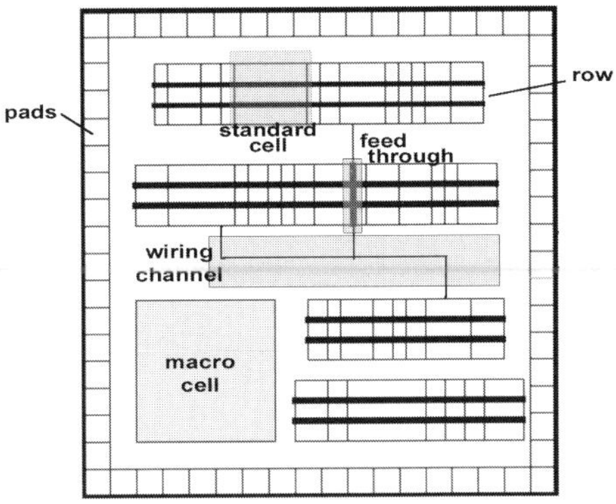

Fig. 16.7 Top layout of a circuit which consists of standard cells and a macro cell

16.2.3 Macro Cells

The term macro cell or mega cell signifies a **predefined block**, which shows a much higher complexity than a standard cell. Examples of macro cells are:

- RAM cell;
- ROM cell;
- serial interfaces;
- timer;
- arithmetic logic units (ALUs);
- processor cores.

Figure 16.7 shows a macro cell combined with standard cells. The form of a macro cell is flexible. In general no row structure is formed. Macro cells can be placed like standard cells at will. All masks are specific for a certain ASIC.

Macro cells are traded as so called **intellectual properties (IP)**, which are discussed in detail in chapter 7. **Hard macro** and **soft macro** have to be distinguished. A **hard macro** has a fixed layout and is described by a simulation, which includes layout influences. The embedding of a hard macro is equivalent to placing of a component.

A **soft macro** is typically defined by a netlist or by VHDL code. Layout and simulation data still have to be generated.

The design style with macro cells shows the **advantages**:

- Fast design with high productivity;
- Short time to market;
- Economical with medium number of parts;
- Low design risk because of use of libraries.

The **disadvantages** are:

- insertion of a macro cell into an individual design demands a certain effort;
- the price of IPs can be remarkable;
- the design style can be uneconomical with a small number of parts as all masks have to be produced.

16.2.4 Gate Array

On a gate array (GA) transistors; and possibly other components; are connected to form **predefined base cells which are arranged in a matrix**. Each of the squares of the inner section of the ASIC shown in fig. 16.8 represents a base cell. The designer defines only those masks, which signify contacts and connection lines between base cells. This procedure is called **personalization** of a gate array and this kind is called a **masked gate array (MGA)**. The manufacturer keeps prefabricated wafers in stock. Hence only the masks for the metal connections have to be produced for customization which reduces production time and cost. 16

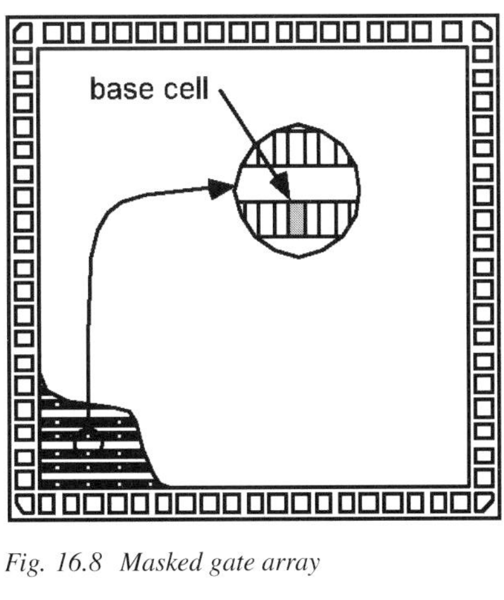

Fig. 16.8 Masked gate array

There are three types of MGAs:
- MGAs with wiring channels;
- Channelless MGAs;
- Structured MGAs.

Figure 16.8 shows a MGA with wiring channels where the base cells of the array are grouped in rows as in a standard cell design. Contrary to the latter the wiring channels cannot be chosen at will but are fixed geometrically. The fixed arrangement of base cells and wiring channels reduces the gate density remarkably compared to a standard cell design.

Figure 16.9 shows a **channelless MGA** which is also called a **Sea of Gates**. There are no special wiring channels between the cells. Instead the wiring is arranged across the base cells. This is possible as those base cells which are crossed by wires are not used and have no contact with the wires. Channelless MGAs allow a higher gate density than MGAs with wiring channels because the wiring can be configured freely.

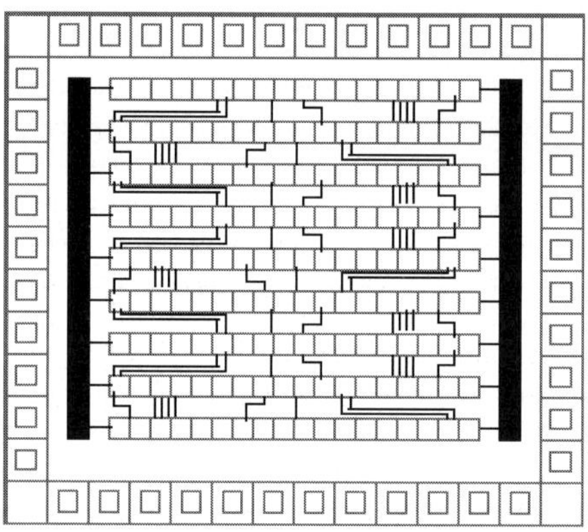

Fig. 16.9 Channelless gate array or Sea-of-Gates

Figure 16.10 shows a so called **structured MGA**. In addition to matrix-like arranged base cells which cover the main part of the area there are ranges with macro cells which contain a RAM or a processor core for instance. With this approach the possibilities of gate arrays and macro cells can be combined.

The great **advantage** of gate arrays lies in that only a few masks have to be produced custom specifically. Compared to the previously discussed design styles this leads to shorter design times and to higher economy when only a small number of parts is required. Compared with standard cell design gate arrays show the **disadvantage** of a lower equivalent gate density and therefore a higher price per part. In addition analog circuits can only be realized on gate arrays with great difficulty.

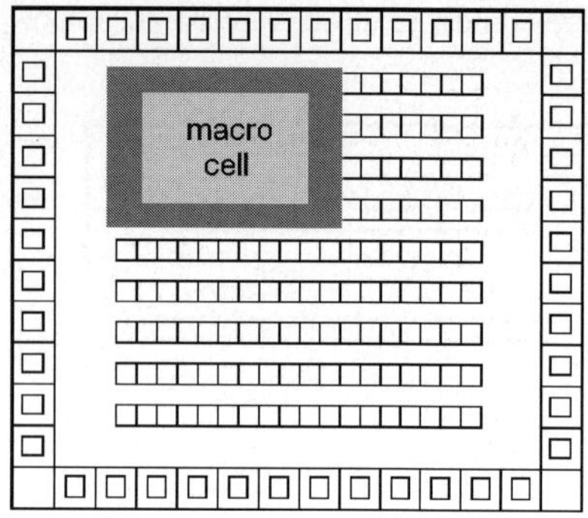

Fig. 16.10 Structured gate array

In the following paragraph the **Gate Forest Technology of IMS** will be discussed as an example of the sea of gates technology. This technology provides masters with 2,000 to 120,000 gate equivalents, which can use up to 408 I/0 connections. In addition to digital master cells there are also analog cells. The technology is therefore suitable for mixed signal applications.

The right hand side of fig. 16.11 shows two non-personalized master base cells which contain two NMOS and two PMOS transistors respectively. This shows that a master base cell forms a gate equivalent. The transistor gate wires in polysilicon are already provided yet the cells themselves are unwired.

On the left hand side of fig. 16.11 one cell is wired as an inverter and another cell is wired as a NAND gate with two inputs. In order to personalize the cells two layers 'metal 1' and 'metal 2' are provided. In addition contacts and vias are needed.

Figure 16.12 shows part of the layout of an A/D-converter described in [16.6] as an example of a personalized gate array. For reasons of clarity only the layers 'metal 1' and 'metal 2' are shown. Bond pads related to I/O connection and interface circuits can be recognized as also some gates belonging to the successive approximation register. Normally the gate density is higher than shown

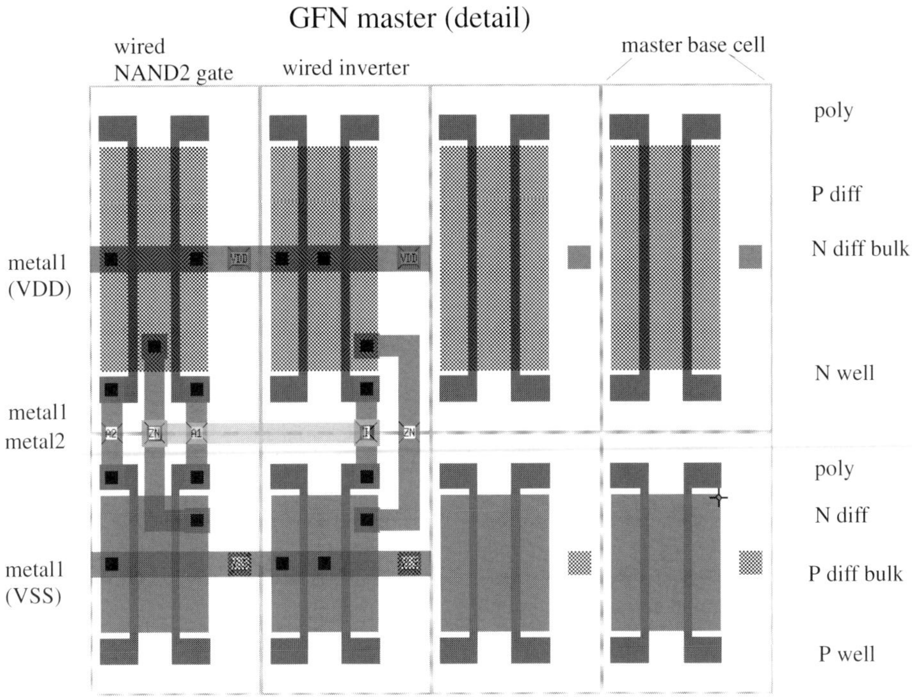

Fig. 16.11 Detail of a master in 'Gate-Forest' technology

in the section, but then the different circuit blocks could hardly be recognized.

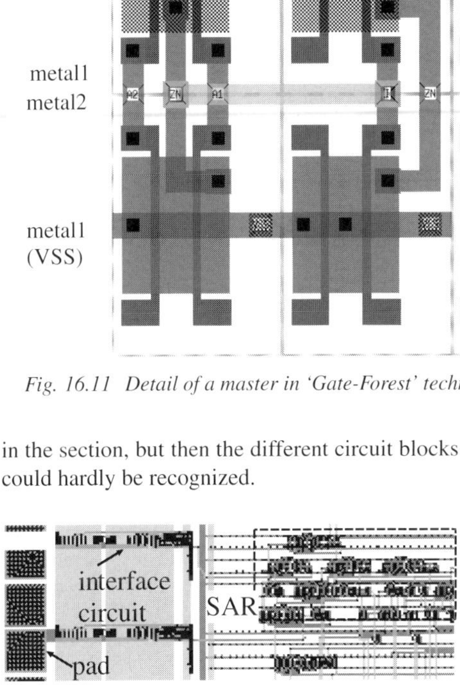

Fig. 16.12 Detail of a personalized gate array in 'Gate-Forest' technology

16.2.5 Programmable Logic: FPGA

The term **Field Programmable Gate Array (FPGA)** signifies programmable logic devices (PLDs) with complex logic cells. As with all PLDs the logic function can be configured custom-specifically by software. Here the FPGAs will be discussed only briefly in order to compare them with other design styles. For a more detailed discussion see chapter 18.

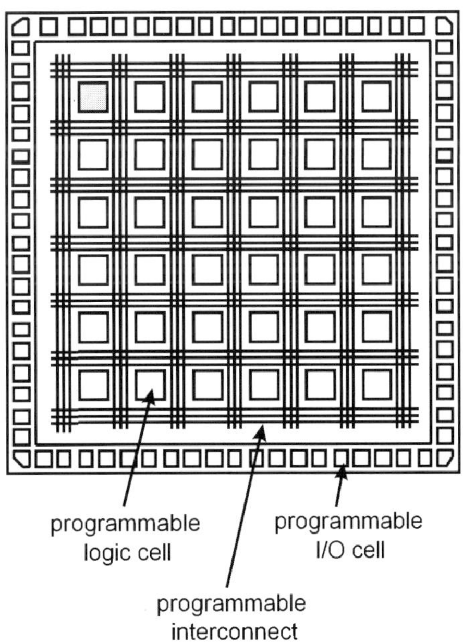

Fig. 16.13 Field programmable gate array (FPGA)

Table 16.2 Comparison between design styles

	full custom	standard cell	macro cell	gate array	FPGA
custom specific masks	all	all	all	max. 50 % of all masks	none
cell measures	variable	fixed	variable	fixed	fixed
cell placement	free	rows	free	matrix	matrix
function of a cell	variable	selectable	fixed	fixed	programmable
wiring	variable	variable channels	variable	variable or fixed channels	programmable
area utilization	excellent	good	good	moderate	poor
performance	excellent	high	high	moderate	low
design productivity [gates/day]	50	200	400	350	500
design time	man years	man months	man months	man weeks	man weeks
production time	12 weeks	12 weeks	12 weeks	3 weeks	hours
economical at number of parts/year	> 100,000	10,000 ... 100,000	10,000 ... 100,000	100 ... 10,000	< 1,000

Figure 16.13 shows that logic cells, interconnect and I/O cells are configurable. Typically the logic cells are arranged as in gate-arrays.

The main **advantage** of FPGAs lies in the extremely short design time. This is because the configuration can be done by software without technological steps involved. The designer himself can program the FPGA and sometimes reprogram it. This makes FPGAs attractive for a small number of parts or for a prototype design of an ASIC still to be designed. Compared to real ASICs FPGAs have the **disadvantages** of higher cost per part and usually lower performance.

16.2.6 Comparison of the Design Styles

Table 16.2 shows the different design styles with respect to properties, advantages and disadvantages for comparison. It is difficult in some cases to attribute a certain feature to a certain design style as the actual ASIC project is of influence. Therefore the statements can only be interpreted as typical. The information was partly taken from [16.11], [16.10], [16.13] and [16.3].

Some properties and features require the following brief explanations:

- **custom specific masks** supplies information on the necessary number of masks and technological steps. In the case of FPGA the actual IC is ready and no masks are needed;
- **cell measures** signifies the cell form (fixed or variable);
- **cell placement** signifies the grouping of cells;
- **wiring** signifies how the cells can be wired;
- **area utilization** signifies the gate density reached;
- **performance** is related to speed and power consumption;
- **design productivity** gives a typical value for the number of gates a designer is able to add to the design per day;
- **design time** results from the number of the gates needed to produce the ASIC;
- **production time** refers to the time used for technological steps. In the case of FPGA production is equivalent to a programming procedure;
- **economical at number of parts/year** provides a range where a certain design style appears economically sensible. This feature is discussed in detail in the next paragraph.

16.3 Economical Aspects

In this section ASIC design styles will be discussed with respect to economical aspects. The costs of ASIC manufacturing, which generally alter rapidly, are not often published by the semiconductor manufacturer. Thus the following values can only be considered as approximate.

16.3.1 ASIC as Product

At first some terms will be introduced which are related to the evaluation of costs and profits of ASICs.

'**Fixed costs**' (FC) are those costs of a product, that are independent of the number of parts sold. With ASICs there are, for instance, the costs of masks or of the CAD installation. The fixed costs have to be taken into account in proportion to every IC sold.

'**Variable costs**' (VC) are those costs of a product, which are proportional to the number of parts sold. With ICs there are, for instance, the costs of a die or a case. Generally the lifetime of a product on the market is restricted and superseded by better successor products. This lifetime is called '**product lifetime**' (PLT). Within PLT a certain **number of parts** of a product will be sold. This is called '**sales volume**' (SV).

The costs per IC are found by

$$ICcost = VC + \frac{FC}{SV} \qquad (16.1)$$

The 'production costs' (PC) are found by

$$PC = FC + VC \cdot SV \qquad (16.2)$$

16.3.2 Fixed Costs

Training comprises acquaintance of the technology related **design kit**, which contains the design rules and the libraries and, in addition, getting to know EDA tools, computers, and system organization are taken into account. The more specific the tools, the higher the training costs.

The costs for computer and software include the costs for buying a workstation and the license costs for EDA tools. The proportion of these costs is dependent on the project time, which differs with each of the design styles.

The **design costs** comprise the costs for design staff and costs for generating the layout of the ASIC. The starting point for table 16.3 is a medium sized ASIC with 20,000 gate equivalents. In order to estimate the number of days required for the design it is necessary to have an idea of the **productivity**, i.e., the design speed (gates/day) which is dependent on the design style. In addition, the experience of the designer, the design method, and the EDA tools play a role.

Part of the design time is needed to design the testability of the IC and to generate the so called **test vectors**.

In addition to the design time of the original circuit the time for **evaluating the IC** samples and for a possible **redesign** has to be taken into account. This effort regarded to be proportional to the effort of the original design.

The non-recurring costs for production of an ASIC are called **Non-Recurring Engineering Costs (NRE)**. These comprise the generation of custom specific masks the number of which is dependent on the design style. The information about masks cost is related to a technology with a feature size of 0.35 µm. In addition, the cost of the production of some ASIC samples and the costs of test related preparation and programming were added to the NRE. Not taken into account were costs for the so called industrialization which qualification and investigation of production methods under different conditions.

Other fixed costs comprise costs for documentation, marketing and advertising. Here the figures vary so much that it does not make sense to insert concrete values.

16

If the evaluation of the ASIC samples show that the performance of the IC does not fulfil the expectations, the design has to be improved. Especially with full custom design the probability of a redesign is high, because untested cells are used. With semi custom ICs the test results are normally in accordance with the simulation results found during the design. Every redesign causes costs for design and NRE. More important, however, is the resulting time loss and the lower product lifetime of number of parts.

The **total fixed costs** are the sum of training costs, computer and software costs, design costs, mask costs, cost for test preparation, and other costs.

16.3.3 Variable Costs

Table 16.4 gives an overview of the variable costs of an ASIC related to different design styles where some data were taken from [16.10].

Table 16.3 Fixed cost for an exemplary ASIC

	Full custom	Standard cell	Macro cell	Gate array	FPGA
Trainig					
days	10	5	5	5	2
cost/day	500 €	500 €	500 €	500 €	500 €
cost trainig	5,000 €	2,500 €	2,500 €	2,500 €	1,000 €
Hardware	5,000 €	2,500 €	2,500 €	1,000 €	500 €
Software	100,000 €	25,000 €	25,000 €	5,000 €	2,500 €
Design					
size (gate equivalent)	20,000	20,000	20,000	20,000	20,000
design productivity (gate equ./day)	50	200	400	350	500
days design	400	100	50	57	40
days evaluation	80	20	10	11	8
days redesign	40	0	0	0	0
cost/day	500 €	500 €	500 €	500 €	500 €
design cost	260,000 €	60,000 €	30,000 €	34,285.71 €	24,000 €
NRE					
no. masks	15	15	15	4	0
cost/mask	2,500 €	2,500 €	2,500 €	2,500 €	5,000 €
cost masks	37,500 €	37,500 €	37,500 €	10,000 €	0 €
cost samples	10,000 €	10,000 €	10,000 €	1,500 €	0 €
days test preparation	10	5	5	5	
cost/day	500 €	500 €	500 €	500 €	500 €
cost test preparation	5,000 €	2,500 €	2,500 €	2,500 €	0 €
Other fixed costs					
documentation					
marketing					
total fixeed cost	422,500 €	140,000 €	110,000 €	56,785.71 €	28 000 €

Table 16.4 Variable cost for an exemplary ASIC

	Full custom	Standard cell	Macro cell	Gate array	FPGA
Die cost					
no. Gate equivalents	20,000	20,000	20,000	20,000	20,000
gate density/cm^2	40,000	35,000	35,000	25,000	20,000
density factor	0.90	0.65	0.65	0.50	0.30
die area/cm^2	0.56	0.88	0.88	1.60	3.33
wafer diameter/inch	8	8	8	8	8
complete dies/wafer	498	301	301	152	62
defect density/cm^2	0.50	0.50	0.50	0.50	0.50
yield	0.77	0.67	0.67	0.51	0.30
good dies/wafer	384	202	202	78	19
cost wafer	1,000 €	1,000 €	1,000 €	1,000 €	1,000 €
die cost	2.61 €	4.95 €	4.95 €	12.88 €	53.96 €
Package cost					
no. pins	68	68	68	68	68
cost/pin	0.02 €	0.02 €	0.02 €	0.02 €	0.02 €
package cost	1.02 €	1.02 €	1.02 €	1.02 €	1.02 €
Test cost	0.25 €	0.25 €	0.25 €	0.25 €	0.25 €
Total variable cost	3.88 €	6.22 €	6.22 €	14.15 €	55.23 €

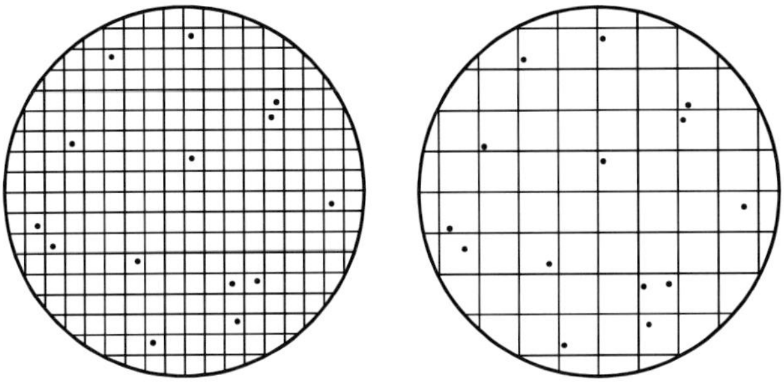

Fig. 16.14 Wafer including defects with dies of different areas

In order to determine the **cost of a die** the required **area** of the die is calculated first. For this purpose the **number of gates** is divided by the **gate density** and multiplied by the **density factor**. The gate density and the density factor of the layout are dependent on technology and design style.

Figure 16.14 shows a wafer with medium sized dies (left) and large dies (right). The dots represent defects caused by errors during production. The dies near the periphery are incomplete and cannot be used. The number of **complete dies** is given by

$$\text{complete dies} = \frac{\text{wafer area}}{\text{die area}} - \frac{\text{wafer circumference}}{\text{die edge length}} \quad (16.3)$$

The **defect density** is a quality feature of the technology. A singlee defect on a die will generally make the die unusable. In order to avoid too high a failure rate with a 0.5 µm technology a defect density of less than $1/\text{cm}^2$ is necessary.

The **yield** is dependent on defect density (Dd) and die area (DA) [16.4]:

$$Yield = \left(\frac{1}{1 + \frac{Dd \cdot DA}{2}}\right)^2 \quad (16.4)$$

For an established technology a yield of more than 80 % is expected. The number of intact dies is calculated by multiplication of the number of whole dies with the yield.

The **costs per processed wafer** refer to an 0.5 µm technology. The costs per die are finally determined from the costs per wafer divided by the number of intact dies.

The **package costs** are approximately proportional to the number of pins but also depend on form and material of course. The numbers given refer to rather cheap SMD plastic cases.

The last row in table 16.4 shows the **test costs** per IC. Every single IC is tested by means of test vectors.

The **total variable costs** of an IC are equal to the costs per die, case costs and package costs.

16.3.4 Design Style Comparison

With the data provided in section 16.3 the IC costs can now be determined according to equation (16.1). Figure 16.15 shows the IC costs as function of the sales volume during the lifetime of an IC for the different design styles, in which the fixed costs were taken from table 16.3 and the variable costs from table 16.4.

With increasing sales volume the fixed costs are less significant and the IC costs decrease. At a sales volume of approximately 1,000 the **break even volume (BEV)** between the FPGA and the gate array design style is to be found. Referring to the above data, the BEV between gate array and standard cells or macro cells lies at 10,000 parts. Only above 100,000 parts does the full custom solution become optimal.

There are a few ways of lowering the IC costs. The fixed costs are mainly determined by the **design productivity**. For this trained designers, modern design methodology, and good EDA tools are necessary. The costs of masks and of sample production can be lowered if the concept of a so called **multi-project wafer** (MPW) is applied. Here several different designs can be processed on one wafer. Also for production of a low number of parts or in the starting phase of a product the MPW can be an attractive approach.

16

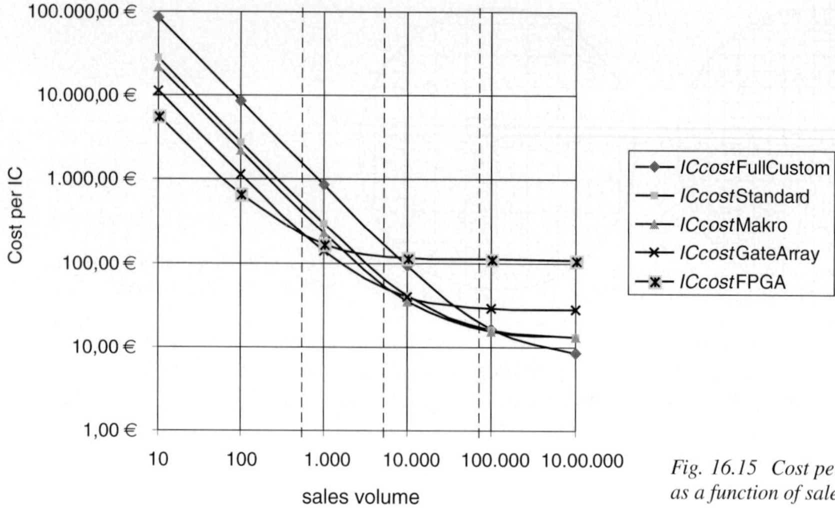

Fig. 16.15 Cost per IC as a function of sales volume

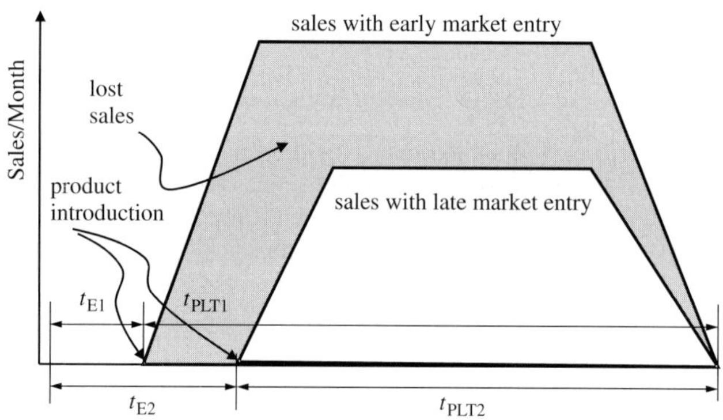

Fig. 16.16 Monthly sales of a product

The price of an IC is determined by the IC costs plus the required **profit per part**. The achievable price mainly depends on competition, which in turn depends on when the innovative product is put to market. The following consideration highlights the design speed as the decisive factor in the success of an IC.

Figure 16.16 shows the monthly sales of a product first under the assumption of an early introduction to the market, second with later introduction [16.2]. If the product is introduced later the product lifetime will be shortened from t_{PLT1} to t_{PLT2}. In addition this product will have a lower market share. Both factors lead to a loss of turnover and lower the sales volume.

The design styles presented differ in design and production time t_E: for the exemplary ASIC the FPGA solution required a design time of 48 days and no time for production whereas the full custom design solution required 520 days design time and 50 days production time. Of course, distributing the workload to several designers can shorten the time required for the development of an IC.

It becomes clear that a combination of different design styles is of great advantage. In order to achieve a fast product introduction a FPGA or gate array solution could be chosen, whereas for mass production a solution with lower variable costs is appropriate.

The design methodology has to be chosen in such a way that a change in design style does not lead to too high a development effort.

16.4 References

[16.1] *Blaud, P.; Fiebig, N.; Ipek, M.*: 'A Direct Conversion IC for Digital Satellite TV'. – Southampton, UK: ESSIRC'97 Intern. Conf., 1997

[16.2] Fa. Chip Express: 'ASIC Industry Outlook'. http://www.chipexpress.com/Docs/outlook.pdf, 2002

[16.3] *Hoppe, B.*: 'ASIC-Design: Realisierung von VLSI-Systemen mit Mentor V8'. – Berlin; Heidelberg u. a.: Springer-Verlag, 1999

[16.4] *Murphy, B.*: 'Cost-size optima of monolithic integrated circuits'. Proc. IEEE., 52, Dec 1964

[16.5] *Reifschneider, N.*: 'CAE-gestützte IC-Entwurfsmethoden'. – München [i. e.]; Haar; London u. a.: Prentice Hall, 1998

[16.6] *Rieger, M.*: 'Einfache Lösungen für Digital-Analog-Wandler und für Analog-Digital-Wandler'. – Albstadt-Sigmaringen: MPC-Workshop, Januar 1998

[16.7] *Sautter, D.; Weinerth, H.* (Hrsg.): 'Lexikon Elektronik und Mikroelektronik'. – Düsseldorf: VDI-Verlag, 1990

[16.8] *Schwanenberger, T.*: 'Design eines flächen- und leistungsoptimierten I2C-Bus-Interfaces in volldifferenzieller ECL-Technik'. Fern-Universität Hagen, Fachbereich Elektrotechnik: Diplomarbeit 1995.

[16.9] SIA Semiconductor Industrie Association: THE NATIONAL TECHNOLOGY ROADMAP FOR SEMICONDUCTORS: TECHNOLOGY NEEDS, 1997
edition: http://notes.sematech.org/ntrs/PublNTRS.nsf

[16.10] *Smith, M.*: 'Application-specific integrated circuits'. – Reading; Massachusetts; Harlow, England u. a.: Addison-Wesley, 1997

[16.11] *Spiro, H.*: 'CAD der Mikroelektronik'. – München; Wien: Oldenbourg-Verlag, 1997

[16.12] Thomson Multimedia: 'CD player with 'one Chip''. PRODUCT FUNCTIONAL SPECIFICATION TCM 121-2/-3 10541940, 1998

[16.13] *Timmermann, D.*: 'Grundlagen der VLSI-Technik':
http://www-md.e-technik.uni-rostock.de/lehre/vlsi_i/folien.htm

17 Library Design

Dirk Jansen

The basic logic elements of an IC design, gates, and flip flops, are not made by the designer himself but mostly taken from libraries. The libraries provide a set of common and frequently used basic building blocks. In the area of digital design full custom design at the transistor level can be avoided. The risk of the design process, as well as the effort of verification, is thus reduced significantly. For analogue integrated circuits, when components are available, analogue libraries are used. Digital circuits are composed nearly entirely of library components, whilst analogue circuits are composed by 80 % or more of library components. These library components are assembled and combined into the intended design. Only cells with special requirements or performances, which are unavailable as stock library components, are custom designed.

Library components are building blocks which are combined to IC designs by interconnection. In the sense of the Gajski Y (see chapter 2), each library component has three views:

- **Symbol** (structure);
- **Model** (behaviour); and
- **Form** (geometry).

See fig 17.1. In digital designs this is a relatively small number of basic components, which are assembled into all further complex circuits.

This concept must not be limited to the design of integrated VLSI circuits. Discrete components like transistors, resistors, and capacitors, arranged on a printed circuit board, have similar symbol views, behaviour description, and housing geometry. In the case of discrete circuits the number of components used which sets up the library may be remarkably large.

The manner in which the library components are connected is described in the **netlist**. **Netlists** and the **library** together describe the entire design. With the arrangement of the cells on the silicon surface – the so called *placement of the cells* which are listed in the netlist – lay out design starts. The electrical connections between the cell pins that are used by the routing are again listed in the netlist. Placement and routing is governed by the **process** defining the *design rules* and layer usage, and which is used for DRC verification.

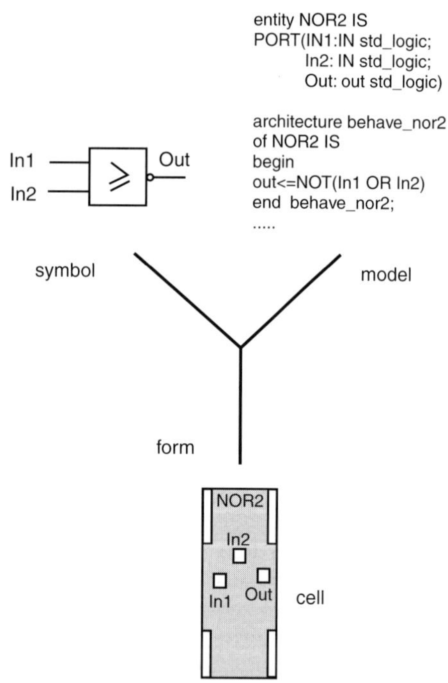

Fig. 17.1 The three views of a library component: Symbol, Model, and Form

For the purpose of simulation the *models* contained in the library are used together with the netlist. In this way netlist and library are the center of the design, where the library of components combine in itself the three arms of the Gajski Y. A further simplification of the design is not needed. The library components build the lowest level of abstraction in the hierarchy.

Choosing the library, the further development process is defined. FPGA libraries allow one to design FPGA, ASIC libraries the design of standard cell or gate array designs, discrete libraries of e.g. 74xx TTL components, discrete board designs. If there are common symbols and functions in different libraries, a one to one mapping of netlist components may be possible. If a direct replacement or mapping is not possible, modern synthesis software can deliver functionally equivalent combinations of basic gates. Basic logic functions AND, OR and INV are available in all digital libraries and can be used directly; more complex AOI gates (AND-OR-INV gates) may not always be available, and may be simplified and reconstituted by the program.

Because of the close connection with the IC manufacturing process, libraries are typically provided by the ASIC manufacturer. This means that detailed design information of the cell is often kept secret by the company Only the physical *shapes* of the cells, the *external interface* contour, and the positions of the pins are published. In most cases this is sufficient to design the circuit in a standard cell design style. The internal structures are inserted into the shapes of the design before mask generation at the manufacturer.

In reality the CMOS manufacturing processes of most plants are not very different. For this reason there are companies which specialize in providing *manufacturer independent libraries*, designed for a superset of process rules. Such libraries (e.g. those by Compass Design Automation Inc., and now Avant! Inc.) can be mapped with little effort onto many CMOS processes. Such 'passport' Libraries allow one to change the manufacturer without major redesign of the IC. Process independent libraries ease the maintenance and further development of library cells and the porting onto new advanced technologies.

In gate arrays and FPGAs the physical or geometrical view is not included. Place and route is completely carried out by the manufacturer. Placement and routing with FPGAs is generally not transparent to the user and programming of the interconnections is carried out by automatic tools.

The libraries provided for FPGA are very similar to ASIC libraries, so that there is an easy mapping process from one technology to another.

In this chapter the typical content of concrete libraries is demonstrated, and how it can be used to design chips.

17.1 Digital Libraries

Digital libraries require a compromise between specialization and standardization. Basically, only a few gates are sufficient for building up all logic constructs. Logic equations can be written in **normal forms** (e.g., the three gate types AND, OR and INV are fully sufficient for mapping every logical equation). It can be shown that even one gate type, NAND or NOR, would allow complete coverage.

Such extreme concepts have applications in the gate array scene, for the ordinary user they are not of interest. He needs a broad set of NAND, NOR, AND, OR, etc. gates, and further memory elements such as flip flops in different types and configurations. A minimal library for digital designs should have about 30 elements (listed in table 17.1).

In the first class basic combinational gates INV, AND, OR are listed. In order to drive a certain number of interconnected gate inputs, buffers and inverters should be available with large drive capabilities. NAND and NOR should be available with more than two inputs to avoid complex logic statements with many levels of internal nodes. For arithmetic applications and data path structures a *full adder* should be included in the library, as well as multiplexers and gates of the AND-OR-INV type. Busses may be build up with *tristate buffers*. These buffers should have sufficient drive capability to drive a large number of inputs to get a good frequency performance.

Memory must be provided in the form of latches and flip flops, some with asynchronous reset/set. For test purposes flip flops with scan input are needed. For the busses further *bus keeper flip flops* are provided, keeping the bus in a defined state when all tristate outputs are in the 'off' state.

The abovedefined minimum digital library contains 30 cells, even fewer are sufficient. Figure 17.2 shows some statistics on numbers of cell types

Table 17.1 Minimal Library for Symbolic Design (30 cells)

Class	Cells	Description
Combinatorial (Gates)	INV, INV_2X, INV_4X	Inverter
	BUF_4X, BUF_16X, BUF_32X	Buffer
	TRI_BUF_8X	Tristate buffer
	NAND2, NAND3, NAND4, NAND8	NAND gate
	NOR2, NOR3, NOR4	NOR gate
	AND2	AND gate
	OR2	OR gate
	EXOR2	EXOR gate
	MUX21, MUX41	Multiplexer 2 : 1, 4 : 1
	FADD	Full adder
	AOI32	AND-OR-INV gate 3 : 2
Memory (Flip flops)	LD_R	Latch with reset, positive strobe
	LD_R_SC	Latch with reset and scan input
	D_FF_R	D flip flop with asynchronous reset
	D_FF_RS	D flip flop with asynchronous reset/set
	D_FF_R_SC	D flip flop with reset and scan input
	JK_FF_R	JK flip flop with asynchronous reset
	JK_FF_R_SC	JK flipflop with reset and scan input
	BK	Bus keeper flip flop

used for about 150 ASIC designs [17.10]. Smith demonstrates that 80 % of the designs analysed use only 20 % of the available cell library.

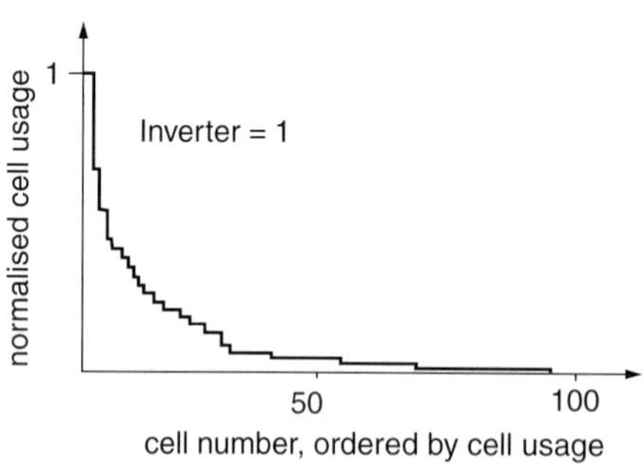

Fig. 17.2 Normalised cell usage statistics of ASIC designs ordered by the usage frequency of the cell type, for about 150 gate array designs analysed (taken from [17.10], p. 143)

The arrangement of the optimal set of basic components depends on:

- design stile (manual or synthesis); and the
- design object (control dominated or data dominated).

Design and maintenance of libraries needs much effort because every cell has to be simulated at the lowest transistor level (e.g., with SPICE) and has to be characterized. Prototypes have to be manufactured, verified, and qualified for a certain process technology. Modelling has to be carried out in several simulator dialects. The life time of a process technology has decreased to less than two years, so modifications to new processes are made continuously. This effort can be justified by the manufacturer, who provides the cells for many applications. The cost of the library maintenance may be distributed over many customers.

As a result of the competition between manufacturers and the ability to map designs using different libraries and processes, a certain standardisation has been established. Today about 300 library components are provided for the IC core, another 100 for the periphery pad cells.

Table 17.2 shows an overview of available libraries, and in which only newer technologies are listed. The libraries offered open the actual technology field for designers down to 0.11 µm

Table 17.2 Libraries and processes from important ASIC manufacturers, showing only newer processes and without any respect for completeness, after [17.11], GA: Gate Array, SC: Standard Cell

Manufacturer	Library	Geometry	Supply Voltage	Type
AMI	AMI6S5	0.6	3...5	SC
	AMI6G6	0.6	3...5	GA
	AMI3HS3	0.35	3.3	SC
ALCATEL-MIETEC	MTC55000	0.25	1.8...2.5	SC
	MTC45000	0.35	3	SC
	MTC35000	0.5	3.3	SC
AMS	CSD	0.35	3.3	SC
	CUB	0.6	3.3	SC
	CXB	0.8	3...5	SC
ATMEL	ATL35	0.35	—	SC
	ATC50	0.5	—	GA
	ATL50	0.5	—	SC
Chip Express	CX3000	0.35	3...5	GA
	CX2000	0.6	3...5	GA
Faraday Technolgy	FG8000A	0.35	3.3...5	GA
	FS8000A	0.35	3.3...5	SC
	FS7000A	0.5	3...5	SC
Fujitsu Ltd.	CC60	0.35	3.3	SC
	CG51	0.5	3.3...5	GA
Hitachi	HG73C	0.35	3.3	SC
	HG72C	0.5	3.3...5	SC
Honeywell (GaAs-Prozess)	HX2000V	0.55	3.3	GA
	BCHFET	0.6	—	SC
Hyundai Electronics	HCB60	0.35	3.3	SC
	HCB50	0.5	3.3	SC
	HSG50K	0.5	3.3	GA
IBM	SA-27E	0.11	1.65	SC
	SA-27	0.12	1.65	SC
	SA-12	0.18	2.3	SC
	SX	0.25	2.3	SC
	5S	0.36	3.0	SC
Kawasaki LSI	KS4000H	0.35	—	SC
	KS400EH	0.35	—	GA
	KZ300GH	0.5	—	SC
	KZ300EH	0.5	—	GA
LSI-Logic	G11P	0.18	2.3	SC
	G10	0.25	3	SC
Lucent Technologies	HL350CDE	0.35	3.3	SC
	HL300C	0.5	3.3	SC
	HL500C	0.5	5	SC

Table 17.2 Libraries and processes from important ASIC manufacturers, showing only newer processes and without any respect for completeness, after [17.11], GA: Gate Array, SC: Standard Cell

Manufacturer	Library	Geometry	Supply Voltage	Type
Macronix	MX05SC	0.5	3...5	SC
	MX06SC	0.6	5	SC
MATRA TEMIC	MG2	0.5	—	GA
	MG1	0.6	—	GA
Matsushita Electronics	MN7F000	0.25	—	SC
	MN7E000	0.35	—	SC
	MN5BA00	0.5	3...5	SC
MITEL Semiconductors	GCS200	0.35	2...3.3	SC
	MVA200	0.35	2...3.3	GA
	CX08	0.8	1.5...5	SC
Mitsubishi	0.25J5	0.25	2.5	GA
	0.5A8LP	0.5	3.3	GA
	0.35E8A	0.35	3.3	SC
	0.18N9	0.18	1.5...1.8	GA
Motorola	M5C	0.5	—	GA
	H4C	0.7	—	GA
NEC	CMOS-9	0.35	—	GA
	CBC9	0.35	—	SC
	QB8	0.44	—	GA
	CBC8VX	0.5	—	SC
OKI Electric Industry	MG113P00	0.25	—	GA
	MG103P00	0.35	—	GA
	MSM32R00	0.5	—	GA
Ricoh	RSC05H5	0.5	—	SC
	RSC08	0.8	—	SC
ROHM	Bu25S	0.5	—	SC
	BU16K	0.6	—	GA
S-MOS	SLA50K	0.35	—	GA
	SLA40k	0.45	—	GA
	SLA35k	0.6	—	GA
STMicroelectronics	CB55000	0.25	2.5	SC
	CB45000	0.35	3.3	SC
	CB36000	0.6	5	SC
SAMSUNG	KG90	0.35	—	GA
	STDM90	0.35	—	SC
	KG70	0.65	—	GA
	STDM80	0.5	—	SC
SANYO	LC27200	0.35	—	GA
	LC98500	0.35	—	SC
	LC22000	0.6	—	GA
	LC98300	0.6	—	SC

Table 17.2 Libraries and processes from important ASIC manufacturers, showing only newer processes and without any respect for completeness, after [17.11], GA: Gate Array, SC: Standard Cell

Manufacturer	Library	Geometry	Supply Voltage	Type
SEIKO EPSON	SLA50000	0.35	—	GA
	SLA40000	0.45	—	GA
	SLA30000	0.6	—	GA
SHARP Corp.	LZ9J	0.5	—	SC
	CMOSA	0.8	—	SC
	LZ9A	0.8	—	GA
SONY	C56	0.4	—	GA
TSMC	TCA753	0.35	—	SC
	TCB650	0.5	—	SC
	TGT650	0.5	—	GA
Texas Instruments	TSC6000	0.18	—	SC
	TGC6000	0.18	—	GA
	TSC5000	0.25	—	SC
	TSC4000	0.35	—	SC
	TGT4000	0.35	—	GA
	TSC2000	0.5	—	SC
	TGT2000	0.5	—	GA
THESYS	THE1008	0.8	—	GA
Thomson-TCS	TSCB45K	0.35	—	SC
	TSCB35K	0.5	—	SC
Toshiba	TC200C	0.4	—	SC
	TC203G	0.4	—	GA
	TC183C	0.4	—	SC
	TC180C	0.5	—	GA
VLSI Technologies	VGC720	0.5	—	GA
	VGC753	0.5	—	SC
VITESSE (GaAs)	FX/Viper	0.6	3.3/−2V	GA
	SLX	0.4	2...3.3	SC

CMOS processes. Many manufacturers offer both standard cell libraries and gate array libraries for the same process and several masters. Far eastern manufacturers and smaller companies primarily offer gate array processes. FPGA manufacturers are not included in table 17.2, the FPGA libraries offered are very similar to gate array libraries. The following discussions are focussed on standard cell libraries.

As an example for the content of such a digital library, table 17.3 gives a summarized view of the CMOS library of ALCATEL MIETEC [17.8].

A large number of different cells arises from:

- all important cells are offered with up to four different driver capabilities;
- more than $1/3$ of all cells are complex gates, e.g., AOI gates.

The diversification of basic cells to cells with larger driver capability is relatively simple by replacing the driver stage, so these cells are designed quickly. At the system level these cells save an additional buffer as well as the interconnection in between.

Table 17.3 Sketch of a typical 0.5 μm CMOS-Standard Cell Core Library [17.8]

Klasse/Typ	Anzahl Zellen	Fläche		Eingänge	Treiberleistung
		min. in μm^2	max. in μm^2		in SL
INV	6	110	440	1	Low ... 16x
BUF	5	165	495	1	2x ... 16x
TRI_BUF	5	440	1 430	1	2x ... 32x
NAND	14	165	715	1 ... 8	Low ... 4x
NOR	12	165	715	1 ... 8	Low ... 4x
AND	12	220	605	1 ... 5	Low ... 4x
OR	12	220	605	1 ... 5	Low ... 4x
MUX	8	330	1 100	2 ... 4	Low ... 4x
XOR	7	330	880	1 ... 3	Low ... 4x
XNOR	7	440	880	1 ... 3	Low ... 4x
ADDER	1	1 045	1 045	3	1x
AOI	106	275	770	3 ... 6	1x
LD	32	495	1 100	2 ... 4	Low ... 4x
D_FF	44	715	1 595	2 ... 4	Low ... 4x

Table 17.4 Technical data of selected representative cells of a 0.5 μm CMOS-Library [17.8], simplified (worst case data)

Cell	Driver SL	Area μm^2	T_{intr} ns	T_{ramp} ns/pF	Power $\mu W/MHz$	Description
IV	1	110	0.09	3.0	0.53	Inverter
IV4	4	165	0.07	0.7	1.1	Inverter 4x
BF2T8	8	330	0.3	0.3	2.9	Buffer 8x
BTSX16	16	715	0.7	0.2	15.6	Tristate buffer
ND2	1	165	0.2	4.0	0.62	Nand gate
NR2	1	165	0.2	4.6	0.55	Nor gate
AN2	1	220	0.4	2.7	0.92	And gate
OR2	1	220	0.4	3.0	0.79	Or gate
EO	1	330	0.6	6.0	0.91	Xor gate
MUX21	1	385	0.6	2.7	0.93	Multiplexer
AO5AN	1	440	0.9	3.0	1.55	And or invert
FADD	1	1045	1.5	2.7	2.7	Full adder
LD1	1	605	1.1	2.7	1.82	Latch
FD2M	1	880	1.6	3.8	1.88	D flip flop
FD2QS	1	1265	1.5	3.0	1.80	D flip flop (Scan)
FJK1	1	1265	1.7	3.0	2.32	JK flip flop

Remarks: Timing simplified, only worst case path and worst case rise/fall, 1 SL = 0,019 pF.

The same is true for AOI gates, which are area optimal, save interconnections and ease routing of logic composed of several levels. Modern logic synthesis supports AOI gates effectively. A disadvantage is the requirement of supplying and support a zoo of AOI gates; this is taken into account.

Technical data of selected cells, shorted and simplified, shows table 17.4.

The cells have, as is typical for standard cells, the same height (25 μm) and vary only in width. In many libraries it is permitted to place the cells horizontally, mirrored, or vertically. Voltage is supplied via VDD and VSS supply busses which are run horizontally through the cell with sufficient width in metal_1. Signal ports or 'pins' are located on metal_1 or metal_2 and generally routed vertically through the cell. During the design phase of the cells internal connections using metal_2 are avoided. In more complex cells, flip flops, metal_2 may be used and blockage layers are defined which tell the automatic routing tool where no connections are allowed. With three or more metal layers it may be routed over the cells, allowing high density channel free routing.

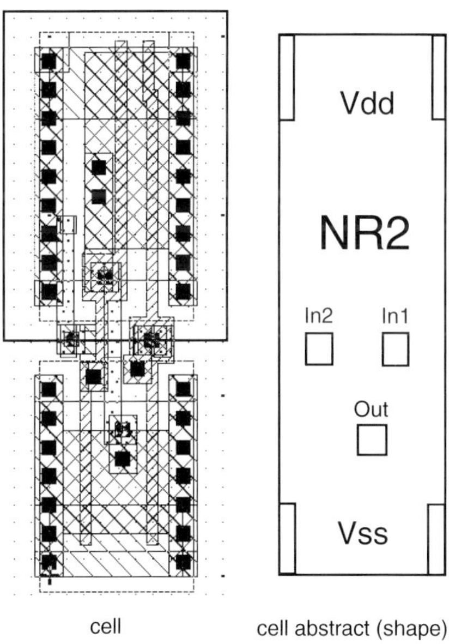

Fig. 17.3 Internal mask-structure of a typical NR2 standard cell and the cell abstract (shape) used for EDA purposes (The N-well may be larger than the shape of the abstract.)

As an example for the internals of a standard cell see fig. 17.3.

In addition to the assembly of standard cells in rows, there is another concept of an array like assembly of gates using specialized libraries. These libraries optimise the lay out of data paths, adders with wide inputs, as well as multipliers and similar systolic structures. The libraries are used together with data path generators. The cells of these libraries are characterised by the signal input being one side, the output on the other, and the supply rails and clock lines passing horizontally and/or vertically through the cell. The cells are designed for direct interconnection by abutment with the neighbouring cells.

Figure 17.4 shows the external interface *(abstract)* of a cell for data path applications. The cell is designed in such a way that electrical interconnections are made by direct abutment. With synthesis gaining more importance today, data path generators are less used, sometimes only for generation of block cells, which are placed in block design style into the IC core. Block design has the disadvantage of inefficient area usage, so the improvement of a design by using high density data path generators may be balanced or outperformed by the block placement.

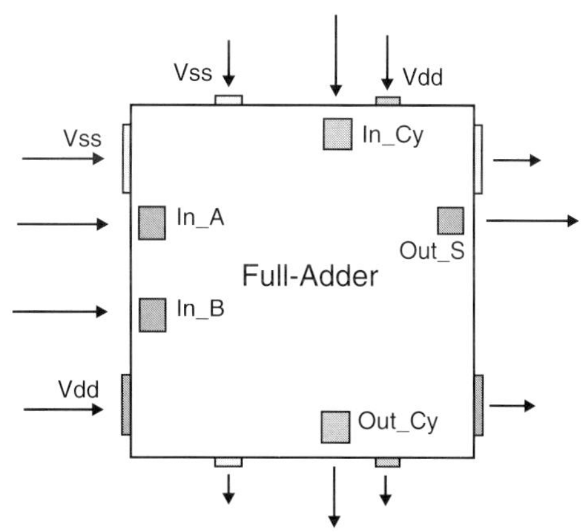

Fig. 17.4 Abstract of a cell for data path designs

The usable set of cells which may be used by data path generators is quite small and limited on a few basic gates and so called *systolic cells* for multiplier and similar structures. In many cases the cells are defined in a lay out language and are

17

designed in detail with help of the design rules of the process by automatic generator tools [17.1].

Table 17.4 provides the geometry and area data as well as the driving capability, given in **standard loads (SL)**. One *standard load* is defined as the input capacity of the smallest inverter in the library. The data is related to an abstract behaviour model of the cell which has been found by **characterization**, and which is simulated at the transistor level with all parasitic effects. Characterization is a technical area on its own and will not be discussed further here.

The propagation delay times measured from one input port to an output port are modelled in many libraries by the following linear timing equation:

$$Delay = T_{\text{intr}} + R_{\text{ramp}} \cdot C_{\text{load}} \qquad (17.1)$$

Delay means here the time for the propagation of a signal's transition slope (the rising and falling edges may have different values) from the chosen input to the assigned output. T_{intr} is the *intrinsic delay,* or basic delay, of the element and R_{ramp} a delay proportional to the load with the dimension nsec/pF. R_{ramp} models the increase of delay with the number of gates connected.

The intrinsic delay is the basic delay of the cell without any external load. It results from internal nodes and loads which are not transparent for the user. R_{ramp} depends on the load capability and design of the driver stage, and represents indirectly the output resistance (pull up or pull down resistance). Because of the pull up resistance formed by a p-channel MOS transistor, and the pull down resistance by a n-channel MOS transistor, there is a different timing behaviour for the rising and falling edges of the output signal, even in a so called *balanced design* the concept of which is to size the output transistors in such a way that delay performance is similar for both slopes.

In the modelling of gates with more than one input and output, time delays for each path and for each type of slope have to be evaluated, which delivers a large number of values even for a simple gate. Gates with larger driver capability, or *fan out* (IV4), show significantly lower values for R_{ramp} than standard gates (IV). But the larger output transistors increase the cell area and the power consumption. The equation (17.1) describes the measured behaviour shown in fig. 17.5 only simplified. Instead of the simple linear equation newer libraries and simulators use tables which are interpolated. Loads, sometimes called *fan in*, are calculated from the netlist and library only once, just before simulation, and are not dynamically changed during the digital simulation. Loads are combined with some wire load modelling, estimating the capacitive load of the interconnection wiring.

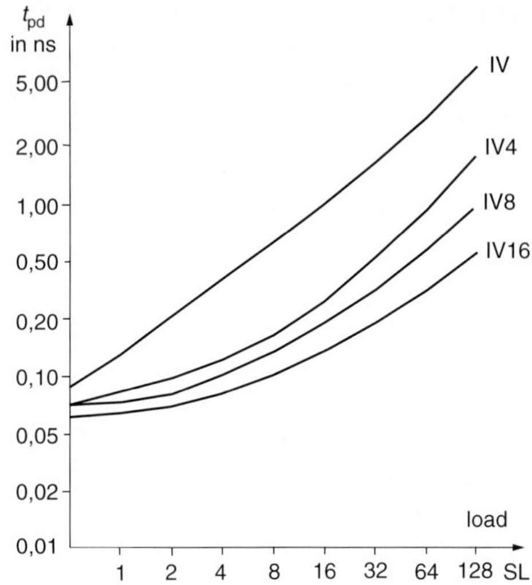

Fig. 17.5 Signal Propagation Delay of Inverter with different Driver Capability, after [17.8]

The current drive performance of an output transistor depends, besides upon the load, upon:

- the temperature of the crystal;
- the driving voltage; and
- the transconductance of the transistor.

With increasing temperature the mobility of the charge carriers in the channel decreases (negative temperature dependency), the resistance, and as a consequence of this the time constant increases. The gate works more slowly. With lower supply voltage the threshold voltage, which is more or less fixed in the process, is reached later because the rising slope of the voltage is less steep. The gate reacts more slowly. Last, but not least, in all switching performance the transconductance of the transistor is included, depending mainly on the thickness of the gate oxide. With a thickness of below 5 nm today, there may be significant tolerances in thickness, so there are *fast* gates for low thickness, *typical gates* for normal thickness and *slow* gates for increased thickness of the gate oxide, depending on the manufacturing lot. Of course there may be other effects, too, which are not of interest here. Manufacturing tolerances are strongly correlated for all transistors in a lot and on one chip, because the thickness of the oxide layer is about the same everywhere.

Switching performance depends strongly on the slope rate of the input signal, because the input capacitance varies non-linearly over the voltage and there are many internal effects [17.10]. Slow input slope rates have to be avoided in each case. The current passing through the driver stage is high when both transistors are in the 'on' condition. For all input signals, a minimum steepness of the slope is required. This is a critical requirement for clock signals because of the internal dynamical structures of clock controlled master slave flip flops. The signals slopes steepness is guaranteed by the way in which the load is compensated by choosing an output stage with sufficient driving capability. The listed delay times are only valid for input signals' rise times of less than 1 nsec.

It is common to list all timing data for the **worst case conditions**:

- 85 degree crystal temperature (highest operating temperature);
- slowest process (slow);
- lowest specified supply voltage (2.7 V).

For the application used, the timing behaviour must be calculated by using *de-rating factors*:

$$\begin{aligned}\text{delay\_real} = {} & \text{delay\_table} \\ & \times \text{de-rating\_process} \\ & \times \text{de-rating\_temperature} \\ & \times \text{de-rating\_voltage}\end{aligned}$$

The de-rating factors may be taken from tables, see table 17.5 to 17.7 as examples.

The tables show the degree of influence which the parameters may possess. The process may have a typical tolerance of ±20 %. The effect temperature has is even larger given a constant supply voltage. To keep the possible combinations of extremes as small as possible during verification it is common to simulate with the *worst case conditions*, which is a conservative approach. The delay times for a normal environment of 25 °C, 3 V and a typical process are only 60 % of the worst case values. Minimal delay times are only seldom requested. A design requiring minimum timing should, in principle, always be avoided. In bipolar and BiCMOS processes additional factors need to be considered, but this will not be discussed further.

Table 17.5 Typical de-rating factor for process tolerance [17.8]

Process	De-rating factor
Fast	0,6
Normal	0,78
Slow	1,00

Table 17.6 Typical de-rating factor for temperature variation [17.8]

Temperature °C	De-rating factor
−55	0,63
0	0,77
25	0,84
40	0,88
70	0,96
85	1,00
125	1,11

Table 17.7 Typical de-rating factor for variation of supply voltage [17.8]

supply voltage	De-rating factor
2,7	1,0
3,0	0,93
3,3	0,86
3,6	0,78

In the next pages two example designs made by the author in a 0.5 μm CMOS technology, mapped on the library of table 17.3, will be discussed in detail some more. The designs differ in content and design style.

Table 17.8 lists the statistics of cells used for a 16 bit microprocessor core design with micro-code ROM, which was originally designed by hand (schematics) and was only optimised by synthesis programs in sub-blocks. The design uses only 20 different cells. The only complex gates used are a multiplexer and a full adder. AOI gates were not originally available. With 420 cells the design is very compact, together with the micro-code ROM it is only 1.5 mm^2, and it is an example for a typical *hard macro (hard IP)*.

Table 17.8 Statistics of cell usage for FHOP V1.1, generated by synthesis, mapped on a library without AOI gates (without micro-code ROM)

Cell class	Type	Number
Buffer	BF1T1	7
	BF1T2	26
	BF1T4	1
Basic gate	IV	30
	AN2	65
	AN3	28
	AN4	1
	OR2	28
	OR3	8
	NR2	7
	NR3	6
	EO	2
	EN	65
Complex gates	MUX21L	1
	MUX21	33
	MUX41	35
	FA1A	16
Flip flops	FD1	45
	FD1S	16
	FD2	2
Sum	20 Types	420

The same processor core with some extended capabilities and an integrated 8bx8b multiplier, but without micro code ROM, is represented as a *soft macro (soft IP)*, designed in VHDL and then synthesized using *Synopsys Design Compiler*, and is listed in table 17.9 for comparison. The core now uses 4 258 cells of 108 different types. The synthesized logic contains many complex gates of AOI type and several buffers. During synthesis the operator library from Synopsys for addition and multiplication was used. The multiplier itself amounts to about one third of the entire design. The size of the core in the same technology is only 10 % larger in area but has higher performance, and may now be mapped to any technology and size requirement.

Designs synthesized from VHDL use broad libraries (via synthesis) much better than manual designs, but must not be optimal for area and power in each case.

Table 17.9 Statistics of cell usage of a complete design synthesized from VHDL, using operator libraries (Synopsys), of a processor core FHOP V2.0 with multiplier

Class	Cells	Number
Buffer	Buffer, Tristate buffer	553
Basic gates	INV, NAND, NOR, AND, OR, XOR, XNOR	2 619
Complex gates	MUX, AOI (41 Typen)	886
Flip flops	FJK, FD	220
Sum	108 Types	4 258

17.2 Pad Cell Libraries

Pad cells, the connection to the surrounding world, are collected in special *pad cell libraries*. They are very important. Pad cells perform the following tasks for a chip:

- Driving of external loads up to 24 mA;
- Buffering of input signals;
- Connection to supply voltage for core and pad area;
- Shielding of all chip internal structures against electro static discharge (ESD), transients and wrong poling;
- Adaptation of logic levels to external circuitry.

Table 17.10 gives a view of a typical pad cell library. Some manufacturers provide pad cells with different pitch for *pad limited designs* or for *core limited designs*. Cells for *pad limited* design are long and narrow, so they may be placed on a grid as small as 100 μm. Pad cells for *core limited* design are short but wide and need a pitch larger than 100 μm, but have a smaller shape of the peripheral ring.

In the library there are input and output cells for CMOS and for TTL signal levels at the output ports. There are simple driver cells with normal load capability and cells with increased driver capability (more *fan out*) and cells with tristate outputs and bi-directional input/outputs. Available input cells are standard input cells with input driver and guard structures, Schmitt trigger cells with hysteresis.

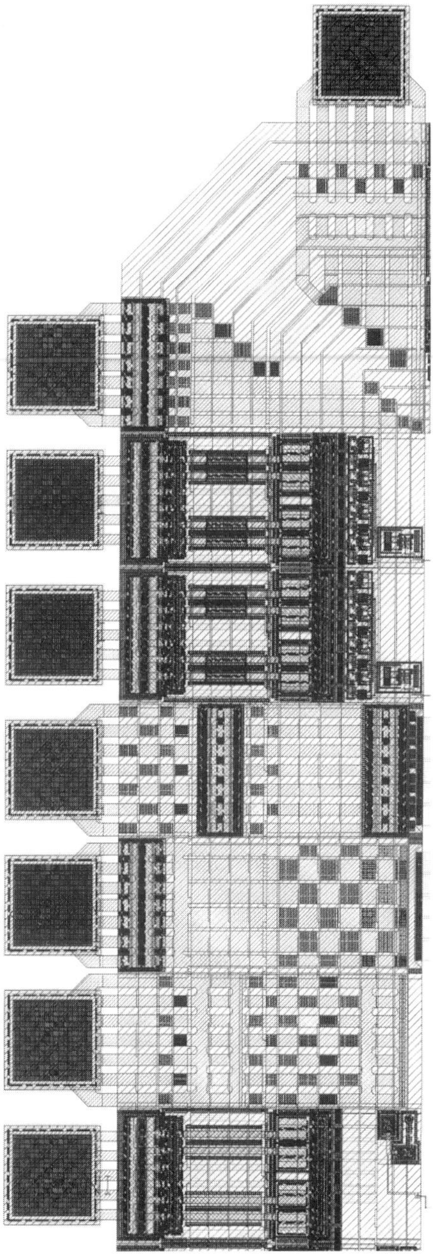

Fig. 17.6 Layout of pad cells (Shown is a part of a chip design. There are output cells, input cells as well as pad cells for supply connection.)

Schmitt trigger cells shall be used with signals with low slew rates. There are further cells with integrated pull up and pull down resistors with selectable resistance. Output cells may be single ended, so there are open drain n-channel outputs as well as open drain p-channel outputs. For analogue designs there are pad cells without any buffers or input drivers, only with protection structures against ESD. See fig. 17.6 for lay out examples.

Output pad cells are classified for driving capability, sometimes measured in mA, and for *slew rate*, the signal slope steepness. A signal with high slew rate and strong driving capability may emit large levels of electromagnetic radiation and give a rise to interference and coupling to other signals on a printed circuit board. With rise times below 1 nsec, there are emissions in the GHz range, which will be emitted by each wire connected to the pad acting as an antenna. So it makes sense to limit or lower the *slew rate* of the signal to avoid noisy emissions. Cells with defined *slew rate*, specially designed for this purpose, contain a miller capacitance in the driver, which generates a trapezoidal signal transition from one logic level to the other. This cell should be used when only slow output signals are needed, e.g., to switch a LED display. Cells with programmable slew rate are available today in all FPGA families.

Pad cells are used for the power supply of the core VDD_{core} and VSS_{core} as well as for the peripheral ring of the pad cells themselves VDD_{pad} and VSS_{pad}. Both power rails are not connected directly on the chip but on the board in order to improve the decoupling of the power supply systems. The corners of the periphery area are build from special corner cells which may have additional pad and guard structures and close the external power ring. For analogue cells, there are additional power pads, defining a complete separate analogue power supply domain, which will not be coupled to the digital supply. With chips today having several hundreds of I/O pads, more than one pair of supply pads are placed on each side of the chip.

All pad cells are provided with protecting guard structures (see fig. 17.7) which consist of integrated diodes to the pad supply rings. Input pads contain an additional input series resistance which forms an effective filter together with the input capacitance, against electrostatic discharge pulses

17

Table 17.10 Typical pad cell library, 731 mm height and 100 µm width for pad limited design [17.8])

Cell/Class	Number of types	Driver capability in mA	Description
OUTPUT_BUFF (CMOS)	7	2...8	Output buffer for CMOS level
OUTPUT_BUFF (TTL)	20	2...24	Output buffer for TTL level
TRI_BUFF (CMOS)	7	2...8	Tristate output buffer for CMOS level
TRI_BUFF (TTL)	16	2...24	Tristate output buffer for TTL level
BIDIR_BUFF (CMOS)	6	2...8	Bidirectional output buffer for CMOS level
BIDIR_BUFF (TTL)	38	2...24	Bidirectional output buffer for TTL level
INPUT (CMOS)	16	—	Input cell for CMOS level
INPUT (TTL)	19	—	Input cell for TTL level
SCHMITT_INP (CMOS)	11	—	Schmitt trigger input cell, CMOS level
SCHMITT_INP (TTL)	14	—	Schmitt trigger input cell, TTL level

(ESD). The lay out is made in such a way that the requirements of a test with the human body model fig. 17.8 are met. This model contains a capacity of 100 pF charged up to 2000 V, and discharged via a resistance of 1500 Ohms to the pad.

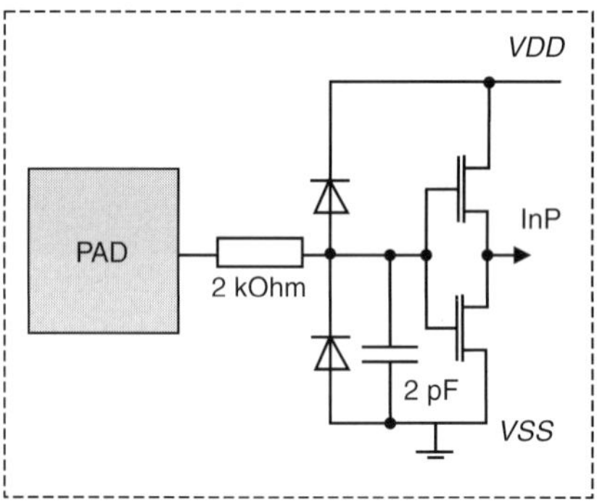

Fig. 17.7 protecting guard structures

For many technical applications the ESD protection after fig. 17.8 is not sufficient; for example in the automobile as well as in the military sector significant higher levels are requested. Many manufacturers design their pad cells to a test voltage of 4...5 kV, which is about four times better than the human body's model.

Output cells should be protected against short circuit, this can be done by series resistances, or better, by limiting the current of the driver.

Beside these standard I/O cells there may be special cells for clock oscillators, allowing a direct connection of a quartz crystal to the chip.

The design of one's own pad cells in a full custom style is very risky because of the ESD problems and the required guard and protection structures. It should be avoided and library cells preferred wherever possible.

Pad cells are described by models in the same way as core cells and are characterised by their timing behaviour. During simulation the external loads (the capacitances of the pins) as well as the board structures may be taken into account. Compared to the core cells, the delay times are remarkably large, the power consumption too. Partitioning of functions over several chips which are connected by a printed circuit board is ineffective for delay and power consumption, in contrast to an entire solution on a single chip.

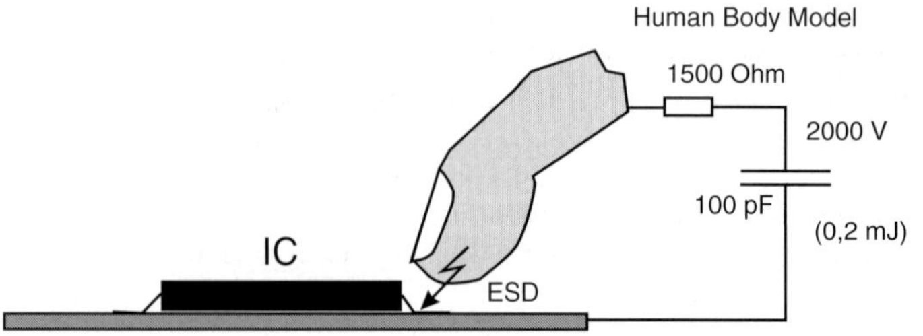

Fig. 17.8 Prüfung der ESD-Festigkeit mit dem Human-Body-Modell

17.3 Library Standards (VITAL)

A consistent and reliable model of digital gates and memory elements is not trivial. In the past several different approaches and modelling languages have been developed for this purpose. Each company which offered a digital simulator developed its own proprietary modelling language and proprietary concepts for capture and definition of delays. ASIC manufacturers are forced to describe their libraries in different dialects, appropriate to the most important EDA sellers. Exchangeability, even a direct comparison of the results, is difficult. So the results of a simulation with a MENTOR GRAPHICS QUICKSIM simulator give significantly different results from one obtained from CADENCE or VIEWLOGIC. Even the logic value systems (12-value logic, 9-value logic, etc.) are not compatible if a detailed analysis was required. Which was right and which company simulate best the real behaviour?

With the definition and standardization of VHDL, a fundamental standard for description of logic circuits was set up. The accompanying packages *IEEE 1164 standard_logic* define a unified 9-level logic value system [17.6]. One of the most important problems in using VHDL in the beginning of the 1990s was libraries which missed the standardized description of delays. This led to the foundation of **VITAL**, the **VHDL Initiative Towards ASIC Libraries**, in 1992. The goal of this initiative is to accelerate the development of 'sign off' quality ASIC libraries in VHDL. The best available procedures and methods for modelling of macro components should be included.

'Sign off' quality means that the manufacturer guarantees the functionality of the circuit which has been simulated using these VITAL libraries. VITAL is now a loosely coupled association of ASIC manufacturers, design houses, the EDA industry, and interested users. It is still the goal of this organization to develop specifications and standards concerning this topic.

The first and may be most important result of this work is the IEEE standard 1076.4 with the title: '**VITAL ASIC Modelling Specification**' [17.12], the last issue 4/99. This standard refers to methods and procedures of the *Standard Delay Format* (SDF) [17.9], to elements of the *std_timing_package* of VHDL Technology Group, to the IEEE 1164 standard concerning unified multi value logic systems [17.6], and is last but not least, based on the huge experience gained from existing modelling languages, software tools, and libraries. VITAL is supported by more than 60 companies working in this field. The standardization is not yet finished. The latest supplement, released in 1999, standardizes methods and procedures to describe memories (RAM, ROM).

A direct result of this standardization initiative are VHDL packages which may be bound to the code of the model description and provide declarations and functions for simulation. These packages are:

- **IEEE.vital_timing**,
- **IEEE.vital_primitives**.

The intentions of VITAL are going further than only for unification of modelling. VITAL packages should help to perform all other simulations procedures using delay times:

- Definition of timing constraints for synthesis;
- Simulation of real time behaviour on gate level using library elements;
- Pre-setting of timing constraints for place and route;
- Back annotation of timing data after place and route for post simulation;
- Evaluation and assessment of error conditions like set-up/hold violations, glitches, etc., and generation of standardized warnings and error messages.

As a medium for transfer of timing data the *generic* construct of the VHDL language is used. These values are defined with standardized naming conventions and predefined types. The *generics* are constants and in this way pre-calculated during compilation of the code. Constants cannot be modified during run time. By this load dependent, temperature dependent, and supply voltage dependent delays are fixed with simulation start. This accelerates simulation significantly.

List 17.1 gives a *VITAL compliant* description of a NOR gate with 2 inputs (NR2). Library descriptions are typically written not by hand but generated by specialized tools (VITAL_MODELER). These tools add the attributes needed for further synthesis with, e.g., SYNOPSYS automatically (as a commentary, see list 17.1).

List 17.1 Vital VHDL description of a NOR_2 cell

```
- Filename          :  nr2.vhd
- Description       :  2 input NOR
- Library           :  Alcatel Mietec - mtc_kp1
- Generated by      :  vital_modeler version 2.0I (25-JUN-96)
- Copyright         :  (c) 1994 VEDA Design Automation Ltd.

LIBRARY ieee;              USE ieee.std_logic_1164.ALL;
-synopsys translate_off
                           USE ieee.vital_timing.ALL;
                           USE ieee.vital_primitives.ALL;
-synopsys translate_on
ENTITY nr2 IS
-synopsys translate_off
    GENERIC
    (
    tpd_a_z         :  VitalDelayType01 := (449 ps, 433 ps);
    tpd_b_z         :  VitalDelayType01 := (477 ps, 497 ps);
    tipd_a          :  VitalDelayType01Z := (0 ps,0 ps,0 ps,0 ps,0 ps,0 ps);
    tipd_b          :  VitalDelayType01Z := (0 ps,0 ps,0 ps,0 ps,0 ps,0 ps);
    TimingChecksOn  :  Boolean := FALSE;
    XOn             :  Boolean := TRUE;
    MsgOn           :  Boolean := FALSE;
    InstancePath    :  string := ""
    );
-synopsys translate_on
    PORT
    (
    a               :  IN    std_logic;
    b               :  IN    std_logic;
    z               :  OUT   std_logic
    );
-synopsys translate_off
ATTRIBUTE VITAL_LEVEL0 OF nr2 :  ENTITY IS TRUE;
-synopsys translate_on
END nr2;

ARCHITECTURE behavioral OF nr2 IS
-synopsys translate_off

ATTRIBUTE VITAL_LEVEL1 OF behavioral :  ARCHITECTURE IS TRUE;
SIGNAL a_ipd        :  std_logic :='X';
SIGNAL b_ipd        :  std_logic :='X';
-synopsys translate_on
BEGIN
-synopsys translate_off

WIREDELAY : BLOCK
BEGIN
    VitalWireDelay ( a_ipd, a, tipd_a );
    VitalWireDelay ( b_ipd, b, tipd_b );
END BLOCK;

VITALBehavior :  PROCESS (a_ipd,b_ipd)

VARIABLE z_GlitchData :  VitalGlitchDataType;
VARIABLE z_zd :  std_ulogic;
```

```
BEGIN

z_zd := not ( b_ipd or a_ipd );

-------------
-Path Delay Section
-------------
VitalPathDelay01Z (z, z_GlitchData, "z", z_zd,
    Paths=>( 0=>(  a_ipd'LAST_EVENT,VitalExtendToFillDelay(tpd_a_z),TRUE),
                 1=>(b_ipd'LAST_EVENT,VitalExtendToFillDelay(tpd_b_z),TRUE)),
    DefaultDelay=>VitalZeroDelay01Z,
    Mode=>OnDetect,
    XOn=>XOn,
    MsgOn=>MsgOn );
END PROCESS;
-synopsys translate_on
END behavioral;
```

Models may be type *level 0* or *level 1*. *Level 0* models are relatively simple descriptions. *Level 1* models may be quite complicated and use the powerful modelling capabilities of the VITAL definitions (more or less). Figure 17.9 shows a structure of level 1 models.

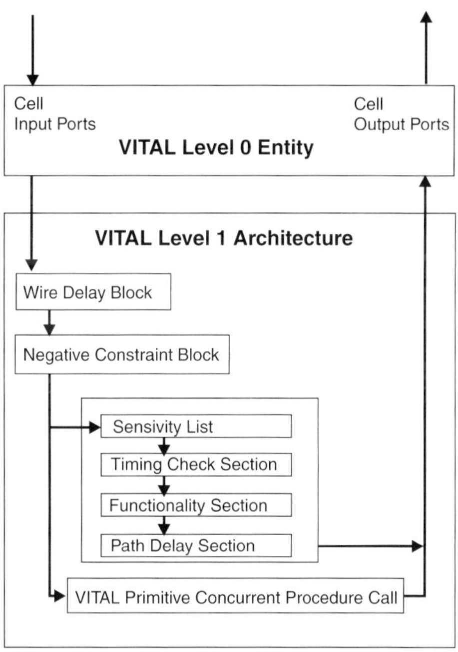

Fig. 17.9 Structure of VITAL-Level 1 models

Input and output signals are declared in the entity (as common in VHDL). Here we find the *generics* which describe the *default behaviour* of the gate. The *architecture* may be of *level 0* or *level 1*.

The first *timing block* models *wire delay*. This is the delay of the interconnection structures between cells, not the internal delay. The related *generics* are set to zero, they will be redefined by *back annotation* with concrete values after the place and route operation.

The following *negative constraint block* may contain preset values for logic synthesis and does not exist in all cases.

Now follows a detailed description of the cells behaviour. There are several possibilities. If a *VITAL primitive*, a cell with the desired function, already exists, it can be called here. Otherwise there must be a behaviour description of the desired functionality in the process section of the VHDL description. The VITAL process contains further a *test section* for the timing check and a *path delay section* which describes the signal propagation and by this the behaviour of the output if there are *glitches* or *spikes* on the signals. The process construct is a very powerful VHDL statement, able to encapsulate several and complicated functions, tests, and behaviour descriptions.

The process contains in addition some switches which allow to control the deepness of the test mechanisms (*Xon*), if error checks should be done (*TimingChecksOn*) or if error messages should be generated at all (*MsgOn*). These generics may be overwritten and allow to control the model behaviour during simulation.

Table 17.11 VITAL timing generic prefixes

Prefix	Interpretation
tpd_	*propagation delay*, delay time between referenced ports
tsetup_	*setup time* between input port and reference port, e.g., between clock and d input of a d flip flop
thold_	*hold time* between input port and reference port, e.g., at flip flops between clock and d input
trecovery_	*recovery time*, for flip flops with active reset (similar to set up time)
tremoval_	*input removal time*, e.g., for flip flops after reset (similar to hold time)
tperiod_	minimal duration of period
tpw_	minimal pulse width
tskew_	time difference between pulse slopes
tncsetup_	*no change setup time*, for back annotation
tnchold_	*no change hold time*, for back annotation
tipd_	*interconnection path delay*, internal delay of the interconnection structure
tdevice_	*device delay*, for back annotation
ticd_	*internal clock delay*, internal delay of clock signal
tisd_	*internal signal delay*, internal signal delay
tbpd_	*biased propagation delay*, delay time for compensation of negative preset times

Table 17.12 VITAL_Primitives: Predefined Basic Gates

VitalAND	VitalAND2	VitalAND3	VitalAND4
VitalOR	VitalOR2	VitalOR3	VitalOR4
VitalXOR	VitalXOR2	VitalXOR3	VitalXOR4
VitalNAND	VitalNAND2	VitalNAND3	VitalNAND4
VitalNOR	VitalNOR2	VitalNOR3	VitalNOR4
VitalXNOR	VitalXNOR2	VitalXNOR3	VitalXNOR4
VitalBUF	VitalBUFf0	VitalBUFf1	VitalIDENT
VitalINV	VitalInv1f0	VitalInv1f1	—
VitalMux	VitalMux2	VitalMux3	VitalMux4
VitalDecoder	VitalDecoder2	VitalDecoder4	VitalDecoder8

VITAL timing generics are a combination of *prefixes*, the name of the input pin and the name of the output pin:

tpd_a_z : VITALdelaytype01 := (449 ps, 443 ps)

has to be interpreted as: delay time between input a and output z. The kind of transition is taken from the type of the variable with the suffix ...01. So 01 stands for a transition from the strongly driven output 0 to strongly driven 1, the delay for this is 449 psec, for the opposite transition 1 to 0 a delay of 443 psec (see also list 17.1).

In table 17.11 the VITAL predefined *timing generics prefixes* are listed.

The *VITAL_primitive package* contains predefined tables and basic gates (primitives), which can be used to describe complex cells very easily.

- *Vital_Primitives*.
- *VitalTruthTable*,
- *VitalStateTable*.

Table 17.12 shows a list of primitives which are predefined. The most important gates like AND, OR, NAND, NOR, as well as multiplexers and decoders are available.

More complex functions (e.g., AOI gates) may be defined using the *VitalTruthTable*. Flip flops are not predefined, they must be modelled using the generalized VitalStateTable function. This function generates *Moore automata*.

It must be remembered that VITAL does not define a canon of library cells, but provides tools and mechanisms to define consistent, powerful and standardized gate models. The IC manufacturer must describe their library elements with the help of VITAL Packages, the deepness of description may be their own decision.

VITAL gains more and more importance today, not least for standardization of the back annotation mechanism. With modern deep sub-micron processes a significant part of the delay is in the interconnection, and so post layout simulation after place and route gains importance. The SDF format for unified description of time data plays here the central role. The relation between the SDF data and the VITAL generics is defined in IEEE 1076, so there is a direct mapping of values.

VITAL, SDF and IEEE 1164 *multi-value logic* standardize and unify methods and procedures for measuring and definition of timing data and allows timing behaviour to be captured in a quantitative way.

One result of the VITAL initiative is that today company proprietary digital simulators are completely replaced by VHDL simulators with same or better functionality.

17.4 Analogue Libraries

Analogue circuits are, when compared to digital circuits, much more difficult to standardize. One reason for that is that many circuit elements must be varied in their parameters. A resistor may take every value between some ohms and some megaohms. Low ohmic resistances are very different in lay out as well as in electrical behaviour from high ohmic resistances. Further requests on these elements like 'matching' are unknown in the digital domain.

Although there are fundamental differences, the advantage of a set of proofed and characterised basic cells is so big that libraries are used also in the analogue domain. There may be a smaller set of different types. Compared to standard cells these analogue blocks are not same size and must be placed in a block style methodology. Table 17.13 gives a summary of a typical analogue library for mixed analogue/digital ASIC design.

All cells are available as shapes (*abstracts*) or in the form of a complete *lay out description* in GDS II format. Additional there are *model descriptions* in SPICE format and sometimes in VHDL-AMS. In many cases the *schematics* are available. The cells are accompanied by a data sheet with the most important technical characteristics. All cells are specific to a certain process and it needs a lot of effort to port them to a new process. Cells as a/d converter and as d/a converter are usually typical IP (see chapter 7) and may be acquired by licensing. They are not included in basic analogue libraries any longer.

Resistors, *capacitor* and standard *transistors* are not part of the libraries. They are not drafted by hand but generated by *lay out generators*. Figure 17.10 shows the screen of the IC Station tool (MENTOR GRAPHICS) with a transistor generator in action designing a NAND gate from schematics. This is called *schematics driven lay out*. The transistor dimensions 'L' and 'W' are specified as properties in the schematics. The transistors lay outs are generated from this in all mask levels. Electrical connections are displayed as *overflows* (this is a symbolic display of connectivity in the form of rubber bands). The transistors may be moved and placed everywhere on the silicon, the overflows being routed by hand or automatically. 17

In the same way resistors and capacitors are generated. The final size and form depends on the parameters taken from the schematics. This idea of using generators can be applied to more complex circuits as operational amplifiers, and comparators too, but this belongs more to the IP domain. Tools like IC Station allow one to edit and manipulate lay out data of analogue cells in a convenient way.

A further, more automatic, generation of analogue lay outs is unusual and makes no sense. There are too many different geometrical limitations, requirements for matching of elements, and many other conditions to follow, which prohibits further automatic lay out generation. Analogue lay outs, mostly full custom and limited to some few cells, will still require the specific VLSI design skills of the engineer.

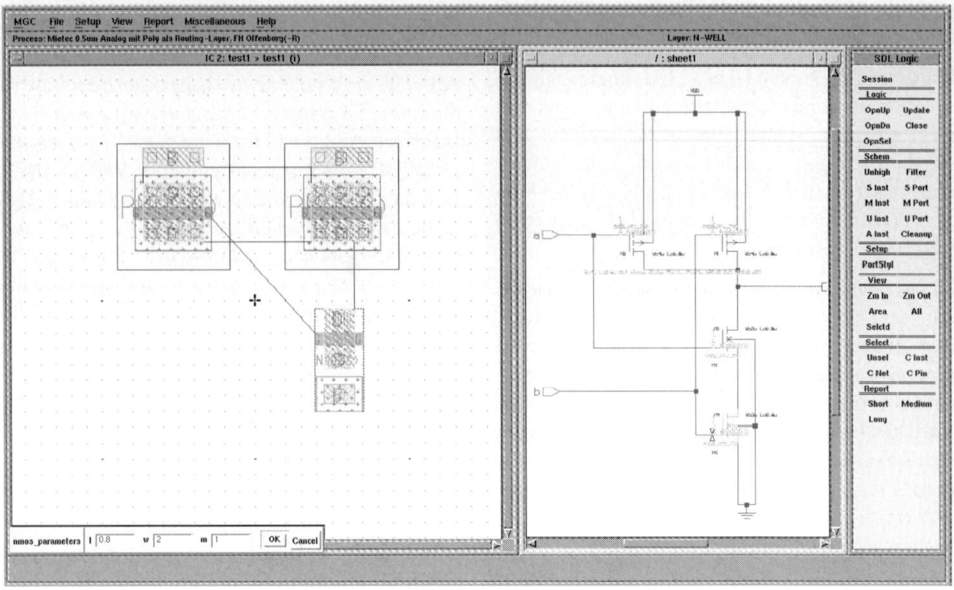

Fig. 17.10 Schematic driven lay out with IC station, using generators for transistor lay out

Table 17.13 Summary of a typical library for mixed analogue/digital ASIC design

Class	Description
Operational amplifier	Operational amplifiers with different bandwidth, driver capability and input stages
Trans impedance amplifier	Trans impedance amplifier for high frequency or low power applications
Comparator	Comparator with different specifications concerning input and output
Band Gap Reference	Cells for band gap references and current distribution
Oscillator	Oscillator for quartz oscillator, RC oscillator, VCO
Analogue/Digital Converter	Flash, successive approximation, and sigma delta A/D converter with different resolution, and concept
Digital/Analogue Converter	D/A converter with different resolution and sample rate

17.5 Macro Libraries

Macros are combinations of basic cells in a defined manner to a new function. There are:

- **Hard Macros**; and
- **Soft Macros**.

A **hard macro** is a complete new cell which is build up hierarchically from basic cells and which is completely placed and routed, and as a direct result is ready for placement. Examples of these kinds of hard macros are ALU, adder, or entire processor cores (see before). Hard macros are handled like normal cells, placed by the cell abstract and integrated into the cell interconnection. Today hard macros are called IP and are the subject of commercial trading (see chapter 7).

The characteristics of **Soft macros** is that they consists of available basic cells, but the interconnection is defined in a netlist. Placement and routing is not pre defined and still open for further manipulation. Examples are counters, registers, and several other building blocks, which are used several times in digital circuits. Classical soft macros are the building blocks of the MSI TTL series 74xx, which are well known to the designer from older discrete designs and which in arrangement and grouping are always a good selection. Soft macro libraries based on the 74xx-family can be found by many FPGA vendors, because many designs were a direct translation of existing discrete designs in the beginning. The importance of these libraries has decreased with the use of high level design languages as VHDL. Larger *soft macros* are today traded as *soft IP*, they should no longer be called *libraries*.

The 74xxx family libraries are more or less replaced by the **LPM standard**, a **L**ibrary of **P**arameterised **M**odules. LPM is approved by EIA as an intermediate standard and is seen as a supplement of the EDIF standard (**E**lectronic **D**esign **I**nterchange **F**ormat). The LPM standard describes functions built up of basic cells which are combined in a netlist to more complex functions. Examples are adders, multiplexers, etc. Because of standardization, LPM functions may be transferred from one vendor to the other. They can be placed in schematics as graphical symbols and do not require further details, only parameters as bus width, type, etc. Mapping to the specific technology may be carried out by the technology provider in an optimised way. LPM functions are supported in the form of generators by every EDA provider today: Cadence, Mentor, Exemplar, Summit Design, Synopsys, Veribest etc.

The primary goal of the LPM standard is a technology independent design with high efficiency in the design tools. The designer may use the LPM modules without knowing all the details of the technology. Mapping and fitting to the target technology is carried out by the synthesis tools. LPM today describes 25 functions, amongst them arithmetic functions, adder and multiplier, multiplexer, decoder but also sequential circuits as counters and register files. Because the modules may be parameterised by their architecture style, bit number, and input combinations, the number of functions which may be generated is very large.

LPM functions are also available as VHDL libraries, generated in a structural VHDL style automatically. The VHDL language supports this idea by the *generate* statement directly. All EDA providers support this kind of design style today.

Design with LPM modules still belongs to the *bottom up design style* using schematic block diagrams and loses importance in the future with a more system oriented design flow.

17.6 Libraries for Printed Circuit Design (IBIS)

For the large domain of printed circuit design, which can only touched upon here, libraries are of central importance. There are four basic libraries:

- symbol libraries;
- model libraries;
- packages libraries;
- pad libraries;

to mention the most important. Symbol libraries are discussed in chapter 3 in detail, there are the IEC libraries and the older, but still widely used, ANSI libraries. Each electrical block has a symbol, and the ports or pins are numerated in a definite way. The geometry of the block is found in the package library, here are standards like JEDEC with different housings (DIL, SO, TSOP, ...) common. Symbol and form are referenced through a mapping in such a way that each symbol port belongs to a geometric pin of the housing. The form of the pin, the geometric footprint on the board (pad), may be again varied, so there may be housings which require holes in the board, other may be soldered to the surface (SMD devices).

Model libraries are offered by nearly all vendors, sometimes as SPICE model description. Simulation of the board functions may then be done with programs like PSPICE or ELDO, also in a mixed digital/analogue style. Simulation of complex circuits is possible by these methods, but the results are often not satisfactory, the effort high, and not good enough in critical points. One reason for that is that there are effects coming from the spatial distribution of components on the printed circuit board. Signals show reflections and couplings, which are not included in the netlist and are not simulated. With thousands of connections, programs like SPICE are not well suited, even if all parasitic effects were modelled.

To verify complex, high frequency printed circuit boards with regards to their *interconnection behaviour* and *signal integrity*, a different approach is required. It has to be focussed on the signal integrity of the widely geometric distributed board structure, not on the general function of the board. The development of modern *motherboards* with operating frequencies above 100 MHz would not have been impossible without these new tools.

A detailed description of the input/output structures of the connected components in a standardized EDA format is important for simulation of the interconnect and signal integrity. The **IBIS standard** ANSI/EIA-656 [17.3] sets up the foundation for all further developments. IBIS stands for I/O Buffer Information Specification. The actual version 3.2 allows one to describe the electrical behaviour of input and output structures of integrated circuits, secondary boards, connectors, etc. in a machine readable format. *IBIS models* are available from many electronic manufacturers [17.4]. The standard is distributed, maintained, and further developed by EIA IBIS Open Forum, an association with more than 35 manufacturers. *IBIS models* are administered in libraries and can be read and processed by specialized programs.

IBIS models describe only the input/output structures, not the functions of the circuits. The functionality of the circuit is not released by the provider and, unfortunately, may be very complex. Classical applications are bus models (e.g., for simulation of PCI bus or processor boards with high clock frequencies). IBIS development and standardization was mostly initiated and driven by INTEL. More information can be found in [17.4].

A company association in Japan is actually developing a similar standard 'IMIC' for analogue integrated circuits: '*Standard for electronic behavioural specification of integrated circuit analogue characteristics*'.

17.7 Maintenance and Porting of Libraries

CMOS processes are still in development to higher resolutions and smaller structure size (see chapter 1). One outcome is a usable process lifetime which is described by design rules lasting only a few years. As the process technology changes they are replaced by new sets of design rules. The designed library of standard cells, which are designed with a certain set of design rules, has to be fitted to each new process and may be partly re-established. So there may be totally new concepts when changing from a process with only two metal layers allowing only channel routing, to a process with more than two metal interconnection layers, allowing area routing without any channels. Simple shrinking of cells without changing details and structures is not sufficient. Further qualitative improvements of processes, as staggering of vias between metal layers, allowing denser routing and cell geometries, have to be taken into account.

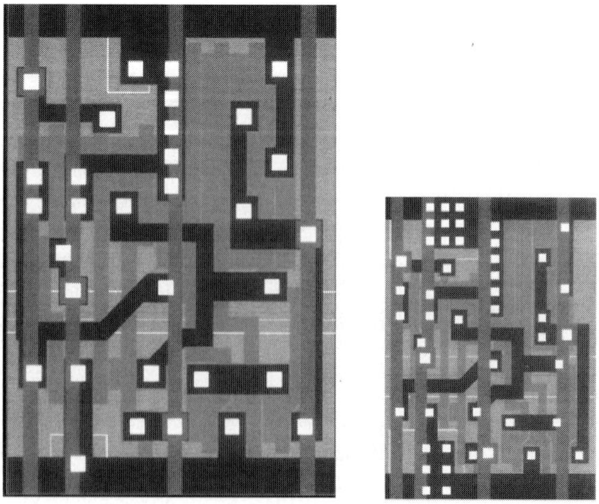

Fig. 17.11 Library cell, ported with a library maintenance program (DREAM© by SAGANTEC)

Maintenance of libraries requires a continuous development of the cells for production yield, which results in small changes of design rules, of which the designer is sometimes not aware. Library maintenance is an important task for the ASIC manufacturer and a key to commercial success.

The transfer of a library to a new process is known as '*porting*' of the library. This may be needed if a certain successful design should be again manufactured on a new process, without going back to the design level. Porting may also be done to *high performance* or *low power libraries* in the same process. In each case a one to one replacement of old cells by an equivalent new cell is required. This has to be differentiated from a redesign at the synthesis level. The development effort for porting a library may be one dimension smaller than a complete re-design if specialized programs (e.g., DREAM© from SAGANTEC) are used.

Porting needs the following steps:
1. Shrinking and adoption to the new design rules;
2. Quality improvements concerning production yield;
3. Extraction of parasitic capacitances and resistances as well as transistor geometries;
4. Simulation of the cell with SPICE at the transistor level for all relevant voltages, temperatures and process tolerances;
5. Abstraction of delay times in a predefined way and setting up of description models (VITAL models);
6. Generation of cell abstracts (shapes with external interface and properties);
7. Qualification of the new cells in the process, capture of cell behaviour by measurements, and adjustment of the models concerning the results.

Figure 17.11 show a library cell ported with DREAM©. The main arrangement is preserved, but in detail there are many changes. Other programs for library generation and porting are provided by specialized companies like Library Technologies Inc. [17.7], AVANT! Inc., and some other important EDA suppliers. There are further companies, which are designing libraries for a 'superset' of CMOS design rules and offer these to ASIC manufacturers. AVANT! offers the library LIBRA PASSPORT [17.2], which is designed with general design rules and shall be a '*multi-foundry access*' library. From the same source the more rule adopted LIBRA VISA is available. Design porting between manufacturers, both using passport libraries, is very simple and needs only low effort.

Library porting requires detailed knowledge and an extensive effort. A totally different approach is to *generate cells automatically from an abstract* description using the design rules. This is possible with simple gates and flip flops without to much area overhead (about 10...20 %). In ALLIANCE, a powerful EDA system developed by Université Pierre et Marie Curie in Paris, the cell library is specified in a metalanguage and must be mapped using the design rules of the technology [17.1]. The method is similar to the 'stick diagram' method used in former times for the description of a lay out concept. The area overhead is in the order of 20 %. In many university and prototype runs this is acceptable; however, for industrial designs this may not be enough area effective, but this may change in future. The area wasted by ineffective synthesis may be high today. As device dimensions shrink, the designer has less available core area, and *pad limited designs* are becoming more and more common.

17

17.8 References

There are only few publicly available references, and with quickly advancing technology there are many fundamental changes. The internet is one of the best modern sources for information on this topic: all companies and associations working in this field are represented with useful information.

[17.1] *EDA System for VLSI design*, free Software from Université Pierre et Marie Curie, ParisCedex 05, ftp://ftp-asi.lip6.fr/pub/alliance, Homepage: http://www-asim.lip6.fr/alliance/

[17.2] Website of EDA supplier AVANT! http://www.avanticorp.com/

[17.3] ANSI/EIA 656: *I/O Buffer Information Specification*, Version 3.2 8/99. http://www.eia.org./EIG/IBIS/ibis.htm and from the ANSI Organisation.

[17.4] *Library of IBIS Models*, Internet Server: http://eda.org/pub/ibis

[17.5] IEEE Std 1076 1993, *IEEE Standard VHDL Language Reference Manual*. Institute of Electrical and Electronics Engineers, Piscataway , NJ, USA

[17.6] IEEE Std 1164 1993, *Standard Multi value Logic System for VHDL Model Interoperability*. Institute of Electrical and Electronics Engineers, Piscataway, NJ, USA

[17.7] *LTI Tools for Cell Design*, ASIC Libraries, Power Simulation, Block Verification and Chip Timing. Company provided information of Library Technologies, Inc. 19959 Lanark Lane, Saratoga, Ca 95070, http://www.libtech.com/

[17.8] ALCATEL MIETEC: *Standard Cell Design Data Book*, 0.5 µm CMOS, Library MTC35xxx Rev. 1.2. Company provided information from ALCATEL Mietec. Oudenaarde, Belgium 1996

[17.9] *Standard Delay Format Specification*, Version 2.1. May be requested from: Open Verilog International (OVI), Los Gatos, CA or http://www.ovi.org

[17.10] Michael John Sebastian Smith: *Application Specific Integrated Circuits*. Addison Wesley Longman, Inc., 1997

[17.11] Internet Website of Synopsys: http://www.synopsys.com/cgi-bin/svp/lib/

[17.12] IEEE std. 1076.4-1995. IEEE-Standard for VITAL Application Specific Integrated Circuits ASIC. IEEE, Piscataway, NJ 08855-1391, USA or http://www.ieee.org/

18 Programmable Logic Devices

UWE PAUL

Programmable logic devices (PLDs) belong to the integrated circuits (IC). They divide up into immediately available standard ICs and 'application specific' integrated circuits (ASIC), which are yet to be made in a rather time consuming way. From the manufacturer's point of view PLDs are standard products for a maximum clientele, from the customer's point of view, however, they are application specific after programming. Thus the PLDs combine the advantage of immediate availability with application specific functionality which explains their attractiveness.

18.1 The Basic Concept of a Programmable Logic Device

How can an already complete circuit still be changed? The forerunners show the close relationship to the programmable memory which reveals the secret.

18.1.1 Historical Milestones

In the middle of the 1960s, Harris Semiconductor developed the first PLD, namely, a **fuse configurable diode matrix**. The diodes connected horizontal lines with vertical lines. These connections could be interrupted by a high current phase during programming.

In 1970 the first programmable read-only memory (**PROM**) developed from it. Such a memory is presented in the following section.

In 1975 Intersil and Signetics presented the first **FPLAs (Field Programmable Logic Array)** at nearly the same time. FPLAs are an obvious generalization of the memory structure. They had, however, no commercial success owing to the lack of computer support for programming.

It was only the simplification to **PAL (Programmable Array Logic)** by Monolithic Memories Inc. (MMI) in 1978 which succeeded on the market. Thanks to good customer support by a PAL manual and the computer support by the PAL assembler (PALASM), which is a program for the translation of Boolean equations into the program data, the demand rapidly exceeded the production capacity. After the takeover of MMI by Advanced Micro Devices (AMD), its daughter Vantis Corp. still offers 5 families of PALs today. This unbroken popularity justifies the detailed study of these components in the sections 18.2 to 18.4 according to the documents [18.1], [18.2] and [18.9].

The development to increasingly more complex programmable logic devices (**CPLD**), respectively, **Field Programmable Gate Array (FPGA)** is dealt with in section 18.5.

18.1.2 From The Memory To The Programmable Logic Device

A memory assigns the associated data to the allocated address. This can also be used for the implementation of logic. Figure 18.1 shows the logic AND operation for the input signals $e1$ and $e0$ which result in the output signal a. The truth table in 18.1 shows the values of 'a' for all possible input combinations of $e1$ and $e0$. This table can also be taken as an assignment of a 4×1 bit memory. Thus the programmable memory is a programmable logic device at the same time.

Table 18.1 AND operation

$e1$	$e0$	a
0	0	0
0	1	0
1	0	0
1	1	1

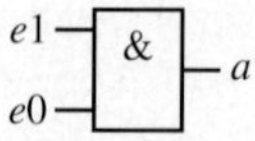

Fig. 18.1 Symbol of an AND gate

Figure 18.2 shows the general realization of a 4×2 bit PROM. The two address inputs are fed into a diode matrix as true and inverted inputs. The vertical lines at the pull up resistors realize positive wired ANDs. A 'low' (0) at an effective diode is sufficient to set the AND line to 'low'. In order to a 'high' (1) at the AND line, all diodes have to be stimulated with 'high'. Only one of the 4 AND lines can have the value 'high'. This line selects the memory word to be addressed. This is the reason why this circuit part is called the address decoder. It is simply an AND matrix.

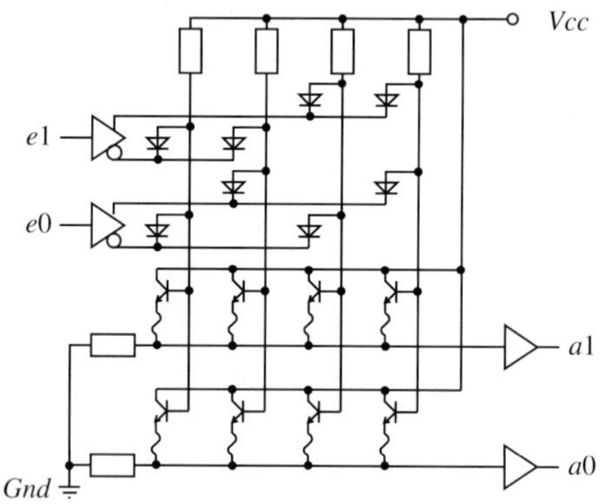

Fig. 18.2 Circuit diagramm of a 4 × 2 bit PROM

The second part of the circuit diagram in figure 18.2 is a transistor matrix. The horizontal lines with the pull down resistors can be pulled up to 'high' by any of the 4 connected transistors. They realize a positive OR operation (a wired OR). Each transistor is connected to the OR line by a fusible link. If this fusible link is blown during programming, the isolated transistor does not operate and the line remains at 'low'. Each AND line of the address decoder selects a pair of transistors showing a 2 bit wide data word at the memory output. This part of the circuit is the data memory and a simple OR matrix.

The fixed AND matrix followed by the programmable OR matrix is shown in a slightly simplified way in figure 18.3a). An '×' shown at the crossing of vertical and horizontal lines represents a diode and/or an effective transistor. Several logic AND operations grouped by an OR operation constitute the **Sum Of Products (SOP)**. There will be an ideal hardware for the realization of logical expressions when the AND matrix is fully covered, too, as shown in figure 18.3b). It is sufficient to have one diode at all the crossings which can also be made ineffective by a fusible link. This development of the memory was called **FPLA (Field Programmable Logic Array)**. Both AND and OR matrixes are programmable.

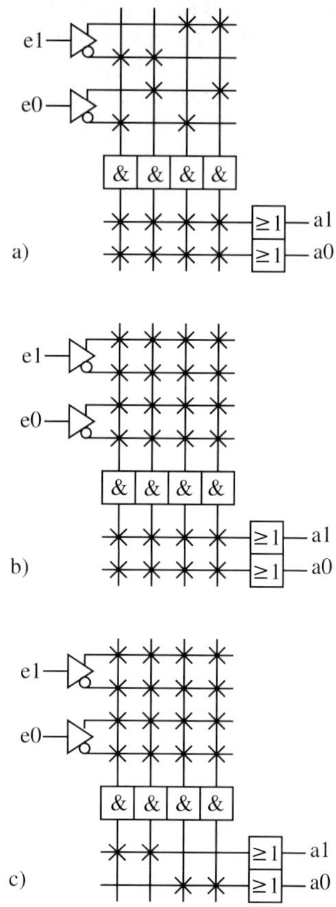

Fig. 18.3
a) Simplified figure of a PROM,
b) of an FPLA, and c) of a PAL

The third version in figure 18.3c) does without the programmability of the OR matrix and merely combines 2 AND lines each with one OR line. This decreases the universality; however, it still meets most of the practical requirements and makes programming much easier. This reversal of the memory, namely a programmable AND matrix followed by a fixed OR matrix, is called **PAL (Programmable Array Logic)**. These PALs started the invasion of programmable logic devices.

18.1.3 Realization Possibilities of Programmable Elements

Apart from the idea of a programmable AND-OR structure, the programmable memory also brought about the production technology and the realization of the programmable elements. In the PROM a thin wire, being a fusible link, connects every single transistor with the OR line. This fusible link can be destroyed by a current pulse. This interruption is irreversible. Components with fusible links can only be programmed once (*One Time Programmable – OTP*). A development style of trial and error is impossible with respect to these components.

The desire to have a possibility of correction led to the **EPLD (Erasable Programmable Logic Device)**. The fusible link is replaced by an N channel field effect transistor with isolated gate as shown in fig. 18.4.

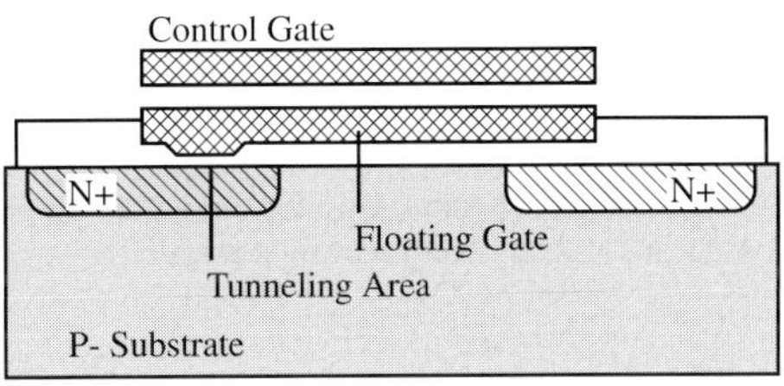

Fig. 18.4 Field effect transistor with isolated gate

In the case of a positive gate potential the FET conducts similarly to an unblown fusible link. The isolated gate is separated from the source diffusion by means of a very thin oxide layer only. When the programming voltage at the control gate is high and positive, negative charge carriers from the source diffusion will penetrate this weak point and will charge the isolated gate with a negative voltage. The field effect transistor will now block similarly to a blown fusible link. The isolated gate will be discharged again by an external energy supply by means of ultraviolet light, making the field effect transistor conductive again. Thus programming can be erased within 20 minutes. An advanced version can be erased electrically (**electrically erasable programmable logic device – EEPLD**) which takes place within milliseconds. Without erase measures the charge is guaranteed to remain in the isolated gate for 10 years, which means that this kind of programming is long lasting.

Another possibility of electrically operating switches is the static random access memory – **SRAM**. The bi-stable circuit made of two over-cross connected inverters stores the information for opening or blocking a field effect transistor as shown in fig. 18.5. This information is lost when the voltage is disconnected, thus the programming is volatile. Therefore programming has to be carried out first after each switching on of the voltage. In addition, reprogramming is possible even during operation. Such a component finds its programming data in an external non-volatile memory.

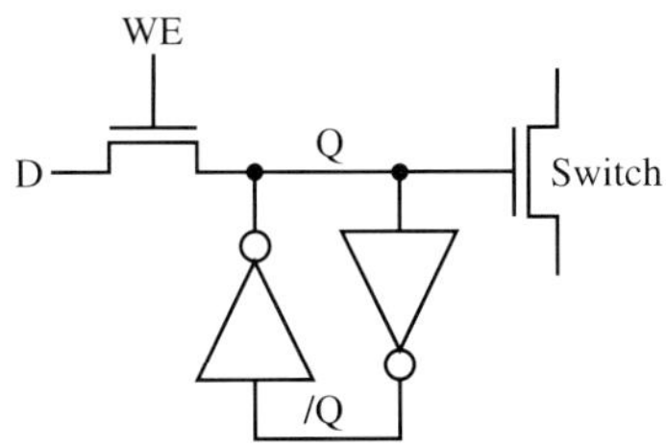

Fig. 18.5 Static memory cell (SRAM) controls field-effect transistors

The latest kind of programming memory cells reverses the principle of fusible links (therefore antifuse). Instead of interrupting a connection the separating insulation at the crossing of two lines is stripped off, thus creating a connection. Figure 18.6 shows two versions of such programming memory cells which hardly need any additional space. The destruction of a special insulating layer can, of course, be made only once (OTP), it is, however, durable instead.

The planned application of a programmed logic device may have an influence on the choice of the programming type. In the case of an unsafe voltage supply with frequent voltage drops, an SRAM-PLD will not work because of the constant reprogramming process. If there are still changes of wiring logic to be expected, an OTP-PLD is not to be considered.

18

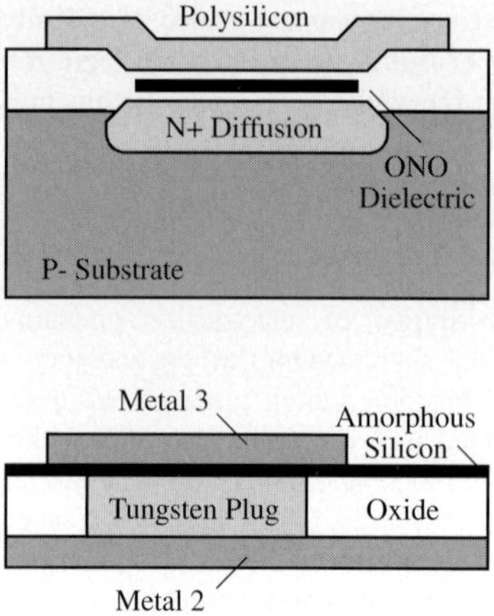

Fig. 18.6 Programmable connections: a) PLICE, b) ViaLink

18.2 Simple Combinatorial Programmable Logic Devices

18.2.1 The Basic Version Of The PAL

This basic type shown in figure 18.8 is directly derived from the PROM.

The component offers 10 inputs and 8 outputs which are active high (Active High – H). These three characteristics are responsible for the additional name 10H8. The 10 inputs feed in true and inverted mode the vertical input lines which are crossing the horizontal AND lines. This is the programmable AND matrix. Thus any of the 16 AND lines disposes of 20 physical inputs; however, only a maximum of 10 inputs can be used. Otherwise a true signal and an inverted signal would be conjunctively linked together, which always results in the low value. The 16 AND lines are combined in pairs with OR gates feeding the 8 output signals. A compact form of all information about the device is also included in the functional block diagram as shown in figure 18.7.

The possibilities offered by the PAL 10H8 are illustrated by means of three examples. In a microprocessor system an enable signal f for a port has to be produced with respect to a given address A:

$$A = \{e9, e8, e7, e6, e5, e4, e3, e2, e1, e0\}$$
$$= 2A5_{\text{hex}} = 10\ 1010\ 0101_{\text{bin}}$$

The binary representation of the address provides the logic operation of the 10 address bits.

$$f = e9*/e8*e7*/e6*e5*/e4*/e3*e2*/e1*e0$$

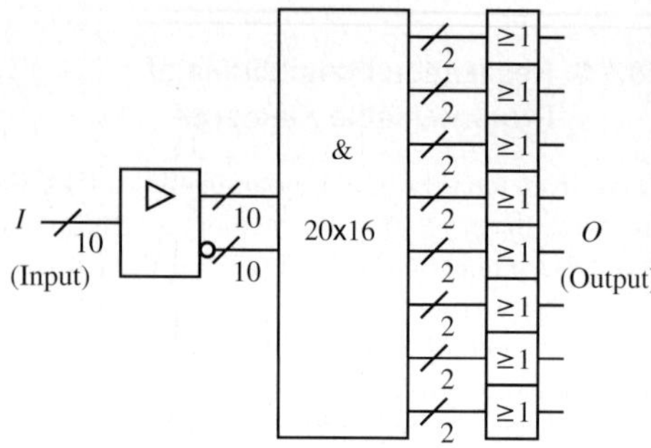

Fig. 18.7 Functional block diagram of PAL 10H8

This wide AND operation is 'programmed' at pin 19 in fig. 18.8: those programmable memory cells which are marked by an 'x' in the upper AND term remain intact, all others will be destroyed during the programming process. The second AND term of pin 19 is not needed, it must always supply a 'low'. For this purpose, all programmable memory cells remain intact. An address decoder realized in this way generates the intended enable signal within typical 0.5 ns.

In addition to this, 4 event lines have to be monitored and flagged with graded priority in the same microprocessor system. In table 18.2 the status signals are allocated to the events, respectively.

Table 18.2 Events $\{e4 \ldots e1\}$ with the respective status signals $\{m2 \ldots m0\}$

$e4$	$e3$	$e2$	$e1$	$m2$	$m1$	$m0$
1	—	—	—	1	0	0
0	1	—	—	0	1	1
0	0	1	—	0	1	0
0	0	0	1	0	0	1
0	0	0	0	0	0	0

In the case of the dashes in the left part of the table, the respective input signal can be ignored. (*don't care*). The equations for the three status signals can directly be seen in table 18.2.

$$m2 = e4$$
$$m1 = /e4 * e3 + /e4 * /e3 * e2$$
$$m0 = /e4 * e3 + /e4 * /e3 * /e2 * e1$$

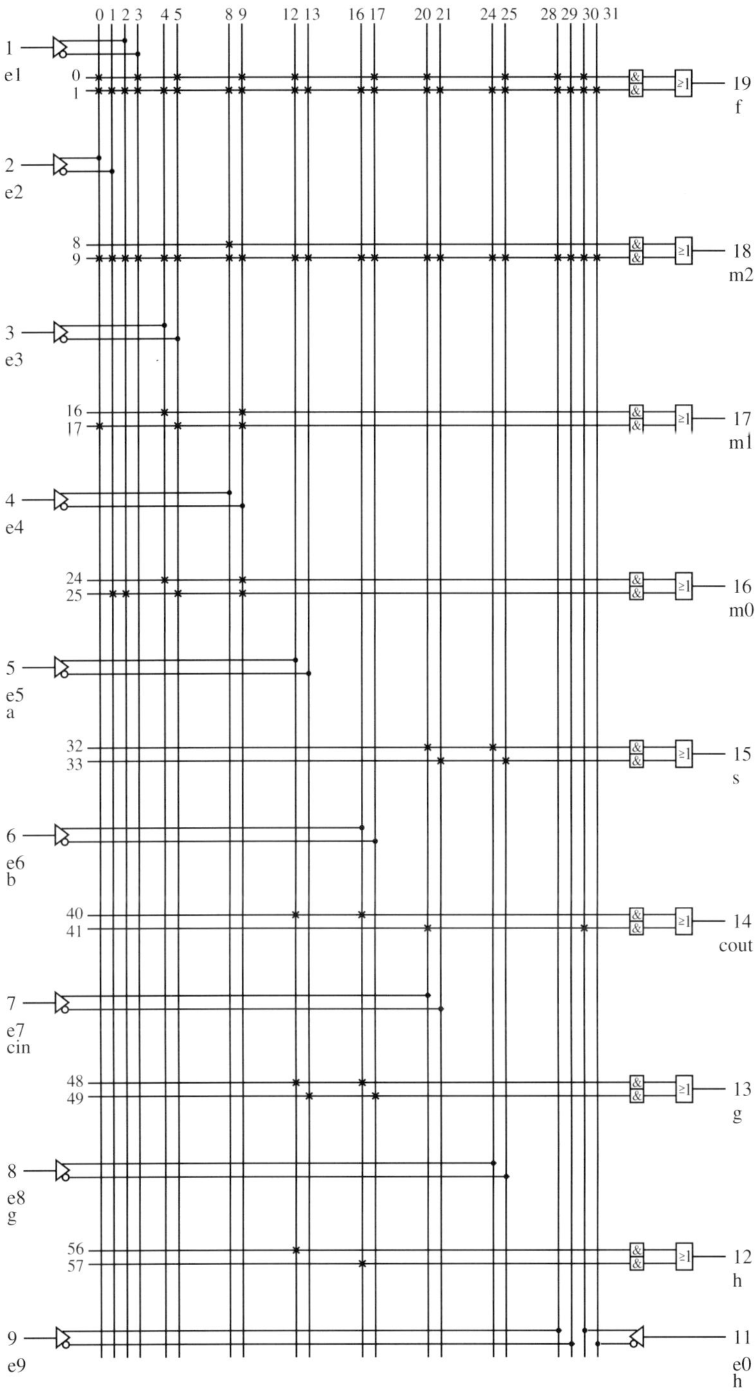

Fig. 18.8 Logic diagram of PAL 10H8

In fig. 18.8 these logic operations are also 'programmed' at the pins 18 to 16. Finally a 1-bit full adder according to table 18.3 illustrates the limits of PAL 10H8.

Table 18.3 Truth table of a full adder

a	b	c_{IN}	s	c_{OUT}
0	0	0	0	0
0	0	1	1	0
0	1	0	1	0
0	1	1	0	1
1	0	0	1	0
1	0	1	0	1
1	1	0	0	1
1	1	1	1	1

The equations for the sum s and for the carry indicator c_{OUT} are as follows:

$$s = a*b*c_{IN} + a*/b*/c_{IN} + /a*b*/c_{IN} + /a*/b*c_{IN}$$

$$c_{OUT} = a*b + a*c_{IN} + b*c_{IN}$$

These equations are too much for the PAL 10H8 because they need 4 and/or 3 AND terms. However, it is possible to reduce the required number of terms to 2 by factoring out and by introducing auxiliary functions.

$$s = (a*b+/a*/b)*c_{IN}+(a*/b+/a*b)*/c_{IN}$$

The reduction $g = a*b + /a*/b$ results in:

$$s = g*c_{IN} + /g*/c_{IN}$$

$$c_{OUT} = a*b + (a+b)*c_{IN}$$

The reduction $h = a + b$ results in:

$$c_{OUT} = a*b + h*c_{IN}$$

In figure 18.8 the 'programming' is carried out at pin 15 and pin 14. In addition, the outputs of pin 13 (g) and pin 12 (h) with the inputs of pin 8 and pin 11 have to be connected externally. This results in a cascaded AND-OR operation. The component is used twice and the delay is doubled. For the 1-bit full adder a component with at least 4 AND terms per OR gate is of course more suitable.

18.2.2 Additional Internal Feedback and Switchable Output Drivers

The logic diagram of PAL 10H8 in fig. 18.8 is already complex enough. In order to concentrate on the essential issue, the following sections deal with simple examples. However, the indications given follow the real models which are shown in brackets.

The example of the full adder shows that a cascade of AND-OR logic is sometimes necessary. For this purpose, some PALs are provided with an internal feedback as shown in fig. 18.9.

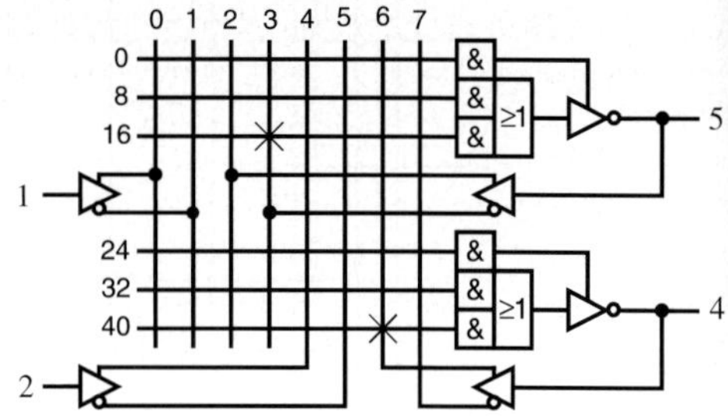

Fig. 18.9 Logic diagram of PAL 4L2

In addition to this, a controllable output driver inverts the signal of the OR gate. Thus the output becomes low active (*active low – L*) which leads to the 'L' in the suffix of PAL 4L2 (*of PAL 16L8*). The upper AND term with the abbreviation of OE (*output enable*) serves as a control of the output driver. Three different connection types can be programmed in this way.

a) OE = 1: all programmable memory cells are destroyed, the pull-up resistor is active: The output driver is constantly enabled and the connection works as an output.
b) OE = 0: all programmable memory cells are intact, true signals and inverted signals result in low: The output driver is constantly disabled and therefore highly resistive. The connection can now be used as an additional input.
c) OE is changed by the logic operation while operating. In this case the connection operates in both directions as input/output.

The internal feedback allows the immediate programming of an output signal in the AND matrix. If it is used in adjacent AND terms, there will be a multi-stage 'transparent' logic. If the output signal is, however, used in its own AND terms, there will be a feedback. Two possibilities have to be distinguished:

There will be a positive feedback if the re-feeding driver/inverter relinquishes the inversion of the output driver. This is programmed in the upper half of fig. 18.9. As long as pin 5 indicates 'low', the positive feedback term supplies a 'high'. However, if pin 5 becomes 'high' it will be confirmed by a 'low' of the positive feedback term and will be maintained until the voltage is cut off. Such a circuit catches the high.

There will be an inverse feedback if the inversion is maintained in the feedback. As shown in the lower part of fig. 18.9, the programmed circuit is never satisfied and constantly changes the output value at pin 4. There is a ring oscillator which could be switched on and off via the middle AND term. It should actually be handled with caution, as it does not have any guaranteed frequency.

The ring oscillator may be useful for test purposes. The delay time of the complete feedback loop, e.g., of the input driver/inverter, the AND line, the OR gate, and the output driver occurs between two counter-sense clock edges. This is identical with the gate sequence of an input pin to an output pin. Thus the delay time t_{PD} between the pins can be determined by the measured oscillator frequency f.

Numerical example:

frequency $f = 66.7$ MHz (measured)
time of oscillation $T = 1/f = 15$ ns
delay time $t_{PD} = T/2 = 7.5$ ns.

In the data sheet for PAL 16L8 the total delay time for the measured frequency should read 7.5 ns.

18.2.3 Programmable Polarity

The two PALs dealt with above both have a defined output polarity, either active high or active low. An additional exclusive OR (EXOR) placed before each output allows the individual programming of the polarity. Figure 18.10 shows an EXOR gate with a programmable input.

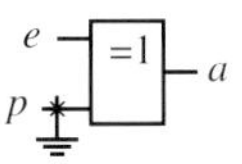

Fig. 18.10 Symbol of the EXOR gate

In the case of an in tact programmable memory cell the input p shows 'low', and in case of a destroyed fusible link there will be 'high' by means of pull up. As shown in the truth table 18.4, the output a for $p = 0$ is identical with the input e. For $p = 1$, however, 'e' will be inverted to become 'a'.

Table 18.4 EXOR operation

e	p	a	
0	0	0	
1	0	1	$a = e$
0	1	1	
1	1	0	$a = /e$

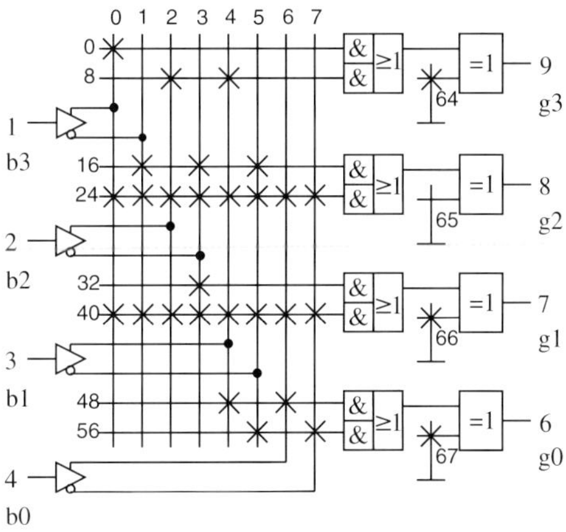

Fig. 18.11 Logic diagram of PAL 4P4

The possibility of inverting may be very helpful. A code transcriber from **BCD Code (Binary Coded Decimal)** to the Excess 3 Gray Code according to table 18.5 is to be realized by means of the PAL 4P4 (of PAL 10P8).

Table 18.5 Truth table BCD to Excess 3 Gray Code

dec.	$b3$	$b2$	$b1$	$b0$	$g3$	$g2$	$g1$	$g0$
0	0	0	0	0	0	0	1	1
1	0	0	0	1	0	0	1	0
2	0	0	1	0	0	1	1	0
3	0	0	1	1	0	1	1	1
4	0	1	0	0	0	1	0	1
5	0	1	0	1	0	1	0	0
6	0	1	1	0	1	1	0	0
7	0	1	1	1	1	1	0	1
8	1	0	0	0	1	1	1	1
9	1	0	0	1	1	1	1	0

The 6 missing input combinations do not apply. The comparison between input signals and output signals directly supplies the equations required:

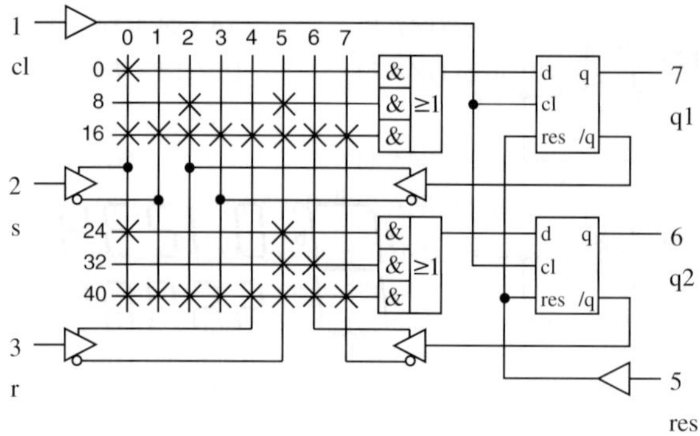

Fig. 18.15
Logic diagram of PAL 4R2

Table 18.7 Truth table of the conflict-free SR flip flops

s	r	$q1$	$d1$	s	r	$q2$	$d2$
0	0	—	$q1$	0	0	—	$q2$
0	1	—	0	0	1	—	0
1	0	—	1	1	0	—	1
1	1	—	1	1	1	—	0

The connection of the input signals s and r with the stored value q defines the next value at the D input of the flip flop, which will be stored with the next clock edge.

$$d1 = s + /r * q1$$

$$d2 = s * /r + /r * q2$$

The equations can be realized with 2 AND terms and are programmed in figure 18.15. The JK flip flop according to table 18.8 offers another way of solving the problem. For $J = K = 1$ it inverts the stored value, and toggles to the opposite. The T flip flop according to table 18.8 combines J and K to a T input.

Table 18.8 Truth table of JK flip flop and of T flip flop

j	k	$q3$	$d3$	t	$q4$	$d4$
0	0	—	$q3$	0	—	$q4$
0	1	—	0			
1	0	—	1			
1	1	—	$/q3$	1	—	$/q4$

$d3 = j * /q3 + /k * q3 \quad d4 = /t * q4 + t * /q4$

Again, the respective equations only need 2 AND terms each and could immediately be programmed into PAL 4R2. Thus the preceding AND-OR structure allows the implementation of other flip flop types, too.

Very often the D flip flop is directly used when designing sequential devices. Starting from the actual state, a subsequent state is prepared depending on the input signals. Figure 18.16 shows a variable divider as an example of a **finite state machine (FSM)**. The circles represent the states, and the arrows show state transitions which are triggered by a clock edge.

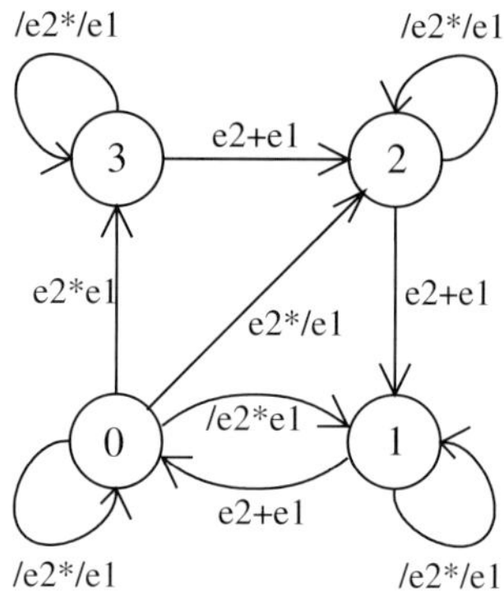

Fig. 18.16 Transition graph (state diagram) of a variable divisor

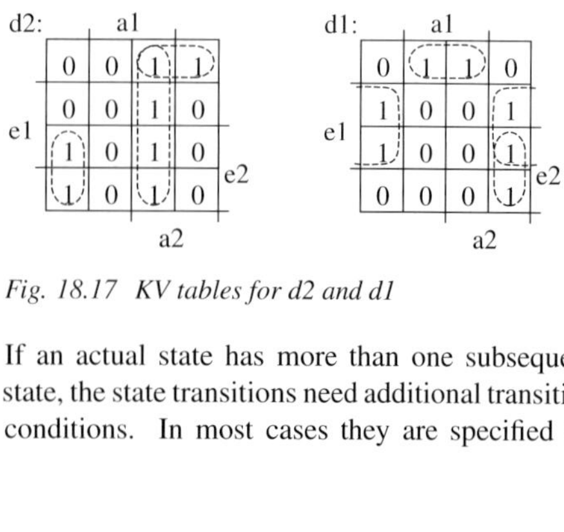

Fig. 18.17 KV tables for d2 and d1

If an actual state has more than one subsequent state, the state transitions need additional transition conditions. In most cases they are specified by

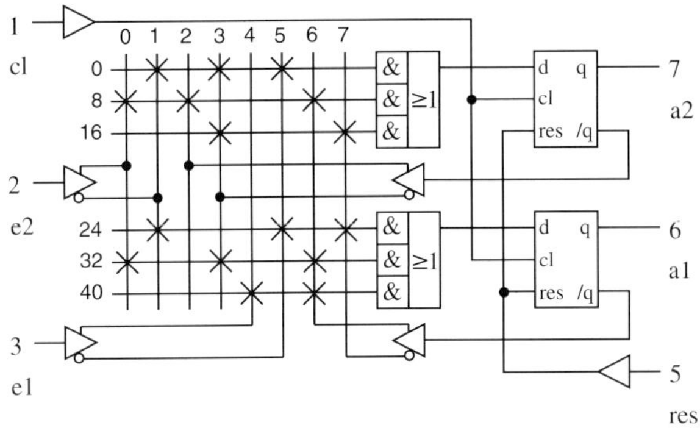

Fig. 18.18 Logic diagram of PAL 4R2 with programmed variable divisor

the input signals and are indicated at the respective arrows. Each state transition can also be found in the functional table 18.9.

Table 18.9 Functional table of the variable divider

$e2$	$e1$	$a2$	$a1$	$d2$	$d1$
0	0	—	—	$a2$	$a1$
0	1	0	0	0	1
1	0	0	0	1	0
1	1	0	0	1	1
$e2 + e1$		0	1	0	0
$e2 + e1$		1	0	0	1
$e2 + e1$		1	1	1	0

The KV tables for the d inputs in fig. 18.17 are equivalent to the above function table 18.9. They result in minimized Boolean equations for the flip flop control:

$$d2 = /e2 * /e1 * a2 + e2 * /a2 * /a1 + a2 * a1$$

(3 AND terms)

$$d1 = /e2 * /e1 * a1 + e2 * a2 * /a1 + e1 * /a1$$

(3 AND terms)

In fig. 18.18 these Boolean equations are programmed into PAL 4R2.

The example of variable dividers clearly illustrates the various methods of description for a sequential device. The transition graph in fig. 18.16, its representation in form of the table shown in 18.9, and the KV tables given in fig. 18.17 are still independent of the realization. The Sum of Products of the Boolean equations already takes the possibilities of the target hardware into account.

The logic diagram in fig. 18.18 with the × marks at the effective inputs provides the wiring diagram and thus it serves as a production document.

18.3.3 EXOR Gate Preceding The Register Inputs

In case of the present PAL 4X2 an EXOR gate supplements the programmable AND-OR structure of the D flip flop as shown in fig. 18.20. Unlike in case of PAL 4P4 with first programmable and then defined polarity, it is now also possible to invert dynamically.

According to fig. 18.19 a T flip flop gate is realized by using the EXOR gate. With 'low' at the T input, the T flip flop takes its former q value with every new clock, thus maintaining its value. A 'high' at the T input inverts the q value, and the flip flop toggles.

18

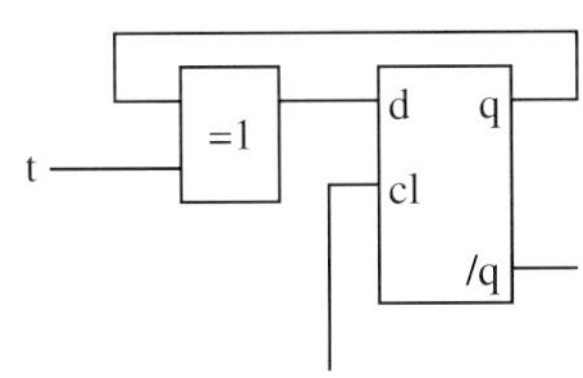

Fig. 18.19 T flip flop existing of EXOR gate and D flip flop

The aforesaid T flip flop is especially useful for the design of counter circuits as is shown by the example of a four digit binary upward counter. Table 18.10 shows the actual state $\{q3, q2, q1, q0\}$, the subsequent state $\{d3, d2, d1, d0\}$ and the toggle conditions $\{t3, t2, t1, t0\}$, too.

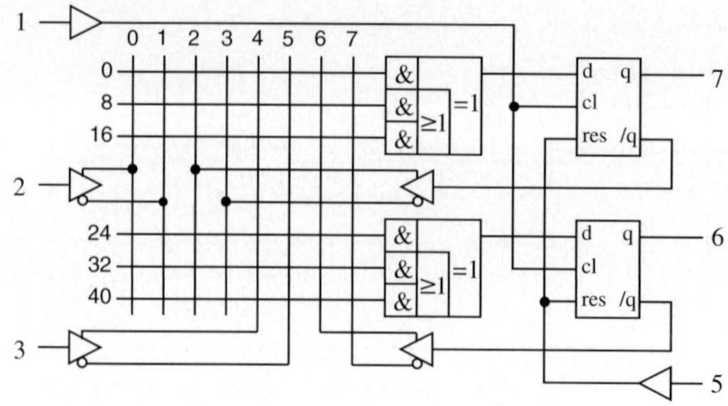

Fig. 18.20
Logic diagram of PAL 4X2

Table 18.10 4 digit binary upward counter

$q3$	$q2$	$q1$	$q0$	$d3$	$d2$	$d1$	$d0$	$t3$	$t2$	$t1$	$t0$
0	0	0	0	0	0	0	1	0	0	0	1
0	0	0	1	0	0	1	0	0	0	1	1
0	0	1	0	0	0	1	1	0	0	0	1
0	0	1	1	0	1	0	0	0	1	1	1
0	1	0	0	0	1	0	1	0	0	0	1
0	1	0	1	0	1	1	0	0	0	1	1
0	1	1	0	0	1	1	1	0	0	0	1
0	1	1	1	1	0	0	0	1	1	1	1
1	0	0	0	1	0	0	1	0	0	0	1
1	0	0	1	1	0	1	0	0	0	1	1
1	0	1	0	1	0	1	1	0	0	0	1
1	0	1	1	1	1	0	0	0	1	1	1
1	1	0	0	1	1	0	1	0	0	0	1
1	1	0	1	1	1	1	0	0	0	1	1
1	1	1	0	1	1	1	1	0	0	0	1
1	1	1	1	0	0	0	0	1	1	1	1

The respective Boolean equations for the d inputs and/or t inputs read as follows:

$$d3 = q3 * /q2 + q3 * /q1 + q3 * /q0$$
$$\quad + /q3 * q2 * q1 * q0$$
$$d2 = q2 * /q1 + q2 * /q0 + /q2 * q1 * q0$$
$$d1 = q1 * /q0 + /q1 * q0$$
$$d0 = /q0$$
$$t3 = q2 * q1 * q0$$
$$t2 = q1 * q0$$
$$t1 = q0$$
$$t0 = 1$$

This clearly shows that the number of required AND terms for the d inputs increases with the number of digits. The toggle equations, however, can be realized with one single AND term. Moreover, it follows a very simple construction law: 'the n-th digit has to toggle if all lower digits are at high, e.g., '1'!'. Considering the EXOR gate in fig. 18.19, the equation for the highest digit of a 10 digit binary upward counter would read:

$$d9 = q9 \oplus q8 * q7 * q6 * q5 * q4 * q3 * q2 * q1 * q0$$

It can hardly be easier than this! As to the equally useful downward counter, 'at maximum' means that all lower digits show the value '0'.

The conversion of the storage element into a T flip flop is certainly not the only possibility of using the EXOR gate. It also supports general logical expressions in the AND-OR-EXOR version.

18.3.3.1 Reed Muller Standard Form

Apart from the two best known standard forms for Boolean expressions, namely, the *Sum of Products (SOP)* and the *Product of Sums (POS)*, there are other standard forms which use for example exclusively the NAND gate or exclusively the NOR gate. An *EXOR Sum of Products (ESOP)* is named after its fathers **Reed** and **Muller**. It contains **only true signals** (not inverted). Any logical expression can be represented by the Reed Muller format. A generalization also allows inverted signals; however it does not allow a signal in both polarities. It is called '**generalized Reed Muller (GRM)**'. If there are signals in both polarities in an AND-OR-EXOR expression this is called '**mixed GRM**'. The last mentioned Sum of Products is supported by the PAL 4X2 because of the input drivers.

How is a given Boolean expression transferred into a Reed Muller format? The inventors proved the existence and definite expression which guaranteed the success of the search, but they were at a loss to indicate a simple guide. A possibility is provided by the '**EXOR Factoring**' according to [18.9], which uses the following identity equations:

$$F \oplus F \oplus A = (F \oplus F) \oplus A = 0 \oplus A = A$$

If a logical expression A is EXOR combined twice with the identical expression F, they cancel out and A remains unchanged. In the case of 'EXOR factoring' one of the EXOR operations is carried out, the other one maintained as the EXOR factor. The equation for the $d3$ input of the upward counter are taken as an example. As all AND terms include the variable $q3$ the latter is taken as an EXOR factor:

$$\begin{aligned}
d3 = q3 \oplus q3 \oplus (&q3 * /q2 + q3 * /q1 \\
&+ q3 * /q0 + /q3 * q2 * q1 * q0) \\
d3 = q3 \oplus [/q3 * (&q3 * /q2 + q3 * /q1 \\
&+ q3 * /q0 + /q3 * q2 * q1 * q0) \\
&+ q3 * (/q3 + q2) * (/q3 + q1) \\
&*(/q3 + q0) \\
&*(q3 + /q2 + /q1 + /q0)] \\
d3 = q3 \oplus [/q3 &* q2 * q1 * q0 \\
&+ (/q3 + q2 * q1 * q0) * (q3)] \\
d3 = q3 \oplus [/q3 &* q2 * q1 * q0 \\
&+ q3 * q2 * q1 * q0] \\
d3 = q3 \oplus q2 &* q1 * q0
\end{aligned}$$

The result is no longer surprising because it is already known as the **toggle condition**. In the **Reed Muller form** $d3$ requires only one AND gate and one EXOR gate compared to four AND gates and one OR gate in the *Sum of Products*.

18.3.4 Arithmetic Combinations Of Inputs With Outputs

The buffers/inverters for the inputs and for fed back outputs are now replaced by four different OR gates. In fig. 18.21 the vertical input lines already represent arithmetic combinations of an input/output signal pair respectively.

Table 18.11 Programming possibilities of a combined signal pair

Fuses	Link	Logic	Arithmetic
---	1 (Pullup)	TRUE	
x--	$e + a$	OR	
-x-	$/e + a$		$e \leqq a$
-x-	$e + /a$		$e \geqq a$
--x	$/e + /e$	NAND	
xx-	a	BUF	
x-x-	e	BUF	
x-x	$e \oplus e$	XOR	$e \neq a$
-xx-	$e = a$	XNOR	$e = a$
-x-x	$/e$	INV	
-xx	$/a$	INV	
xxx-	$e * a$	AND	
xx-x	$/e * a$		$e < a$
x-xx	$e * /a$		$e > a$
-xxx	$/e * /a$	NOR	
xxxx	$a * /a = 0$	FALSE	

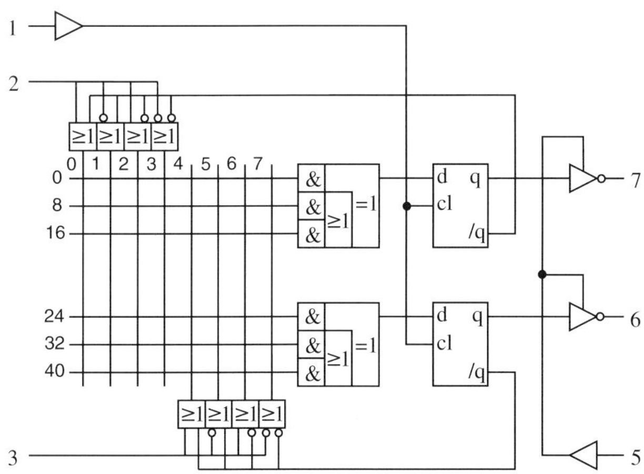

Fig. 18.21 Logic diagram of PAL 4A2

In this way it is possible to program 16 individual functions within the AND matrix. They are shown in table 18.11 in the form of a programming specification. For some functions the arithmetic denotation is added.

The above table shows all theoretically possible functions with up to two variables. The abnormal functions with a constant value high (1) or low (0) may cause astonishment, the practice shows however, that they do occur in the case of defective components.

The aforesaid PLD type is preferentially applied to **fast arithmetic parallel operations**. It can demonstrate its efficiency by searching a maximum value for example. It will compare the 4-digit binary numbers $\{e3, e2, e1, e0\}$ with its stored output values $\{a3, a2, a1, a0\}$. If the actual input value is higher than the one stored, this value shall be stored as the new maximum value.

A complete function table taking all possible combinations of numbers into account requires $2^4 \times 2^4 = 256$ lines. The comparison is, however, only made by positions starting with the highest position. In addition there are only three possible comparison results per position, e.g., higher, lower, or equal. In the latter case the next lower position has to be queried. The result is the compressed function table 18.12 with only nine lines.

In the Sum of Products the Boolean equations for the d0 input would therefore require nine AND terms. But a change is realized in only four lines of the function table. A T flip flop which is satisfied by the toggle conditions can be programmed by means of the EXOR gate. The existing Reed Muller standard format requires a maximum of only four AND terms for each position:

$$
\begin{aligned}
d3 &= a3 \oplus \quad (e3 > a3) \\
d2 &= a2 \oplus [\ (e3 > a3) * (e2 \oplus a2) \\
&\qquad + (e3 = a3) * (e2 > a2)\ ] \\
d1 &= a1 \oplus [\ (e3 > a3) \\
&\qquad * (e1 \oplus a1) \\
&\qquad + (e3 = a3) * (e2 > a2) \\
&\qquad * (e1 \oplus a1) \\
&\qquad + (e3 = a3) * (e2 = a2) \\
&\qquad * (e1 > a1)\ ] \\
d0 &= a0 \oplus [\ (e3 > a3) \\
&\qquad * (e0 \oplus a0) \\
&\qquad + (e3 = a3) * (e2 > a2) \\
&\qquad * (e0 \oplus a0) \\
&\qquad + (e3 = a3) * (e2 = a2) \\
&\qquad * (e1 > a1) * (e0 \oplus a0) \\
&\qquad + (e3 = a3) * (e2 = a2) \\
&\qquad * (e1 = a1) * (e0 > a0)\ ]
\end{aligned}
$$

These equations follow directly from the function table 18.12. A position needs to change only when it does not match. This explains the additional unequal conditions at the lower positions. Because of the programming specification in table 18.11 the arithmetic terms can be programmed directly. The importance of the said components demands the highest respect when you try to solve the task, for example, with the PAL 8R4.

Table 18.12 Function table for the d inputs of a maximum value finder

3rd position	2nd position	1st position	0 position	*d3*	*d2*	*d1*	*d0*
$e3 < a3$	–	–	–	$a3$	$a2$	$a1$	$a0$
$e3 > a3$	–	–	–	$e3$	$e2$	$e1$	$e0$
$e3 = a3$	$e2 < a2$	–	–	$a3$	$a2$	$a1$	$a0$
$e3 = e3$	$e2 > a2$	–	–	$a3$	$e2$	$e1$	$e0$
$e3 = a3$	$e2 = a2$	$e1 < a1$	–	$a3$	$a2$	$a1$	$a0$
$e3 = a3$	$e2 = a2$	$e1 > a1$	–	$a3$	$a2$	$e1$	$e0$
$e3 = a3$	$e2 = a2$	$e1 = a1$	$e0 < a0$	$a3$	$a2$	$a1$	$a0$
$e3 = a3$	$e2 = a2$	$e1 = a1$	$e0 > a0$	$a3$	$a2$	$a1$	$e0$
$e3 = a3$	$e2 = a2$	$e1 = a1$	$e0 = a0$	$a3$	$a2$	$a1$	$a0$

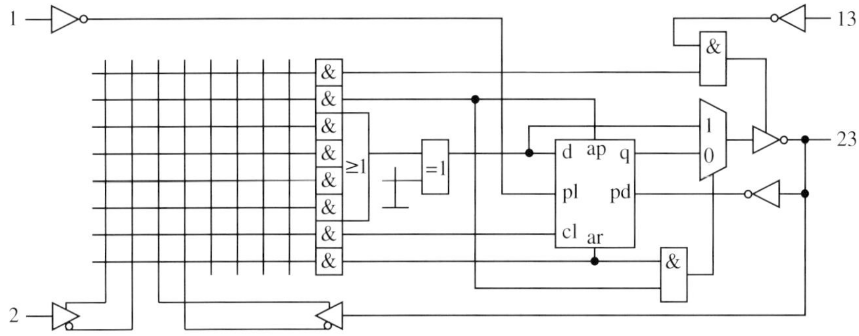

Fig. 18.22 Output macro cell of PAL 20RA10 (Registered Asynchronous)

18.3.5 Asynchronous Register Functions

All sequential logic devices dealt with so far are provided with a clock input which controls all storage elements simultaneously. The outputs can only change synchronously with this clock. Additional possibilities for set or reset are required for some applications. Such possibilities are, for example, provided by PAL 20RA10. Its outputs are provided with a macro cell each, as shown in fig. 18.22.

Out of the eight AND terms the four outer ones each serve a specific task:

- control of the output driver;
- asynchronous set (ap); and
- asynchronous reset (ar); as well as
- clock (cl) of flip-flop.

The four middle AND terms are collected by the OR gate. The polarity of the d input can be selected by means of an EXOR. The resolution of the set/reset conflict is very interesting: If both signals ap and ar are active the register is avoided, making the output combinatorial.

The preload function (PL) facilitates the test of complex state machines. It allows, by using output pins, to charge arbitrary, even forbidden, states into the flip flops. Following a clock edge it is possible to test a subsequent state. If several subsequent states are possible, depending on the input conditions, they can directly and successively be tested without running through long sequences.

The combination of combinatorial and sequential logic devices had a stimulating effect on imitations.

18.3.6 Generic-Array-Logic GAL 16V8 (versatile – V)

The GAL components of the Lattice company use the EEPROM technology and are electrically programmable and electrically erasable. Such a possibility of correction during a development is very welcome, especially for prototypes. GAL components are simply a temptation to try other circuit realizations [18.5].

The logic diagram in fig. 18.23 shows the well known structure with eight programmable AND terms for each of the eight output logic macro-cells (*OLMC*).

Figure 18.24 shows an output logic macro-cell OLMC. Four multiplexers allow different configurations. The output multiplexer (OMUX) selects the type of output. The switchable output driver drives the TSMUX (tri-state multiplexer), one of its four inputs being an AND term. If the AND term is selected the Product Term Multiplexer (PT-MUX) will replace it by a low (0) signal at the OR gate. Finally, the Feedback Multiplexer (FMUX) is responsible for the desired feedback. Theoretically it may be possible to program $2\times4\times2\times4 = 64$ different combinations, but in practice the selection is limited to only six variants. In table 18.13 they are even subdivided into three operation modes.

In '*registered mode*' the pins 1 and 11 serve as a clock input (CLK) and as control (OE) of the output driver. In '*simple mode*' they feed, as standard inputs, the AND matrix via the feedback. As far as the pins 19 and 12 are concerned, being inputs, they have to back up to the neighboring output cells. In such a competition of suppression

18

18.4.1 Programming of the GAL 16V8

The GAL 16V8 offers in its logic diagram according to fig. 18.23 a programmable AND matrix with 2 × 16 inputs and 8 × 8 AND terms. These 2048 crossings must be addressed individually in order to be able to interrupt the connection. In addition to this further programming positions define the desired architecture of the output cells.

The GAL 16V8 offers an integrated selection logic to address the total of 2196 programming positions. This logic has to be driven with the existing pins. For programming purposes the pins necessarily have another task, as is shown in fig. 18.25.

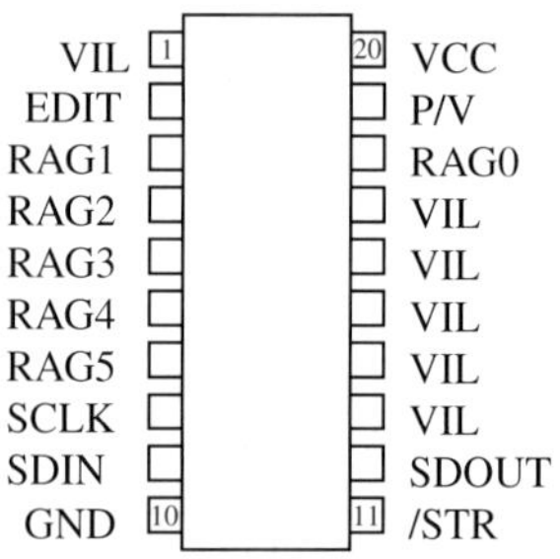

Fig. 18.25 Pin allocation of GAL 16V8 for programming

First, the programming data are latched to a shift register via the serial data input (SDIN). Each position of the shift register is allocated to an AND term (product term, PT). The row address (RAG5 to RAG0) indicates the input within the AND terms. Figure 18.26 shows the row address map block diagram.

Only 36 different row addresses are used:

- RA0 to RA31 cover the AND matrix area;
- RA 32 receives the 'electronic signature', e.g., a field with 64 bits for arbitrary user data such as circuit name, version identification or inventory identification;
- RA 60 is used for the definition of the architecture of the output cell; and
- RA 61 is for the security cell. It provides a copying protection against unauthorized read-out of programming data. After being programmed it does not allow access to the AND matrix area;
- RA 63 drives the erase cell. Its programming releases a complete bulk erase. The component is afterwards in the original state again.

A programming voltage of approximately 16.5 V at pin EDIT puts the GAL 16V8 in its programming mode. With each clock at SCLK the shift register accepts the new value at SDIN. A clock pulse of the length of 10 ms at pin /STR starts the actual programming. The signal P/V changes from programming mode to verification mode. For control purposes the state of any programming position can be read out at SDOUT by means of the shift register.

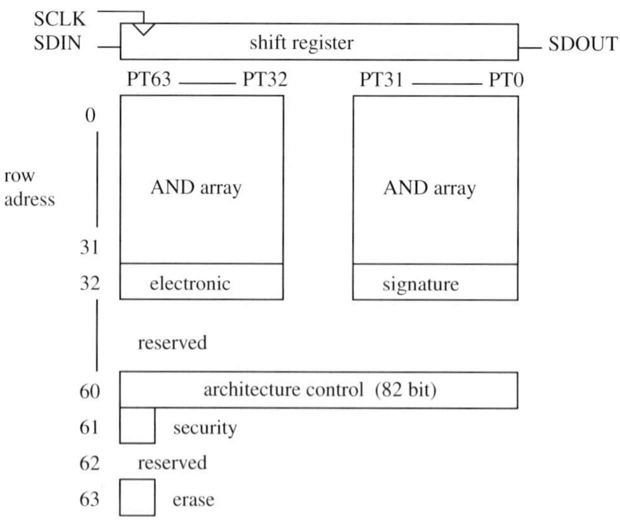

Fig. 18.26 Row address map block diagram of GAL 16V8

The attempt to program the GAL 16V8 with simple laboratory means such as power supplies and switches is obviously doomed to failure. Better computer supported tools are needed.

18.4.2 Computer Support for Programming

Starting from the design input (see chapters 3 to 8) three steps lead to the programmed logic device:

- First of all, the circuit specification has to be transferred to rather simple logical relations between the input signals and the output signals. Moreover, the polarity, and if necessary, a storage element has to be selected. This equation is still independent with respect to realization.
- In the second step the equations are harmonized with the resources of the chosen component. The × marks in the logic diagram (*Fuse Plot*) promise a success and represent the programming instruction.
- The actual programming is carried out by a programming unit, e.g., the programmer. It transmits the programming data to the logic component and supplies the required voltage pulses.

As promised, the variable divider mentioned in section 18.3.2 shall be taken as an example to demonstrate these steps. The listing in 18.1 shows the design input with the names of the input signals and output signals, their relationship in form of Boolean equations and the desired pin allocation. The sign ':=' assigns the value of the logic operation to a storage element, the sign '&' represents the wired AND.

List 18.1 Design input for the variable divider

```
*IDENTIFICATION
TITLE: VARIABLE DIVIDER

*XNAMES
A2, A1, E2, E1, CLK, OE;

*YNAMES
A2, A1;

*BOOLEAN-EQUATIONS
A2 := /E2 & /E1 & A2 + E2 & /A2 & /A1
        + A2 & A1;
A1 := /E2 & /E1 & A1 + E2 &  A2 & /A1
        + E1 & /A1;

*PLD
TYPE = GAL16V8 ;

*PINS
CLK=1, E2=2, E1=3, OE=11, A1=18, A2=19;

*END
```

The minimized Boolean equations in list 18.2 are realized after the first working step. Those equations which had been entered were obviously already minimized.

List 18.2 Boolean equations of the variable divider

```
**********************************
******** BOOLEAN EQUATIONS ********
**********************************
A2  :=  A2    &  A1
     + /A2    & /A1    &  E2
     +  A2    & /E2    & /E1     ;
A1  := /A1    &  E1
     +  A2    & /A1    &  E2
     +  A1    & /E2    & /E1     ;
```

Each line in list 18.2 requires an AND term for its realization. A maximum of three AND terms, which are wired OR, is necessary for each output. This is therefore the Sum of Products which is supported by the target component GAL 16V8.

The second step, namely, the adaptation to the resources of the logic component, does not require any transformation of the equations found. List 18.3 only shows the necessary AND terms of the logic diagram with the × marks of the still functioning connections.

List 18.3 Reduced logic diagram of GAL 16V8 with × marks of still functioning connections as well as programming positions for selection of architecture (Configuration Fuses)

```
***********************************************
****************** FUSE PLOT ******************
***********************************************
                              @ @  @ @  @ @  @ @  @ @  @ @
                              0 0  0 0  0 0  0 0  0 0  0 0
          E A  E A  0 1  0 1  0 1  0 1  0 1  0 1
          2 2  1 1  4 7  5 6  6 5  7 4  8 3  9 2
          --------------------------
          0 0  0 0  0 1  1 1  1 1  2 2  2 2  2 3
          0 2  4 6  8 0  2 4  6 8  0 2  4 6  8 0
-------------------------------
                                                  A2
000000  -X- -X- --- --- --- --- --- ---
000032  X-X --X --- --- --- --- --- ---
000064  -XX- -X- --- --- --- --- --- ---
                                                  A1
```

18

```
000256  --- X-X --- --- --- --- --- ---
000288  X-X- --X --- --- --- --- --- ---
000320  -X- -XX- --- --- --- --- --- ---

CONFIGURATION FUSES:
002048  -XXXXXX
002056  XXXXXXXX XXXXXXXX XXXXXXXX XXXXXXXX
002088  XXXXXXXX XXXXXXXX XXXXXXXX XXXXXXXX
002120  XX----
002128  --XXXXX --XXXXX XXXXXXXX XXXXXXXX
002160  XXXXXXXX XXXXXXXX XXXXXXXX XXXXXXXX
002192  X -

6 OF 64 PRODUCT TERMS USED = 10  %
DEVICE CODE:   3655
FUSE CHECKSUM: 168D
```

18.4.3 The JEDEC Form

The last step, e.g., the transmission of the programming specification into the target component is realized by the **programmer**. This is a variable voltage generator and analyser for a great number of pins which are contacted via a ZIF adapter (*Zero Insertion Force*). The *programmer* receives the programming data in the international standard JEDEC form *(Joint Electron Device Engineering Council*).

This ensures the understanding between arbitrary computers and *programmers*. The list 18.4 shows the programming data of the given example in the JEDEC form.

List 18.4 Programming data of the given example in JEDEC form

```
$DEVICE
  GAL16V8 ;
$PIN
    1 = CLK ;
    2 = E2 ;
    3 = E1 ;
   11 = OE ;
   18 = A1 ;
   19 = A2 ;
$END
*
D3655*
G0*
QP020*
QF02194*
F0*
L00000 11011101111111111111111111111111*
L00032 01101110111111111111111111111111*
L00064 10011011111111111111111111111111*
L00256 11110110111111111111111111111111*
L00288 01011110111111111111111111111111*
L00320 10111001111111111111111111111111*
L02048 11000000*
L02056 00000000000000000000000000000000*
L02088 00000000000000000000000000000000*
L02120 00111111*
L02128 11100000111000000000000000000000*
L02160 00000000000000000000000000000000*
L02192 01*
C168D*
```

All information given prior to the first '*' are comments for the operator of the *programmer*. The actual programming data starting with a characteristic letter are followed by numbers and are finished by '*':

- D represents **D**EVICE and it opens the field of the component code. It also allows the programmer to control whether the proper component has been used;
- G represents **G**UARD applies for security cell (not programmed in the example given);
- QP represents **Q**UANTITY of **P**ins indicates the number of pins; and
- QF represents **Q**UANTITY of **F**uses indicates the number of programming positions;
- F indicates the 'De**F**ault' value for the programming positions which are not listed;
- L for **L**INK defines the desired connection state (1 means interruption, 0 keeps the connection in tact) starting from the indicated starting address;
- C for **C**hecksum presents the 4 digits of the check sum.

The addresses of the programming positions are realized by mere consecutive numbering from left to right and from top to bottom, as shown in fig. 18.23. Such a clear allocation allows the conversion of the two-dimensional '*fuse plot*' according to list 18.3 into the one-dimensional JEDEC file in list 18.4. The × marks are replaced by 'low (0)' and the minus signs are replaced by 'high (1)'. Knowing this and knowing the pin allocation the path can also be traced back; the JEDEC file results in a '*fuse plot*', revealing the Boolean equations. Thus a JEDEC file can also be accepted by a logic simulator.

After receipt of an error-free complete JEDEC file by the *programmer*, it will find in its archive, amongst its component codes, the exact voltage values and pulse widths for the programming of the component. This information is missing in today's data sheets, recommending *programmers* instead.

Only manufacturers of programming devices (see section 18.6) are furnished by the semiconductor companies with their respective latest data. This guarantees a careful and yet long lasting programming.

18.5 Complex Programmable Logic Devices

The progressive semiconductor technology allows structures on the chip which are becomming smaller and smaller, even realizing more and more complex logic circuits. Three different basic architectures will be presented in the following. [18.8], [18.10], [18.11], and [18.12] supply detailed information. An up to date outline is given by [18.3]. They are even the main subjects in the special editions [18.4] and/or [18.6].

18.5.1 Multiple Array Matrix (MAX) of ALTERA

ALTERA is the inventor of *erasable programmable logic devices* (EPLD). With seven types of devices they offer solutions for nearly any development requirement today. In the MAX components ALTERA combine several PAL structures to **complex programmable logic devices (CPLD)**. Figure 18.27 shows the architecture in principle of a MAX component.

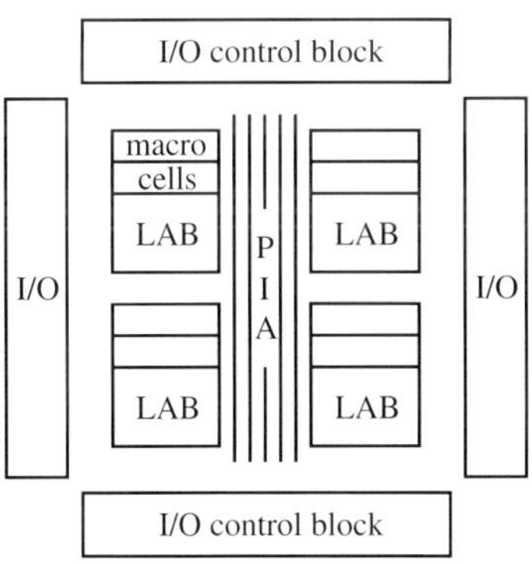

Fig. 18.27 Architecture of a MAX component

The functional blocks are called **LAB** (*Logic Array Block*) and consist of the well known *macro cells* (MC) shown in fig. 18.28 which are supplemented by several *Expander Product Terms*. A *Programmable Interconnect Array* (PIA) provides the interconnection amongst the functional blocks. It allows all kinds of connections among all macro cells. The I/O Control Blocks drive the links to the outside.

Such an uncomplicated wiring scheme does not require high standards from the design tools and ensures short predictable signal propagation times. Thus a given clock rate requires only the selection of the appropriate MAX component. There are no losses of velocity to be expected by allocation and wiring.

Each macro cell consists of a D flip flop. Its data input is driven by four programmable AND terms via an OR EXOR structure. The clock input can be switched between a global system clock and an AND term. Two other AND terms influence the storage element directly via the set and/or reset input. The remaining AND term drives the switchable output driver containing the D input or the Q output of the flip flop.

The vertical input lines of the programmable AND matrix are provided with 4 different sources. In addition to the input pins and the signal feedback of the macro cell they are also driven by the Programmable Interconnect Array and by extension AND terms. The latter allow the extension of the AND terms for programming complex logical expressions.

The ALTERA component, which is at present the most complex one, provides 560 of such macro cells (EPM 9560). A comparable component belonging to the MACH family of VANTIS consists of 512 MCs (M5-512/256). The third component from the ispLSI 80000 family made by LATTICE consists even of 840 MCs (ispLSI8840). This corresponds to a complexity of 45,000 gate equivalents. Such a vast application example would exceed the scope of the present chapter and has to be dropped.

It is difficult for the number of pins to keep pace with the increasing number of macro cells. The identification of the MACH component shows that the 512 MCs must share only 256 I/O pins. Every second MC is buried, e.g., it can provide its output value only internally. In practice this does not result in a disadvantage, as the outside often needs some few important signals only.

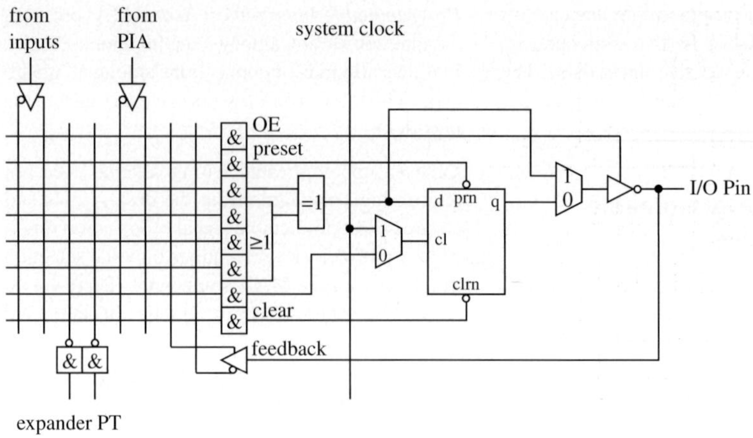

Fig. 18.28 Logic of a macro cell (MC)

The CPLD chip is often supplied with a different number of pins for different packages. This means that connection signals are abandoned once again. On the other hand, the user can choose amongst different complex CPLDs in one and the same package. If a system modification requires a more powerful component, it matches the foot prints of its predecessor. The printed circuit board remains unchanged, it only gets another assembly.

18.5.2 Logic Cell Arrays (LCA) of XILINX

Probably inspired by the regular structures of the gate arrays, XILINX adopted a chequered design of logic cells for their LCAs as shown in fig. 18.29. The logic cells are connected between each other and between the input/output blocks by means of horizontal and vertical connecting paths. Such an architecture is also called a Field Programmable Gate Array (FPGA).

Each logic cell consists of a *configurable logic block* (CLB) according to fig. 18.30.

Unlike the macro cell of the CPLD a *Look Up Table* (LUT) replaces the programmable AND/OR structure in the configurable logic block. This Look Up Table is a mere static random access memory (SRAM) with seven inputs and two outputs, corresponding to a storage organization of 128×2 bits. This allows the programming of two individual combinatorial functions with up to five variables. For sequential tasks the flip flop outputs are also available as the 6th and 7th variable.

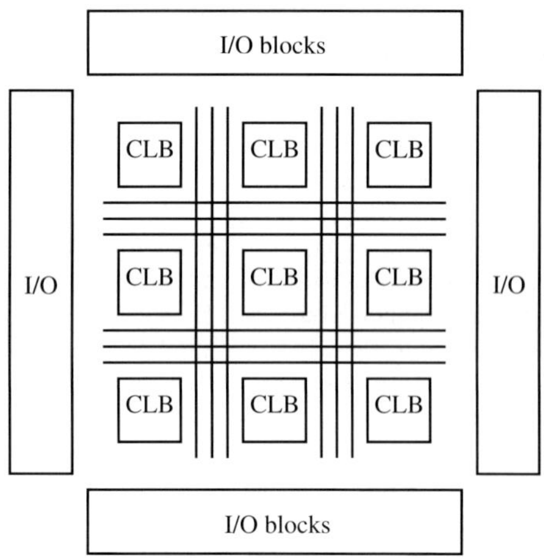

Fig. 18.29 Architecture of the LCA components

Four signals can be selected for the data input of the flip flops. In addition to the two outputs of the Look Up Table and to the flip flop output there is also a direct input 'data in' of the logic cell. This output makes the realization of shift registers easier. A clock signal and a reset signal are available separately. The output signals of the logic cell are defined by the Look Up Table or the flip flops.

A configurable logic block corresponds to approximately two macro cells, it has, however, significantly less input signals. For example, if wider AND operations have to be realized, several CLBs have to be cascaded. This is not a problem, as there are sufficient CLBs in most cases. The

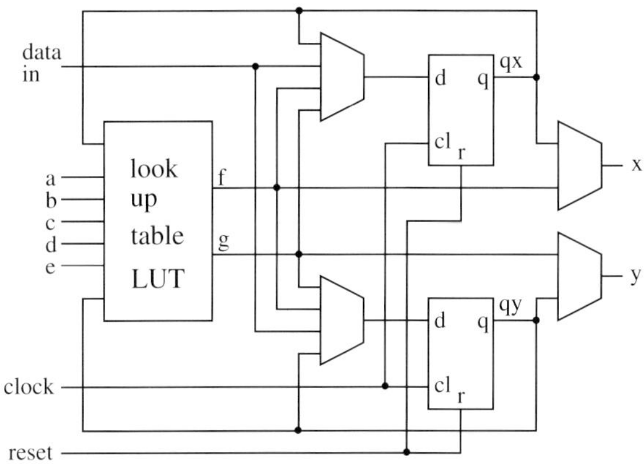

Fig. 18.30 Configurable logic block (CLB)

LCA (XC40250XV), which is the most complex at present, offers a matrix of $92 \times 92 = 8{,}464$ CLBs. This corresponds to a complexity of 250,000 gate equivalents.

18.5.3 ACT Components, the FPGA of ACTEL

ACTEL were the first to succeed in a reliable realization of anti-fuses. ACTEL selected a line by line order of so called logic modules (LM) for their FPGAs as shown in fig. 18.31. The horizontal routing channels are positioned between the lines.

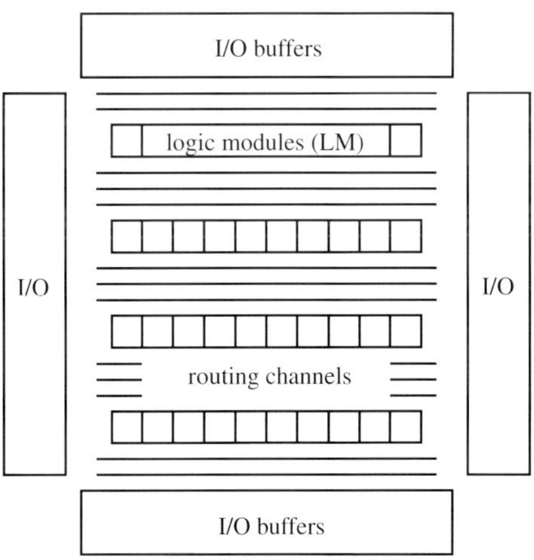

Fig. 18.31 Architecture of ACT components

The architecture of the ACT components shown in figure fig. 18.31 is also known from the *Gate Arrays*. The logic module is new and surprisingly simple. According to fig. 18.32 it consists of three multiplexers only.

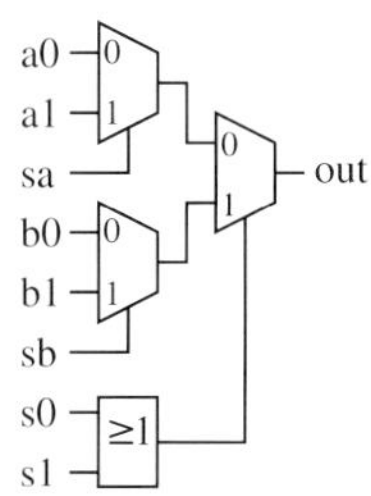

Fig. 18.32 ACT logic module (LM)

The above mentioned logic module allows the realization of all logic combinations of up to four variables. At the first glance this seems to be impossible, as there is no inverter available. As a proof, fig. 18.33 shows the three basic functions, e.g., inverter, AND gate, and OR gate.

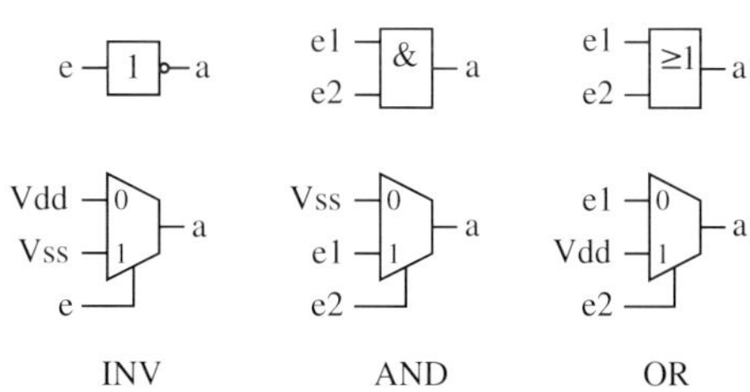

Fig. 18.33 Realization of the basic functions with a multiplexer

In all three examples an input signal selects which one of the constants or variables may define the output. More complex functions require several of these logic modules, whereas flip flops need only two. As complex circuits mostly require a great number of flip flops, ACTEL expand every third logic module with an additional flip flop. At present the most complex ACT component (A54SX64) provides 3,600 combinatorial logic modules and 2,160 storage logic modules. This corresponds to a complexity of 64,000 gate equivalents.

18.5.4 Design steps for highly complex CPLDs or FPGAs

A circuit design with many thousands of gates cannot be realized with Boolean equations only. A possibility of a more complex definition for digital circuits is needed, namely, a hardware definition language such as VHDL. It allows the definition of the future circuit into the design entry according to fig. 18.34.

It is possible to simulate the behavior of the VHDL definition using the EDA support tool. By this means it is possible to detect logical errors and to correct them in the VHDL definition. If the simulation proves the desired behavior, the circuit synthesis will be carried out in the next step. This results in a circuit made of logic gates and flip flops in the form of an Electronic Design Interchange Format (EDIF) netlist and in the form of a VHDL structure description at the same time. The latter is, however, understood by the simulator allows the subsequent simulation which follows the synthesis.

The EDIF netlist may be used by a graphic program, not shown in fig. 18.34, which is used to draw the wiring diagram. In this case it is the definition for the target technology mapping. The gates and flip flops have to be replaced by the macro cells of the CPLD or by the logic modules of the FPGA. This necessary conversion and optimization of the circuit can be simulated and controlled with respect to the function by the revised VHDL structure description. In this phase the EDA tool also monitors the fan out of each gate as the number of the drivable subsequent inputs. Especially in the case of clock signals the driver load often has to be divided into several parallel drivers. It is also possible to try several CPLDs or FPGAs during this design step. Finally, it is a fact that the selected logic circuit is provided with sufficient logic resources.

This is a necessary requirement which is, however, not sufficient for allocating and connecting. All logic elements are now allocated to a physical place on the chip. In the case of a CPLD allocating is not critical, as the connection is always possible because of the great number of inputs into the programmable AND terms. Moreover, the signal delay is defined by connection lines from the beginning. In case of an FPGA there is only a limited number of long, medium, and short connecting lines available in the connecting channels. If these lines are not sufficient the connection remains incomplete. The allocation has to be modified now or even the next larger FPGA has to be taken. There will be varying signal paths with the respective delay times, which will be calculated and provided in the *Standard Delay Format* (SDF). Together with the VHDL structure description the delay times allow a precise simulation of the timing behaviour. This is the last possibility of control for the developer prior to programming. A second EDIF netlist is set up in parallel containing all topological information, too. It is used by a graphic program to draw a symbolic layout. In a critical situation the allocations and the connections can still be changed here with a layout editor. Afterwards the delay times have to be calculated and simulated once again, of course.

The bit pattern for the programming is realized in the last design step. There are many different ways in which to feed the programming data into the target hardware:

a) EPROM memories, CPLDs in EPROM technique, and non-erasable programmable FPGAs require a programming device;
b) FLASH memories and CPLDs in FLASH technique are *In System Programmable* (ISP) or *In Circuit Reconfigurable* (ICR). The JTAG interface which belongs to the '*Boundary Scan Test* – BST' is misused for a short time for this purpose;
c) After each power on of the system, the FPGAs in SRAM technique have to be programmed anew. In 'slave mode' they are written like a memory. In 'master mode' they read their programming data independently from a non-volatile memory device, usually a serial EPROM.

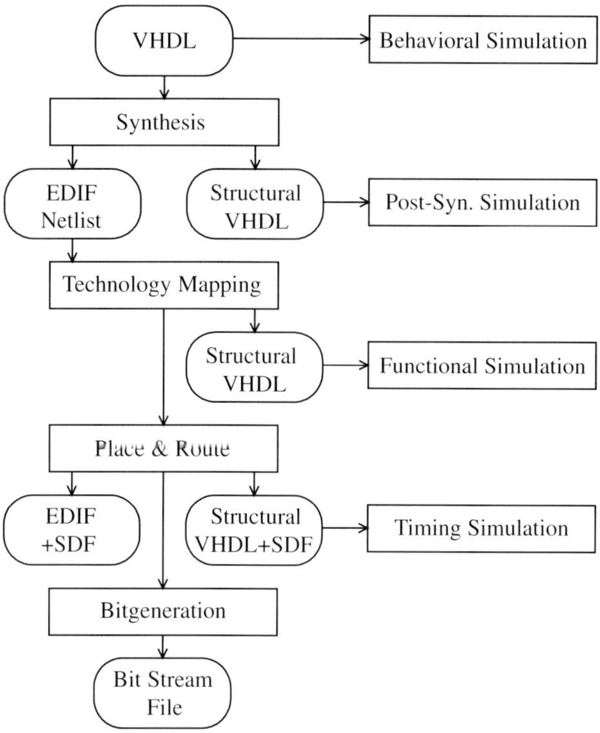

Fig. 18.34 Design entry for CPLDs or FPGAs

Then they leave their programming status and start to operate.

The correct functioning of the programmed logic device is finally controlled by logic analysis. If errors are detected which had not been revealed in any other preceding simulation, there will be a difficult investigation for the possible reasons. ACTEL offers a unique help with their '*Silicon Explorer*', which is able to monitor two arbitrary internal nodes via the '*ActionProbe*' circuit. This allows the back-tracing of a signal path up to the error cause.

The outlined design steps illustrate the need of an extensive EDA software. Such a software is provided by manufacturers of programmable circuit devices (see section 18.6) and by independent software companies (see section 18.6). The recent proposal made by XILINX is called '*SiliconXpresso*'. It promises to translate a circuit specification received via Internet by means of its 'WebFITTER' into the programming data for an XC9500 CPLD. This results in the design method of the 21st century, e.g., the '*Internet Reconfigurable Logig*' (**IRL**).

18.5.5 Comparison and Prospects

Comparing the basic elements of the CPLDs (fig. 18.28) and FPGAs (figures 18.30 and 18.32) their varying sizes are remarkable. A macro cell of the CPLD looks coarse grain compared to the fine grain logic module of the FPGA by ACTEL. The latter can more easily be adapted to the various logic requirements; they can either connect or store. However, if a macro cell is only supposed to connect, the flip flop will remain unused. Typically only approximately 50 % of the available gates of a CPLD are usable gates. With respect to the FPGAs the limited connection resources only allow an efficiency of 70 to 90 %. Any comparison is therefore open.

Another criterion can be found in the programmable elements. Long lasting programmable fusible links by ACTEL or field effect transistors

18

with isolated gate of the CPLD compete with the volatile programming of SRAM FPGAs. The latter keep their configuration in an always readable non-volatile memory. An unfair product pirate can also read and copy the circuit without hindrance. A copy protection is only guaranteed by durable programmable logic devices; they refuse reading. Advantage CPLD and anti-fuse FPGA.

The SRAM FPGA characteristics to forget what has been programmed becomes a virtue when hardware has to be reconfigured during operation. This requires programming within the system. For example, an FPGA being an interface to a magnetic disk can be configured for writing and for reading and vice versa. An obvious generalization of such an idea results in a universal programmable hardware made of several SRAM FPGAs. It allows, for example, the emulation of highly complex ASIC designs or the realization of task-specific computers. It is no longer necessary to compile the problem in a computer-specific way, but the computer receives its problem-specific configuration.

All advantages of the programmable logic devices lead to their impressive **economic success**. According to table 18.14 the above mentioned suppliers cover more than 80 % of the programmable logic devices on the world market. An annual increase of more than 30 % to 7.9 billion dollars is expected for the year 2002. With respect to the ASIC market, however, an annual total increase of only 20 % is estimated [18.7].

Table 18.14 Companies and their respective market share

XILINX	ALTERA	VANTIS	LATTICE	ACTEL
28 %	27.5 %	13.4 %	10 %	7.5 %

According to the above table XILINX and ALTERA are both on top. As to the products, the FPGA are ahead of the CPLDs, as is shown in table 18.15. The simple programmable logic devices (SPLD) will lose, and in 2002 they will only hold 3 % of the market.

Table 18.15 Products and their market shares (prognosis 2002)

FPGA	CPLD	SPLD
46 %	38 %	16 %
(53 %)	(44 %)	(3 %)

There is very strong competition in the component market between the FPGAs and CPLDs. For example, XILINX, who are the market leader in FPGA, are also competing with the (XC9500) as far as CPLDs are concerned. In contrast to this, ALTERA who are the market leader in CPLDs, win market shares with respect to the FPGAs by offering their FLEX components. LATTICE takes over VANTIS.

The above mentioned increases illustrate the attractiveness and great acceptance of programmable logic devices.

18.6 References

[18.1] *Auer, A.*: 'Programmierbare Logik-IC'. Hüthig, 1994

[18.2] *Bolton, M.*: 'DigitalSystems Design with Programmable Logic'. Addison-Wesley, 1990

[18.3] *Bursky, D.*: 'High-Density FPGAs Go Toe-To-Toe With Gate Arrays'. Electronic Design, April 20, 1998

[18.4] *Cong, J.; Ebeling, C.*: 'Special Section on Field Programmable Gate Arrays. IEEE Trans'. VLSI Systems, Vol. 6 No. 2, June 1998

[18.5] *Hack, U.; Hoffmann, M.*: 'Das GAL-Buch'. Elektor-Verlag, 1993

[18.6] *Hauck, S.*: 'The Roles of FPGAs in Reprogrammable Systems'. Proc. IEEE, Vol. 86 No. 4, April 1998 (with 148 sources!)

[18.7] nach ICE: 'Wachstumsraten von mehr als 30 Prozent'. Markt & Technik Nr. 6, 1998

[18.8] *Jenkins, J. H.*: 'Designing with FPGAs and CPLDs'. Prentice-Hall, 1994

[18.9] *Pellerin, D.; Holley, M.*: 'Practical Design Using Programmable Logic'. Prentice-Hall, 1991

[18.10] *Schwabe, I.*: 'Begleittexte zum Entwicklerforum Programmierbare Logik'. Design & Elektronik, 1997

[18.11] *Tischler, M.; Oertel, K.*: 'FPGAs und CPLDs – Technologie und Anwendungen'. Hüthig, 1998

[18.12] *Wannemacher, M.*: 'Das FPGA-Kochbuch'. Internat. Thomson Publishing, 1998

Suppliers of Programmable Logic Devices

ACTEL Corporation, 955 East Arques Avenue, Sunnyvale, CA 94086-4533, USA (www.actel.com)

ALTERA Corporation, 101 Innovation Drive, San Jose, CA 95134, USA (www.ALTERA.com)

Atmel Corporation, 2325 Orchard Parkway, San Jose, CA 95131, USA (www.atmel.com)

Cypress Semiconductor Corporation, 3901 North Fist Street, San Jose, CA 95134 (www.cypress.com)

DynaChip Corporation, 1255 Oakmead Parkway, Sunnyvale, CA 94086, USA (www.dyna.com)

GateField Corporation, 47100 Bayside Parkway, Fremont, CA 94539-9942, USA (www.gatefield.com)

LATTICE Semiconductor Corporation, 5555 Northeast Moore Street, Hillsboro, OR 97124-6421 (www.latticesemi.com)

Lucent Technologies, 555 Union Boulevard, Allentown, PA 18103, USA (www.lucent.com)

Motorola Incorporated, 2100 East Elliot Road, MD EL563, Tempe, AZ 85284, USA (sps.motorola.com/fpga)

Philips Semiconductors, Hammerbrookstraße 69, D-20097 Hamburg (www.semiconductors.philips.com)

QuickLogic Corporation, 1277 Orleans Drive, Sunnyvale, CA 94089-1138, USA (www.quicklogic.com)

Texas Instruments Incorporated, Semiconductor Group, P.O. Box 809066, Dallas, TX 75330, USA (www.ti.com/sc/docs/schome.htm)

VANTIS Corporation, 995 Steward Drive, P.O. Box 3755, Sunnyvale, CA 94088, USA (www.vantis.com)

XILINX Incorporated, 2100 Logic Drive, San Jose, CA 95124-3450, USA (www.xilinx.com)

Suppliers of Manufacturer-neutral Design Tools

Cadence Design Systems (www.cadence.com)

Exemplar Logic (www.exemplar.com)

IST International Semiconductor Technologies (www.ist.com)

Mentor Graphics Corporation (www.mentorg.com)

MINC (www.minc.com)

Model Technology (www.model.com)

OrCAD (www.orcad.com)

Synopsys (www.synopsys.com)

Synplicity (www.synplicity.com)

VeriBest (www.veribest.com)

Viewlogic (www.viewlogic.com)

19 Semiconductor Process Technologies

HERMANN CLAUSS

Approximately up to the year 1970 integrated circuits were realized mainly in bipolar technology. Though the basics of MOS technology already were known, the reproducibility of the threshold voltages of MOS-Field Effect Transistors (MOS-FET) was insufficient. The problem could be solved, however, with the breakthrough of the ion-implantation technology. Since that time MOS is the dominating technology for the production of integrated circuits.

For analog circuits the most important technology still is the bipolar technology, because the transconductance of Bipolar Junction Transistors (BJT) carrying a comparable amount of current is much higher than that of MOS FET [19.8]. Additionally for high output currents the chip area needed for BJTs is less than that for MOS FETs.

A combination of both technologies is achieved on the one hand by the BiCMOS technology (Bipolar-CMOS) [19.21] combining CMOS circuits with BJTs (NPN and PNP types) and on the other hand by the BCDMOS technology [19.10] which combines on a single chip bipolar transistors with CMOS circuits and additionally (for high voltages) with DMOS transistors. But both technologies are very costly and are implemented only if all their advantages are required.

First of all, the following paragraphs present the silicon planar technology and its processing steps. Thereafter the processing steps for bipolar and MOS circuits are presented. These presentations are completed by the description of components, which can be integrated on a chip.

19.1 Basics of the Silicon Planar Technology (SPT)

19.1.1 Introduction

The realization of modern monolithic integrated circuits is achieved mainly by the SPT. Integrated circuits for very high frequencies are realized most often by the GaAs technology because of the high electron mobility of this material.

Using SPT the active regions of the IC are generated close to the surface of the mono-crystalline silicon wafer. To achieve that, doping materials (B, P, As, Sb) are introduced by solid-state diffusion or by ion-implantation into the surface of the wafer thereby changing the concentration and the type of the majority charge carriers (either electrons or holes). Silicon oxide can be easily grown by thermal oxidation of silicon on the surface of the wafer. This material (SiO_2) is nearly impermeable for the doping materials normally used in silicon. The atoms of the doping material can pass only through windows in the silicon oxide layer created by a photolithographic process using a photo mask. The process steps and their sequences for the production of bipolar and CMOS circuits are presented in detail in the following paragraphs.

19.1.2 Oxidation

For the oxidation of the silicon surface the wafers are exposed in a tube of quartz glass within an oxidation furnace at temperatures from 900 to 1200 degrees centigrade. The oxidizing atmosphere consists either of oxygen (dry oxidation) or of water vapor (wet oxidation).

In the latter case the water vapor is created by controlled burning of hydrogen and oxygen inside the oxidation tube.

The rate of oxidation rises with the temperature (see fig. 19.1) and is dependent on the atmosphere in the oxidation tube. Dry oxidation results in low oxidation rates with high quality of the oxide. With wet oxidation the oxidation rate rises rapidly but the quality of the oxide and of the silicon-silicon dioxide interface is lower than of dry oxidation.

To increase the gas pressure within the oxidation tube results in raising the oxidation rate as well. Examples for oxide growth using High Pressure Oxidation are shown in fig. 19.1 as well.

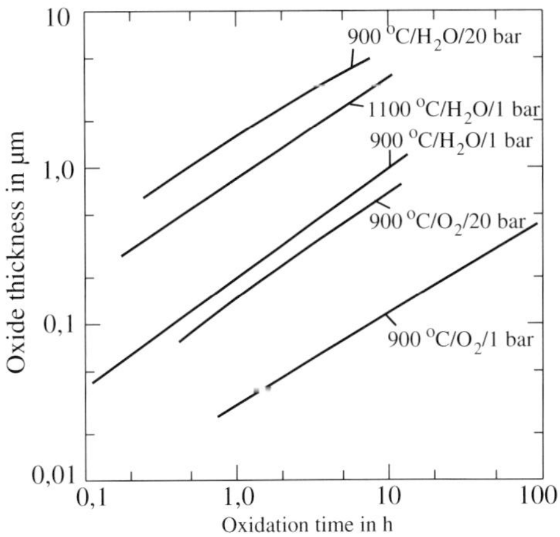

Fig. 19.1 Oxide thickness as a function of the oxidation time for different oxidation conditions (Results at 900 °C according to [19.9])

During the oxidation all silicon areas are oxidized, either they are pure or are coated by a SiO_2 layer. Selective oxidation [19.22] of distinct regions is achieved by coating the areas not to be oxidized with silicon nitride which is impermeable for oxygen and thus prevents the underlying silicon areas from being oxidized.

Figure 19.2 shows schematically the processing steps for a local oxidation (LOCOS) without (left hand side) and with (right hand side) pre-etching of the crystal silicon.

First of all a thin protection oxide layer is grown. Above that silicon nitride (Si_3N_4) is deposited which has been structured by a photo lithography process (see fig. 19.2b).

During the following oxidation process the Si_3N_4 layer prevents the oxidation of the silicon surface (see fig. 19.2b). The oxide grows a little bit under the Si_3N_4 layer (see fig. 19.2c) thereby creating the so-called bird's beak structures. Later on the nitride and the protection oxide are etched off and the gate oxide is grown (see fig. 19.2d).

By the use of silicon pre-etching the height of oxide steps may be decreased. But a totally planar surface cannot be achieved by this method.

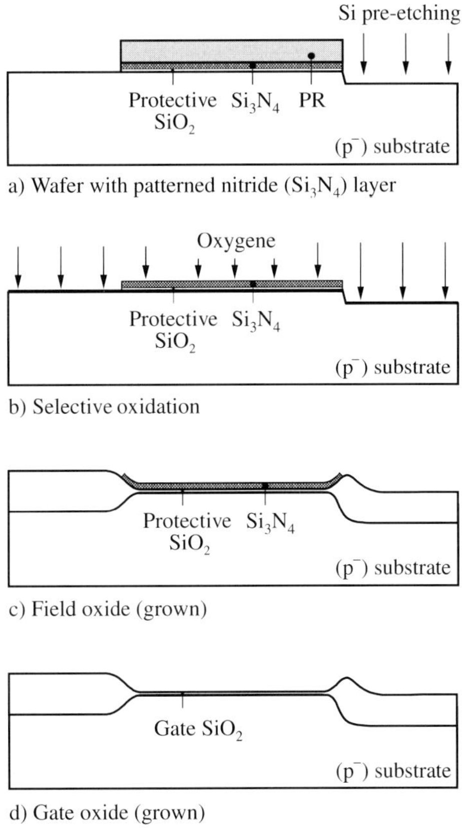

Fig. 19.2 Processing steps for local oxidation without (a, c) and with (b, d) pre-etching of the crystal silicon

19.1.3 Photo Lithography

For the structures to be transferred onto the wafers a photo lithography process (mask process) is used. The photo mask consists of a quartz plate, which is coated by a thin chromium layer. The desired structures are transferred to this layer by electron beam writing. Figure 19.3 shows the steps of the mask process for the transfer of structures into an oxide layer.

First of all a thin layer of photoresist (PR) is put onto the oxidized wafer (see fig. 19.3a). This is done by putting some drops of the PR onto the wafer. These are then equally spread over the complete wafer by fast rotation (spinning). In the next step the PR is exposed to ultraviolet light using a photo mask. The UV light changes the inner structure of the PR.

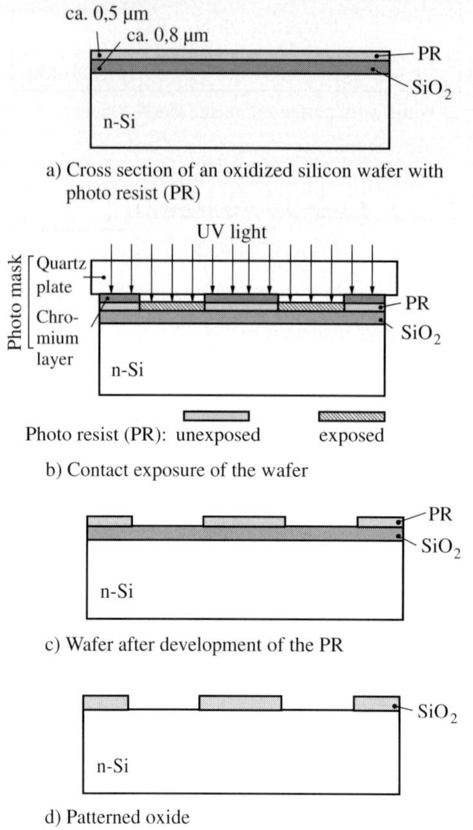

Fig. 19.3 Sequence of a photo mask process

By development of the PR during the next process step the exposed parts of the PR are removed from the wafer (see fig. 19.3c). On the wafer remain only the areas of the PR, which were hidden by the mask (unexposed areas) (positive PR). Alternatively with the negative PR procedure the unexposed areas are removed from the wafer.

By means of this the PR keeps a positive or a negative image of the photo mask structures.

Thereafter the PR is hardened by heating. With the PR as a mask the desired patterns are then etched into the oxide using a liquid containing hydrofluoric acid. Figure 19.3d shows a cross section of the wafer after the removal ('stripping') of the PR.

The contact printing lithography depicted in fig. 19.3b has the disadvantage that the mask touches the PR coated wafer during exposure causing rapid contamination of the mask by sticking particles of the PR. This is the reason why in modern processes the mask is projected optically onto the wafer [19.15].

During the projection exposure the whole mask is put as an image, scaled 1 : 1, onto the wafer. The disadvantages of this method are the low depth of focus and the low optical resolution because of the large image area, especially when large wafer diameters are used.

For very small structures ($< 1\ \mu m$) the projection exposure has been replaced by the use of wafer steppers. Applying this technique, only a small part of the wafer (about 14 mm × 14 mm) is exposed with a much better resolution. Thereafter the exposing field is stepped under laser control to the next section and focused. Then this next section is exposed and so on. By the application of wafer steppers even uneven surfaces may be compensated for and the resolution is much better than with the 1 : 1 projection exposure.

The masks are adjusted automatically with reference to the existing structures with tolerances $<$ 0,2 µm. The photo masks are patterned by electron beam writing. Targets are chromium plated quartz plates, coated with resist layer which changes its structure under the exposure to the electron beam. The electron beam is electromagnetically deflected under the control of the mask data processed by the electron beam writer. After the electron beam writing procedure the parts of the resist not touched by the electron beam are removed. The resist is hardened by heat and with the resist as a mask the underlying chromium layer is etched. The resolution of this process is $<$ 80 nm.

19.1.4 Doping

Bulk doping of the silicon wafers is accomplished already during the growth of the silicon monocrystals. Important doping substances for silicon are boron (B) as an acceptor for p-conduction and phosphorus (P), arsenic (As), and antimony (Sb) as donors for n-conduction. Figure 19.4 shows the resistivity as a function of the doping concentration in n-type and p-type silicon [19.12].

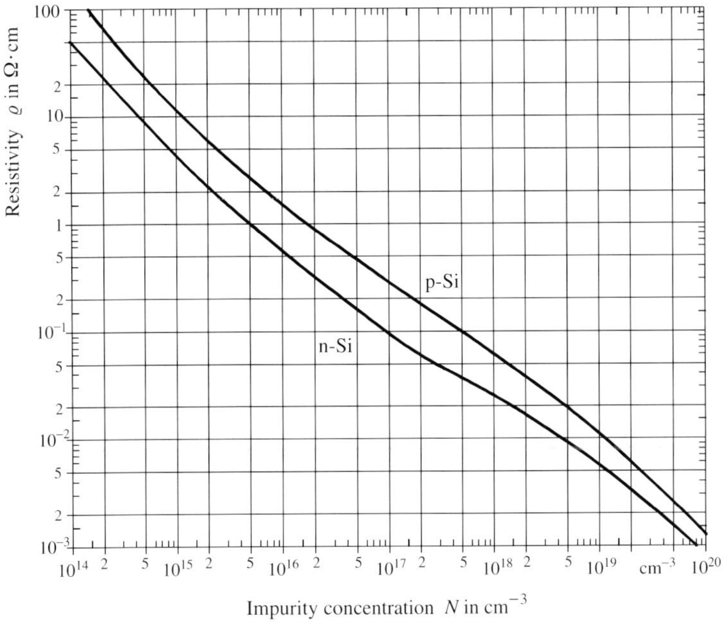

Fig. 19.4 Resistivity of n-type and p-type silicon as a function of impurity concentration

The following IC production steps change the doping locally, e.g., for the creation of PN-junctions. Doping atoms may be brought into silicon either by solid state diffusion or by ion implantation. Both are presented in the following paragraphs.

Solid State Diffusion

At high temperatures (900 . . . 1 200 °C) doping atoms may move from the wafer surface into the silicon supported by voids in the silicon lattice. This movement is called solid state diffusion. Thereby the doping concentration is highest at the surface and decreases with the distance from it. Figure 19.5 shows as an example the concentration of boron in an initially n-type wafer doped with arsenic.

In cases of predominant boron concentration silicon is p-conducting. In cases of predominant arsenic concentration silicon is n-conducting. The pn-junction is formed just at the location where the boron and arsenic concentrations have the same value. The corresponding value of the penetration depth is the 'junction depth' x_j of the pn-junction.

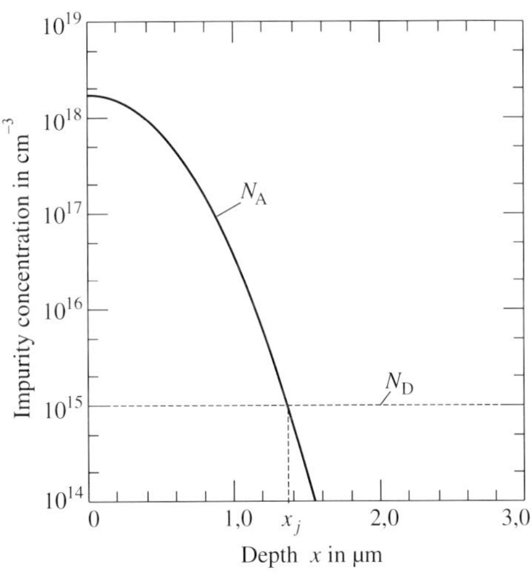

Fig. 19.5 Doping profile of a diffused p-zone in n-type bulk material

The introduction and distribution of the doping substances in a solid state diffusion process is achieved mostly in two steps. First of all with the **pre-deposition diffusion** ('Pre-dep') the doping atoms are introduced into the wafer. Many wafers are exposed at the same time to an atmosphere

Table 19.1 Mean penetration depth and standard deviation for important doping substances in silicon as functions of the acceleration voltage

Acceleration voltage	Impurity Type Boron MPD/STDEV	Phosphorus MPD/ STDEV	Arsenic MPD/STDEV
20 kV	80 nm/30 nm	30 nm/11 nm	20 nm/7 nm
50 kV	200 nm/60 nm	75 nm/25 nm	40 nm/14 nm
100 kV	350 nm/100 nm	150 nm/45 nm	75 nm/25 nm

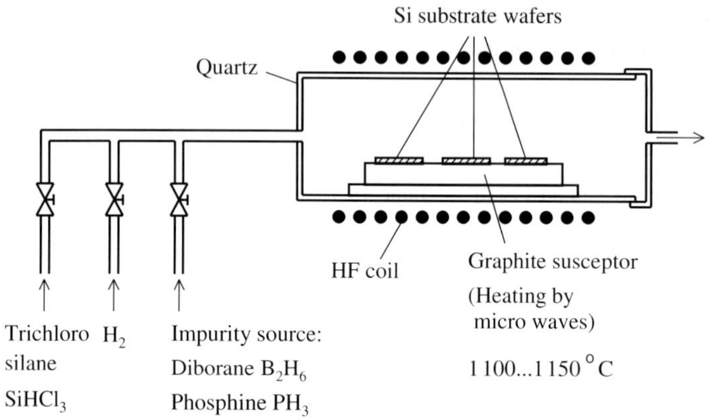

Fig. 19.10 Scheme of an epitaxy reactor

the substrate. To dope the epi layer a carrier gas containing the desired doping atoms is added to the atmosphere, which is decomposed at the wafer surface setting free the doping atoms. During this process the doping atoms are directly built into the lattice structure of the epitaxial layer. The doping concentration is adjusted by the composition of the gas.

By this means even weakly doped layers may be grown on highly doped substrates.

19.1.5.2 Metallization

The conductors on IC's must fulfill different requirements listed below:

- High conductivity;
- Low contact resistance metal silicon;
- Good adhesion on silicon dioxide;
- Good covering of the oxid steps;
- Must be suited for multi layer metallization;
- Low electro migration.

The wiring of integrated components on a chip is mostly accomplished by aluminum conductors. For each of the requirements mentioned above, there are particular materials suited better than pure aluminum. But aluminum is a good compromise for the requirements altogether.

Aluminum has a low resistivity ($2.7\ \mu\Omega \cdot \text{cm} \Rightarrow 50\ \text{m}\Omega/\square$ with a layer thickness of $0.5\ \mu\text{m}$). Additionally the contact resistance to the (n^+) and (p^+) silicon is sufficiently low ($R_c \approx 10^{-7}\ \Omega \cdot \text{cm}^2$) [19.16].

By modern magnetron sputtering processes for layer deposition the step coverage is much better than with the vapor deposition process used earlier. Unwanted layers may be removed by back-sputtering techniques before new layers are deposited. This causes very low contact resistances between the metal layers in multi-layer metallization.

Very small contact windows and/or shallow pn-junctions may cause problems resulting from micro alloying [19.1], [19.17]: during annealing, a process step to decrease the contact resistances, at temperatures of about 420 °C, crystal silicon diffuses into the conductor aluminum. The problem may be reduced by adding 0.8 . . . 1 % silicon to the aluminum because then the aluminum is always saturated while it is deposited and therefore only very little aluminum diffuses out of the wafer surface. The silicon is added to the targets for the sputtering process and is transferred together with the aluminum onto the wafers.

Problems with the metal lines caused by electromigration can emerge from shrinking structures. High current densities and high temperatures in the metal lines may cause a transport of line material in the direction of the electron movement. Inhomogenities at this transport may cause interruptions of the lines and thus failing of the device (Attrition of integrated circuits) [19.2]. Figure 19.11 shows an aluminum line without the covering oxide after the application of a high current density (10^6 A/cm^2) and of a raised temperature. On the one hand hillocks were created, on the other hand an interruption of the line occurred.

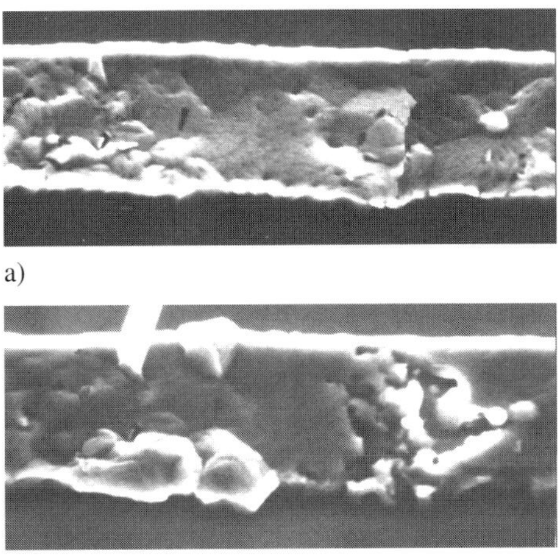

a)

b)

Fig. 19.11 Electromigration in aluminum [19.3]
a) Metal line before the test, b) Metal line after the application of 10,000 A/mm^2 *at* 200 °C *(A whisker, small hillocks and an interruption were created.)*

Electromigration may be reduced substantially by alloying 1 ... 4 % copper into the aluminum. By means of this the mean lifetime of the metal lines may be increased up to a factor of 1,000.

To prevent electromigration the maximum permissible current density is limited to about 1,000 A/mm^2 corresponding with 1 mA/µm^2. The actual value depends on the type of the manufacturing process, the composition of the material, and the operating temperature of the lines.

The limit above must be checked during the design process, which causes a lot of computing time for the design rule check of the layout.

Recently copper lines are in use for ICs as well because of the lower specific resistance and the better electromigration properties as compared to aluminum. The application of copper allows smaller cross sections of the lines in combination with shrinking structures.

19.1.5.3 Polysilicon and Polycides

In modern CMOS technologies polycrystalline silicon layers (Polysilicon) are used for the gates of MOS transistors (Silicon Gate Technology). Polysilicon is deposited by decomposition of Silan (SiH_4) at the surface of the hot wafer (600 ... 650 °C) in a CVD process (Chemical Vapor Deposition) [19.18]. The sheet resistance is then adjusted by ion implantation. Gates and short interconnection lines should have low layer resistances. With a layer thickness of 0.5 µm a sheet resistances of 20 Ω/□ may be achieved. For long lines polysilicon is not suited because of its in comparison with aluminum high sheet resistance.

The sheet resistance may be decreased by the creation of silicides on the polysilicon composed of titanium, tantalum, wolfram or molybdenum. For this purpose the particular metal is deposited on the polysilicon and subsequently sintered into its surface. In this way the desired heavy metal silicide is created with specific resistances ranging from 20 Ω · cm with titanium silicide ($TiSi_2$) to 100 Ω · cm with $MoSi_2$ [19.19]. For $TiSi_2$ a sheet resistance of 0.8 Ω/□ is calculated for a layer thickness of 0.25 µm. The sandwich structure Metal Silicide on Polysilicon is called Polycide as well. For resistors with high resistances there is used low-doped polysilicon too.

19.2 Bipolar Technology

19.2.1 Process Description

Bipolar technology uses mainly the processing steps required for the production of insulated NPN Bipolar Junction Transistors (BJT) on a chip. Figure 19.12 shows in perspective the cross section

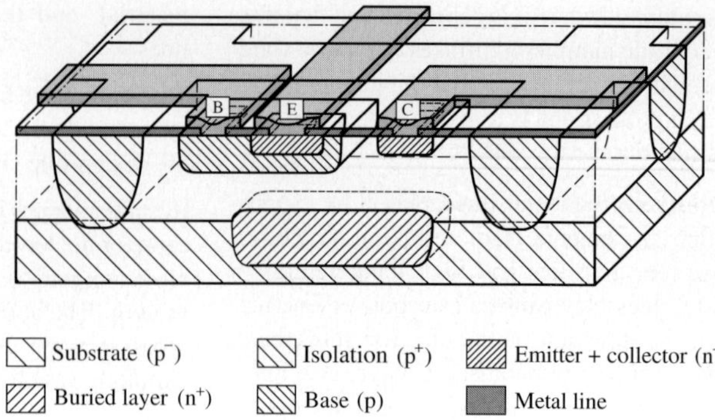

Fig. 19.12 Cross section of an integrated NPN BJT

of an integrated NPN BJT in Standard Bipolar Technology.

The components of an IC are insulated by reverse biased PN-junctions surrounding the particular device. To achieve this a (n^-) layer is deposited on a (p^-) substrate wafer by epitaxy ('Epilayer'). Additionally (p^+) zones ('Isolation' zones) are diffused from the surface through the epilayer down to the substrate. By this means the n-zone is bordered in all directions by a pn-junction ('n-well').

Using planar technology the transistor zones are contacted at the silicon surface. Therefore the n-well, which forms the collector zone, is contacted by a (n^+) diffusion zone.

The doping concentration and the thickness of the epilayer are determined by the required breakdown voltages of the devices within the n-wells. To achieve a low resistive path for the collector current, a highly doped (n^+) layer is diffused into the substrate, the so-called buried layer, before the deposition of the epi layer. Its effect on the component properties is discussed later on in the context of the properties of the integrated devices.

Standard Buried Collector Process (SBC-Process)

The process is designed to manufacture integrated transistors with $V_{CEo} > 20$ V.

The sequence of production steps is as shown in figs. 19.13 to 19.19:

- Substrate wafer P type (boron), $5 \ldots 10\ \Omega \cdot \text{cm}$
- Thermal oxidation: 1,100 °C, 2 h, oxide thickness $t_{ox} = 1\ \mu\text{m}$

a) Mask M1: Buried Layer (see fig. 19.13):

(n^+)-diffusion (arsenic or antimony), 1,200 °C, 10 h, $x_j = 5\ \mu\text{m}$, $R_s = 10 \ldots 20\ \Omega/\Box$, then stripping off the oxide. The atoms of arsenic or antimony diffusing slowly are used to prevent the unwanted redistribution of the buried layer into the weakly doped epi layer during the following high temperature processes.

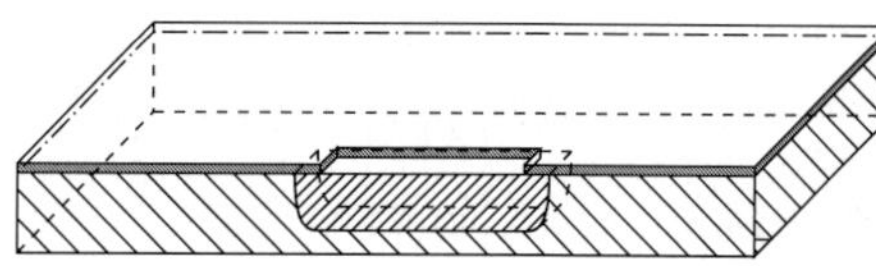

Fig. 19.13 Wafer, oxidized, Mask 1: Buried Layer, diffused

b) Deposition of the epitaxial layer (see fig. 19.14):

1 100 °C, 1 h, $d_{epi} = 10\ \mu\text{m}$;
$N_D = 0{,}5 \ldots 1 \cdot 10^{16}\ \text{cm}^{-3}$ (phosphorus)

- Thermal oxidation: 1,100 °C, 2 h, oxide thickness $t_{ox} = 0.7\ \mu\text{m}$.

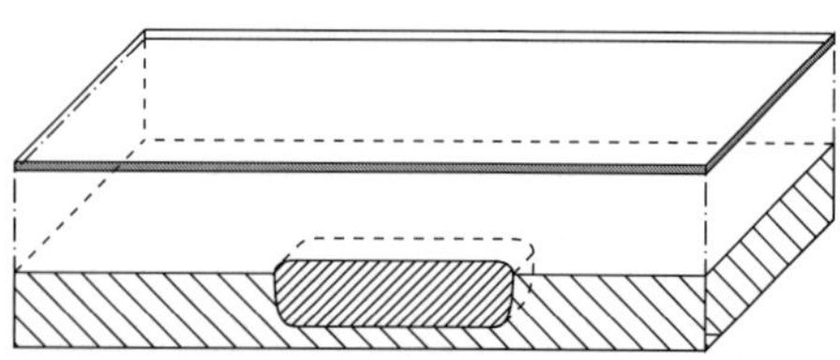

Fig. 19.14 Deposited epitaxial layer and thermal oxide after epitaxy

c) Mask M2: Isolation (see fig. 19.15):

Diffusion: p^+ (boron), 1,100 °C, 3 h, $x_j = 10\ \mu\text{m}$, $R_s = 20\ \Omega/\Box$; afterwards oxide stripping to decrease the height of oxide steps on the surface;

- Thermal oxidation: 1,050 °C, 50 min, oxide thickness $t_{ox} = 0.5\ \mu m$.

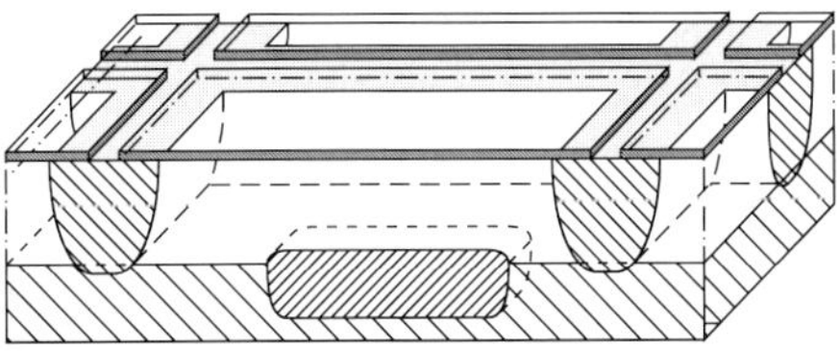

Fig. 19.15 Mask 2: Isolation, p^+ zones diffused

d) Mask M3: Base (see fig. 19.16):

Diffusion: p^+ (boron), 1,050 °C, 1 h, $x_j = 2\ \mu m$, $R_s = 200\ \Omega/\square$

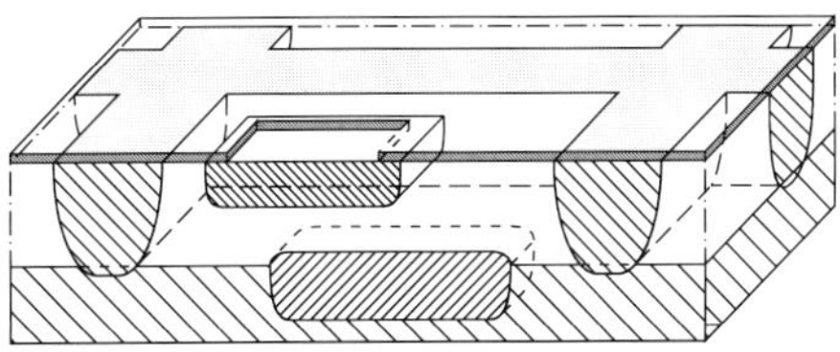

Fig. 19.16 Mask 3: NPN BJT base, diffused

e) Mask M4: Emitter and Collector (see fig. 19.17):

Diffusion: n^+ (phosphorus), 1,000 °C, 0.5 h, $x_j = 1.6\ \mu m$, $R_s = 10\ \Omega/\square$

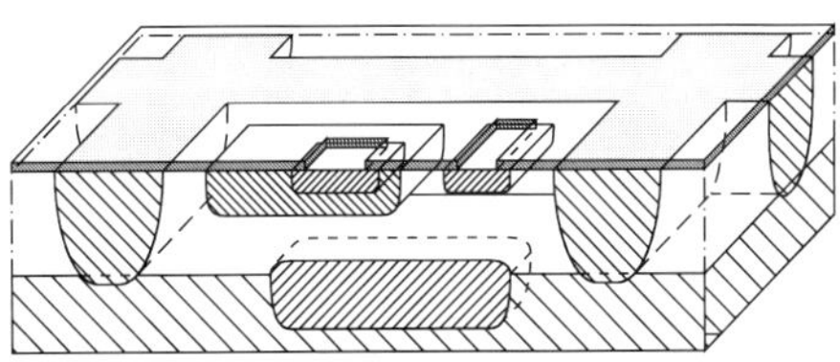

Fig. 19.17 Mask 4: Emitter and Collector, zones diffused

f) Mask M5: Contact Windows (see fig. 19.18):

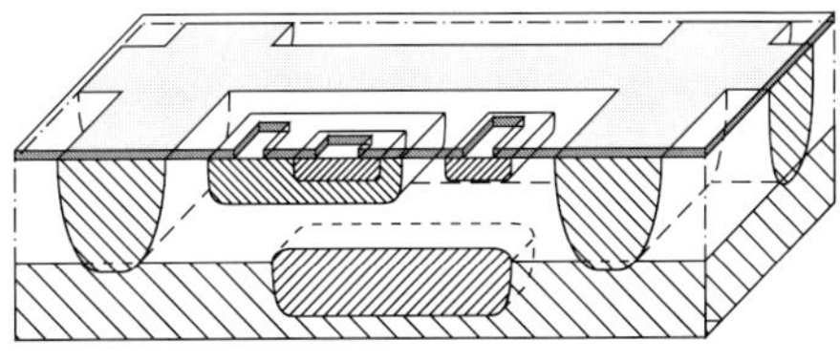

Fig. 19.18 Mask 5: Contact Windows, contacts for base, emitter, and collector, etched

g) Mask M6: Metal interconnection lines (see fig. 19.19):

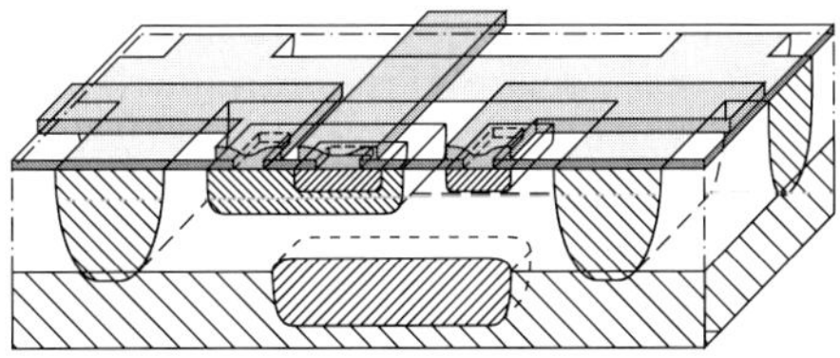

Fig. 19.19 Mask 6: Metal interconnection lines, etched

Annealing the metal lines at about 400 ... 450 °C to minimize contact resistances. Deposition of the protecting oxide over the metal lines.

h) Mask M7: Pad Window

Oxide removal from the areas for the placement of the bonding wires for the connection of the metal lines on the chip to the case contacts.

Testing and Housing

The finished dies (chips) are tested on the wafer and the faulty ones are marked. Afterwards the wafers are put onto a thin film, fixed by adhesion. Then the dies on the wafer are separated by sawing up the wafers in the so called scribe lanes using a diamond saw. Meanwhile the chips remain in order attached to the film and may be picked up for housing or to be mounted on a metal strip.

The Standard Bipolar Process consists of a minimum number of seven masking steps. Frequently it is expanded by two additional mask processes to realize highly doped (low resistive) connections between collector contact and buried layer (mask M2a between M2 and M3) and high resistive implanted resistors (mask M3a between M3 and M4).

19

19.2.2 Integrated Components

As presented in the preceding paragraph the processes for the production of integrated NPN BJTs are the basic technology of the Bipolar Process. If required, it may be extended by some additional process steps. In the following paragraphs the structures and properties of integratable components using the Bipolar Process are explained.

19.2.2.1 NPN Bipolar Junction Transistor (NPN BJT)

The NPN BJT is created in the weakly doped epitaxial layer as a double diffused transistor. Figure 19.20 shows the perspective view of the structure of a double diffused NPN BJT in Standard Technology. The base width is less than 0.5 μm, for transistors with transit frequencies $f_T > 10$ GHz even less than 0.1 μm.

The BJT function, i.e., the movement of electrons from the emitter through the base to the collector, takes place in the region below the emitter zone ('Inner transistor'). The emitter area is the effectively the area of the transistor.

The Buried Layer decreases the series resistance between the inner collector and the collector contact and reduces the current gain of the parasitic Substrate PNP BJT (SPNP, see 19.2.2.3). This parasitic transistor is activated during the saturation of the NPN BJT.

The maximum permissible reverse voltage is given by the breakdown voltage of the plane collector junction and is determined therefore by the doping concentration and thickness of the Epi layer, which covers the Buried Layer. The process as described before yields breakdown voltages of the plane collector PN junction of about 90 V. This corresponds to V_{CEo} values > 20 V when the DC current gain is $B = 70 \ldots 150$. But the measured breakdown voltage of the collector junction is only about 60 V owing to the junction rim curvature determined by penetration depth of the PN junction. The breakdown voltage of the emitter junction (owing to the higher doping concentration) is less than that of the collector junction and is about 7 V.

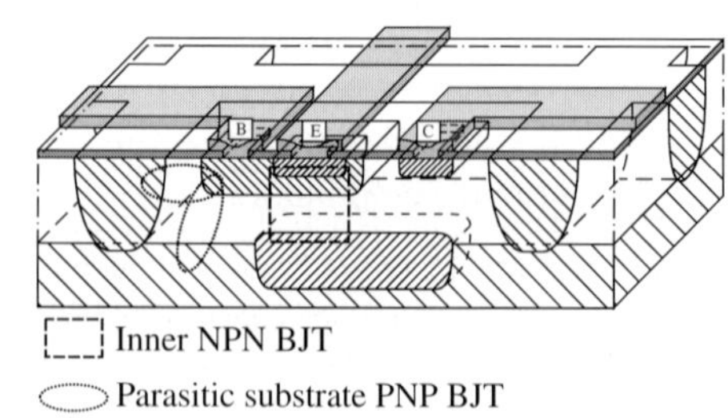

Inner NPN BJT

Parasitic substrate PNP BJT

Fig. 19.20 Perspective view of an integrated NPN-Bipolar Junction Transistor (BJT) in standard technology

Power transistors are produced with multiple emitters in parallel to increase the lengths of the outlines of the emitters. Additionally they are surrounded by a closed ring of (n^+)-doped material, which extends down to the buried layer in order to decrease both the collector series ohmic resistance and the current gain of the parasitic SPNP. Both effects reduce the saturation voltage, which results in a higher maximum permissible collector current.

In the base zone created by diffusion or ion implantation the impurity concentration decreases from the emitter to the collector edge. The concentration dependent rearrangement of the majority carriers in the base causes a 'Drift Field' which shortens the transit time of the electrons through the base and thus increases the transit frequency f_T.

19.2.2.2 Lateral PNP-Transistors

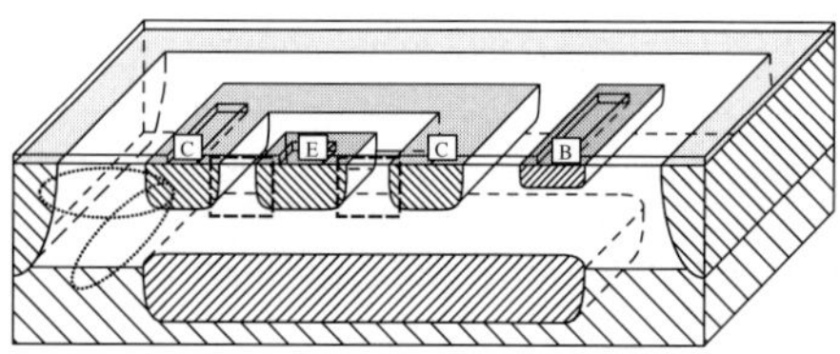

Inner LPNP BJT

Parasitic substrate PNP BJT

Fig. 19.21 Perspective view of a lateral PNP BJT (LPNP); (metal lines omitted)

Figure 19.21 shows in perspective view the structure of a lateral PNP BJT (LPNP). The (p^+) emitter zone is surrounded closely by a (p^+) doped collector. The base is formed by the epi layer, which is connected via the buried layer with the (n^+) base contact. In fig. 19.21 the inner LPNP transistor is marked by dotted lines. The distance between both (p^+) zones, i.e., the base width W_B, depends on the breakdown voltage required for the LPNP because the collector depletion layer extends into the base. This is the reason why $W_{B\text{-}LPNP} \gg W_{B\text{-}NPN}$. Nevertheless the current gain $B_{LPNP} \approx B_{NPN}$. A large disadvantage of the LPNP is, that having the same overall chip area its maximum collector current is only 1/10 of the value for NPN BJT. Furthermore, the transit frequency of the LPNP is essentially lower than that of the NPN BJT because of the large base width and because of the absence of a drift field in its base zone (like in the SPNP). There holds

$$f_{\text{T-LPNP}} \approx \frac{1}{50} f_{\text{T-NPN}}$$

19.2.2.3 Substrate PNP Bipolar Junction Transistor (SPNP BJT)

Figure 19.22 shows the substrate PNP transistor (SPNP) in perspective. The inner transistor is marked by dotted lines. The emitter is the diffused (p^+) zone. The base is formed by the n-doped epi layer and is contacted by the (n^+) zone. The collector is formed by the p-substrate and by the p^+ isolation zones. The collector is consequently connected to the substrate and not available as an insulated contact. In general this transistor is used only for special applications, e.g., emitter follower stages.

The maximum permissible current of the SPNP is comparable with that of an NPN transistor as are the current gains. But because the base zone is essentially wider and has no drift field (unlike the NPN transistor) the transit frequency is also essentially lower than that of an NPN transistor. It holds

$$f_{\text{T-SPNP}} \approx \frac{1}{50} f_{\text{T-NPN}}$$

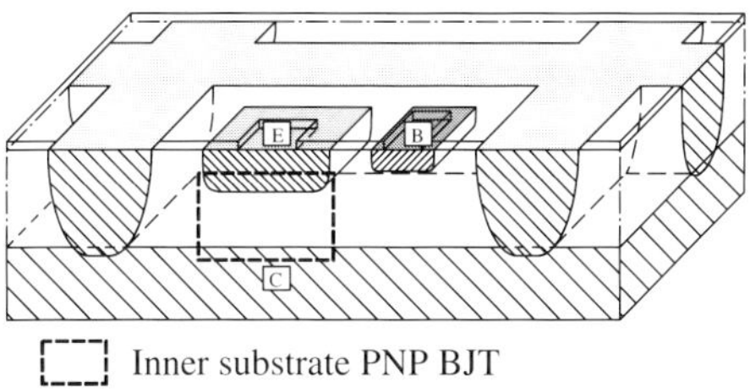

Fig. 19.22 Substrate PNP transistor (SPNP) (metal lines omitted)

Structures comparable with those of the SPNP are inevitably created between the base and isolation zones of the NPN BJT and between collector and isolation zone of the LPNP BJT. These Parasitic Substrate PNP transistors get into the active mode only if the collector junction of the NPN or LPNP transistors is forward biased, i.e., if these transistors are saturated. A part of the base current flows then through the parasitic SPNP transistor directly to the substrate, which is the most negative point of the supply voltage.

19.2.2.4 Integrated Diodes

Most PN diodes are created by the use of NPN transistors having a short circuit between base and collector. Their forward voltage drop equals the forward voltage V_{BE} of the transistor. This configuration is advantageous in that the transistor is operated in the active region and therefore the parasitic SPNP cannot be activated. The disadvantage of this configuration is the low break voltage of the diode which equals the breakdown voltage of the emitter junction which is about $6 \ldots 7$ V. For diodes with higher break voltages LPNP transistors may be used with short-circuited base and collector as well and a supplementary deep (n^+)-diffusion around the collector to minimize the effect of the parasitic SPNP transistor.

Z-diodes may be realized with NPN transistors, having short-circuited base and collector and reverse biased base-emitter junction. The Z-voltage is $6 \ldots 7$ V, depends on the technology, and is of very good reproducibility.

Schottky diodes are created at the contact between aluminum and the low doped n^--type epi layer. The depletion layer of this junction allows breakdown voltages > 20 V.

To create non blocking ohmic contacts between aluminum and n-type silicon the contact area must be (n^+)-doped. By means of this the thickness of the depletion layer becomes so thin that it may be tunneled in both directions by electrons and the current is independent of its direction.

19.2.2.5 Integrated Resistors

Integrated resistors in bipolar technology are most often realized as p-zones in an n-well (see fig. 19.23). The PN junction between resistor body and n-well is reverse biased by connecting the well contact to a voltage level which is equal to or higher than the most positive voltage occurring in the resistor body. The sheet resistance of the resistor body ($R_{\text{sR}} = 1 \ldots 5$ k$\Omega/\square$) is most often higher than the base layer sheet resistance ($R_{\text{sB}} = 100 \ldots 300\ \Omega/\square$). During the layout of the resistors, first of all the resistor body is drawn as a path which is characterized by the width and length of the center line, frequently as meander to save area on the chip. Then the resistor heads and the

Table 19.2 Threshold Voltage V_{T0} vs. oxide thickness and bulk doping

Oxide Thickness	50 nm	100 nm	200 nm	500 nm	1 000 nm	Remarks
V_{T0}/V	−0.30	−0.31	−0.33	−0.40	−0.50	1; NMOS; $N_B = 3 \cdot 10^{14}$ cm^{-3}
V_{T0}/V	+0.4	+1.0	+2.2	+5.8	+11.8	2; NMOS; $N_B = 1 \cdot 10^{16}$ cm^{-3}
V_{T0}/V	−1.0	−1.3	−1.7	−3.1	−5.3	3; PMOS; $N_B = 3 \cdot 10^{14}$ cm^{-3}
V_{T0}/V	−1.7	−2.6	−4.2	−9.2	−17.7	4; PMOS; $N_B = 1 \cdot 10^{16}$ cm^{-3}

Source and drain are bordered through a thick **Field Oxide** created by local oxidation (LOCOS process). Table 19.2 shows the threshold voltage V_{T0} of NMOS and PMOS structures as a function of bulk doping and oxide thickness.

The negative threshold voltage of the NMOS structure on weakly doped substrate material shows that there occurs every time an inversion of the surface. This is the reason why below field oxide regions the doping concentration is increased so far by ion implantation of boron (field implantation) until the unwanted inversion is avoided reliably (see fig. 19.25).

The necessary thickness of the field oxide depends on the supply voltage of the circuit. Higher supply voltages require thicker field oxides and/or higher ion implantation doses during the field implantation. The thickness of the field oxide is about 0.5 . . . 1 µm.

Principally transistors in MOS circuits are self-insulating because source, drain, and channel in normal operation are reverse biased with respect to the substrate (exception: Latch-up in CMOS structures see section 21.1.3). Unlike in bipolar technology, special structures for the insulation of the components are unnecessary. This results in the very high packaging density of active elements in silicon.

The sequence of processing steps of the NMOS silicon gate technology is shown in Figures 19.28 to 19.36. A detailed presentation of the particular steps of photo resist deposition, patterning, resist stripping and cleaning was omitted in these drawings.

Basic materials are weakly doped (p^-) silicon wafers. During the first processing cycle the field oxide is grown on the wafers by local oxidation excluding the active areas where the MOS FET will be placed. For local oxidation (see section 19.1.2), a very thin SiO_2 layer is grown on the wafers (protecting oxide) and then a Si_3N_4 layer is deposited above it. By means of a photo resist process the nitride layer is structured in such a way that the Si_3N_4 is preserved over the active areas (Mask M1, active area).

To increase the field threshold voltage a field implantation (boron) is applied. Figure 19.28 shows the cross section of the transistors active area at this time.

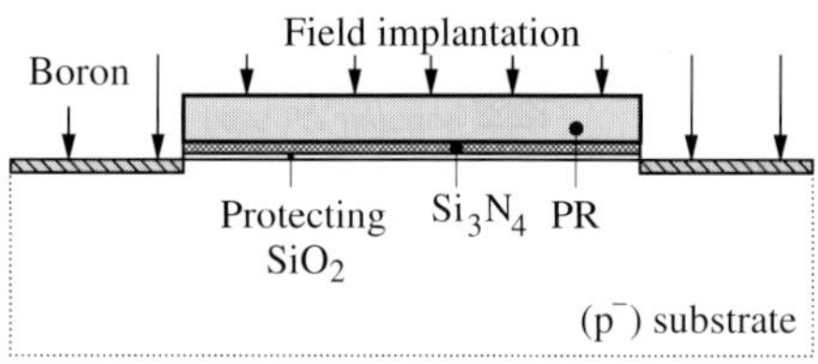

Fig. 19.28 Field implantation

After photo resist stripping the field oxide is grown by local oxidation (see fig. 19.29).

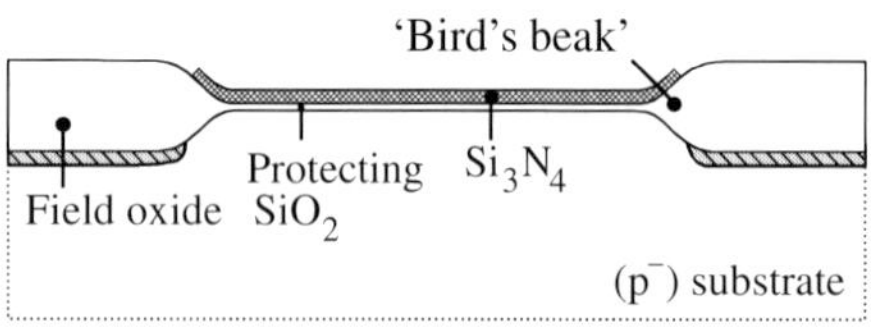

Fig. 19.29 Local oxidation of the field oxide

After that the Si_3N_4 and the protecting oxide are removed and the gate oxide is thermally grown.

For the adjustment of the threshold voltage V_T of the enhancement transistors the V_T-implantation is applied over the whole surface (see fig. 19.30).

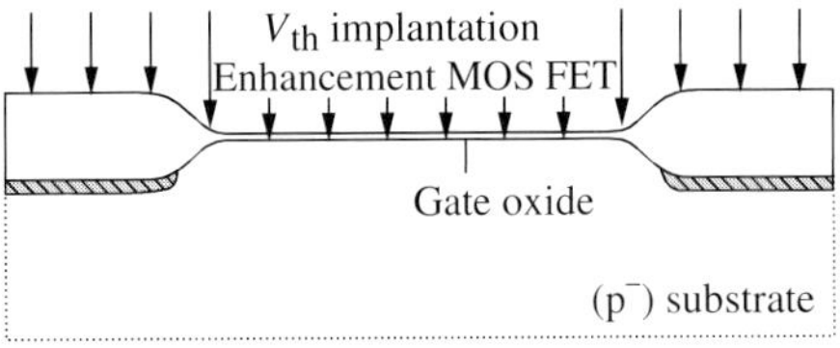

Fig. 19.30 V_T implantation for the depletion mode MOS FET

Subsequently the channel regions of the depletion mode transistors are implanted using a resist mask (M2, depletion-implantation) which has windows only for the depletion mode transistors active area (see fig. 19.31).

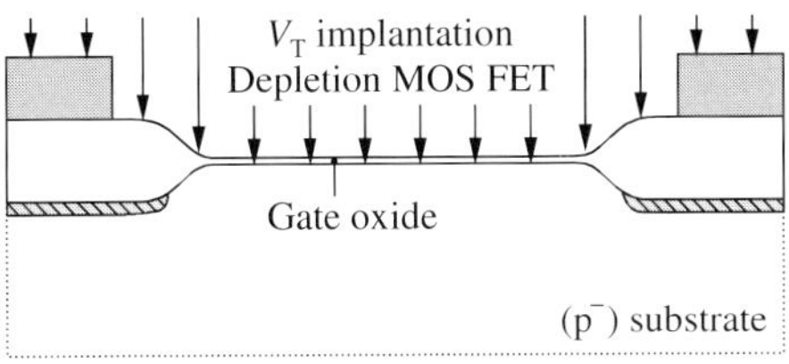

Fig. 19.31 Kanalimplantation der selbstleitenden Transistoren

After stripping the photo resist phosphorus doped polysilicon is deposited on the wafer. The next mask (M3, polysilicon) defines the gates. Polysilicon stripes remain over the active area and form the gates (see fig. 19.32).

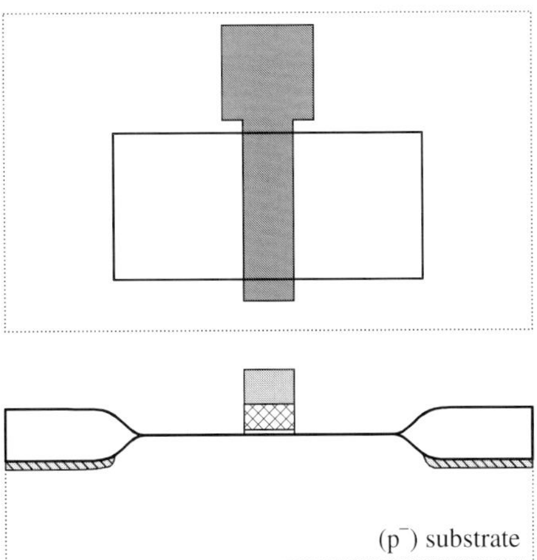

Fig. 19.32 Structuring of polysilicon

Subsequently phosphorus or arsenic is implanted over the whole wafer. By this process the source and drain regions of the MOS transistors are created (see fig. 19.33). The channel length is determined by the width of the polysilicon gates and thus self-aligned. By this the capacitances resulting from the overlapping of gate source and gate drain are minimized.

In the field oxide regions phosphorus is implanted into the oxide and consequently not electrically activated. Thus the channel width is self-aligned as well and equals the width of the active area.

Now a thermal oxidation is applied, during which the open monosilicon and polysilicon areas are covered by a thin protective oxide (about 0.1 µm), whose function is explained later on. At this state of processing the wafers are covered by relatively thick oxide layers with steep edges of oxide and polysilicon. These are very disadvantageous for the deposition and structuring of the metal interconnect lines. It may even happen that the thickness of the metal layers is significantly reduced at the oxide edges.

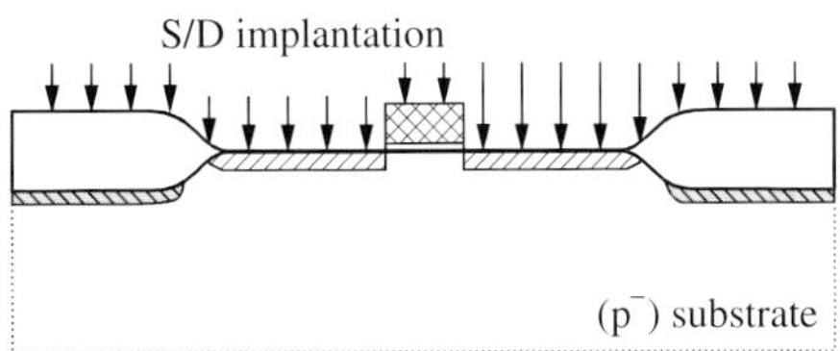

Fig. 19.33 Source drain implantation

To prevent this, a layer consisting of Phosphorus Silicate Glass (PSG) is now deposited. This is a SiO_2 layer highly doped with phosphorus. The property of the PSG layer is that its consistency is soft even at low temperatures of about 1,000 °C, thus causing a planarization of the wafer surface by a reflow process [19.20].

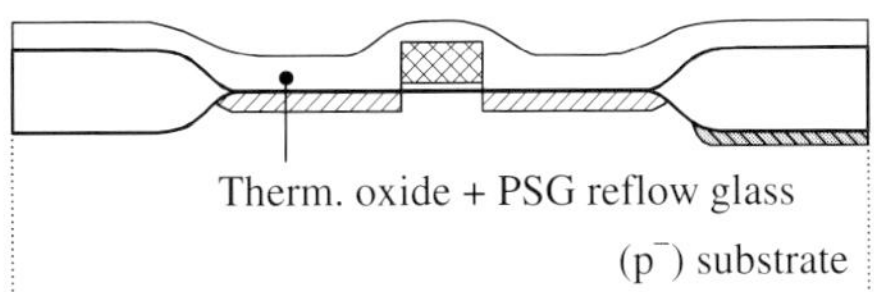

Fig. 19.34 PSG deposition and reflow process

By adding of boron to the PSG (Boron phosphorus silicate glass, BPSG) the processing temperature of the reflow process may be further reduced (about 900 °C). The thin protecting oxide, mentioned above, has to be created before the deposition of the silicate glasses, because during this high temperature process unwanted doping atoms could diffuse out of the glass into the silicon beneath it.

The mask following now (M4, contact windows) opens the contact windows to the poly- and mono-silicon (see fig. 19.35). At the same time the edges of the oxidation steps are rounded by an appropriate technology. In addition, after the etching procedure the oxidation steps are rounded again by a further reflow process.

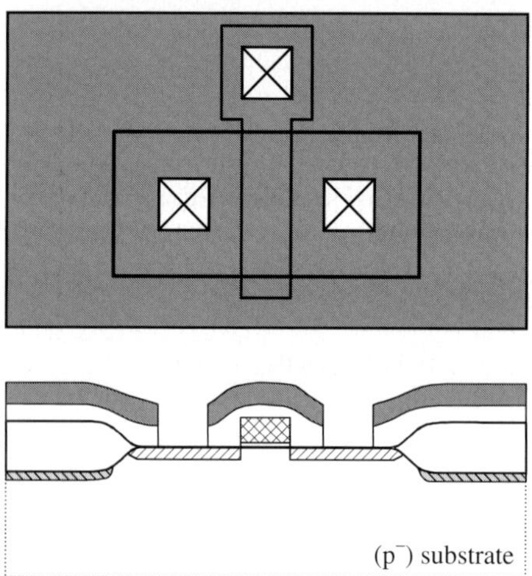

Fig. 19.35 Contact window mask

After the structuring of the contact windows, aluminum with added silicon (ca. 1 %) and copper (ca. 2 . . . 4 %) is deposited on the wafer by sputtering and structured by a further mask (M5, metal mask) (see fig. 19.36). After tempering of the wafer for the reduction of the contact resistance between aluminum and silicon a protecting oxide is deposited over the metal lines.

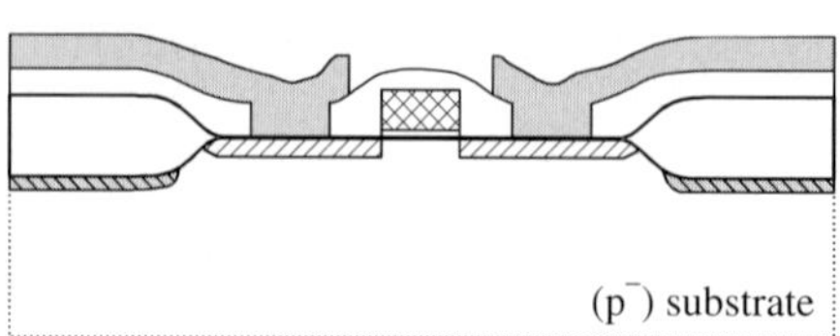

Fig. 19.36 Metal interconnection lines

By means of the final mask process (M6, Pad windows) the areas (pads) for the bonding wires, which connect the chip with the case are opened.

The minimum number of mask processes steps is six and thus is well suited for the production of complex VLSI circuits. This number increases if other types of structures are needed as, e.g., direct contacts between poly- and mono-silicon or for further metallization layers.

A modification of the NMOS process, the LDD technology (Lightly Doped Drain, see fig. 19.37) [19.7], is used especially for modern components with very short channel lengths (< 1 µm). The goal is to realize a flat doping profile of the PN junction at the drain side of the channel instead of an abrupt one. This helps to avoid the shift of the threshold voltage due to hot carrier degradation in transistors with very short channel lengths.

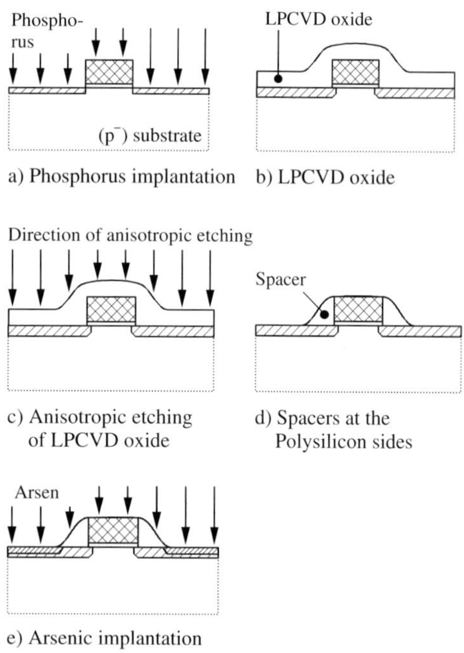

Fig. 19.37 LDD technology

Using the LDD technology the wafers are processed including polysilicon patterning. Then phosphorus is implanted with a much lower concentration than normally used for source and drain (see fig. 19.37a). Then with a Low Pressure Chemical Vapor Deposition process (LPCVD) an oxide with a thickness of about 0.25 μm is deposited. The oxide covers the oxide step on the wafer very smoothly (see fig. 19.37b). Subsequently the oxide is stripped in a anisotropic dry etching process. That means the vertical etch rate is much higher than the lateral etch rate. The oxide thickness is bigger at the sides of the polysilicon stripes than on horizontal areas. Oxide stripping is done in such a manner that the thinner oxide is removed whereas the thicker oxide partly remains in areas at the sides of the poly silicon stripes. These oxide residues called 'Spacer' act as distinct broadening of the polysilicon stripe during subsequent high dose source drain arsenic implantation. That means that between the rim of the channel and the highly doped arsenic zones there are low phosphorus doped stripes. In this way the unwanted high doping concentration just at the side of the channel can be avoided.

The following process steps are the same as with the normal NMOS technology.

In complex circuits the number of components per chip is limited by the maximum permissible power, which can be dissipated by the case. By the use of CMOS circuits essentially less power dissipation is possible as with NMOS circuits. The technology needed for CMOS is presented in the following section.

19.3.2 CMOS Technology

CMOS technology allows realizing NMOS FET as well as PMOS FET on the same chip. Figure 19.38 shows the cross section of a CMOS inverter. The process used, is derived from the NMOS technology. For the P-channel FET there is created a weakly n-doped area ('n-well') into which the PMOS structures are inserted.

Besides the n-well process there are still two further variants of the CMOS process. The 'p-well process' starts from an n-doped substrate wafer and uses a 'p-well'. This technology is not directly compatible with the NMOS technology and shows some electrical disadvantages. It is used rarely. Much more frequently used is a process with two wells (n-well and p-well), called 'Twin Tub Process' which is discussed shortly at the end of this paragraph.

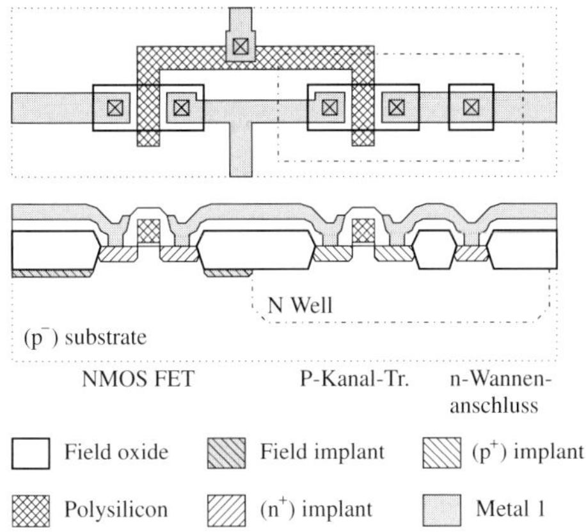

Fig. 19.38 CMOS inverter (top view and cross section)

In the n-well CMOS process the n-well is connected by a (n^+)-contact with the source of the PMOS FET. The field oxide regions are created by local oxidation. A field implantation is only applied to the (p^-)-surfaces. Frequently additional guard rings are used to prevent Latch Up. The function of this measure is explained in detail in section 21.1.3. Because of that these guard rings are not mentioned while the CMOS technology is described. They may be created by the process steps discussed subsequently and require no additional process steps.

Sequence of Process Steps of the N-Well CMOS Technology

Substrate material is a (p^-)-doped silicon wafer. The wafer is oxidized and then the windows for the phosphorus implantation of the n-well are opened by mask 1 (M1, n-well) (see fig. 19.39).

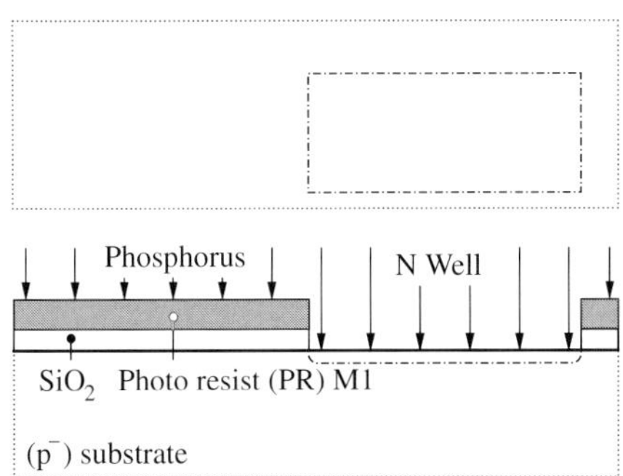

Fig. 19.39 n-well implantation

For the local oxidation of the field oxide zones the oxide is removed from the wafer surface. Then there is grown a thin protection oxide and subsequently Si_3N_4 is deposited on the whole wafer.

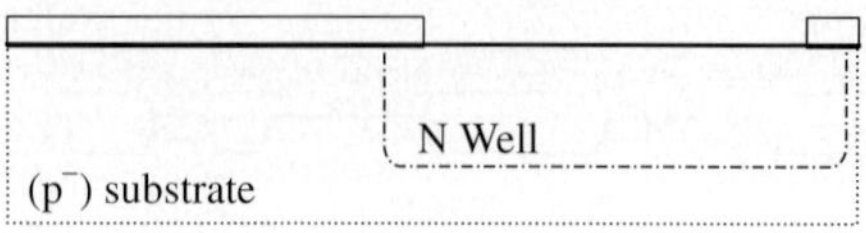

Fig. 19.40 n-well after 'drive in'

By the following mask (M2, active regions) the nitride is structured in such a way that the areas where the field oxide is to be grown become uncovered (see fig. 19.41). The photo resist remains on the nitride regions and is hardened in such a manner that this photo resist remains unchanged during the following mask process.

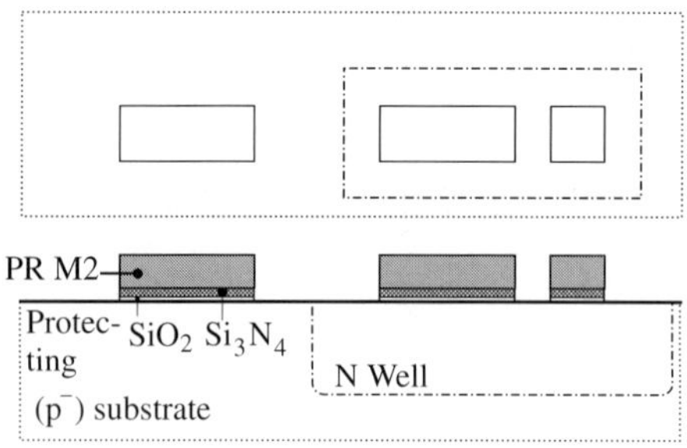

Fig. 19.41 Definition of the active areas

During the next mask process (M3, field implantation) the n-well areas are coated by photo resist. Then the field implantation is applied to the (p^-)-doped areas of the surface. The photo resist of M2 covers the active areas of the NMOS FETs. Figure 19.42 shows the structures at this state.

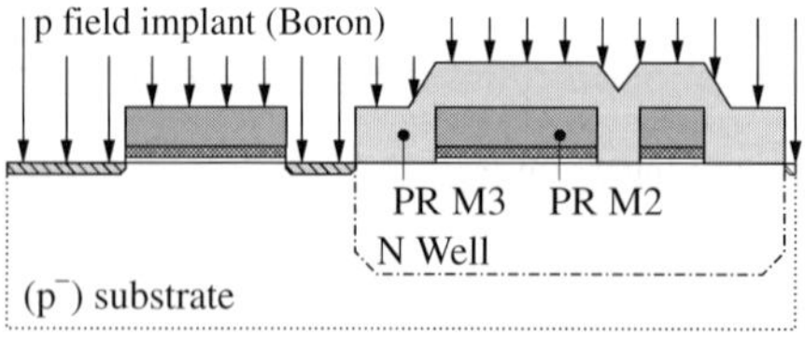

Fig. 19.42 Field implantation of the p^--doped surfaces

After the field implantation all photo resist is removed from the wafer and the field oxide is grown selectively using the silicon nitride as a mask (see fig. 19.43). Then the nitride and the protective oxide are removed by etching and then the thin gate oxide is grown.

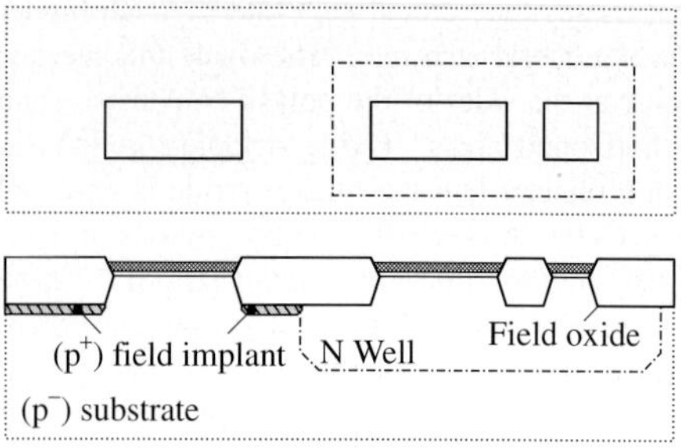

Fig. 19.43 Local oxidation

Using the following Photo resist masks:

M3a, n-channel V_T implantation

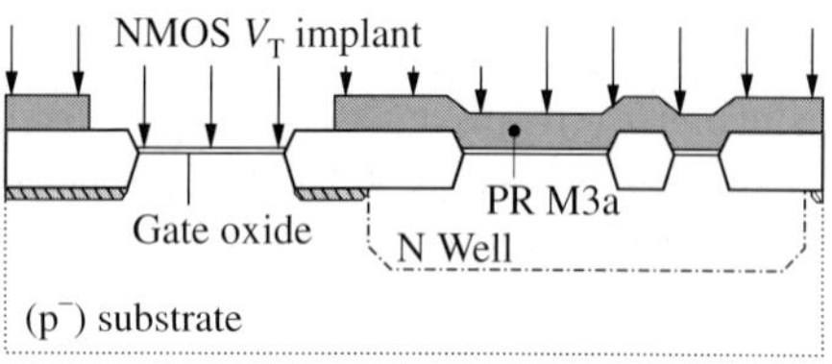

Fig. 19.44 V_T implantation n-channel

and M3b, p-channel V_T implantation implantations are applied to adjust the threshold voltages of the NMOS FET and PMOS FET (see Figures 19.44 and 19.45) respectively. By special control of the process, one of the masks may be omitted eventually.

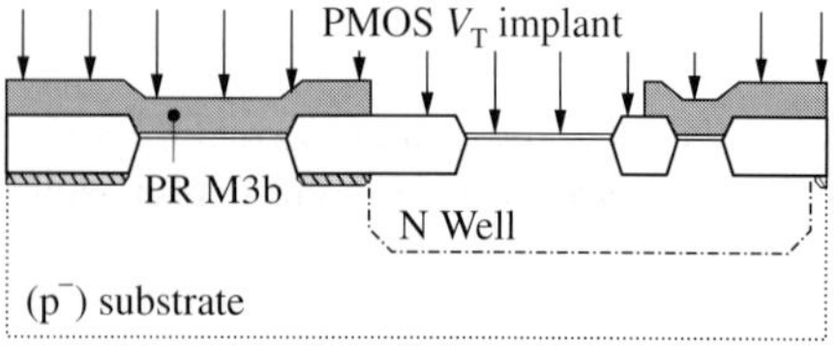

Fig. 19.45 V_T implantation p-channel

Polysilicon doped with phosphorus is deposited and structured by the next mask (M4, polysilicon) (see fig. 19.46).

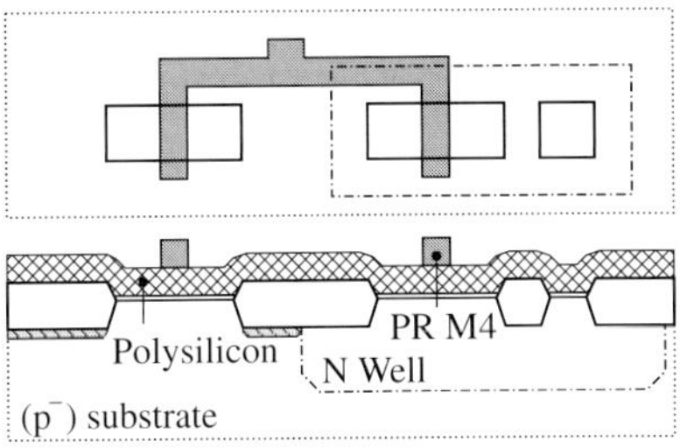

Fig. 19.46 Polysilicon mask

After polysilicon etching the Photo resist remains on the wafer (see fig. 19.47).

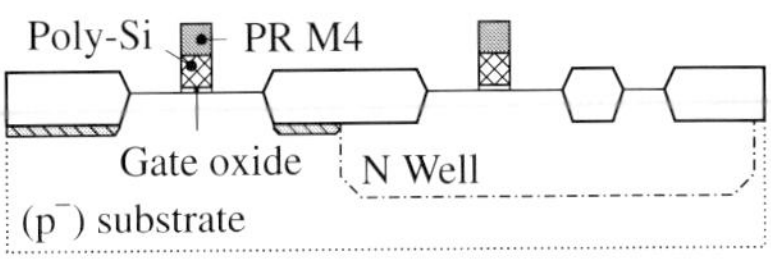

Fig. 19.47 Polysilicon structured

In a further photo process a new resist mask is applied (M5, p-channel S/D Implantation), which opens only the PMOS active areas for source/drain implantation with boron (see fig. 19.48). The photo resist which was applied by M4 and which is still on the wafer protects the polysilicon stripes against the implantation of boron.

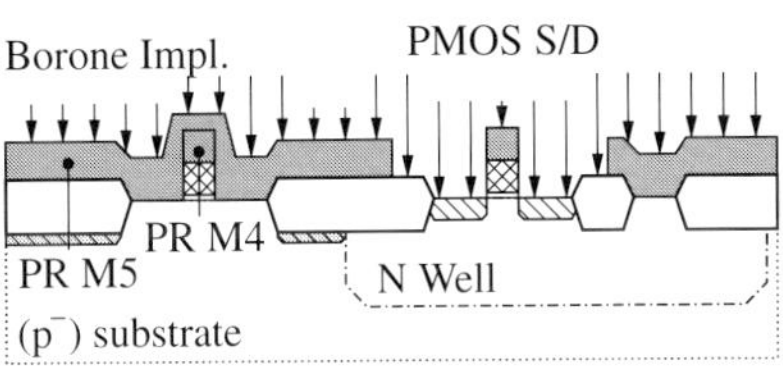

Fig. 19.48 P-channel S/D implantation

After the implantation of boron the resist layer is stripped. Then in a further mask process (M6, n-channel S/D implantation) the PMOS active regions are covered with photo resist (see fig. 19.49) and the NMOS source/drain areas and the n-well contacts are created by phosphorus implantation.

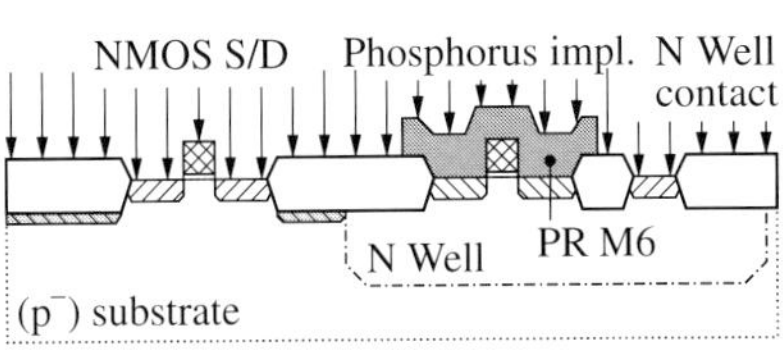

Fig. 19.49 N-channel S/D implantation

The subsequent process steps correspond with the NMOS process described before. I.e., or the reflow process following, a thin thermal protection oxide is grown and a PSG or BPSG layer is deposited.

During a reflow process the steps on the wafer surface become smaller and the edges are rounded (see fig. 19.50)

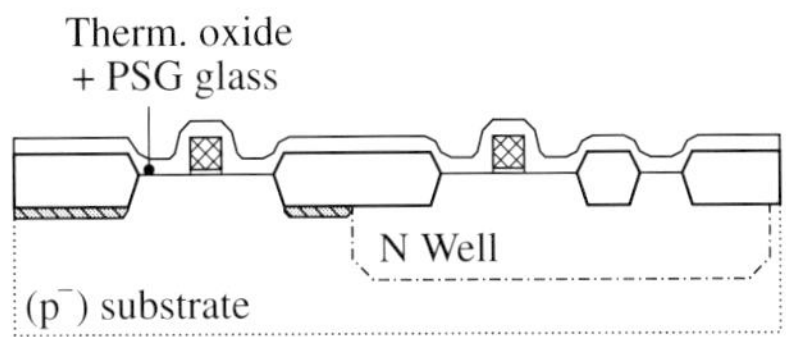

Fig. 19.50 PSG reflow process

By the next mask (M7, contact windows) the contact windows are opened to the mono- and polysilicon (see fig. 19.51) and the edges of the contact windows are again rounded by a reflow process of the existing PSG/BPSG layer.

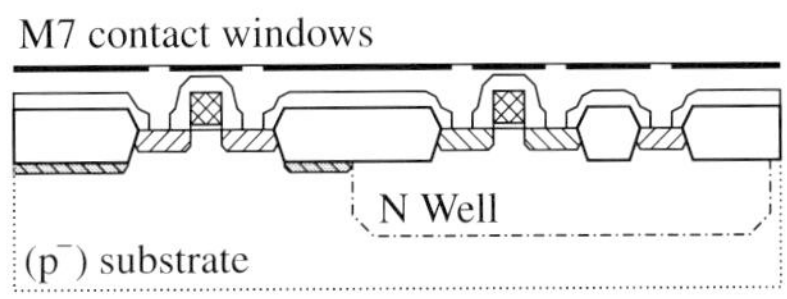

Fig. 19.51 Contact window mask

Then aluminum with additives of Si and Cu is sputtered onto the wafer and structured by a photo mask (M8, metal 1) (see Figures 19.52 and 19.53).

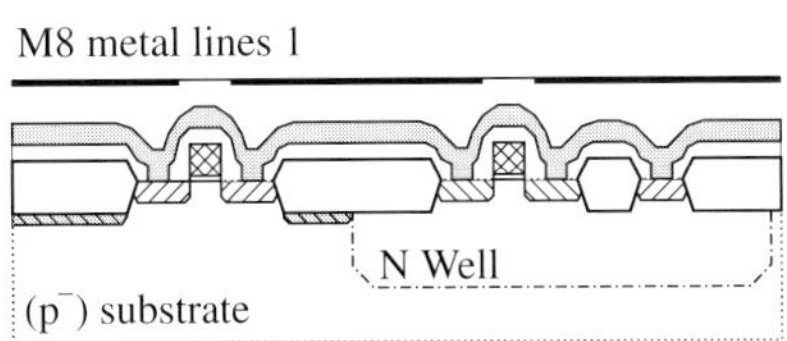

Fig. 19.52 metal 1 mask

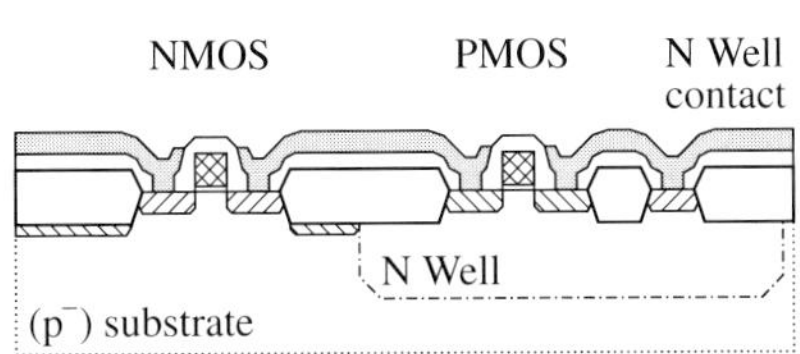

Fig. 19.53 Metal 1 lines structured

Finally, there is a pyrolytic deposition of the protecting oxide and the opening of the pad windows by a mask (M9, pad mask).

As can be seen, the CMOS process requires nine masks at minimum plus a maximum of two masks for the adjustment of the threshold voltage of the transistors. For each of additional metal layers two additional mask processes are required.

Twin Tub Process

Alternatively to the n-well process described above there is increasingly used a CMOS process using two types of wells (**Twin Tub Process**). Using local oxidation the structure shown in fig. 19.54 requires only one photo mask process to implement the two ion implantation steps for the two wells.

Fig. 19.54 Twin Tub Structure

The Twin tub process [19.5] starts with a highly (n^+)-doped substrate on which a lowly doped epitaxial n^- layer is deposited. Thereon is deposited a SiO_2 Si_3N_4 double layer, which is structured in such a way that the Si_3N_4 is removed over the p-well areas (see fig. 19.55).

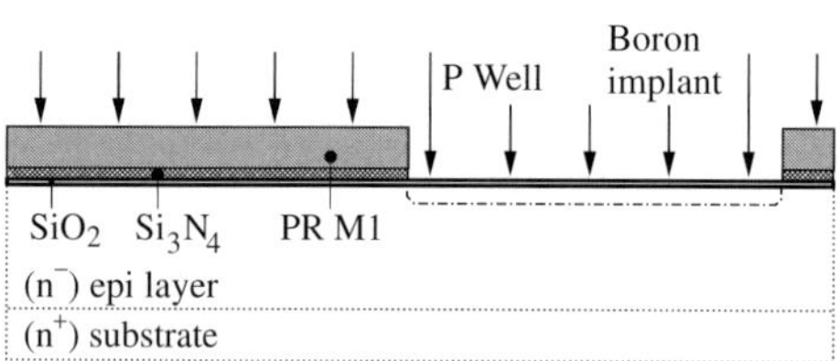

Fig. 19.55 Implantation of the p-well

After implantation of boron and photoresist stripping follows a first p-well drive in and simultaneously an oxide is grown selectively over the p-well areas (see fig. 19.56). Thereafter the nitride is removed and (without a further photo process) phosphorus is implanted for the n-wells (see fig. 19.57). In an additional drive in diffusion process both wells are diffused into the silicon until the required junction depth is achieved (see fig. 19.58). The oxide is removed then from the surface of the wafer.

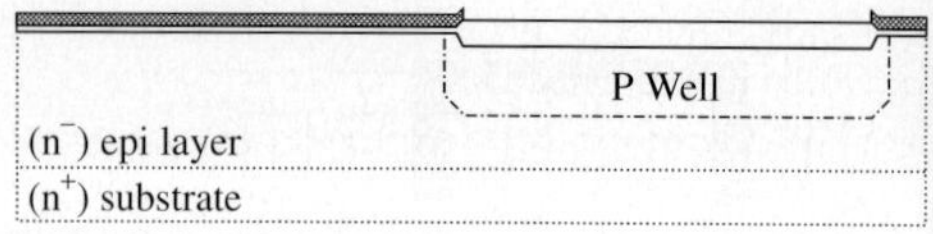

Fig. 19.56 Selective oxidation of the p-well surface

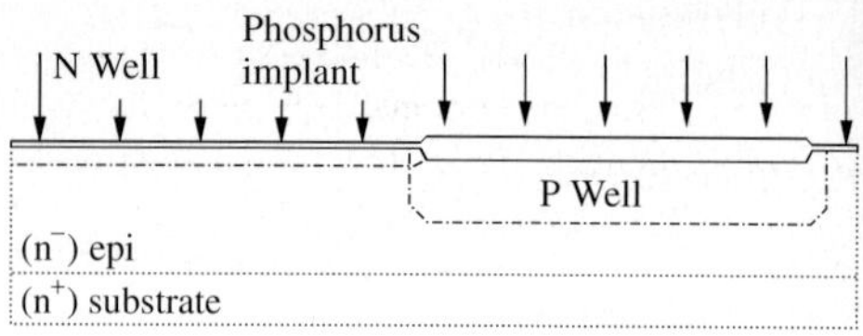

Fig. 19.57 Implantation of the n-well

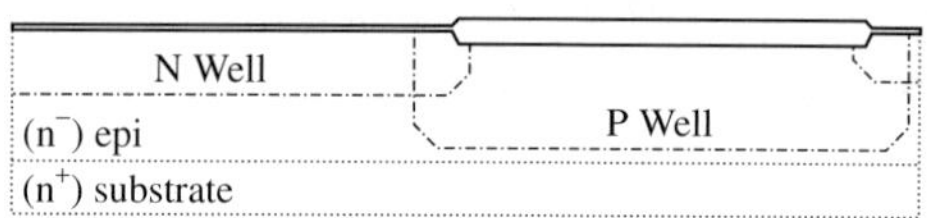

Fig. 19.58 Drive in diffusion of both wells

The further process steps correspond with the n-well process after implementation of the n-well. The surface impurity concentrations of the well regions may be adjusted for the required threshold voltages of the NMOS and PMOS transistors. The highly doped substrate improves the stability against latch-up of the CMOS structures (see section21.1.2).

19.4 Further Enhancements of Technology

The technologies discussed above for bipolar and CMOS circuits offer good compromises with respect to the requirements of the circuits. Many different ASICS may be realized by these technologies. It should be mentioned that the separate development of the two technologies (Bipolar and CMOS) prevailing in the past, is no longer continued and that they merge into combined technologies. Advancements of the MOS technology are adapted for bipolar circuits as well (e.g., local oxidation). On the other, hand typical techniques used in the bipolar technology (e.g. buried layer zones and epitaxial layers) are increasingly introduced into the CMOS technology (e.g., Twin Tub process).

For further enhancements of the technology there are three main targets:

- Shrinking of the structures for higher complexity of the ICs;
- Higher transit frequencies;
- Extension of the operating range of the supply voltages.

Examples of advanced technologies are presented in the following paragraphs. It should be kept in mind that the improvement of one parameter may cause a degradation of other parameters. Most often the improvement of some properties causes a higher number of process steps.

19.4.1 Bipolar Process for High Frequencies

To achieve smaller structures in bipolar technology, on the one hand progresses in the photo lithography are applied. On the other hand, the diffused isolation zones are replaced by oxide zones which extend down to the substrate and are created by local oxidation. This requires a considerable reduction of the epi- layer thickness and of the junction depth of base and emitter thus reducing the breakdown voltage of the devices.

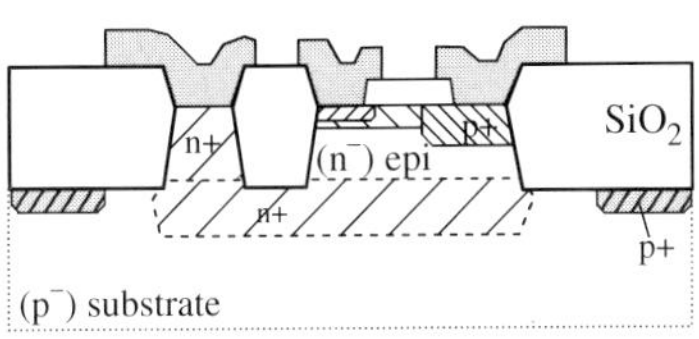

Fig. 19.59 Bipolar transistor with oxide isolation

Figure 19.59 shows the cross section of an NPN BJT for high frequencies. Caused by the elimination of the diffused (p^+)-isolation zones a considerable shrinking of the structures and reductions of the junction capacitances are achieved. Moreover the reduction of the junction depth and the use of arsenic emitters lead to steeper doping profiles in the base zones of the transistors. This causes a stronger drift field and consequently a higher transit frequency. All together transit frequencies of greater than 20 GHz may be achieved. As mentioned before the smaller junction depths cause a reduction of the breakdown voltage and thus demands smaller operating voltages of the IC's.

19.4.2 BiCMOS

Additional to the improvements in the shrinking of structures in MOS technology there are efforts to combine digital and analog circuits on a single chip. For analog circuits, e.g., bipolar transistors are advantageous because of their much higher transconductance as compared to MOS FETs carrying the same current. Therefore by the BiCMOS technology CMOS structures and additional bipolar transistors may be realized on the same chip. Figure 19.60 shows cross sections of transistors manufactured by this technology [19.23]. Not only NMOS and PMOS FET but also bipolar junction transistors may be realized. The example shows an NPN BJT but also Lateral PNP BJTs may be realized by this technique.

The CMOS n-well process has been extended in this case by three mask processes (Buried Layer, Deep (n^+)-Diffusion, P-Base for the NPN BJT) and by the corresponding implantation and diffusion processes as well as by a (p^-)-epitaxial layer.

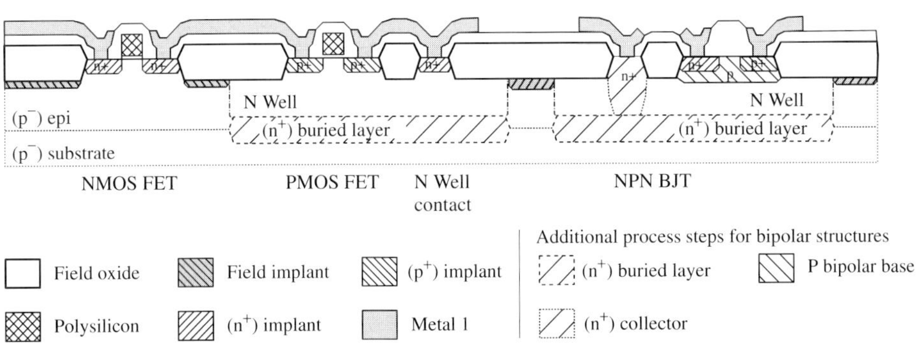

Fig. 19.60 BiCMOS technology: schematic cross section

Besides the BiCMOS process discussed here, process sequences were developed allowing the realization of Bipolar, CMOS, and DMOS transistors simultaneously. This BCDMOS technology is very costly and is used for Application Specific Circuits (Asics), e.g., in automotive applications. Details of this technology may be found, e.g., in [19.10].

19.5 References

[19.1] *Clauss, H.*: 'Probleme der Aluminium-Metallisierung bei Halbleiterbauelementen'. In 'Hochvakuum und Dünne Schichten', IEZ, München, 1973

[19.2] *Clauss, H.*: 'Physik der Fehlerursachen an Halbleiterbauelementen'. AEG-Telefunken, BMFT-Forschungsbericht T 76–77 (1976), pp. 26–55

[19.3] see [19.2], p. 40

[19.4] *Goser, K.*: 'Großintegrationstechnik'. Bd. 1. – Heidelberg: Hüthig Buchverlag, 1990, p. 86

[19.5] *Hoffmann, K.*: 'VLSI-Entwurf'. – München; Wien: R. Oldenbourg Verlag, 1998, p. 313

[19.6] *Hofker, W. K.*: Philips Res. Rep. Suppl. 8(1975)

[19.7] *Laker, K. R.; Sansen, W. M. C.*: 'Design of Analog Integrated Circuits and Systems'. – New York: McGraw-Hill Inc., 1994, pp. 156–162

[19.8] *Klar, H.*: 'Integrierte Digitale Schaltungen MOS/BICMOS'. – Berlin: Springer-Verlag, 1996, p. 42

[19.9] *Lie, L. N.; Razouk, R. R.; Deal, B. E.*: 'High Pressure Oxidation of Silicon in Dry Oxygen'. J. Electrochem. Soc. 129 (1982), pp. 2828–2834

[19.10] *Murari, B.; Bertotti, F.; Vignola, G. A.*: 'Smart Power ICs'. – Berlin: Springer-Verlag, 1996, p. 13 ff.

[19.11] *Parillo, L. C.*: 'VLSI Process Integration'. In S. M. Sze, Ed., VLSI Technology. – New York: McGraw-Hill, 1989

[19.12] *Rein, H.-M.; Ranft, R.*: 'Integrierte Bipolarschaltungen'. – Berlin: Springer-Verlag, 1980, p. 58

[19.13] see [19.12], p. 38

[19.14] see [19.12], p. 75

[19.15] *Schumicki, G.; Seegebrecht, P.*: Prozesstechnologie. – Berlin: Springer-Verlag, 1991, p. 56 ff.

[19.16] see [19.15], p. 245

[19.17] see [19.15], p. 249

[19.18] see [19.15], p. 219

[19.19] see [19.15], p. 251

[19.20] see [19.15], p. 231 ff.

[19.21] *Widmann, D.; Mader, H.; Friedrich, H.*: 'Technologie hochintegrierter Schaltungen'. – Berlin: Springer-Verlag, 1988, p. 301 ff.

[19.22] see [19.21], p. 67 ff.

[19.23] see [19.21], p. 301

[19.24] *Ziegler, J. F.; Biersack, J. P.; Littmark, J.*: 'The Stopping and Range of Ions in Solids'. – Pergamon Press, 1985

20 Integrated Circuit Techniques

GERHARD ALBERT

20.1 General Comments

Circuit technology is the basic level of the design hierarchy, but is the most important for the design process. Alone can the selection of the correct realization technology be responsible for the economic success of the circuit. The degree of complexity which originates through numerous parasitic elements is too high, even with the simplest circuits, in order to be able to find closed solutions for the design process. Through the description of active and nonlinear transistors by very primitive models, a 'closed' approximation can be calculated in simple circuits. The principal features of the circuits can be extracted by an analysis of the given accuracy of the model as well as being able to start optimizations. In many cases, however, one must consult even small signal equivalent model examinations for the simplification purposes.

Because of the differences of the technologies and design rules only principles can be represented. If one produces new functional modules from the given circuit diagrams, then the actual designs of the circuit parameters are to be accomplished with the desired process data. After that a fine optimization should be performed, supported by simulations (with exact models).

Layout examples are to be interpreted only symbolically because concrete design rules are lacking. However, it should be emphasized that analog circuits respond very critically on the implementation of the layout. Whilst it is produced sufficiently successfully for digital circuits in the automatic operation mode of the CAD programs, it is still an 'artwork' for analog circuits and frequently is still produced today with only inferior computer support. Bases for automation exist. The system [20.18], for example, composes a layout from given tested macro cells and performs a subsequent compaction step. [20.5] is a layout description language which contains the fundamental building blocks needed for analog circuits, efficiently places the blocks, and generates a layout. However, the breakthrough to the general utilization has not yet occurred.

In the following sections an attempt is made to represent the basic principles of the integrated circuit design of all current technologies on the narrowest area. First, general relations, model equations, parasitic elements, noise issues, and scaling are treated. After that a separation takes place into analog and digital circuits as well as A/D and D/A converters, which perform the transition between the digital and analog worlds.

Simplified DIN symbols are applied to represent the circuits. For MOS transistors only enhancement types are used in modern processes, so in the symbols only P and N channel must be distinguished. According to general practice the P channel type is marked with a curl at the gate symbol. The substrate connections are left out for clarity and are represented only in problem cases, for example if the control of a transistor can take place also over the backgate (body effect). In formulas there is no distinction by different spelling between whether it is a DC value, a complex value, or a time function. In cases of doubt the type of the formula is pointed out in the text. According to international practice incremental resistances and conductances are expressed in small signal operations by small letters. The symbol $+V_B$ is used uniformly for the power supply voltage of the circuits for MOS and bipolar circuits. This means a positive potential above ground (GND).

20.1.1 Application Areas of the Individual Technologies

Historically the bipolar process delivered the basis for precise analog functions with very good matching, high transconductance, fast ECL circuits (**E**mitter **C**oupled **L**ogic), and efficient drivers; however, there was also considerable power dissipation. PMOS, NMOS, and later CMOS, could not excite the pampered circuit designers at first, because all data were noticeably worse except drastically reduced power dissipation. Meanwhile CMOS became generally accepted in many areas through ist fine structures and reproducible process technology. Only in the area of very fast logic and distinct high frequency technology is the bipolar transistor still dominant.

BICMOS, the most recent technology, can take the advantages of MOS and bipolar technology and therefore is generally applicable. However, the technological expenditure is very high. Since the MOS gate length entered the submicron region the MOS transistor has become more and more useless for analog applications owing to distinctive short channel effects. In this case the bipolar transistor in the BICMOS technology has to replace the MOS transistor. A summary of these advantages and disadvantages of the CMOS and bipolar technology is listed in table 20.1. The BICMOS designer can choose the most favorable qualities according to requirements, as well as forming combinations to obtain even better results.

Table 20.1 Features of Bipolar and CMOS Transistors in Comparision

	Bipolar transistor	CMOS transistor
Matching	very good	good
Transconductance	very good, $\sim I_C$	bad, $\sim \sqrt{I_D}$
Speed	very high	high
Quiescent current	very high	≈ 0
Driver capability	very good	bad

20.1.2 Model Equations for the Bipolar Transistor

The most important relationships of the model equations are shown in condensed form in table 20.2 and in the following formulas. The equations are based on the *Ebers Moll* device model; however, all terms of the inverse transistor are deleted for simplification (the base and collector current are assumed positive entering, the emitter current leaving the transistor). For the small signal parameters the additional index 'A' indicates that the value of the parameter is meant in the working point.

Ebers Moll Device Model for the NPN Transistor in active Operation

(PNP transistor: Change the sign of currents and voltages accordingly):

- Emitter current:

$$I_E = I_{SE} \cdot \left(e^{\frac{V_{BE}}{V_T}} - 1 \right) \cdot \left(1 + \frac{V_{CE}}{V_A} \right) \approx I_{SE} \cdot e^{\frac{V_{BE}}{V_T}} \cdot \left(1 + \frac{V_{CE}}{V_A} \right) \tag{20.3}$$

V_A Early voltage

- Collector current:

$$I_C = A \cdot I_{SE} \cdot \left(e^{\frac{V_{BE}}{V_T}} - 1 \right) \cdot \left(1 + \frac{V_{CE}}{V_A} \right) \approx I_{SE} \cdot e^{\frac{V_{BE}}{V_T}} \cdot \left(1 + \frac{V_{CE}}{V_A} \right) \tag{20.4}$$

- Current gain:

$$A = \frac{I_C}{I_E} \approx 1, \qquad I_B = I_E - I_C,$$

$$B = \frac{I_C}{I_B} = \frac{A}{1-A}$$

Table 20.2 Some constants and general relationships

Bandgap energy (Si)	$E_g = 1.2\ \mathrm{eV}$
Boltzmann's constant	$k = 1.38 \cdot 10^{-23}\ \mathrm{W \cdot s/K}$
Dielectric constants	$\varepsilon_0 = 8.86 \cdot 10^{-12}\ \mathrm{F/m}$ $\varepsilon_{r(Si)} = 11.7; \quad \varepsilon_{r(SiO_2)} = 3.9$
Electronic charge	$q = 1.602 \cdot 10^{-19}\ \mathrm{A \cdot s}$
Thermal voltage, about 26 mV at room temperature	$V_T = kT/q$
Intrinsic conduction	$n_i^2 \sim T^3 \cdot e^{-\frac{E_g}{kT}}$ (20.1)
Mobility of the carriers	$\mu(T) = \mu(T_0) \cdot (T/T_0)^{-r}$ (20.2) with $r = 1.5 \ldots 2.7$ at normal doping $r \approx 2$

- Base current:

$$I_B = \frac{1}{1-A} \cdot I_{SE} \cdot \left(e^{\frac{V_{BE}}{V_T}} - 1 \right) \approx \frac{I_{SE}}{B} \cdot e^{\frac{V_{BE}}{V_T}} \quad (20.5)$$

- Emitter saturation current:

$$I_{SE} = A_E \cdot n_i^2 \cdot \mu(T) \cdot V_T \cdot f_E \quad (20.6)$$

A_E emitter area;
f_E technological constant

- Base emitter voltage:

$$V_{BE} = \frac{T}{T_0} \cdot V_{BE}(T_0) + V_{G0} \cdot \left(1 - \frac{T}{T_0}\right) - (4-r) \cdot V_T \cdot \ln\left(\frac{T}{T_0}\right) + V_T \cdot \ln\left(\frac{I_C(T)}{I_C(T_0)}\right) \quad (20.7)$$

Small Signal Parameters

- Transconductance:

$$g_m = \frac{I_{CA}}{V_T}$$

- Small signal current gain:

$$\beta = \left.\frac{\partial I_C}{\partial I_B}\right|_{\text{Arbeitspunkt}} \approx B$$

- Small signal diffusion capacitance:

$$C_{BE} = \tau_B \cdot g_m$$

τ_B base transit time

- Incremental base emitter resistance:

$$r_{BE} = \frac{V_T}{I_{BA}} = \frac{\beta}{g_m}$$

- Output impedance:

$$r_{CE} = \frac{V_A}{I_{CA}}$$

20.1.3 Simplified Device Model Equations for the MOS Transistors

The most important relationships of the model equations are shown in condensed form for MOS N channel transistors. (The drain current is assumed positive entering, the source current leaving the transistor; P channel transistor: change the sign of currents and voltages accordingly; the gate currents are assumed to be negligible.) The equations are based on the simple charge model with substrate control and channel shortening in the saturation region. This channel shortening can be taken into account in the computation through the constant λ or through L_{eff}. For the small signal parameters, the additional index 'A' indicates that the value of the parameter is meant in the operating point.

Strong Inversion

- Cut off region: for $V_{GS} < V_{Th}$

$$I_D = 0$$

- Drain current in the triode region: for $V_{GS} > V_{Th}$ and $V_{GS} > V_{DS} + V_{Th}$

$$I_D = K \cdot \left[(V_{GS} - V_{Th}) \cdot V_{DS} - \frac{V_{DS}^2}{2} \right] \quad (20.8)$$

- Drain current in the saturation region: for $V_{GS} > V_{Th}$ and $V_{GS} < V_{DS} + V_{Th}$

$$I_D = \frac{K}{2} \cdot (V_{GS} - V_{Th})^2 \cdot (1 + \lambda \cdot V_{DS}) \quad (20.9)$$

- Transconductance parameter (in A/V^2):

$$K = C_{ox} \cdot \mu(T) \cdot \frac{W}{L} \quad (20.10)$$

W/L Width/length of gate;
C_{ox} Specific gate capacitance (C/area)
V_{Th} Threshold voltage
λ Channel length modulation parameter (1/V)

- Threshold voltage of the N channel FET:

$$V_{Th} = V_{Th0} + \gamma \cdot \left(\sqrt{2 \cdot \Phi_F - V_{BS}} - \sqrt{2 \cdot \Phi_F} \right)$$

- Threshold voltage of the P channel FET:

$$V_{Th} = V_{Th0} - \gamma \cdot \left(\sqrt{2 \cdot \Phi_F + V_{BS}} - \sqrt{2 \cdot \Phi_F} \right)$$

with

$$\gamma = \frac{1}{C_{ox}} \cdot \sqrt{2 \cdot q \cdot \varepsilon_0 \cdot \varepsilon_{rSi} \cdot N_S}$$

$$\Phi_F = V_T \cdot \ln\left(\frac{N_S}{n_i}\right)$$

V_{BS} Bulk-source voltage;
$\Phi_F \approx 0{,}3$ V Fermi potential;
N_S Substrate doping

- Effective channel length in saturation

$$L_{eff} = L - \sqrt{2\frac{\varepsilon_0 \cdot \varepsilon_{rSi}}{q \cdot N_S} \cdot [V_{DS} - (V_{GS} - V_{Th})]}$$

Small Signal Parameters in the Saturation Region

- Transconductance:

$$g_m = K \cdot (V_{GSA} - V_{Th}) \cdot (1 + \lambda \cdot V_{DSA}) = \frac{2 \cdot I_{DA}}{V_{GSA} - V_{Th}} = \sqrt{2 \cdot I_{DA} \cdot K}$$

20

- Back gate transconductance:

$$g_{mB} = \frac{\gamma \cdot g_m}{2 \cdot \sqrt{2 \cdot \Phi_F - V_{BSA}}}$$

- Output conductance:

$$g_{DS} = \frac{\lambda \cdot I_{DA}}{1 + \lambda \cdot V_{DSA}} = \frac{K \cdot \lambda}{2} \cdot (V_{GSA} - V_{Th})^2 = \frac{1}{r_{DS}}$$

- Gate source capacitance:

$$C_{GS} \approx \frac{2}{3} \cdot W \cdot L \cdot C_{ox}$$

Subthreshold [20.11]

for $V_{GS} < V_{Th}$, but $V_{GS} \approx V_{Th}$ at very small currents

- Drain current:

$$I_D = I_{D0} \cdot \frac{W}{L} \cdot \left(1 - e^{-\frac{V_{DS}}{V_T}}\right) \cdot e^{\frac{V_{GS} - V_{Th}}{m \cdot V_T}} \quad (20.11)$$

with $m \approx 1{,}5 \ldots 2$ gate effectivity and $I_{D0} \approx 20$ nA.

20.1.4 Parasitic Elements, Transistors, Transmission Lines

The above model equations comprise only the non-linear relationship between input voltage and the output current for very low frequencies. Usually, simple relationships are used for a few parasitic elements as well. The Figures 20.1 and 20.2 show the simplified equivalent models for the bipolar and MOS transistors. The controlled current generator is given by the corresponding model equation from the previous chapters or the small signal term g_m. The above models and also the simulation models of the simulators don't contain the parasitic bipolar transistors. The most important cases are shown in principle in the Figures 20.3 to 20.6. The parasitic substrate PNP transistor of the vertical NPN transistor in the bipolar process only becomes conducting, if the actual transistor gets into saturation; thus the parasitic transistor only rarely disturbs. Things look differently with the lateral transistors of the bipolar and MOS technologies: In each case there is a parasitic substrate transistor that is always conducting, if the actual transistor also conducts; an undesirable current always flows into the substrate. A route with very low resitivity (substrate contact) must necessarily be provided for this current in order to avoid undesirable potential movements in the substrate.

The two bipolar transistors in the CMOS process that can be responsible for the latch up effect, are not shown because the latch up danger should be banned through relevant design rules.

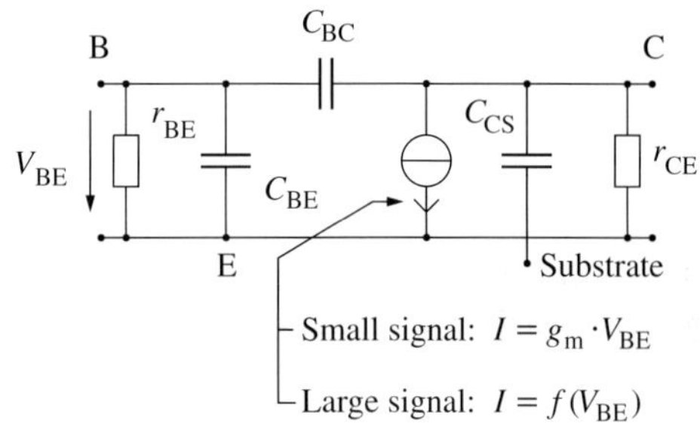

Fig. 20.1 Simplified equivalent model of the bipolar transistors

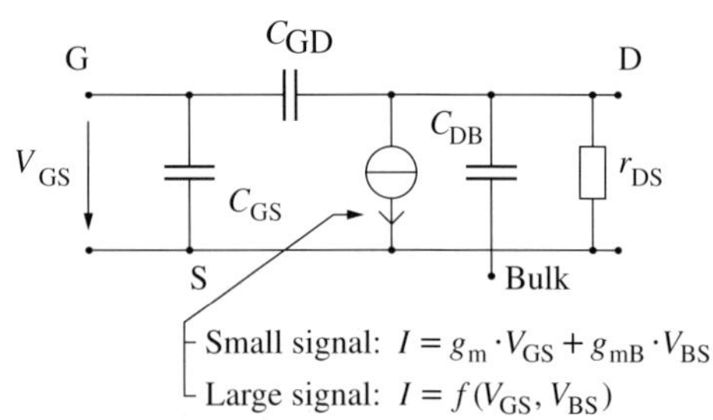

Fig. 20.2 Simplified equivalent model of the MOS transistor

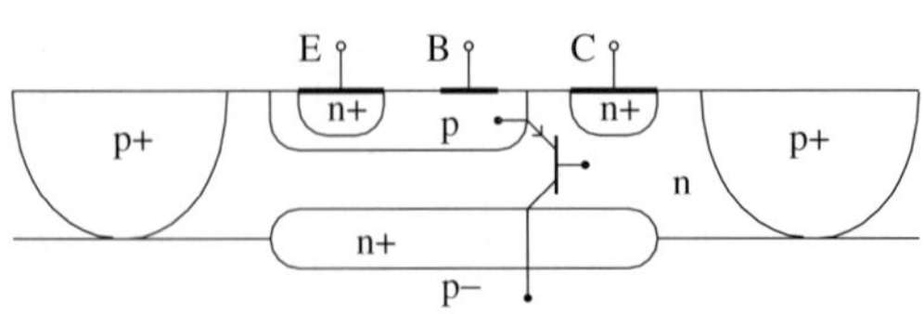

Fig. 20.3 Parasitic transistor of the vertical NPN transistor in the bipolar process

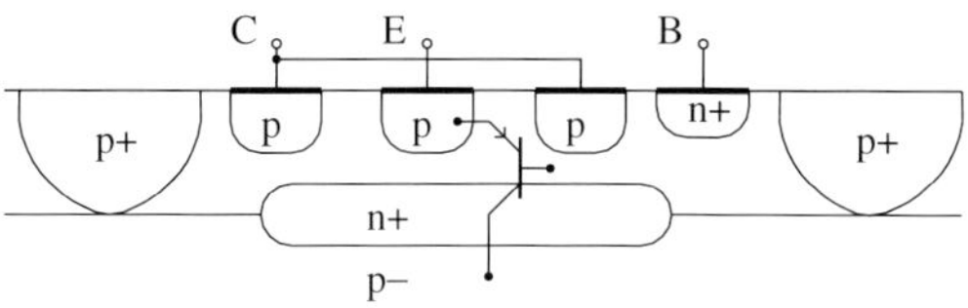

Fig. 20.4 Parasitic transistor of the lateral PNP transistor in the bipolar process

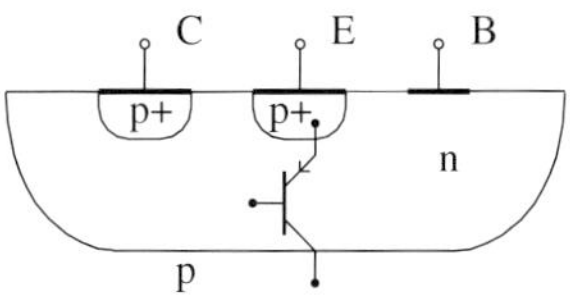

Fig. 20.5 Parasitic transistor of the lateral PNP transistor in the N well CMOS process

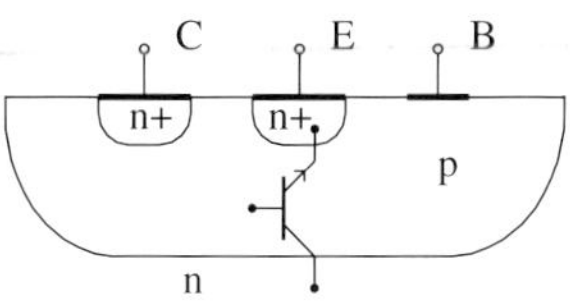

Fig. 20.6 Parasitic transistor of the lateral NPN transistor in the P well CMOS process

The connection wires provide further parasitic elements. When their lengths are short the delays are essentially smaller than the gate delays, and the connection can be taken into account sufficiently by the line capacity alone.

$$C_{\text{Load}} = C' \cdot W \cdot L$$

After [20.27] the delay time of a *RC*-line is $t_{\text{delay}} = (R \cdot C)/2$. R is the total resistance of the line and C is given by the total capacity of the wire. High frequency issues can usually be neglected with low frequencies. However with a clock frequency of 500 MHz (under the assumption that the 5th harmonic is still necessary for a proper clock form, and that transmission line phenomenas appear from approximately one eighth of the wave length) which is the case with line lengths from approximately 1.5 cm.

20.1.5 Noise

A good circuit design requires that the designer is familiar with the noise sources which appear. Therefore, the components can be designed by observing the required quiescent currents, frequency response, and noise demands with coarse hand calculations, and that the circuit simulations can be started as goal-oriented from good initial values. The noise behavior of MOS transistors is usually very much worse than that of bipolar transistors. The main noise sources are the thermal noise, the shot noise, and the $1/f$ noise. The latter is dominant from low to middle frequencies.

The noise behavior of bipolar transistors can approximately be described through a noise voltage source and a noise current source:

- Thermal noise:

$$\overline{V_{\text{r}}^2} = 4 \cdot k \cdot T \cdot r_{\text{BB}} \cdot \Delta f \tag{20.12}$$

- Shot noise and $1/f$ noise:

$$\overline{I_{\text{r}}^2} = 2 \cdot q \cdot I_{\text{B}} \cdot \left(1 + \frac{K_1}{f}\right) \cdot \Delta f \tag{20.13}$$

Here r_{BB} is the neglected base resistance in fig. 20.1, I_{B} the base current and K_1 a process dependent noise constant. By reducing the base resistance and the base current both noise sources can be reduced; however, the transconductance also decreases with it.

Because of the absence of a gate current (at low frequencies) the MOS transistor can be described by a noise voltage source alone (simplified SPICE model):

- Thermal noise and $1/f$ noise:

$$\overline{V_{\text{r}}^2} = \left(4 \cdot k \cdot T \cdot \frac{2}{3} \cdot \frac{1}{g_{\text{m}}} + \frac{K_2}{W \cdot L \cdot C_{\text{ox}} \cdot f}\right) \cdot \Delta f \tag{20.14}$$

K_2 another process-dependent noise constant

In particular, the critical $1/f$ term becomes reduced by a large transistor area $W \cdot L$. The constant K_2 is approximately 50 times bigger for NMOS transistors than for PMOS transistors at the present time [20.19], therefore PMOS input transistors are mainly used. The principle location of the noise sources is recognizable in fig. 20.7.

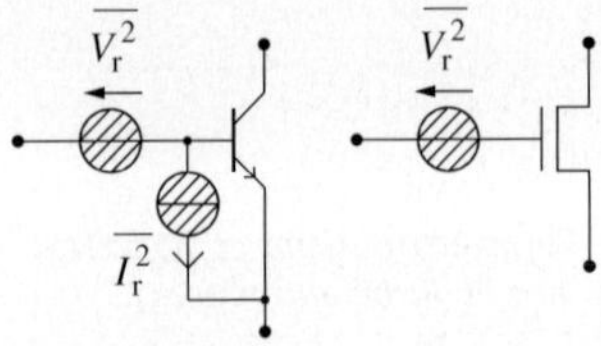

Fig. 20.7 Noise sources of the bipolar and MOS transistor

20.1.6 Scaling

Driven by the rapid advance in process technology, and in circuit technique one has to find procedures with which existing libraries can be adapted without a very large expenditure. All dimensions are divided linear by a factor $a > 1$, so the module areas shrink $\sim a^{-2}$. For the further treatment of the element's dimensions there are two recipes. First, one tries to perform the scaling in a way that the emerging electric field strengths remains approximately constant [20.27]. The second direction has the goal of retaining the supply voltages. The second procedure admittedly has the advantage that the peripheral components can work on traditional supply voltages, that delays decrease $\sim a^{-2}$, but also the power dissipation density increases $\sim a^3$. However, an increase of the power dissipation density is hardly possible anymore; the heat drainage of the chip has already arrived at the uppermost border and cannot be increased anymore without new methods. Therefore a further decreasing of the supply voltage will be expected. In table 20.3 some effects of the scaling are presented. A procedure for the adaptation of digital libraries is described in [20.2].

Table 20.3 Effects of the scaling

Procedure	Effect	
	constant field strength	constant supply voltages
Surface	$\sim a^{-2}$	$\sim a^{-2}$
Power dissipation	$\sim a^{-2}$	$\sim a$
Power dissipation density	1	$\sim a^3$
Delay	$\sim a^{-1}$	$\sim a^{-2}$

20.2 Analog Circuits

The principles of circuits are quite similar in all technologies. Therefore first the circuits are discussed and then, if necessary, a distinction between MOS or bipolar realization is made. Most circuits that work in the sub-threshold region can be described sufficiently exactly by the equations valid for bipolar transistors.

The basic function blocks for the automated system design must show constant characteristics as independently as possible from the application. Therefore the main focus lies on temperature and power supply independence of the circuits. Still, specifical problems are addressed which result from modern requirements, such as decreasing power supply voltages, less power dissipation, as well as the growing influence of the parasitic elements.

20.2.1 Matching Principle

The matching principle of the integrated circuit technology is based on the absolute values of components being strongly dependent on statistical process variations (up to ± 35 %); the relative deviations, however, are much less concerned ($< \pm 5$ %). With good layout preparation, relative tolerances are attainable in practice far below 0.1 %. However, the aforementioned values can only be achieved if thermal influences and other environmental conditions for the components are the same. The relative tolerance plays the leading role when ratios of values are involved. Thus an essential difference of the integrated technology from the discrete circuit technology is that as many as possible of the relevant circuit items depend only on resistance, respectively capacity, respectively transistor ratios.

From table 20.4 it is evident that the matching values of the capacities are by far the best, also the temperature and voltage coefficient are outstanding. For capacities with $C \sim W \cdot L$ and for bipolar transistors with $I_{SE} \sim W \cdot L$ the area $W \cdot L$ plays the most important role, for MOS transistors it is the relationship W/L. The matching behavior of resistances shall be studied. The resistance value is given through:

$$R = R_{Sheet} \cdot \frac{L}{W} \cdot (1 + \alpha_T \cdot \Delta T)$$

Table 20.4 *Matching values, temperature and voltage coefficient [20.14]*

Element	Matching	Temperature coefficient	Voltage coefficient
Resistance, diffused	±0.4 %	+2 000 ppm/C	200 ppm/V
Resistance, implanted	±0.12 %	+ 400 ppm/C	800 ppm/V
Capacity, MOS	±0.06 %	+ 26 ppm/C	10 ppm/V

where R_{Sheet} is the sheet resistance of the semiconductor material at the temperature T_0.

If two resistances are arranged locally very close together (equivalent temperature), then the sheet resistance and the temperature coefficient can be assumed to be the same as a first approximation. Thus the resistance ratio results to

$$\frac{R_1}{R_2} = \frac{R_{\text{Sheet}} \cdot \frac{L_1}{W_1} \cdot (1 + \alpha_{\text{T}} \cdot \Delta T)}{R_{\text{Sheet}} \cdot \frac{L_2}{W_2} \cdot (1 + \alpha_{\text{T}} \cdot \Delta T)} = \frac{L_1}{L_2} \cdot \frac{W_2}{W_1}$$

α_{T} temperature coefficient;
ΔT temperature difference from the reference temperature T_0

Thus the quotient depends only on two geometrical ratios. These length and width ratios are, however, geometrically very exactly specified, so that only technological inaccuracy enters the quotient. Even this inaccuracy can be eliminated nearly completely from the result if both resistances can be replaced by series connections respectively parallel connections of resistances with the same W/L.

Example:

$$\frac{R_1}{R_2} = 4$$

Choice:
$R_1 = R + R = 2 \cdot R$ and $R_2 = R||R = R/2$.

Now the quotient can only be falsified by different technological production of the geometrically identical elements. The technological sizes are dependent to a minor extent on the environment of the elements. Therefore the environment of all resistances should be similar. This can be achieved, for example, by means of adding so called dummies. Locally uneven diffusion, respectively implantation, cause gradients in the sheet resistance on the wafer. For small ranges the gradient can be accepted as linear. The effect of gradients can be eliminated by skilful arrangement of the resistances if the individual elements possess a common geometrical center (common centroid).

Figure 20.8 shows in principle an optimized arrangement with dummies. With capacitance ratios one builds the individual capacities up from parallel connections of many unit capacities, usually in common centroid arrangement.

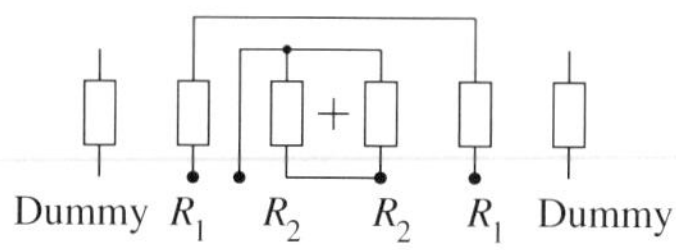

Fig. 20.8 Common centroid matching with dummies (center '+')

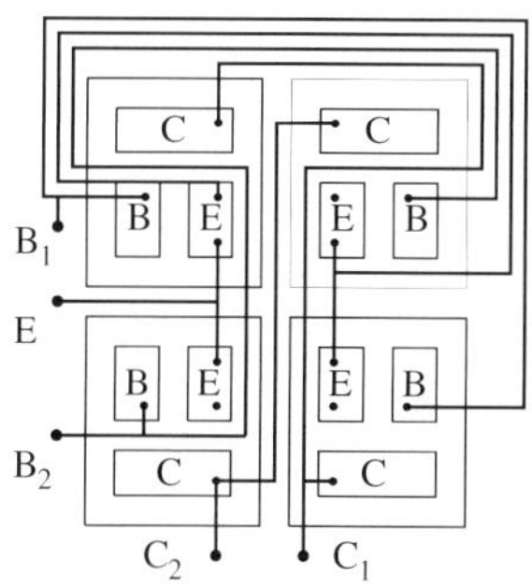

Fig. 20.9 Common centroid matching of two bipolar transistors

The matching principle with transistors is far more common in differential amplifiers and current mirrors. The design principles for this hardly differ from the aforementioned ones. Usually parallel connections of two smaller transistors are used as the replacement for one larger transistor, because otherwise the wiring is too complicated. In fig. 20.9 the principle of a common centroid arrangement of bipolar transistors is represented, the wiring is only symbolically shown for reasons of clarity. The temperature has a serious influence on many parameters and concomitantly on matching parameters, thus the intrinsic conduction density occurs in many semiconductor parameters. Since silicon, however, possesses a rather high thermal conductivity, almost comparable with aluminium, the thermal coupling is very good with closely adjoining elements. However, it becomes more difficult to keep the temperature's influence negligible on the matching as the chip area is decreasing and the power dissipation is increasing.

Rules for good matching:

- similar elements, even better: same elements and series respectively parallel connected circuits. (Series circuits only with resistors);
- interlaced elements (common centroid);
- parallel orientation of the elements on the chip;
- equal number of corners of resistances;
- perimeter and area must be both in the ratio of the capacitors (with capacitors);
- arrangement of the elements on isotherms.

20.2.2 Temperature Dependencies

As already mentioned, practically all semiconductor parameters are temperature-dependent. However, only the most important dependences can be taken into account in the design process owing to the complexity. Therefore a temperature simulation with exact models must be started later. An interesting function of the bipolar transistor is $I_{SE}(T)$, see equation (20.6), which shows a very steep increase with the temperature and the known temperature dependence of the voltage $V_{BE}(T)$. The latter nearly possesses a linear decrease as a function of the temperature with a slope of approximately $-2\ \mathrm{mV/K}$, see equation (20.7). With the MOS transistor the quantity K is a function of the temperature over the mobility $\mu(T)$, as indicated in equation (20.10). The change of the threshold voltage with the temperature is usually disregarded with the design because it will be dropped at symmetrical construction in the first approximation anyway. Resistances usually have a positive temperature coefficient, depending upon doping and technological processing. Oxide capacities have a weak temperature dependence through the dielectric constant of the oxide material.

20.2.3 Basic Elements

20.2.3.1 Resistances, Active Resistances

For the realization of resistances different types are available depending upon process. The selection has to consider the requirements of temperature and voltage dependence. In bipolar processes one usually uses resistances that are produced with the base or emitter diffusion (small sheet resistance), respectively, with the epitaxial layer. In more distinguished processes implanted resistances are available as well with outstanding characteristics. The above resistance types are separated from each other by means of PN junctions, i.e. it must be guaranteed that the PN junctions are reverse biased. Therefore the respective background material has to be connected over a corresponding contact with the required voltage. In pure MOS processes only polysilicon resistances and resistances of the respective well material are available. Table 20.5 shows a short overview of the different sheet resistances and tolerances.

Table 20.5 Properties for different types of resistors (R_{Sheet} = Sheet resistance)

Type	from R_{Sheet}	to R_{Sheet}	Absolute tolerance
Diffused resistor	$5\ \Omega/\square$	$200\ \Omega/\square$	±30 %
Implanted resistor	$200\ \Omega/\square$	$5\,000\ \Omega/\square$	± 1 %
Polysilicon resistor	$10\ \Omega/\square$	$100\ \Omega/\square$	±30 %

The term 'active' resistor means the use of active elements as resistors which follow Ohm's law in the small signal operation. Figure 20.10 shows an example for the application of a bipolar transistor.

For the active resistor one gets:

$$r = \frac{\mathrm{d}V}{\mathrm{d}I} = \frac{1}{g_m + \dfrac{1}{r_{BE} \| r_{CE}}} \approx \frac{1}{g_m}$$

For the MOS transistor there are three possibilities:

- Voltage controlled resistor in the triode region with
$$r = \frac{\mathrm{d}V}{\mathrm{d}I} = \frac{1}{K \cdot (V_{GSA} - V_{Th})}$$
see fig. 20.11
- Resistor in the saturation region with
$$r = \frac{\mathrm{d}V}{\mathrm{d}I} = r_{DS}$$

- According to fig. 20.10 only in MOS implementation, fig. 20.12 shows with
$$r = \frac{dV}{dI} = r_{DS} \| \frac{1}{g_m} \approx \frac{1}{g_m}$$
the last realization.

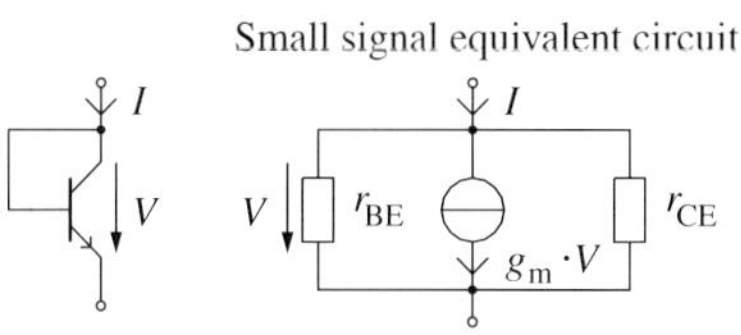

Fig. 20.10 Bipolar transistor as 'active' resistor

It should still be noticed that the 'active' resistor with the bipolar transistor actually works in the active region, even if its collector base voltage is equal to zero.

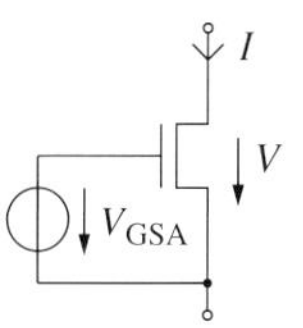

Fig. 20.11 MOS transistor in the triode region as voltage controlled 'active' resistor

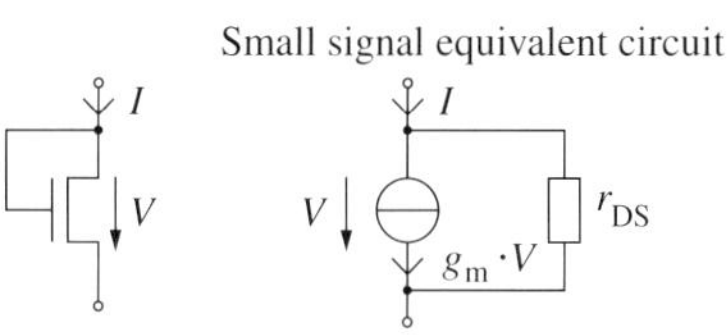

Fig. 20.12 MOS transistor as 'active' resistor in the saturation region

20.2.3.2 Capacitances

Usually capacitances are undesirable, especially with transistors and wiring. However, if they are not, then one desires voltage and temperature-independent capacitances. In the bipolar process most capacitances team up with PN junctions. Unfortunately these junction capacitances are very voltage dependent and are therefore useful only for blocking purposes; also their quality factor is very temperature dependent because the reverse biased current is $\sim n_i^2$. For more exact purposes, the capacity 'aluminium SiO_2 diffusion' must be used. However, one has to pay attention to the unavoidable parasitic capacity to the substrate.

In the MOS process the desirable capacity is 'polysilicon SiO_2 polysilicon'. Also the capacity polysilicon is usable as well as aluminium with SiO_2 as dielectric upon diffused silicon, but through the semiconductor layer the capacity is a little bit voltage dependent. All capacities mentioned here have an unavoidable parasitic capacity to the substrate. According to table 20.4 the matching accuracies of capacities are outstanding. They are therefore used extensively, for example, with switched-capacitor filters.

20.2.3.3 Switches

In bipolar circuit design switches were used almost only for digital circuits. In analog circuits the offset voltages and currents are prohibitive. The MOS technology provides essentially better switches. The attainable resistance values are $R_{On} \approx 100\ \Omega$ and $R_{Off} > 5\ \text{M}\Omega$. Figure 20.13 shows the principle circuit of a NMOS transistor as switch.

Under the assumption that the control voltage is larger than V_E and V_A the output voltage can follow the input voltage, but is in principle restricted to maximally $V_{control} - V_{Th}$. In the digital operation $V_{control} = V_{E\,max} = V_B$ and therefore a high level ('1') is reduced by about the threshold voltage. The NMOS transistor as switch can transfer the low level precisely and the high level inaccurately.

If one looks accordingly at a PMOS transistor one finds, exactly vice-versa, that it precisely transfers the high level and the low level ('0') inaccurately. If one now connects one NMOS and one PMOS transistor in parallel, see fig. 20.14, one gets a good switch which transfers signals in the range between zero and the control voltage well. However, both switching transistors need complementary control voltages. Such a combined switch is called a transmission gate. In this form the switch system is also applicable for analog signals if the input and output voltages remain smaller than the control voltage. However, the R_{On} resistance has a complicated dependence on the input and output voltage. Therefore the load must remain very small at the output side in order to avoid a different voltage distribution at the R_{On} resistance and the load resistance. During the switching process the capacitors between source or drain and the gate couple over charges to the input and output. Figure 20.15 shows these parasitic capacitances.

With the CMOS switch the clock over coupling can be minimized by complementary clock signals affecting the two gates while switching. With optimal matching of the transistor capacities the brought in charges are consequently equal to the drained charges. Of course, the CMOS switch has higher capacities compared to a switch with only one transistor. The single transistor switch can be improved by introducing a dummy transistor with complementary control voltage. However, this increases the total capacitance, so that the transmission gate should be preferred if a complementary transistor is available.

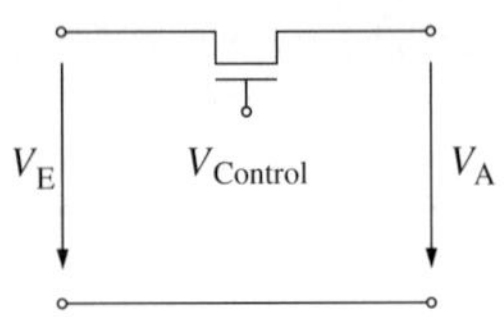

Fig. 20.13 Principle of a NMOS transistor as switch

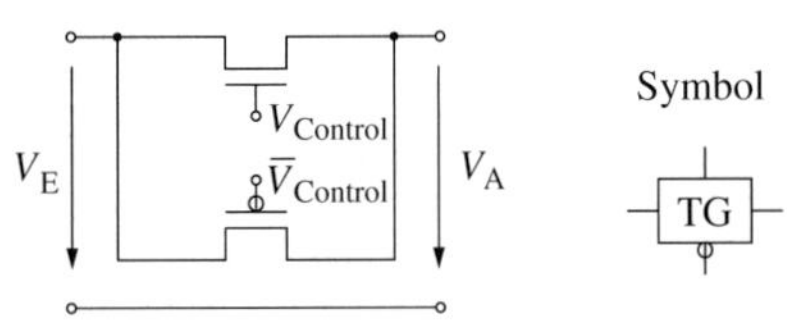

Fig. 20.14 CMOS transmission gate

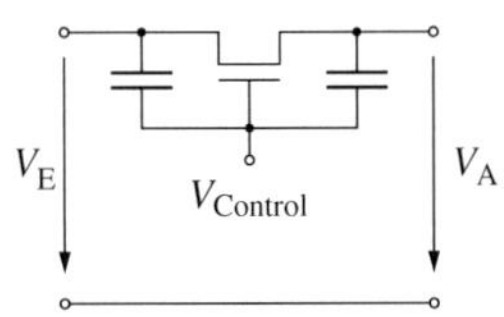

Fig. 20.15 Clock over coupling at the NMOS switch

20.2.3.4 Current Sources and Current Mirror Circuits

Current sources are, in fact, amongst the most important basic modules of the integrated circuit technology because they use the matching principle in an efficient manner and bring results some what independent of technological variations. Additionally, current sources for operating point purposes often result in substantial space savings compared with solutions, which use the high impedance of resistances. The background for the almost exclusive use of current sources for the operating point definition is that one hopes that all current sources admittedly also show a variation under matching conditions on a chip if there are process variations; however, all sources together change into one direction if all sources depend on a common reference current. The most important requirements are: in every sense stable currents, as high as possible output resistances with the ability of a high voltage swing and a good frequency response. In the course of process improvements, the supply voltages will decrease on and on and there will appear ever stronger short channel effects through finer structures, compare section 20.1.6. The requirement points, output resistance and high voltage swing ability, become more and more important. With the following circuits in each case the DC behavior is regarded. However, the circuits have parasitic capacitances, thus the stability must be checked carefully and individually, especially in cases with feedback.

As an example all circuits are here introduced with NPN as well as NMOS transistors, the corresponding PNP as well as PMOS realizations work accordingly.

Current Mirror Circuits with Bipolar Transistors

Figure 20.16 shows the circuit of a discrete current source with a bipolar transistor and the current versus voltage characteristic. One recognizes a current increase with increasing collector to emitter voltage caused by the Early effect. The behavior of the circuit for voltages above $V_{CE\,sat}$ can be regarded as a current source with output resistance.

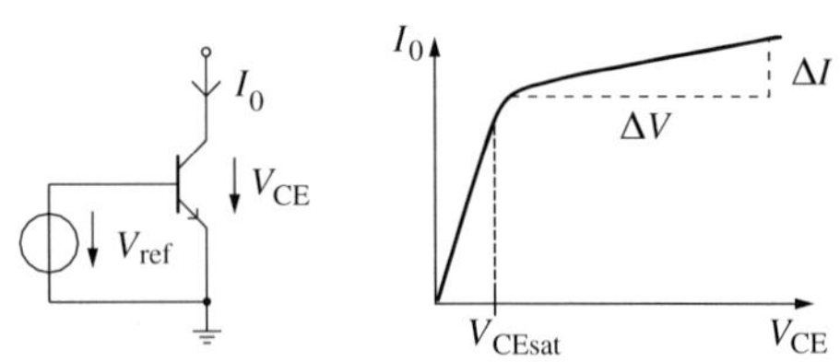

Fig. 20.16 Circuit of a discrete current source

This circuit has the deficiency that it shows very bad temperature behavior in integrated technology. By applying a constant base emitter voltage the collector current rises strongly with the temperature. By inserting of a resistor in the base and in the emitter line, the behavior can be improved, but Widlar proposed a circuit better adapted to the integrated technology; it is represented in fig. 20.17, [20.29].

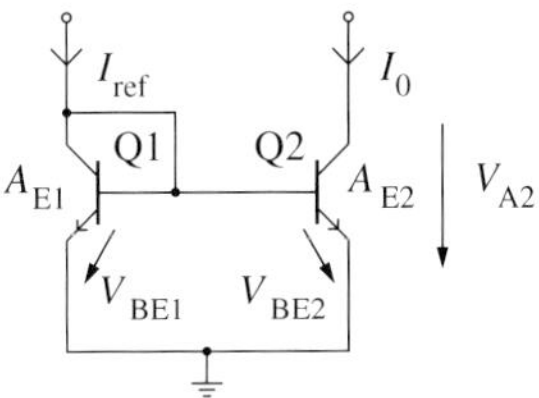

Fig. 20.17 Current mirror circuit from Widlar

Here, the reference voltage is replaced by a transistor with collector-base short circuit ('active' resistance) and a reference current. First, one could assume a step backward in comparison with the previous circuit, because now one already needs a reference current for a current source. This circuit just mirrors one current and therefore is called current mirror. The advantage of this circuit becomes evident after a rough analysis. The collector current of a bipolar transistor is due to equation (20.4):

$$I_C \approx I_{SE} \cdot e^{\frac{V_{BE}}{V_T}} \cdot \left(1 + \frac{V_{CE}}{V_A}\right) \Rightarrow$$

$$V_{BE} = V_T \cdot \ln\left[\frac{I_C}{I_{ES} \cdot \left(1 + \frac{V_{CE}}{V_A}\right)}\right]$$

The condition $V_{BE1} = V_{BE2}$ and neglecting the base currents within the above equation for both transistors provides:

$$\ln\left[\frac{I_{ref}}{I_{ES1} \cdot \left(1 + \frac{V_{CE1}}{V_{A1}}\right)}\right] = \ln\left[\frac{I_0}{I_{ES2} \cdot \left(1 + \frac{V_{CE2}}{V_{A2}}\right)}\right]$$

$$\Rightarrow$$

$$I_0 = I_{ref} \cdot \frac{I_{SE2} \cdot \left(1 + \frac{V_{CE2}}{V_{A2}}\right)}{I_{SE1} \cdot \left(1 + \frac{V_{CE1}}{V_{A1}}\right)}$$

If one assumes matching conditions for both transistors, which means $T_1 = T_2$, then the ratio of the two reverse saturation currents is immediately equal to the ratio of the emitter areas

$$\frac{A_{E1}}{A_{E2}} = \frac{I_{SE1}}{I_{SE2}}$$

If one still neglects the terms which describe the Early effect then one gets:

$$I_0 \approx I_{ref} \cdot \frac{A_{E2}}{A_{E1}} \tag{20.15}$$

The output current is therefore related to the reference current by the ratio of the emitter areas without any temperature dependence. If one does not neglect the base currents some other expression arises for the ratio of the emitter areas, namely:

$$\frac{A_{E2}}{A_{E1}} = \frac{X \cdot \left(1 + \frac{1}{\beta}\right)}{1 - \frac{X}{\beta}} \tag{20.16}$$

$X = I_0/I_{ref}$ is the desired ratio of the currents.

For low current gains, for example with lateral transistors, the deviations are not negligible. There is an improved circuit, fig. 20.18, especially for more than one current source, where the base current of the current source transistors is fed in via an emitter follower. Then one can substitute β in equation (20.16) in approximation by $\beta \cdot (\beta + 1)$.

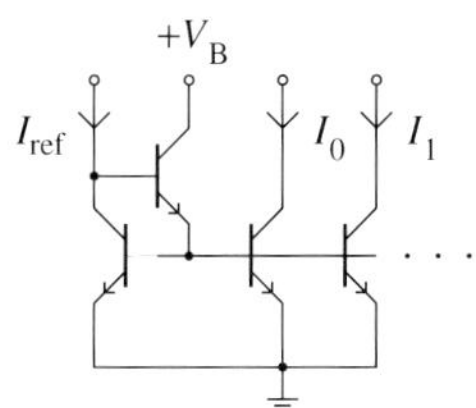

Fig. 20.18 Current mirror circuit for more than one current source

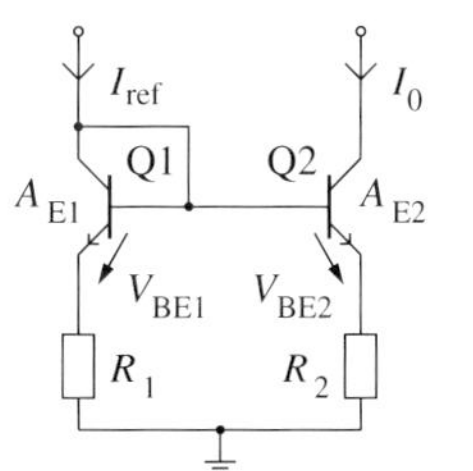

Fig. 20.19 Current mirror circuit with negative feedback

The output resistance of the circuit in fig. 20.17 at the current output is $r_{Out} = V_{A2}/I_0$, therefore this is immediately the output resistance of the transistor Q2 alone. This output resistance is too low for some applications; a negative feedback (emitter degeneration) provides improvement. Figure 20.19 shows the already known current mirror with the feedback resistances R_1 and R_2; where R_1 hardly has an influence on the output resistance. However, there is a very flexible possibility of modifying the temperature coefficient of the circuit.

For the calculation of the output current the base currents and the Early effect still should be neglected. The sum of voltages in the base circuit of the transistors delivers

$$V_{BE1} + I_{ref} \cdot R_1 = V_{BE2} + R_2 \cdot I_0$$

If one introduces the corresponding expressions for the base emitter voltages under matching conditions, one gets the result

$$I_0 = \frac{R_1}{R_2} \cdot I_{ref} + \frac{V_T}{R_2} \cdot \ln\left(\frac{I_{ref}}{I_0} \cdot \frac{A_{E2}}{A_{E1}}\right) \qquad (20.17)$$

One recognizes that the output current is composed of two terms, in which the one is temperature independent and the second term is proportional to the absolute temperature. The temperature dependent part is dependent on the absolute value R_2 of the resistance and therefore is not easily reproducible.

Because the expression in the argument of the logarithm can be bigger or smaller than one, even the sign of the temperature coefficient of the output current is adjustable. The special case of temperature independence is achieved if the argument of the logarithm is equal to one. The current transfer relationship then still follows equation (20.15) as well as (20.16). For the calculation of the output resistance the small signal equivalent circuit is shown in fig. 20.20. The transistor Q1 serves as the 'active' resistance $1/g_{m1}$.

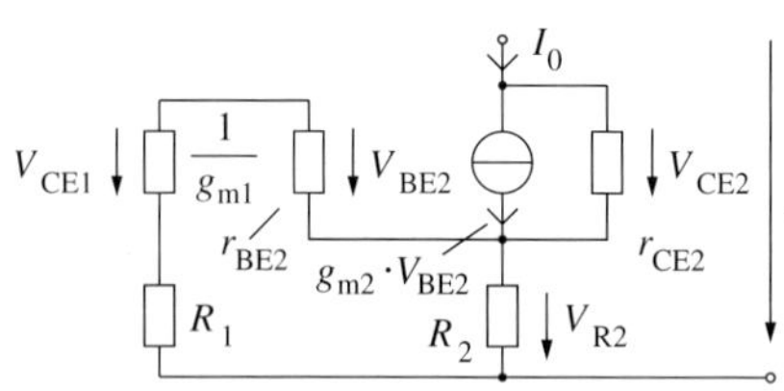

Fig. 20.20 Small signal equivalent circuit of the current mirror with feedback

The equation for the determination of the output resistance of the circuit shall be deduced briefly, since the procedure is similar with all current sources. It applies:

$$I_0 = g_{m2} \cdot V_{BE2} + \frac{V - V_{R2}}{r_{CE2}}$$

$$V_{R2} = I_0 \cdot (R_2 \| R); \quad V_{BE2} = -V_{R2} \cdot \frac{r_{BE2}}{R}$$

with

$$R = r_{BE2} + \frac{1}{g_{m1}} + R_1$$

One gets the dynamic output resistance by differentiation to

$$r_{Out} = \frac{dV}{dI_0} \qquad (20.18)$$

$$= r_{CE2} \cdot \left[1 + \frac{R_2 \cdot r_{BE2}}{R_2 + R} \cdot \left(g_{m2} + \frac{R}{r_{CE2} \cdot r_{BE2}}\right)\right]$$

$$r_{Out} \approx r_{CE2} \cdot (1 + R_2 \cdot g_{m2})$$

for $r_{BE2} \approx R \gg R_2$ and $g_{m2} \gg 1/r_{CE2}$.

One thus receives a dynamic output resistance which is higher by approximately the feedback factor $R_2 \cdot g_{m2} = V_{R2}/V_T$ than the output resistance of the transistor without feedback. The feedback unfortunately reduces the ability of a high voltage swing of the current source by the voltage drop across R_2. One nevertheless receives an increase of the output resistance of around the factor 11 with a voltage drop of approx. 250 mV.

Another possibility of increasing the output resistance with a feedback is the Wilson current mirror, whose circuit is shown in fig. 20.21. The mode of action is that for both transistors (Q1 and Q2), which determine the current ratio, the Early effect has practically no effect, because they are driven with virtually constant collector emitter voltages. Since the condition of same base emitter voltages is applicable to the output current, the current ratio still is given by equation (20.15) as well as (20.16). The associated small signal equivalent circuit is represented in fig. 20.22a and in fig. 20.22b with simplified names.

$$R_2 = r_{BE1} \| r_{BE2} \| r_{CE2} \| \frac{1}{g_{m2}} \approx \frac{1}{g_{m2}}$$

The equivalent models shown in the figures 20.20 and 20.22b are equivalent. Therefore the output resistance is given by equation (20.18) directly, if the appropriate substitutions are used.

The equivalent resistance R, used in equation (20.18), can be determined from the left circuit part of fig. 20.22:

$$R = \frac{r_{BE3} + r_{CE1}}{1 + g_{m1} \cdot r_{CE1}} \approx \frac{1}{g_{m1}}$$

Thus the dynamic output resistance results in:

$$r_{Out} = r_{CE3} \cdot \left(1 + \frac{g_{m1} \cdot r_{BE3} \cdot g_{m3}}{g_{m1} + g_{m2}}\right) = r_{CE3} \cdot \left(1 + \frac{g_{m1} \cdot \beta_3}{g_{m1} + g_{m2}}\right) \approx r_{CE3} \cdot \left(1 + \frac{\beta_3}{2}\right)$$

The minimum voltage is

$$V_{min} = V_{CE\,sat} + V_{BE}$$

One can increase the output resistance also with a cascode circuit. The circuit uses two pairs of transistors in a series connection. The two transistors Q1 and Q2, which determine the current ratio, show virtually constant collector emitter voltages and the Early effect has hardly any effect. Figure 20.23 shows the corresponding circuit. Since the condition of same base emitter voltages is applicable to the output current, the current ratio is still given by equation (20.15) as well as (20.16). The simplified small signal model is shown in fig. 20.24. The two transistors Q1 and Q4 are 'active' resistances and do not play any role for the output resistance in the first approximation. Their resistance is assumed small compared with the input resistances. With appropriate substitution of the corresponding values to equation (20.18), the dynamic output resistance is given

$$r_{Out} = r_{CE3} \cdot \left(1 + \frac{r_{CE2} \cdot r_{BE3} \cdot g_{m3}}{r_{CE2} + r_{BE3}}\right) = r_{CE3} \cdot \left(1 + \frac{r_{CE2} \cdot \beta_3}{r_{CE2} + r_{BE3}}\right) \approx r_{CE3} \cdot (1 + \beta_3)$$

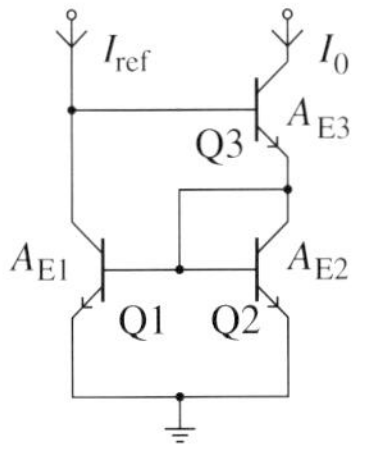

Fig. 20.21 Wilson current mirror circuit

The minimum voltage is

$$V_{min} = V_{CE\,sat} + V_{BE}$$

Finally table 20.6 gives a summary of the presented bipolar current mirror circuits.

Table 20.6 Properties of bipolar current mirror circuits

Circuit type	Output resistance	Minimum voltage
Widlar circuit	r_{CE}	$V_{CE\,sat}$
Widlar circuit with feedback	$r_{CE} \cdot \left(1 + \frac{V_R}{V_T}\right)$	$V_{CE\,sat} + V_R$
Wilson circuit	$r_{CE} \cdot (1 + \beta/2)$	$V_{CE\,sat} + V_{BE}$
Cascode circuit	$r_{CE} \cdot (1 + \beta)$	$V_{CE\,sat} + V_{BE}$

Current Mirror Circuits with MOS Transistors

With all current mirror circuits in MOS technology, the attainable transconductance usually is essentially smaller than with bipolar transistors; thus the output resistances are smaller as well.

The Widlar current mirror circuit with MOS transistors is shown in fig. 20.25. One uses equation (20.9) for the calculation of the current transfer ratio, under matching conditions and neglecting the channel-length modulation. One assumes that the transistors work in the saturation region.

$$I_D = \frac{K}{2} \cdot (V_{GS} - V_{Th})^2 \Rightarrow V_{GS} = \sqrt{\frac{I_D \cdot 2}{K}} + V_{Th}$$

Both transistors share the same gate source voltage

$$V_{GS1} = V_{GS2}$$

$$\sqrt{\frac{I_{ref} \cdot 2}{K_1}} = \sqrt{\frac{I_0 \cdot 2}{K_2}} \Rightarrow I_0 = I_{ref} \cdot \frac{L_1}{W_1} \cdot \frac{W_2}{L_2}$$

With the MOS current mirrors the W/L ratio corresponds to the ratio of the emitter areas. So the minimum voltage across the current source transistor must be at least $V_{DS\,min} = V_{GS} - V_{Th} = V_{DS\,sat}$ for it to remain in saturation.

With the following current mirrors the body effect (change of the threshold voltage) is neglected, because this effect increases the output resistance and one is always with the given values on the safe side.

The feedback variant of the Widlar current mirror is uncommon in MOS technology, because in most pure MOS processes no qualified resistances are available. Also the expression for the current transfer ratio is very difficult to survey. There is a common circuit which has a feedback resistance only on the output side. This circuit is not recommended, however, because of the unbalanced temperature dependence there enters the current

20

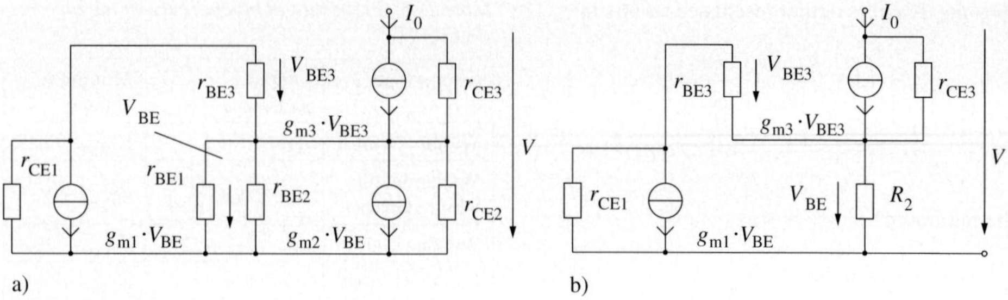

Fig. 20.22 a) Small signal equivalent circuit of the Wilson current mirror; b) Simplified small signal equivalent circuit of the Wilson current mirror

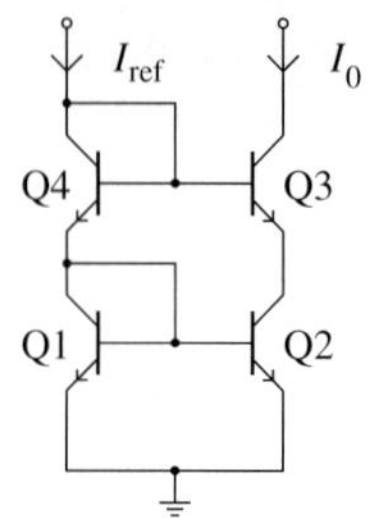

Fig. 20.23 Cascode current mirror circuit

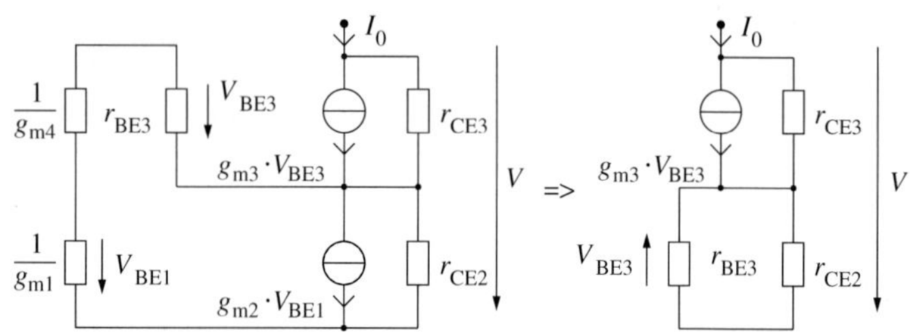

Fig. 20.24 Small signal equivalent model of a cascode current mirror circuit

transfer ratio via the parameter K. It is better to scale the W/L ratio as well as the resistances, then each temperature dependence is removed:

$$\frac{I_0}{I_{ref}} = \frac{R_1}{R_2} = \frac{\frac{W_2}{L_2}}{\frac{W_1}{L_1}}$$

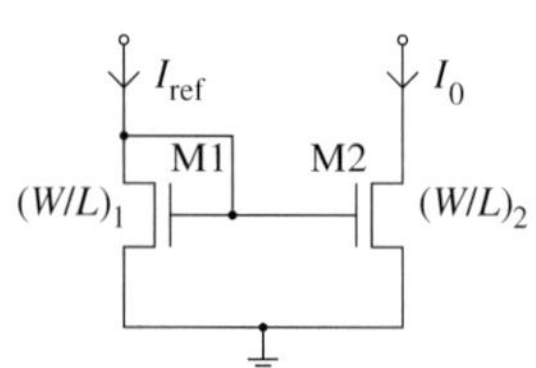

Fig. 20.25 Widlar current mirror with MOS transistors

One can very simply determine the output resistance of the circuit from the equivalent model shown in fig. 20.20 and equation (20.18) through substitution of the corresponding parameters. A simplification happens because there is no finite input resistance with the MOS transistor, one can therefore assume $r_{BE} \to \infty$:

$$r_{Out} \approx r_{DS2} \cdot \left[1 + R_2 \cdot \left(g_{m2} + \frac{1}{r_{DS2}}\right)\right]$$
$$\approx r_{DS2} \cdot (1 + R_2 \cdot g_{m2})$$

The minimum voltage across the current source must be at least:

$$V_{DS\,min} = V_{DS\,sat} + I_0 \cdot R_2$$

The circuit with MOS transistors in fig. 20.27 corresponds to the Wilson current mirror in fig. 20.21. The fig. 20.28 shows the small signal equivalent circuit. The transistor M2 works as 'active' resistance; furthermore, the output resistance r_{Out} attracts attention. It is the output resistance of the reference source I_{ref} that loads the transistor M1 additionally. One gets for the output resistance

$$r_{Out} \approx \left\{1 + \frac{g_{m3}}{g_{m2}} \cdot \left[g_{m1} \cdot (r_{DS1} \| r_{in}) + 1\right] + \frac{1}{r_{DS3} \cdot g_{m2}}\right\}$$

The minimum voltage across the current source must be at least:

$$V_{DS\,min} = V_{DS\,sat} + V_{GS}$$

betragen.

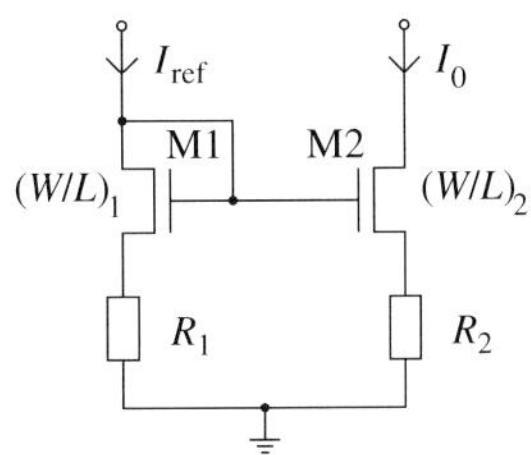

Fig. 20.26 Widlar current mirror with feedback and MOS transistors

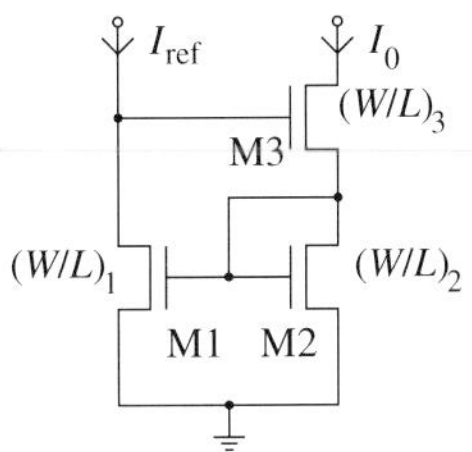

Fig. 20.27 Wilson current mirror with MOS transistors

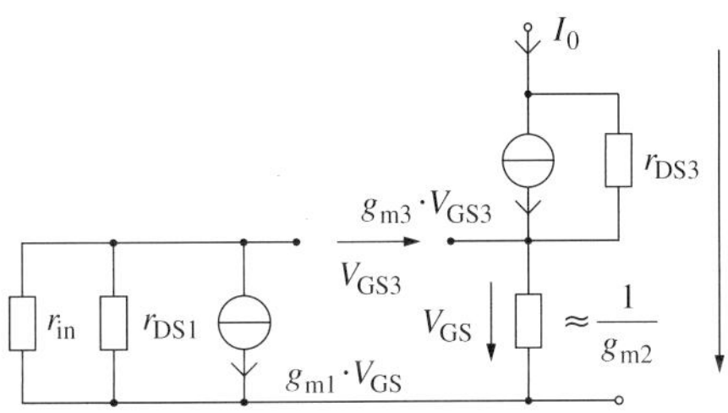

Fig. 20.28 Small signal equivalent circuit of the Wilson current mirror with MOS transistors

In fig. 20.29 the MOS variant of the cascode circuit of fig. 20.23 is to be seen. The associated small signal equivalent model is shown in fig. 20.30. The two transistors M1 and M4 work as 'active' resistances; however, the small signal input current is zero because of the infinite input impedance at the gate of M3. One gets as output resistance

$$r_{Out} \approx r_{DS3} \cdot \left(1 + g_{m3} \cdot r_{DS2} + \frac{r_{DS2}}{r_{DS3}}\right)$$

The minimum voltage across the current source must be at least:

$$V_{DS\,min} = 2 \cdot V_{DS\,sat}$$

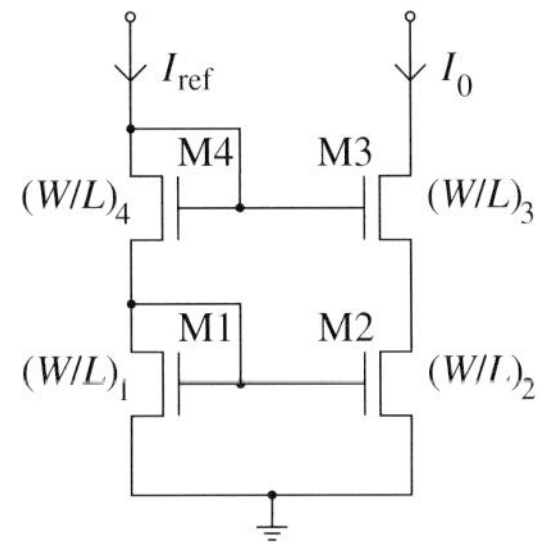

Fig. 20.29 Cascode current mirror circuit with MOS transistors

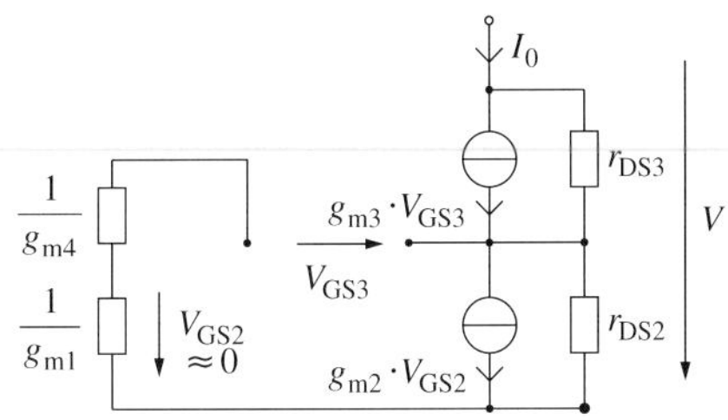

Fig. 20.30 Small signal equivalent model of a cascode current mirror circuit with MOS transistors

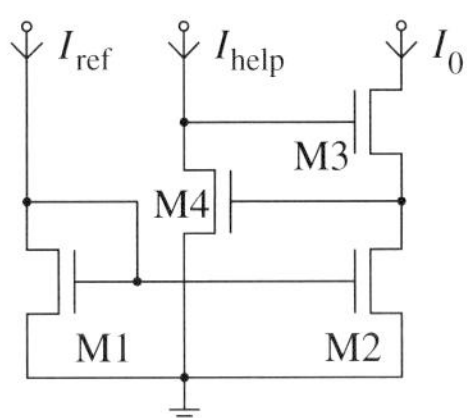

Fig. 20.31 Regulated cascode current mirror circuit with MOS transistors

One can obtain a circuit with very large output resistance with the so called regulated cascode circuit [20.25]. Figure 20.31 shows the circuit and fig. 20.32 the equivalent small signal circuit. The term r_{Out} represents the output resistance of the auxiliary current source, transistor M1 works as an 'active' resistance, however, without a small signal current, so that $V_{GS2} = 0$ and also the current source does not work at M2. For the output resistance one can find the following expression from the equivalent small signal circuit:

$$\begin{aligned} r_{Out} &\approx r_{DS3} \cdot \Big\{1 + g_{m3} \cdot r_{DS2} \times \\ &\quad \times [g_{m1} \cdot (r_{DS1} \| r_{in}) + 1] + \frac{r_{DS2}}{r_{DS3}}\Big\} \\ &\approx r_{DS3} \cdot g_{m3} \cdot r_{DS2} \cdot g_{m1} \cdot (r_{DS1} \| r_{in}) \end{aligned}$$

20

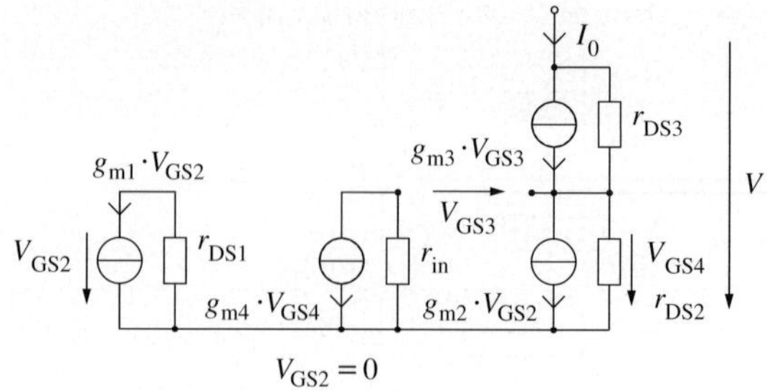

Fig. 20.32 Small signal equivalent model of a regulated cascode current mirror circuit with MOS transistors

Table 20.7 Properties of MOS current mirror circuits

Circuit type	Output resistance	Minimum voltage
Widlar circuit	r_{DS}	$V_{DS\,sat}$
Widlar circuit with feedback	$r_{DS} \cdot (1 + g_m \cdot R_2)$	$V_{DS\,sat} + I_0 \cdot R_2$
Wilson circuit	$r_{DS} \cdot (1 + g_m \cdot r_{DS}/2)$	$V_{DS\,sat} + V_{GS}$
Cascode circuit	$r_{DS} \cdot (1 + g_m \cdot r_{DS})$	$2 \cdot V_{DS\,sat}$
Regulated cascode circuit	$r_{DS} \cdot (1 + (g_m \cdot r_{DS})^2)$	$V_{DS\,sat}$

The value of the output resistance is at least one order of magnitude higher than with the preceding arrangements. Also the minimum voltage is quite small, caused by the action of the control amplifier M4, the current source transistor M3 may be controlled even a little bit into the ohmic range, without noticeable change of output current.

The minimum voltage across the current source should be at least:

$$V_{DS\,min} = V_{DS\,sat}$$

Finally table 20.7 gives a summary of the presented MOS current mirror circuits.

20.2.4 Basic Amplifier Circuits

Because the basic circuits for bipolar and for MOS transistors can be derived from the same structure and also in the simplified small signal models by simple substitution from each other, the equivalent small signal circuits are to be regarded only once. First the bipolar model is used and then for the appropriate MOS circuit $C_{BE} \rightarrow C_{GS}$, $C_{BC} \rightarrow C_{GD}$, $C_{CS} \rightarrow C_{DB}$, $r_{CE} \rightarrow r_{DS}$ and $r_{BE} \rightarrow \infty$ have to be substituted. For circuits with back gate control their quantities are added. The bipolar transistors work in the active region, the MOS transistors in the saturation region.

While the DC quantities, such as input and output impedances and transfer functions, are usually definable without further approximations, the frequency responses are very complex and difficult to survey even with the derived simple equivalent small signal models. Only the 3 dB cutoff frequencies are usually computed approximately according to the method of open circuit time constants [20.20], [20.12].

As results the complex input resistance Z_S (does not contain R_S) and the complex output resistance Z_L are shown, as well as the voltage transfer factor A for very low frequencies (inclusive the load resistors R_S and R_L), and approximately the 3dB cutoff frequency $\omega_{3\,dB}$. The values of R_S and R_L can be very different, so that in many cases an approximation is almost impossible.

Common Emitter and Common Source Circuit

Probably the most frequently applied circuits of the discrete and integrated technology are the common emitter respectively common source circuits. They deliver high gain and are 'good natured' in their operational behaviour. The input impedance at low frequencies is very high for MOS amplifiers and in a middle range for bipolar amplifiers. The cut off frequency is mainly caused by the base/gate to collector/drain capacitance transformed to the input side (the Miller capacity).

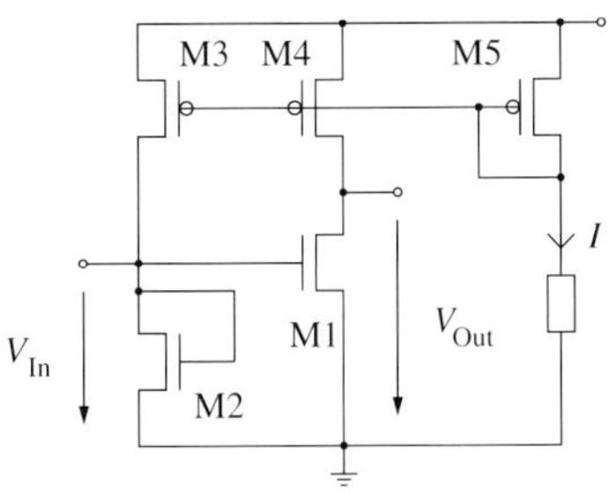

Fig. 20.33 Common source circuit with bias

In fig. 20.33 a complete circuit is shown as an example with an operating point given by current mirrors. The transistors M2 to M5 adjust the biasing point; the actual amplifier transistor is M1. The gate bias for M1 is created by the current mirror with the transistors M5 and M3; they produce a suitable voltage drop across the 'active' resistance M2. The current source transistor M4 injects the drain current for the amplifier transistor. For the principal contemplations a simplified circuit will be analyzed, see fig. 20.34. There, for example, the resistance R_L represents the parallel connection of the output resistance of the current source M4 and an assumed load resistance. V_S and R_S are the elements of an equivalent voltage source with internal resistance.

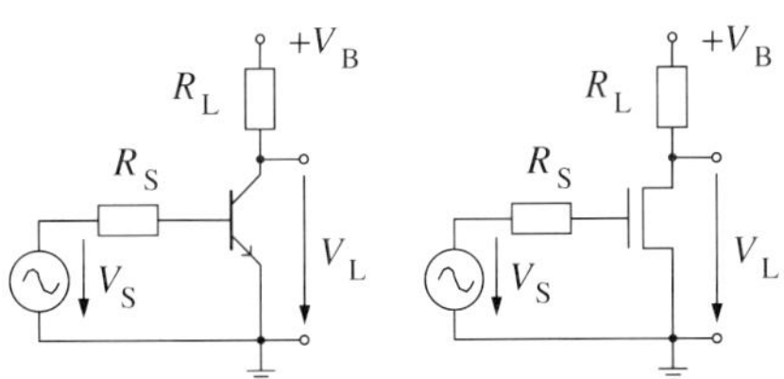

Fig. 20.34 Principle of a common emitter/source circuit

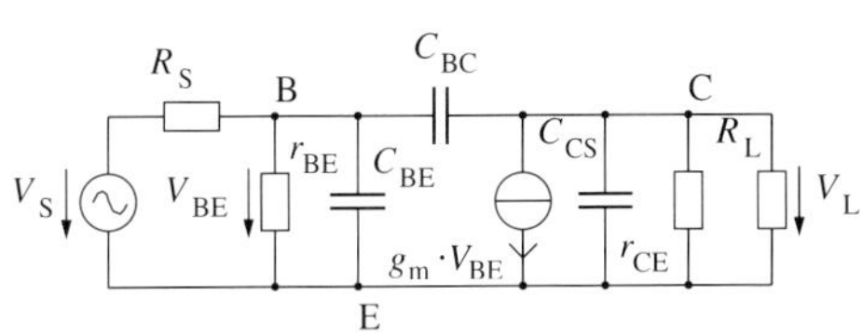

Fig. 20.35 Equivalent small signal model of the common emitter/source circuit

Results for the Bipolar Amplifier

- Voltage gain at $f = 0$:
$$A = \frac{V_L}{V_S} = -\frac{r_{BE}}{r_{BE} + R_S} \cdot g_m \cdot \frac{R_L \cdot r_{CE}}{R_L + r_{CE}}$$
- Input impedance:
$$Z_S \approx \frac{1}{\frac{1}{r_{BE}} + j\omega \cdot \left\{ C_{BE} + C_{BC} \cdot \left[1 + g_m \cdot (r_{CE} \| R_L) \right] \right\}}$$
- Output impedance:
$$Z_L \approx \frac{1}{\frac{1}{R_L} + \frac{1}{r_{CE}} + j\omega \cdot C_{CS}}$$
- 3 dB cut off frequency:
$$\omega_{3\,dB} \approx \frac{1}{(R_S \| r_{BE}) \left\{ C_{BE} + C_{BC} \left[1 + g_m (r_{CE} \| R_L) \right] \right\} + (r_{CE} \| R_L) C_{CS}}$$

Results for the MOS Amplifier

- Voltage gain at $f = 0$:
$$A = \frac{V_L}{V_S} = -g_m \cdot \frac{R_L \cdot r_{DS}}{R_L + r_{DS}}$$
- Input impedance:
$$Z_S \approx \frac{1}{j\omega \left\{ C_{GS} + C_{GD} \cdot \left[1 + g_m \cdot (r_{DS} \| R_L) \right] \right\}}$$
- Output impedance:
$$Z_L \approx \frac{1}{\frac{1}{R_L} + \frac{1}{r_{DS}} + j\omega \cdot C_{DB}}$$
- 3 dB cut off frequency:
$$\omega_{3\,dB} \approx \frac{1}{R_S \left\{ C_{GS} + C_{GD} \left[1 + g_m (r_{DS} \| R_L) \right] \right\} + (r_{DS} \| R_L) C_{DB}}$$

Common Collector or Common Drain Circuit

This basic circuit also called an emitter/source-follower is frequently used as a buffer circuit or as an impedance converter. The input has high impedance, the output low impedance. It has also very good high frequency characteristics; however, caution is required if the load contains capacitive components, then the input impedance can become negative, which usually causes instability. An exact analysis of the problem and a compensation circuit are found in [20.16]. With the MOS amplifier, the substrate control is taken into account through the back gate conductance g_{mB}.

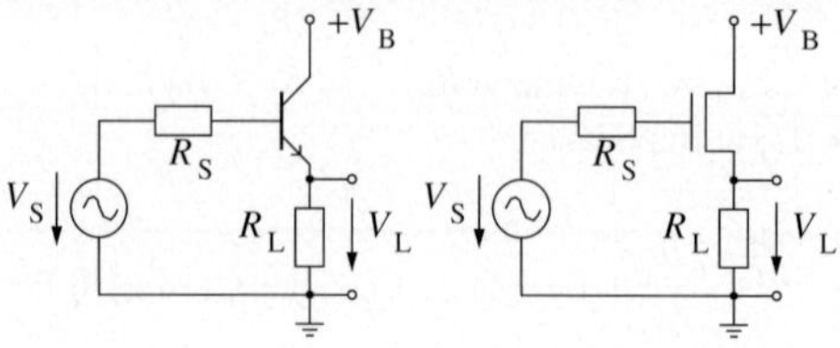

Fig. 20.36 Principle of a common collector/drain circuit

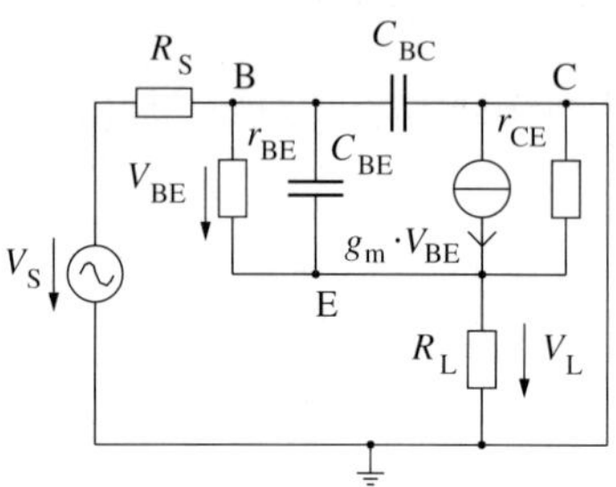

Fig. 20.37 Equivalent small signal model of the common collector/drain circuit

Results for the Bipolar Amplifier

With $r_{CE} \gg R_L$:

- Voltage gain at $f = 0$:

$$A = \frac{V_L}{V_S} = \frac{R_L \cdot (1 + r_{BE} \cdot g_m)}{r_{BE} + R_S + R_L \cdot (1 + r_{BE} \cdot g_m)}$$

- Input impedance:

$$Z_S \approx \frac{1}{\dfrac{1 + j\omega \cdot r_{BE} C_{BE}}{r_{BE} + R_L(1 + g_m r_{BE}) + j\omega \cdot r_{BE} C_{BE} R_L} + j\omega C_{BC}}$$

at $f = 0$:

$$Z_S = r_{BE} + R_L \cdot (1 + g_m \cdot r_{BE}) = r_{BE} + R_L \cdot (1 + \beta)$$

- Output impedance for $C_{BC} \approx 0$:

$$Z_L \approx \frac{1}{\dfrac{1}{R_L} + \dfrac{1 + g_m \cdot r_{BE} + j\omega \cdot r_{BE} \cdot C_{BE}}{r_{BE} + R_S \cdot (1 + j\omega \cdot r_{BE} \cdot C_{BE})}}$$

at $f = 0$:

$$Z_L = \frac{1}{\dfrac{1}{R_L} + \dfrac{1 + g_m \cdot r_{BE}}{r_{BE} + R_S}}$$

- 3 dB cut off frequency:

$$\omega_{3\,dB} \approx \frac{1}{C_{BE} \cdot A_1 + C_{BC} \cdot A_2}$$

with:

$$A_1 = \frac{R_L + R_S}{1 + R_L \cdot g_m + \dfrac{R_L + R_S}{r_{BE}}}$$

$$A_2 = \frac{R_S\,[R_L + r_{BE} \cdot (1 + g_m \cdot R_L)]}{R_S + R_L + r_{BE} \cdot (1 + g_m \cdot R_L)}$$

Results for the MOS Amplifier

With $r_{DS} \gg R_L$:

- Voltage gain at $f = 0$:

$$A = \frac{V_L}{V_S} = \frac{R_L \cdot g_m}{1 + R_L \cdot (g_m + g_{mB})}$$

- Input impedance:

$$Z_S \approx \frac{1}{\dfrac{j\omega \cdot C_{GS}(1 + g_{mB} R_L)}{R_L(g_m + g_{mB}) + j\omega \cdot C_{GS} R_L} + j\omega \cdot C_{GD}}$$

- Output impedance for $C_{GD} \approx 0$:

$$Z_L \approx \frac{1}{\dfrac{1}{R_L} + g_{mB} + \dfrac{g_m + j\omega \cdot C_{GS}}{1 + j\omega \cdot R_S \cdot C_{GS}}}$$

at $f = 0$:

$$Z_L = \frac{1}{\dfrac{1}{R_L} + g_{mB} + g_m}$$

- 3dB cut off frequency:

$$\omega_{3\,dB} \approx \frac{1}{C_{GS}\left[\dfrac{R_L + R_S(1 + R_L g_{mB})}{1 + R_L(g_m + g_{mB})}\right] + C_{GD} R_S}$$

Common Base or Common Gate Circuit

The common base/gate circuit provides substantially higher cut off frequencies than the other basic circuits; however, this type of circuit has stability problems in the discrete circuit technology, because a base/gate parasitic inductance produces negative input impedance. In integrated circuit technology a careful layout can prevent this characteristic. The input impedance is rather small ($\approx 1/g_m$), the output resistance high. Preferred applications are high frequency LNA circuits and within cascode circuits as second stage (see section 20.2.7). The figures 20.38 and 20.39 show the principle and the equivalent small signal circuit.

With the MOS amplifier the substrate control is taken into account through the back gate transconductance g_{mB}.

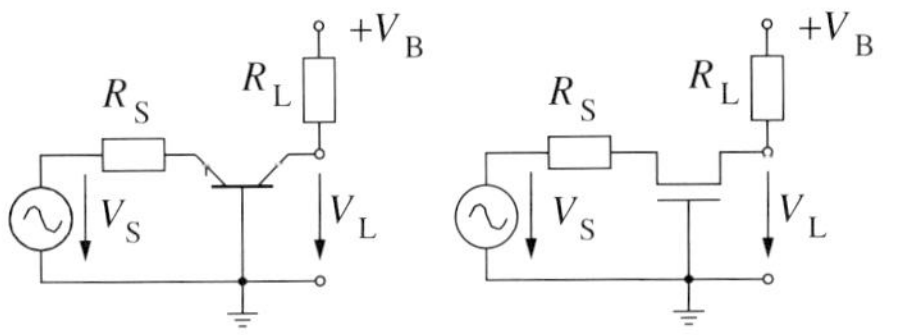

Fig. 20.38 Principle of a common base/gate circuit

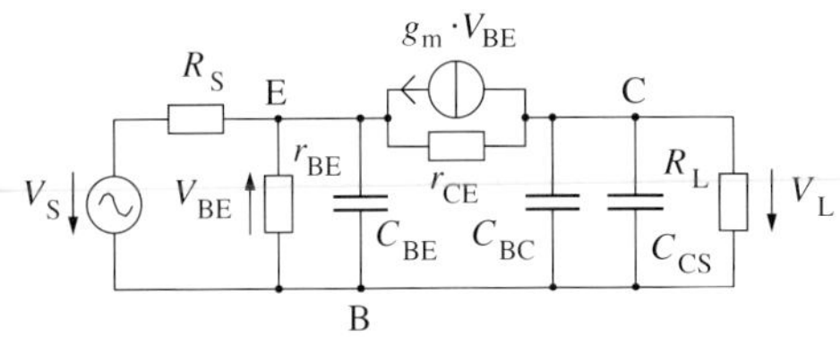

Fig. 20.39 Equivalent small signal model of the common base/gate circuit

Results for the bipolar amplifier

With $r_{CE} \gg R_L$:

- Voltage gain at $f = 0$:

$$A = \frac{V_L}{V_S} = \frac{R_L \cdot r_{BE} \cdot g_m}{r_{BE} + R_S \cdot (1 + r_{BE} \cdot g_m)}$$

- Input impedance:

$$Z_S \approx \frac{r_{BE}}{1 + g_m \cdot r_{BE} + j\omega \cdot r_{BE} \cdot C_{BE}}$$

- Output impedance:

$$Z_L \approx \frac{1}{j\omega \cdot (C_{BC} + C_{CS})} || R_L$$

- 3 dB cut off frequency:

$$\omega_{3\,dB} \approx \frac{1}{C_{BE}\left[\frac{R_S \cdot r_{BE}}{r_{BE}+R_S(1+r_{BE}g_m)}\right]+(C_{BC}+C_{CS})R_L}$$

Results for the MOS amplifier

With $r_{DS} \gg R_L$:

- Voltage gain at $f = 0$:

$$A = \frac{V_L}{V_S} = \frac{R_L \cdot (g_m + g_{mB})}{1 + R_S \cdot (g_m + g_{mB})}$$

- Input impedance:

$$Z_S \approx \frac{1}{g_m + g_{mB} + j\omega \cdot C_{GS}}$$

- Output impedance:

$$Z_L \approx \frac{1}{j\omega \cdot (C_{GD} + C_{DB})} || R_L$$

- 3 dB cut off frequency:

$$\omega_{3\,dB} \approx \frac{1}{C_{GS}\left[\frac{R_S}{1 + R_S(g_m + g_{mB})}\right] + (C_{GD} + C_{DB})R_L}$$

20.2.5 Differential Amplifier

This kind of amplifier is typical of the integrated technology, because it develops only under matching conditions its excellent qualities, especially the amplification of small voltages. Since on an integrated circuit capacitances larger than approximately 30 pF can be produced only uneconomically, galvanic coupling must be used when coupling of several amplifier stages. Thus a temperature influence very easily arises, since many semiconductor parameters are temperature dependent. A solution is to split the input signal into two voltages, the voltage difference between them carrying the information. With the differential amplifier only the differences of the amplifications have an effect. Owing to the high symmetry of transistors placed close together with almost identical temperatures, the error in the differential amplifications is very small. To improve the matching of the transistors, the two transistors are split up into four 'half transistors' and then parallel paired crosswise; see, in addition, section 20.2.1 as well as fig. 20.9. The principle of a differential amplifier is presented in fig. 20.40. Assuming identical transistors, operating in the active region with identical collector resistance and at the same temperatures and neglecting the influences of the Early effect, equation (20.4) helps to derive five equations, in which V_D is the differential input voltage:

$$\begin{aligned}
I_{C1} &\approx I_{SE} \cdot e^{\frac{V_{BE1}}{V_T}} \\
I_{C2} &\approx I_{SE} \cdot e^{\frac{V_{BE2}}{V_T}} \\
I_0 &= I_{C1} + I_{C2} \\
V_D &= V_{In1} - V_{In2} = V_{BE1} - V_{BE2} \\
V_{Out} &= R_C \cdot (I_{C1} - I_{C2})
\end{aligned}$$

By elimination of intermediate quantities one finally receives as the result:

$$V_{\text{Out}} = R_{\text{C}} \cdot I_0 \cdot \tanh\left[\frac{V_{\text{D}}}{2 \cdot V_{\text{T}}}\right] \approx \frac{R_{\text{C}} \cdot I_0}{2 \cdot V_{\text{T}}} \cdot V_{\text{D}}$$

for $|V_{\text{D}}| < V_{\text{T}}$.

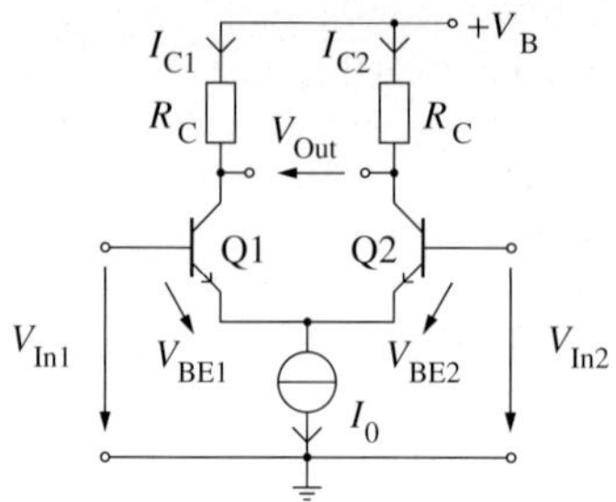

Fig. 20.40 Principle of a differential amplifier

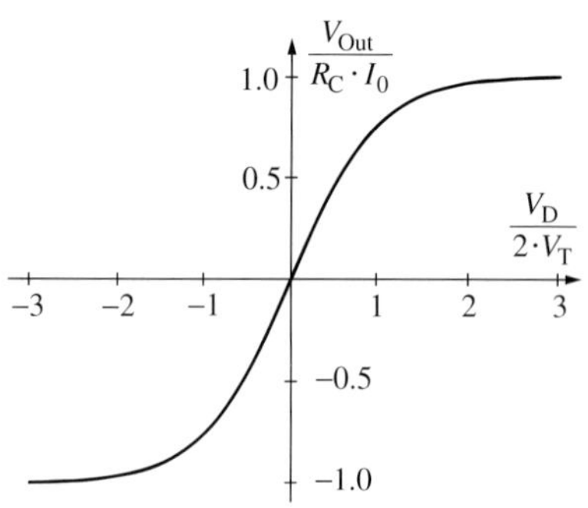

Fig. 20.41 Transfer characteristic of the differential amplifier

A diagram of the above equation in normalized form is shown in Figure 20.41. The output voltage has a temperature dependence owed to V_{T}, which can be compensated, however, by keeping the current I_0 linear with the temperature (PTAT current: **P**roportional **T**o **A**bsolute **T**emperature); review in addition the current mirror, whose temperature behaviour is adjustable (section 20.2.3.4).

One recognises however that the linear range for the differential voltage is limited to about V_{T}, thus to about 25 mV at room temperature. This is often not sufficient, and therefore one inserts feedback resistors into both emitter lines or uses a circuit after [20.16, page 592 ff], which connects in parallel two differential amplifiers with different emitter areas. One obtains thereby a smaller slope, which means lower gain at the origin; however, the range of the input voltage is increased. The principal course does not change. If one does not want a linear gain but a signal compression or limitation, then the differential amplifier offers a solution, i.e., for input voltages $|V_{\text{D}}| > 5 \cdot V_{\text{T}}$ the output voltage remains about constant. In the delimitation range one of the transistors is not conducting and the other transistor conducts the total current.

In the above contemplation both exact symmetry and an ideal current source were assumed. In practice the circumstances are not that ideal, i.e., owing to different emitter areas one gets an offset voltage as well as temperature dependence, and owing to a finite output resistance of the current source I_0 a non-zero common mode gain. If one replaces in fig. 20.40 the two transistors with NMOS transistors one gets the transfer equation for the MOS differential amplifier

$$\begin{aligned} V_{\text{Out}} &= R_{\text{C}} \cdot (I_{\text{D1}} - I_{\text{D2}}) \\ &= R_{\text{C}} \cdot I_0 \cdot V_{\text{D}} \cdot \sqrt{\frac{K}{I_0} - \frac{K^2 \cdot V_{\text{D}}^2}{4 \cdot I_0^2}} \end{aligned}$$

This equation applies until one of the transistors is reaching the cutoff region that is the case, if V_{D} times the square root is equal to one:

$$V_{\text{D}\max} = \pm\sqrt{\frac{2 \cdot I_0}{K}}$$

Qualitatively the shape of the transfer function is very similar to that in fig. 20.41.

One important criterion with the differential amplifier is, apart from the offset voltage, the range of the permissible common mode voltage at the differential input without which the differential mode gain suffers. One finds additional computations in section 20.2.7 treating operational amplifiers. The differential amplifier is versatilely applicable, e.g., as input stage of an operational amplifier, as transconductance amplifier, as voltage to current converter or as amplifier in differentially implemented circuit parts.

20.2.6 Transconductance Amplifier

The main application field of the transconductance amplifier is the integrated filter technology, e.g., the g_{m} C filter [20.26]. In principle a transconductance amplifier is a voltage controlled current source $I_{Out} = V_{In} \cdot g$. The transconductance g must be adjustable to compensate for variations and must have a wide range for low or high frequency applications. Usually the input signal is present as a differential voltage. The differential stage described in the previous chapter supplies a current difference; if one does not change these via both collector resistors into voltages, but combines them by a current mirror into one output current, one has a simple transconductance amplifier [20.22], see fig. 20.42. With the bias voltage of transistor M0 the quiescent current I_0 is determined. Under the assumption of complete symmetry, identical transistors and input voltages, so that all transistors are in saturated operation, the output current disappears and both drain currents are a half of I_0.

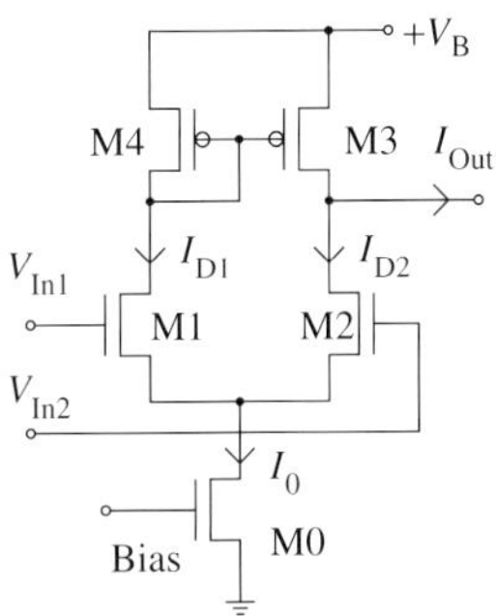

Fig. 20.42 Simple transconductance amplifier circuit

If one keeps V_{In2} at this potential and if the input voltage V_{In1} increases by a small amount, then I_{D1} increases to $I_0/2 + \Delta I$ and V_{In1} decreases to $I_0/2 - \Delta I$. The current I_{D1} is reflected at the current mirror, so that the output current adds to $2 \cdot \Delta I$. Thus the current mirror, as an active load, just adds the currents of both differential transistors. This circuit has been well known for a long time with bipolar operational amplifiers as a phase adding circuit. The above conductance g is given here as the transconductance g_m of the differential transistors.

An improved version, working almost over the entire supply voltage range, is also described in [20.22] and presented in fig. 20.43.

Apart from the mentioned improved operating voltage range, this circuit still has the advantage that the differential amplifier transistors operate on low impedance symmetrical loads (M3 and M5 are both 'active' resistors). Thus both inputs are less afflicted with the Miller effect. The cut off frequency is practically determined just by the time constant of the output node:

$$\omega_{3\,dB} \approx \frac{1}{(r_{DS4} \| r_{DS8}) \cdot (C_{DB4} + C_{DB8})}$$

Further transconductance amplifiers are to be found in [20.4] and [20.16]. The circuits shown above are implemented in MOS technology. The finite base currents of the bipolar transistors mostly cause trouble. Also frequently only very small transconductances are needed, which can be achieved easily with the smaller transconductances of the MOS transistors.

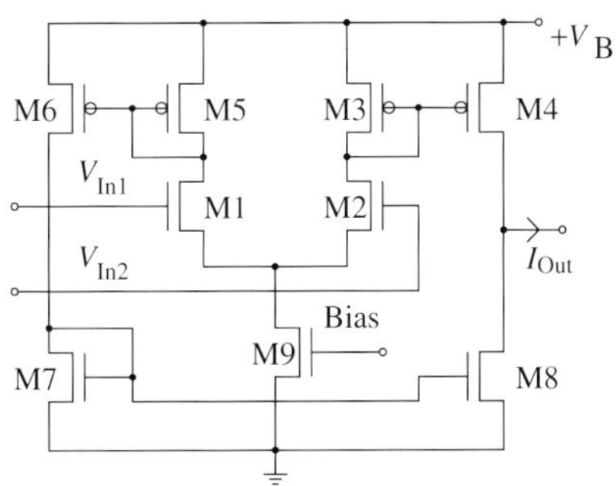

Fig. 20.43 Improved transconductance amplifier circuit

20.2.7 Operational Amplifiers

The requirements for operational amplifiers are very different; this led in the discrete technology to 'Jacks of all trades' controlling the market. In the integrated technology this procedure is not recommended, since power consumption and chip area become unnecessarily high. Usually only very simple types are necessary, often even without a special output stage if only capacitive loads are present. If the amplification factor is well known in feedback operation, then the compensation can be individually accomplished. Thus substantial energy dissipation can be saved and/or other characteristic values can be improved. Fundamental work has been done on the bipolar range [20.29] and on the MOS sector [20.13]. If the actual process leaves the choice of the transistor types to

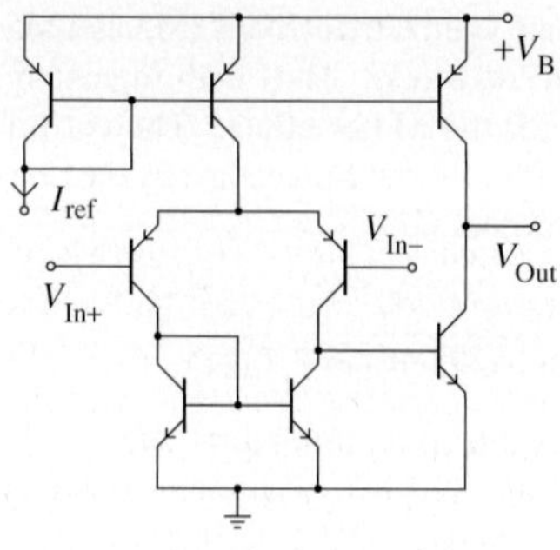

Fig. 20.44 Simple bipolar operational amplifiers

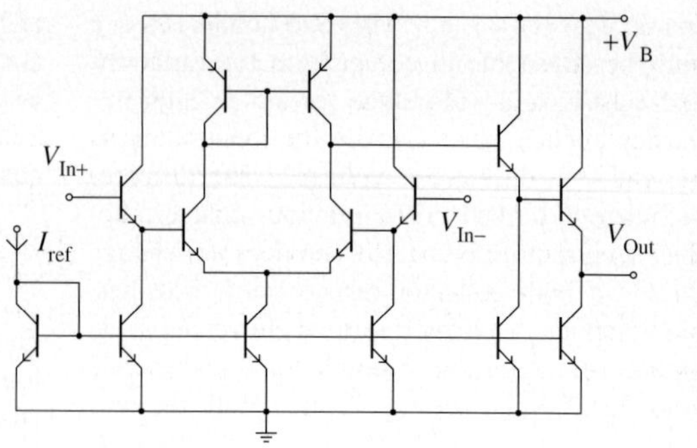

Fig. 20.45 Simplified circuit of a video operational amplifier

Table 20.8 Properties of transistor input stages

	NMOS transistor	PMOS transistor	Bipolar transistor
Thermal noise	−	−	+
$1/f$ noise	−	+	++
Slew rate	−	+	−
Input impedance	++	++	−
Voltage gain	−	−	++
Bandwidth	−	−	+
Matching, offset	−	−	++

the developer, then the following basic principles are recommended:

- Input stage: see table 20.8,
- Second stage: if possible NPN transistors, because of high transconductance;
- Output stage: preferably bipolar transistors, with capacitive load: omit the output stage.

A two-stage circuit, found very frequently, will therefore serve as the fundamental circuit and is presented in fig. 20.44 in bipolar technology. With NPN transistors in the input stage the circuit would not perform well, since the second stage would require a lateral PNP transistor, which is a very bad transistor. With bipolar circuits usually also an ohmic load is involved. Thus an output stage is necessary in any form which separates the load from the high impedance node of the circuit. With NPN Darlington transistors in the input stage and a Darlington emitter follower as output stage, e.g., a video operational amplifier with only one amplifier stage arises, after [20.21] or [20.1]. Figure 20.45 shows the simplified circuit. The PNP current mirror with lateral transistors operates above the rather low cut off frequency as a passive circuit, therefore the open loop gain (approximately 60 dB) decreases back to half. This circuit admittedly has a systematic offset error, which, however, does not disturb in this application.

With help of the CMOS circuit in fig. 20.46, which is very frequently used, some important parameters will be studied. The two transistors M1 and M2 form, together with the current source transistor M7, a differential stage whose drain current steers the second amplifier transistor M5 by the current mirror M3 and M4. One takes the output voltage without a special output stage at the working resistance of the transistor M5, which means the current source transistor M6. The amplifier is compensated through C_C and R_C, in which R_C is usually realized by a transistor in the ohmic range. The two differential amplifier transistors are affected, according to process, by the body effect or not. For symmetry reasons M1 and M2 as well as M3 and M4 must be dimensioned equal. The drain current of the transistors M5 and M6 must be the same for the state $V_{In1} = V_{In2} = V_B/2$ to achieve $V_{Out} = V_B/2$.

In addition one can implement M5, M6 and M7 with the same W/L ratio and M3 and M4 with half of the W/L ratio. The factor two arises from the partition of the current I_0 into the two transistors M1 and M2. The open loop gain can be expressed sufficiently exact simply by the multiplication of the amplification factors of the two stages:

$$A_0 \approx g_{m1} \cdot (r_{DS2} \| r_{DS4}) \cdot g_{m5} \cdot (r_{DS5} \| r_{DS6})$$

The cut off frequency of the open loop gain is calculated in the first approximation, first without compensation:

$$\omega_{3\,dB\,open\,loop} \approx \frac{1}{\{C_{DB2}+C_{DB4}+C_{GD5}[1+g_{m5}(r_{DS5}\|r_{DS6})]\}(r_{DS2}\|r_{DS4})}$$

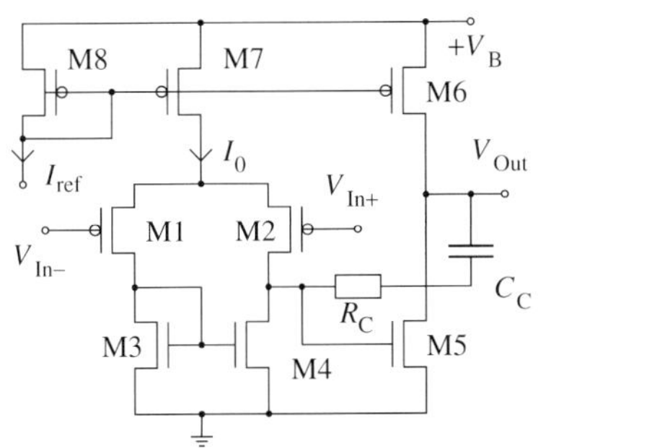

Fig. 20.46 Simple CMOS operational amplifier

In the case of compensation this cut off frequency is given almost just by the first amplifiers stage output resistance multiplied by the 'Miller multiplied' compensation capacity; that means the cut off frequency drops significantly:

$$\omega_{3\,dB\,comp} \approx \frac{1}{\{C_C[1+g_{m5}(r_{DS5}\|r_{DS6})]\}(r_{DS2}\|r_{DS4})}$$

Regarding the expressions for the cut off frequencies it is noticeable that they do not depend on the voltage gain of the first stage but on its output resistance. Therefore one can increase the open loop gain by enlargement of g_m (increase of the W/L of the input transistors) without substantial changes in the frequency response. By [20.13] an increase of the open loop gain is also possible just by decreasing I_0, since thereby the output resistances increase super-proportionally to the decrease of g_m. However, this deteriorates the frequency response together with the slew rate, since this is given through

$$S_R = \frac{I_0}{C_L} \approx \frac{I_0}{C_C}$$

The quantity C_L is thereby the entire effective capacity at the drain of M2 and thus is essentially determined by C_C. Together with the slew rate the linear settling time has to be taken into account. It gives information on how long after an input voltage step the system needs to settle in a certain approximation to the steady state condition. Here linear means that the voltage step does not force the operational amplifier into the slew condition. For a residual error of 0.1 %, for instance, approximately a duration of 7-fold the time constant of the system has to elapse

$$t_{settl} \approx 7 \cdot A_g \cdot \frac{C_C}{g_{m1}}$$

where A_g is the voltage gain of the operational amplifier in feedback mode.

For the noise behavior of the operational amplifier mainly the input stage is responsible if its gain is larger than the gain of the current mirror consisting of the transistors M3 and M4. The squared averaged noise voltage of the current mirror transistors enters only with about

$$\left(\frac{g_{m\,Current\,mirror}}{g_{m\,Input}}\right)^2$$

of the total noise.

With operational amplifiers within the low frequency range almost only the $1/f$ noise plays a role, so that after equation (20.14) as only accessible variable the transistor area $W \cdot L$ is at the disposal to the designer for optimisation. Wide input transistors are recommended also for the suppression of statistical matching errors. The frequency response suffers, however, because of a larger capacitive load with wide transistors.

There are still further, increasingly more important parameters available for the evaluation of operational amplifiers, i.e., the common mode range (CMR) and the common mode rejection ratio (CMRR). Both parameters have substantial effects with reduced operating voltages. First consider the common mode range of the CMOS operational amplifier according to fig. 20.46. The maximum input voltage is reached if M7 is operated at the border of the saturation region, and the voltage $V_{GS1} = V_{GS2}$ just generates the desired current:

$$V_{Min+} = -V_{DS7} - V_{GS1} = \sqrt{\frac{2 \cdot I_0}{K_7}} + \sqrt{\frac{I_0}{K_1}} - V_{Thp}$$

The lower limiting level is reached, if the transistors M3, M4 and M5 touch the border of the saturation region, and the differential amplifier transistors just generate the desired current:

$$V_{Min-} = V_{GS3} - V_{DG1} = \sqrt{\frac{I_0}{K_3}} + V_{Thn} + V_{Thp} \approx \sqrt{\frac{I_0}{K_3}}$$

For a threshold voltage of 0.75 V the sum of the unusable voltages is approximately 1.5 V, which means half of the operating voltage range is already unusable with today's usual voltages. One can achieve a substantial improvement if one connects the regarded circuit and its complementary differential amplifier circuit at the input in parallel and adds the output currents accordingly to the output. With this circuit the common mode range reaches to within approximately 0.4 V of the supply voltage borders. Problems arise because the transconductance of the total differential amplifier is not constant over the dynamic range, and thus compensation becomes difficult [20.4, page 620 ff].

The common mode rejection is not infinite, nevertheless it is very high in real operational amplifiers. The suppression can take place only in the differential amplifier stage, because all following stages are no longer differentially implemented and all signals are treated equally. Common mode gain is mainly caused by the common current source and their finite output resistance. The remedy is an improved current source with higher output resistance. Any unbalance in the differential amplifier causes common mode gain. Furthermore, the dependence of the drain current on the drain source voltage moves the operating points and thereby produces unintended changes by the common mode gain.

The circuit of a modern operational amplifier with folded cascode is shown in fig. 20.47. The generation of the various bias voltages is not drawn for better simplicity of presentation. The differential amplifier stage is implemented as in the previous Figure, only the source current source is improved by means of an additional cascode transistor M11 to increase the common mode rejection. The two transistors M3 and M4 only adjust the operating point. The way of the signal leads over the differential stage with the transistors M1 and M2, then over the folding stage into the common gate stage with the transistors M5 and M6.

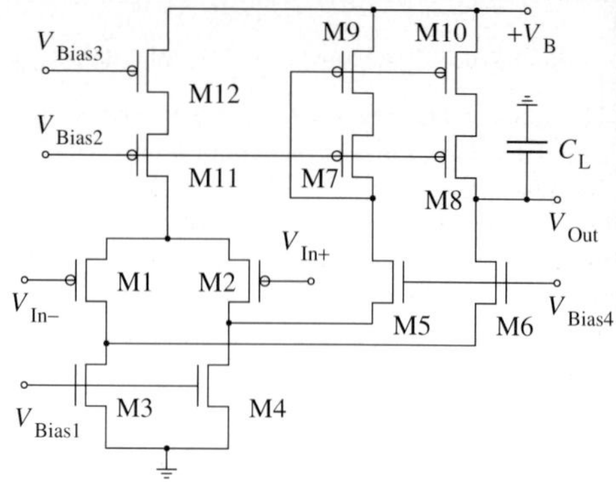

Fig. 20.47 A CMOS operational amplifier with folded cascode

The output of the cascode stage leads to a wide swing current mirror with the transistors M7 to M10. The output is attached to the high impedance node between the cascode output and the current mirror output. If necessary a frequency response compensation, respectively an output stage, can also be attached here.

There are many advantages of this circuit compared to the preceding: first the permissible input common mode range is somewhat increased, thus the differential amplifier transistors operate no more on a current mirror but on two current sources (which provides an improvement of about one threshold voltage). However, the voltage over the cascode transistors is slightly increased (a degradation of about $V_{DS\,sat}$). By the 'folding' the range of the output voltage is expanded to approximate $V_B - 3 \cdot V_{DS\,sat}$, because the cascode stage operates, as seen from the supply voltage, in parallel to the input differential amplifier. The last advantage is that the circuit has only one high impedance node, and thus the frequency response compensation, if at all necessary, can very simply be achieved by the capacitor C_L indicated in the diagram. The output of the differential stage operates on the very small input impedance of the common gate stage, and thus the influence of the parallel capacities can be neglected. Assuming the two amplifier stages without internal feedback, the open loop gain can approximately be calculated by multiplying the voltage gain of the differential stage with the cascode stage. Furthermore the splitting and the later combining of the branches of the amplifier can be ignored.

$$A_0 \approx g_{m1} \cdot \left(r_{DS1} \| r_{DS3} \| \frac{1}{g_{m6}} \right) \cdot g_{m6} \cdot (r_{DS6} \| r_{iM8})$$
$$\approx g_{m1} \cdot (r_{DS6} \| r_{iM8})$$

The effective output resistance is very high, since both stages involved are cascode stages. This means that the open loop gain is, for instance, approximately between 500 and 2000. Also useful operational amplifiers can be built with only one amplifier stage. An amplifier as just described, but in a complementary implementation (P and N channel transistors exchanged) delivers around approximately a factor of $\sqrt{3}$ higher amplification factor with the same supply current and around the same factor higher transit frequency. The noise behavior becomes worse, however. These operational amplifier principles can naturally be transferred also to bipolar transistors, but then an output stage is necessary in addition, for example with an emitter follower. One finds special circuit technologies for bipolar operational amplifiers with very small supply voltages in [20.10]. Increasing complexity and the influences of noise within combined digital as well as analog functions often require completely differentially implemented operational amplifiers. This is, however, a substantial additional expenditure of circuit technology and power dissipation. Furthermore, the individual components often present a smaller bandwidth because of increased additional capacitive loads. A procedure in [20.9] permits, by simple doubling and modification of existing components, the introduction of differential circuit technology.

20.2.8 Current/Voltage References

Almost all analog circuits use the matching principle, thus it is possible to make practically all relevant currents and voltages of a circuit linear dependent on one reference source. Excepting very simple circuits, which do not need high quality stabilization, the band gap principle suggested by Widlar [20.28] became generally accepted: One voltage with positive and one with negative temperature coefficients are added in the correct ratio, thus the sum of both voltages is temperature independent. The base emitter voltage of a bipolar transistor is used (approximately -2 mV/K) for a voltage with negative temperature coefficient and the voltage drop of an asymmetrical Widlar current mirror (see section 20.2.3.4) across a resistor serves as voltage with positive temperature coefficient. More frequently current references are needed, since many operating points are adjusted by current mirrors.

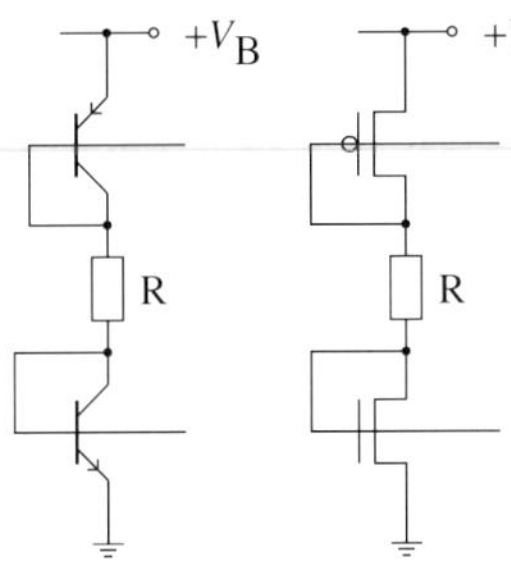

Fig. 20.48 Operating point adjustment with resistor

First to a very simple generation of reference voltages shown in fig. 20.48. The circuit produces near the lower and upper supply voltage lines a voltage drop across an 'active resistance', which also can be converted via a current mirror into a current. The reference voltage depends on temperature, resistance, and operating voltage. For the realization of low currents one needs large resistance values and thus much chip area. The circuit of fig. 20.49 gives smaller resistance values. The circuit part on the left side with the transistors M5 to M9 serves as starting circuit, because many circuits of this kind possess two stable operating points, one almost without current and one with the desired operating point. In operation the starting circuit is disconnected by the switched off transistor M6. When switching on, the transistor chain M7 to M9 provides a very high resistance to ground, so that M6 becomes conducting and drives a current to M1. Thus the circuit starts. The necessary resistance is

$$R = \sqrt{\frac{2}{I}} \cdot \left[\sqrt{\frac{1}{K_1}} - \sqrt{\frac{1}{K_2}} \right]; \quad K_2 > K_1$$

if the transistors M3 and M4 have the same sizes.

20

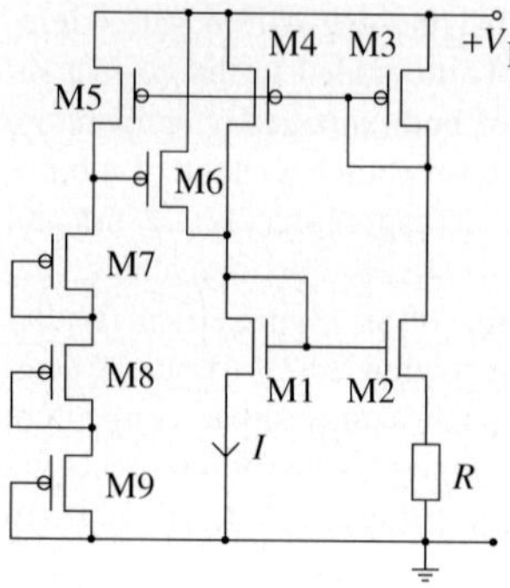

Fig. 20.49 Operating point adjustment with current mirror

The current I is temperature dependent owing to the resistor R and the transconductances K_1 and K_2 of the transistors M1 and M2. If in the above circuit the MOS Transistors M1 and M2 were replaced by bipolar transistors a PTAT current is achieved, which can be adjusted by the resistor R:

$$I = \frac{V_T}{R} \cdot \ln\left(\frac{A_{E2}}{A_{E1}}\right)$$

The emitter area of transistor Q2 must be larger than the emitter area of Q1; otherwise the circuit is not realizable. This reference source is ideal for the supply of bipolar transistors, since these have a decreasing transconductance with rising temperature (in addition see section 20.1.2). Usually, however, the resistance also has a considerable temperature coefficient, so that the current rising linearly with the absolute temperature is not perfectly achieved.

One can extend the two Widlar current mirrors into types with larger output resistance, in order to increase the independence from the supply voltage. The above temperature dependence also can be achieved with the CMOS process using the parasitic substrate transistors. As an example fig. 20.50 shows this for an N well process without the starting circuit. Again the above current equation results if M1 and M2 as well as M3 and M4 are dimensioned equal. The temperature dependence can be almost completely eliminated by adding only one resistor; fig. 20.51 shows the extended circuit. This circuit is the application of an old principle [20.6], which was intended for the injector current supply of I²L circuits, and still suitable for today's needs (see fig. 20.52). The design equation can be taken over from there and reads as follows (using $V_{BE}(T)$ from equation (20.7)):

$$\frac{1}{z-1} \cdot \frac{\ln N + \ln z}{1 + \ln N + \ln z} = \frac{1}{1-\alpha_T T_0}\left[\frac{V_{G0} - V_{BE0} - V_{T0}(r-4)}{V_{BE0}} + \alpha_T T_0\right]$$

with

$$N = \frac{A_{E2}}{A_{E1}}, \quad z = \frac{I_1}{I_2}\bigg|_{T=T_0} = \frac{I_{10}}{I_{20}},$$
$$V_{BE0} = V_{BE}|_{T=T_0} \quad \text{and} \quad V_{T0} = V_T|_{T=T_0}$$

Where α_T is the temperature coefficient of the used resistors at the reference temperature T_0. The values of the resistors arise as a result of the following equations:

$$R_1 = \frac{V_{BE0} \cdot z}{I_{ref} \cdot (z-1)} \quad \text{and}$$
$$R_2 = \frac{V_{T0} \cdot z}{I_{ref}} \cdot (\ln z + \ln N)$$

The compensation succeeds accurately only at one temperature, but the range of very good compensation is very wide, approximately ±50 K.

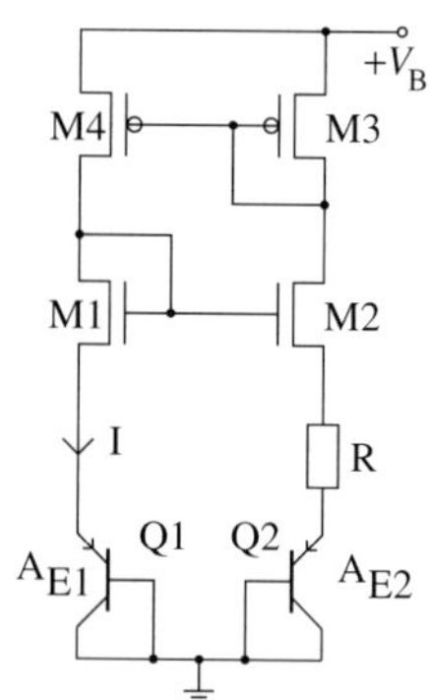

Fig. 20.50 Operating point adjustment with current mirror and parasitic transistors

The aforementioned circuits supplied mainly reference currents, which one can convert, e.g., with resistors into reference voltages. Large resistances are necessary for small currents. The direct design of a band gap reference with a voltage of $V_{G0} \approx$ 1.2 V is difficult with increasingly lower supply voltages. A circuit diagram is shown in fig. 20.53.

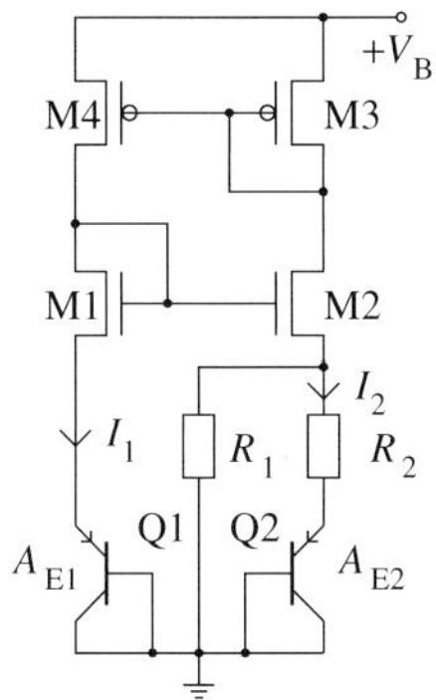

Fig. 20.51 Operating point adjustment with current mirror and parasitic transistors with temperature compensation

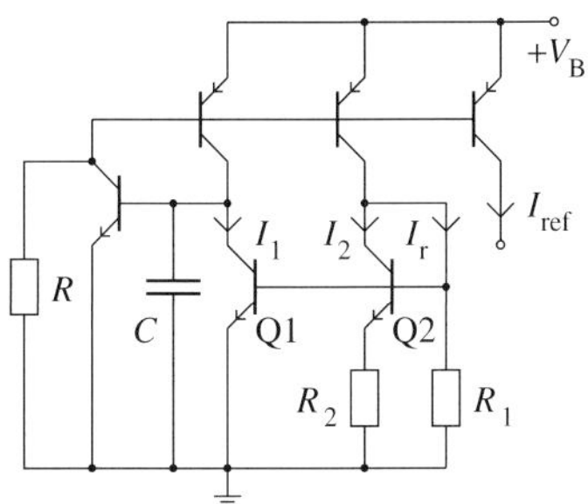

Fig. 20.52 Injector circuit [20.6]

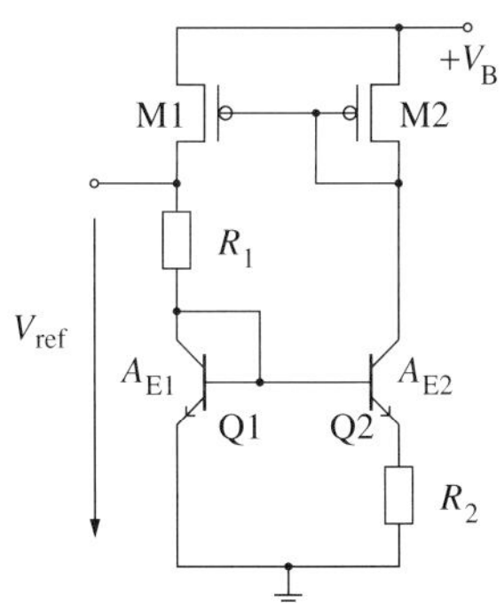

Fig. 20.53 Band gap reference with bipolar and MOS transistors

The current mirror with the NPN transistors can be replaced only very badly by lateral transistors, because these have parasitic substrate transistors causing temperature dependent current divisions and putting the function of the circuit in question. The NPN transistors are therefore realizable only in a genuine BICMOS processes. The principle operation is easily recognisable: the bipolar current mirror supplies a current with positive temperature coefficient by the feedback action of R_2 via the PMOS current mirror back to the input transistor Q1. The reference voltage consists of the voltage drop across Q1 with negative temperature coefficient and the voltage drop across R_1 with a positive temperature coefficient. In [20.20, p. 234] a Brokaw band gap reference is described, modified for CMOS realization. In fig. 20.54 the circuit diagram without a starting circuit is represented. In each case M3 and M4 as well as M5 and M6 need an identical layout. The temperature compensation should be dimensioned using equation (20.7):

$$V_{T0} \cdot \frac{R_2}{R_2 - R_1} \cdot \ln(N)$$
$$= V_{G0} - V_{BE0} - V_{T0} \cdot (r - 4)$$

and

$$I = \frac{V_{T0}}{R_2 - R_1} \cdot \ln(N)$$

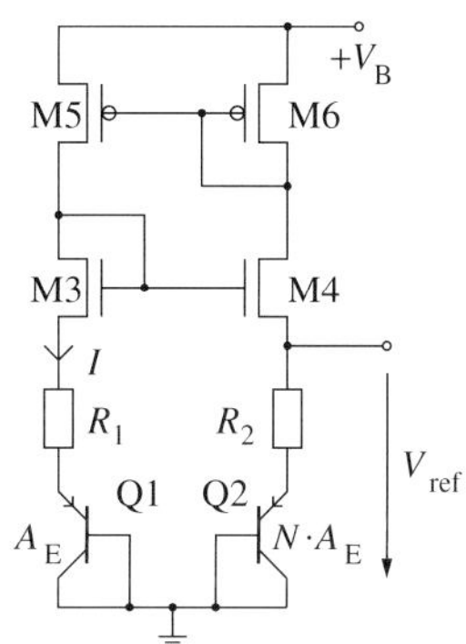

Fig. 20.54 Band gap reference with CMOS transistors

The temperature coefficient of the resistances, if for both resistances equal, does not enter the circuit's behavior. The current I of the circuit is a *PTAT* current if the resistances are assumed as temperature independent.

If a reference source is required not only to adjust the operating point, then circuits with operational amplifiers deliver more precise results. However, that needs a resistance and an offset alignment or a chopper operational amplifier, since otherwise the offset problems prevent a tidy function of the circuit. An example of a realization after [20.3] is shown in fig. 20.55. The operational amplifier employed is equipped with a *chopper*, which however, is not drawn.

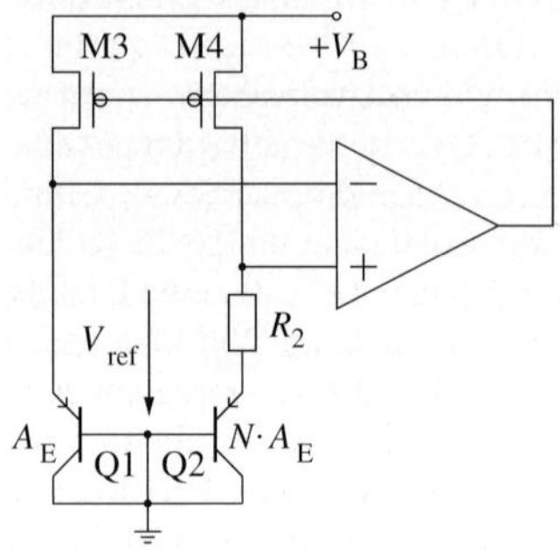

Fig. 20.55 Band gap reference [20.3]

M3 and M4 have the same size. The dimension of the temperature compensation follows the equations of the preceding circuit with $R_1 = 0$. An improvement of the temperature compensation towards wider ranges is achieved by better adapting the voltage with positive temperature coefficient to the base emitter voltage. An example of that is to be found in [20.24]. In all circuits the current mirrors should be designed with high output resistances to become as independent as possible of the supply voltage.

20.2.9 Oscillators

Depending upon the application the requirements are different for an oscillator: for digital applications phase noise and frequency stability plays practically no role, with analog applications this is the main problem as well as the suppression of disturbing influences, such as temperature and the coupling of disturbances. A controllability of the frequency is often necessary; a common goal is small power dissipation and operation with small supply voltages.

The simplest oscillator is the ring oscillator (see fig. 20.56). It consists only of a ring connection of n inverters. Under normal conditions n must be odd to start oscillations. If the inverters are, however, differentially laid out then an oscillation can be achieved also with an even number of inverters by exchanging two outputs. This detail is important with high frequency circuits if two oscillations with 90 degrees of phase shift are needed for special mixers (quadrature mixers). The ring oscillator is used frequently as a test for delay times of the technology, because the frequency of oscillation f_{Osc} informs us of the mean delay time τ_{Delay} of an inverter:

$$f_{\mathrm{Osc}} = \frac{1}{2 \cdot n \cdot \tau_{\mathrm{Delay}}}$$

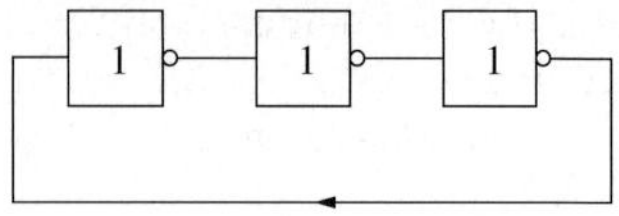

Fig. 20.56 Ring oscillator with three inverters

A voltage controlled version of a CMOS ring oscillator is shown in fig. 20.57. The frequency is controlled by the transistors M_S working in the ohmic range; the delay is changed via the adjusted resistors. Another kind of control of the delay is done by capacitors coupled with transistors in the ohmic range. Also the frequency of oscillation can be changed roughly by switching on or off delay stages. With CMOS realizations the shortening of the ring even saves current. With all types, however, the delay time depends on the supply voltage, unless special inverter types independent of supply voltage are employed. ECL inverters are quite constant in their delay time owing to the predefined signal swing. Relaxation oscillators have advantages compared with ring oscillators where phase jitter and current consumption are concerned. Also operating voltage independence is rather good within symmetrical circuit design.

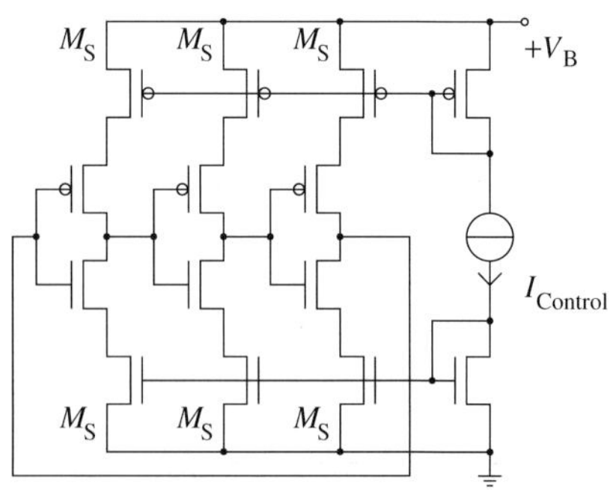

Fig. 20.57 Current controlled ring oscillator with inverters ($n = 3$)

Figure 20.58 shows the circuit of a current controlled oscillator and in fig. 20.59 the time functions of the voltages V_1, V_2 and V_3 are charted. The realization of the load elements was accomplished by two transistors each. The transistors M3 and M5 work in the ohmic range and serve as pull up resistors. The transistors M4 and M6 are connected as 'active' resistors with large W/L ratio; they limit the voltage drop to approximately V_{Th} when the control current $I_{Control}$ varies. Considering a symmetrical circuit the pulse length is equal to the pulse pause. The constant current I provided by the transistors M7 and/or M8 has to charge the capacitor in one phase up to $2 \cdot V_{Th}$, thus the necessary time is t.

$$t = C \cdot \frac{2 \cdot V_{Th}}{I_{Control}}$$

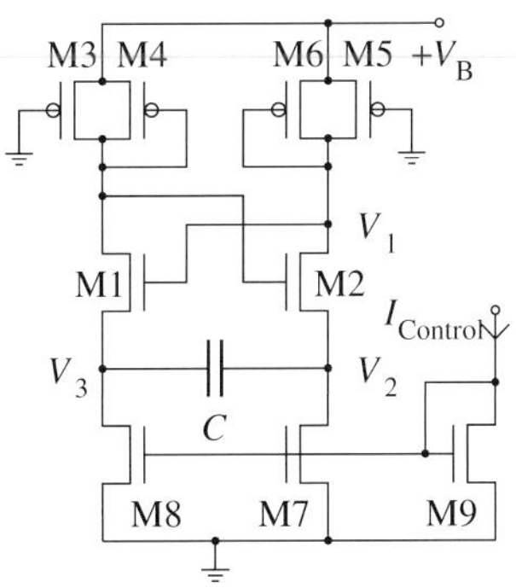

Fig. 20.58 Circuit diagram of a current controlled CMOS relaxation oscillator

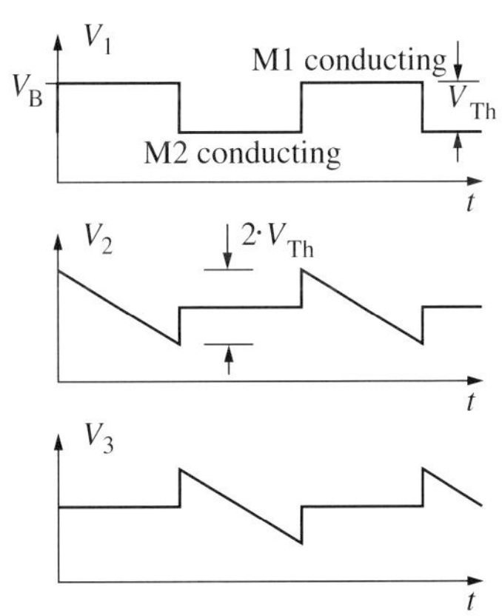

Fig. 20.59 Voltage waveforms of the current controlled CMOS relaxation oscillator

The frequency of oscillation is then

$$f_{Osc} = \frac{1}{2 \cdot t} = \frac{I_{Control}}{4 \cdot C \cdot V_{Th}}$$

With this circuit the body effect must be considered. In the above equation the increased threshold voltage must be inserted. In first approximation the frequency is proportional to the charging current and independent of the supply voltage. A moderate voltage dependence of the capacitor may be neglected. However, the parasitic capacitances to the substrate at both connections have to be dimensioned carefully to be identical, otherwise the rise times on both sides of the circuit become asymmetrical. The circuit shown here has an ancestor within the bipolar circuit technology, there, however, mostly two further emitter followers serve as level shifters. The phase noise of this kind of circuit is not suitable for more ambitious systems of the communication technology, after measured data in [20.7] approximately -73 dBc/Hz at 10 kHz distance from the carrier frequency (dBc: relative to the power of the carrier).

Oscillators for applications in communications engineering devices have two principal claims, i.e., frequency stability and small phase noise. The noise spectrum results from the active and passive elements of an oscillator producing thermal and $1/f$ noise. In particular, the $1/f$ noise, which reaches high values just at very low frequencies, is most disturbing.

This noise spectrum modulates the oscillator frequency in its phase, and because the spectral distribution rises very rapidly with low frequencies, the spectrum of the phase noise also lies close to the oscillator frequency. With today's very closely occupied frequency bands a receiver must be able to receive a weak signal trouble-free even if in the adjacent channel a very much stronger signal (e.g., around 80 dB more strongly) is present. If the oscillator in a receiver is now afflicted with larger phase noise then its noise, as well as the strong signal from the adjacent channel, mixes into the desired channel and covers the information signal. The requirements on small phase noise are very stringent and can be achieved only with large effort. The GSM specifications (global system for mobile communications) require a one-sided spectral noise density smaller than -100 dBc/Hz within 10 kHz distance from the carrier frequency. The amplitude noise usually is not very strong, because harmonic oscillators (sinusoidal oscillators) always contain some amplitude limiting functions.

In [20.20, p.530 ff] and [20.8] criteria are formulated which lead to good phase noise characteristics:

$$\overline{V_r^2} \sim \frac{1}{\text{Output power of oscillator}} \cdot \left(\frac{1}{\text{loaded Q}}\right)^2 \cdot \Delta f$$

→ a signal amplitude as large as possible;
→ a high quality factor Q of the resonator.

The outstanding noise characteristic of quartz oscillators compared with oscillators with LC resonant circuits is not only owed to the very high quality of the quartz resonators, but is additionally caused by the high voltage transformation which the internal node in a quartz (fig. 20.60) gains [20.8]. The equivalent model combines all possible losses in the resistance R_S. The capacity C_0, which essentially results from the quartz mounting plate, exceeds C_S by several orders of magnitude. Up to now all efforts have failed to integrate a quartz resonator on a silicon crystal. Also surface wave resonators (SAW, Surface Acoustic Wave), suitable for higher frequencies, have not been able to be integrated so far. They have the same equivalent model as the quartz. For very high frequencies, starting from approximately 1 GHz, inductors can be built in the form of planar spirals of aluminium normally used for wiring. The attainable quality values Q are usually below ten because the resistance afflicted substrate material has a distance of only about 1 µm from the coil. Somewhat better characteristics can be achieved with inductors made from bonding wires; however, the reproducibility is not very good.

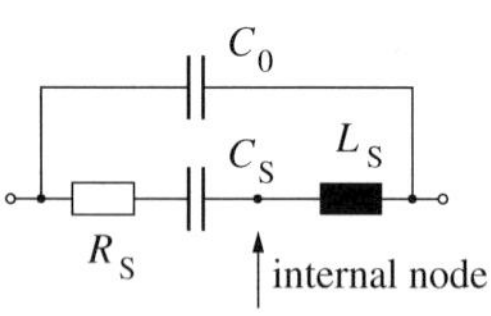

Fig. 20.60 Electrical equivalent model of a quartz resonator

For reasons of better phase noise values the following circuits are all equipped with a resonator. For the analysis of the condition for the start of oscillation the method of the negative resistance can be used. One separates the resonator at a suitable point from the amplifier and calculates the input impedance of the amplifier at this point. The condition of the start of oscillation is given if the negative internal resistance just compensates the real part of the resonator's impedance. A very frequently used oscillator circuit, see fig. 20.61, is the Pierce circuit, here realized by a CMOS inverter with feedback. The feedback resistor between input and output of the amplifier should be as high as possible, and can be achieved by the leakage current of a turned off transistor. The transconductance, necessary in the small signal operation for the start of oscillation, can be determined from the small signal equivalent model to

$$g_m \approx \frac{C_1 \cdot C_2}{R_S \cdot C_0^2} \quad \text{with} \quad C_1, C_2 \gg C_0 \gg C_S$$

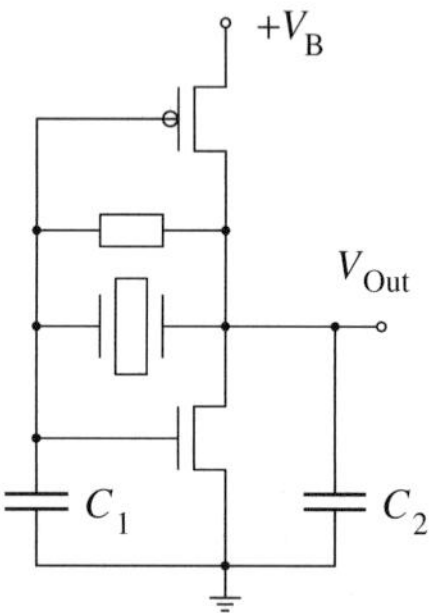

Fig. 20.61 Schematic of a Pierce oscillator with CMOS transistors

The calculated transconductance is the sum of the transconductances of the two transistors under operating conditions. According to [20.20] one should start a design with five times the starting transconductance and then continue an optimisation from this value. For analog purposes the transconductance may not be chosen too large, otherwise the oscillation is too strongly limited and, apart from the usable oscillation contains too many harmonics.

A voltage controlled oscillator in bipolar technology is shown in fig. 20.62. The two cross-coupled transistors generate a negative resistance of a value of approximately $2/g_m$ between the collectors, which can be changed very simply by adjusting the quiescent current. As a result of the differential amplifier a symmetrical delimitation of the oscillator amplitude arises and thus a stable amplitude. The oscillator signal can be coupled to other stages inductively or with galvanic differential coupling at the collector nodes. The oscillation frequency may be controlled with two symmetrically coupled varactor diodes. By stringent symmetry of the circuit and in the layout too very good uncoupling of disturbances from other parts of the circuit can be achieved.

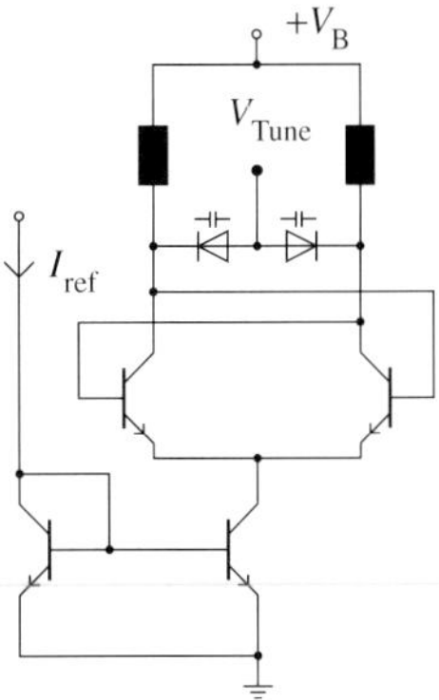

Fig. 20.62 Schematic of a voltage-controlled oscillator with bipolar transistors

20.3 Digital Circuits

About 90 % of the integrated circuits belong to digital circuits; however, this is true only if counting the chips but not the integrated transistors. Contrary to the case of analog circuits, deep knowledge of the actual semiconductor processes is hardly any longer necessary for a successful design, since the corresponding design tools can not only select the required basic cells, but can place and route them too. The basic cells are normally offered by the semiconductor manufacturers in sufficient quantity, thus only in exceptional cases do the required cells have to be designed and tested. This is also the reason why here the basic elements in their different realizations are treated only very superficially. In addition, dynamic principles are not considered (except the dynamic memory) because they are not suitable in many cases for designs in an automated procedure, but should be used only in complete tested cells. Comment of an experienced designer: 'Dynamic circuits are the simplest method of creating a chaotic circuit behavior'.

20.3.1 Basic Elements

The kind of realization of the basic elements has substantial influence on the switching speed and the power dissipation. Whereas the switching speed is determined by the signal swing, the available charging current and the capacity to be charged, the power dissipation still needs to be separated into dynamic and static power dissipation. Pure CMOS circuits have almost no static power dissipation apart from very small leakage currents of the source and drain PN junctions. The dynamic power dissipation is $\sim f_{CLK} \cdot C_{Load} \cdot V_B^2$, where f_{CLK} is the clock frequency and C_{Load} the load capacity. 'Standard' BICMOS circuits follow the same conditions as CMOS circuits. Bipolar circuits, which do not operate in the saturation region (ECL and/or CML), have no dynamic, but not always an insignificant static power dissipation.

20.3.1.1 Inverter

The CMOS inverter's basic form is shown in fig. 20.63. Its transfer characteristic is traced in fig. 20.64 for the practically important case that $K_n = K_p$ along with $(W/L)_p \approx 3 \cdot (W/L)_n$ and $V_{Thn} = -V_{Thp} = V_{Th}$.

The point with $V_{In} = V_{Out}$ then lies at $V_B/2$. The points with slope $= -1$ of the transfer characteristic are given after [20.27, p. 506] as

$$V_{IL} = \frac{3}{8} \cdot V_B \cdot \left(1 + \frac{2}{3} \cdot \frac{V_{Th}}{V_B}\right)$$

and

$$V_{IH} = \frac{5}{8} \cdot V_B \cdot \left(1 - \frac{2}{5} \cdot \frac{V_{Th}}{V_B}\right)$$

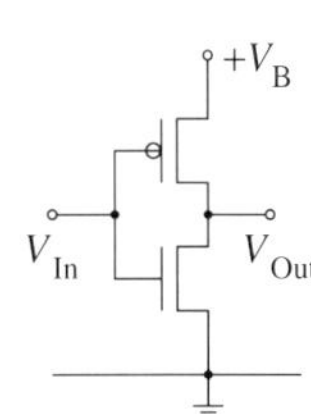

Fig. 20.63 Schematic of a CMOS inverter

The active switching range is thus independent of the threshold voltage and equal to $V_B/4$. During the input passing through the switching range, a drain current flows through both transistors. This current and the additional charging and load currents must be provided by the power supply source during switching.

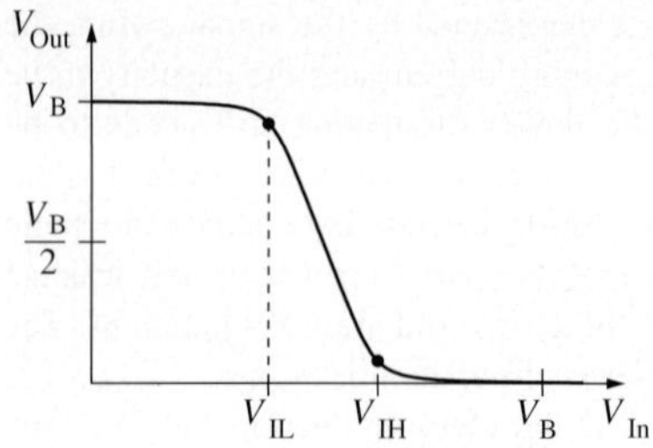

Fig. 20.64 Transfer characteristic of a CMOS inverter

For BICMOS circuits there are still no standard circuits developed. This is probably because the processes generate very differently fast MOS, respectively bipolar transistors, and thus depending upon the process there is a different optimization necessary. BICMOS inverters (totem pole) in the simplest form do not reach the full output level; the voltage drop across a base emitter diode is missing at each bound of the supply voltage. That restricts the use with supply voltages becoming smaller. Even the switching time is not as shortened as expected by the additional charging, respectively discharging current supplied by the bipolar buffer transistors (only approximately $1/2$ instead of $1/\beta$ [20.15] and [20.17]). An enlargement of the output level is achieved by connecting resistors in parallel between base and emitter at each of the two bipolar driver transistors, see fig. 20.65. The current through these resistors additionally must be supplied by the CMOS transistors. With most BICMOS circuits the logic functions are realized with MOS transistors and the driver functions with bipolar transistors.

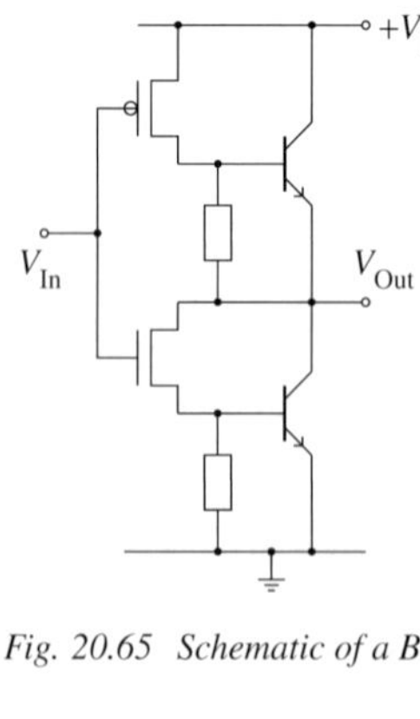

Fig. 20.65 Schematic of a BICMOS inverter

The maximum switching speed of bipolar transistors can be achieved only in the unsaturated operation, which means the circuit is almost an analog circuit. This becomes apparent in the circuit design. Thus fig. 20.66 shows an CML inverter (**C**urrent **M**ode **L**ogic). Notice that all logic voltages are related to the most positive point of the supply voltage, thus their levels do not depend on the supply voltage. It is a slightly limiting differential amplifier with the emitter current source replaced by a resistor, since no common mode problems exist. The circuit design of this logic differs from the classical ECL logic (**E**mitter **C**oupled **L**ogic) that between two cascaded gates no emitter followers are necessary for coupling. The operating conditions are adjusted to a logic swing of, for instance, only 0.3 V, thus the transistors just do not reach the saturation. The reference voltage has to be the average between the logic low and logic high levels.

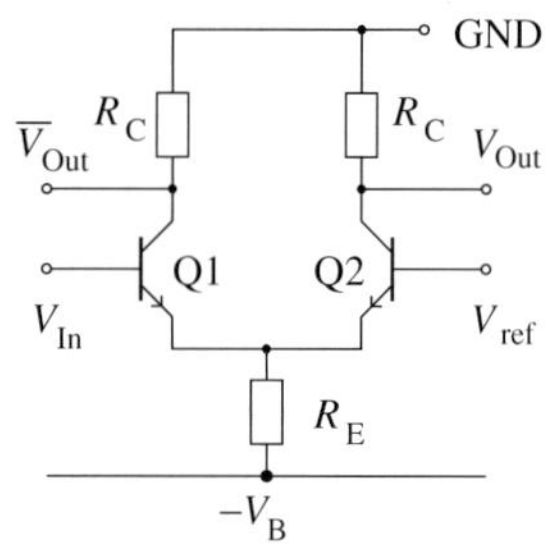

Fig. 20.66 Schematic of a CML inverter

20.3.1.2 NAND Gate

The schematic of a CMOS NAND gate is shown in fig. 20.67. To achieve a switching point at $V_B/2$, the W/L ratio of the N channel and P channel transistors must be selected as follows:

$$(W/L)_p = \frac{1}{4} \cdot \frac{\mu_n}{\mu_p} \cdot (W/L)_n \approx \frac{3}{4} \cdot (W/L)_n$$

for $V_{Thn} = -V_{Thp} = V_{Th}$.

The body effect of the transistors connected in series was neglected.

Within the BICMOS NAND gate, see fig. 20.68, according to section 20.3.1.1 the logic function is realized with MOS transistors and the driver function with bipolar transistors. The CML logic does not directly provide a simple realization of a NAND gate. A series connection of current switches, as usual in ECL logic, is quite complicated, because with the differential amplifier circuits in series several bias sources and potential shifting circuits become necessary. Thus the desired result is achieved by connecting two inverters to the inputs of an OR gate.

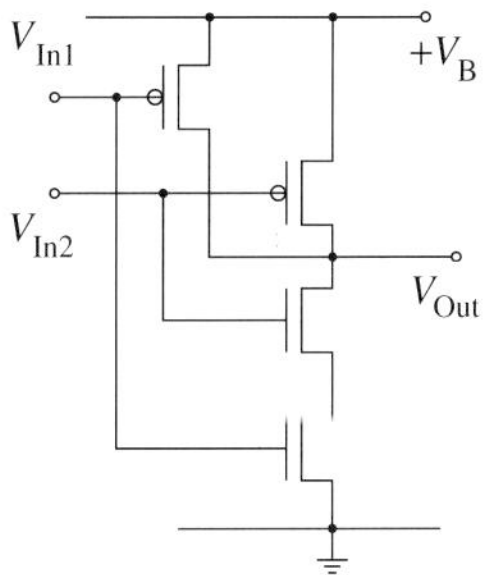

Fig. 20.67 Schematic of a CMOS NAND gate

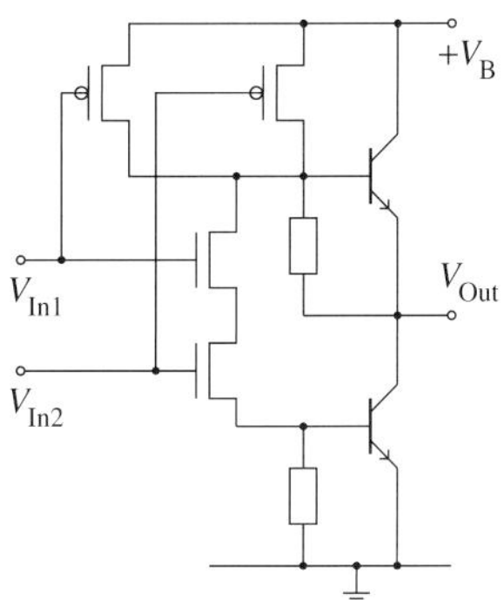

Fig. 20.68 Schematic of a BICMOS NAND gate

20.3.1.3 NOR Gates

The schematic of a CMOS NOR gate is shown in fig. 20.69. To adjust the switching point at $V_B/2$, the W/L ratios of the N channel and P channel transistors must be selected as follows:

$$(W/L)_p = 4 \cdot \frac{\mu_n}{\mu_p} \cdot (W/L)_n \approx 12 \cdot (W/L)_n$$

for $V_{Thn} = -V_{Thp} = V_{Th}$.

The body effect of the transistors connected in series was neglected. The very high differences in the W/L ratio provide large capacities allied with the P channel transistors, thus in CMOS technology NOR gates are slower than NAND gates.

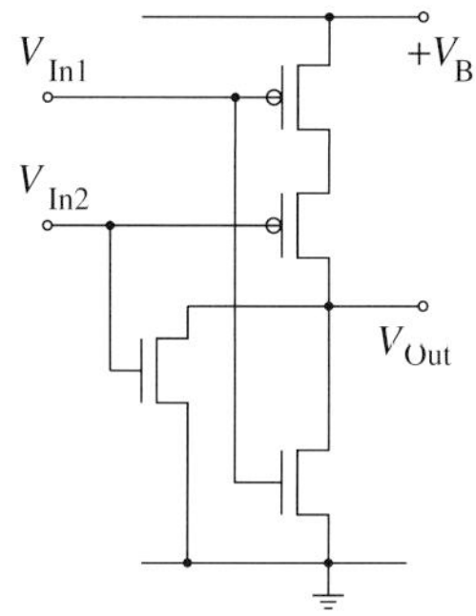

Fig. 20.69 Schematic of a CMOS NOR gate

The realization of the NOR gates in BICMOS technology is shown in fig. 20.70, the realization with bipolar transistors in fig. 20.71.

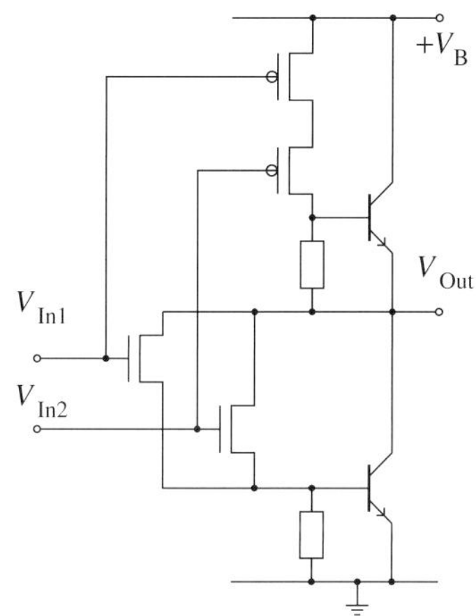

Fig. 20.70 Schematic of a BICMOS NOR gate

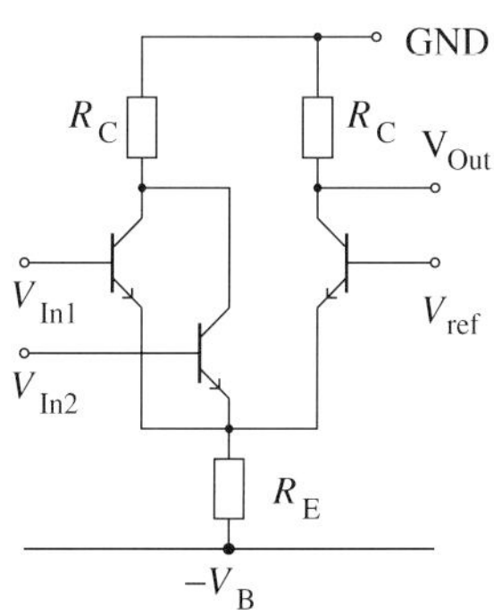

Fig. 20.71 Schematic of a CML NOR gate

20.3.1.4 Transmission Gates

As already described in section 20.2.3.3, transmission gates are useful in analog as well as in digital operations. They are used in logic gates, flip flops, decoders and multiplexers, respectively demultiplexers. Within some applications both transistors of the transmission gate are not necessary, thus only half the number of transistors is required. If, e.g., NMOS transistors were used the high level can not accurately be transferred but only decreased by V_{Thn}, thus the noise immunity suffers somewhat. Lowering the switching point at the following gate, respectively inverter, by slightly changing the W/L ratio will decrease this effect a little bit. As an example Figure 20.72 shows the schematic of a 4 to 1 multiplexer. This type of circuit can evidently also be used in the reverse direction as a demultiplexer. One recognizes in the schematic that in each signal path from the input to the output two transmission gates are connected in series. This AND gate can be eliminated by shifting this AND into the control of the transmission gates.

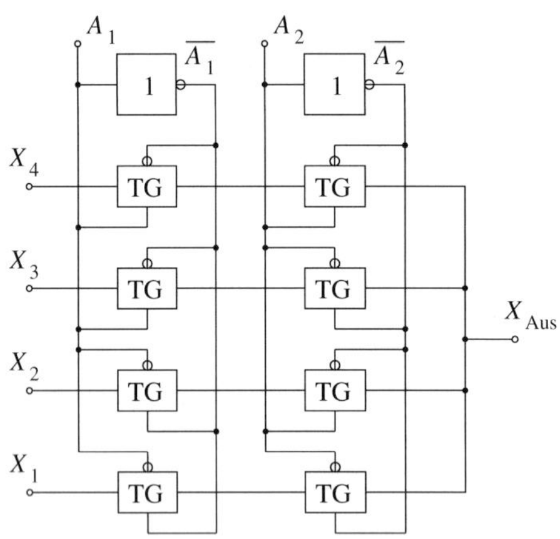

Fig. 20.72 Schematic of a CMOS 4 to1 multiplexer

20.3.2 Flip Flops

The many different flip flops are realized within bipolar technology (CML and ECL) by cross-coupled gates, within (BI)CMOS usually by inverters and transmission gates. Only the somewhat unusual realization with transmission gates will be described here. It became generally accepted because it requires a smaller number of transistors than the gate realizations. As an example a clocked D flip flop is considered. Figure 20.73 shows the schematic with gate symbols, and in parenthesis the number of transistors necessary for the realization in CMOS technology are indicated. Figure 20.74 shows the realization with transmission gates. The second inverter in the clock line should not be eliminated by common use of the first inverter, because a short delay is necessary for the transmission gate, which connects the output and input of the two inverters for the data storage. Otherwise the D input and the Q output work shortly against each other. By designing one of the transmission gates 'more weakly' respectively 'more strongly' (by reduction, respectively enlargement, of the W/L ratio) this behavior can be made tolerable however.

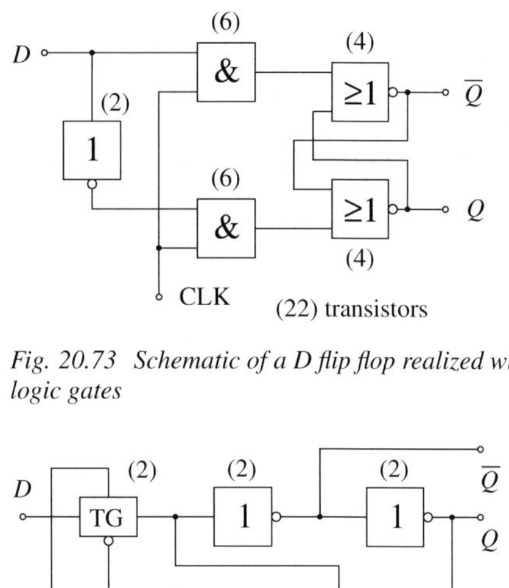

Fig. 20.73 Schematic of a D flip flop realized with logic gates

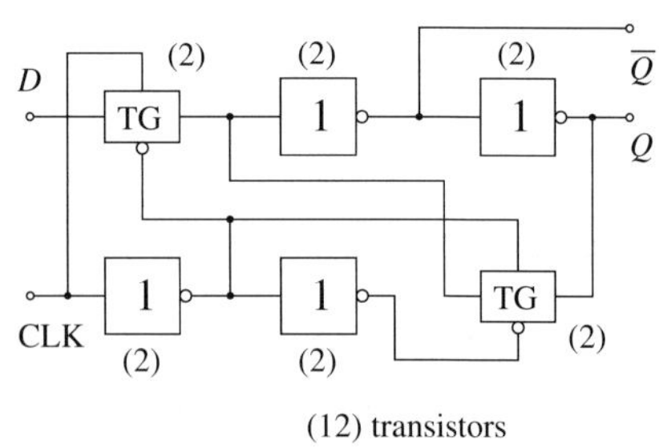

Fig. 20.74 Schematic of a D flip flop, realized with transmission gates

20.3.3 Random Access Memory (RAM)

In the course of time the request for storage size increased tremendously, memory sizes are rarely below 1 kilobytes even with simple applications. In order to achieve small sized address decoders a matrix arrangement with line and column addressing (row/column address) is used. Thus the size of the decoder is divided into two with half the bit number each. The above memory needs 10 address bits to be demultiplexed to 1024 outputs without division in line and column addressing. With splitting two decoders are necessary, each with 5 address bits together decoded to 64 outputs. Figure 20.75 shows the principle of the matrix arrangement of a read/write memory with the splitting into row and column decoder. The decoder is, in principle, a demultiplexer see fig. 20.72, with $X_{Out} = V_B$ as input and the outputs of the decoder are the signals X_1 to X_4. To increase the speed only the NMOS transistor of the transmission gates are used. At each crossing of a row line (word line) and a column line (bit line) a memory cell is placed. The column line consists, however, physically of two lines, the *Bit* and $\overline{Bit}$ lines. Via these lines the read or write information are transferred from or to the memory cell. The design of memory cells and the affiliated peripheral circuits is very sophisticated, therefore many semiconductor manufacturers provide the necessary module generators.

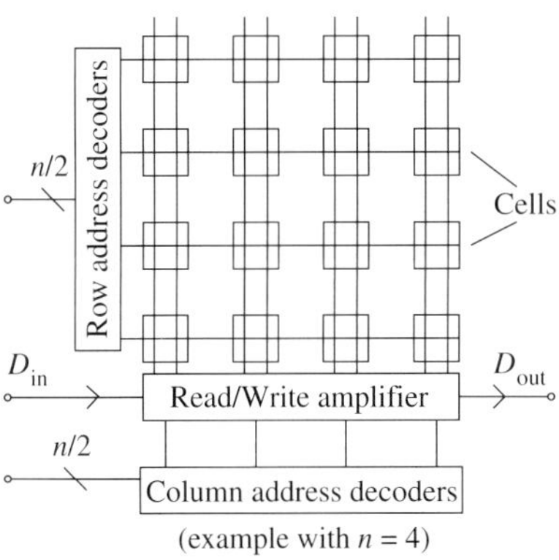

Fig. 20.75 Principle of the memory array of a random access memory

20.3.3.1 Static Random Access Memory (SRAM)

Two basic circuits became generally accepted as the static memory cell. Both differ only in the load elements. Figure 20.76 shows the cell with two PMOS load transistors, thus the cell has 6 transistors and is therefore called a 6 transistor cell.

The core cell, the actual memory, is built from two cross-coupled inverters. The selection and writing takes place via the two switch transistors which are activated by the word line. In order to write to the cell, first the appropriate information is switched to the *Bit* and $\overline{Bit}$ lines and then passed to the cell by setting the row selection. For reading, the *Bit* lines are precharged to a common voltage, then after activation of the word line one of the *Bit* lines is conditionally discharged and read as data information with a differential amplifier.

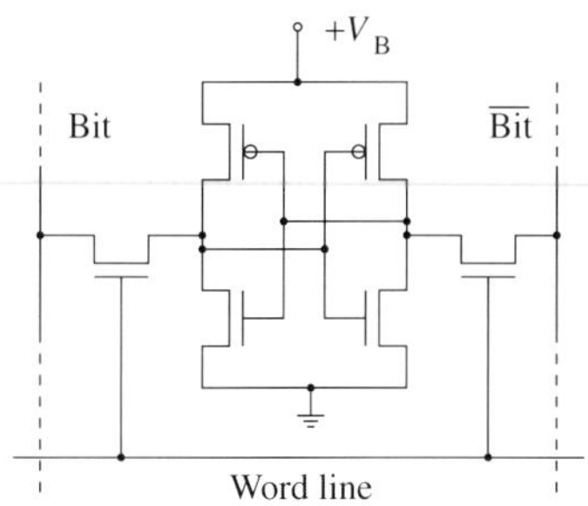

Fig. 20.76 Schematic of the static 6 transistor cell

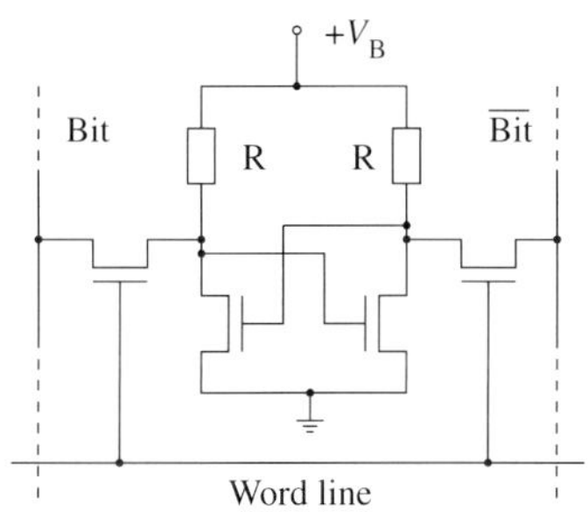

Fig. 20.77 Schematic of the static 4 transistor cell

The read amplifier for the static cell is quite uncritical, since the cell supplies large level changes. Figure 20.77 shows the 4 transistor cell with the load elements replaced by very high impedance polysilicon resistors (> 10 MΩ). Whilst the 6 transistor cell does not dissipate static power, the 4 transistor cell does, because in both resistors a current flows, although a very small one. The reason for the use of the resistors is that they hardly need any chip area, because the resistors can be placed over otherwise unused surfaces. The use of bipolar transistors from a BICMOS process in the column decoder and the read amplifier will approximately double the speed compared to a pure CMOS realization.

20.3.3.2 Dynamic Random Access Memory (DRAM)

Dynamic memory cells keep the information by charge storing. There are different realizations of the storage, usually the gate area of a MOS transistor with today's structure sizes is not sufficient. For the circuit design the realization of the capacity is not important. Figure 20.78 shows the 1 transistor cell which is not really complex. The writing of the information is done as with the static memory. The reading of the information causes problems, because the capacity of the memory cell is relatively small compared to the large capacity of the Bit line. Therefore first the Bit line is precharged, then after activation of the row selection line the potential of the Bit line is read as data information with a special read amplifier. The read signal is the potential, changed by connecting the charged cell. The change of potential is very small, thus the read evaluation amplifier represents a critical component within the dynamic memory.

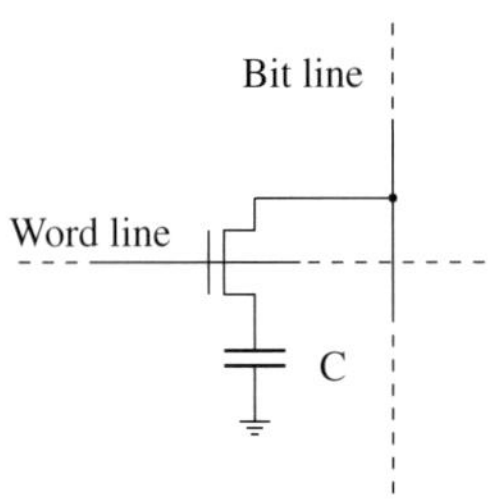

Fig. 20.78 Schematic of the dynamic 1 transistor cell

20.4 Digital to Analog and Analog to Digital Converter

The converter components as mediators between the world of pure analog technique and pure digital technique receive a strong impetus. Several reasons are responsible for that. One is that the technology of the circuits has progressed so far, that even fastidious analog functions are realizable with only small changes, if at all, in the processing of digital circuits. Then the trend continues to ever further digitization of the once analog signal processing. Thus the clear boundary between analog and digital circuits is disappearing. That means pure analog circuits and pure digital circuits increasingly gain a rarity value. The working of the analog and digital components together on a chip is very problematic for the analog part. The noise spectrum of the digital circuits is transferred via the substrate, capacitively, inductively, and via changes of temperature to the analog parts. Thus the analog part of the circuit must be protected by relevant shielding measures against that noise of the digital circuits.

The conditions of accuracy request high qualification of the circuits; for example, the small resolution of 8-bit of a converter already needs an entire accuracy of approximately 0.4 %. Thus the limit of the accuracy attainable by matching alone is almost reached. 16-bit resolution, today usual within some applications, requires approximately 0.0015 % accuracy. Such values are achieveable only by the exhaustion of all means and alignment with LASER, respectively, with similar methods. In principle the conversions generate always a difference between the desired value and the value reached, since the quantization into n bits can represent only 2^n different voltage or current values. The error signal is called quantization noise. The signal to noise ratio (SNR), measured in dB, achieveable with a quantization into n bits is

$$SNR \approx (6 \cdot n + 1.7)\,\mathrm{dB}$$

It is assumed that the input signal is sinusoidal and reaches the limiting bounds of the converter. Important criteria for the choice are resolution, accuracy, and speed, in which the accuracy of a converter is impaired with rising speed. The binary number represents an integer value, whose decimal equivalent X is given as

$$X = \sum_{i=0}^{n-1} S_i \cdot 2^i$$

n is the number of digits of the binary word, S_i are the binary coefficients with the value 0 or 1, S_{n-1} is thus the most significant bit (MSB).

20.4.1 Digital to Analog Converter

20.4.1.1 Converter with Resistor Chain

The oldest converter principle is the voltage divider principle. The voltage drop across a number X of partial resistors ($X < 2^n$), within a total chain of 2^n identical partial resistors, represents the analog value via a network of switches. Figure 20.79 shows the principle for $n = 2$. Thereby the converter output arises according to the voltage divider rule to

$$V_{Out} = \frac{X \cdot R}{2^n \cdot R} \cdot V_{ref} = \frac{X}{2^n} \cdot V_{ref}$$
$$= V_{ref} \cdot \sum_{i=0}^{n-1} S_i \cdot 2^{i-n}$$

The number X controls a 1 from n decoder, which in turn activates the X-th switch. Here NMOS transistors serve as switches, thus the reference voltage V_{ref} always has to be smaller by at least one threshold voltage than the gate potential of the selection transistor. In addition the output should be provided with a buffer amplifier with very high input impedance, thus the 'on resistance' of the selection transistor does not have an influence on the accuracy.

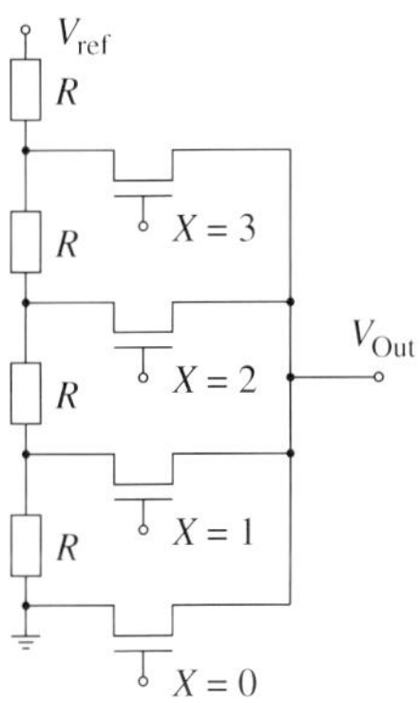

Fig. 20.79 Principle of a digital to analog converter with resistor chain

The single stage switch net shown here provides the disadvantage of a high capacity at the output node, which increases the conversion time. It can be replaced also by a n stage switch net without decoder. Then high longitudinal resistances with reduced capacities are achieved (compare section 20.3.1.4). A current flows continuously in the resistor chain, thus the static power dissipation can be reduced by increasing the resistance value according to the feasibility. The accuracy of the procedure directly depends on the accuracy of the resistances, thus a good matching and layout are important. The procedure also permits nonlinear conversions by implementing resistances with different values.

20.4.1.2 *R 2R* Converter

In the previous chapter a resolution of n bits required an expenditure of 2^n resistors and at least the same number of switches . A converter with a so called R 2R ladder network has only $2 \cdot (n-1)+1$ resistors and n switches. A substantial reduction of the expenditure for large n results. Figure 20.80 shows the principle of a 4 bit converter with *R 2R* ladder network as example. Both ends of the network have to be terminated according to the condition that from each internal node of the network in each direction the impedance $2 \cdot R$ arises. At the two end nodes that is not the case; however one can make sure that the correct current transfer characteristic is achieved. The *R 2R* ladder network is ideally suited for an integrated realization, because voltage dividers can be manufactured very precisely using the matching principle. The network has the characteristic that the node potentials and the currents flowing off the nodes are halved in each step to the next node. Thus, for example, the current I_0 reaches the input of the operational amplifier only with 1/8 of its original strength. The output voltage is

$$V_{Out} = R \cdot I_{ref} \cdot \sum_{i=0}^{n-1} S_i \cdot 2^{(i-n+1)}$$

The switches S_i, which are controlled by the binary word, are realized by two transistors operating differentially. Figure 20.81 shows the bipolar and MOS realization.

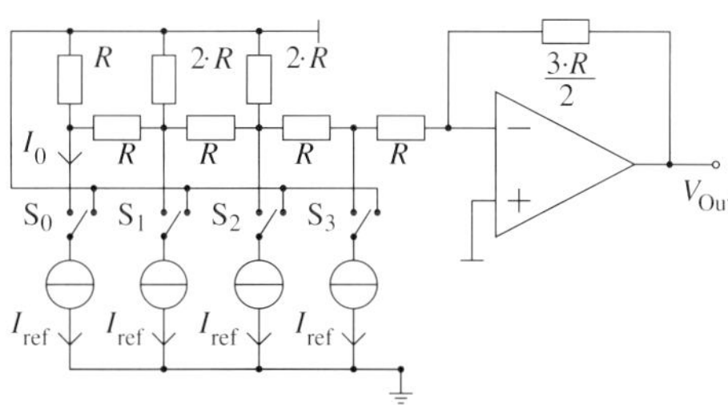

Fig. 20.80 Principle of a digital to analog converter with R 2R resistor chain

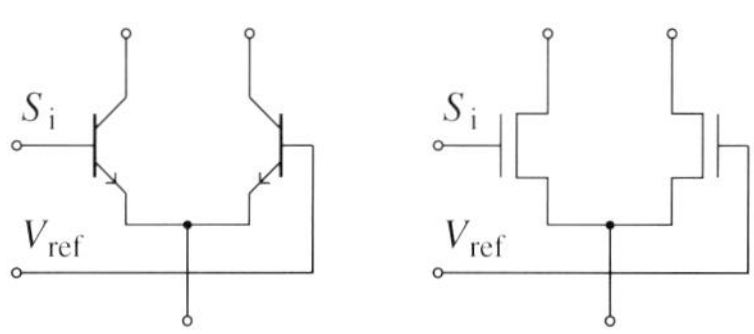

Fig. 20.81 Realization of the switches

21 Geometric Layout

HARALD TOEPFER

21.1 The Layout of CMOS circuits

21.1.1 Introduction

The layout of an integrated circuit consists of a great number of polygons attached to distinct layers. During the design process the layers are generated according to the needs of the foundry.

Derived from these layout data the mask data are created by a software that is provided mostly by the chip manufacturer as well. During the generation of the data for the masks some of the original layout data are left unchanged whilst others are distorted or even newly computed on the base of the original layout data. The original design layers are fewer in number than the layers for the masks. Therefore they are easier to handle.

The layout of CMOS circuits is nowadays generated automatically by place and route tools in most cases, whereas analog circuits are created mostly manually by layout editors.

Layout editors are software programs for the creation and checking of chip layouts. Such editors are e.g.: **Virtuoso** from Cadence, **IC-Station** from Mentor Graphics, **L-Edit** from Tanner EDA and **Layed** from Catena.

A typical layout editor provides besides others the following functions:

- import and export of GDSII data;
- display of the layout and selectable layers;
- generation of transistors based on parameterizable base cells;
- wiring in different layers with predefined contacts;
- design Rule Check;
- saving of the design to an own data base.

In many cases place and route tools and simulators can be started from the workspace of the layout editor.

21.1.2 The Layers of the CMOS Layout

The layout of a CMOS circuit consists of about 10 to 20 layers. Table 21.1 shows the usual stack of a simple CMOS technology with two metal layers. The layers marked with the word drawn are the **layers of the original design**. On the other hand, the layers P Well and N Implant are generated automatically with the mask data.

The layers according to table 21.1 are not standardized. In another process (design kit) the layer N Implant can be drawn as the layer P Implant and will be generated automatically. Even the names of the layers may vary from chip manufacturer to chip manufacturer.

The drain and the source electrodes of the MOS transistors are formed by the highly doped (N+) and (P+) regions, which are called **diffusion regions** as well. The name diffusion region originated historically. Previously these regions were formed by diffusion of highly doped oxides, nowadays exclusively by **ion implantation**.

Some Design Kits use the N+ and the P+ Region as originally drawn Layer. Other Design Kits use for these two highly doped regions the common layer **active**, called active region as well. To differ between (N+) and (P+) regions there are the two layers **N Implant** (sometimes n_diff, n_select) and **P Implant** (p_diff, p_select). If an active region is surrounded by a **P Implant** layer, the result is a highly doped (P+) region. If it is surrounded by an **N Implant** layer the result is a (N+) region. Because all active regions are either (N+) or (P+) doped, in most cases only one of the implant layers (e.g., P **Implant**) is drawn. The **N Implant** layer is computed together with the mask data by subtracting the **P Implant** layer from the overall chip area.

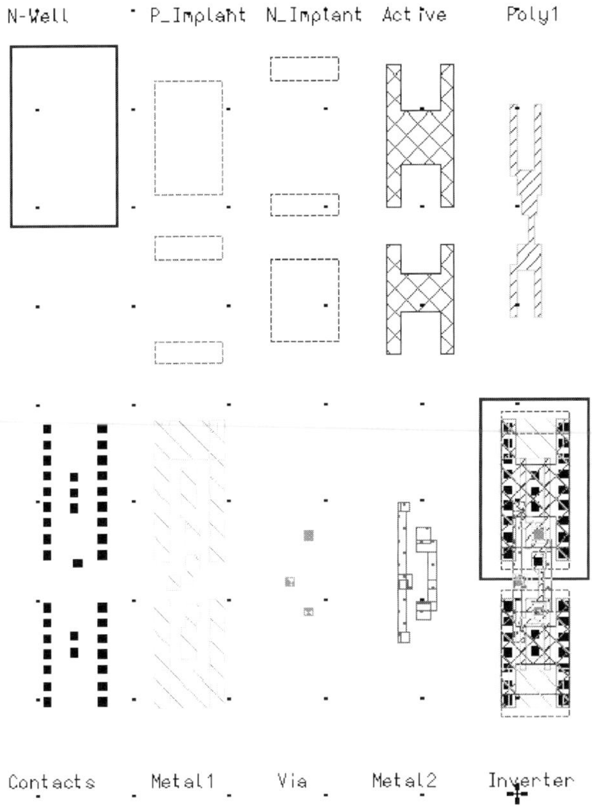

Fig. 21.4 Layout of the standard cell Inverter with its layers

The P substrate together with the P Well must be connected to ground potential, whereas the N Well must be pulled up to the positive supply voltage. In order to do this substrate and well contacts should be used as often as possible. A P substrate contact is formed by a (P+) doped active region which is connected to ground via Contact and Metal_1. An N Well contact is formed by a (N+) doped active region.

Substrate and **well contacts** prevent variations of the threshold owed to local substrate and well voltage variations (body effect). Furthermore, they bypass parasitic currents and improve resistance against latch up effects. They should therefore be placed as often as possible. The maximum distance allowed is about 100 μm. Better are distances from 10 to 50 μm.

Figure 21.4 depicts the layout of the standard cell Inverter. The logical equations for the parts of the CMOS technology using the layers of Table 21.1 are:

Field oxide:	not Active
(P+) regions:	Active and P_Implant and not Poly
(N+) regions:	Active and not P_Implant and not Poly
NMOS transistor:	Active and not P_Implant and Poly and not N_Well
PMOS transistor:	Active and P_Implant and Poly and N_Well
Source/Drain (NMOS):	Active and not P_Implant and not Poly and not N_Well
Source/Drain (PMOS):	Active and P_Implant and not Poly and N_Well
N Well contact:	Active and not P_Implant and not Poly and N_Well

P Substrate contact: Active and P_Implant and not Poly and not N_Well

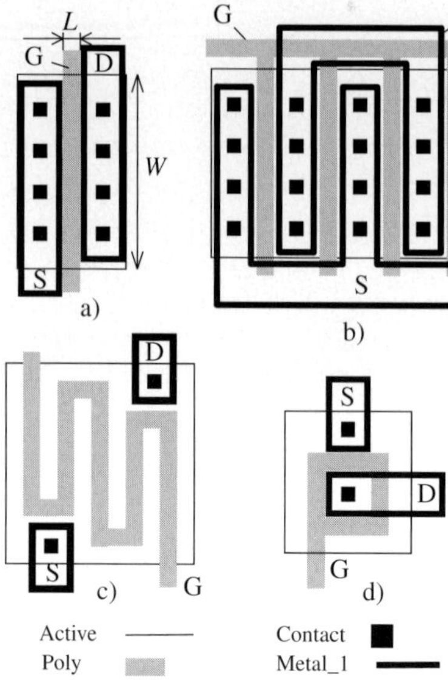

Fig. 21.5 Transistor layouts
a) linear transistor, b) comb transistor,
c) meander transistor, d) ring transistor

Figure 21.5 depicts different transistor layouts. The great number of source and drain contacts of the linear (a) and the comb (b) transistor provide low resistance between the source/drain diffusion regions and the metal stripes. These layouts are especially suited for analog circuits and power mosfets.

The meander transistor (c) has a large W, although the series resistance of source and drain are relatively large owing to the long diffusion areas. In **salicide processes** this is not a problem because of the low area resistance of the active areas. The ring shaped transistor (d) has a small drain area. The amount of leak current of the drain bulk diode is small, which is important at high temperatures. The small drain area results also in low junction capacity.

At the edges of the drain and source active regions there are placed **LDD layers** into the chip (LDD: Lightly Doped drain). The LDD areas are doped lower than the active regions. They take over some amount of the drain voltage and relieve the transistor (see fig. 21.6). The mask for the LDD is mostly derived from the Implant layers. For ESD protection devices the NLDD implant may be switched off by a special layer (Nldd_Prot).

The 11 layers of table 21.1 form a simple CMOS base technology for digital circuits with two metal layers. Today's CMOS technologies comprise up to 7 metal layers for the simplification of wiring. Each metal layer needs a Via layer for the connections to the metal layer lying below it. The lowest metal layer in many cases has a finer pitch than the higher ones. The upmost layers preferably are used for the supply voltages.

Figure 21.6 shows schematically a cross section of a CMOS process with 1 poly and 3 metal layers. The active areas and the Poly are covered by Silicide which reduces the area resistance of these layers to 2 . . . 4 Ω/square. **Silicides** are metal silicon compounds.

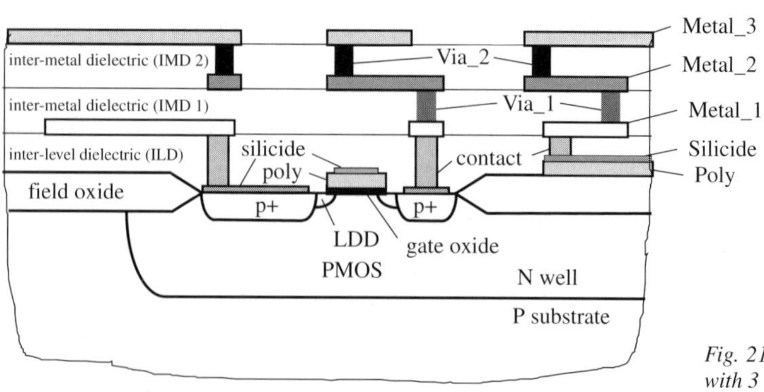

Fig. 21.6 Cross section CMOS with 3 metal layers and Silicide

To form a silicide layer the active areas and the Poly are etched free. Then the metal (tungsten, titanium, cobalt, or platinum) is sputtered. The silicide is then generated by thermal processing. Where there is no Poly and no active region no Silicide emerges and the sputtered metal layer is left unchanged and is etched away later. The silicide remains [21.14]. If active regions as well as Poly are silicidized, one calls that **salicide** (self **ali**gn sili**cide**).

The Silicide can be switched off by a special mask (Sal_Protect) which is important for ESD structures, because silicidized areas may break through from surges [21.2].

Modern CMOS technologies use some further layers as there are:

- Low VT layer for single transistors with low threshold voltages;
- Injector Oxide for EEPROM cells, thin Tunnel oxide (about 5 nm).

21.1.3 CMOS Layout and Latch up

A problem with CMOS circuits is the so called latch up. This means the ignition of a parasitic thyristor (see fig. 21.7). Caused by latch up the current consumption of a chip raises rapidly. At the same time the voltage at the chip falls down to the hold voltage of the parasitic thyristor. If the hold voltage is lower than the supply voltage a continuous current flows through the chip which may destroy it thermally.

In the circuit of fig. 21.7 a latch up is initiated if substrate and well currents cause voltage drops over the collector resistors R_p or R_n which are high enough to forward bias the base emitter diode of a parasitic transistor and when at the same time the loop gain of the circuit is higher than unity [21.16], [21.24], [21.17], [21.43].

The substrate or well currents may by caused by:

- exceeding VDD or falling below ground level of the voltages at the I/O pads;
- surges on the supply voltages;
- capacitive currents caused by fast rising voltages;
- internally induced currents into the thyristor by forward biased diodes.

The danger of a latch up is high in big output inverters and at switching inductive loads.

The EIA/JEDEC standard No. 78 [21.16] defines **two test criteria** for the stability against latch up with which every chip must comply:

- the supply pins are connected to 1.5 times VDD;
- into all I/O pads currents of $\pm 100\,\text{mA}$ are driven.

To prevent latch up, substrate and well currents have to be bypassed by means of many substrate and well contacts. For a 0.5 µm CMOS process a maximum distance of 40 µm between such contacts is requested [21.17]. This rule is very important for *standard cell layouts*. Therefore each cell must contain one substrate and one well contact as a minimum (fig. 21.3). Furthermore, for latch up stability it is recommended to provide a large distance L (fig. 21.7) between the source regions that form the emitters of the parasitic transistors. This will lower the current gain of the parasitic transistors.

Critical is the **latch up protection of pad cells**, because of over or under shooting transients of the voltages at the pads high currents may be induced into the thyristor structure. Therefore the latch up protection in these structures is designed very carefully. The substrate and well contacts in these cells are mostly designed as guard rings.

Figure 21.8 depicts the position of the guard rings in well and substrate. From an electric point of view the guard rings are base contacts of the parasitic PNP and NPN transistors. They are therefore called base guard rings as well. They decrease the resistors R_p and R_n in the schematic of fig. 21.7. The guard rings should be covered by a metal ring that is connected directly to the supply voltage or ground respectively and that has many connections to the active layer below it. Frequently such guard rings are not only installed within pad cells but also between the ring of pad cells and the inner core of the chip.

According to [21.17] the hold voltage of the parasitic thyristor with $L = 20$ µm (fig. 21.7) without guard ring is 3 V and with guard ring 9 V. A hold voltage of 9 V means that with a supply voltage of 5 V the thyristor is opened immediately after a latch up. These values may be improved by

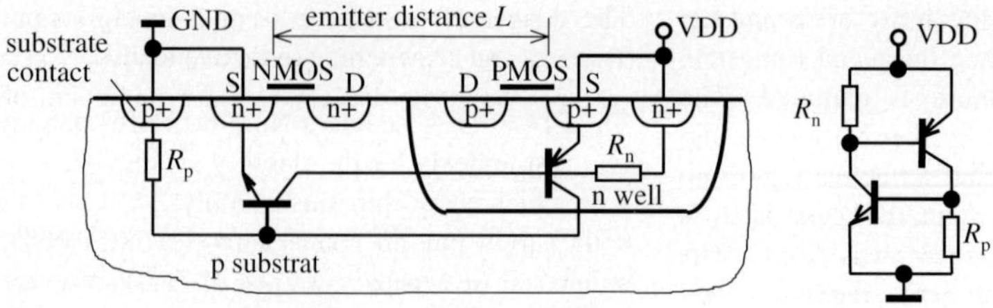

Fig. 21.7 CMOS circuit with parasitic Bipolar Transistors

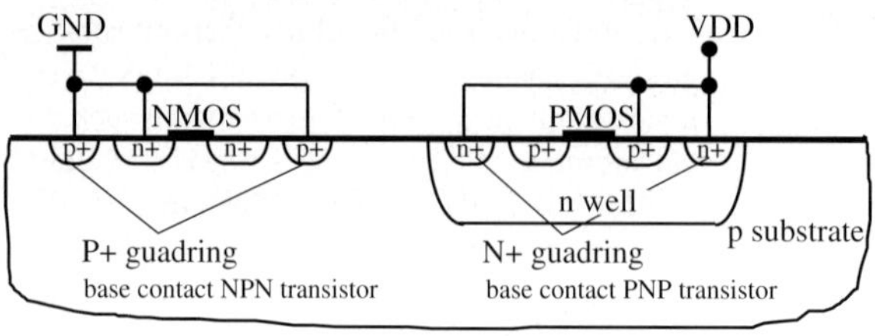

Fig. 21.8 latch up protection with base guard rings

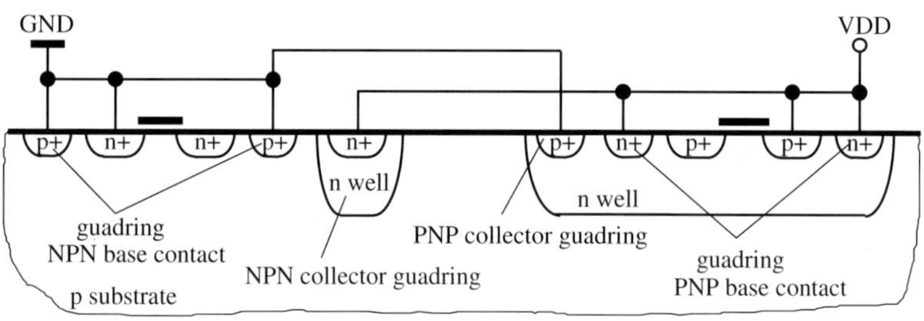

Fig. 21.9 latch up protection with base and collector guard rings

increasing the distance L. As a disadvantage guard rings require a lot of chip area.

In addition to the **base guard rings** frequently **collector guard rings** are used. These collector guard rings operate from an electrical point of view like additional collectors which sink the collector current out of the base region of the parasitic transistors and pass them directly to ground and supply voltage. The collector guard rings bypass R_p and R_n. The currents through R_p and R_n are decreased and the probability of triggering the parasitic thyristor is reduced.

Test results of Ker [21.17] suggest that the latch up sensitivity through the additional placement of collector guard rings as compared to the exclusive use of base guard rings is not substantially improved.

In many cases even large power transistors are surrounded by guard rings. Besides improved latch up protection the implementation of such guard rings mainly provides a better shielding of the remaining components of the circuit against interfering pulses which are created and radiated during the switching transients of these devices.

For the improvement of the latch up sensitivity for digital CMOS ICs epi substrates with highly doped and low resistive bulk are often used. With these types of substrate the actual circuit is implemented in a weakly doped epitactic layer at the surface. Only for specialized HF circuits uniformly doped substrates are used [21.15]. Figure 21.1 shows the typical cross sections of a uniformly doped and of an epi substrate.

Retrograde wells improve the latch up stability as well. The doping of the well is performed by a high energy implanter (4 . . . 10 MeV). Owing to this high energy a highly doped region is created at the bottom of the well causing low collector resistances.

21.1.4 Resistors in CMOS

Table 21.2 presents an overview of the usual resistances in Ω/square of the most important CMOS layers (each resistor is composed of a series connection of squares with the corresponding resistance value in Ω/square). The exact values particularly those of the (N+)/(P+) and Poly regions are extremely dependent on the technology.

Table 21.2 Resistance per square of different layers

Layer	Resistance/square
Metal (Aluminium)	40 . . . 100 mΩ
N+, P+ without Silizide	50 . . . 60 Ω/□
N+, P+ with Silizide	2 . . . 4 Ω/□
Poly without Silizide	30 . . . 400 Ω/□
Poly with Silizide	3 Ω/□
N Well	1 . . . 1.5 kΩ/□
Contact	1 . . . 10 Ω/□
Via	0.5 . . . 2 Ω/□

Analog circuits as A/D converters, filters, and oscillators frequently depend on reproducible and stable resistors. Integrated resistors vary by 20 . . . 50 % in their absolute values. More important is the matching precision of resistors with the same geometry produced in the same lot that is as close as 0.2 . . . 2 %.

All integrated resistors are semiconductor resistors. E.g., they heavily depend on temperature and voltage variations. The temperature coefficient of resistors in wells and diffusion regions is about 0.05 %/K [21.44].

Additionally the value of the resistance depends on the voltage difference between the resistor layer and the substrate.

In CMOS technologies resistors are implemented in the poly, well and active regions.

Poly Resistor

Poly resistors are used frequently. They have no common junction with the substrate and therefore are relatively independent against voltage variations (about 0.01 %/V). The width of poly resistors is 1 . . . 10 µm. Resistors with a width of 2 µm reach matching ratios of about 1 %, those with a width of 8µm reach 0.3 %. The matching accuracy is directly proportional to the area used for the resistors. More detailed information concerning the matching of poly resistors can be found in [21.49].

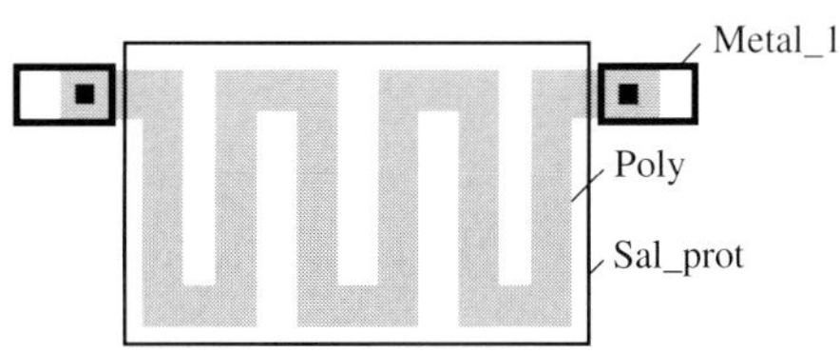

Fig. 21.10 Poly resistor without Salicide

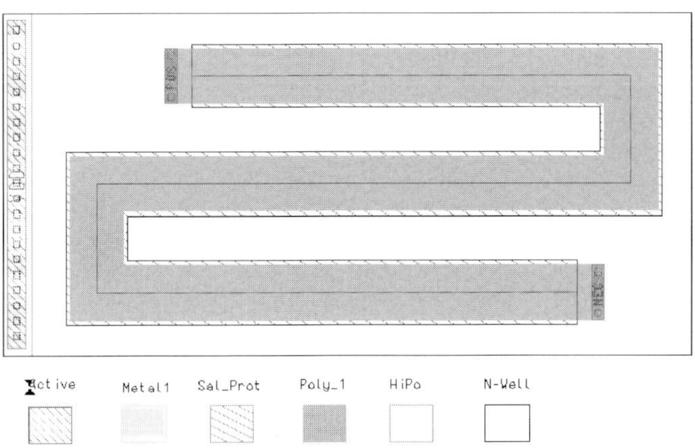

Fig. 21.11 Layout of a 30 KΩ Poly resistor with HiPo layer

Due to the low square resistance values of the poly layer the resistors are implemented as large meanders.

Polysilicon with Silicide is suited only for low resistance values up to 100 Ω. The total resistance value is highly influenced by the metal poly contact. Therefore these resistors should not be interrupted by contacts and metal. At the terminals of such resistors however multiple contacts are installed in parallel.

For resistors in the KΩ range Poly without Silicide can be used. In this case Poly is to be covered with the Sal_Protect layer. Figure 21.10 depicts such a resistor. Poly resistors are used, e.g., in $R/2R$ ladders in A/D and D/A converters [21.11].

Some analog CMOS processes additionally use a special lower doped polysilicon. To implement this the polysilicon has to be covered by an additional High Resistive Poly (HiPo). This low doped and consecutively highly resistive Poly has a resistance per square of about 2 KΩ.

Figure 21.11 shows the layout of a HiPo resistor with 30 KΩ. It is covered by Sal_Prot and HiPo.

Well Resistor

In digital processes where there is no high resistive Poly in use frequently well resistors replace the Poly resistors [21.44]. Owing to the high square resistance of 1 . . . 1.5 KΩ values of the resistances reach up to the MΩ range. Well resistors should be wide (5 . . . 10 μm) because of the deep diffusion. Figure 21.12 shows a well resistor of about 5 KΩ. Well resistors, owing to their junction, exhibit a relatively strong dependence on the voltage difference between well and substrate (about 1 %/V). Their temperature coefficient is about 0.6 %/K [21.44].

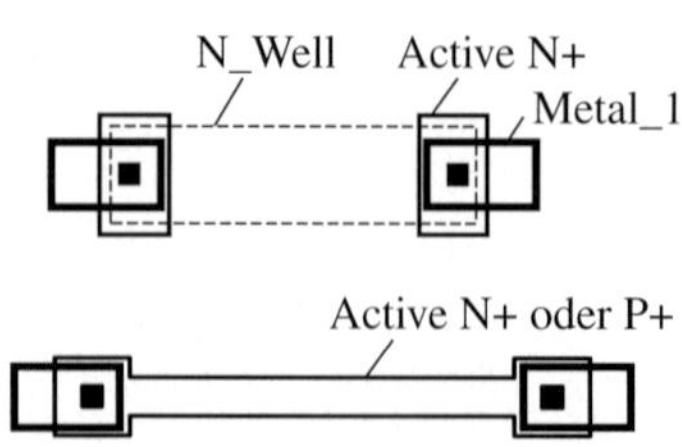

Fig. 21.12 Well and diffusion resistors

By means of a (P+) region placed over the well the square resistance of the well can be raised. But this decreases the voltage stability further.

Diffusion Resistor

Resistors can be implemented in diffusion areas (N+ or P+) as well. Owing to the low square resistance of the diffusion areas diffusion resistors can be used only for low resistance values up to a few KΩ. They are used less frequently because their voltage dependency (about 0.4 %/V) is higher than that of Poly resistors.

21.1.5 Capacitors in CMOS Circuits

Capacitors in CMOS circuits are used frequently in oscillators and filters. The value of the capacitance is the sum of the area capacitance and the side wall capacitance. Table 21.3 shows the approximate values of the capacity for some layers of a submicron technology.

In a digital CMOS process capacitors can be realized in different manners. The active regions (P+ and N+) are pure junction capacitors [21.15]. Therefore they are strongly dependent on voltage.

So they can used as variable capacitors in oscillators [21.27].

Voltage independent capacitors in a digital CMOS process can be implemented only between Poly and metal or between metal layers. But the area capacitance of a Poly metal capacitor is only 0.05 Femtofarad/μm^2. A capacitor of 1 picofarad would need such a large area as 20,000 μm^2. Therefore gate capacitors are preferred although they do not have such a high voltage stability.

Double Poly Capacitor

Analog CMOS processes frequently comprise an additional Poly layer (Poly_2). Poly_2 is placed over Poly_1 and is separated from the latter by a relatively thin dielectric layer of silicon oxide or silicon nitride. The 0.3 μm analog digital CMOS process presented in [21.25] the capacity per area of the Poly_2 layer is about 1.7 femtofarad/μm^2. By means of a particular **Kapa Implant Layer** that has to be placed over the Poly, the implantation of both poly layers is equalized in such a manner that the voltage coefficient of the capacity becomes very low. In [21.25] the voltage coefficient was

Table 21.3 Typical area and side wall capacitances of a CMOS submicron technology

Layer	Area capacitance in fF/μm^2	Sidewall capacitance in fF/μm
Gate Capacitor, Oxide thickness 15 nm (max. C at Enhancement)	2.3	
N+, P+ (junction capacitor at $V = 0$)	0.5 . . . 1	0.3 . . . 0.5
Well	0.07 . . . 0.15	0.2 . . . 0.7
Poly Substrate	0.07	0.05
Metal_1 Substrate	0.03	0.04
Metal_2 Substrate	0.02	0.03
Metal_1-Metal_2	0.05	0.05

decreased by this implantation from 350 ppm/V to 38 ppm/V.

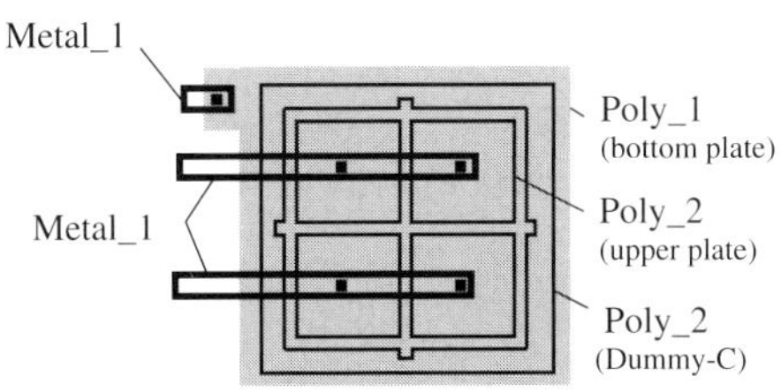

Fig. 21.13 Double Poly capacitor array with dummy capacitor

Fig. 21.14 Layout of a double Poly capacitor

Double poly capacitors are frequently used in Switched Capacitor Filters and in A/D converters [21.19], [21.41]. Mostly an array of equally valued capacitor tiles (about 0.5 pF) is implemented. By parallel connection of the individual capacitors integer ratios of capacitance values can be realized (see fig. 21.13). A dummy capacitor at the edge of the array provides an equal environment for all capacitors. Capacitors in these arrays achieve very high matching accuracies of about 0.5 % [21.44], [21.1], [21.34], [21.33], taking into account the wiring capacities even up to 0.1 % [21.23].

The advantage of Poly Poly capacitors is that the lower Poly electrode has only a low capacity against the substrate. By placement of the capacitor in a well the Poly_1 layer can be shielded against the substrate.

Implanted Gate Capacity

Usually a gate capacity (Layers Active and Poly) is driven into depletion mode. In this state the gate capacity is strongly dependent on voltage and frequency. Some analog CMOS processes use an additional Implant layer that creates a high doping in the gate region below the gate oxide. The layout of such a capacitor is similar to fig. 21.15. Merely an additional Implant layer is placed over the Poly electrode.

By means of this an excellent conducting (N+) capacitor plate is generated. The high capacity per area of the gate oxide can be used then as a high quality and voltage independent capacitor. But the lower plate of this capacitor is not insulated from the substrate.

21

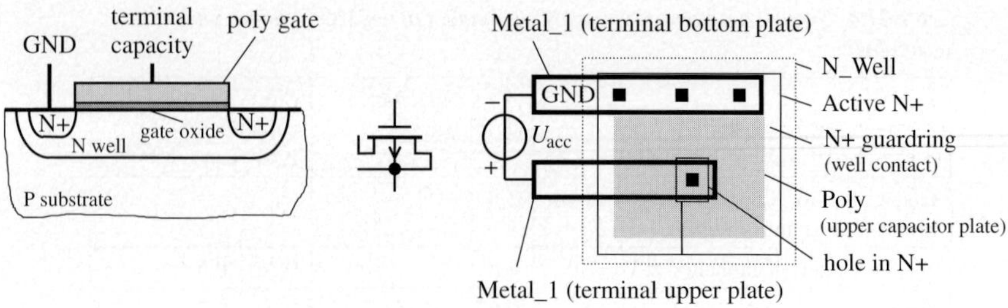

Fig. 21.15 Cross section, schematic symbol, and layout of a well gate capacitor in enhancement mode

Gate Capacity in Enhancement Mode

In digital CMOS processes there is no special implantation step for high quality capacitors. In spite of this capacitors are needed on the chip. Frequently gate well capacities are then used, which are driven into enhancement mode. The function of such a capacitor may be performed in principle by a PMOS transistor the gate voltage of which is reversed. Such a capacitor therefore is called MOSCAP [21.42].

Figure 21.15 shows the layout of such a capacitor. The capacitor is located in an N well that is mostly connected with GND. The upper electrode of the capacitor is formed by a poly rectangle.

It is not allowed to place a contact hole in the gate oxide region. So for the connection of the poly electrode the designer must make a hole in the active layer, which creates field oxide below the Poly contact. To achieve low series resistances and high Q factor the capacitor should be surrounded by a (N+) guard ring that is connected to ground.

Caused by the voltage Vacc between the upper Poly plate and the N well the charge carriers are enhanced (accumulation) in the well gate oxide junction, thereby forming a low resistive lower capacitor plate. The higher Vacc the more the well is driven into enhancement mode thereby enlarging the actual capacity and its quality.

Figure 21.16 shows the voltage dependency of the gate well capacity for Vacc (voltage between gate and well) between −3 V and +3 V at a frequency of 1.3 GHz [21.40]. According to this the capacity at $V_{acc} = -2$ V is only half of the value at $V_{acc} = +1$ V.

If the lower plate of the capacitor, e.g., the well, is not at ground potential but at a higher level, the accumulation voltage V_{acc} decreases then also the Q factor and the capacity will decrease. In spite of enhancement gate capacitors never reach the voltage stability of double poly capacitors. At $V_{acc} = 2$ V the voltage dependency is not better than 800 ppm/V [21.42]. By series connection of two Poly well capacitors (with the poly plates connected) the voltage dependency can be partly compensated [21.42].

Capacitors of this kind are often used as frequency controlling devices in *RC* or quartz oscillators.

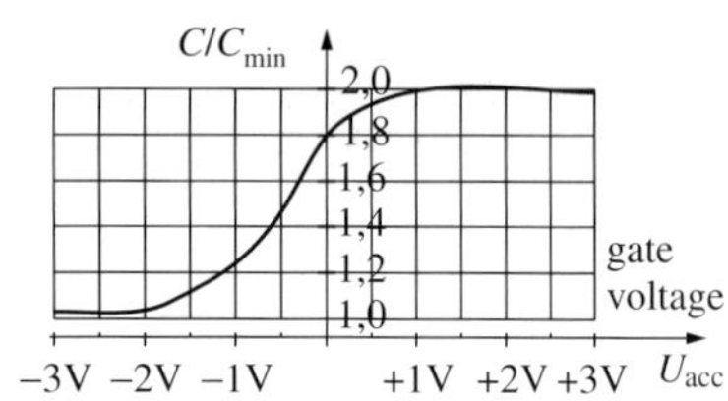

Fig. 21.16 Voltage dependence of a gate well capacitor at 1.3 GHz [21.40]

21.1.6 Diodes and Bipolar Transistors in CMOS

Diodes are used in CMOS circuits preferably as voltage references and as rectifiers. In the standard N well CMOS process there is no diode with two free terminals but only a (N+) P substrate diode. The anode of this device is connected to GND.

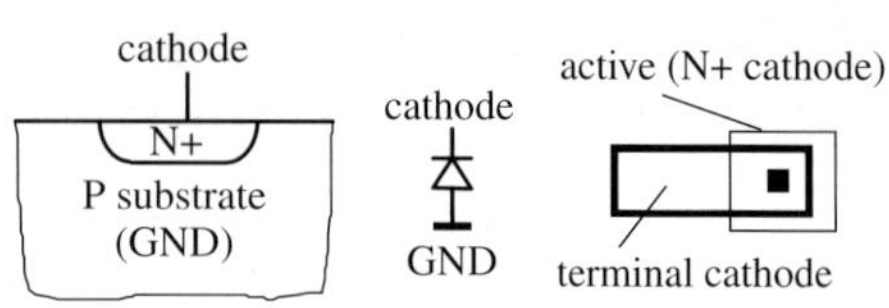

Fig. 21.17 (N+)-P substrate diode: Cross section, symbol, and layout

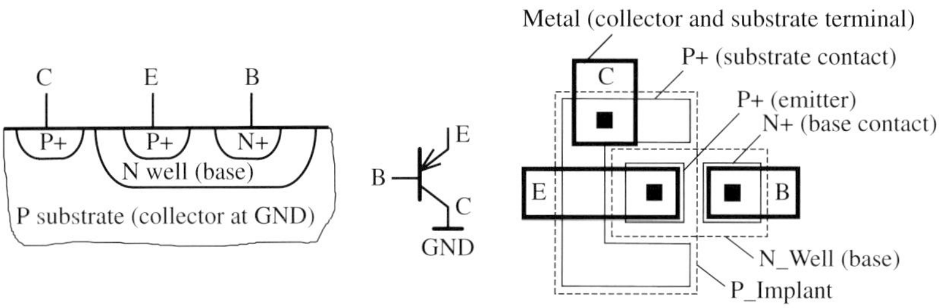

Fig. 21.18 Vertical PNP transistor: Cross section, symbol, and layout

Through such a diode the supply current for the chip can be applied to the P substrate. Figure 21.17 depicts cross section, symbol, and layout of such a diode.

In an N Well CMOS technology vertical and with some limitations even lateral PNP bipolar transistors can be realized.

Vertical PNP bipolar transistors are frequently used. Their collector is fixed to GND. Figure 21.18 shows the cross section through a vertical PNP bipolar transistor, called **substrate transistor** as well.

The collector is formed by the substrate, the base by the well and the emitter by a (P+) region. The U-formed (P+) region surrounding the base well is a substrate contact which sinks the collector current from the substrate.

Because of the fixed connection of the collector to GND only an emitter follower (common collector circuit) can be realized with such a transistor. So a voltage gain is impossible.

Vertical bipolar transistors yield current gains from 20 to 60. At a vertical PNP transistor in a 0.7 µm CMOS technology built by the author the base emitter voltage was $V_{BE} = -660$ mV (emitter area 3 µm × 3 µm, $I_E = 1$ µA) and the emission coefficient of the BE diode was $NF = 1.02$.

Vertical bipolar transistors are frequently used in band gap sources [21.44], [21.45], [21.37].

In a standard CMOS technology there are no vertical bipolar transistors available with a free collector. For a complete vertical bipolar transistor at minimum a base implantation and a buried layer are needed in addition to the CMOS layers [21.44].

In a standard CMOS technology **lateral PNP bipolar transistors** can be realized with a free collector [21.8], [21.48]. Such transistors are used in band gap circuits and audio amplifiers [21.38].

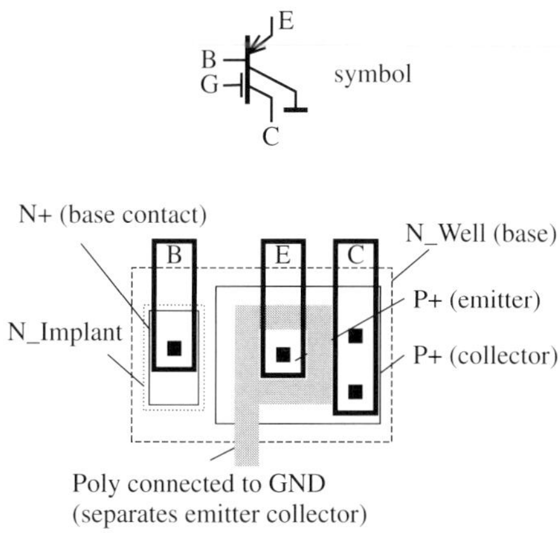

Fig. 21.19 Lateral PNP bipolar transistor in standard CMOS

Figure 21.9 shows symbol and layout of a lateral PNP transistor. The small emitter area in the center is separated from the surrounding collector by a Poly ring. But, caused by the Poly ring, a parasitic MOS transistor emerges in parallel with the PNP transistor (pin G in the symbol). To cut off the MOS transistor the Poly ring has to be connected with VCC. Part of the collector current flows vertically into the substrate whereas the rest of the collector current flows horizontally to the outer collector ring. In the symbol this fact is visualized by the two collectors. The collector connected with GND symbolizes the collector current into the substrate.

Intelligent design of the emitter (small area) causes less than half of the emitter current to flow into the substrate. A lateral PNP transistor realized in 0.7 µm technology for testing purposes had a relatively high current gain of about 90. But for collector currents higher than 3 μA the current gain will decrease.

In conclusion, lateral PNP transistors can be seen as less than ideal solutions. In modern CMOS technologies the implant depth is becoming smaller and smaller, the areas of the emitters and collectors are diminished, and therewith the current yield as well.

21.1.7 Special Requirements for Layout of CMOS Analog Circuits

Often analog circuits are integrated together with digital circuits on a common CMOS chip. In many cases the layout of the analog cells is designed manually. Automatic design tools exist for particular circuits like Switched Capacitor Filters or Operational Amplifiers. The program LAYLA [21.20], e.g., is such a tool for the design of OP Amps. But there is no general tool for the design of all analog circuits [21.31].

The transconductance of CMOS transistors is lower than that of bipolar transistors, the offset voltage and the $1/f$ noise are higher than those of bipolar transistors. These disadvantages can be partially compensated by a careful design.

Figure 21.20 shows a typical layout of an analog cell in MOS technology (band gap cell). Standing out are the large MOS transistors. By means of the large dimensions of these transistors a good matching and a low output conductance g_{DS} will be achieved. Additionally the generation of hot electrons is suppressed which could deteriorate the transistor data [21.6], [21.28].

Critical features in analog CMOS layout are:
- crosstalk between analog and digital parts of the chip;
- matching of resistors and capacitors;
- offset voltages and thermal noise in differential amplifiers;
- matching of current mirrors.

Matching stands for the concurrence of electrical data for congruent geometries, e.g., the ohm values of two resistors with the same geometry.

Crosstalk between Analog and Digital Circuits

The switching level of digital circuits is equal to the total supply voltage, whereas analog circuits work with much lower signal levels. Consequently analog circuits can be influenced by unwanted couplings from the digital part of the chip.

Couplings can result from:
- Capacitive coupling, e.g., by closely spaced conductors;
- common conductors for ground and supply voltages (ground loops);
- coupling caused by the substrate (see next paragraph).

Capacitive couplings between conductors carrying analog and digital signals can be reduced by intelligent routing. No conductors carrying digital signals should be routed over analog parts of the circuit. Analog and digital supply voltages should be routed on separate conductors. Ground and supply rails should be wide enough to avoid voltage drops. The supply voltages for the analog and digital parts of the circuit should be applied over separate pads [21.44], [21.22].

Capacitors and resistors can be shielded against the substrate by placement in wells. A conductor can be shielded by two conductors on the left and right that are connected to a clean ground. Conductors which are sensitive against interference should be as short as possible.

Matching of Resistors

Resistor layouts should be wide enough for good matching. The wider the dimensions of the resistors the less is the influence of variations of the width on the resistance [21.44], [21.22], [21.48], [21.49].

Exactly matching resistors have to be composed from the same base cells. If a 1 KΩ and a 10 KΩ resistor are needed, then the 10 KΩ resistor should be composed of 10 single 1 KΩ resistor elements.

Matching resistors with low tolerances should not be covered by metallic conductors. They should be closely spaced to avoid the influence of surface gradients.

The resistors should have the same environment and the same direction. This is the reason why resistor arrays are surrounded by dummy resistors

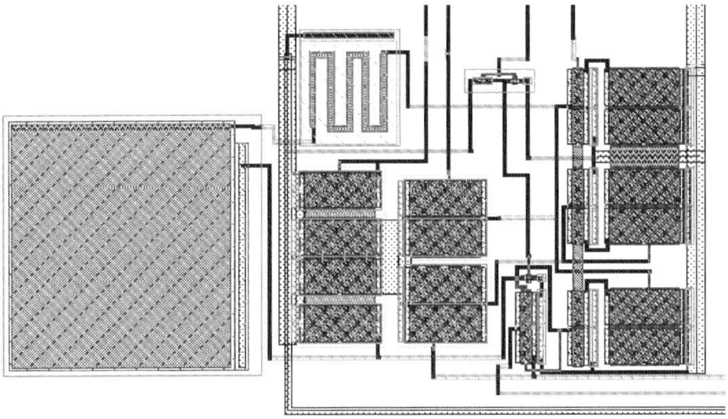

Fig. 21.20 Layout of a CMOS analog cell (band gap)

causing the same environment even for the devices at the edges of the active array.

Figure 21.21 depicts the layout of two resistors with the same value of the resistance with additional dummy resistors. The surface gradients of the resistance per unit length are minimized by the spatial formation (array) of the segments for the resistors R_1 and R_2.

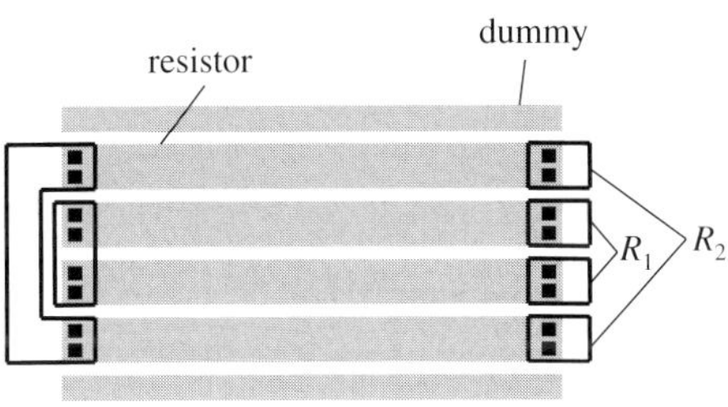

Fig. 21.21 Layout of matched resistors with dummy resistor

The same principles should be applied for closely matching capacitors.

Matching of Differential Amplifiers and Current Mirrors

A differential amplifier consists of two transistors, the matching of which determines the offset voltage. The drain current of a MOS transistor is:

$$I_D = \frac{\beta}{2}(U_{GS} - V_T)^2 \tag{21.1}$$

The threshold voltage V_T and the transistor parameter β are characterized by particular tolerances. According to [21.26] the mean square variation $\sigma^2(V_T)$ and $\sigma^2(\beta)/\beta^2$ are given by:

$$\sigma^2(V_T) \approx \frac{A_{VT}^2}{W \cdot L} + S_{VT}^2 \cdot D^2 \tag{21.2}$$

$$\frac{\sigma^2(\beta)}{\beta^2} \approx \frac{A_\beta^2}{W \cdot L} + S_\beta^2 \cdot D^2 \tag{21.3}$$

D	Distance between transistors;
$W \cdot L$	Transistor area (Product width × Length);
A_{VT}, A_β	Constants describing the influence of the transistor area on variations of V_T and β;
S_{VT}, S_β	Constants describing the influence of the transistor distance on variations of V_T and β.

The variation constants A_{VT}, A_β, S_{VT}, S_β are only valid for a specific technology. Approximate values for a 25 nm gate oxide are [21.21]:

$$A_{VT} \approx 10 \ldots 30\ \text{mV} \cdot \mu\text{m}, \qquad A_\beta \approx 2\ \%/\mu\text{m},$$
$$S_{VT} \approx 4\ \mu\text{V}/\mu\text{m}, \qquad S_\beta \approx 2 \cdot 10^{-6}\ 1/\mu\text{m}$$

A_{VT} will be decrease at thinner gate oxides. Equation (21.2) and (21.3) shows that the standard deviation of mismatch is inversely proportional to the square root of the effective channel area $W \cdot L$. Therefore exactly matching transistors should be large in general. The cut off frequency of the $1/f$ noise is diminished by a **large gate area** as well.

Lovett [21.21] found a good correspondence with the formulae mentioned above during his tests with a 0.8 µm CMOS technology. Although the transistors characterized by him with length's of $L = 6.6$, 4, 1.3, and 0.9 µm. The last two ones exhibited a large variation of β. Consequently the transistors must have a distinct minimum length so that the $1/(WL)$ formulation is valid for the β variation as well.

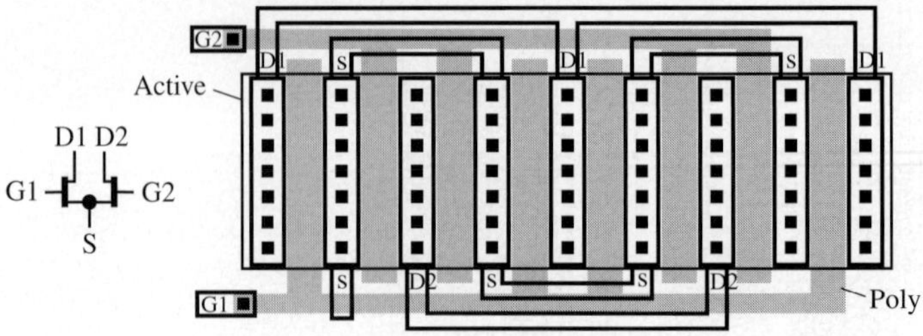

Fig. 21.22 Layout of a MOS differential pair

NMOS transistors have a better matching than PMOS transistors.

Exactly matching transistors should be in a **close neighborhood** and should have the **same direction**. They should also have the **same environment** (possibly dummy transistors should be provided) [21.10]. They should **not be covered by metal** conductors.

The variance of the **offset voltage of differential pairs** will be:

$$\sigma^2(V_{GS}) = \sigma^2(V_T) + \frac{(V_{GS}-V_T)^2}{4} \cdot \frac{\sigma^2(\beta)}{\beta^2} \quad (21.4)$$

For differential amplifiers with a gate voltages close to V_T the second term of (21.4) can be neglected. Then the variation of V_T cause the mismatch.

Figure 21.22 shows the layout of a differential pair. By the arrangement of the 8 transistor segments the surface gradients and the directional variations of the drain currents are compensated for [21.44], [21.22]. This arrangement is called common centroid, e.g., the segments of the two matched transistors are in equal distances from the center [21.3], [21.22].

In analog circuits **current mirrors** are frequently used and their currents have to match exactly. The variation of the drain current is given by:

$$\frac{\sigma^2(I_D)}{I_D^2} = \frac{4\sigma^2(V_T)}{(V_{GS}-V_T)^2} + \frac{\sigma^2(\beta)}{\beta^2} \quad (21.5)$$

The first term describes the influence of V_T variations. To diminish the influence of V_T variations the denominator $(V_{GS} - V_T)$ of the first term must be large, e.g., V_{GS} should be high. This is achieved by dimensioning the current mirror transistors in such a manner that their gate voltage is 1 . . . 2 Volts higher than V_T. So the matching of the currents is improved essentially [21.21], [21.26], [21.34].

Therefore current mirrors contain preferably long and narrow transistors with a W/L ratio below unity are used [21.26]. Close to exactly matching transistors large substrate and well contacts should be placed to stabilize the substrate and well potentials.

21.1.8 Substrate Noise

A particular problem in mixed analog and digital circuits is the substrate coupling [21.44], [21.12], [21.35], [21.46], [21.4], [21.13], [21.20]. It causes digital signals to be coupled into the analog part of the circuit as noise. Substrate coupling can be reduced by guard rings. On the other hand, guard rings and substrate contacts may induce additional noise in the substrate.

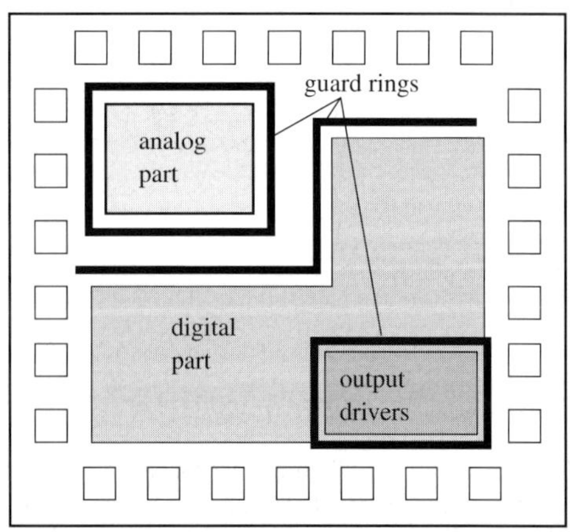

Fig. 21.23 Floorplan of an mixed analog digital chip

Figure 21.23 shows an example for the implementation of guard rings on a chip. The guard rings are (P+) substrate contacts connected to GND [21.44], [21.22]. Besides the (P+) guard rings N Well guard rings may be placed which mainly interrupt sur-

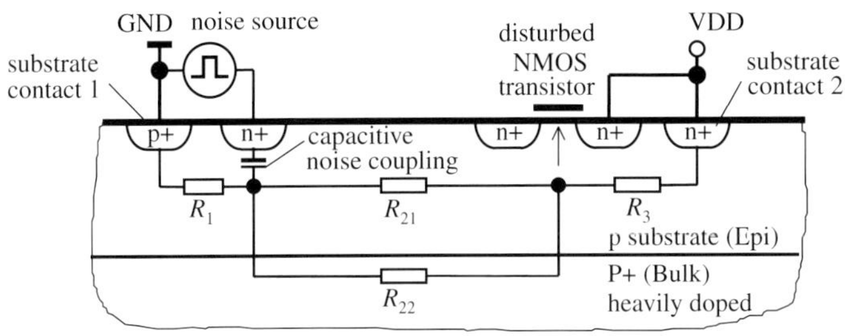

Fig. 21.24 Model for capacitive coupling of noise in an epi substrate

face currents in the low resistive channel stoppers [21.22].

The main causes for **substrate noise** are:

- Capacitive Coupling by gate and drain regions, conductors and bond pads;
- Coupling from substrate contacts with non perfect ground potential;
- Ionization currents caused by hot carriers [21.4].

Charge noise currents are coupled into the substrate by gate and drain regions by large conductors and bond pads. Digital circuits generate large substrate currents because of their short transients. The same is valid for power transistors because of their large gate and drain areas.

Substrate contacts may induce noise currents by galvanic coupling. Long ground conductors with high inductance or resistance are dangerous. Current peaks flowing in these conductors cause voltage peaks. These voltage peaks may be galvanic coupled into the substrate via substrate contacts.

Highly sensitive analog circuits therefore should be placed far away from digital circuits and from driver transistors.

Figure 21.24 shows the model for capacitive coupling of substrate noise in an epi substrate on a heavily doped bulk [21.44], [21.39], [21.13].

The noise is caused by the junction capacitance of an (N+) region connected to the noise source. This (N+) region, e.g., could be the drain of a power transistor. The substrate noise current modulates the right hand MOS transistor and its drain current as well, e.g., the transistor will be disturbed by the substrate noise.

Some part of the noise current may be diverted through R_1. The closer the substrate contact 1 is placed with respect to the noise source the lower is the resistance of R_1 and the more of the noise current may be diverted through R_1. The distance between the noise source and substrate contact 1 should be smaller than the thickness of the epi layer (about 10 µm) to divert the major part of the noise current into substrate contact 1.

The noise voltage is coupled via R_{21} and R_{22} to the disturbed NMOS transistor. The resistance of the epi layer is much higher than the resistance of the highly doped (P+) bulk ($R_{21} \gg R_{22}$). The interfering voltage influencing the NMOS transistor is applied through the voltage divider formed by R_{22} and R_3.

Therefore the substrate contact 2 should be placed as close as possible to the disturbed NMOS transistor. The distance should be less than the thickness of the epi layer. The closer the guard ring is with respect to the noise source the better its effect.

Generally the effect of guard rings in epi substrates is limited because of the low resistance of the highly doped bulk [21.39]. As soon as an interfering signal has been coupled into the bulk it spreads over the whole bulk region and is hardly to suppress.

If the distance between noise source and the disturbed transistor is higher than the thickness of the epi layer then $R_{21} \gg R_{22}$. The noise current flows completely through the substrate. Enlarging the distance do not improve the noise suppression.

In epi substrates it is extremely important to **keep clean the bulk**. If the noise is coupled in the heavily doped (P+) bulk then it will disturb the whole chip. The bulk should be connected to PCB ground with low resistance and low inductance by multiple bond pads. The more bond pads used for GND the cleaner will be the P+ bulk.

21.2 Standard Cell Layout

21.2.1 Introduction

Almost all digital circuits on chips are nowadays designed in standard cell technology with automatic routing. This is partly true for analog circuits as well, whereas the layout of analog cells is frequently carried out manually.

The advantage of the standard cell approach is that the designer does not need to go into details of the transistor level and is able to create the layout at a higher level.

Figure 21.25 shows the basic structure of a chip in standard cell technology. At the edges of the chip the pad frame with the pad cells are located. The pad cells contains the pad and ESD protection structures. Some pad cells contains also output buffer.

The protection structure consists of MOS transistors, bipolar transistors, field transistors, and resistors. A pad cell design including the ESD protection structure is often supplied by the chip manufacturer. By means of that a particular pulse voltage of about 1.500 . . . 4.000 V according to the Human Body Model is guaranteed without destroying the chip.

At each corner of the chip a so called corner cell is placed. It provides the VCC/GND ring that passes through pad cells to be continued round the corner. This VCC/GND ring through the pad cells is necessary because the protection circuits should have a low impedance connection with these potentials.

If the pad frame is not completely filled by the core cells the chip is called Pad Limited. Otherwise Core Limited.

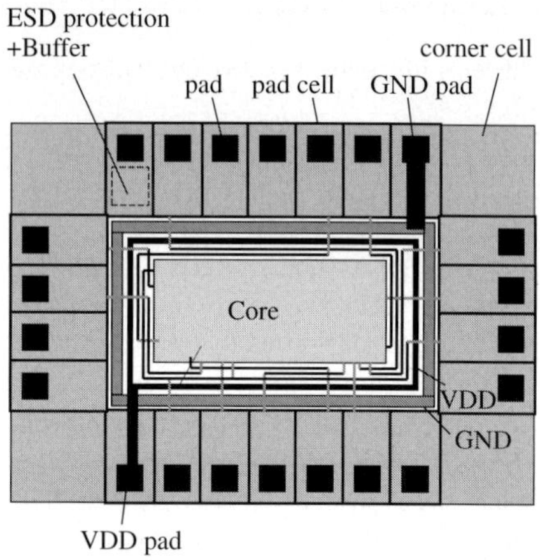

Fig. 21.25 Basic structure of an IC in standard cell technology

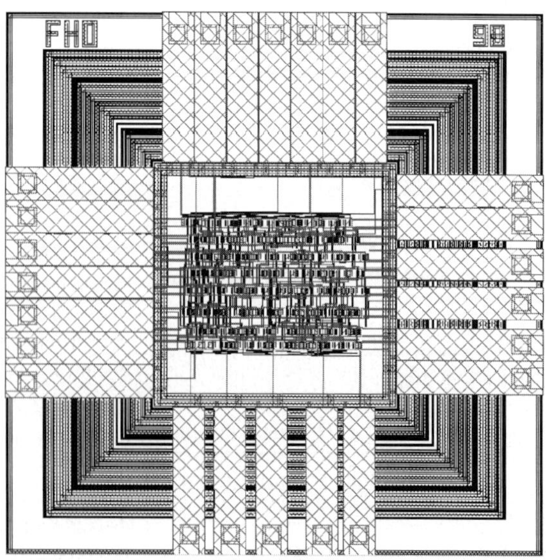

Fig. 21.26 Layout of a chip in standard cell technology

21.2.2 Abstract of a Standard Cell

For every standard cell there are several views: **Schematic, Symbol, Layout and Abstract view**.

The Layout view of a standard cell consists of a rectangle with standardized Site height. At the upper and lower edge of the cell there are continuous conductors carrying VDD and GND, called rails. Besides the actual circuit layout the cell contains substrate and well contacts.

Examples of standard cell layout are presented in figs. 21.27 and 21.28.

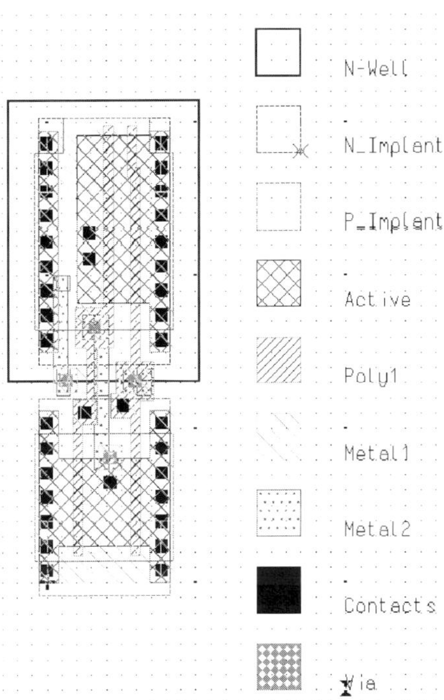

Fig. 21.27 Layout of the standard cell 'nor2'

Figure 21.29 shows the Abstract view of a simple DL (D Latch) with 3 ports CLK, D, and Q. The complete layout is not shown in the Abstract view, only the outline and the ports. The horizontal supply rail as well as the inner wiring of the cell are carried out in Metal_1. The ports are Metal_2. The ports are small lands located on the Metal_2 layer, like CLK and D in fig. 21.29. These lands are then connectedwith Metal_2 (fig. 21.30).

The ports can even consist of a complete conductor (Full Pin Model) the Q port in fig. 21.29. And they can even consist of multiple polygons in different layers.

The Abstract view can be generated in the layout editor on the base of the complete Layout view of the cell. Some abstracts contains blockages called obstructions as well. These are inner conductors being not connected outside and which are not allowed to be crossed by conductors of the same layer. The Metal_1 Blockage of fig. 21.29 is not allowed to be crossed by a Metal_1 conductor.

Figure 21.30 shows the interconnection of 3 cells in a row. The power supply and the horizontal wiring are executed in Metal_1, the vertical wiring in Metal_2. The cells can be placed without or with space between them. If there is some space left between cells the VDD and the GND conductors have to be bridged from one cell in a row to the next one in order to provide the power supply for the cells in a row. Rows of any length can be built by this scheme. Placing and routing of these cells is done automatically by suitable software.

In fig. 21.30 the signal wiring is in a special routing channel. This kind of wiring is called **Channel Routing**. It needs a large part of the chip area.

In fig. 21.30 Metal_1 routing inside the row is impossible because of the large vertical Metal_1 blockage. If a further metal layer is available for the routing then the signals can be routed inside the channel and no additional routing channels are needed. This approach therefore is called **Over the Channel Routing**.

Figure 21.31 depicts the differences between both methods: the over the cell routed layout occupies less chip area than the channel routed one but it needs one more metal layer. Figures 21.32 and 21.33 show examples of standard cell layouts. For technologies with two metal layers the Channel router is used. For more than two metal layers the Over the Cell router is used. Nowadays more than 5 metal layers are in common use, therefore the latter is applied.

Place and route tools are complicated and expensive. Main suppliers for this tools are companies like Cadence, Mentor Graphics,and Avanti. place and route tools are even available for PCs from Tanner EDA.

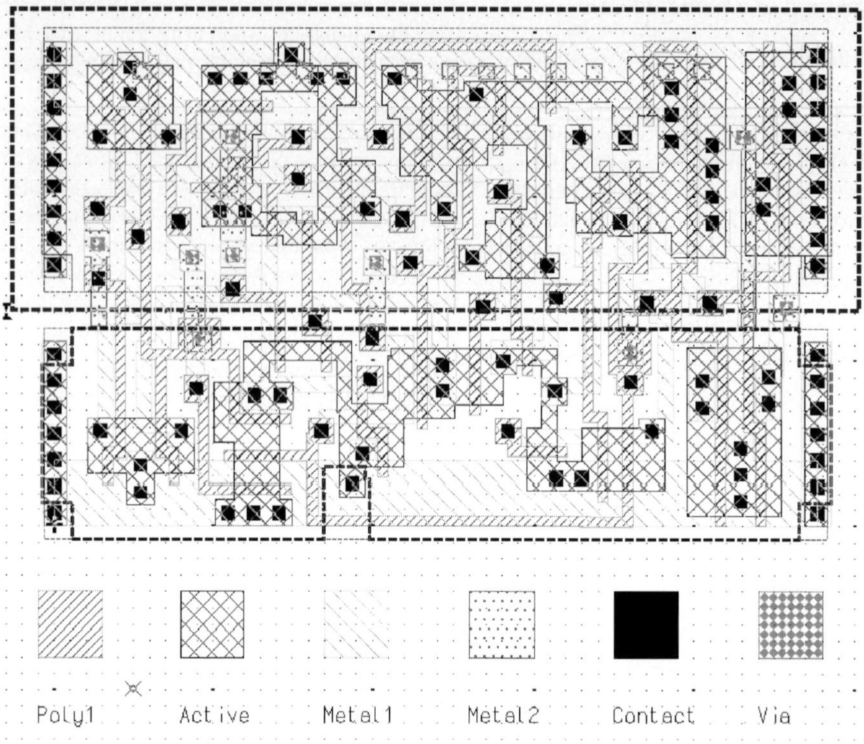

Fig. 21.28 Layout of the D Flip Flop standard cell 'dffrs'

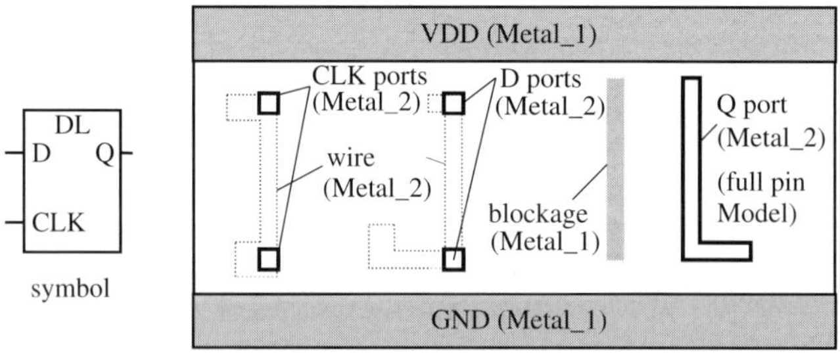

Fig. 21.29 Schematic symbol and abstract of a standard cell

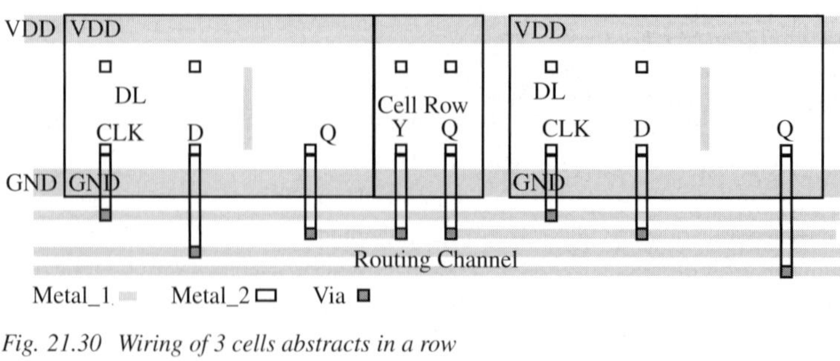

Fig. 21.30 Wiring of 3 cells abstracts in a row

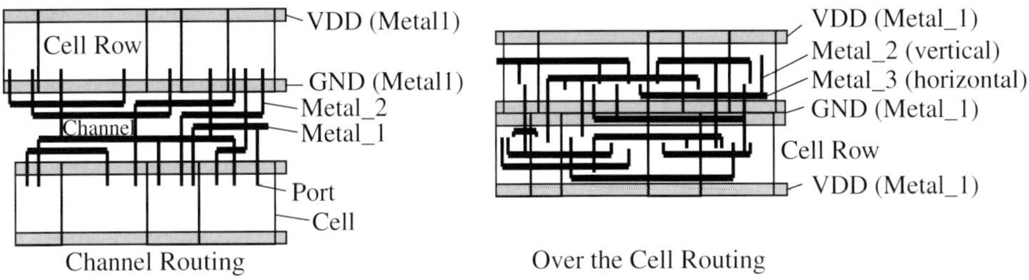

Fig. 21.31 Routing modes: channel routing and over the cell routing

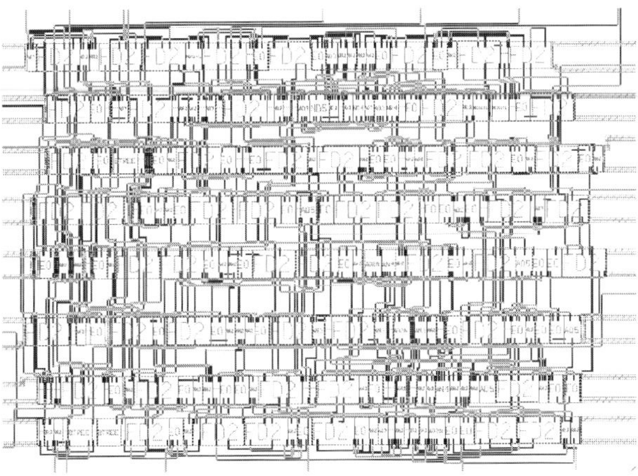

Fig. 21.32 Standard cell layout with channel routing

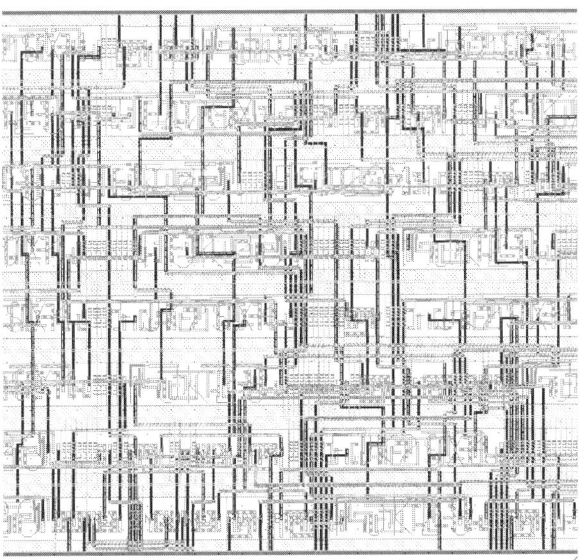

Fig. 21.33 Standard cell layout with over the cell routing

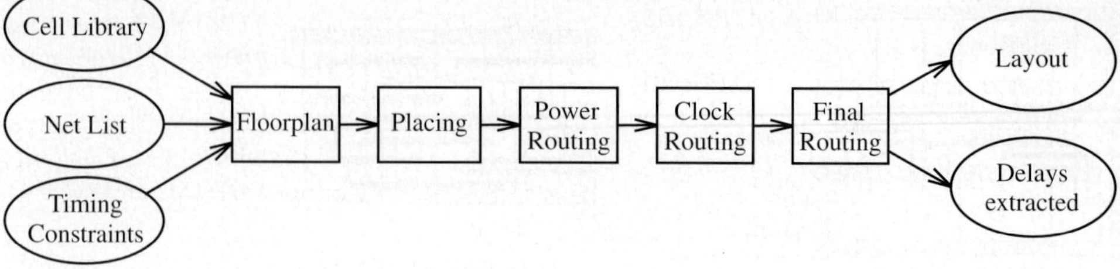

Fig. 21.34 Flowchart place and route

In the following paragraph place and route with the tool Silicon Ensemble from Cadence is explained [21.5]. We do not explain the theory here which is presented in detail in [21.32] and [21.36]. For our practical work with these tools it is not so important.

21.2.3 Floor Planing

Figure 21.34 shows a principal flow chart of lace and route with the (software) program Silicon Ensemble. The tool requires the following inputs: the netlist, a cell library with the macros of the netlist, and possibly timing constraints for the chip.

These input data are imported into the SE database (SE: Silicon Ensemble). Afterwards the layout is created in three steps: Floorplan, placing and routing.

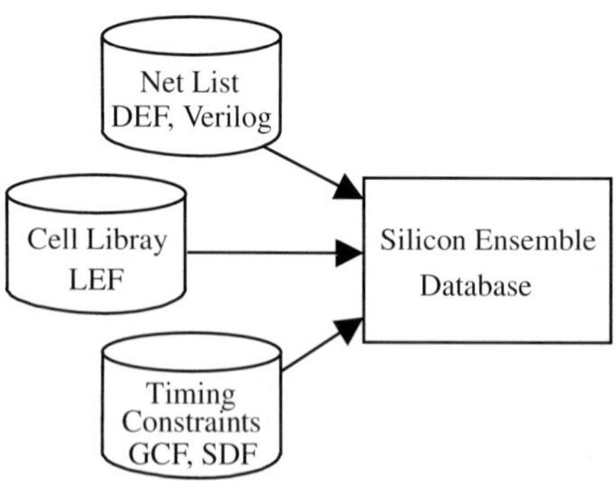

Fig. 21.35 Input data for Silicon Ensemble

Figure 21.35 shows one more time the input data for Silicon Ensemble. The netlist of the circuit to be routed can be extracted directly from the schematic editor Composer as VERILOG netlist or it can be imported from a DEF file.

DEF (Design Exchange Format) is a data format from Cadence. 'xxx.def' files are in ASCII and might contain the following information:

- **Design Section:** Technology, Site extension, Routing Grid, default capacities for place, list of used components, and groups with placing;
- **Netlist:** List of interconnections, delays, max. capacities, net weight for routing, layout data from routing;
- **Timing constraints:** Maximum delays for particular paths.

The DEF format is suited as well for netlists as for the routed cell layouts including time constraints.

All the design data of the SE database can be imported and exported from and to a DEF file during the whole place and route process.

A place and route tool uses the cells (macros) from a Cell Library, which usually is supplied by the chip manufacturer.

For the description of the cell library there is the LEF format (Library Exchange Format) from Cadence. A detailed description of the LEF format can be found in the section 21.3. Besides the design rules LEF comprises information about logic, delays and abstract layout of a cell. The LEF file is imported into the SE database.

Timing constraints can be added as a SDF or GCF file [21.36].

The floor planner does the following jobs:

- creation of core rows, placement of blocks like RAM, ROM, and ALU;
- creation of the I/O rows;
- computation of the routing grid.

A row is a sequence of equally oriented cells. The row is composed of an integer multiple of sites. All cells placed in a row must use an integer multiple of a site.

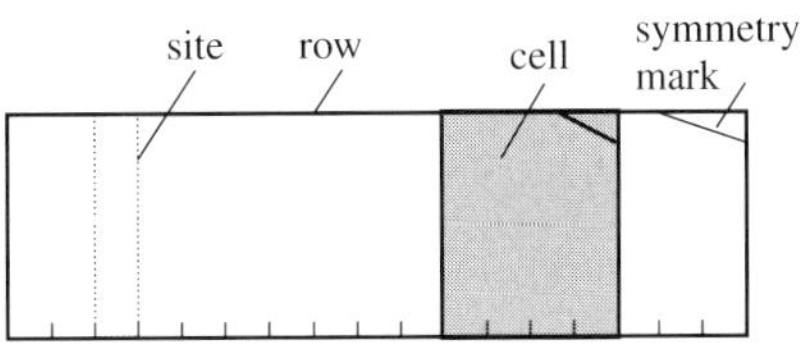

Fig. 21.36 Row, Cell, Site

Figure 21.36 shows a row with the corresponding site. The cell in this example is 4 sites wide. The triangle in the upper right corner marks the direction of the orientation (Symmetry). For each wiring layer there is defined a routing grid. The routing grid corresponds with the distance from center to center of two conductors one of which includes a via (see fig. 21.37).

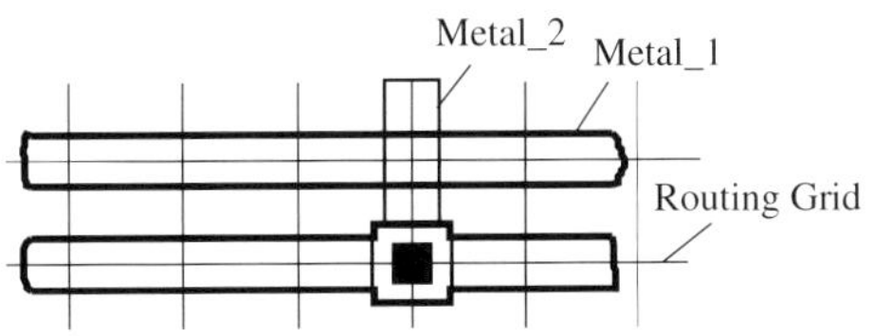

Fig. 21.37 Routing Grid

From the routing grid a global routing grid is computed (GGrid) which is about 10 times as large as the routing grid.

Figure 21.38 shows a typical floorplan. The chip consists of five core rows, a block, and four I/O rows with the pad cells. The floor planner may be influenced by the following steering inputs: required chip area, ratio of height/width, distance of the rows, distance of I/O to core, and distance of blocks to rows.

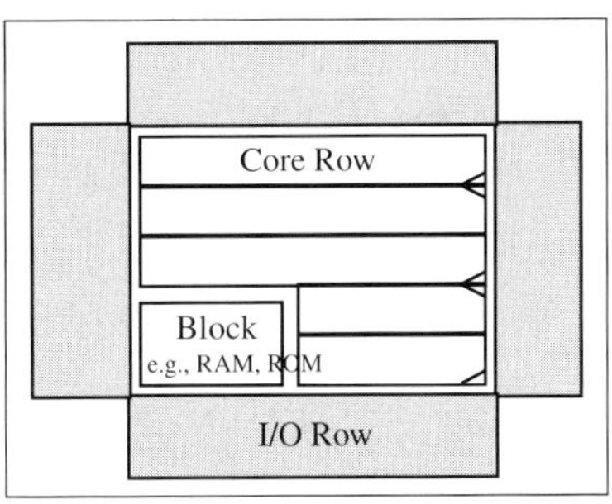

Fig. 21.38 Floorplan with Core, I/O Row, and Block

The blocks may be dragged by the mouse directly. Rows may be deleted or new rows may be inserted. The floorplanner calculates a row utilization using the actual floor plan, e.g., how many percent of the core rows are actually occupied by cells. Recommended are 80 . . . 90 %. The floorplanning may be repeated several times in order to optimize the layout.

The I/O cells (Pads) should be placed during the floorplanning process as well. The pad cells may be placed automatically or by hand. If marked by the attribute fixed or placed they cannot be moved in the succeeding placing. Blocks (as RAM, ROM) should be fixed as well.

21.2.4 Placement

After floorplanning the cells may be placed in the prepared rows. In the **Silicon Ensemble** software package for the placement there are the following tools: **Q Place** (Quick Place) and **Ultra Placer**.

Placement with the Q Place tool is executed in 3 steps:

- **Global placement:** hierarchical partitioning of the netlist, initial cell placement;
- **Detailed placement:** Shortening of the wiring (conductors) by movement of the cells;
- **Annealing:** final placement of the cells in the rows, removal of overlaps.

The placer optimizes for minimum wire lengths. Alternatively there is a kind of placement possible that optimizes the placement for timing (with the switch Timing.Mode set to on).

The placer then tries to fulfill the Timing Constraints given in the SE database.

From the result of the placement it may be estimated whether there are too narrow paths for the routing. If necessary the floorplan may be revised. After the placement it is possible to estimate the wiring lengths and the delays as well. A variable Timing.Weight determines how tight the timing constraints are taken into account by the placement tool.

To measure the quality of the placement the log file of the placer contains the following data: Total Wire Length, HMAX, VMAX. The Total Wire Length is the total length of the wiring which is estimated by the Steiner Tree [21.32]. It should

21

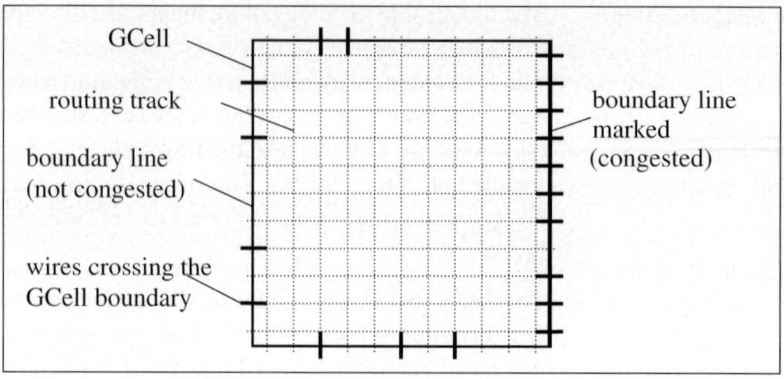

Fig. 21.39 Congestion Map of a GCell

be as small as possible. The parameter HMAX represents the maximum number of intersections of a horizontal line from one end of the placement area to the other end with wiring paths. The parameter VMAX is the same for a vertical line. The ratio HMAX/VMAX is nearly proportional to the density of vertical and horizontal wiring.

Exact information about so called Hot Spots with too dense wiring can be achieved from the **congestion map**.

During the floorplanning the chip area is covered by a routing grid. The lines of this grid are the routing tracks. Wires must lie on the routing tracks. About 10 routing tracks form a **GCell**. The congestion map indicates for each of the four boundary lines of a GCell by how many wires they are crossed. Fig. 21.39 shows the formation of a congestion map for only one GCell. The complete congestion map contains all GCells of the chip.

If the number of wires crossing a boundary line of a GCell is higher than a particular number represented by a variable (e.g., 8) then this line is congested (crowded). There are different colors to mark a congested line depending on the overshooting number of wires. In fig. 21.39 the right hand line is congested. There are too many wires crossing it. It will be marked on the display by a color while the other lines remain dark.

After the placement conductor capacities and delay times may be extracted. The delay times can be compared with the timing constraints. The Cadence tool Pearl allows one to read out the delays from the schematic and highlight the corresponding locations in the layout.

21.2.5 Routing

Routing starts with the special routing and power routing. Power routing connects all cells with the VDD and GND pads. Power routing generates the Power/GND Rings, power strips and power connects to the rows, blocks and pads (see fig. 21.40). The design rules for the power lines are contained in the LEF file.

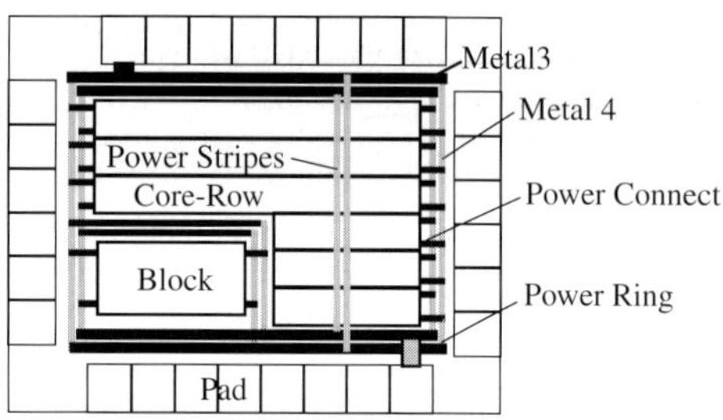

Fig. 21.40 Chip with power routing

The power rings are generated in the power router by the command Add Rings. Additional power stripes may be built in by the command Add Stripes.

With the commands Connect Ring or Follow Pins the power/GND rails of the cells will be connected to the power/GND ring.

A clock tree with definable clock skew is generated by means of the tool CTGen. CTGen uses the capacities and delays which were estimated after the placement. CTGen inserts additional clock buffers if needed and places them on the chip. Thereafter the tool calculates the expected clock skews and clock delays. The clock tree can be routed separately before the other signals are routed, in order to balance the individual clock paths.

After the power routing the floorplan can be improved. A measure for the quality of routing is the congestion map. By the use of VSize hot spots in the congestion map can be removed. VSize allows one to change the row width and the channel height, the stretching and compacting of the floorplan.

Following the application of VSize the routing is completed by the tools Global Route/Final Route or Warp Route (the warp router combines Global and Final Route).

Global Route generates a routing plan for the following application of Final Route. It associates individual conductor paths with routing tracks.

Global route is executed in two steps, the Initial step and the final step. It creates the congestion map and displays the number of congested GCells and the total number of GCells. The ratio of congested cells to the total number of cells should not exceed a particular value. Otherwise the final routing is impossible.

The final router completes the routing and creates the finished layout. It subdivides the routing area in SBoxes. The SBoxes are located in the GGrid raster but their borders extend to about eight times the length of the GGrid. The SBoxes overlap each other.

The router deals with the SBoxes sequentially. Every SBox is routed locally. Caused by the overlapping of the SBoxes the connections to the next SBox are achieved.

After the first routing pass there is a search and repair operation. Faulty routings are searched for and corrected by new local routings. It is possible to do this even manually by dragging the mouse over a faulty SBox and repeating the routing of this box.

As soon as all routing errors are eliminated a Final Cleanup is executed for shortening the wire lengths and diminishing the number of vias. The completely routed layout may be saved to the database as a GDSII or DEF file.

As an alternative to the Global/Final Router Cadence offers the Warp Router (Envisia Ultra Router). The Warp Router is faster and allows Timing Driven Routing and is able to solve congestion problems in a better way.

21.3 The LEF Data Format

LEF (Library Exchange Format) is a data format from Cadence. It is used to describe a standard cell library. LEF is an ASCII format.

It includes the design rules for the routing and the Abstract Layout of the cells (Macros) in the library. The LEF data format contains no information about the internal netlist of the cells.

The xxx.lef file consists of the following sections:

- **Technology**: layer, design rules, via definition;
- **Site:** Site extension (raster of the rows);
- **Macros:** cell description, cell dimensions, layout of the pins and blockages, capacities, and delays.

The technology is described by the Layer and Via statements. To each layer the following attributes may be associated: type, width/pitch/spacing rules, direction, resistance and capacitance per unit square, antenna Factor.

Layer types may be: routing, cut (contact, via), masterslice (active, poly, no routing), overlap.

The **direction statement** defines whether vertical or horizontal routing is associated with the particular layer.

The Antenna Area Factor statement allows to define factors for the weight of the layer area by which the antenna size is calculated (see pin statement).

The following example describes the metal layers M1, M2, and the associated Via CUT12:

```
LAYER M1
  TYPE ROUTING ;
  WIDTH 1.5 ;
  SPACING 1.2 ;
  DIRECTION VERTICAL ;
END M1 ;

LAYER CUT12
  TYPE CUT ;
END CUT12 ;

LAYER M2
  TYPE ROUTING ;
  WIDTH 1.5 ;
  SPACING 1.2 ;
  DIRECTION Horizontal ;
END M2 ;
```

21

Besides the above layer commands there are there commands for the power routing.

The via statement describes the layout of a standardized via consisting of a lower metal rectangle, an upper metal rectangle, and a via rectangle in between. The via is given a particular name. For the power routing special contact holes may be defined.

The following example defines the via VM12 from Metal_1 to Metal_2. The coordinates are the diagonal locations of the corners of the rectangles:

```
VIA VM12 DEFAULT
  LAYER M2
    RECT -1.2 -1.2 1.2 1.2 ;
  LAYER CUT12
    RECT -0.6 -0.6 0.6 0.6 ;
  LAYER M1
    RECT -1.2 -1.2 1.2 1.2 ;
END VM12 ;
```

The Site statement defines a Site (see fig. 21.36). The Site receives a name and a size. The site may be one of a class Pad or Core:

```
SITE CORE11
  CLASS CORE
  SIZE 2 BY 25;
END CORE11 ;
```

The Macro statement describes the standard cell. Each macro is given a name. The macro contains mainly the following data: size, symmetry (allowed rotations and mirror images of the macro), site name, pin, blockages, class (Pad, Core, Ring, Block), timing.

For every pin of the circuit there is a PIN statement. Every pin may have multiple ports. As an example the pin of fig. 21.41 has two ports. Within the Pin statements there are the Port statements for these purposes. The two ports of pin A may be written into a Port statement. Alternatively, for each port a separate Port statement may be written.

The Pin may be associated by an Antenna Size statement. This defines the maximum permissible antenna area. The following listing is a shorthand description of the macro Inv (# is the symbol for comments):

```
MACRO INV
  SIZE 14 BY 25 ;
  # Cell size
  SITE CORE11 ;
  # Site "CORE11" is used
  Symmetry y ;
  # Cell may be y-mirrored

PIN A
  DIRECTION INPUT ;
  # Input A
  USE SIGNAL ;
  # A is signal, not Power
  CAPACITANCE 0.1
  # input capacity of A
  PORT
  LAYER M1
  # Port in layer Metal 1
    RECT 3 7 5 9
    # Coordinats xo yo x1 y1, lower Port
    RECT 3 16 5 18
    # Coordinats xo yo x1 y1, upper Port
  END ;
END A ;
PIN Y...
END Y ;
PIN VSS
  DIRECTION INOUT ;
  SHAPE ABUTMENT ;
  # Power router may route over VSS Pin
  USE GROUND ;    # Pin is GND
  PORT VSS
  LAYER M2;
    RECT 0 0 14 5 ;
  END ;
END VSS ;
PIN VDD ...
END VDD ;
  OBS    # Blockage in Metal 1
  LAYER M1 ;
  RECT 7 0 9 10 ;
  RECT 7 15 9 25 ;
END ;

TIMING
FROMPIN A ; TOPIN Y ;    # Delay A -> Y
  RISE INTRINSIC 0.2 0.3
      VARIABLE 0.11 0.14 ;    # 40
  FALL INTRINSIC 0.2 0.3
      VARIABLE 0.11 0.14 ;
  UATENESS INVERT ;
END TIMING ;
END INV;
```

In the **timing statement** the delay times of the macro are defined. The path for which the delay is valid is defined by FROMPIN, TOPIN. The two time values behind RISE INTRINSIC are the minimum and maximum intrinsic delay (delay at load capacity zero) for the rising slope at the output pin Y.

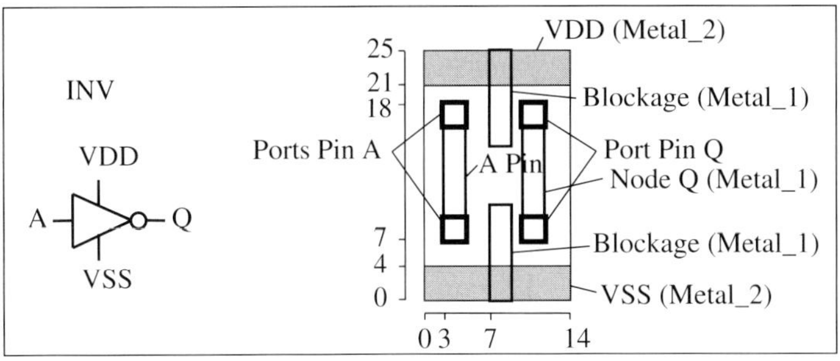

Fig. 21.41 Symbol and Abstract of INV

The two time values behind VARIABLE represent the minimum and maximum variable delay (additional delay time per unit of load capacity). The line starting with FALL contains the delay values for the falling slope.

The statements UATENESS may be followed by INVERT, NONINVERT, and NONUNATE. These define whether the signal between FROMPIN and TOPIN is inverted or not inverted. NONUNATE defines whether the output signal not only is dependent on the input FROMPIN but also from other inputs.

21.4 The GDSII Format

Different formats for the exchange and storage of layout data are in use:

CIF (Caltech Intermediate Format), Applicon, Mebes and GDSII (originally Calma Stream Format). The GDSII data format from Cadence for exchange and storage is the most used standard format today. It can be used for import and export of the data in all layout programs.

GDSII stores Numbers not in ASCII but as data bytes in the signed integer or real format. Only cell and lib names are stored in ASCII. Therefore a simple text editor cannot be used to read the data of a GDSII file.

For checking purposes some layout editors allow to save the GDSII file additionally as a text file as well (Dump).

Listing 21.1 shows the GDSII text dump of the cell m_test (from the library top) created by the layout editor Virtuoso (Cadence). This cell consists of only one polygon and a rectangle in layer 9.

There are different versions of the GDSII format. Today versions 3, 4, 5, and 6 are in use.

From the view of the data structure the GDSII file is a library containing multiple layouts with any number of elements.

The GDSII file starts with the following data: Library Name (ASCII text), Version Number, time of last modification, length of a layout unit.

These data are followed by the cell abstracts.The cell name is in plain text as well in the GDSII file. The cells themselves contain any number of elements as there can be:

- boundary (image element with locations of corners);
- box (rectangle);
- path (conductor);
- sref (call of an instance with reference location);
- aref (array of cells);
- text (sequence of xy-strings, label);
- node (node name).

Every layout element contains a layer number from 0 to 63.

Every element has a number for the data type, that characterizes the type of the element (e.g., data type 0: normal layout; data type 30: special layout; data type 23: pintext).

Further details concerning GDSII may be found in [21.29]. All layout programs in use today are able to import and export GDSII data. But there are special programs for GDSII which can improve the handling of the data and make transparent the contents of the files. An example is GDS Display from Dolphin [21.9].

List 21.1 GDSII-Text-Dump

```
Begin Library
Library Name:top, DB per unit user unit:1000 User Unit:  Micron
Begin Cell Definition
Cell Name :  M_TEST, View Name :  Layout
Polygon - Layer :  9 Data Type :  0 No of points :  8
(5400,6100) (5400,4600) (3500,4600) (3500,3600)
(2100,3600) (2100,5000) (-1600,5000) (-1600,6100)
Rectangle - Layer :  9 Data Type :  0 BBOX : (-6300, -7700) (-2300,6400)
End Library
```

21.5 References

[21.1] *Allen, P. E; Holberg, D. R.*: 'CMOS Analog Circuit Design'. – New York: Holt, Rinehardt and Winston, 1987

[21.2] Amerasekera: 'ESD in Silicon integrated Circuits'. – John Wiley & Sons, 1995

[21.3] *Baker, R. J.; Li, H. W.; Boyse, D. E.*: 'CMOS Circuit Design, Layout and Simulation'. – IEEE-Press, 1998

[21.4] *Briare, J.; Krisch, K. S.*: 'Substrate Injection and Crosstalk in CMOS Circuits'. Custom Integrated Circuits Conference, 1999, pp. 483–486

[21.5] 'Silicon Ensemble – Version 5.2'. Programmunterlagen Fa. Cadence

[21.6] *Chan, V.-H.; Chung, J. E.*: 'The Impact of NMOSFET Hot-Carrier Degeneration on CMOS Analog Subcircuit Performance'. IEEE Journal of Solid State Circuits, Vol. 30.6, 1995, pp. 644–649

[21.7] *Chen, J. Y.*: 'CMOS Devices and Technology for VLSI'. – Prentice Hall, 1990

[21.8] *Degrauwe, M. R.*: 'CMOS Voltage References Using Lateral Bipolar Transistors'. IEEE Journal of Solid State Circuits, Vol. 20, 1985, pp. 1151–1156

[21.9] 'GDS Display'. Programmunterlagen Fa. Dolphin, www.dolphin.fr

[21.10] *Fiez, T. S.; Naiknaware, R.*: 'Automated Hierachical CMOS-Analog Circuit Generation with Intramodule Cennectivity and Matching Conciderations'. IEEE Journal of Solid State Circuits, Vol. 34.3, 1999, pp. 304–317

[21.11] *Gärtner, P.; Grube, R.; Engelhardt*: 'Mixed-Signal Gate Array'. Abschlussbericht, IMS Stuttgart 1998

[21.12] *Gharpurey, R.; Meyer, R. G.*: 'Modelling and Analysis of Substrate Coupling in Integrated Circuits'. IEEE Journal of Solid State Circuits, Vol. 31, 1996, pp. 344–353

[21.13] *Gharpurey, R.; Meyer, R. G.*: 'Modelling and Analysis of Substrate Coupling in Integrated Circuits'. IEEE Journal of Solid State Circuits, Vol. 31.3, 1996, pp. 344–353

[21.14] *Hilleringman, U.*: 'Silicium Halbleitertechnologie'. – Stuttgart: Teubner-Verlag, 1996

[21.15] *Hoffmann, K.*: 'VLSI-Entwurf'. – München: Oldenbourg Verlag, 1990

[21.16] IC Latch-up Test, EIA/JEDEC Standard No. 78, Electronic Industries Association 1997

[21.17] *Ming-Dou Ker, Wen-Yu Lo, Chung-Yu Wu*: 'New Experimental Methodology to Extract Compact Layout Rules for Latchup Prevention in Bulk CMOS IC's'. Custom Integrated Circuits Conference, 1999, pp. 143–146

[21.18] *Lakshmikumar, K. R., et al.*: 'Characterization and Modelling of Mismatch in MOS Transistors for Precition Analog Design'. IEEE Journal of Solid State Circuits, vol. 21, 1986, pp. 1057–1066

[21.19] *Laker, K. E.; Sanson, W.*: 'Design of analog integrated Circuits and Systems'.– New York: Mc Graw-Hill, 1994

[21.20] *Lampaert, K.; Gielen, G.; Sansen, W.*: 'Analog Layout Generation for Performance and Manufacturability'. – Kluwer, 1999

[21.21] *Lovett, S., et al.*: 'Optimizing Transistor Mismatch. IEEE Journal of Solid State Circuits'. vol. 33.1, 1998, pp. 147–150

[21.22] *Maloberti, F.*: 'Layout of Analog and Mixed Analog-Digital Circuits'. In: *Franca, J. E.; Tsividis, Y.* (Herausgeber): Design of Analog-Digital VLSI Circuits for Telecommunication and Signal Processing. – Prentice Hall, 1994

[21.23] *McNutt, M. J., et al.; Shyu; Temes, G. C.; Yao, K.*: 'Random Error Effects in MOS Capacitors'. IEEE Journal of Solid State Circuits, Vol. 29.5, 1994, pp. 611–616

[21.24] *Menozzi et al.*: 'Layout Dependence of CMOS Latchup'. IEEE Transaction of Electron Devices, Vol. 36, 1989, pp. 1892–1901

[21.25] *Miyamoto, M., et al.*: '0,3μm Mixed Analog/Digital CMOS Technology for Low-Voltage Operation'. Custom Integrated Circuits Conference 1993, pp. 24.4.1–24.4.4

[21.26] *Pelgrom, M.*: 'Matching Properties of MOS Transistors'. IEEE Journal of Solid State Circuits, Vol. 24.5, 1989, pp. 1433–1439

[21.27] *Porret, A.; Melly, T.; Enz, C.*: 'Design of High-Q Varactors for Low-Power Wireless Applications using Standard CMOS Process'. Custom Integrated Circuits Conference 1999, pp. 641–644

[21.28] *Quader, K. N.*: 'Hot-Carrier-Reliability Design Guidelines for CMOS Logic Circuits'. IEEE Journal of Solid State Circuits, Vol. 29.3, 1994, pp. 253–262

[21.29] *Rai-Choudhury, P. (Editor)*: 'Spie Handbook of Microlithography, Micromachining, and Microfabrication'. Volume 1, see www.cnf.cornell.edu/SPIEBook/order.htm

[21.30] *Reifschneider, N.*: 'CAE-gestützte IC-Entwurfsmethoden'. – Prentice Hall, 1998

[21.31] *Rutenbar, R.*: 'Analog Design Automation: Where we are? Where we are Going?' Custom Integrated Circuits Conference 1993, pp. 13.1.1–13.1.7

[21.32] *Shervwani, N.*: 'Algorithms for VLSI Physical Design Automation'. – Kluwer, 1995

[21.33] *Shyu, J. B.; Temes, G. C.; Yao, K.*: 'Random Error Effects in MOS Capacitors'. IEEE Journal of Solid State Circuits, Vol. 17.6, 1982, pp. 1070–1076

[21.34] *Shyu, J. B.; Temes, G. C.; Krummenacher, F.*: 'Random Error Effects in Matched MOS Capacitors and Current Sources'. IEEE Journal of Solid State Circuits, Vol. 19.6, 1984, pp. 948–955

[21.35] *Singh, R.*: 'A Review of Substrate Coupling Issuses and Modeling Strategies'. Custom Integrated Circuits Conference 1999, pp. 491–498

[21.36] *Smith, M. J. S.*: 'Application-Specific Integrated Circuits'. – Addison-Wesley, 1997

[21.37] *Song, B. S.; Gray, P. R.*: 'A precision curvature-compensated CMOS Bandgap Reference'. IEEE Journal of Solid State Circuits, Vol. 18.6, 1983, pp. 634–643

[21.38] *Stefanelli, B., et al.*: 'A very low noise CMOS-Preamplifier for Capacitive Sensors'. IEEE Journal of Solid State Circuits, Vol. 28.9, 1993, pp. 971–978

[21.39] *Su, D. K., et al.*: 'Experimental Results and Modelling Techniques for Substrate Noise in Mixed-Signal Integrated Circuits'. IEEE Journal of Solid State Circuits, Vol. 28.4, 1993, pp. 420–430

[21.40] *Svelto, F.; Deantoni, S.; Castello, R.*: 'A 1.3 GHz CMOS VCO with 28 % frequency tuning'. Custom Integrated Circuits Conference 1999, pp. 645–652

[21.41] *Taylor, R. C.*: 'Switch Capcitor Filters'. In: *van der Plasche, R.; Sanson, W.; Huijsing, J.* (Herausgeber): Analog Circuit Design – Low Power, Low Voltage IC. – Boston: Kluwer, 1995

[21.42] *Temes, G. C., et. al.*: 'MOSFET-Only Switched-Capacitor Circuits in Digital CMOS-Technology'. IEEE Journal of Solid State Circuits, Vol. 34.6, 1999, pp. 734–747

[21.43] *Troutman, R. R.*: 'Latchup in CMOS Technology'. – Kluwer, 1986

[21.44] *Tsividis, Y.*: 'Mixed Analog-Digital VLSI Devices and Technology'. – Mc Graw Hill, 1995

[21.45] *Tsividis, Y.; Ye, R.*: 'Bandgap voltage reference sources in CMOS technology'. Electron. Letters, Vol. 18, 1982, pp. 24–25

[21.46] *Verghese, N. K.; Allstod, D. J.; Wolfe, M. A.*: 'Verification Techniques for Substarte Coupling and their Application to Mixed Signal IC-Design'. IEEE Journal of Solid State Circuits, Vol. 31.3, 1996, pp. 354–365

[21.47] *Vittoz, E. A.*: 'MOS Transistors Operated in the Lateral Bipolar Mode and their Application in CMOS Technology'. IEEE Journal of Solid State Circuits, Vol. 18, 1983, pp. 273–279

[21.48] *Vittoz, E. A.*: 'The Design of High-Perfornance Analog Circuits on Digital CMOS Chips'. IEEE Journal of Solid State Circuits, Vol. 10, 1985, pp. 657–665

[21.49] *Wittman, R., et al.*: 'Trimless high precision ratioed resistors in D/A and A/D converters'. IEEE Journal of Solid State Circuits, Vol. 30.8, 1995, pp. 935–939

22 Geometric Verification

HARALD TOEPFER

22.1 Introduction

The term **Geometric Verification** subsumes different checks executing at the finished layout or during the design of the layout. Most important is the DesignRuleCheck. It verifies the minimum spaces and widths in the layout prescribed by the manufacturer.

The recovery of the electric schematic from the layout is called **Extract**.

The **LVS** (Layout versus Schematic) tool compares the extracted schematic against the original netlist. This proves finally that the layout corresponds exactly with the simulated schematic. A successful LVS guarantees the correspondence Schematic Layout.

The ERC (Electrical Rule Check) checks on basic electric faults like opens, shorts and floating nets.

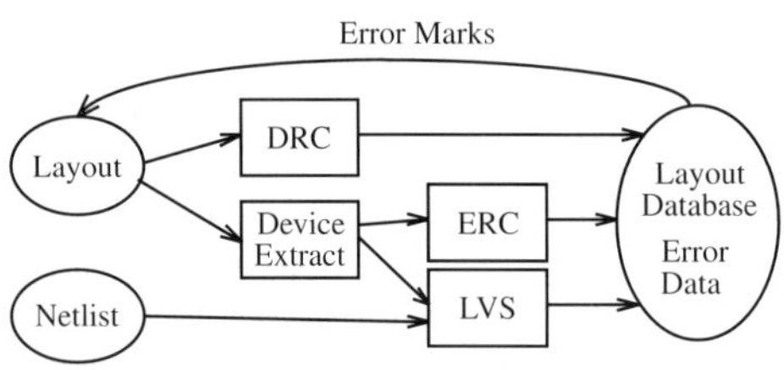

Fig. 22.1 Workflow of the Geometric Verification

The rules for DRC, Extract, ERC and LVS are contained in a technology file. In the tool DRACULA from Cadence this command file consists of:

Description Block:
- library and cell name;
- names of input, output, report and graphic files;
- scale and resolution of the layout data;
- input layer block;
- assignment of the internal DRACULA layer names to the input data;
- assignment of label strings to the layers.

Operation Block:
- Layer preprocessing and interconnect definition;
- Commands for DRC, Extraction, LVS.

In the following paragraphs some commands of the Operation Block will be explained. The program DRACULA comprises a great number of Operation Block commands. Only the most important ones are presented. Detailed information is to be found in the DRACULA reference [22.2].

22.2 Layer Preprocessing

These commands generate new layers. In order to do this DRACULA accepts the following commands.

22.2.1 Logic Combinations

Logic associations are important operations, e.g., for the definition of Gate, Source, and Drain contacts. Through the association of *layer_a* and *layer_b* the new layer_out is created. The following associations may be applied: *and, or, xor, not* (see fig. 22.2).

The areas of the layer_out polygons created by the *not* combination are equivalent to *layer_a* minus *layer_b*.

Syntax:
Operator layer_a layer_b layer_out

Operators:
and, or, xor, not

All DRACULA layout commands permit alternatively to save the output polygons not only in layer_out but also directly in a file.

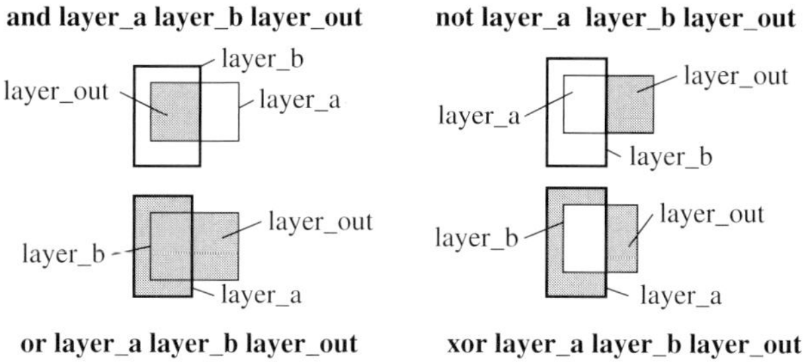

Fig. 22.2 Logic Combinations for and, not, or; and xor

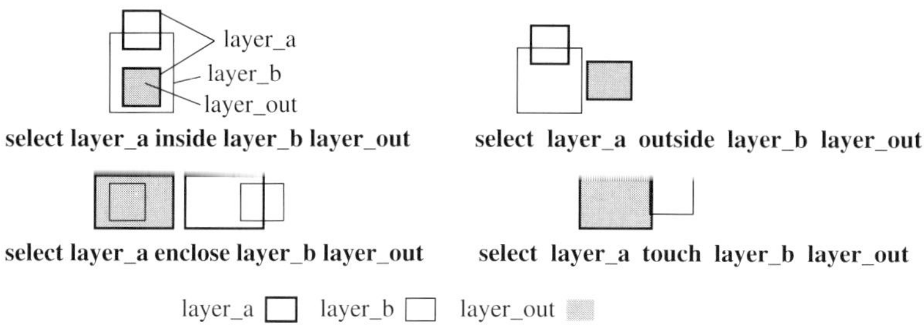

Fig. 22.3 Select Commands: inside, outside, enclose, touch, cut; and overlap

22.2.2 Select Commands

Only the polygons fulfilling particular criteria are selected by a *selectcommand*. The polygons selected are copied into the *outlayer*.

Syntax:
select layer_a <relation> layer_b layer_out

Relation:
inside, outside, enclose, vertex, cut, touch, hole, label, overlap

inside selects polygons from *layer_a* that are completely covered by *layer_b*.

outside selects *layer_a* polygons that are completely uncovered by *layer_b*.

enclosose selects *layer_a* polygons that cover a *layer_b* polygon completely.

touch selects *layer_a* polygons that touch *layer_b* polygons.

cut selects *layer_a* polygons that are partly covered by *layer_b* polygons.

hole selects *layer_a* polygons that surround a *layer_b* polygon completely.

label selects a *layer_a* polygon containing a label input with the command.

Figure 22.3 demonstrates the effects of the select commands.

22.2.3 Sizing Commands

Sizing commands grow and shrink polygons. These commands are needed, e.g., for the generation of masks. The program DRACULA from Cadence comprises the commands size, grow, and shrink for these purposes. The command *size* grows and shrinks the polygons independently of the direction, the commands *grow* and *shrink* modify the sizes in selected directions.

For the *size* command there are numerous options determining, e.g., the mode of fusion or the treatment of closely neighbored polygons.

Syntax for the *sizing* commands:

size layer_a [whitin Layer_b] by value layer_out
grow layer_a x_value y_value layer_out
shrink layer_a x_value y_value layer_out

Positive values cause growing and negative values shrinking of the polygons. Applying the option *within layer_b* a layer_a polygon is not grown to an extent larger than the borders of layer_b.

22

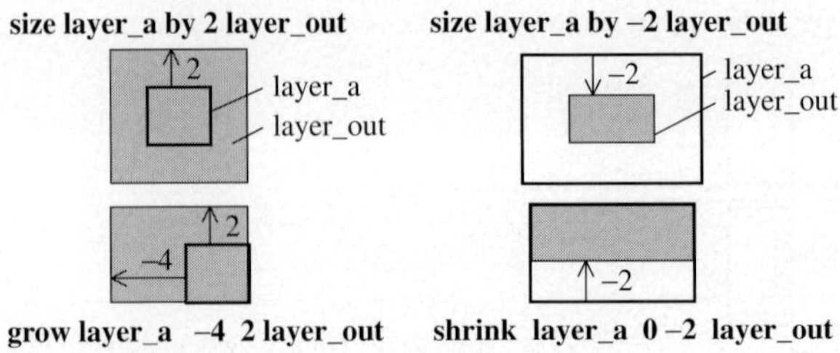

Fig. 22.4 Commands: size, grow; and shrink

With the grow and shrink commands a positive x_value cause a shift of the right hand border of the polygon, a negative x_value a shift of the left hand border. Positive y_values cause a shift of the upper border, negative y_values a shift of the lower border (see fig. 22.4).

22.3 Design Rule Check, DRC

Every chip manufacturer supplies Design Rules for their technology; that is, a table with layer combinations and the associated geometric measures. For a CMOS technology this table comprises about 100 rules which fix the minimum distances and widths of the particular layers and layer combinations. Table 22.1 and fig. 22.5 show the rules of a 0.7 µm technology for the polysilicon layer. For all the other layers there is a similar number of rules.

Table 22.1 Design Rules of a 0.7 µm CMOS Technology for the Layer Poly (dimensions in µm)

No.	Design Rule	Size
1	Minimum polysilcon width for interconnect	0.7
2	Minimum polysilcon width for transistors	0.7
3	Minimum polysilcon width for low VT PMOS transistors	1.2
4	Minimum polysilicon spacing on field oxide	1.0
5	Minimum polysilicon spacing on thin oxide	1.0
6	Minimum polysilicon extension beyond gate, on field oxide	0.9
7	Minimum polysilicon spacing to unrelated active area	0.3
8	Minimum polysilicon spacing to related active area	0.6

The actual DRC is accomplished by a checking tool that is part of the layout editor. Well known checking tools are DIVA and DRACULA from Cadence and ICrules from Mentor Graphics. The errors found by the DRC Checker are listed but they may be highlighted directly in the layout editor as well. Figure 22.6 shows the display window of the IC Station from Mentor Graphics. The transistor displayed has a DRC error in the poly layer, in addition the marked edges have a spacing that is too low.

The *Design Rules* have to be written into the technology file in a syntax which is readable by the DRC tool. With respect to the numerous layer combinations being possible the set up of the technology file is not an easy task. Such files are mostly written by experienced specialists. Errors difficult to detect require complicated error criteria. Under some circumstances these criteria cause the output of errors that are not in real. Normally technology files supplied by the manufacturer are preferred.

Within the program DRACULA there are a lot of commands for the error detection. Every error is named and dumped in an error text file. Additionally, locations of error are dumped to a graphic error file and may be visualized together with the layout.

The most important DRC commands are width, ext for the measurement of spaces, and enc for the measurement of the width of frames, e.g., around a contact hole.

Syntax of the DRC commands:

width [[options]] Layer_a measure d [layer_out] [output name n]

ext [[options]] layer_a [layer_b] measure d [layer_out] [output name n]

enc [[options]] layer_a layer_b measure d [layer_out] [output name n]

int [[options]] layer_a [layer_b] measure d

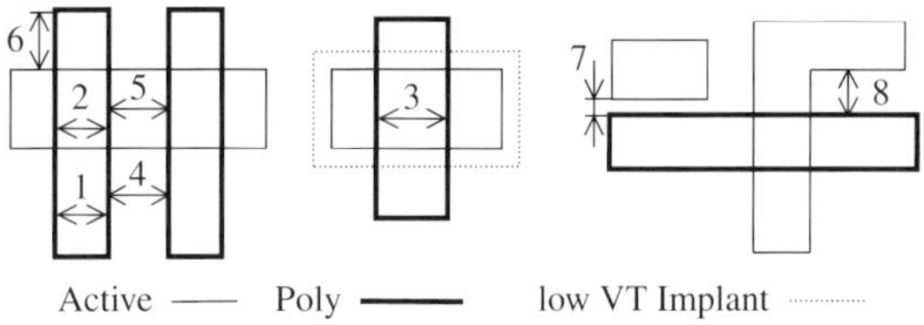

Fig. 22.5 DRC Rules from Table 22.1 with Rule Numbers

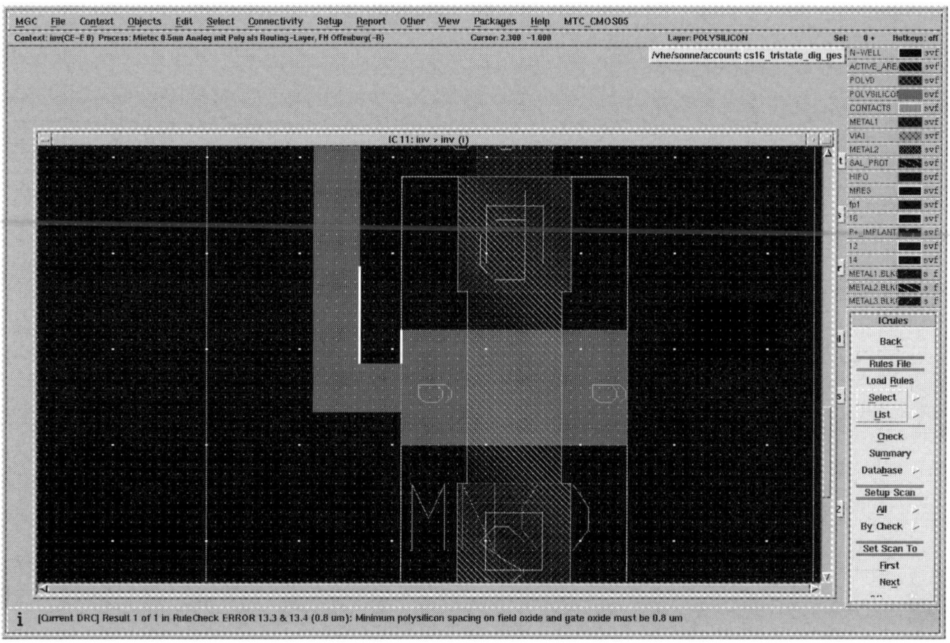

Fig. 22.6 IC Station Display Window (Mentor) with DRC Fault in Poly (thick white line)

[layer_out] [output name n]

area layer_a range/ne/eq d1 d2 [layer_out] [output name n]

length layer_a measure d [output name n]

For *measure* can be inserted:

lt	*less than*
le	*less than or equal*
gt	*greater than*
ge	*greater than or equal*
range d1 d2	value to be checked must be between *d1* and *d2*

width measures the width of a layer_a polygon.

ext measures the distance between two layer_a polygons or between a layer_a and a layer_b polygon.

enc measures the width of the frame of a *layer_a* polygon through a *layer_b* polygon.

int measures the depth of the intersection between *layer_a* and *layer_b*.

area measures the area of the *layer_a* polygon.

length measures the length of the borderline of a *layer_a* polygon.

Multiple rules may be combined by an &.

Details of distance measuring and of the generation of the error polygons as well as the references to the electric nodes may be controlled by options.

Options are enclosed in square brackets in DRACULA. More than one option may be combined in one bracket.

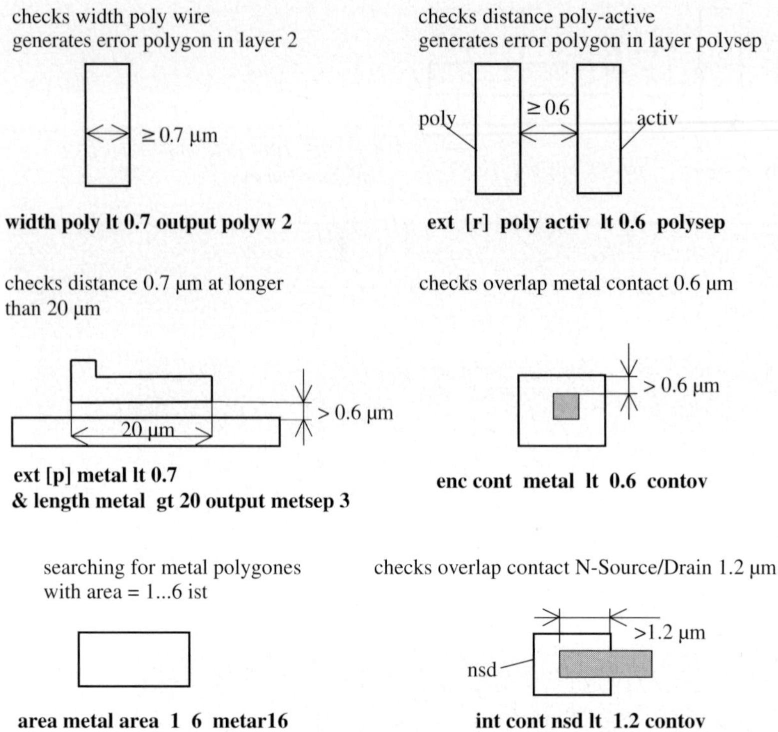

Fig. 22.7 Examples for the DRC Commands: width, ext, enc, area; and int

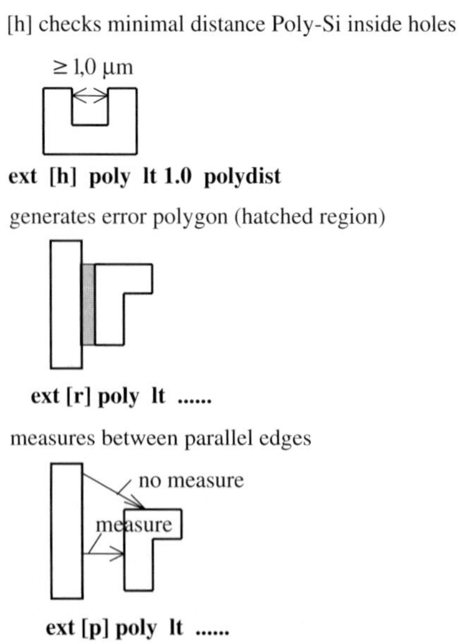

Fig. 22.8 Options: [h], [r], and [p]

Here are some examples for **Options**:

[h] by *ext [h]* the distance is checked inside the polygon as well;

[r] in the error region an *error shape* is created;

[p] during the measurement of distances only parallel lines are taken into account;

[n] checks only polygons belonging to the same node (opposite: [n']).

22.4 Extract

Extract recovers the schematic from the layout. Extract needs two definitions:

1. How are the particular layers of the layout electrically connected (*connect* commands)?
2. Which combination of layers forms a device and how is that connected (*element* commands)?

Ad 1: In DRACULA there are the following commands for the recovery of the conducting paths:

Syntax of the *Connect* commands:

connect-layer = layer_a layer_b ...
connect layer_a layer_b by contact_layer
sconnect layer_a layer_b by contact_layer
stamp layer_a layer_b

The command *connect layer* connects overlapping shapes of layer_a and layer_b.

The command *connect* defines the *contact_layer* as a contact layer between *layer_a* and *layer_b*.

The command *sconnect* (soft connect, high resistive connection) connects in a similar mode *layer_a* and *layer_b* by the contact layer. It is often used for the description of the substrate/well contacts. Then, e.g., *layer_b* is the well and *layer_a* the diffusion region that establishes the contact with the well.

By means of the *stamp* command *layer_a* is given the node name of *layer_b*.

The extraction of a device starts from a *recognition layer*. This is an *auxiliary layer* created by logic combination that exists only together with the corresponding device. For the MOS transistor the *recognition layer* is its channel, in fig. 22.9 the layer *ngate*.

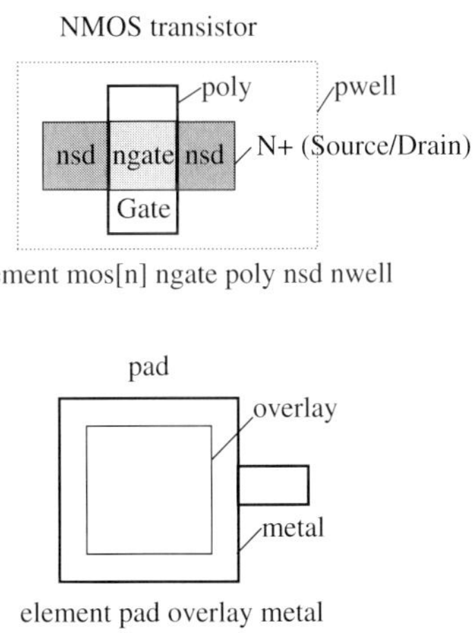

Fig. 22.9 Element Commands for NMOS and Pad

Devices are defined by the *element* command. This command contains the *recognition layer* and the layers of the terminals of the devices. There exist *element commands* for the components MOS, bipolar transistor, resistor, Diode, capacitor, and pad. The connection layers must exist in the *connect layer* statement as well.

Syntax of the element commands:

element mos [[type]] recog_layer gate source/ drain [substrate]

element bjt [[type]] recog_layer collector base emitter [substrate]

element res [[type]] recog_layer res_term [substrate]

element dio [[type]] recog_layer anode cathode [substrate]

element cap [[type]] recog_layer pos neg [substrate]

element pad vapox_layer metal_layer [substrate]

In the following example the *connect* commands will be demonstrated using an inverter (see fig. 22.10). The inverter is realized in a simple two well technology with one metal layer.

Layer:

Active regions:	*active*
(P+) Implant:	*p_implant*
Polysilicon:	*poly*
Contact:	*cont*
Metal:	*metal*
N Well:	*nwell*
P Well:	*pwell*

Generation of auxiliary layers in the Operation Block:

```
not active p_implant nact
  ; (N+) diffusion regions (active regions)
and active p_implant pact
  ; (P+) diffusion regions (active regions)
and nact pwell ndiff
  ; (N+) diffusion regions in P Well
and pact nwell pdiff
  ; (P+) diffusion ranges in N Well
and ndiff poly ngate
  ; Gate region, recognition layer for NMOS
and pdiff poly pgate
  ; Gate region, recognition layer for PMOS
not ndiff poly nsd
  ; Source/Drain NMOS
not pdiff poly psd
  ; Source/Drain PMOS
and nact nwell ntap
  ; Well contacts for N Well
and pact pwell ptap
  ; Well contacts for P Well
```

22

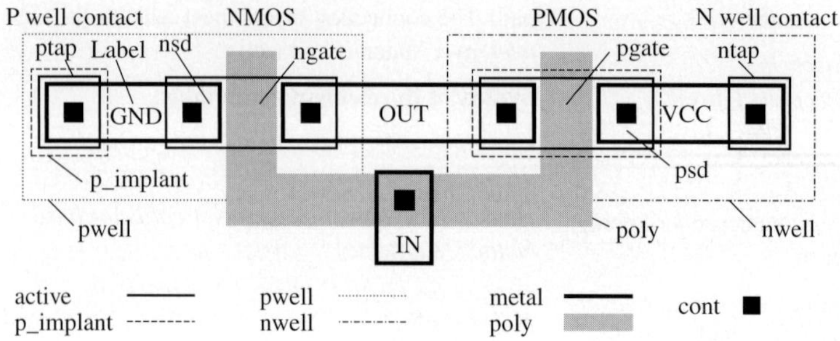

Fig. 22.10
Layout Inverter

Connect layer command in the input layer block:

```
connect layer = nwell pwell nsd psd poly
metal
  ; connect layers in bottom up order
```

Connect commands in the Operation Block:

```
connect metal poly by cont
  ; contact Metal Poly Si
connect metal nsd by cont
  ; contact Metal N+
connect metal psd by cont
  ; contact Metal P+
sconnect nsd nwell by ntap
  ; high resistive contact N Well N+
sconnect psd pwell by ptap
  ; high resistive contact P Well P+
```

Extract commands for the NMOS transistors in the Operation Block:

```
element mos[n] ngate poly nsd nwell
  ; extraction NMOS
element mos[p] pgate poly psd pwell
  ; extraction PMOS
```

22.5 Extraction of Parasitic Capacitors and Resistors

In CMOS technology the delay time of an inverter is mainly determined by an *RC* low pass filter consisting of the drain source resistance of the transistors and the load capacity at the inverter output. For the computation of the delay time it is very important to extract the effective load capacity from the layout. The load capacity is composed of the capacity of the drain regions, the gate capacities of the inputs and the capacity of the wiring. In larger chips the capacity of the wiring is dominant.

All conductors being located on top of each other or side by side create parasitic capacities. The extraction of these capacities is a rather complicated process. To do this in DRACULA there are a lot of commands with numerous options. We present only some typical examples:

For the extraction of parasitic capacities there are the commands *parasitic* and *fringe*. The command *parasitic cap* extracts area and edge capacitance. The command *fringe cap* extracts edge capacities only. The syntax of both commands is similar to the *element command*.

Syntax:

parasitic cap [[Type]] recog_layer terminal_1 terminal_2

fringe cap [[Type]] recog_layer terminal_1 terminal_2

The *recog_layer* defines the capacitor and determines the area. *terminal_1/2* are the contact layers of the capacitor respectively. The capacity per unit square of the particular capacitor is inputted by the command attribute.

Referencing the example Inverter of section 22.1 the parasitic capacities are extracted by the following commands:

```
and metal poly cmp
  ; Recognition Layer capacity metal poly
parasitic cap [mp] cmp metal poly
  ; extraction metal poly C
attribute cap [mp] 6E-17
  ; Capacity per unit square
  ; Cmp = 6E-17 F/µm²
fringe cap [fm] met met
  ; extraction edge capacity metal metal
attribute cap [fm] 3 0.003
```

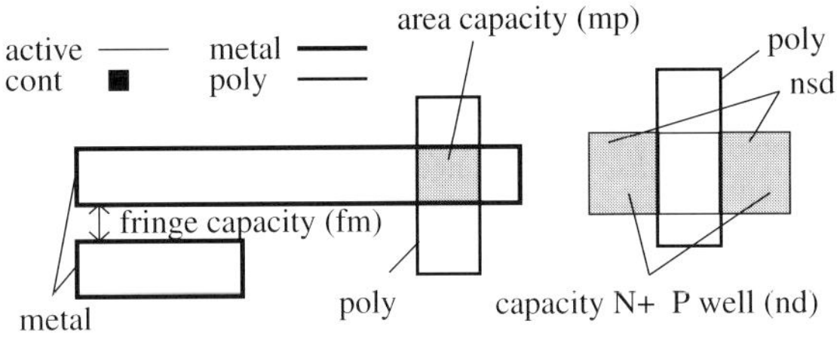

Fig. 22.11
Parasitic capacities

```
  ; max.  distance and capacity
  ; per unit square of Cfm
parasitic cap [nd] nsd nsd pwell
  ; extraction N+ Drain capacity
attribute cap [nd] 5E-16 3E-16
  ; capacity per unit square
  ; Cnd = 5E-16 F/µm²
  ; capacity per unit border
  ; Cnd = 3E-16 F/µm
```

The command *attribute cap [nd] 5E-1 6 3E-16* defines for the (N+) regions a area-capacity per unit square of 5E-16 F/μm^2 and a fringe capacity of 3E-16 F/μm.

During the extraction a huge number of parasitic capacities is created. These capacities are dumped to a text file or saved as components in a SPICE file whereby the minimum capacity may be defined to limit their number. There is an option allowing one to display each single capacity or to group all capacities of a node together to one load capacity the second electrode of which is pulled to ground (lumped capacitor). With the knowledge of all extracted capacities a post layout simulation may be executed.

By the command *lextract* the area and circumference of the Source/Drain regions may be extracted and inserted via attach into the SPICE statement of the particular transistor.

By the command *lextract* an **Antenna Check** is possible as well. The gate oxide used nowadays is extremely thin. During the ion implantation, on a long conductor connected to the Gate Poly, a voltage may be generated which can damage the gate oxide.

Therefore the ratio of the conductor area connected to the Gate to the Gate area itself (Antenna Ratio, AR) needs not to exceed a distinct value. The Antenna Ratio is given a defined value within the design rules (e.g., 100). The conductor area comprises all poly and metal layers except the top metal layer. If there is connected in parallel with the gate a substrate diode there will be no antenna problems, because the Gate charges are bypassed immediately.

In the absence of diodes long conductors must be interrupted by contacting the top metal layer (close to the Gate if possible).

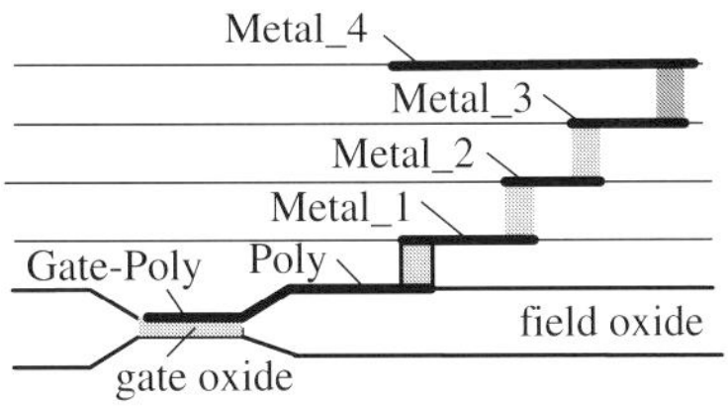

Fig. 22.12 Antenna Rules

In the example of fig. 22.12 with 4 metal layers the Antenna Ratio (AR) would be:

$$AR = \frac{A_{\text{Poly}} + A_{\text{Metal\_1}} + A_{\text{Metal\_2}} + A_{\text{Metal\_3}}}{A_{\text{Gate}}}$$

Metal_4 is not taken into account because it is the top layer.

By the command l*extract* the area of the Gate and of the conductors is extracted. By compute the ratio AR is computed and by chkpar compared with the reference value. The extraction of parasitic conductor resistances is much more complicated than the extraction of capacities because the (resistive) conductor path in the netlist has to be interrupted for the insertion of extracted resistors.

The effect of the command *cut term recog_layer contact_layer res_layer term_layer* in DRACULA" is that the conductor is interrupted by the extracted resistive layer recog_layer; the command creates the new layers res_layer and term_layer" which represent the body and the terminals of the parasitic resistor.

23 Assembly- and Packaging Methods

GÜNTHER SCHUSTER

23.1 Die Assembly

In the context of IC processing the term *die assembly* is used for the mechanical attachment of the die to the substrate, and the thermal binding of the circuit to the substrate. With power semiconductors the electrical connection gains special importance. During assembly (*pick and place*) the die is picked up with a vacuum pipette first, then aligned if necessary, and then placed accurately on the substrate. In the simplest case the pick and place device can be a semi-automaton, manually steered for position and movement. For the assembly of larger die numbers fully automatic machines are used. In mass production the sawed dies are arranged on a self-adhesive stretched foil (*blue tape*) suspended on a frame by which they are supplied directly to the subsequent treatment by the automaton. With small numbers of dies and in manual assembly the *dies* can be kept in a *waffle pack* (tray).

23.1.1 Assembly by Gluing

The most usual attachment method is gluing. Adhesive connections between IC dies and substrates are made with adhesives based on **epoxy resins**. Epoxy resins are macromolecular materials composed from diane and epichlorhydrine by poly addition. For adhesives the following substances are used as hardeners:

- *aliphatic polyamine*: leads to a hard binding;
- *aromatic amines*, *polyamides* or *polyaminoamide*: condensates of resins with polyamine: results in flexible binding;
- *dizyandiamide*: hardened by thermal processing.

The adhesive must be dispersed evenly and in constant quantity on the substrate. Application techniques used are:

- manually with a spatula, or a pipette;
- manually with a dispenser;
- by screen printing;
- in by stamping technology.

Dispensers are syringes with which adhesives, solder pastes, and masking compounds are applied. Dispensers can be used for small as well as for large production volumes. The process can be performed by hand or with the help of dosing equipment.

The **screen printing technology** is particularly suitable for mass production, since all substrates are coated at the same time.

In addition **stamping technology** is used mainly for the production of housed semiconductors, and is, furthermore suitable for *hybrid technology*. The adhesive is dispersed in a storage vessel, a flat bowl, with a stripper onto a layer of constant thickness. A metal stamp is lowered into the adhesive. When pulled out a certain quantity of the adhesive remains on the stamp. Subsequently the stamp is pressed onto the substrate and leaves there a part of the adhesive on the surface. The quantity of the adhesive and the intended thickness of the layer can be adjusted by variation of the adhesive film's thickness in the storage vessel, the stamp form, and by the pressure.

After applying the adhesive the parts may be assembled on the substrate. In order to achieve a good connection the chip will be pressed into the adhesive until all sides are fully covered with glue. The adhesive must not rise over the edges of the chip until the die's surface is wetted, which would cause electrical shorts in the metallization, see fig. 23.1. The thickness of the adhesive layer commonly used is about 25 μm.

The adhesives are hardened on a hotplate by infrared radiation in a temperature cabinet, or in a vapour phase soldering equipment. Hardening temperature and curing time depend on the type of adhesive. They can be selected from different combinations of hardening temperature and curing time. High temperature hardening saves time and affects the mechanical characteristics positively, since the resin interlaces more strongly. The **glass transition temperature** increases likewise, but the elasticity of the adhesive decreases. In practice the temperature is limited by temperature sensitive components. For *adhesives containing a solvent* a hardening in two steps is recommended, in which in the first step the solvent is driven out before the interlacing of the resin begins. During hardening various reaction products escape, sometimes together with the solvents.

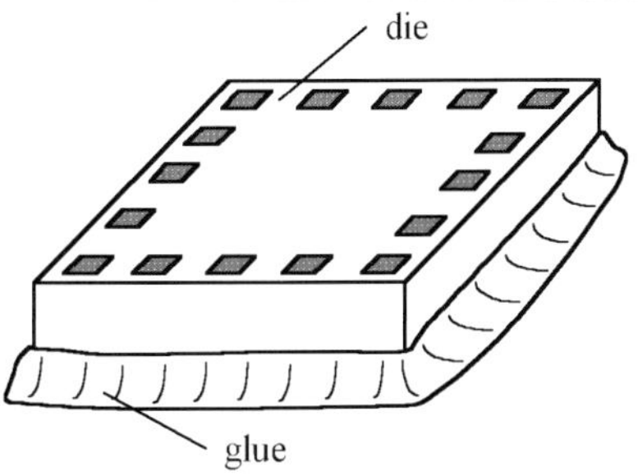

Fig. 23.1 Well processed die assembly glued to the substrate with an adhesive

Adhesives with one or two components are available on the market:

Adhesives with two components must be mixed before processing in a relationship prescribed by the manufacturer. After mixing the resin with the hardener the adhesive must be used in a relatively short time. The so called *pot lives* are from some hours to some days, depending on the product and manufacturer. Two component adhesives can be stored up to 16 months at room temperature, and for this they are recommended if only small quantities are processed.

Adhesives with one component are ready for direct use. They can be applied during their entire life span. Equipment used and the dispenser must not be cleaned after usage or at the end of each day. But storage life is very limited and amounts only 6 month under cooled storage conditions.

The mechanical strength of an adhesive is indicated in the data sheets by the parameter's *shearing strength*. Typical values of mechanical strength lie within the range of 10 N/mm^2. The shearing strength depends on the thickness of the adhesive layer and the hardening temperature. High hardening temperatures strengthen the adhesive joints.

Adhesives are organic compounds which decompose at high temperatures. The permissible continuous operating temperature is about 100 °C to a maximum of 160 °C. Going beyond these limits for a short time, e.g., during thermosonic bonding and welding, is possible, however, without degradation. *Polyimid* based adhesives can be used up to an operating temperature of 350 °C. A further upper temperature limit is the *glass transition temperature*, which usually lies below the maximum continuous operating temperature. At this temperature the adhesive begins to soften.

If **flexible connections** have to be manufactured, silicone adhesives (poly-condensation products) can also be used. This connection technique is used with large elements, e.g., solar cells and micro-mechanical components. Silicone adhesives have high coherence and adhesive strength, small dielectric losses, and are suitable for high temperature applications.

The **advantages** of a chip assembly using adhesives based on epoxy resin can be summarized as follows:

- good stability against stress during thermal shock;
- low temperature stress during hardening for semiconductors and substrates;
- continuous operating temperature may be higher than the processing temperature;
- simple processing procedure;
- easy to repair.

The **disadvantages** of adhesives are:

- electrical and thermal conductivity is clearly worse than with alloys and solder joints.

Inorganic fillers affect the electrical and thermal conductivity as well as the mechanical strength of the adhesive. Fillers are firm, non-volatile substances which lead to exactly defined mechanical, physical, and chemical characteristics in the glue layers. Fillers are built into the resin matrix as chemically neutral particles. Fillers may extend the operating temperature range, strengthen the adhesive layer, decrease shrinking, and improve all

23

Table 23.1 Comparison of adhesives with some material characteristics

Material	Thermal Conductivity $W \cdot m^{-1} \cdot K^{-1}$	Thermal Expansion 10^{-6} K^{-1}	Specific Conductivity $\Omega \cdot cm$
Epoxy, pure	0.3	50/150	$> 10^{15}$
Epoxy, Ag-filled	1.5...3.8	50/150	$10^{-4} \ldots 10^{-5}$
Epoxy, Al_2O_3-filled	0.7...1.5	50/150	$> 10^{15}$
Aluminia	200	23	$2.9 \cdot 10^{-6}$
Kupfer	400	18	$1.7 \cdot 10^{-6}$
Aluminiaoxide	30	5.5	$> 10^{15}$
Berilliumoxide	220	6	$> 10^{15}$
Silicon	145	2.3	$10^{-3} \ldots 10^{3}$

manufacturing properties, and, last but not least, extend pot life. Semiconductor chips which do not require an electrically conducting back contact can be glued with these mechanically strong and heat conducting adhesives. Suitable fillers are *aluminium oxide*, *aluminium nitride*, and *boron nitride*. To achieve a good thermal conducting joint a thin layer of adhesive with filler is recommended.

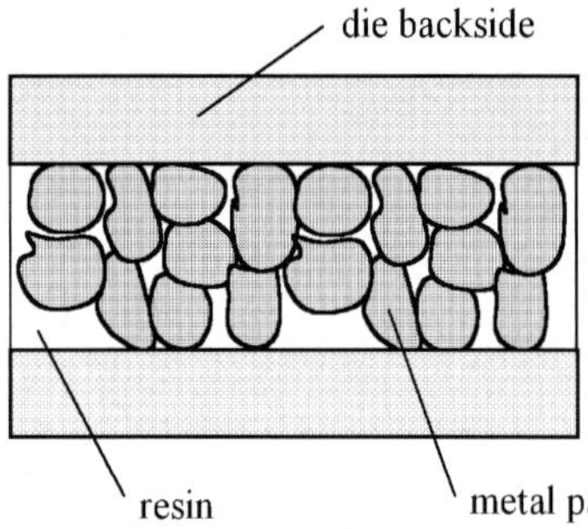

Fig. 23.2 Glue connection of a die with an electrical conducting epoxy adhesive

Fillers for **electrically conducting adhesives** are *gold, palladium, silver, nickel*, and *copper* in powder form, see fig. 23.2. The content of filler needed for improved electrical conduction is relatively high at 30 % precious metal. Silver-filled adhesives are widely used in semiconductor technology. They are preferred against the gold-filled because of the lower price. The lower conductivity of the silver filled adhesives is insignificant for most applications. Gluing a silicon semiconductor with an electrically conducting epoxy resin, a non-ohmic contact may be formed. The connection semiconductor/metal may form a p/n junction which behaves like a diode. To avoid that happening, the back side of the semiconductor must be metallized or highly doped. Suitable are, e.g., *nickel* and *gold* metallizations.

Table 23.1 compares some widely used adhesives with metals, ceramics, and silicone for specific thermal conductivity, thermal coefficients of expansion, and specific electrical conductivity.

In applying electrically conducting adhesives, the characteristics of the surfaces involved are to be considered. Well suited are: *covar, gold, palladium, nickel, platinum, rhodium*, and *silver*.

The following materials have to be considered as **unsuitable** or only usable with attendant problems: *aluminium, copper, bronze, silicon, tin* and *lead*.

The **anisotropic conducting adhesives** earn a special treatment. These adhesives are filled with spherical conducting particles. The adhesives are manufactured with a filling grade of a volume of only about 5 %. The junction serves, on the one hand, as a mechanical fixing; on the other hand, the electrically conducting balls create a local contact through the film, but isolate each contact from the other. Figure 23.3 describes the principle of operation.

The main application field today is the electrical connection to LCD modules which are connected via a flexible band (*flex print*) to the printed circuit board. In future the assembly and contacting of *grid array packages* will also be considered.

Anisotropic conducting adhesives have the following **advantages**:

- the adhesive may be applied without any sophisticated adjustment, except that the components must be placed and adjusted with care;
- Electrical connection to many contacts is made in a simple one step operation.

The **disadvantages** are:

- one of the two parts which are connected must be elastic;
- the maximum operating temperature is approx 80 °C;
- the principle implies a certain probability of short-circuiting between the contacts or non-contacting of desired connections.

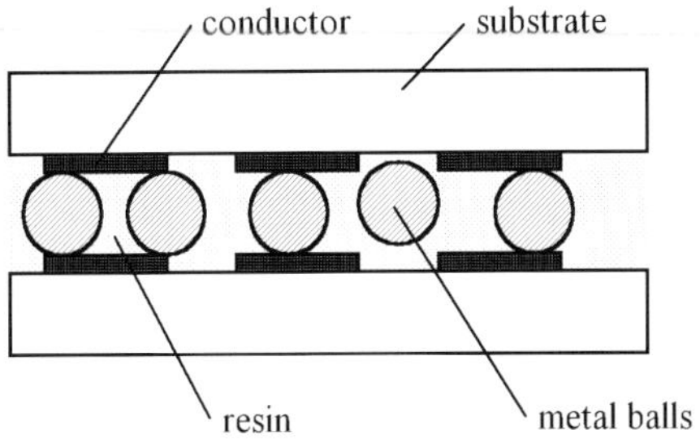

Fig. 23.3 Glue connection with an anisotropic conducting adhesive

23.1.2 Assembly by Soldering

The re-flow technique is used, in particular, for the mounting of dies on arbitrary carriers. For a good solder connection the control of the solder film is important. Solder pastes, which are applied by screen printing or a dispenser, as well as pre-forms of tin solder, are suitable for this process. Silicon itself is not suitable for soldering, it must be sputtered beforehand with a solderable metallization. Several layers are applied in a sputter or vapour phase process. Applicable materials are *chrome*, *titanium*, *nickel*, *silver*, *palladium* and as a final cover layer *gold*. The semiconductors must be stored in an oxygen-free atmosphere to keep the surfaces free from oxidation.

Widely used soldering alloys may be used for the soldering. Selection follows the requirements for melting temperature and for mechanical strength. In low temperature melting solders pure colophon may be used as flux. Better results are achieved with soldering in a reducing atmosphere.

The **advantages** and **disadvantages** of soldering used for die mounting are:

- the mechanical characteristics of a metallic solder connection are better than an adhesive joint;
- electrical conductivity and the heat conductivity is superior;
- The solder material is ductile and forms an intermediate layer between the silicon crystal and the carrier. This prevents mechanical stress with the different expansion under temperature variation. But with very large dies this may not be sufficient, so these differences in temperature expanding coefficients may be the reason for crystal cracks or mechanical tensions;
- The electrical, thermal, and mechanical characteristics decrease under temperature variations and may cause material fatigue;
- The high temperature during the soldering process may impact or damage the component;
- Oxidation of the solder material may cause acceleration of material fatigue. A hermetically closed housing is preferred. The semiconductors must be protected against oxidation during storage.

23.1.3 Eutectic Die Attachment

A widely used method for die attachment is *eutectic bonding*, using the material combination of *silicon* with *gold*. The process is also known as *eutectic die bonding*. Silicon forms an alloy with gold, which now behaves like a solder. The condition for successful eutectic bonding is the existence of such an eutectic alloy between the partner materials. The phase diagram for gold and silicon shows an eutectic point for an alloy with a fraction of 30 % silicon (in atomic numbers) at 350 °C. The used process temperature about 400 °C is only a little higher than this eutectic temperature.

During bonding the silicon die is pressed with a predefined pressure on the metallized carrier, a rubbing supports the wetting of the phase borders and accelerates the bonding process. After bonding the junction is annealed in a temperature process for some hours.

A gold metallization of the silicon crystal improves processing, pure silicon builds up an oxidation film in open air very quickly. This oxide prevents

23

the bonding in the beginning and is removed by friction when moving the die under pressure, as described before, so that the alloy may be formed.

For using eutectic die bonding on printed circuit boards metallic pre-forms of thin gold foils may be used. They are placed between the die and the conducting surface of the board.

In a **hybrid circuit** the complete substrate has to be heated to process temperature. If more than one die has to be bonded the temperature stress lasts until the last die is mounted. The number of chips on one substrate is limited by this stress because the chips should not be damaged by this procedure. For hybrid circuit die mounting some special equipment has been developed which allows a local heating of the substrate for bonding purpose. Focused infrared radiation or hot gas is used as a source of heat.

The special **advantages** of eutectic die bonding are:

- creation of low impedance contacts showing no p/n junction effects;
- very good thermal connection with high thermal conductivity;
- high mechanical strength of the connection;
- no additional soldering or adhesives are needed.

The **disadvantages** of eutectic bonding are:

- high processing temperatures;
- large induced material stress by thermal mismatch which may result in mechanical tensions or damage of the semiconductor die;
- only a limited number of dies are mountable on a hybrid circuit;
- the required annealing process needs time and increases the costs;
- expensive and critical process.

23.1.4 Measuring and Inspection Techniques

In manufacturing a simple visual inspection is suitable for quality control. Stereo microscopes or electronic image processing systems can be used as aids. Instructions for the execution of visual control are indicated, e.g., in MIL Std. 883 [23.4]. The subject of the investigations are:

- damage of the semiconductor crystal and/or the metallization;
- correct orientation of the chips;
- appearance of adhesive films or solder of the joints;
- contamination of the semiconductor or circuit by precipitation of solvents or fluxing agents.

The adhesive strength of a connection is examined in the *shear test*, see fig. 23.4. A shear test is a destructive test and is used on samples during series production. The carrier with the mounted chip is placed in a fixture. The shearing tool now attacks at one side of the die and stresses the connection. The shear force is now increased until the junction is destroyed. The maximum force is recorded. The locality of the break and the visual appearance of the break is of importance. The shear test of a good bonding leads normally to a **break of the silicon crystal**. Limits for minimal shear resistance can be defined by oneself or will be taken from MIL Std 883, Method 219, [23.4]. Because shear resistance depends on the chosen hardening temperature of the adhesive or the soldering temperature of the solder connections, this test may be used to optimise the manufacturing process parameters.

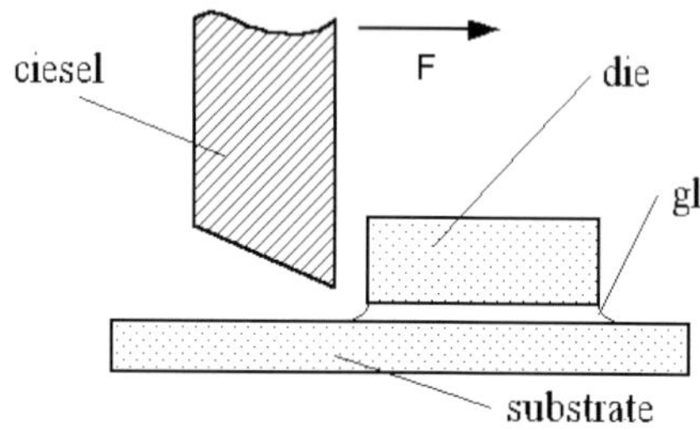

Fig. 23.4 Shear test for inspection of bond quality and strength

Further examinations can be:

- peel test;
- pull test;
- measurement of thermal resistance;
- measurement of the electrical resistance, and
- accelerated aging by temperature cycling.

23.2 Electrical Connections

After the mechanical attachment the die must be connected electrically to the contacts of the carrier or the package.

23.2.1 Wire Bonding

The electrical connection technique most frequently used is wire bonding. In particular, wires of gold and aluminium are used. During the bonding process thin metal wires are welded with the contacts. Wire bonding is a so called *micro-joining technology*. This is the creation of indissolvable connections by welding, soldering or gluing when the dimensions of the connections are in the range of those of the material's structure. Connections created by wire bonding are solid welds. The bonding between the partner materials is based on the cohesive forces of the solid. The difficulty in the welding process is to achieve an optimal approximation of the lattices of the materials over a large area, so that the atomic coherence forces become effective. At least one of the two partners must be ductile, therefore. In the case of wire bonding this is always the wire.

In wire bonding technology the welding methods used are:

- Thermosonic Bonding; and
- Ultrasonic Bonding.

Gold Lead Bonding

Thermosonic bonding is the most important wire bonding procedure in semiconductor technology. In principle this procedure represents a combination of **thermo-compression welding** and **ultrasonic welding**. The welded joint is created under the effect of pressure, heat, and ultrasonic power. On *ball wedge bonding* machines today almost exclusively gold wire is processed. The standard wire for making contact with semiconductors has a diameter of 25 μm.

The process cycle, see fig. 23.5, can be described as follows: the wire is led by a capillary, whose hole diameter is adapted to the wire. The supernatant end of the wire, underneath the capillary, is melted by the energy of a spark discharge. The melt forms a ball by its surface tension, which exhibits, e.g., a 2.5–3 times larger diameter in the rigid condition than the wire itself, see fig. 23.6.

Subsequently the capillary is lowered onto the substrate and the weld is made under pressure, heat and ultrasonic effect. The ball is centred and formed in the capillary opening. Certain capillaries give the bond a nail-head-like appearance, this is why the procedure is called *nail head bonding*. The capillary now is raised and moved to the second contact position. The wire is pulled out of the capillary by doing so. Because of the circular symmetry of the ball bond this movement may take place in each direction. When lowering the capillary again, the wire forms a loop. The form of this loop has influence on the reliability of the bond.

The second contact bonding is created in the same way under pressure, heating and ultrasonic effect. In contrast to the first bonding contact, the edge of the capillary works here as a wedge-shaped welding tool, similar to the bond wedge used in pure *ultrasonic wedge bonding*.

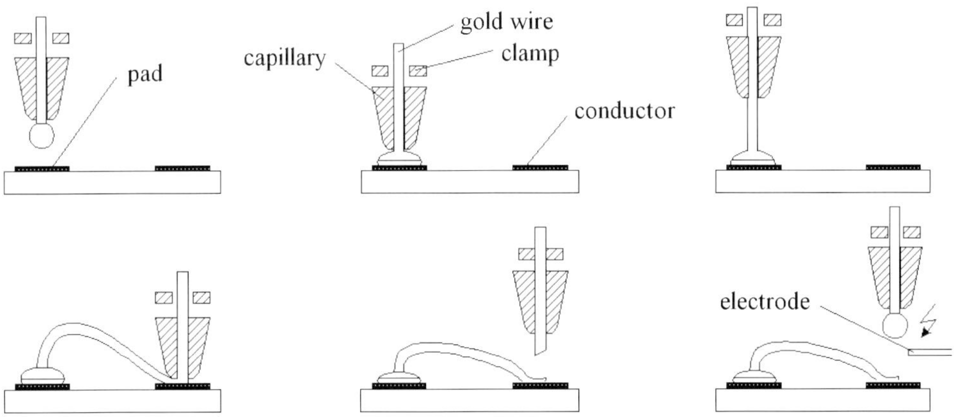

Fig. 23.5 The thermosonic bonding process

23

- With longer welding times there is a risk of material cold solidification;
- Wedge-wedge bonding is more difficult to automate because of more mechanical control of the bonding direction which reduces the operating speed;
- no bonding of substrates sensitive to ultrasound is possible.

Capillaries are used for bonding tools, in particular for bond wedges. Material and geometry of the bonding tool must be adapted to the requirements of the bonding procedure. Bond wedges for wedge-wedge bonding are usually manufactured from *tungsten carbide*. They show a service life of more than 1 million weld bonds. For welding of **gold wires**, which is not so often used, bond wedges made of *titanium carbide* or *osmium alloys* are used, since *tungsten carbide* forms an alloy with *gold*. Wire guidance and bond wedge are normally integrated in the tool. The wire is fed into the tool at an angle of 30 °C to the horizontal. Tear off load and wire ductility of aluminium wires are affected by cold material solidification caused by deformation, and an additional thermal treatment during production. One must distinguish between soft, middling, and hard wires. The material **AlSi1** with 25 μm diameter shows a tear off load of 90...110 mN for soft wires and 150...180 mN for hard wires.

23.2.2 Materials Suitable for Bonding

The metallization of semiconductors, housings, and film circuits can be welded with *gold* or *aluminium* wires in a similar way. The characteristics of the welded joints created deviate, however, from each other. Connections between similar material combinations (wire/metallization) are relatively uncritical in their behaviour. *Gold/gold* and *aluminium/aluminium* bonds are thermally and mechanically very stable, their long term behaviour is very reliable. A small decrease in strength (pull test) may be observed and has to be attributed to re-crystallizing of the wire's material. These kind of bonds are used for highly reliable circuit interconnections.

Bonds between different material combinations are found in many circuits. An inter-metallic phase is created at the boundary between the different materials. Depending on the material combination, these bonds may be very reliable or may lead to problems. The often used interconnection of gold with aluminium has to be mentioned here. If these two materials are welded with one another the **inter-metallic phases** $AuAl_2$ on the *gold* side and $AuAl_2$ on the *aluminium* side are formed at the boundary surface. The existence of these two phases does not yet damage the contact. The strength and hardness is higher than those of the two base metals. On the other hand cracks in the brittle phases, or microscopically small cavities, can under certain conditions destroy the contact by the **'Kirkendall' effect**. These cavities are formed along the phase boundary of the inter-metallic phase to the pure metal during the growth of the inter-metallic phases owed to the different diffusion constants of aluminium and gold. If their growth is not blocked these defects fuse, and the contact becomes detached. The purple colour of the gold/aluminium inter-metallic phase gave the name '**purple plague**' to this fault mechanism.

Thick film paste has a completely different structure and its bonding characteristics cannot be compared with those of semiconductors and housing interconnections. The bond characteristics of a thick film paste is affected by:

- **paste composition**: kind of the metals used, how many metals are included and additives used;
- **screen printing procedure**: surface conditions, thickness of conductor film;
- **burn in procedure**: temperatures used, processing time;
- **conductor width**;
- **long time storage** behaviour of the burned in conductors.

The surface finish of the thick film conductor is an important factor for a high quality bonding. It depends on the contents of the paste, the burn in procedure, the thickness and width of the conductor. In order to create surfaces which may be bonded easily it has to be as even and as dense as possible and should posses small roughness. Thick film conductors made from gold are easy to bond. Since gold is stable against oxidation, it changes its characteristics little even after longer storage or after several temperature processes during hybrid manufacturing. Using gold wires to bond to printed gold conductors leads to very reliable interconnections.

23.2.3 Flip Chip Technique

In the 1970s considerations were undertaken to replace the sequential and sometimes unreliable wire bonding technology by another interconnection method. Flip chip bonding is one of the oldest **simultaneous contacting processes**. This technology allows the highest component density possible, since the space requirement is reduced nearly to the area of the semiconductor dies. A semiconductor component is called a 'flip chip' whose bond pads are provided with mushroom-shaped bumps, see fig. 23.9.

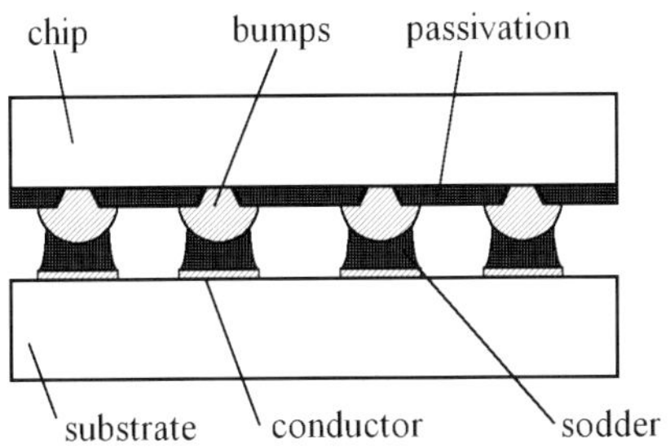

Fig. 23.9 Die, mounted by flip chip technique

Bumps consists of aluminium, gold, copper, or solder alloys, like lead/tin, lead/indium, etc., depending on the connection technique used. Their function is similar to that of a wire in wire bonding. The bumps are applied during the normal manufacturing process of a semiconductor chip. First, in a sputtering or an evaporating process step intermediate layers (chrome, titanium, nickel, platinum, copper, titanium/tungsten) are deposited. They give a better adhesion, work as a diffusion barrier, and serve as a basis for the galvanic process which follows. These layers cover the whole disk and form the cathode in the galvanic bath. In a photo-lithographic step the bump positions are defined and the bumps created by galvanic deposition. Now the photoresist and the unused metallization is removed by a solvent and acerbic step. In this way mushroom shaped gold bumps are produced.

A production process for mushroom-shaped bumps is, e.g., the **Controlled Collapse Process** of IBM. The basis for this technology is the strong selective ability of solders to wet different materials. On the connection pads of the semiconductor chips lead and tin are successively vapour deposited. The alloy relationship is exactly controlled thereby. During the following melting process the actual solder alloy is developed. It pulls together over the pad immediately to a bump, since it cannot wet the surrounding passivation layers.

Other processes for bump manufacturing are:
- mask and screen printing;
- deposition of nickel/gold;
- stud bumping.

For interconnection of the flip chip to the substrate the following procedures are used, depending on the kind of bumps:
- Thermo-compression welding (gold bumps);
- Ultrasonic welding (aluminium bumps);
- Re-flow Techniques (Pb/Sn bumps);
- Conductive adhesives.

The chip is placed face down on the carrier. Adjustment is critical between the die and the substrate metallization. To assist mounting semi permeable mirrors and IR optics are used. The connection mask on the substrate must be an exact mirror-symmetric image of the die pads. The geometric deformation of the substrate must stay within a fraction of the height of the bumps. The average height of a bump is only 20 . . . 70 μm.

Tape Automated Bonding (TAB) or film bonding, see fig. 23.10, is another simultaneous interconnection method. It combines two technologies:
- *flip chip bonding*; and
- flexible printed circuit board technology.

A flexible printed circuit board thereby serves either as an intermediate substrate (chip on tape), or it is already the final circuit (e.g., wrist watch applications, cheque card applications), where the die is bonded to. Different kinds of TAB are used, which differ mostly by the number of interconnection layers. The three layer system, which consists of polyamide foils (*Kapton*) bonded to a copper foil with epoxy resin, is the most common. A five layer system with two conductive layers is very similar to a flexible printed circuit board.

If the tape serves only as an intermediate carrier there are now two processing steps. With

A possible simple classification is that of distinguishing between **THT designs** (*Through Hole Technology*) and **SMT designs** (*Surface Mount Technology*). SMT designs have today the largest market share, for this technology has clear advantages in automation, miniaturization, and number of interconnections possible compared with THT.

The triumphant progress of SMD packages (SMD stands for Surface Mounted Devices) started in the 1970s. At that time the semiconductor components used were mostly discrete devices. First, package development concentrates on passive components such as resistors and capacitors which may be directly soldered to the printed circuit board surface. This was the birth hour of the SMD packages; later, additional packages for single discrete semiconductor devices and integrated circuits followed. The well known DIPs (Dual in Line Packages) for ICs were recognized as improvable and are now replaced by SOT and PLCC (Plastic Leaded Chip Carrier) packages. Today there are many different packages for ICs on the market, the smallest are only a tiny bit larger than the chip itself (chip size packages).

The **advantages of SMD packages** are obvious: small outline, low profile, higher interconnection density, mounting on the front *and* the backside of the board, and last but not least, smaller weight and suited for full automated production.

New innovation for packages are coming from memory and processor developments for communication engineering and from ASICs. The *ultra thin packages* were first developed for dynamic memory chips (DRAM) and later applied to other ICs with success. There is a trend to higher numbers of leads, forced by ASICs, microprocessors, and gate arrays. Best market acceptance is forecast for metric P-QFP packages (MQFP), which are available with pin number of up to 376, larger pin numbers will be available in future. Since 1991 the two most important industry associations **EIAJ** (Electronic Industry Association Japan) and **JEDEC** (Joint Electronic Devices Engineering Council, USA) are cooperating more and more, where JEDEC is applying now metric standards to new package designs, following the long term trend to metric standardization.

Very important for the market share of certain package standards is the safe handling of the processing steps by the customer. In consumer electronics today the mounting of high pin number packages with a pitch of 0.5 mm, 0.4 mm, and 0.3 mm is usual. One main problem is the prevention of any bending of the fine pitch package leads during handling at the manufacturer as well as at the customer. The requirements for co-planarity are increasing. One failure in this point may lead to a failure of the board. In Japan we have the trend towards providing fine pitch packages on a belt, coiled into a reel. In the USA more so called **guard ring quad flat packs** are offered, from which the customer has to punch out the packages himself.

23.3.1 Metal Packages

In the 1950s semiconductor components were predominantly packaged in metal or glass packages. In communications metal housings have nearly completely lost any importance, they were replaced by plastic and ceramic packages. For power semiconductors metal packages are still in use, because the thermal conductivity of metal is much superior to ceramics and others.

The most known representatives of metal packages for semiconductors are the TO package (Transistor Outline). Rectangular forms are still in use in the hybrid technology (see fig. 23.12). Concerning position and ordering of the leads one has to distinguish between **plug in packages** with leads on the package's bottom, and **flat packages** whose leads are arranged at one or all sides of the package.

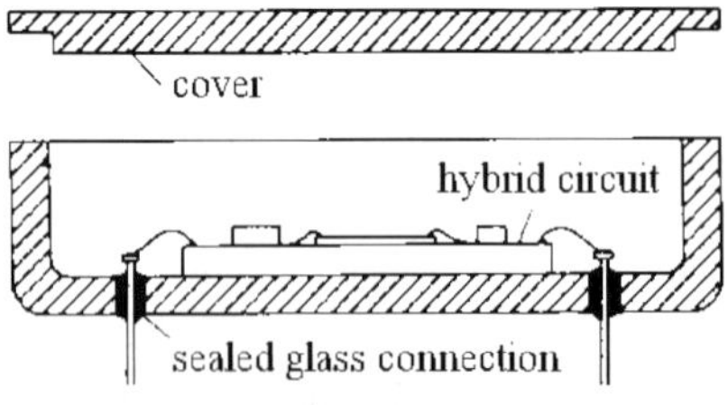

Fig. 23.12 Metal plug-in housing for hybrid electronics

Materials for metal packages are, in particular, *covar* (54 % iron, 29 % nickel, 17 % cobalt) with a thermal expansion coefficient of $4.6 \cdot 10^{-6}\ K^{-1}$. Further in use are NiFe47Cr and NiFe45, sometimes normal steel is used. In many cases the surface of the metal is protected against corrosion by a thin nickel or gold layer.

The leads are realized by means of *metal-glass through connections* (seals). Concerning these lead seals one must distinguish between glass melting lead seals and lead seals in which the glass is under constant internal tension. In both cases special glasses with adapted thermal expansion coefficients have to be used. Ready to use packages are provided by specialized companies.

Metal packages are sealed by **impulse welding** or **roll welding**. Using *impulse welding* the housing and the cover are fixed under pressure between electrodes which have the shape of the cover. A short high current impulse induces a melting of material or welding at the interconnection border between the cover material and the housing material. Welding is done in one step over the whole cover in one single moment. If the metal housing is very large and the sealing edge is longer than 100 mm impulse welding is no longer applicable and *roll welding* must be used. Today, metal housings may be sealed with laser welding, electron beam welding (in vacuum), and soldering too. Welding allows a long term hermetic sealing of the electronics housed.

Metal packages are very expensive and only used today with high quality and high reliability electronics (space electronics, aerospace applications, military electronics).

23.3.2 Ceramic Packages

Packages made from ceramics, glass ceramics, or glass can be sealed hermetically too. They represent a high performance package for high quality, long term use electronics. In contrast to metal packages they provide no protection against electromagnetic radiation. Simple glass packages are coated with films protecting against light; multi-pin ceramic packages may be protected with a shielding metal layer at the bottom and cover. A particular advantage of ceramics is that the thermal expansion coefficient may be adjusted to the expansion coefficient of silicon. Packages with many pins are able to be manufactured.

The simple glass housings are used for two pole components, e.g., diodes. They are produced in millions, are very cheap, and there are many forms suited for THT as well as for SMT. The multi-pin ceramic packages are more complex in manufacturing and for this reason are more expensive.

Figure 23.13 shows the cross-section of a typical multi-pin ceramic package (PGA, Pin Grid Array), how it is used for housing of high performance microprocessors. The main body is assembled from thin ceramic layers (so called green sheets), printed with conductors. This compound is pressed and burned in at high temperatures. The leads are applied in a further assembly step by welding or soldering.

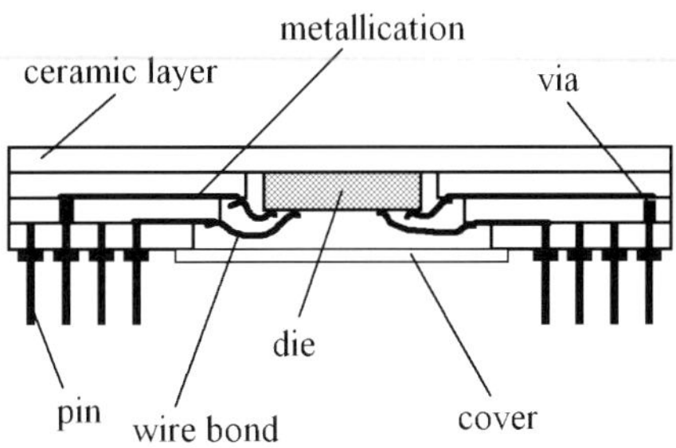

Fig. 23.13 Cross-section of a PGA ceramic package

The die is attached to the package by eutectic bonding and the electrical connections bonded to the conducting layers of the substrate. The package is sealed hermetically with a metal cover which is done with glass solder, precious metal solder, and sometimes solder pre-forms as the intermediate material.

Smaller housings are available as so called **leadless carriers**. The large contacts are optimised for making contact with printed solder deposits.

In particular, ceramic packages have the following **advantages**:
- high mechanical and thermal stability;
- high interconnection density, sometimes arranged in an array format (PGA);
- improved heat dissipation in comparison with plastic packages;
- may be hermetically sealed.

Disadvantages are:
- expensive in production and handling;
- no or bad electromagnetic shielding.

23.3.3 Plastic Packages

The predominant number of semiconductors today are packaged in plastic packages, fig. 23.14, since this is the most economical and most flexible kind of packaging. For the production of plastic packages **lead frames** are needed first. Lead frames are metallic carriers which build the platform for the bonded die and the electrical bond connections. They are the basis for the production of many plastic packages. Lead frames are usually manufactured in smaller numbers by lithography, in large numbers by punching. The materials used are CuFe or NiFe alloys. The thickness of the lead frames is typically 0.25 mm with older housings, newer packages with small pitch have only 0.1 mm sheet thickness.

The lead frame must fulfil different requirements in mechanical as well as in electrical ways. One must distinguish between

- The inner frame, or island, where the die is attached;
- The area where the bonds are made; and
- The lead area.

The manufacturers prefer lead frames which do not have to be re-worked. Outgoing leads have to be coated by tin in a galvanic bath for better solderability.

As material for the plastics, mostly epoxy resins and silicon resins, more seldom diallylphthalate, phenol resins, and polyurethane resins are used.

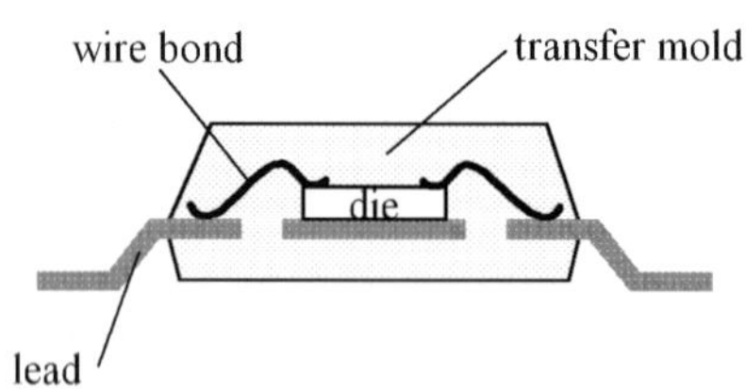

Fig. 23.14 Cross section of a plastic package

Usually the plastics are combined with inorganic fillers as there are silicon dioxide, aluminium oxide, glass fibres, and other materials. These fillers **improve the functional characteristics** of the plastic significantly. In particular these lead to:

- A better mechanical and thermal strength;
- A smaller coefficient of thermal expansion;
- A better thermal conductivity.

There are mainly three kinds of manufacturing processes used:

- Transfer moulding;
- Assembly by adhesives from pre-forms;
- Casting.

Transfer moulding is the process which is used mostly for plastic packaging of semiconductors. Before these steps the die is attached to the lead frame and electrically connected via bonding. The lead frames will be arranged in a heated steel form, normally more than one at a time, depending on the size of the frame. The plastic is melted in an extruder and injected into the moulding form under high pressure. The hardening of the plastics takes place during 1–3 minutes at high temperature and pressure. Then the packages are removed from the form and sometimes a post-hardening process is performed. The plastic materials used are types of thermosetting polymers, usually epoxy resin based. To achieve reproducible results the process parameters have to be controlled carefully.

During moulding, care has to ensure that the bond wires are not affected by the injection process. It may be useful to protect the wires with a *globe top* before injection moulding. The requirements on the quality, strength, and shape accuracy of the steel form are quite high; they are very expensive and subject to wear. Sometimes a de-flashing is needed, when remaining plastic particles have to be removed. Components mounted on lead frames require that the leads must be cut free and bent to the final shape. This manufacturing step is called 'trim and form'.

The **assembly of the housing** by pre-formed plastic parts is used together with sensor packages. Adhesives are mostly used for assembly, and in a few cases thermal melting or ultrasonic welding is applied.

Casting in pre-produced forms is widely used in hybrid technology, but also for electronic assembly groups. Sometimes the inner cover of the housing is used as a 'lost form'. If silicone forms are used the form may be used for more than one casting. Silicone resins, PU, and epoxy resins are used as casting materials. Because the casting is quite thick and not even everywhere, there may be large material tensions and shrinking after hardening, which may effect the functions of the electronic circuit or may even destroy it. To avoid this, elastic casting materials are preferred; in addition, fillers are very effective.

The **advantages** of plastic packages are summarized here:

- most economical kind of packaging;
- extremely flexibly in form and lead numbers;
- small and easy to handle.

The **disadvantages** are:

- reduced thermal strength;
- reduced chemical stability;
- low thermal conductivity;
- no shielding against electromagnetic radiation;
- not hermetically sealed;
- admission of humidity possible over long term;
- relatively high thermal coefficient of expansion.

Despite this large list of disadvantages plastic packaging is the widely used way of packaging semiconductors, not least because of the very low price of the packages compared with all other technologies.

23.3.4 Other Packaging Methods

In hybrid technology where open semiconductor chips (*naked chips*) are attached in a *chip and wire* style, a protection for the semiconductor die and the bonding wire is very important. For this a casting compound, known as **globe top**, which is applied with a dispenser on the sensitive parts. *Globe tops* are special high performance casting resins based on of epoxy and/or silicone resins with fillings. A very low viscosity prevents fluxing the globe top must stick at the dispensed place. To prevent unwanted fluxing, a shield ring of plastic may be used in addition, see fig. fig. 23.15. The globe top is black in colour for light protection and must be totally inert to the silicon die. Hardening is done by tempering in a furnace (about 150 °C for some hours).

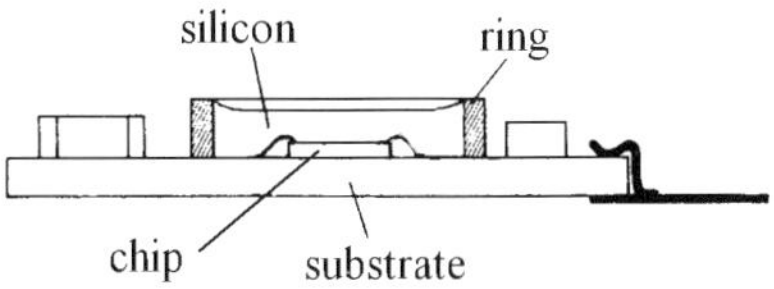

Fig. 23.15 Globe top used to protect open dies and bond wires in hybrid circuits

Larger hybrid assemblies may be protected, in addition, by plastic sintering or dipping in organic protection materials if no classical metallic housings are used.

23.3.5 Measuring and Inspection Techniques

Fully packaged components can be optically inspected, usually only by taking a sample from the lot. With a microscope or a magnifying glass the outside appearance is examined for damages, bends or other defects. The inspection may be automated with an image recognition system.

The most important test is the **final electrical test** which is accomplished in 100 % of all components produced. Many specified parameters can be tested in the packaged condition only. This test is usually done with automatic test equipment, including the handling, at ambient temperatures. The extent of the final test, the test time, and by this the test costs, are dependent on the functionality of the circuit. Further sample tests cover the measurement of electrical parameters over the full temperature range, in most cases at the corner temperatures. Components relevant to safety are fully tested against their test specifications.

Further tests may verify specified characteristics as climatic stress and temperature shock survival. Depending on the application, further tests may cover corrosion stability and estimated life span. For example, a salt spray test may be performed, stability against several solvents, oil, and fuels may be validated. For hermitically sealed housings coarse and fine leakage tests are demanded. These tests are very sensitive. In the fine leakage test the component is place in a high pressure helium atmosphere for a certain time. After removal from the pressure vessel a helium leakage detector will trace every small leaking helium molecule emitted by the package.

Testing is a significant cost factor in electronic production and packaging. The test requirements are standardized in many documents, see [23.4] for high reliability components.

23.3.6 Identification and Handling

After successful testing of the devices they are marked with identification numbers, etc. This may be done by stamping, laser writing, or by applying an identification label.

Stamping, a printing process often used, is not very flexible and mostly used for high production numbers. The wet ink has to be dried in a temperature process. Laser printing is fast, flexible, and stable against wear, but requires a surface material suited to this process. Adhesive labels are only used with very low production lots and large packages (modules).

Further packaging into reels, boxes, waffle packs, etc., depend on the requirements of the customers and the price of the devices. Small, leadless, and cheap devices may be delivered without any further packaging in volumes which are handled by the manufacturer with suitable tools. Larger devices are packaged into plastic stamps which can be stacked. Integrated circuits with many leads, e.g., BGAs, are provided in trays or waffle packs. For large production numbers this kind of packaging is much too expensive.

Packages with leads sensitive to bending and which are designed for further automatic manufacturing processes, are provided in magazines or in belts which are coiled into reels. The majority of SMD devices for mass production are processed from these reels in an automatic pick and place process.

Bulk case is a special form of volume packaging which avoids the large waste of packaging material typical of belt packaging, but is not flexible enough and is used in only few applications.

All packaging of devices has to take care of **ESD protection** (Electrostatic Discharge) which is crucial together with MOS devices. The packaging material has to be made from *low conductivity plastics* to avoid any high voltage accumulation.

23.4 References

[23.1] *Hanke, H. J.* (Editor): 'Baugruppentechnologie der Elektronik-Hybridträger'. – Berlin: Verlag Technik, 1994

[23.2] *Scheel, W.* (Editor): 'Baugruppentechnologie der Elektronik-Montage'. – Berlin: Verlag Technik, 1997

[23.3] *Reichl, H.* (Editor): 'Hybridintegration'. – Heidelberg: Huethig Verlag, 1988

[23.4] 'Test Methods and Procedures for Microelectronics'. MIL-STD 883 C. Department of Defense, USA, 1983

24 Printed Circuit Board Technologies

Andreas Foitzik

24.1 Requirements on Interconnection Technologies

Electronic installations and devices consist of systems or subsystems that can be ordered in a hierarchical manner as follows:

- discrete components and integrated circuits;
- subassemblies and Multi-Chip Modules (MCM);
- boards, e.g., printed circuit boards;
- devices;
- installations.

An electronic interconnection is necessary in between either two of the above mentioned hierarchies ensuring electronic contact. This is called a wiring even though the single interconnects literally may not be wires but, e.g., printed board circuit tracks.

A **printed circuit board has manifold functions**: it carries the electric wiring system, additionally ensuring the mechanical support of the components, and last but not least, it conducts away the heat generated by the components during operation. In HF circuits antenna functions have additionally to be taken into account. The requirements on a printed circuit board are defined by its applications, by its mechanical and thermal environment, and only then by its electrical functions. A typical example for this hierarchy are standardised boards, or the so called mother boards stemming from PC technologies.

As a carrier for components and wiring the printed circuit board dictates the **reliability** of an electric system, especially when mechanical loads are applied, vibrations, shocks, or temperature changes (typical in automotive applications). Since up to 20 % of the added value of a final product stems from the printed circuit board it is of importance for the **economic** success of that product.

The technological possibilities of manufacturing carriers with wiring functions are manifold. Trying to categorise the available technologies one finds thick film and thin film technologies, chemical and physical methods, pure and hybrid technologies, plugged, damped, and soldered interconnects, integrated and assembled circuits. In this article we will focus on printed circuit boards, i.e., card-like 2D substrates made of glass fibre reinforced polymers, and circuit tracks (made of copper) forming the circuit's layout. The characteristics of special technologies and appropriate materials, as well as the demand for low cost production, result in a number of restrictions. Such restrictions are known to the designer as design rules for the layout.

Regarding the routing one has to distinguish between the power routing and the signal routing. Power routing requires plane structures low in induction, nowadays being realised as power layers in 'multi layer' circuits), whilst the demands concerning the signal routing have been increased dramatically owing to the requirements for the transfer of signals:

- highest speed for signal propagation;
- defined electrical properties (wave resistance);
- extreme density of wires;
- small and smallest geometrical dimensions of circuits and packages.

The velocity of propagation of a signal within a wire is defined by the phase velocity c, this being defined by the permittivity ε_r as follows:

$$c = \frac{c_0}{\sqrt{\varepsilon_r \cdot \mu_r}}$$

The signal propagation delay therefore is increased by a factor of 2.2 in the case of a typical material for printed circuit boards being used (FR4 with $\varepsilon_r = 4.8$), in the case of ceramic substrates ($\varepsilon_r = 9$) by a factor of three, whilst the employment of materials appropriate for HF requirements such as Teflon ($\varepsilon_r = 2.06$) or glass ($\varepsilon_r = 2.25$) are restricted to special applications.

Similarly the maximal possible length of the circuit paths is shortened by the same factor, which is proportional to the wavelength of the signal. In the case of digital signals the minimum wavelength can be derived from the rise time of the signal t_f.

$$\lambda = \frac{c_0 \cdot 2 \cdot t_f}{\sqrt{\varepsilon_r \cdot \mu_r}}$$

$$\lambda_f < \frac{1}{10}\lambda \quad \text{without exclusion via wave resistance}$$

Taking a rise time of 1 ns and FR4 as a material one gets a wavelength of approximately 27 cm using printed circuit boards. In order to minimise reflections, radiation, and conducting effects the length of single interconnects should not be larger than 2.7 cm. Thus the requirements for the smallest geometrical dimensions of circuits stem from the demand for shortest circuit tracks, further minimising the signal propagation delay. The shortest circuit tracks nowadays can be realised on ICs in the case of monolithic integration of circuits. Such integration, on the other hand, is not always appropriate or advantageous. Thus ICs and other components will be assembled as electric circuits in order to meet a customer's requirements.

The challenges for shortest signal propagation delays thus are shifted towards assembly technologies, the attachment and the contacting of functional elements onto printed circuit boards or other carriers. This is called **macro integration** (of carriers) instead of **micro integration** (of ICs). The most important parameter describing such macro integration is the packing density defined as the quotient of the sum of all surface areas of assembly and the total surface area of the printed circuit board. Another definition of the packing density is the so called chip packing density, the quotient of the total surface area of ICs and the total surface area of the printed circuit board. Increased packing densities require higher densities of electrical contacts as well as a higher density of wiring D_V.

The maximum density of wiring

$$D_V = \frac{n_{ve}}{b_L + a}$$

is a function of the width of a circuit track b_L as well as a function of the distance a between either two circuit tracks according to fig. 24.1. In the case in which all technological parameters minimising the width and the distance of circuit tracks are optimised, a higher number of wiring layers n_{ve} is necessary. Such three dimensional arrangement of circuit wiring in a printed circuit board is called a multi-layer printed circuit board, while it should be noted that carriers of that kind are more costly, of course. The real density of wiring is lower compared to that expected from the equation for DV, since additional surface areas are necessary for contacting surface mount devices, drilling holes, and pads (in the case of through platings for interlayer connections and/or through mounted devices). Minimising these surface areas further helps to increase the density of wiring.

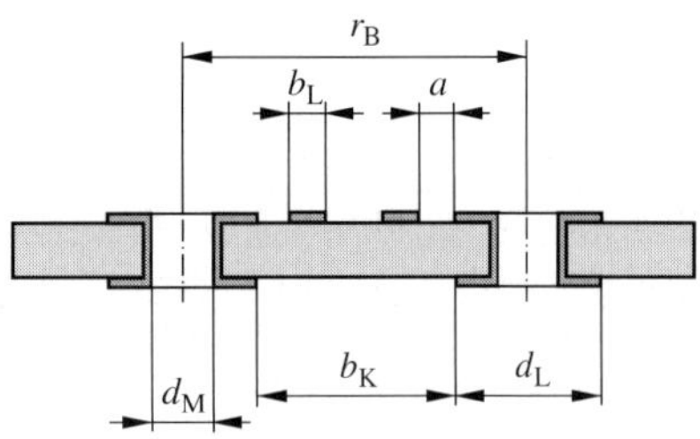

Fig. 24.1 Cross-section of a printed circuit board including through contacts

The density of circuit tracks within a wiring layer only can be further increased if the parameters given in fig. 24.1 can be minimised, given that the appropriate technologies are available. Some technologies may offer such minimisation, other may not. This points to the necessity of an optimisation of all those technological processes governing the density of wiring D_V. Later on we will take a closer look at the problems hindering further macro integration. In the next chapter we will, of course, discuss the different types of construction of printed circuit boards as well as focus on some process routines manufacturing such carriers.

24.2 Manufacture of Printed Circuit Boards

The requirements on manufacturing printed circuit boards are threefold: applying copper as structured wiring systems, assuring the electrical contact be-

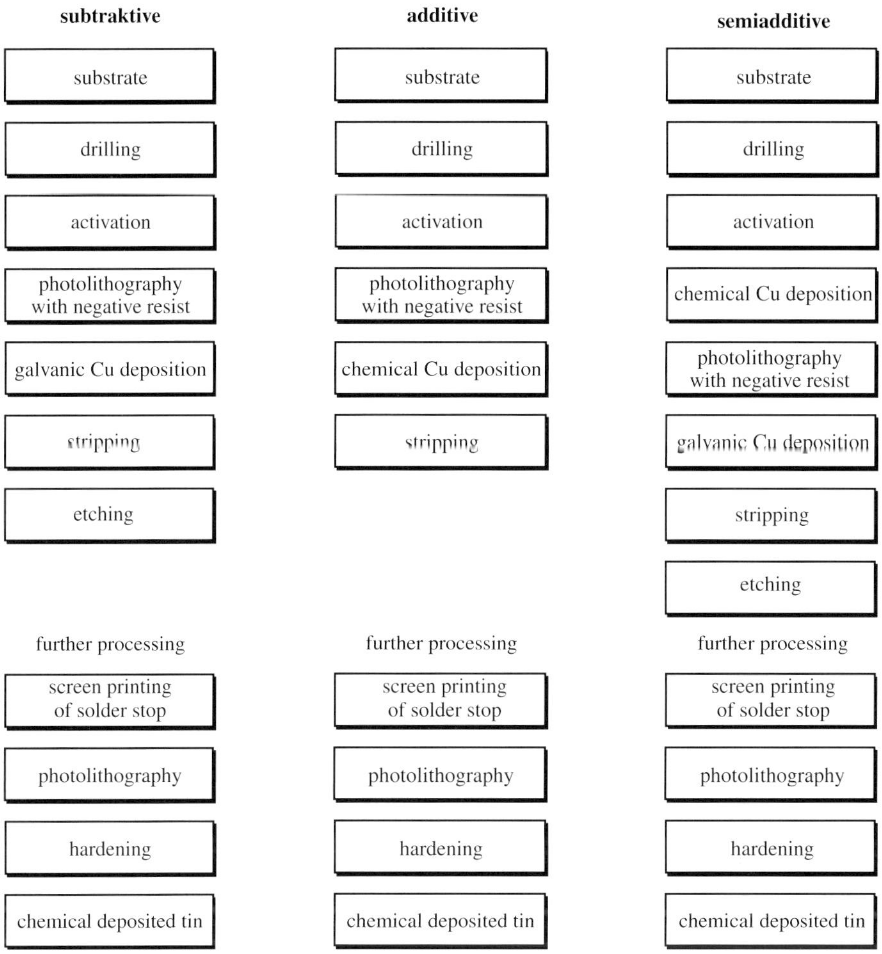

Fig. 24.2 Technological sequences for manufacturing printed circuit board by subtractive, additive and semi-additive processing

tween different layers of wiring, as well as the preparation of the printed circuit board for the accommodation of components.

A detailed description of the flow of manufacture would go beyond this articles's scope aiming for an overview. The reasons are, on the one hand, the huge number of different types of printed circuit boards, and on the other hand the huge number of technological procedures available. There are circuit boards carrying electric functions just on one side, the so called single layer boards, whereas double layer boards carry circuit wiring on both sides, whilst these two sides have no electrical interconnections, which in turn are realised in through hole plated double layer boards. Multiple layer boards carry additional wiring layers within the board, whilst one has to distinguish between such boards that have or do not have drilling holes, the latter allowing for higher densities of wiring and exhibiting higher reliability, but also showing increased costs of manufacture. Becoming more important are, e.g., flexible printed circuit boards, the basic material of these boards allowing quite extensive elastic strains and thus being adapted to non-planar designs. More exotic varieties are multiple wire boards, rigid flexible boards, 3D boards manufactured via injection moulding or boards with inlays made of metal (metal core boards, advantageous for improved heat transfer).

The topmost circuit wiring normally is manufactured onto the surface of the printed circuit board and thus slightly elevated, but also recessed circuit wiring is known.

Through contacts can run through the whole board or extend just to a specific depth (blind holes). The metallisation of these drilled holes can take place utilising physical chemical methods as well as hollow rivets.

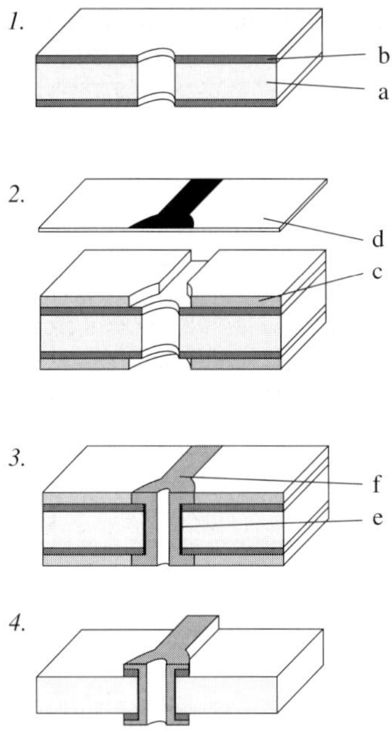

Fig. 24.3 Manufacture of printed circuit boards and through contacts by semi-additive processing: a) substrate; b) conducting starting metallisation; c) photoresist; d) image carrier; e) metallisation of the through contact or the through hole, respectively. (The conducting starting material is etched away after stripping the resist, resulting in the wiring as the only metallic layer on top of the basic material (step 4).)

Despite the huge number of different types as well as of technological procedures, it is possible to identify technological steps of processing being applied to the manufacture of printed circuit boards and determining the maximal density of wiring D_V. These are:

- design and manufacture of layouts;
- drilling and milling;
- lithography;
- etching;
- metal plating.

These technological steps are employed in different sequences when utilised for the so called additive, semi-additive, and subtractive processing routines.

In fig. 24.2 the sequences for an additive and a semiadditive routine are sketched. Besides these two specific routines there are many others realised in industrial production exhibiting differences in the actual work flow.

In the next sections single technological steps are discussed. We will focus on the technological procedures as well as identify the problems which result in deviations between the manufactured structure and the layout or the design, respectively.

24.3 Fabrication of Image Carriers: Film Bases for Reproduction Technologies

Once the circuit is developed and the equivalent topology is generated as a 2D plot the data for the different manufacturing procedures have to be generated. This includes the generation of the data for the circuit paths on the printed circuit board. Subsequently this layout has to be made up as image carriers. The technologies for creating image carriers have been advanced all the time, even though data formats are still used that were developed originally for older techniques. The criteria for using special technologies are:

- availability;
- accuracy and tolerances;
- maximum size of the printed circuit board;
- technology of the printed circuit board (single layer boards, double layer boards, ...);
- time required to realise a specific image carrier.

The layouts are produced via appropriate CAD software. Drawing or splicing (with subsequent photographic reproduction) do not play a significant role anymore. In order to avoid the critical photographic reproduction the layout is transferred directly to the film material of which the image carrier consists.

Table 24.1 Shape accuracy and positioning accuracy of image carriers

method	shape accuracy in μm	positioning accuracy in μm
drawing	50 ... 100	150 ... 200
splicing	30 ... 50	50 ... 150
vector drawing	15	30
laser drawing	30	25
laser writing	< 1	< 0.05

Transferring the design onto the image carrier can be carried out, applying **vector light drawing** utilising **plotters**. Plotters are numerically controlled machines that allow one to transfer the design from the layout onto (or into) the image carrier sequentially. Light from a halogen lamp passes through **apertures** just onto a specific region of the image carrier. Shifting the film and/or the plotter it is possible to 'draw'. For the different sizes of the regions to be drawn (for example narrow and wide circuit paths) the plotter is equipped with a large set of apertures. If just local areas have to be exposed (for example pads) special apertures are available and the light is switched on and off ('**flash**'). While reaching a new starting position the light is switched off, thus avoiding unintended exposure. A deviance from the layout stems from too low exposure rates at the edges of the apertures as well as on drawing circuit paths using quadratic apertures, if the translation is not carried out perpendicular to the edges of the apertures. The dependency of the effective brightness from the speed of drawing mostly can be compensated by increasing the absolute intensity of the halogen lamp.

In the case of utilising **laser light drawing** the digitised circuit layout is transferred line by line onto the image carrier, similarly to the process of laser printing. For that purpose the original data given in terms of vector drawing (for example **Gerber format**) have to be transformed to a matrix format (a grid pattern format). A write head containing a line of laser diodes transfers the pixels to be exposed onto the image carrier. Differences in exposure equivalent to those in vector light drawing do not occur. The time necessary for the manufacture of an image carrier is independent from the fraction of exposed area. Deviations stem from the absolute size of the laser pixel as well as from the raster elements in between the pixels.

In **laser light writing** the image carrier passes a fully focussed beam of, e.g., a HeCd laser with a wavelength of 442 nm. Structures larger than the beam diameter (this is the normal case) are painted in (filled) by the laser beam, a procedure that is known from, e.g., electron beam lithography. The exposure time for each pixel is equal, unlike at the points of return; in the case of line by line writing the prolonged exposure time has to taken into account. Areas not to be exposed are not considered in terms of moving the laser beam. Laser beam writing offers the best lateral resolution and the best accuracy of positioning, but the long times for manufacture are disadvantageous, especially when taking into account the quadratic increase in time with an increased area to be exposed. This technology therefore could not assert itself up to the other technologies, and so far is used for special applications as well as in the IC and MEMS business writing masks. With increasing integration the advantages of this technology will remain profitable. Thus an increased use of direct writing laser beam methods has to be expected in the future.

Evaluating the quality of the single methods we will focus on the **contouring accuracy** (since this is decisive for the realisation of further diminished widths of circuit paths) as well as **positional accuracy** (the accuracy in positioning a special geometry onto the designated area). An overview is given in table 24.1.

Regarding the positional geometry the distortion of the film owed to temperature changes or moisture has to be taken into account, especially when processing large formats (usually multiplex formats are processed). This gives reason for specific technological requirements on buildings and rooms where the lithography takes place. The accuracy of position ultimately defines the size of contact areas sideration of tolerances as well as of wear of the drill. The drill's absolute length is determined, thus ensuring precise depths of drilling holes.

24.4.2 Milling

The contours of the printed circuit boards as well as the breakouts are realised by milling. The milling tools are made of hard metals, typical diameters ranging from 0.8 to 3 mm. A distinction is drawn between tools with diamond tooth formation and spiral formation, the latter being used for glass fibre reinforced materials. The best quality of the cutting surface is produced by milling in the 'opposite direction', milling tools with a right hand spiral proceed leftwards when cutting. Typical parameters are: rotational speeds of 20,000 to 60,000 min^{-1}; feed rates of 0.5 to 1.5 m/min depending on the size of the milling tool; tolerances in the range of 50 µm can be realised; 3 to 4 printed circuit boards are machined at the same time as is known from drilling holes. Thus drilling and milling can be accomplished subsequently in the same machine.

Alternative procedures utilise technologies such as:

- jet cutting, especially when thermal loads as well as dust particles have to be avoided;
- laser beam drilling in the case in which ultra thin holes or blind holes are desired.

24.5 Lithography

Copper has to be structured into a wiring according to the design or the layout, respectively. Thus it is necessary to define those areas where later on the wiring is processed, and subsequently to process appropriate materials onto the printed circuit board according to the specific layout. Depending on the specific technologies used (which are summarised as subtractive, additive, or half-additive technologies) these materials:

- protect the underlying copper layer during etching processes;
- protect those areas where no copper should be added during metallisation processes.

Some very interesting polymers, the so called photoresists, reveal a different solubility in water, bases, or alcohol once they have been exposed to light. Thus if they are exposed to light it is possible to structure them just locally as determined by the layout. This processing routine is called photolithography, or just lithography and is used in printed circuit board technologies as well as in the semiconductor industry for the manucacture of ICs. Lateral dimensions ranging from about 1 m down to about 0.18 µm (the status quo in 2002) can be manufactured utilising photolithography. The fabrication of the Gigabit Chip will require structural dimensions as small as 0.11 µm and the forecasts predict dimensions as small as 0.05 µm.

Photoresists will become either more soluble (positive resist) or less soluble (negative resist) on exposure to light. The removal of the non-exposed resist in case of the negative resists, or the exposed resist in the case of positive resist, is called development, according to the visualisation of exposed and unexposed regions of photographic material. The specific kind of resist determines the application of the resist as a protective layer during etching or metallisation, respectively, as shown in fig. 24.7.

Some important properties of resists are:

- spectral sensitivity;
- thickness after drying;
- homogeneous distribution on the printed circuit board;
- lateral resolution and steepness of the edges after development;
- adhesion on the substrate;
- durability during cleaning, etching, and metallisation;
- costs.

Photoresists are commercially available as fluids or solid materials, each exhibiting advantages regarding the above mentioned properties. It should be mentioned, however, that fluid resists offer better lateral resolution, increased adhesion, and low costs whilst solid resists offer increased processing reliability and allow special applications, such as the bridging of even large holes. Fluid resists are applied by dipping, pouring, spinning, screen processing, or spraying, while solid resists are applied by lamination.

Once the resist is applied, the image carrier is adjusted. Subsequently the exposure is carried out utilising UV light in suitable machines ensuring parallel optical paths. Diffuse light results in

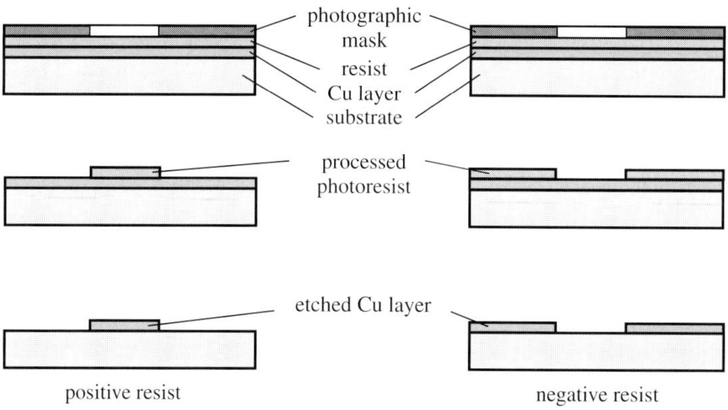

Fig. 24.7 Etching and galvanisation in case of using negative and positive photoresist, respectively

faulty exposure, and thus finally in aberrations of the wiring with respect to the layout (fig. 24.8). Subsequently the resists are developed, then the etching or the metallisation takes place, and finally the resists no longer required are removed.

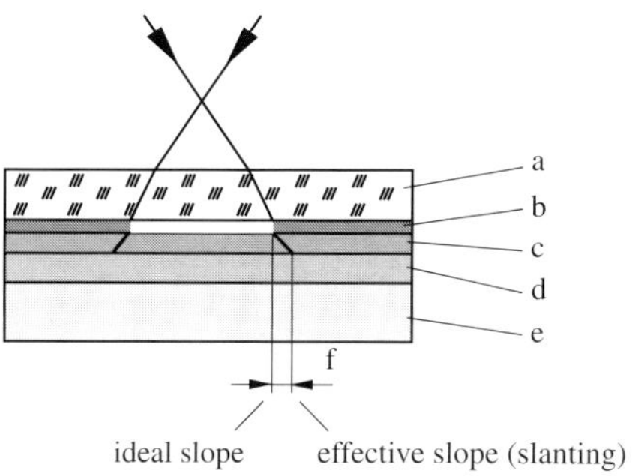

Fig. 24.8 Faulty exposure owed to non-parallel light

24.6 Etching

Almost all materials can be dissolved in a chemical reaction resulting in either a gaseous or fluid substance which can be removed. Thus wet etching as well as dry etching procedures are known. Dry etching procedures are costly in terms of investment as well as operation, therefore dry etching procedures are most commonly used in the semiconductor industry, allowing the processing of lateral dimensions in the order of less than a micrometer. In printed board technologies wet etching procedures have been used up to now. The main application is the manufacture of wiring according to the above mentioned subtractive technology. During etching the copper is removed while those areas later on forming the conducting paths are protected by etch resistant materials, like the resists mentioned above, whilst metals being used in processing may also be used as etch resists, such as lead tin, tin, gold, or nickel.

Etching of metals is a so called redox reaction, a combination of oxidation and reduction. The material to be etched is oxidised (chemically: delivering electrons), the etchant is reduced (chemically: accepting electrons). In materials sciences a distinction is drawn between etching (fast aggressive removal of material) and polishing (slow removal of material, the shining up of surfaces).

Etchants for copper and copper alloys are Fe(III) chloride, ammonium and sodium sulphate, hydrogen peroxide, copper(II) chloride, special copper salts, and chromic acid. Etching technologies can be described as dipping, spinning, spraying, and foam etching techniques (fig. 24.9). For large quantities etching equipment are arranged in modules, the printed circuit boards first being etched, subsequently the etchant being neutralised, and lastly the boards being rinsed and dried.

The chemical reactions at the phase boundary metal/etchant as well as the transport phenomena in the reaction zone determine the speed and efficiency of the etching process, whilst it is very difficult to measure and control these phenomena. In most cases the technician at the etching machine can conclude from the etching results whether the etching parameters are well set or have to be optimised. These parameters are:

- the kind of etchant and concentration;
- duration of etching;
- etching temperature;
- kind of resist and kind of lithography;
- thickness of the metallic layer to be etched.

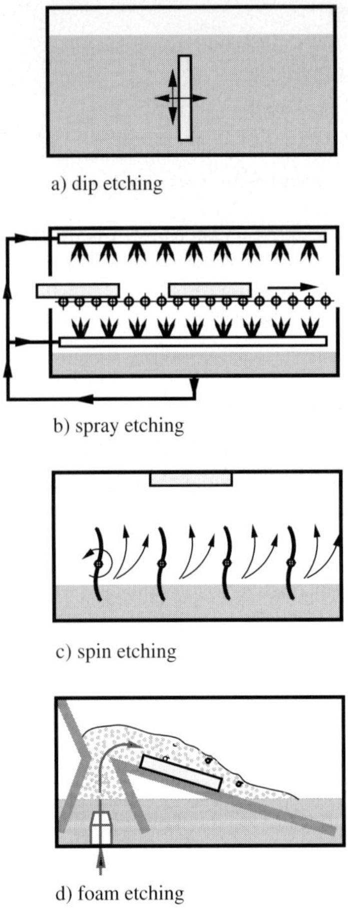

Fig. 24.9 Common etching processes

Economic success and quality of etching are determined by the speed of etching, commonly about 10 to 50 μm/min. The etching rate is a function of time, since the etchant is enriched with products from the chemical reaction. Thus a frequent change of etchant or even a continuous exchange of used etchant for fresh etchant is necessary.

The quality of the etching procedure is further determined by the so called under-etching, the removal of material even below the masking resist. The reason is the isotropic etching behaviour of most chemicals, during etching an axial as well as a lateral etching rate is observed (fig. 24.10). The etching factor is defined as the ration of these etching rates. In dipping the etching factor is just 1 to 2, on spraying the etching factor is 2 to 2.5 (spraying perpendicular to the surface results in some anisotropic etching behaviour), the additional application of protective fluids generating protective layers that are more easily removed from the lower surfaces (owing to the spraying procedure) result in etching factors of about 4.

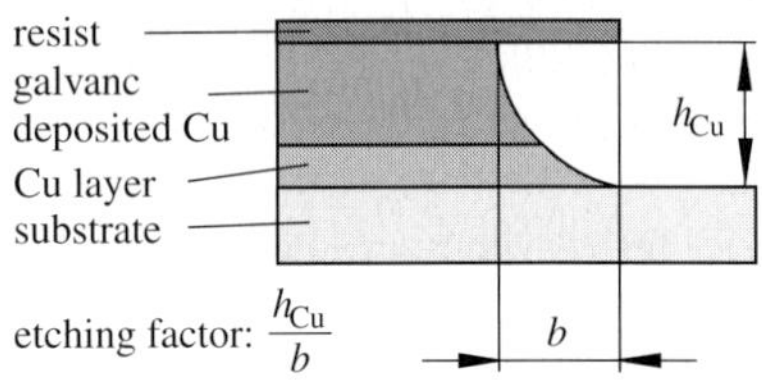

Fig. 24.10 Under-etching, definition of the etching factor

A counter-example for isotropic etching is the so called anisotropic etching of silicon single crystals utilising KOH (potassium hydroxide) as etchant. This results in etching grooves with surfaces along specific crystallographic planes, and these surfaces being almost atomically flat. Anisotropic etching is a basic process in the so called bulk micro-machining manufacturing of MEMS (Micro Electro-Mechanical Systems).

24.7 Deposition of metals

Quite a number of materials can be deposited, attached to other materials either chemically or physically without any connecting technology. As in the case of etching a number of wet and dry procedures are available. In manufacturing printed circuit boards metals are deposited mostly by utilising wet chemical processes, either by electrochemical processes or by electrodeless plating, a purely chemical process.

In the case of galvanic deposition metallic layers grow much faster, therefore this technology is widely distributed for the deposition of copper

Fig. 24.11 Fabrication line for manufacturing printed circuit board (prototypes) (Picture taken in the Goeppingen Micro-Laboratory of the Esslingen University of Applied Sciences)

in all subtractive technologies. A prerequisite is a continuous metal layer that ensures electrical contact.

Though electrodeless deposition is one to two orders of magnitude slower than galvanic deposition, it offers the advantage of plating even nonconducting materials. Thus electrodeless plating is used, e.g., for the metallisation of holes that cannot be metallised otherwise. Additionally there are processes in full additive technology to plate the whole copper wiring on a pure basic nonconducting basic material. In this chapter we cannot discuss all these technologies in detail. Similarly we will not discuss the pre-treatment of the printed circuit boards (activation procedures) which are necessary to ensure the adhesion of a first metal layer on the basic material. Even though the final protection of the copper layers (that are easily covered with an oxide layer) with metallic or non-metallic protective layers is not taken into account in this article.

In the manufacture of prototypes a space-saving integration of different processes, such as activation and galvanisation together with etching and protection, is possible. A fabrication line extending over no more than just 2 m but still containing these processes is shown in fig. 24.11.

24.7.1 Galvanic Deposition

The basic principles of galvanic deposition are described by Faraday's law and the Nernst equation, as discussed in every good textbook on chemistry. A sketch indicating the main principle is shown in fig. 24.12. The cathode accepts electrons, the metal atoms of the anode are reduced and are dissolved in the electrolyte. By diffusion in the electric field the metal ions drift to the cathode, accept electrons, and accumulate as electrically neutral metal atoms into or onto the crystallographic lattice of the cathode. Thus material of the anode is permanently dissolved and is accumulated onto the cathode. Typical growth rates are 0.2 to 0.4 $\mu m/min$ for copper and 0.1 to 0.3 $\mu m/min$ for all other technologically interesting metals (Pb, Sn, Ni, Ag).

Criteria concerning the quality of the galvanic layers are:

- the geometry of the deposition (the goal is a homogeneous thickness without any kind of pores;
- adhesion;
- the microstructure of the material, diameter of grains, distribution of grain diameters, lamellar (unwanted) or globular (wanted) grains;
- chemical purity;

24

- ductility (the feasibility for plastic deformation on mechanical stresses on further processing or during operation);
- solderability;
- resistance regarding wear and corrosion;
- electrical contacts of inner metal layers in case of multiple layer boards.

Critical areas for galvanisation are:
- corners and edges (owing to increased intensities of the electrical field);
- through contacts, especially in the case of large thickness to diameter ratios (thickness of the board to the diameter of the hole).

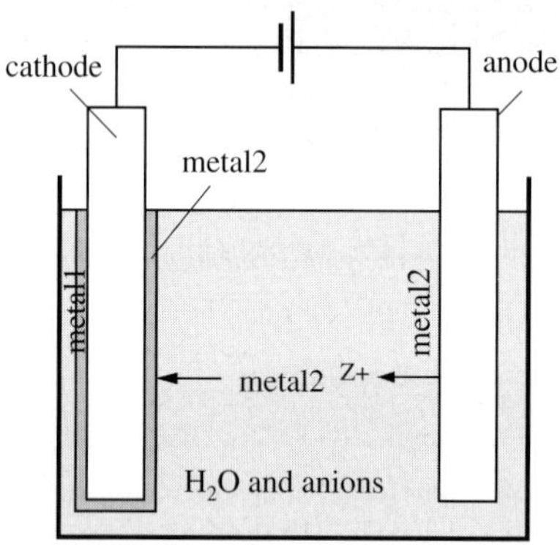

Fig. 24.12 Galvanic metallisation

In resolving these problems the technical potentials of the particular manufacturing line have to have been taken into account whendesigning the printed circuit board.

24.7.2 Electrodeless Metal Plating

Electrodeless metal plating is a so called redox reaction (see section 24.6). Regarding the manufacture of printed circuit boards only the copper deposition is of technological interest (with the excemption of special applications for tin). In an appropriate electrolyte copper ions (e.g., by adding CuSO4) as well as a reduction agent (e.g., formaldehyde, HCOH) are dissolved. The copper ions are reduced at catalytically active metal surfaces or appropriate seeds (i.e., their oxidation number is reduced from +2 to 0, thus being reduced, metallic copper has a defined oxidation number of 0), the reduction agent is oxidised (in the following example the oxidation number of the carbon C in the formaldehyde is raised from 0 to +2):

$$2\,HCHO + Cu^{2+} + 4\,OH^- \rightarrow$$
$$Cu + 2\,HCOO^- + 2\,H_2O + H_2$$

The reaction takes place at 22 to 28 °C and thus is very slow, the growth rate is just 0.01 to 0.03 μm/min. The reason is the nucleation of copper as well as the growth of copper nuclei onto the substrate. For a detailed description of this fascinating topic of metal physics special textbooks may be consulted.

A technician permanently has to control the temperature, the concentration of all chemical constituents as well as the optimised pH value of 12 to 13 (strongly basic). The goals are a homogeneous distribution of thickness of the metal layer on flat surfaces as well as in through holes accompanied with a low porosity.

24.8 Subsequent Processes

Following the galvanic processes a stop resist for the soldering is applied by spraying or screen printing. After drying and curing the resist the surface of the circuit paths are covered with tin and possibly re-melted. Special care is necessary for wrap connections covered with a thin layer of gold for optimised electrical contact.

Subsequently the printing for assembling is carried out and the boards are separated into individual units (commonly multiple printed panels are manufactured). In case of multiple layer boards tests ensuring isolation or electrical contact are necessary. All boards are inspected optically.

All these processes are not discussed here, since they are of minor interest to the designer.

24.9 Mounting and Interconnection

The printed circuit board is equipped with components, that are soldered to the board to ensure both electrical and mechanical contact. Up to now components have been used that are mounted in through holes and soldered to the backside of the board, whilst more and more surface mount devices are used that are mounted and soldered just on one side of the board (surface mount technology). Thus two different methods for soldering are used: wave soldering and re-flow soldering.

For assembling the printed circuit board with SMDs (surface mounted devices) the board has to be covered with lead tin paste at all those places where the components need to be contacted with the board on soldering. This is done using screen printing. In cases in which the contact areas are large enough the sticking strength of the paste is sufficient for holding the components onto the board after assembling, otherwise the board has to be provided with appropriate dots of glue prior to the assembling process.

Assembling takes place using highly specialised machines; the components are taken from magazines and placed onto the board, taking care of their correct orientation as well as high precision in their correct location. A distinction is drawn between Pick and Place, Collect and Place, Shooter and Revolver machines. These machines can assemble up to 100,000 components per hour onto printed circuit boards. For a detailed review specialised textbooks are available.

Components being mounted in through holes are soldered using wave soldering. This method is also used for components in surface mount technology if the density of components is not too high. On wave soldering fluid lead tin is provided by elongated nozzles, thus the observer seems to look at a soldering wave virtually standing still. The bottom side of the board is led over this wave. In fact, there are two waves necessary for faultless solder contacts. The first one uses a turbulent wave and ensures thewetting of all contacts with the solder, whilst the second one uses a laminar wave and ensures avoidance of short circuits that may have been formed in between two contacts by the turbulent wave (this is physically done by the surface tension of the solder, ensuring that only those areas that were intended are wet with solder). Nevertheless extremely dense arrangements of components cannot be soldered using wave soldering, thus re-flow soldering is getting more and more imortant.

Components to be mounted by surface mount technology are soldered by re-flow soldering. The soldering paste is fluidised, as the name indicates. The printed circuit boards are pre-heated by radiation, heated air, or by a combination of both methods, subsequently the temperature is raised just below the melting point of the solder and just very briefly additionally heated up to fluidising the solder. This procedure ensures minimal thermal stresses for the components.

Increasingly also bare chips are mounted on printed circuit boards ensuring the highest possible packaging of electrical components. The bare chips mostly are glued to the board for mechanical contact and electrically contacted with extremely thin wires. This latter technology is called wire bonding. If additionally an electrical contact with the lower surface of the IC is necessary, electrically conducting glues may be used. In this case this lower surface is additionally structured and the single structures have to be electrically contacted to the board individually, one may utilise the so called anisotropic glues. Anisotroic glues contain metal particles of optimised size ensuring electrical contact to the lower surface but no electrical contact between different contacts.

For even higher integration of electrical components on the printed circuit board alternative concepts have been developed such as Flip-Chip (FC) and Ball Grid Arrays (BGA). On top of the equivalent ICs specific areas are covered with lead tin paste (this is called a solder bump). Such components are mounted head first. Novel technologies for heating and soldering such arrangements are necessary, since in applying re-flow soldering the ICs have to be completely heated up to the melting point of the solder before the fluidisation of the solder takes place. Therefore condensation soldering has been developed. Appropriate gases condense at the solder bumps and the condensation heat is used to fluidise them. So far only special applications use these novel technologies.

24.10 MCM – Multiple Chip Modules

As mentioned in the previous chapter, bare chips can be mounted in such a way that almost no additional space is necessary (Chip Size Package, CSP). If designing and manufacturing an integrated circuit is not economical another solution for a highly integrated electric circuit would be the integration of some standard ICs onto a board, thus generating a new kind of component that may be mounted onto a printed circuit board. Such a new component would need more space than a single

25 Printed Circuit Board Design

Bernd Kohlhammer

25.1 Introduction

Performance requirements for electronic devices are constantly growing. The requirements put on printed circuit boards (PCBs) as a basis of these devices are growing in the same way. Without powerful EDA tools being available, the design of modern PCBs with high packing density would no longer be possible. Since each PCB must be optimally fitted to the intended application, many possible variations emerge. Professional EDA tools must therefore offer extreme variability. Manuals with more than 1000 pages are necessary to enable the user to apply these tools. Despite these aids, the tools can actually be used only with a lot of experience and fundamental technical knowledge.

Therefore only some few aspects can be treated in this chapter. The processes which contribute to the understanding of computer aided design tools for PCB design are described. These processes are independent of the particular EDA system. Anyhow, it makes sense to describe these processes concretely on the basis of a distinct PCB layout system. The EDA systems of Mentor Graphics and OrCAD were chosen as widely used professional systems.

The features an EDA system should have for PCB design depend on the complexity of the PCBs. The more features a system offers the higher the expenditure is for the training. In this chapter it should be tried to show the most relevant points for the selection of EDA systems for PCB design. The layout process itself and the interrelated problems with layout are described. It will be explained why the careful definition of design rules is important, or why an unfavorable choice of wire spacing may complicate the routing solution. The back annotation is described in detail, too.

Very important for all EDA systems: The provider of the EDA system must be able to supply an exhaustive and up to date library of electronic components that may be placed on the board.

25.2 Printed Circuit Design Flow

Figure 25.1 shows the design flow of computer aided PCB design in a block diagram.

According to fig. 25.1 the computer aided design can be subdivided into the design steps:

- schematic entry;
- layout design process;
- post processing.

25.3 Schematic Entry for PCB Design

The circuit's schematics is the basis for the description of an electronic circuit. It is largely independent of the technique of PCB design used. In chapter 3 symbolic schematic capture with graphical symbols is described in detail. Figure 25.2 shows an example of a schematics.

The *graphical representation of a schematics* shows the functional connections of the circuit clearly. Schematics of large systems are partitioned into several pages and hierarchies. That leads to a significant increase in clearness.

The schematic serves as a basis for all further logical, circuit related, and constructive design steps. The circuitry includes all information necessary for simulation in order to verify the demanded electrical performance. A library contains all models necessary for simulation. From the graphical representation the EDA system generates the equivalent alphanumeric description called the netlist. The netlist file describes the interconnections using the names of nets, components, and pins. Given this netlist the simulator can calculate the circuit's behaviour. The formats of the netlists (EDIF) used for simulation and further processing are described in section 8.1.

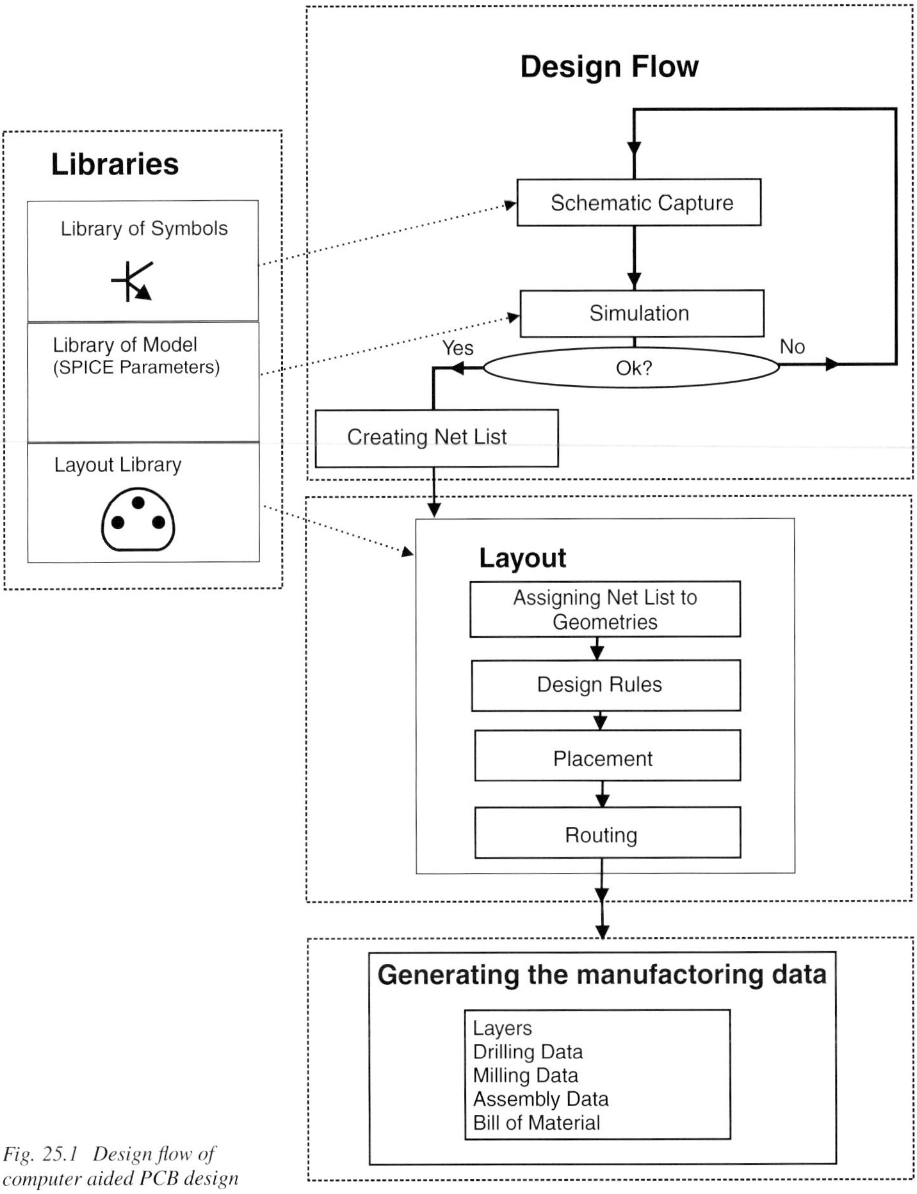

Fig. 25.1 Design flow of computer aided PCB design

A graphical symbol is assigned to each component. The symbolic, electrical, and geometrical *properties* of the components are predefined. To each component names have to be assigned, for example, R6 or U7. These names refer to the electric *properties* as well as the *models for simulation*, as well as for the *geometrical properties for layout*.

Powerful EDA systems offer *Electrical Rule Check* (ERC) tools as an important support for the user. These tools scan schematic designs and check for conformance to basic design rules and electrical rules. The results of these checks are marked on the schematic pages with ERC markers, and are also listed in a report. Checks performed by

25

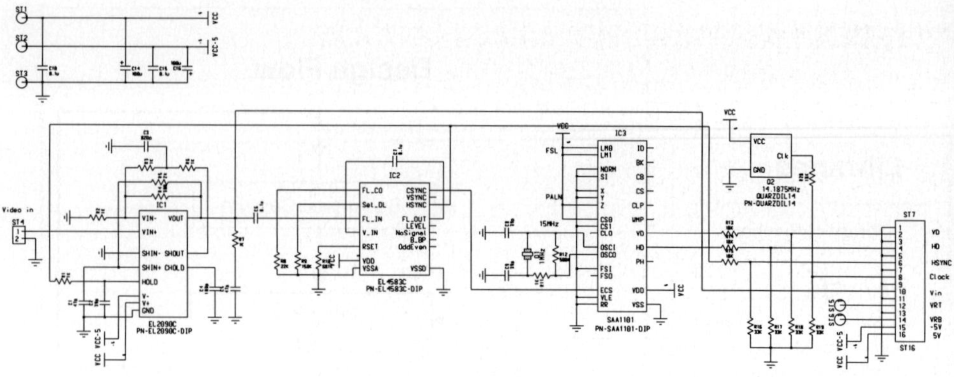

Fig. 25.2 *Example of a circuit's schematics*

the ERC tools, for example, include unconnected nets, pins, ports, or identical part references. The user can control whether electrical rule violations are reported as *errors* or as *warnings*. ERC tools make it easy to locate and fix design errors or electrical errors. Conditions that generate errors can be specified. Critical problems, which have to be fixed, must be defined as *errors*. *Warnings* are intended for situations that may or may not be acceptable and may be ignored. All errors have to be fixed before further simulation verifying electrical behaviour.

High performance EDA systems, however, offer not only the possibility of checking simple errors in the circuitry. Furthermore, one can assign limits for tolerable currents and voltages to the component symbols (see section 25.3.1 Libraries). In the ERC run these limits are checked. Thus the reliability of electric circuits can be improved considerably.

It is shown here with the example of the OrCAD system how the designer is able to control the ERC run after its needs. In OrCAD the conditions for whether electrical rules violations are reported as errors or warnings can be selected in the ERC matrix. Figure 25.3 shows a possible decision matrix.

For example, the ERC matrix in fig. 25.3 configures the ERC run to report an error if two outputs are shorted. A warning is reported if an input stays unconnected.

If all required functions are performed and verified by simulation, the netlist used for layout is generated from the schematics in accordance with fig. 25.1. This netlist contains all listed symbols, the device pins and their interconnection, as well as related properties and is used for further layout processing. The netlist may be written in a system-independent file format (EDIF), but is often processed in a company proprietary format and not visible to the user. In the same way, EDIF files with netlists imported from foreign EDA systems may be further processed. In many cases it is possible to write or edit the netlist directly in alphanumerical form with a text editor, too.

25.3.1 Libraries

Huge and actual libraries are an important prerequirement for the effective use of an EDA system. The libraries contain components with all their properties and their electrical and geometrical data. Properties are, e.g., the model parameters for circuit simulation as well as geometrical data for layout, see, for example, fig. 25.1. A component from the library must be completely defined by the properties for the dedicated EDA tool. If a circuit shall be analysed with a simulator the electrical models must be completely available.

In general, the symbols used for components are based on standard definitions. Sometimes internal company conventions are used. The symbols are maintained in the *symbol library*. Symbols which are not available in the attached library must be generated by the user. The generation of new symbols with all required properties may be very costly. **Therefore it must kept in mind that a**

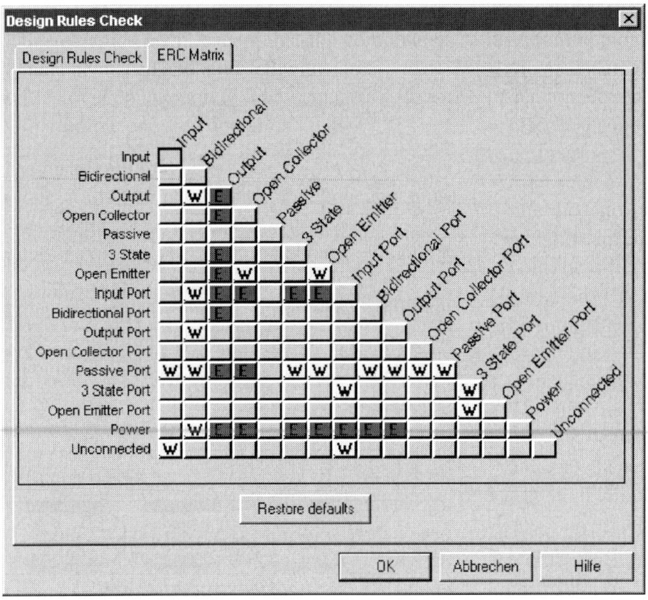

Fig. 25.3 Decision matrix for ERC configuration; ***W*** *stand for* ***Warning****,* ***E*** *stands for* ***Error***

large symbol library must be available with the acquisition of an EDA system. The quality and the capability of an EDA system for PCB design depends decisively on the quality of the component library provided.

The so-called *Board Process Library* of Mentor Graphics contains more than 14,000 digital, analog, passive, electro-mechanical, and opto-electronic components. All digital (and most of the other) components in the library contain data sheet information, electrical models for simulation, thermal model for thermal simulation, and other ERC properties.

If no symbol exists for a specific component in the symbol library it must be created by the user. With the *creation of a symbol object* a relationship must be established between the component symbolic representation, its electrical properties, and its geometrical footprint. When the object is stored there must be consistency between *geometry, symbolic representation, allocation of pins*, the *sequence of pins*, and further properties. The required data for creation of a symbol object for a given component depend on the software used and the component properties.

The Mentor Graphics EDA system uses the following data:

- Graphical symbolic representation;
- Characterization of component type (e.g., IC, transistor, resistor, diode);
- Geometry of the footprint (dimensions, number of pins, layout of pins, power supply connections);
- Allocation between the symbol representation and the physical footprint;
- Information about functional equivalent circuits and swappable pins in a physical package (e.g., in a series 7400 package are four logically equivalent NAND gates, both inputs of a NAND gate show same behaviour and by this they are swappable);
- Geometry of the pin pad (width of wire connection, size and form of the necessary pads);
- Data for automatic assembly support and manufacturing tolerances;
- Cross-references to libraries, which support simulation and test (e.g., model names);
- Limits for components loads (e.g., maximal thermal dissipation power).

25.3.1.1 Graphical Symbols

Graphical symbols characterize the electrical properties of components in a graphical representation. The geometrical properties are described by the package. The symbol object contains all the in-

25

formation which is required for PCB design. This includes the part number, the type of component, number of pins, pin assignment, etc. As simple examples fig. 25.4 shows the graphical symbol of a npn-transitor and of a resistor.

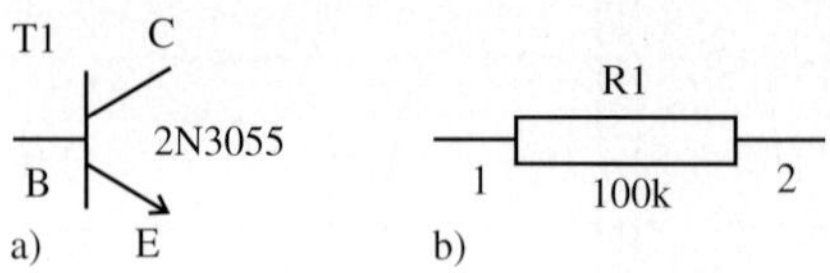

Fig. 25.4 Examples for graphical symbols a) npn transistor, b) 100 kOhm resistor

An EDA software must offer tools for creating *user specific symbols* as well as to modify existing symbol objects and to add and modify related symbol properties. This tool is usually called the *symbol editor*.

For layout creation special properties must be assigned to the symbol objects. With graphical symbols taken from a library, the properties are already defined, only the values must be edited. In addition to standard properties, company specific properties may be added (e.g., part number code, provider code, etc.).

Table 25.1 Examples for properties used for PCB design (Mentor Graphics, Board Station)

INST	Properties named **INST** (instance) must have definite values to give each component an individual, distinctive *registration number*.
COMP	Property named **COMP** (component) specifies the *component type*.
REF	The property named **REF** (reference) specifies the *numbering of symbols*, e.g., resistors R1, R2, etc.
PART_NO	The property named **PART_NO** (part number) is used *for identification of components* in the companies stock management. It is further used for the assignment of the symbol to a specific package geometry. There may be different part numbers for the same symbol existent (e.g., several transistors of the same type share the same symbol but have different packages).
INSTPAR	The property value named **INSTPAR** specifies the electrical behavior or parameters of components, e.g., resistance 10 kOhm.

As an example for properties used in PCB design see table 25.1.

25.3.1.2 Component Geometries

The component geometry defines the physical shape and characteristics of a component. In Mentor Graphics systems these chracteristics are named 'component attributes'. This includes, e.g., the pin numbers, pin location, or the component placement outline. Further component geometry characteristics are component height, restrictions on placement orientation, assignment of a component specific padstack, the reference name, and an addition of drill holes if they are needed for manufacturing.

Professional EDA systems offer tools which use the *thermal outline definition* along with *component height information* to develop a model for **thermal analysis**. The Mentor Graphics programs *AutoTherm* and *AutoThermMCM* accept such models. The component geometry allows multiple thermal outlines. There may be one thermal outline defined that applies when the component is placed on the top of the board and a different thermal outline may be defined that applies when the component is placed on the bottom of the board.

Figure 25.5 shows two examples of component packages.

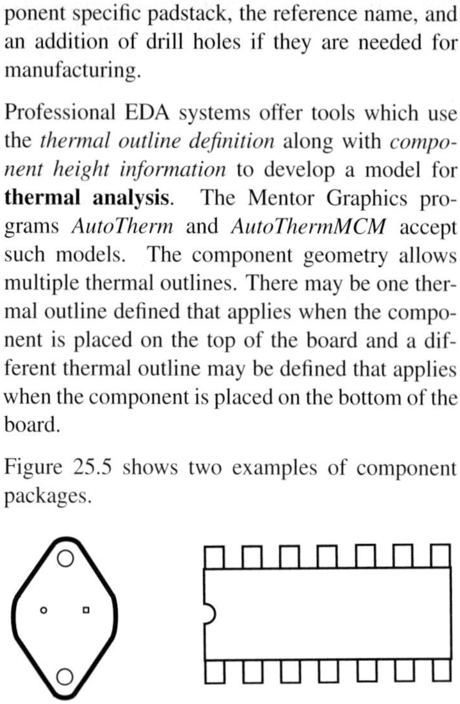

*Fig. 25.5 Examples for component packages: TO3 and DIL package (**D**ual **I**n **L**ine) with 14 pins (DIP 14)*

The layout process flow needs a referral (mapping) of the *physical pins on the component* to the *logical pins of the symbols*. This mapping has to be unique.

There are different possibilities to map the pins:
1. One-to-one pin assignment (fig. 25.6);
2. Assignment by a mapping file (fig. 25.7).

For an explanation of these two different concepts an operational amplifier in a 'dual in line' package with 14 pins is selected. Figure 25.6 shows the first form of assignment. For each pin on the symbol a physical pin on the package is assigned. The physical pin numbers on the component are *identical* with the logic pin numbers of the symbols.

The second form of assignment is shown in fig. 25.7, by the creation of a mapping file which *maps* the physical pins of the package to the logic pins of the symbols. Any logical pin name may be chosen. For instance, in fig. 25.7 the logical pins are named 'IN-, IN+, V-, OUT, V+'. These logical pins names are associated with the physical pin numbers '4, 5, 6, 10, 11'.

The one-to-one assignment (fig. 25.6) has the advantage that the effort of creating a new symbol is low. The disadvantage however is that the assignment must be modified if the package is changed (fig. 25.8).

The disadvantage of a mapping file (fig. 25.7) is that the expenses for creating mapping files are higher than for one-to-one assignments. On the other hand, it is an advantage that a mapping file is independent of the package. The change of a component is possible without additional expenditure, too. Figure 25.8 shows an example: The DIL package with 14 pins (fig. 25.7) is replaced by a TO5 package with 12 pins.

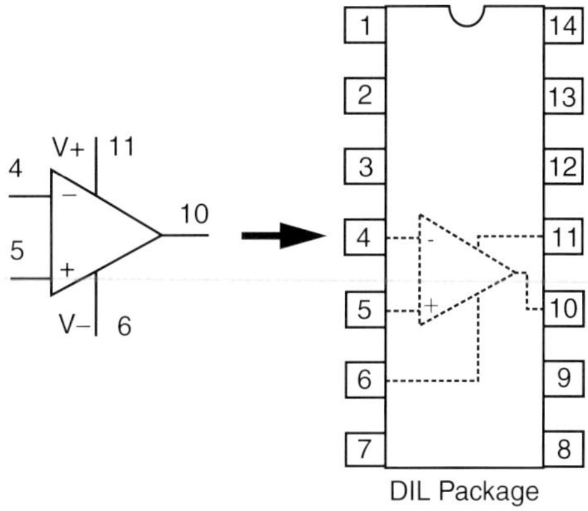

Fig. 25.6 One-to-one pin assignment of an operational amplifier in a DIL-package

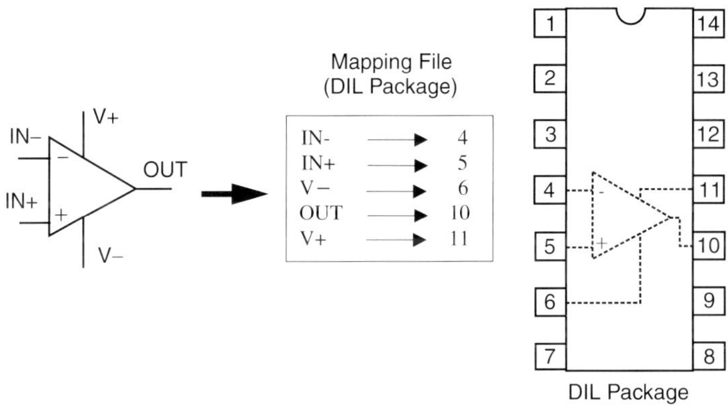

Fig. 25.7 Pin assignment to a DIL package done with a mapping file for the example of fig. 25.6

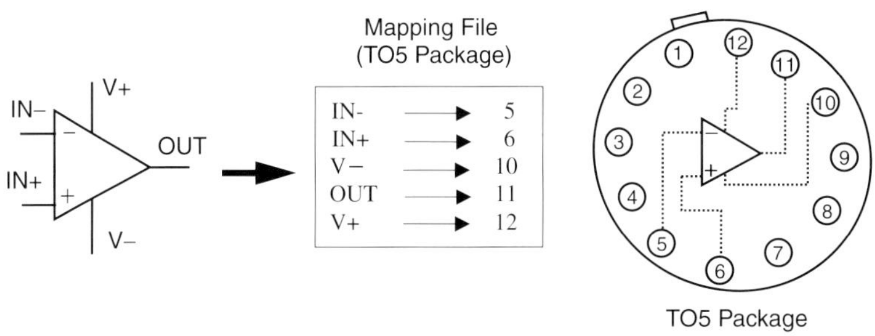

Fig. 25.8 Pin assignment to a TO5 Package with mapping file the same example as in fig. 25.6

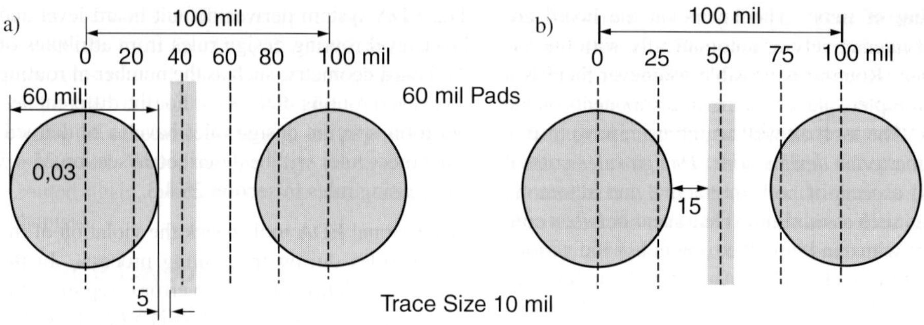

Fig. 25.10 Influence of grid spacing on traces and clearance violations. a) 20 mil grid, no trace allowed between pins because of design rule violation; b) 25 mil grid allows 1 trace between the pins

with a 25 mil grid the minimum space is now 15 mils. Thus a routed trace between the 60 mil pads is DRC conformant.

The example demonstrates the importance of a good choice of grid spacing on routing results.

25.4.4 Defining Board Geometry

The board geometry object defines the *basic shape* of the board. The object properties contain the values for the design rules assigned to the board. A placement of components is only allowed if the so-called *board outline* is defined. The board geometry requires three graphically defined entities:

- the *board outline*;
- the *board placement outline*;
- the *board routing outline*.

The *board outline* represents the shape and size of the board and describes the total physical area of the board. The *board placement* outline defines the area where components may be placed. There should be no components placed outside this line. The *board routing outline* identifies the area where traces may be routed.

Some of the design rules assigned to the board add more graphical objects to the board, such as the shapes of the placement area and of the routing area. Outlines have to be specified for each side of the board and may be specified for each layer. There may be more graphical shapes for special areas, where individual rules are valid or only special actions allowed. For instance, a placement rule area defines a region on the board and specifies a component height constraint to that area. Or an area may be specified as the placement region for a particular circuit.

Figure 25.11 shows a board with the three outlines. There is a slot defined inside the board. This could be a milling area to fit a special application.

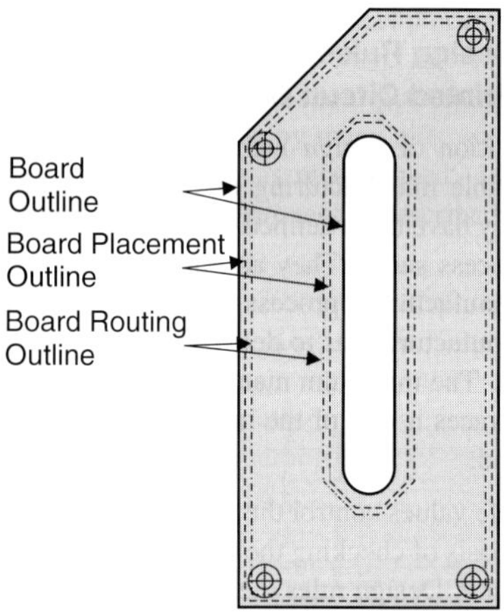

Fig. 25.11 Example of a board geometry with the different outlines

25.4.5 Placement of Components

The purpose of placement is to define positions for components on the board's surface. Modern EDA systems provide both **interactive and automatic placement** capabilities, as well as placement support features to help the user for achieving the intended results.

Interactive and automatic placement has to conform to a set of placement rules to ensure that components are placed in the proper areas of the board and that they are not located too close to each other. These rules are a combination of placement restrictions defined for the board geometry and placement constraints of the used component geometries.

When a board geometry is created attributes for the placement rules have to be assigned. These attributes define the following data set:

- Placement **grid definition**;
- Placement **outline shapes**;
- Placement **keep outs** where placement is not allowed;
- Placement regions where only components with certain **maximum height** are allowed;
- Placement region of components belonging to a **predefined group**;
- Placement region for components **not belonging to predefined groups**;
- **Minimum spacing** between components (fig. 25.12);
- Regions for **valid component orientation** (fig. 25.13).

The placement grid may differ from the routing grid. When the grids are different the user has to define the grids in such a way that some, or all, of its points coincide with each other.

Placement *keep outs* are optional areas within the placement area in which components cannot be validly placed. A *keep out area* can be associated with both top and bottom placement layers, or with only one single placement layer.

A board may contain one or more placement regions that restrict placement in those areas because of component height. Component height is defined by optional attributes assigned to individual component geometries.

Placement clearance is the value of minimum distance allowed between placed components. The *component placement outline* defines the space, the component needs on the board. As shown in fig. 25.12, this clearance value is used during DRC checking to prohibit placing components too close of each other. Components may be placed closer when automatic checking is switched off. In this case, clearance violations are noted in DRC, but overlapping components are reported as errors.

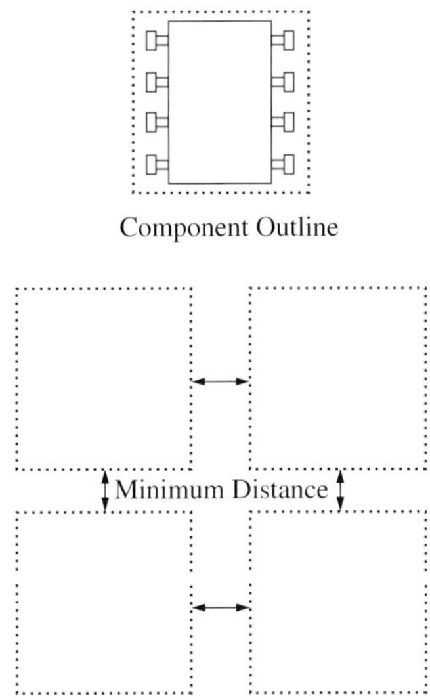

Fig. 25.12 Component outlines and clearances

When a *component placement outline* is created for library components one or more placement outlines can be defined. The outline may be defined for display on all placement layers, or different outlines may be defined for different placement layers. Generally the entire region defined by the component placement outline should be within the board's placement outline. However, when the component is interactively placed its placement outline may overlap the board's outline and still be a valid placement, as long as all the pins are within the board's placement outline.

Generally it should be noted that a **good placement is most important for the following routing process**. The placement step needs in many cases most of the time effort of a PCB design. In the planning phase of board design the user must consider the criteria which should be applied to the placement. Such criteria may be to minimize net lengths, to group analog and digital circuit parts, power supply nets, etc.

Defining component orientation may be useful too. By default, a board geometry allows component placement at any orientation or angle. Directional placement orientation can be restricted to *orthogonal* or *orthogonal and diagonal* by the presence of optional board property values. This allows some regularity for similar components. In the example of fig. 25.13 it is shown that the component columns are arranged with uneven spacing. The designer anticipates a vertical routing problem on the board and leaves space between the component columns for vertical tracks.

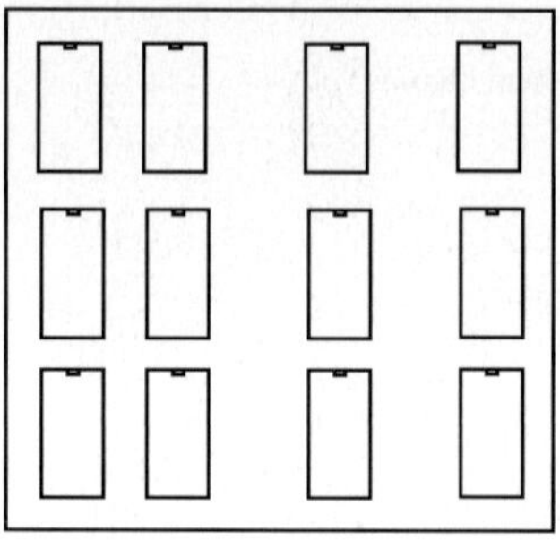

Fig. 25.13 Directional placement of components; all components are placed with same orientation, leaving space for a vertical interconnection channel

Individual components, a component group, or all components can be *protected*. **Protected components** cannot be moved in subsequent placement operations such as placing, moving, pivoting, rotating, etc.

An important criterion for acquisition of EDA software for the PCB design can be that of whether *design variants* are available. *Design variants* are PCBs with the same routing but different components placed, e.g., modified versions for different applications. In variants several netlists share one board. The designer places and routes all components of all variants. Some components are common to all variants, other components belong only to a distinct variant. If two or more components never appear in the same variant, the EDA system considers these components as mutually exclusive. To use this feature the related footprint properties are added to *mutually exclusive components*. Mutually exclusive components can be placed in an overlapping configuration which otherwise would cause placement violation errors. Later, when generating manufacturing output, an assembly drawing and a bill of materials for a single variant are created.

25.4.6 Interactive Placement

Interactive placement procedures allow the designer to place components where he wants to place them. Automatic rule checking in the background prevents invalid placement, e.g., overlapping of footprints, grid violations, directional orientation, outline overlapping, etc. The electrical interconnections are displayed in the form of so called *overflows*, which behave like *rubber bands*, showing which pins are connected by the same net. The designer now *picks and places* each component, arranged by the EDA tool in the beginning outside the board, until all components are validly placed on the board. The overflows are restructured with each placement. The overflow density may be displayed in a colour area coding, helping to estimate routing success. Placed components can be edited, moved, reoriented, or unplaced again interactively. During this *pick and place* process the full experience of the designer is used to find the optimal arrangement of the components. There are a lot of frame conditions in board design which are not captured in rules or are difficult to formulate in rules. Separation of power supply components, shielding, grouping of analog and digital circuits, etc. are examples, as well as positioning of connectors, grouping of components with same height, etc.

Figure 25.14 shows an example of an interactively placed PCB, the main IC connector is placed near the middle of the board, power supply connectors are placed in a group at the board edge, etc.

Interactive placement can be effectively combined with automatic placement, when first all *critical components* are place manually, then protected and then all remaining *noncritical components* are placed automatically. After some re-editing of placement this allows fast and effective designs and is mostly adapted.

25.4.7 Automatic Placement

In automatic placement the components are placed according to *placement algorithms*. The EDA tool finds an optimal placement in the sense of the available rules, predefined criteria, routing density, connectivity, and geometric placement conditions. One of the main criteria is to minimize the sum of all net lengths. To find this minimum several iteration steps are calculated and all the known algorithms for placement are applied. But in contrast to standard cell placement, in which the variance of cell sizes is very restricted and these algorithms

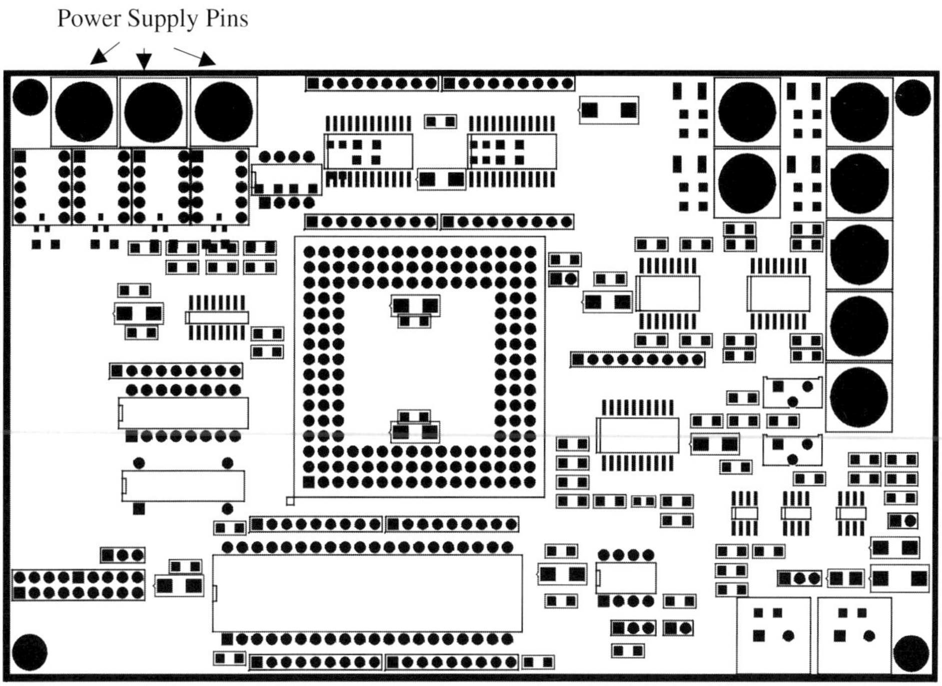

Fig. 25.14 Example of an interactively placed board

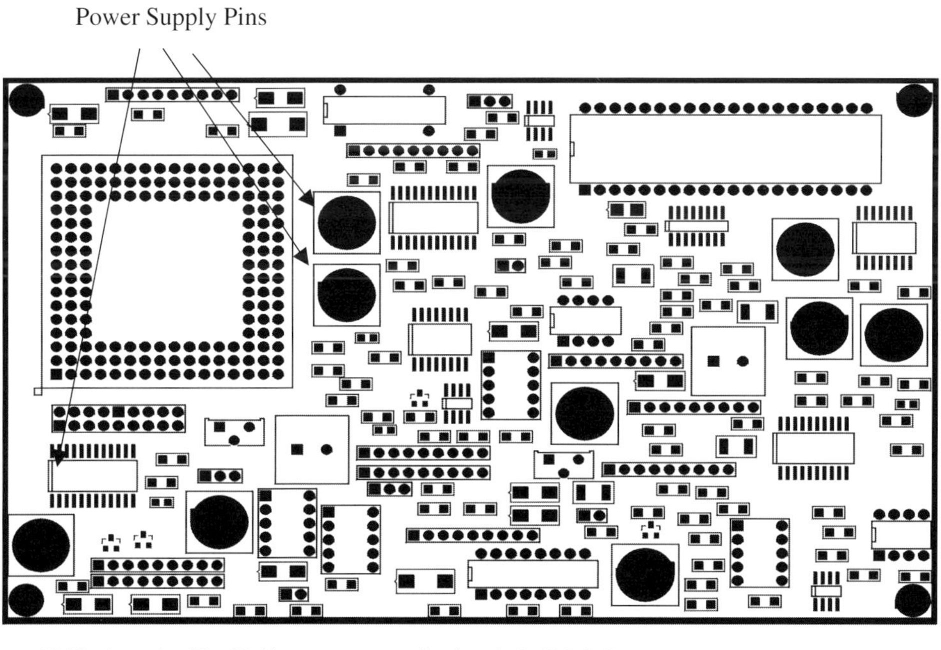

Fig. 25.15 Example of fig. 25.14, now automatically placed (OrCAD Software)

25

are very effective, PCB boards show very different component footprints and are difficult to handle. There are many constraints which are difficult to formulate as a rule, and because of this are not available for the optimisation process.

Figure 25.15 shows the example of fig. 25.14 now placed automatically. It is obvious that this placement is far from optimal because the e.g. connectors for power supply are distributed over the board.

Automatic placement in EDA systems supports placement on both the top and bottom layers of the board. The placement can be built gradually or all at once.

In practice, automatic placement of all components is seldom used. The usual way is to place all critical components interactively, protecting them, and then to use automatic placement for the rest. Adding components to groups by using group properties, e.g., grouping analog components to the group *analog* and digital components to the group *digital*, supports the placement process significantly.

The placement can be improved by using *automatic component swapping* to adjust results of automatic placement. There are several tools for re-editing and improving a valid placement interactively. All these tools rearrange placed components and optimise net connections. Placement may be *fixed*, as it is needed, e.g., for connectors, or *movable* which allows small movement (a few grid steps) and re-arrangement during the routing process.

25.4.8 Routing

Routing tools create a set of traces which connect the component pins on the board. EDA tools allow the user to route traces *interactively*, *automatically*, or *both*. Routing is not possible until all components are validly placed.

Routing layers can be differentiated into *power layers* and *signal layers*. *Power layers* contain a *power* and/or *ground net* that connects components to *power* and/or *ground*. *Signal layers* contain the traces that interconnect the signal pins of the components. *Plated through hole vias*, as well as *additive technology based vias*, are supported to feed routing from layer to layer. *Power net* and *ground net* can also be routed as *signals* on *signal layers*.

Routing is done in a grid of potential *trace paths* that run horizontally or vertically across the board (*grid routing*). Gridless routing is only allowed for special purposes and should be avoided. The user has to specify the grid settings for the design when he sets up the routing rules. The preferred routing direction, vertically or horizontally, is specified in a direction property belonging to the layer.

Just as placement is constrained within a defined area and by varous placement rules, a series of design rules prohibit routing on certain areas of the board (*keep out areas*) and maintain minimum clearances between all traces.

The basis for the routing process is again the netlist which contains all connections between component terminals. These un-routed connections are displayed as *overflows* or *guides* in a predefined colour, easily to distinguish from traces. These *guides* are replaced with *traces* as routing proceeds.

The *board routing outline* (fig. 25.11) limits the area within which traces can be routed. The board routing outline is the same for all routing layers. Inside the routing outline are optional *keep out areas* where routing traces are not allowed, either on a specific layer or on all routing layers. A *routing keep out area* restricts traces and vias, as a *trace keep out area* restricts only traces, and as a *via keep out area* restricts just vias. The routing outline and keep out areas are attributes of the board geometry. Keep outs can also be included in the component specification as part of the geometry definition.

Keep outs may be applied to *critical nets* in analog circuits, for example. An optimal routed layer improves electromagnetic interference (EMI).

Design rules control several aspects of both interactive and automatic routing, such as the minimum clearances between traces and how the router uses the various pin and via pad stacks.

As in manual placement, the user can **interactively route** a design before or after using the autorouter. Generally, interactive routing is used to pre-route critical signal nets and supply nets before activating the *autorouter*. Interactive routing is further used to edit autorouted, but sub optimal, boards or to pre-route traces in difficult areas of high density, e.g., within a grid area or dense connectors with fine pitch. Of course, connections which are left unrouted by the autorouter (shown as *overflows*) are interactively routed by ripping up traces and re-routing of nets. This work is generally time consuming and needs a great deal of experience.

Autorouting can be done whenever there is a valid complete placement on the board. Before using the autorouter the user should embark on a *routing strategy*. He should consider his design from an autorouter's point of view:

- What is the best method of specifying a routing grid and how many layers do we need?
- What technology rules are specified by the manufacturer?
- What are the characteristics of components and connectors on the board?
- What are the general design rules?

The rules have to be carefully defined before starting the autorouter. There are:

- general rules, used in every board of this kind;
- company specific rules, mostly related to quality, soldering and assembly;
- board manufacturer rules;
- application specific rules.

The quality of the design and the costs of manufacturing the board, of later assembly of the components and testing are directly linked to these rules set up. Every bit of manual editing work after routing is time consuming and a high cost factor, besides the risk of introducing new errors. This rule set up is tried to be kept standardized, it will only be slightly changed with different application boards.

The routing algorithms used today are classical *maze runner algorithms* (e.g., the Lee algorithm), which are steered by cost factors, applied to paths found. These cost factors are directly influenced by the design rules. The *Lee algorithm* provides a set of equivalent *shortest paths* for a net. Traces going left to right in a horizontally defined layer are much 'cheaper' than traces going up or down, so they are preferred. Each *bend* in a path has extra costs, *vias* are expensive too, so they are avoided as far as possible. It is useful to configure the router by redefining these cost factors to improve routing performance, but in many cases the default values are sufficient. Routing performance further depends on the sorting of the nets in the netlist. Short distance connections are routed first, long distance connections are routed later. The sorting may be influenced by defining nets as *critical*, and these are routed first. Autorouting is done by several iterations with some *rip up mechanism* (**rip up router**) for difficult nets. The router works multidimensionally in all layers of the board. Autorouting is stopped when all overflows are routed or if there is no improvement possible after a predefined number of rip up tries. If there are overflows left, they have to be routed manually.

The process contains some **heuristics**, so it is not possible to an identical routing twice, although the starting conditions are the same. A good placement leads today, with professional systems, to a 100 % resolution, remaining overflows of up to 10 may be acceptable, if there are more the process has to be started again. If there is again no improvement the placement has to be modified or the rules defined have to be checked.

Figure 25.16 shows an example of fig. 25.14 after the autorouting process. Only some few traces have been corrected.

The fig. 25.1 and fig. 25.9 show that in the design process the *design schematics* and the *netlist* derived from it are the source of all following steps.

Both the simulator, as well as the different layout tools, refer to the schematics. The simulator and the layout tools both use some information from the schematic, for example, the number of components, but not all. Most of the component properties required for simulation as well as for layout are, however, different. For example, geometrical data is normally irrelevant for the simulator, the layout process, however, is essentially based on these data. Therefore the components used in the schematic must contain properties for simulation as well as geometrical data for the layout process.

It is important that changes or corrections are still possible in all phases of the design process. The design must stay consistent. This implies that each design change is administered correctly in a so called *back annotation*. In fig. 25.9 it is shown that the back annotation process registers all changes which are made during the layout process, and re-report these changes to the schematic. The back annotation process is optional and can be

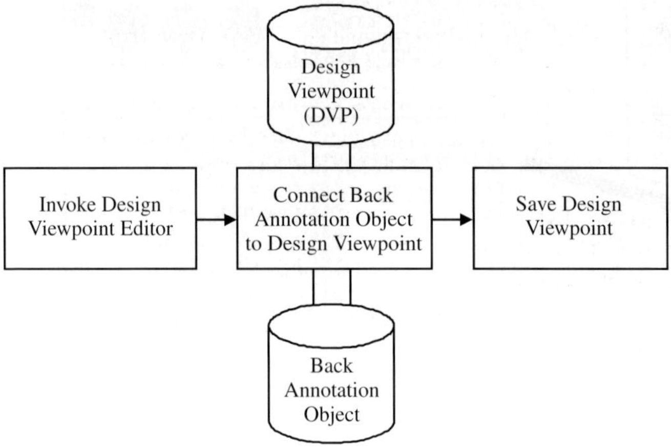

Fig. 25.19 Connection between back annotation objects and viewpoint

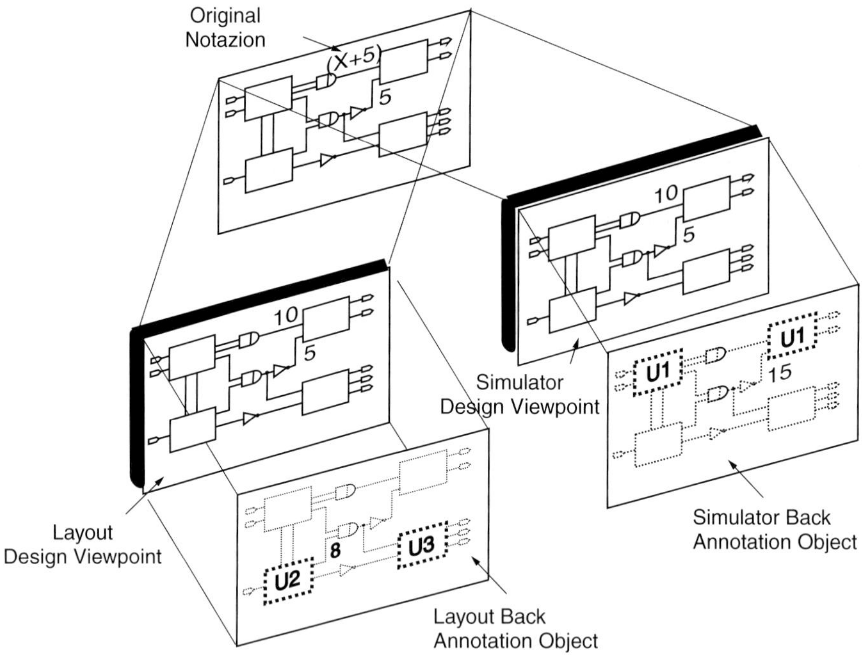

Fig. 25.20 Illustration of concurrent engineering on the same design

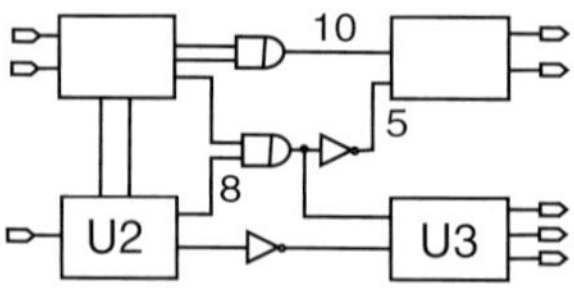

Fig. 25.21 Layout viewpoint after change

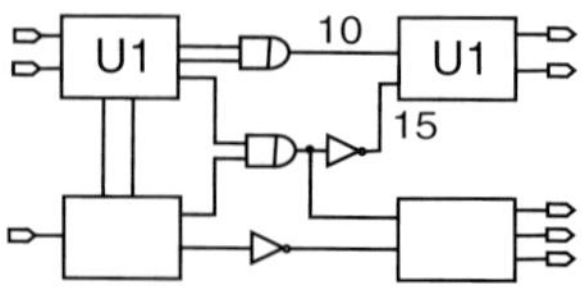

Fig. 25.22 Simulator viewpoint after change

the same design and to start downstream processes sooner. Tasks such as simulation and physical layout may start early even when significant modifications still have to be made to the original design source.

Figure 25.20 illustrates how the concept of the *design viewpoint* makes this possible. The simulator and the layout tool both 'see' the source design in their respective design viewpoint. These downstream tools capture the changes in their respective back annotation objects.

In our example it will be assumed that two designers work on the same project. One uses the simulator tool, the other the layout tool. Figure 25.20 shows their individual view of the design source. All changes made to their viewpoints as results of their design work are stored in the respective *back annotation objects*.

The designer who analyses the circuit behaviour with the simulator has changed the *delay value* from of 5 ns to 15 ns to evaluate its effect on circuit performance. Additionally, in order to ensure a minimum wire length between the two upper blocks he has pre-assigned the reference designator U1 to this blocks, which tells the layout tool to include these blocks in the same physical package.

The second designer, working with the layout tool, makes some changes from a layout perspective. The bottom two blocks U2 and U3 were defined. Reference designators are assigned to these blocks and a timing value of 8 ns is added to the wire. This value may be calculated by the layout tool from the physical wire length.

Figure 25.21 shows the current viewpoint of the simulator, fig. 25.22 the viewpoint of the layout tool.

The changes are saved in the respective back annotation objects. They are included in the original schematics if both designers are satisfied with their results and decide to include these design changes into the design object. This represents an incremental concurrent design style.

25.4.9 Output Files

When the design is completed the board is checked by a DRC run. The most important checks are:

- Checking that all traces in the artwork are the same as in the circuit schematics;
- Check for design rules violations.

If the DRC is passed successfully, it is certified that the design can be produced in the intended manufacturing technology.

Now the **manufacturing documentation** is created. The *postprocessor* creates files for all layers and several additional manufacturing data.

Examples of **manufacturing data** are:
- Connection lists;
- Mask description of each layer;
- Soldering masks;
- Aperture table;
- Drill table;
- Placement drawings;
- Fabrication drawings (milling);
- Manufacturing documentation.

The manufacturing data can be directly used to control a photoplotter for mask creation and numeric controlled (NC) machines for drilling and milling.

The manufacturing documentation includes following data:
- Schematics;
- Netlists (EDIF);
- Bill of materials;
- Drill table;
- All drawings.

An extract of a typical *bill of materials* is shown in table 25.2, sorted according to components (items), and table 25.3 sorted by *reference designation*. These extracts can be easily generated from the inherent component data base in the design.

Manufacturing data is manufacturer specific and must be adapted to the data formats of the production process. Modern PCB design tools are flexible enough to provide data and formats for all relevant production lines.

The basic steps in creating manufacturing data, sometimes called *artwork data* are:
- The artwork order must be defined which determines the association between board layers and film layers;
- The aperture table must be defined which establishes the apertures used by the photoplotter;

Table 25.2 Bill of materials for an example design, sorted after items

# Board Station BOM file					
# date : Thursday May 6, 1999; 13:13:42					
ITEM_NUMBER	COMPANY PART NO.	GEOMETRY	COUNT	DESCRIPTION	REFERENCE
1	PN-C-SMD	RC0805_N	13	CAPACITOR, 0.1u	C1 C2 C3 ... C24
2	PN-C2	C_RM2	3	CAPACITOR, 0.1u	C12 C17 C19
3	PN-CE2	C_gepolt_S_RM2	3	POL_CAPACITOR, 100u	C13 C18 C20
4	PN-LEISTE-16-2	ST16_2	1	ST16	ST15
5	PN-LEISTE-32-1	ST32_1	1	ST32_1	ST11
6	PN-R-SMD	RC0805_N	5	RESISTOR, 10	R1 R3 R4 R5 R6
7	PN-SAA9740H-SMD	SOT314-2	1	SAA9740H	IC2
8	PN-SN74LVC4245-SOL	SOL24	2	SN74LVC4245	IC4 IC5
9	PN-STIFT	loetstift	6	STIFT1	ST1 ST2 ST3 ST4 T14 U1
10	PN-STIFT2	STIFT2	5	STIFT2	ST6 ST7 ST8 ST9ST10
11	PN-TDA8792-SOL	SOL24	1	TDA8792	IC3

Table 25.3 Bill of materials for an example design, sorted after reference designator

REFERENCE	ITEM_NUMBER	COMPANY PART NO.	GEOMETRY	DESCRIPTION
C1	1	PN-C-SMD	RC0805_N	CAPACITOR, 0.1u
	. . .			
C12	2	PN-C2	C_RM2	CAPACITOR, 0.1u
C13	3	PN-CE2	C_gepolt_S_RM2	POL_CAPACITOR, 100u
	. . .			
C24	1	PN-C-SMD	RC0805_N	CAPACITOR, 0.1u
IC2	7	PN-SAA9740H-SMD	SOT314-2	SAA9740H
IC3	11	PN-TDA8792-SOL	SOL24	TDA8792
IC4	8	PN-SN74LVC4245-SOL	SOL24	SN74LVC4245
IC5	8	PN-SN74LVC4245-SOL	SOL24	SN74LVC4245
R1	6	PN-R-SMD	RC0805_N	RESISTOR, 10
	. . .			
R6	6	PN-R-SMD	RC0805_N	RESISTOR, 10
ST1	9	PN-STIFT	loetstift S	TIFT1
	. . .			
ST11	5	PN-LEISTE-32-1	ST32_1	ST32_1
ST14	9	PN-STIFT	loetstift	STIFT1
ST15	4	PN-LEISTE-16-2	ST16_2	ST16
U1	9	PN-STIFT	loetstift	STIFT1

Table 25.4 Aperture table

1	2	3	4	5	6	7	8	9	10
Pos.	Shape	Type	Width(X)	Height(Y)/Diameter	Orientation	Mirror	Power	Dcode	Used Last Run
1	rectangle	flash	0.300000	1.200000	0	false	false	10	false
2	rectangle	flash	1.200000	0.300000	0	false	false	11	false
	...	...	...	...	...	...	...	...	...
6	circle	trace	0.000000	0.254000	0	false	false	15	false
7	circle	trace	0.000000	0.500000	0	false	false	16	false
	...	...	...	...	...	...	...	...	...
10	circle	flash	0.000000	1.600000	0	false	false	19	false
11	rectangle	flash	1.600000	1.600000	0	false	false	70	false
12	rectangle	flash	2.600000	2.600000	0	false	false	71	false
13	circle	flash	0.000000	2.600000	0	false	false	20	false
14	circle	trace	0.000000	0.010000	0	false	false	21	true
15	circle	flash	0.000000	1.300000	0	false	false	22	false
16	circle	flash	0.000000	1.300000	0	false	true	23	false
	...	...	...	...	...	...	...	...	...
21	rectangle	flash	1.700000	1.700000	0	false	false	28	false
	...	...	...	...	...	...	...	...	...
27	rectangle	flash	2.700000	2.700000	0	false	false	127	false

Object Painting Aperture Position Resolution 14 0.008000
Area Fill Aperture Aperture Position Spacing 14 0.040000 0.040000
Thermal tie aperture 14
Power aperture 16 Tie is 0.500000 Air Gap is 0.300000

25

Table 25.5 Example of an artwork format report (Gerber Format)

ARTWORK FORMAT	
Photoplotter coordinate mode is absolute	
Photoplotter does not support circular interpolation Arcs and circles are interpolated by 8 segments Photoplotter supports Gerber G code	
Photoplotter supports millimeter units Scale is 1.000	
Data format 5.3	
Photoplotter modal coordinate is off	
Photoplotter modal shutter open(D01) is off	
Leading zeros are present	
Trailing zeros are present	
Record length of the artwork output file is	80
Film size is(width by height) :	406.400000 by 508.000000
Automatic offset specified will center board artwork on film sheet	
Distance from film(0,0) to board origin is(x,y):	128.000000 , 209.000000
Panel artwork offset is	(0 ,0).
Photoplotter command block end char is	'*'
Photoplotter machine stop code is	'M02'

Table 25.6 Part of a Gerber data file for an artwork photo plotter

```
G90*G71*G01*D02*G54D10*X00149710Y00245560D03*X00150210Y00245560D03*
X00150710Y00245560D03*X00151210Y00245560D03*X00151710Y00245560D03*
X00152210Y00245560D03*X00152710Y00245560D03*X00153210Y00245560D03*
X00153710Y00245560D03*X00154210Y00245560D03*X00154710Y00245560D03*
X00155210Y00245560D03*X00155710Y00245560D03*X00156210Y00245560D03*
X00156710Y00245560D03*X00157210Y00245560D03*X00157210Y00257260D03*
X00156710Y00257260D03*X00156210Y00257260D03*X00155710Y00257260D03*
X00155210Y00257260D03*X00154710Y00257260D03*X00154210Y00257260D03*
X00153710Y00257260D03*X00153210Y00257260D03*X00152710Y00257260D03*
X00152210Y00257260D03*X00151710Y00257260D03*X00151210Y00257260D03*
X00150710Y00257260D03*X00150210Y00257260D03*X00149710Y00257260D03*D02*G54D11*
```

Table 25.7 Example of a drill table

DRILL TABLE

Drill Position	Drill Size	Upper Bound	Lower Bound	Plated	Symbol
1	0.600000	0.600000	0.600000	yes	
2	0.800000	0.800000	0.800000	yes	
3	1.300000	1.300000	1.300000	yes	

Table 25.8 Drill format report

DRILL FORMAT
Drill machine coordinate mode is ABSOLUTE
Drill machine supports MILLIMETER units
Scale is 1.000 Data format is 3.3
Drill machine modal coordinate is OFF
Leading zeros are PRESENT
Trailing zeros are PRESENT
Drill machine origin (x, y) at (0.0, 0.0)
Drill machine command block end char is ''
Drill machine stop code is 'M30'

- The *Gerber format* files for describing the artwork must be generated. These files are used to control the photoplotter.

The *artwork data format* must match the format required by the specific photoplotter. This is generally the so called *Gerber format*, because the manufacturer of a widely used photoplotter is a company called 'Gerber' which defined a widely adopted standard. The photoplotter uses apertures to write the masks on the film. These apertures have to be specified in the aperture table, see table 25.4 as an example.

In the second column of table 25.4 the geometrical shape of the aperture is described, e.g., a rectangle or a circle. The size of the aperture in width and height is defined in columns 4 and 5. Column 9 notes the so called **Dcode**. This value is used for plotting and may be seen again in the plotting file table 25.6. All traces are plotted with a circle aperture which is moved over the film (column 3, *trace*). All pads are made by a flash projection (*flash*).

In table 25.4 one row (pos. 16) is marked in grey. In the table the cell at column 8 (Power) and row (pos. 16) is marked '*true*'. This specific aperture is called *power aperture*. With these apertures pads can be created which are connected to *ground* or *power*. The corresponding aperture on the photoplotter used has a user-defined power aperture *flash shape* which can be used for thermal ties. These shapes may vary. Therefore a number of options exists in defining the final thermal tie shape on artwork. Figure 25.23 shows an example of a power aperture flash shape with the definition of tie and air gap values as chosen in table 25.4. The term 'flash' results from the way these pads are produced: the mask shape is projected on the film with a flash light, giving the complete symbol in one step.

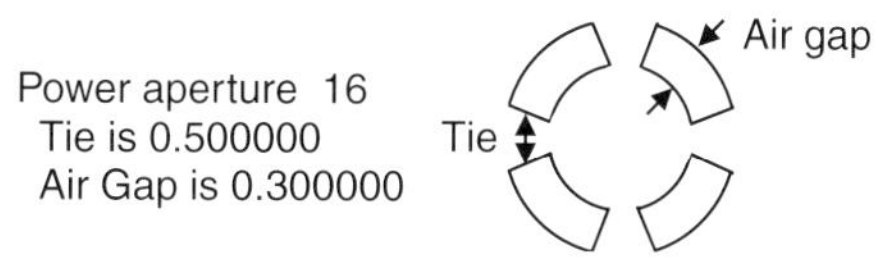

Fig. 25.23 Power aperture flash shape with definition of tie and air gap

The *Gerber artwork format* is a description of how artwork data is written to artwork files. This format must match the settings of the photoplotter.

An example of an artwork format report (in the Gerber format) is printed in table 25.5. In this report all parameters needed by the photoplotter are given. The report shows, e.g., that arcs and circles are interpolated by 8 segments. The scaling factor is set to 1.000, the data format version is 5.3, etc.

How a *Gerber data file* looks is shown in table 25.6, showing a small part of such a file, which is generally in ASCII format.

Drill data and milling data are used by numerically controlled (NC) machines for drilling the through holes and milling the board's outlines. The

26 EDA Tutorial

DIRK JANSEN

This tutorial will demonstrate in an easily understood example what is known as *top down design*. The example selected, a 'rolling dice', was taken from student education in the university of the author. It has been used for many years in the VHDL basics course [26.3] and for first prototype runs in new CMOS technologies and for this reason it exists in several different implementations. It may be traced back to a first implementation [26.1], awarded as the **second best student design** on the international EUROPRACTICE conference in 1994 in Dresden [26.6]. It contains several different techniques for realizing a *Finite State Machine*, tightly coupled and locking with each other, and is far from trivial. The design with about 230 gates may be regarded as small today, but a complete reprint of the schematics generated from the code or the VHDL code itself is not possible in the context of this book, see [26.4] for more details. The design is demonstrated here in all its steps until it reaches the stage of being a product which is used by the author, e.g., as advertising tag for the university, as a visiting card, or similar purposes.

With this design we want to demonstrate modern integrated EDA tools as well as the variability of the results. The example is suitable for being transferred to other kinds of programs or other designs, and it may be taken as a basic guideline for digital designs in general.

26.1 System Design

The specification of the 'rolling dice' may be taken from the following, verbal formulation [26.3]:

- The display of the dice result, which may be one of 1, 2, …, 6, shall be in the familiar form of a physical working dice but using LEDs.
- One button only will activate and operate the circuit.
- If the button is pressed the dice will be seen as 'rolling', this means that all display positions may be shown one after the other in a fast sequence. This sequence must be so fast that a definite stop for a certain result is impossible and the results seems to be random.
- When the button is released, the dice will roll down with continuously decreasing frequency, until it stops on a random number.
- The time it takes to roll down will depend on the time for which the button is pressed. The longer the button is pressed the longer the roll down shall last up to a certain maximum.
- Each change in the display will be marked acoustically by a 'click' sound, which is emitted by a piezo speaker. The 'click' sound should be about 4 kHz in frequency and last about 10 msec (a few oscillations only).
- After roll down ends, a melody of a few notes will be generated, depending on the resulting number, with the 'six' specially accented.
- For the basic clock frequency a clock crystal of 32 kHz shall be used, no other external components are allowed.
- The design must be fully synchronous.

A block diagram of the circuit, already partioned into the most important blocks, is shown in fig. 26.1.

The abovementioned functions may be realized in a few blocks, communicating with each other.

The rolling of the dice is carried out by a FSM [1)] which represents a counter from 1 to 6 with related coding. Figure 26.2 shows the (simplified) state diagram of this FSM.

This state machine is controlled by the *enable* signal which is generated by a NCO (a numerical controlled oscillator). A NCO is a circuit where a parallel input word defines the frequency which is generated at the output. From its principle concept

1) FSM stands for Finite State Machine

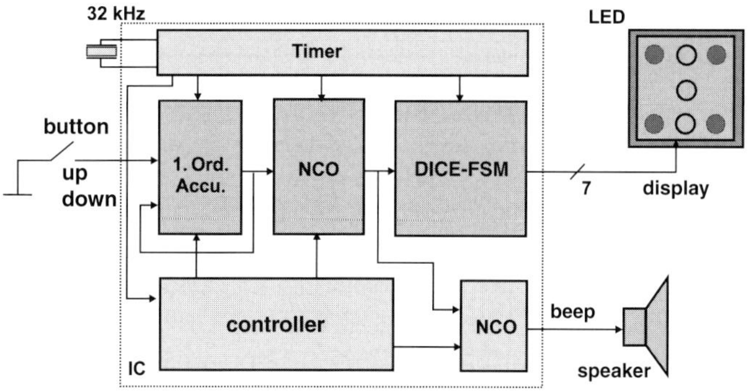

Fig. 26.1 Block diagram of the 'rolling dice' circuit

this is an accumulator, which adds up the input word with each clock pulse. If the accumulator overflows it starts from the beginning again. The most significant bit (MSB) of the accumulator may be interpreted as a phase bit. The frequency of the phase bit's changing depends linearly on the input word. The enable signal, derived from the phase bit and resulting in an one step counting, controls the roll down or roll up of the dice's counter.

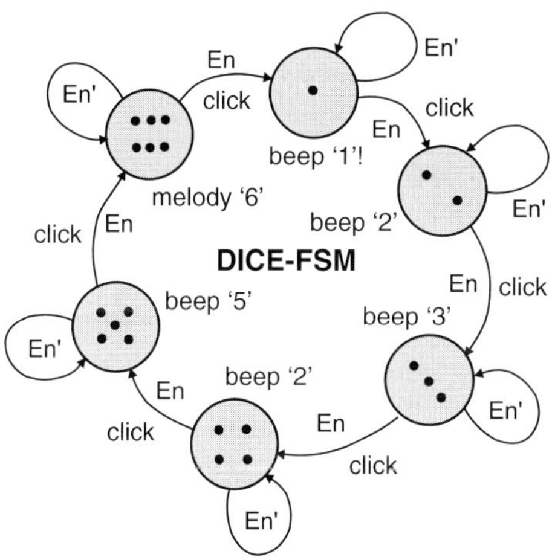

Fig. 26.2 State diagram of the dice's finite state machine

In this way the input of the NCO defines the rolling frequency of the dice. To create the rolling up and rolling down behaviour a second accumulator is used which is combined with a loop back to a first order system, seen from the point of view of control theory, the output of which is the input for the NCO. A button press now results in an exponential increase with an $1 - e^{-t/T}$ behaviour of the counting frequency, releasing the button in a $e^{-t/T}$ behaviour until zero is reached and the dice stops. The time constant T is chosen fixed by ergonomic points of view and depends on the feedback configuration.

During counting, with each enable of the counter a '*click*', a fix frequency sound, must be generated, which is done by a second NCO. This NCO generates the sounds for the melody at the rolling stop too.

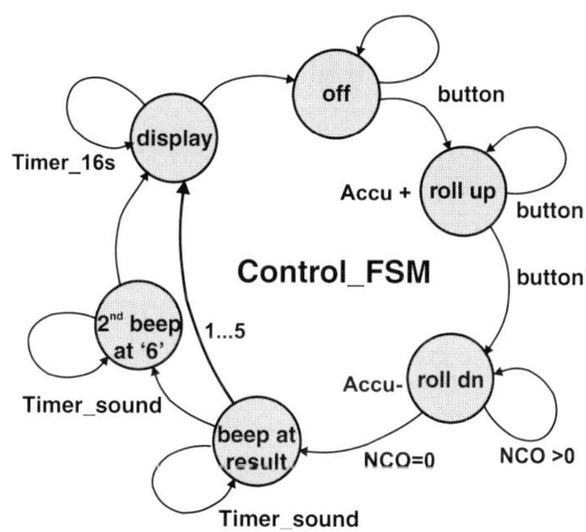

Fig. 26.3 State diagram of the control FSM (simplified)

The sequence of functions is controlled by a central control FSM, the simplified state diagram is shown in fig. 26.3. On pressing the button the first time the FSM goes into '*rolling*' state, the system is fully activated, the display enabled, and the rolling up behaviour initiated by continuous summing in the accumulator. If the button is now released, the system goes into the '*rolling down*' state, the NCO is driven down, and the '*click*' is generated. On stopping (NCO = 0) the melody generation starts with a few notes, which needs several states. At the same time the *timer_16s* starts, which resets the system after 16 seconds to the state '*off*' and shuts all circuitry completely off.

In state '*off*' all circuitry except the button supervising functions is inactive, so power consumption in this state is reduced to less than 1 μA. In this way an additional on/off power switch may be avoided. The electronic dice will be powered by a 3 V lithium battery for more than a year of intermittent operating time.

26.2 Implementation of the Design by Schematic Entry

All schematics are designed and input into EDA using the schematics input tool Design Architect (Mentor Graphics) one after the other. Figure 26.4 shows one sheet with the central dice counter FSM circuit.

Table 26.1 Statistics on gate usage for the example design 'rolling dice', classical bottom up design style

Library Cell	Number
AN2	43
AN3	3
AN4	2
AO1	1
AO2	1
AO3	2
AO5	4
AO6	1
AO7	6
AO2N	3
EO	51
FD2	52
IV	12
MUX21L	3
ND2	17
ND3	9
ND4	1
ND5	1
NR2	8
NR3	1
NR4	1
OR2	5
VDD	1

The logic may be designed with classical logic design methods by hand or derived from the state diagrams using supporting programs, e.g., LogIC. This is done by further dividing the blocks into sub blocks until a design by hand is feasible and well known and proven circuits can be used. Verification of these circuits, sub-blocks and blocks is nowadays done by functional digital simulation.

The result of this classical structured method is 230 gates, see table 26.1 for a statistics of the used gates.

The result is now transformed in a netlist (e.g., EDIF) and available for further processing.

26.3 Implementation of the Design using the High Level Language VHDL

As an alternate design approach to schematics input the design is described block by block in VHDL. In contrast to the schematic style the description is much more abstract, the FSM may be defined in a general formulation, and a detailed dissolution in single gates is not needed. List 26.1 shows an extract from the resulting code, describing the dice counter FSM.

List 26.1 Description of the main dice counter FSM in VHDL code (FSM part only)

```
counter :  PROCESS (mode, enable)

BEGIN
    nxt_mode <= mode;
    CASE mode IS
        WHEN st0 => IF enable='1' THEN
                nxt_mode <= st1;
            ELSE
                nxt_mode <= st0;
            END IF;
        WHEN st1 => IF enable='1' THEN
                nxt_mode <= st2;
            ELSE
                nxt_mode <= st1;
            END IF;
        WHEN st2 => IF enable='1' THEN
                nxt_mode <= st3;
            ELSE
                nxt_mode <= st2;
            END IF;
        WHEN st3 => IF enable='1' THEN
                nxt_mode <= st4;
            ELSE
                nxt_mode <= st3;
```

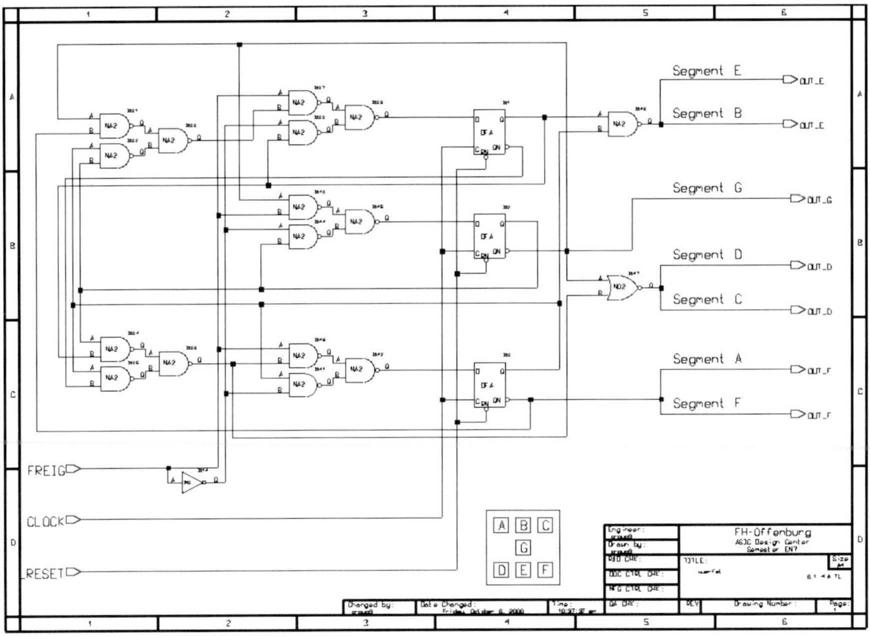

Fig. 26.4 Circuit diagram of the central dice counter FSM

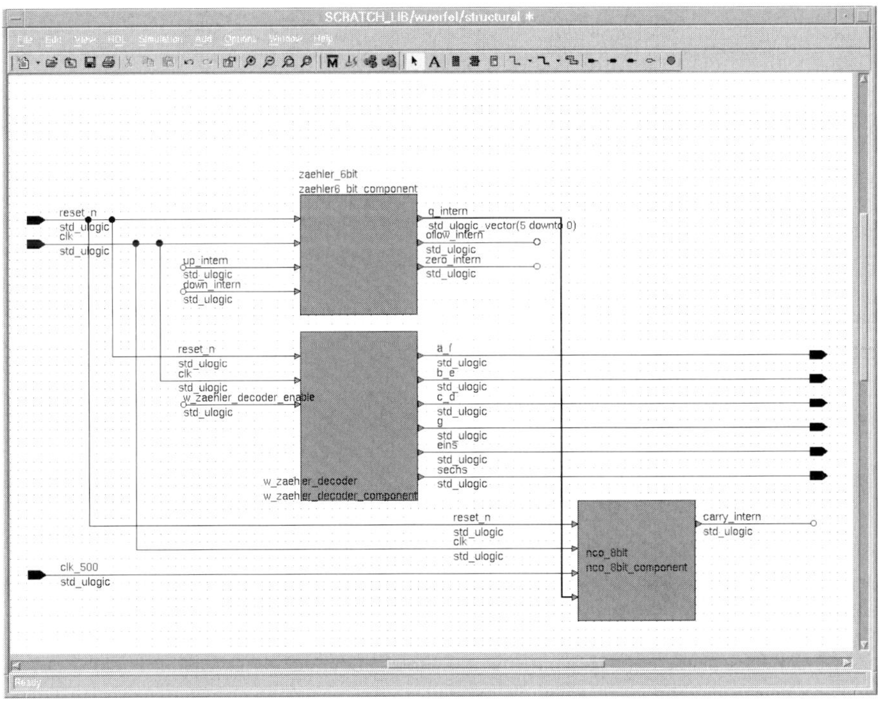

Fig. 26.5 Screen print of RENOIR with the blocks of the dice example, described in VHDL

```
            END IF;
        WHEN st4 => IF enable='1' THEN
                nxt_mode <= st5;
            ELSE
                nxt_mode <= st4;
            END IF;
        WHEN st5 => IF enable='1' THEN
                nxt_mode <= st0;
            ELSE
                nxt_mode <= st5;
            END IF;
    END CASE;
END PROCESS counter;
```

Each block is described by its own code and verified by digital simulation. The blocks are arranged with the help of a tool for VHDL block design style (*RENOIR*, Mentor Graphics) into a complete structural description fig. 26.5.

The programming tool *RENOIR* now generates the structural VHDL code shown in list 26.2, instantiating the components w_counter_decoder, nco_8bit, counter_6_bit. The example is not complete and describes only a part of the entire design to keep transparency. Of course, the code may be written by hand in the same way. The tool *RENOIR* allows one to arrange and administer the building VHDL blocks in a hierarchical way, which eases supervision in complex designs. Also in Mentor *Design Architect*, a tool which is used mainly for graphical symbolic schematic entry, blocks written in VHDL may be combined with schematics blocks build up from gates and flip flops. But *Design Architect* is oriented primary towards symbols and does not have all the special features for VHDL checking like *RENOIR*. Similar tools are available from other EDA providers too.

RENOIR allows one to input and display state diagrams graphically. The main dice counter FSM may be input in this way directly (fig. 26.6).

Finite state machines written in VHDL may displayed too, the description forms *text* or *diagram* are directly related to each other and show different aspects of the same contents.

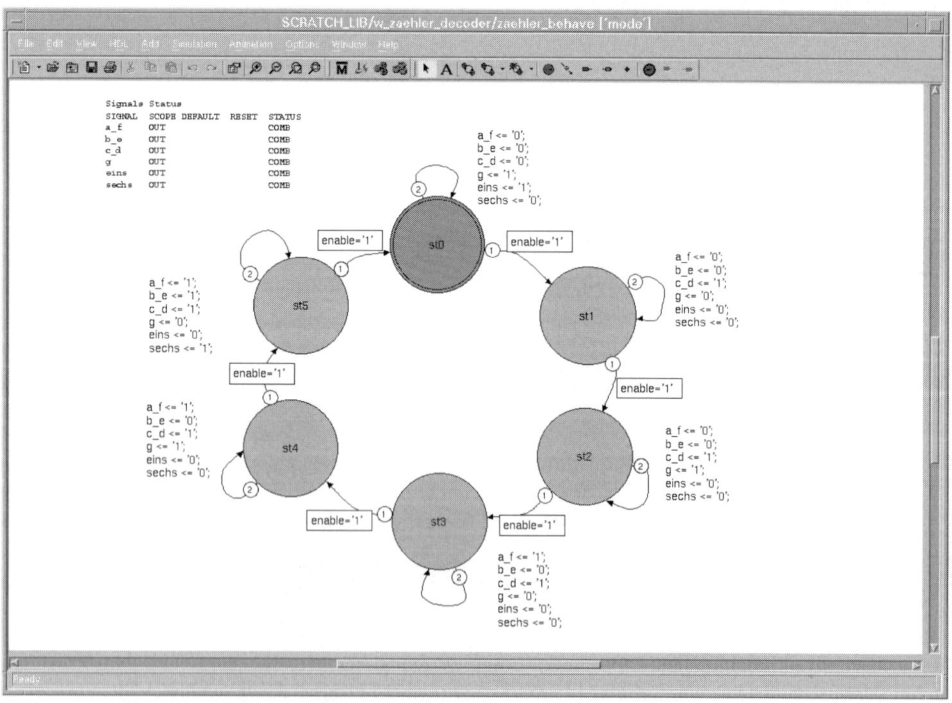

Fig. 26.6 Input display of the dice counter FSM with RENOIR

List 26.2 Structural VHDL description, defining the interconnection of the components (dice example, only a part)

```
ARCHITECTURE structural OF dice IS

COMPONENT w_counter_decoder
       PORT
       ( reset_n              :IN   std_ulogic;
       clk                    :IN   std_ulogic;
       enable                 :IN   std_ulogic;
       a_f                    :OUT  std_ulogic;
       b_e                    :OUT  std_ulogic;
       c_d                    :OUT  std_ulogic;
       g                      :OUT  std_ulogic;
       one                    :OUT  std_ulogic;
       six                    :OUT  std_ulogic);
END COMPONENT;

COMPONENT nco_8bit
       PORT
       ( reset_n              :IN   std_ulogic;
       clk                    :IN   std_ulogic;
       enable                 :IN   std_ulogic;
       a                      :IN   std_ulogic_vector (5 downto 0);
           carry              :OUT  std_ulogic);
END COMPONENT;

COMPONENT counter_6bit
       PORT
       (reset_n               :IN   std_ulogic;
       clk                    :IN   std_ulogic;
       up                     :IN   std_ulogic;
       down                   :IN   std_ulogic;
           q                  :OUT  std_ulogic_vector (5 downto 0);
       oflow                  :OUT  std_ulogic;
       zero                   :OUT  std_ulogic);
END COMPONENT;

FOR ALL:nco_8bit             USE ENTITY work.nco_8bit(nco_8bit_behave);
FOR ALL:counter_6bit         USE ENTITY work.counter_6bit(counter_6bit_behave);
FOR ALL:w_counter_decoder    USE ENTITY work.w_counter_decoder(counter_behave);

SIGNAL carry_intern, up_intern, w_counter_decoder_enable, down_intern, zero_intern,
oflow_intern                 :std_ulogic;
SIGNAL q_intern              :std_ulogic_vector (5 downto 0);
BEGIN
       w_counter_decoder_enable<=carry_intern OR button;
       up_intern<=button AND NOT oflow_intern AND clk_32;
       down_intern<=carry_intern AND NOT button;
       w_counter_decoder_component  :w_counter_decoder
           PORT MAP (reset_n, clk, w_counter_decoder_enable, a_f, b_e, c_d, g, one,
                     six);
       counter6_bit_component       :counter_6bit
           PORT MAP (reset_n, clk, up_intern, down_intern, q_intern, oflow_intern,
                     zero_intern);
       nco_8bit_component           :nco_8bit
           PORT MAP (reset_n, clk, clk_500, q_intern, carry_intern);
       start_click<=down_intern;
       start_melodie<=zero_intern;
END structural;
```

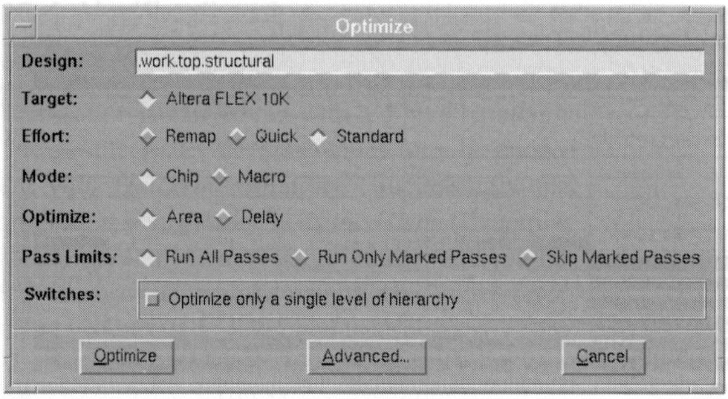

Fig. 26.9 Compile Window of the synthesis tool LEONARDO

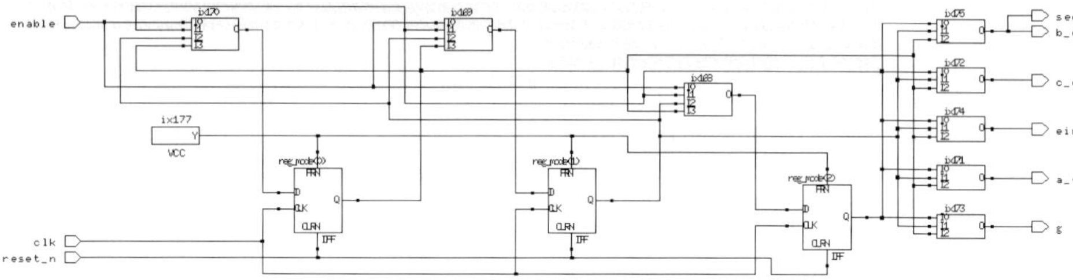

Fig. 26.10 Section of the schematics, re-transformed from the synthesized netlist, showing the 'dice counter FSM'

Synthesis was carried out with *LEONARDO* (Mentor Graphics); fig. 26.9 shows a screen display of the compile window.

Synthesis generates the netlists (EDIF) for the target technology, which may be re-transformed to schematics. Figure 26.10 shows a section with the dice counter FSM, and how it is generated by synthesis of the VHDL code. There are again three flip flops used, but with more complex post decoding. This may result from a certain code for the counter having been pre-defined (binary, gray code, etc.) during the set up of the synthesis.

Some sections from the netlist generated by the synthesis program are shown in list 26.4.

The netlist contains several calls of a LUT function[1)] which reflects the internal structure of the ALTERA FLEX 10K FPGA architecture used. The schematics are not directly readable in this form. The re-transformed schematics is again generated from these LUT functions. To allow a comparison it is mapped onto generic gates as AND, OR and EXOR cells as well as D flip flops. Table 26.2 shows design statistics for the target technology ALTERA FLEX 10K.

List 26.4 Section taken from the synthesized netlist of design 'rolling dice'

```
(cell w_counter_decoder (cellType GENERIC)
  (view counter_behave (viewType NETLIST)
    (interface
      (port reset_n (direction INPUT))
      (port clk (direction INPUT))
      (port enable (direction INPUT))
      (port a_f (direction OUTPUT))
      (port b_e (direction OUTPUT))
      (port c_d (direction OUTPUT))
      (port g (direction OUTPUT))
      (port one (direction OUTPUT))
      (port six (direction OUTPUT)))
```

[1)] LUT stands for Look Up Table, this is a kind of function allowing a truth table to be programmed.

```
    (contents
      (instance reg_mode2 (viewRef NETLIST (cellRef DFF (libraryRef flex10 ))))
      (instance reg_mode1 (viewRef NETLIST (cellRef DFF (libraryRef flex10 ))))
      (instance reg_mode0 (viewRef NETLIST (cellRef DFF (libraryRef flex10 ))))
      (instance ix212 (viewRef NETLIST (cellRef LUT (libraryRef flex10 )))
        (property lut_function (string "((IN1' IN2)+(IN1 IN3 IN4'))"))
        (property area_report (string "0"))
        (property area_units (string "LCs")))
      (instance ix213 (viewRef NETLIST (cellRef LUT (libraryRef flex10 )))
        (property lut_function (string "((IN1' IN2)+(IN2' IN3)+(IN3 IN4')+(IN1 IN2' IN4))
        "))
        .
        .
        .
(net reset_n
    (joined
      (portRef reset_n )
      (portRef CLRN (instanceRef reg_mode2 ))
      (portRef CLRN (instanceRef reg_mode1 ))
      (portRef CLRN (instanceRef reg_mode0 ))))
  (net enable
    (joined
      (portRef enable )
      (portRef IN1 (instanceRef ix212 ))
      (portRef IN1 (instanceRef ix213 ))
      (portRef IN1 (instanceRef ix214 ))))
```

List 26.5 Static timing analysis report of the design 'rolling dice', FPGA version, report from LEONARDO

```
- Critical Path of the entire design

Critical Path Report
NAME                                        GATE        ARRIVAL
--------------------------------------------
enable/                                     0.00 up     0.00
ix66/OUT                                    INBUF       2.40
clock_generator_component_intern_carry(1)/O  F2_LUT      6.50
clock_generator_component_intern_carry(2)/O  F2_LUT      10.60
clock_generator_component_intern_carry(3)/O  F2_LUT      14.70
clock_generator_component_intern_carry(4)/O  F2_LUT      18.80
clk_1000_int/O                              F2_LUT      22.90
clk_500_int/O                               F2_LUT      27.00
clock_generator_component_intern_carry(7)/O  F2_LUT      31.10
clock_generator_component_intern_carry(8)/O  F2_LUT      35.20
clock_generator_component_intern_carry(9)/O  F2_LUT      39.30
clk_32_int/O                                F2_LUT      43.40
clock_generator_component_intern_carry(11)/O F2_LUT      47.50
clk_8_int/O                                 F2_LUT      51.60
clock_generator_component_intern_carry(13)/O F2_LUT      55.70
clock_generator_component_intern_carry(14)/O F2_LUT      59.80
clock_generator_component_intern_carry(15)/O F2_LUT      63.90
clock_generator_component_intern_carry(16)/O F2_LUT      68.00
clock_generator_component_intern_carry(17)/O F2_LUT      72.10
time_out_dup0/O                             F2_LUT      76.20
ix71/OUT                                    OUTBUF      80.00
time_out/                                               80.00
--------------------------------------------
data arrival time                                       80.00
                                                        -------
```

26

It is interesting that the design now needs only 158 elements, 45 of which are d flip flops. The number does not tell us very much, of more interest is the statement that 117 logic cells, which are the primary resources of the FPGA, are used. The smallest FPGA from the FLEX 10K series, the 10K10 with 576 logic elements[1)], will be sufficient for this design. A synthesis on other FPGA families (as there are MAX 7000 or even the older version MAX 5000) may also be successful.

After synthesis a timing analysis is reported, see list 26.5, showing a 'critical path' of 80 nsec. The dice may be clocked at up to 10 MHz, so 32 kHz will work without any problem.

Synthesis and consequent emulation of the design on FPGA improves the trust in the design and is inevitable for circuits with ergonomic interfaces and long timing chains. Misinterpretation of specification and inconsistencies may be detected and eliminated early, the validated VHDL code may then be taken as a '**golden code**' for different target technologies, which now have to be checked only for timing behaviour.

26.6 Synthesis on ASIC Standard Cell Technology

Synthesis with the target technology '*standard cell*' (same is true for *gate array*) is not very different from synthesis with FPGA technology. The only difference is in the target library, which is built up from basic logic gates with ASICs in contrast to the more complex structured FPGA logic cells. So there are AND, NAND, OR, EXOR, etc. available in each ASIC library, but LUT cells are not. New are complex gates like AND OR INV gates (AOI gates) in different forms, which leads to some improvements in area and power consumption compared with basic gates, saving some interconnection wires in between the subgates too. The synthesis tool prefers to use these gates where ever possible. The gate usage statistics of the design mapped to an ASIC library with *LEONARDO* are shown in table 26.3. Area optimised complex gates are used quite often.

Table 26.3 Gate statistics of the design 'rolling dice' after synthesis on a ASIC library

Library Cell	Number	Library Cell	Number
AN2	26	AO16AN	1
AN3	2	AO17	2
AN4	1	AO18	2
AO1	2	AO20	1
AO1AN	2	AO20L	2
AO1N	1	AO20NL	1
AO2	3	AO21	1
AO2A	1	AO21N	1
AO3A	1	AO22	1
AO4A	2	AO22N	1
AO6	1	AO23	4
AO6A	3	FA1A	8
AO6C	1	FD2M	31
AO7	2	FD2SM	8
AO7A	1	FD4M	6
AO7C	1	GND	2
AO7N	1	IV	4
AO8	1	IVL	3
AO8L	2	IVP	38
AO9	1	ND2	2
AO9N	1	ND3	1
AO10	2	ND4	1
AO10L	1	NR2	10
AO11N	3	NR2L	4
AO13	2	NR3	5
AO15	1	MUX21N	19
AO15A	1	OR2L	1

The synthesized design now contains 225 cells (without GND), about the same as the design made by hand in the beginning table 26.1. The design made by hand was done quasi-optimally, or put in another way: modern synthesis tools allow one to obtain design quality comparable with hand drafted designs, but in a **fraction of the design time**.

As a comparison, again the core design of the dice counter FSM as a schematics is extracted in fig. 26.11.

The static timing analysis report of LEONARDO is shown in list 26.6. The critical path is now 16.78 nsec, the design is now five times as fast as in FPGA. Synthesis optimisation was on area, not on timing.

[1)] Logic Element, ALTERA name for 'Logic Cell'

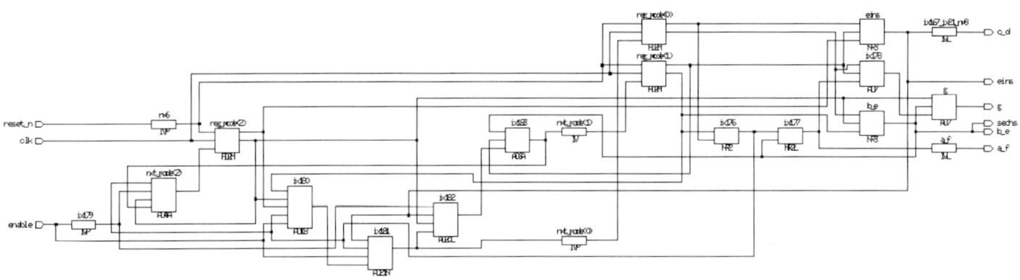

Fig. 26.11 Core of the dice counter FSM after synthesis on standard cells

List 26.6 Static timing analysis report of the critical path in the design 'rolling dice', synthesized on a standard cell library

```
Critical Path des Gesamtdesigns
NAME                                                    GATE    ARRIVAL
-------------------------------------------------
dice_component_counter6_bit_component_ix254_ix21/Q      FD4M    1.80
dice_component_nco_8bit_component_modgen_78_i0/CO       FA1A    3.38
dice_component_nco_8bit_component_modgen_78_i0_dup_310/CO FA1A  4.13
dice_component_nco_8bit_component_modgen_78_i0_dup_311/CO FA1A  4.88
dice_component_nco_8bit_component_modgen_78_i0_dup_312/CO FA1A  5.63
dice_component_nco_8bit_component_modgen_78_i0_dup_313/CO FA1A  6.39
dice_component_nco_8bit_component_modgen_78_i0_dup_314/CO FA1A  7.14
dice_component_nco_8bit_component_modgen_78_i0_dup_315/CO FA1A  7.89
dice_component_nco_8bit_component_modgen_78_i0_dup_316/CO FA1A  8.60
dice_component_nco_8bit_component_carry/Z               AN2     9.16
dice_component_ix73/Z                                   IVP     9.30
dice_component_ix74/Z                                   NR2L    9.99
dice_component_counter6_bit_component_ix314/Z           NR2     10.92
dice_component_counter6_bit_component_ix315/Z           AO6A    11.87
dice_component_counter6_bit_component_ix316/Z           AN2     12.86
dice_component_counter6_bit_component_ix322/Z           NR3     13.91
dice_component_counter6_bit_component_ix338/Z           AO7C    14.59
dice_component_counter6_bit_component_ix339/Z           AO2     15.18
dice_component_counter6_bit_component_ix340/Z           AO15A   15.81
dice_component_counter6_bit_component_ix341/Z           AO10NL  16.78
dice_component_counter6_bit_component_ix254_ix18/D      FD4M    16.78
data arrival time                                               16.78
-------------------------------------------------
data arrival time                                               16.78
                                                                -----
```

With such timing data for the critical path a timing related post simulation at the gate level using the detailed timing data of the cells used, may be omitted. But this is only allowed for a fully synchronous design with one central clock.

26.7 Completion of the design with Pad Cells and Validation of the Entire Design

The circuits developed lack any pad cells for the input and output signals of the chip. Selection of these cells is subject to considerations of the external circuit. This is output loading for driver cells, and for input cells the behaviour of the input signal determines which kind of cell has to be used.

In the dice circuit, a Schmitt Trigger Input cell is used for the input signal derived from the button to obtain a hysteresis behaviour for slowly increasing signals. Bouncing of the button has to be suppressed by the FSM which processes the button signal and which synchronizes the signal with the internal clock. This functionality has to be respected during the design of the control FSM and is

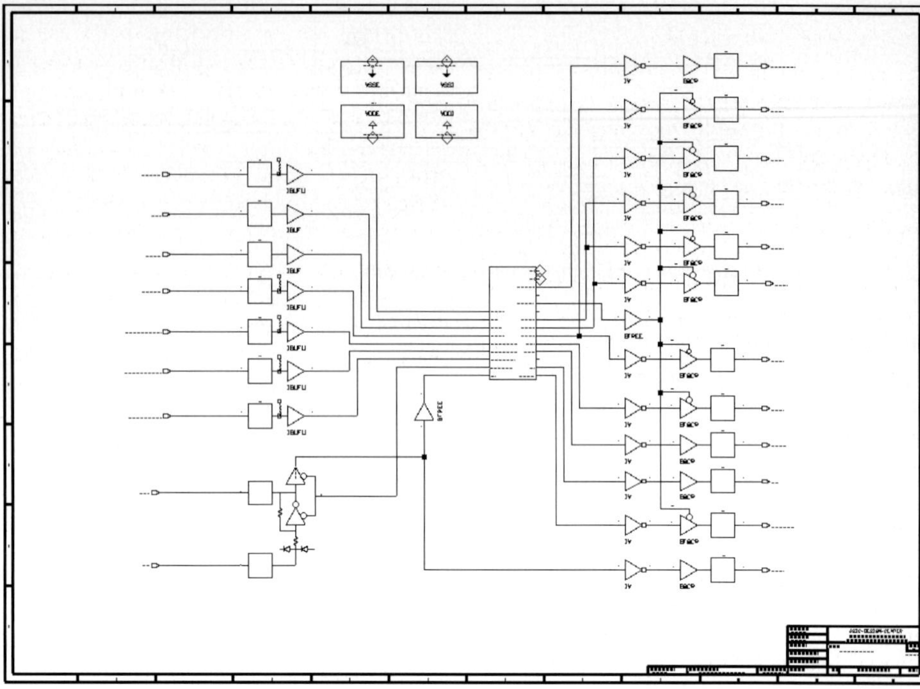

Fig. 26.12 Completion of the 'rolling dice' design with pre-placed pad cells (top sheet)

not part of the pad cell selection considerations. To be brief, there are no pad cells with anti bouncing characteristics.

For all other input cells standard input pad cells are selected, most of them with an internal *Pull Up*[1] resistor. The internal pull up resistors presents a logical '1' to the pad even when there is no connection to the cell at all; in the same way a pull down resistor generates a '0' input. If the pin is connected to a circuit such internal pull up or pull down resistors may be a source of power consumption, and so the later usage scenario has to be considered carefully.

On the output driver side the load on the circuit from external components has to be considered, such as the capacity of the wires on the printed circuit board or of connected cables. In our case these are the LEDs, consuming about 3 mA, so a simple standard output pad cell will do what is required. In many libraries there are pad cells with driver capabilities up to 24 mA; if more load capability is needed pad cells may be used in parallel.

When positioning the pad cells on the peripheral ring the power supply method has to be considered carefully. So there should be no power driver cells in the direct neighbourhood of sensitive analog or oscillator cells. If this happens, an influence on the phase, or even a severe disturbance of the frequency, of the oscillator may take place.

The power supply of an ASIC is provided in two distinct ways for core and pad ring. The supply systems are connected with each other on the printed circuit board, not on the chip. This avoids mutual interference caused by the series resistance and impedance of the bonding connections. But not only the ohmic resistance is of importance, also a low inductance has to be realized for a good power supply. Usually the power supply pad cells are arranged in the middle of the sides of the chip or in the corners, Vdd and Vss are placed on

[1] A pull up resistor is a resistor of about 10 . . . 100 kOhm which connects the input pin to the positive supply voltage and consequently simulate a logic '1' if no other components are connected to this pin.

opposite sides, Vdd_core beside the Vdd_pad and Vss_core beside the Vss_pad.

The relative placement of the pad cells is predefined by placement properties at the schematics level or as attributes in the structured VHDL file. Figure 26.12 shows the schematic of the *dice project* with predefined pad cells.

The delay of the pad cells is usually much greater than the delay of core cells. The complete design with connected pad cells and external circuitry and loads has to be validated in a final and detailed gate simulation. The result of this simulation may then be directly compared to test (measured) results later.

For an industrial design the latest **test structures** have to be introduced, so far such structures have not been included previously during synthesis. In a design of this kind normally a *scan path* and *JTAG* connections are added, both of which may be done automatically. The test patterns needed may be generated with ATPG tools automatically. In this case this was omitted because a functional test is sufficient for these low production numbers and an automatic wafer test is not intended.

26.8 Place and Route in a Standard Cell Design Style

In the design of FPGA circuits, after the synthesis of the EDIF files, the compiling and programming of the FPGA device, everything is defined and the component is completely configured. With ASIC designs, there are two possible scenarios:

- Delivery of the design to an ASIC provider in the form of a so called **simulated netlist** (EDIF);
- Place and Route of the design with one's own tools, verification of the results, and delivering of the design as a geometrically exact **GDS II File**.

For similar designs in industry the netlist version is frequently chosen, but here the second scenario which includes the **backend processing**, the physical design, will be demonstrated in more detail. This needs special software tools, which allow the mask design at the lowest geometry level, and tools for checking the generated geometries against the predefined design rules (*Design Rule Check, DRC*), and allow one to extract parasitics and electrical connections (XTRACT, CALIBRE). These tools are integrated in complex software systems, e.g., *IC station* (Mentor Graphics) or similar from other EDA providers. Because of the high investment needed for these programs, as well as the investment in personal skills in using these tools, such programs are only available at special design houses, universities, or companies engaged in analogue or huge mixed signal IC designs.

The following description demonstrates the steps with Mentor IC station which was installed at the site of the author. The technology used is a fairly old 2 metal layer CMOS technology (ALCATEL MIETEC 0.7). Newer technologies may be used in the same way with more metal layers (up to 6 are usual today), and a channel-less routing style which allows high density function blocks may be adapted. For the small 'rolling dice' example such a high resolution technology would be wasted, because chip size is determined by the number of pad cells, a further miniaturisation of the core would not save any silicon area (so called *pad limited design*).

The first steps with using the *IC Station* are:

- Loading of the process file, which describes the technology;
- Loading of the cell library of the target technology;
- Loading the netlist of the logic which has to be processed.

From the netlist and the geometry of the cells the program generates a floor plan automatically, containing rows for the possible placement of the cells. The length of the rows, which are ordered in a rectangle, considers the number of cells used plus some extent of about 10 %. This floor plan may be edited with the graphical tools, rows may be moved, inserted, or deleted, rows may be mirrored or turned. All rows in the core area have the same height as the core cells. For the periphery area the row size is adopted to the larger pad cells, see fig. 26.13.

A few operations now allow to place the standard cells automatically, the pad cells in their predefined way and in their relative position. The core cells are placed using optimal placement algorithms, so that there is a minimum interconnection length. Figure 26.14 shows placement of the cells for the 'rolling dice' example. The interconnections are

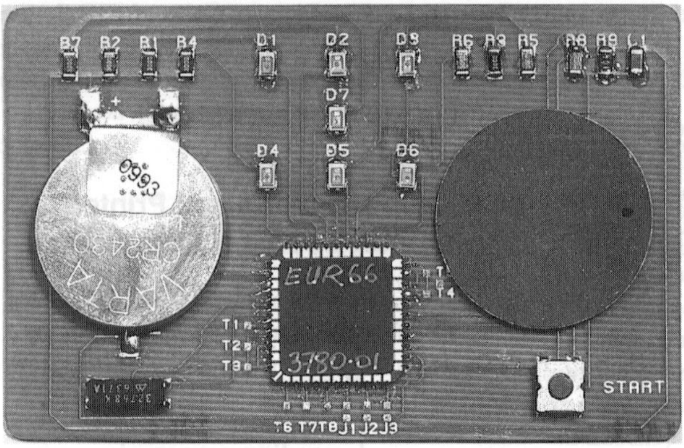

Fig. 26.18 Complete board with mounted 'rolling dice' IC in a CLCC44 package in a chip card format

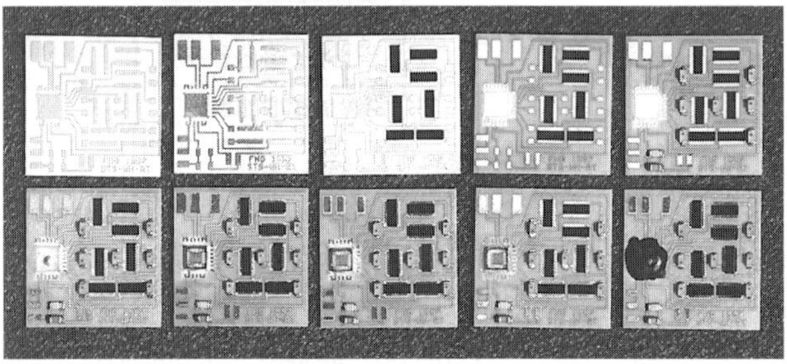

Fig. 26.19 Production steps for manufacturing of the thick film hybrid substrate: conductor printing; burn in; resistor printing; over glaze printing; burn in; mounting of discrete components; die bonding; chip bonding; globe top application

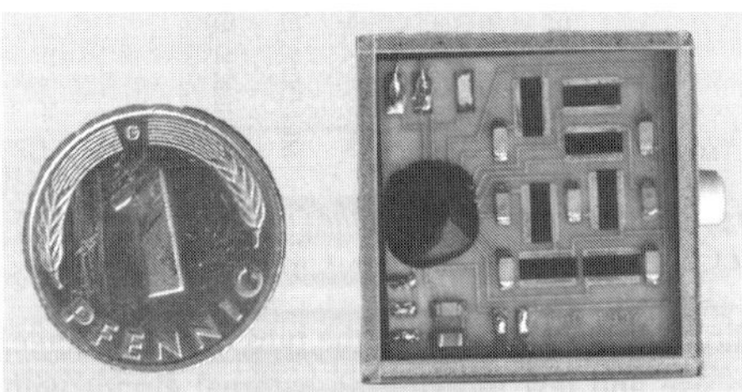

Fig. 26.20 The 'rolling dice' as a hybrid circuit in thick film technology with 'chip on board' in operation, showing the result '5', dimensions $10 \times 10 \times 10$ mm^3

26.10 References

[26.1] Joachim Schweiker: 'Entwicklung eines Einchip Würfels in ES_2 Standardzellentechnologie', Studienarbeit at the Fachhochschule Offenburg, SS 93

[26.2] Roland Tornar, Stefan Schleer, Wolfgang Harter: 'Würfel in Hybridtechik, Aufbau des ausrollenden Würfels als Hybridschaltung auf einem Keramiksubstrat'. Studienarbeit at the Fachhochschule Offenburg, SS 97

[26.3] Wolfgang Vollmer, Dirk Jansen: Lecture notes VHDL Course, Fachhochschule Offenburg, WS 98/99

[26.4] Homepage of the ASIC Design Center, University of Applied Sciences, FH-Offenburg: http://www.asic.fh-offenburg.de

[26.5] Homepage of europractice: http://www.europractice.com/

[26.6] Fifth EUROCHIP WORKSHOP ON VLSI TRAINING, 17.–19. October 2000, Dresden, Germany

Overview EDA
1, 2

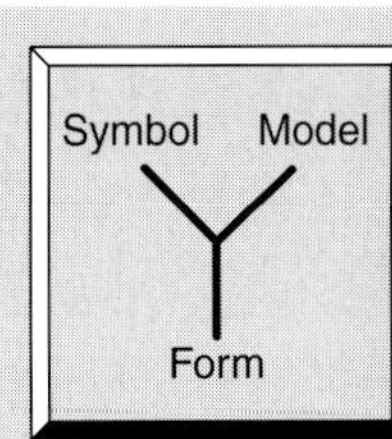

Symbolic Design
3

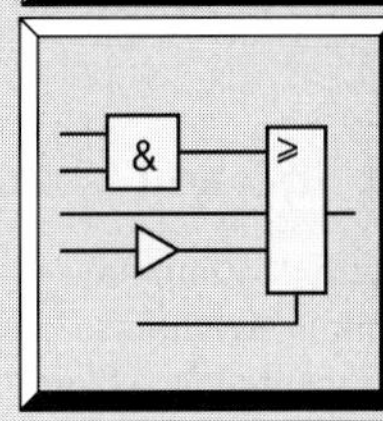

High Level Language Design
4, 5, 6, 7, 8

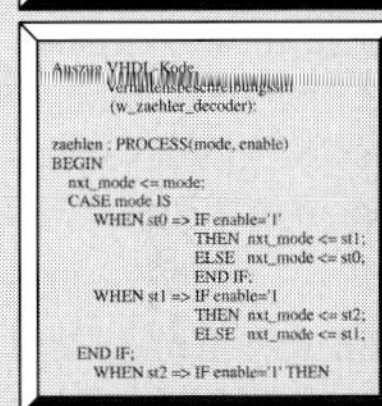

Modelling and Verifications
9, 10, 11, 12, 13, 14, 15

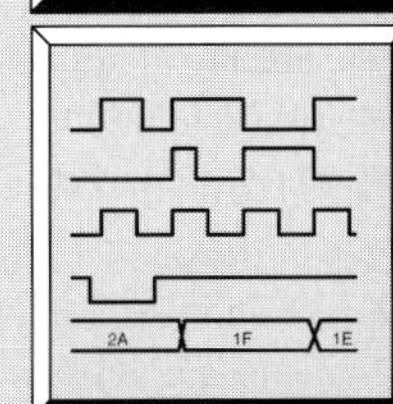

Implementation
16, 17, 18, 19, 20, 21, 22, 23, 24, 25

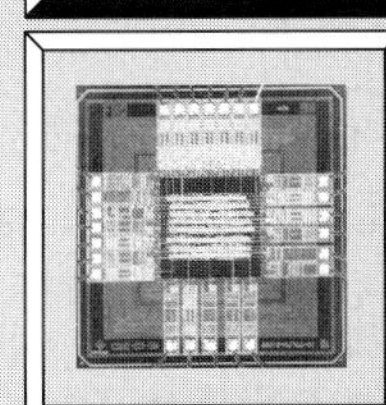

Tutorial
26

Appendix
A, B, C, D, E

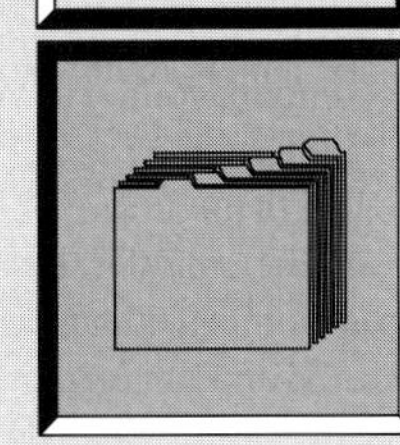

Appendix A Symbols

GERT VOLAND

A.1 IEC and National Standards

In order to facilitate the readability of electronic symbols and the worldwide exchange of electronic documentation, efforts have been made to standardize these symbols. The Technical Committee of the International Electrotechnical Commission (IEC) published documents which have been adopted by local standardization groups in almost all countries of the world [A.1].

The newest edition of the electrical and electronic symbols is published as

- **IEC 60617.**

Currently about 1500 symbols are covered by this standard (Part 2 to Part 13).

Access to the standards documents is normally organized on-line through the Internet, although paper versions are available. It is not free of charge, though, and licenses for subscriptions have to be purchased at the national publishers. There may also be national adaptations to the IEC standards.

In the US, standards are organized by the American National Standards Institute (ANSI) [A.2] where the IEC publications are combined with national standardization efforts. Electrical standards are then made available to the public by the Institute of Electrical and Electronics Engineers (IEEE) [A.3]. The American version of IEC 60617 is published as IEEE Standard No.: 91A/91 [A.4].

In Germany the IEC standards are published by the VDE-Verlag (Verein Deutscher Ingenieure) [A.5], while the national standards are organized by the Deutsches Institut für Normung (DIN) [A.6]. The institute has published the electrical symbols as DIN EN 60617 [A.7].

An earlier system, published in the US as IEEE/ANSI Standard Y32.14-1973, is still in use. It has triangles for inverters and half-rounded shapes for AND and OR. On many CAD systems these familiar shapes are utilized for the library elements.

The main characteristic of the IEC standardization is the graphical structure with element and control blocks, as well as the dependency notation. The standards are especially important for widely used devices such as commercial standard products in the SSI and LSI range. The most important examples are symbols for the 74-series of TTL and CMOS logic devices.

In the field of Application Specific Integrated Circuits (ASICs) and Field Programmable Gate Arrays (FPGAs), the symbol standardization has only very little impact. As a consequence, the libraries keep their CAD system specific character. Additionally, further non-standardized elements are needed in ASIC designs, such as I/O Pads, Bus-Keepers or HI/LO symbols. The generation of block symbols in hierarchical designs does not usually follow the rules of IEC symbol standards either. The older IEEE/ANSI symbols are also found in most FPGA/CPLD systems.

In practical design environments, the systematic approach to symbols often does not seem to be very important. However, symbols of different appearance must be easily identifiable. In the rest of this Appendix, the most important features of symbols for different design environments are therefore described separately:

- elementary logic circuits;
- design of Printed Circuit Boards (PCB);
- design of cells for Integrated Circuits (IC) and
- Cell Based IC designs (CBIC).

A.2 Symbols for Digital Designs

A.2.1 General Shape of Symbols

The symbol composition is described in IEC 60617-12: Binary Elements (available at [A.1]).

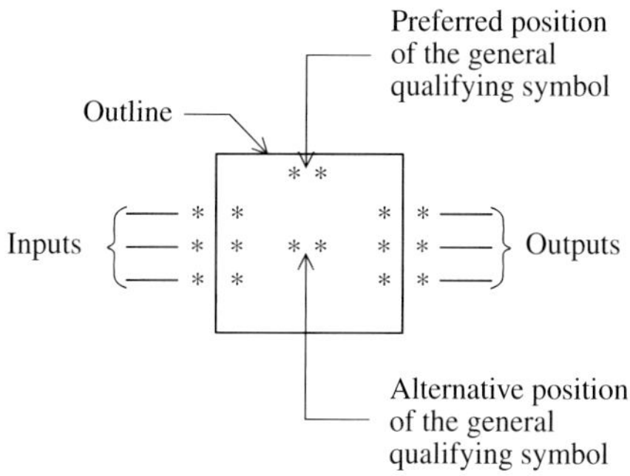

Fig. A.1 Symbol Composition

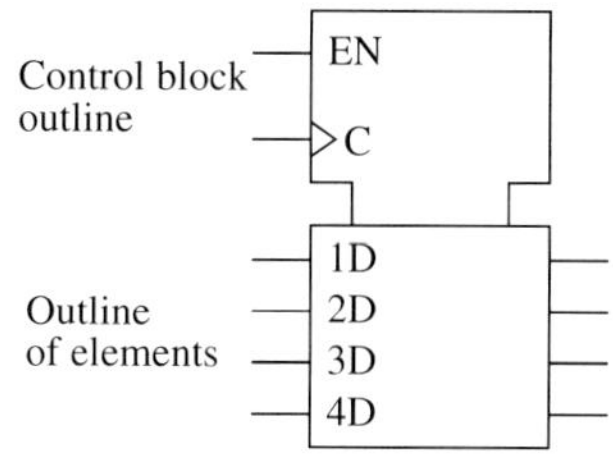

Fig. A.2 Common control block and data structure of a symbol

A symbol comprises an outline or a combination of different outlines together with one or more qualifying symbols. As shown in fig. A.1, general qualifying symbols are used to define exactly what logical operation is performed by this device. It is important to note that the input lines do not carry any descriptive names. The function of the input and output lines is solely determined by the dependency notation.

Table A.1 The most important general qualifying symbols

Symbol	Functional Description
&	AND function
$\geq$ **1**	OR function
= m	*m* of *n* Element
=	EXNOR function
= 1	EXOR function
1	Buffer
X/Y	Code converter
MUX	Multiplexer/data selector
DX	Demultiplexer
Σ	Adder
P – Q	Subtracter
COMP	Magnitude Comparator
ALU	Arithmetic Logical Unit
SRGm	Shift register, *m* stages
CTRm	Counter, cycle length 2^m
CTRDIVm	Counter, cycle length *m*
ROM	Read Only Memory
PROM	Programmable ROM
RAM	Read/write memory (Random Access Memory)

Table A.2 The most important input and output qualifying symbols

Symbol	Qualifying Symbols for Inputs and Outputs
	Signal flow, right to left
	Bidirectional signal flow
	Data, Toggle, J and K input for storage elements (Flipflops)
	Set, Reset, Enable input
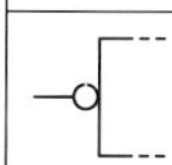	Logic negation at input
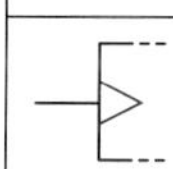	Dynamic (clock) input on rising edge
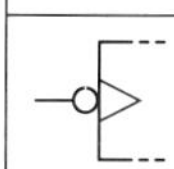	Dynamic (clock) input on falling edge
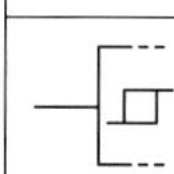	Hysteresis input (Schmitt-Trigger)
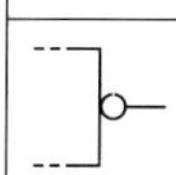	Negation at output
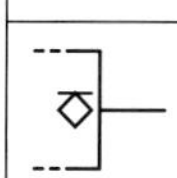	Open-drain, open collector output. Alternative with underscore: open source, open emitter
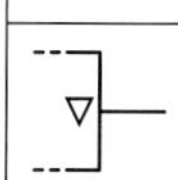	Tri-state output, can be switched to active or high impedance

A

In contrast to using qualifying symbols, pin names and pin numbers are usually used in ASIC and FPGA libraries. The positions next to input and output lines with an '*' are to be used for input and output qualifying symbols, such as negation circles (outside) or clock triangles (inside), as shown in table A.2.

As shown in fig. A.2, a device symbol may be composed of a common control block and an outline containing the functions. The control inputs could be clocks (C), enable inputs (EN) or read/write inputs for memory elements. In fig. A.2 the data inputs (D) are controlled by these inputs. Since there are several D inputs, they are numbered by nD.

Table A.1 shows the most important general qualifying symbols which determine the functionality. They can be placed in either one of two positions within the outline (fig. A.1).

Table A.2 shows some of the most important graphical input and output elements. They may be located inside or outside the symbol outline.

Note: Instead of the negating circles, a triangular shape may be used alternatively to indicate a negation.

Table A.3 Characters for the dependency notation

Character(s)	Dependency	Effect in state 1
A	ADDRESS	Memory element(s) selected
C	CONTROL	Data input(s) enabled
EN	ENABLE	Affects corresponding outputs
G	AND	Inputs ANDed
M	MODE	Mode selected
N	NEGATION	Negates the state (EXOR)
R	RESET	Sets output to 0
S	SET	Sets output to 1
V	OR	Inputs ORed
X	TRANSMISSION	Bidirectional connection
Z	CONNECTION	Logic connection

The dependency notation provides a powerful tool for denoting the relationship between inputs and outputs without actually showing all the elements and interconnections involved. The standardized characters are summarized in table A.3. With this system, complex functional relationships can be symbolized in a compact way. These types of symbols were very successful for the widespread 74-series of logic functions.

In ASIC libraries only some of the characters are used, such as C for clock inputs, EN for enable inputs, S and R for set and reset inputs for latches and flip flops.

A.2.2 Comparison Table for IEC and ANSI symbols

Table A.4 shows some relevant ASIC symbols in IEC and ANSI notation. Gates with multiple inputs are not limited in the number of input lines (at least not by the symbol shape). There are libraries that might have 16 input NAND gates. Such a symbol does not necessarily explain the implementation method because the function may be realized in several stages due to area or delay constraints.

Table A.4 Some of the most important digital symbols in IEC and ANSI notation

Function	IEC	ANSI
Buffer	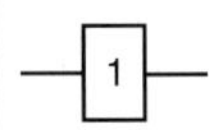	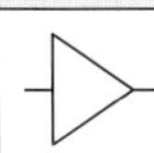
Inverter	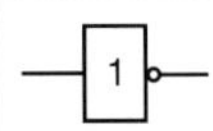	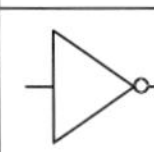
2-Input AND	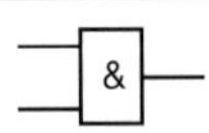	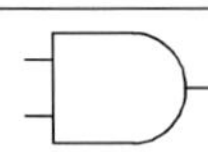
2-Input NAND	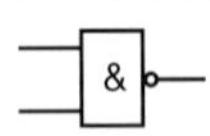	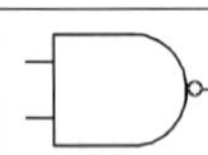
3-Input NAND	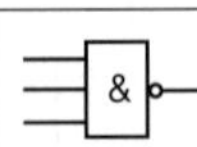	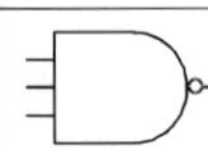
2-Input OR	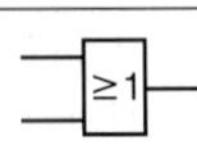	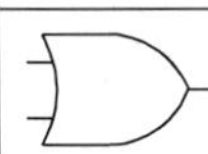
2-Input NOR	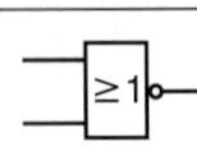	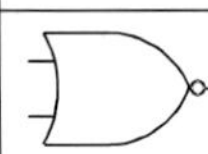

Table A.4 Some of the most important digital symbols in IEC and ANSI notation (continued)

Function	IEC	ANSI
3-Input NOR	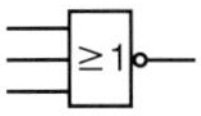	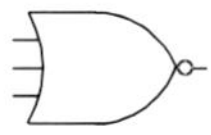
EXOR	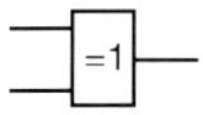	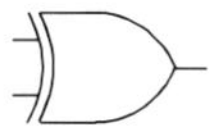
EXNOR	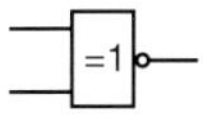	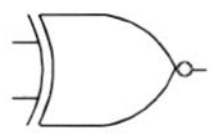
Tri-State Buffer	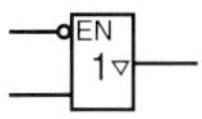	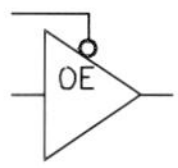
And-Or-Invert	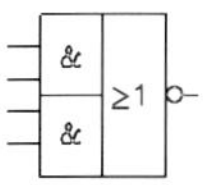	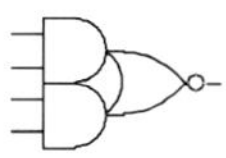
D Latch	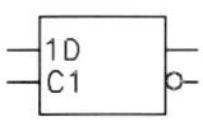	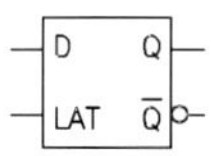
D Flip flop with Reset	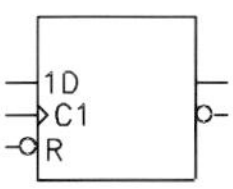	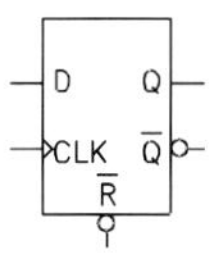
JK Flip flop with Set and Reset	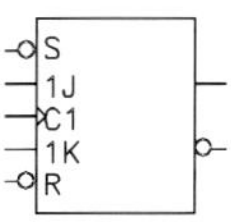	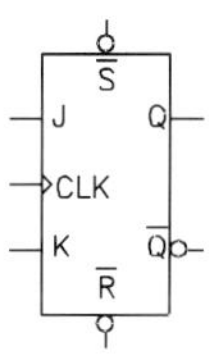
4-Bit Adder, cascadable	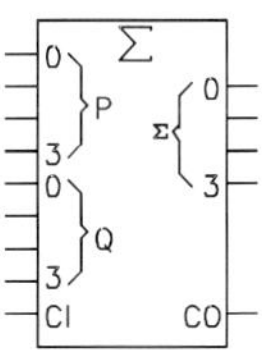	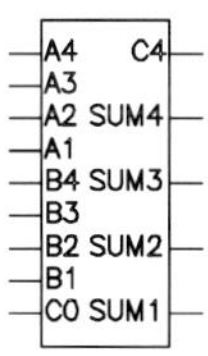
8 : 1 Multi-plexer with Enable and Reset	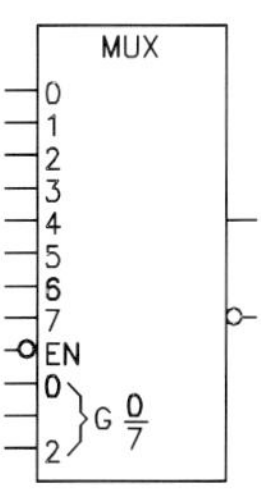	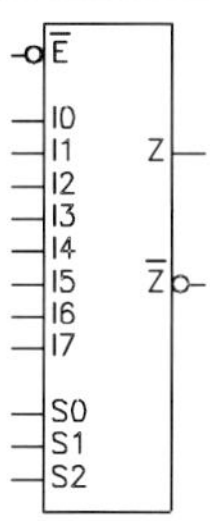

Table A.5 Some symbols for CBIC designs (MIETEC library on MENTOR)

Function	Symbol
Schmitt-Trigger Input-Pad with Pull-Down	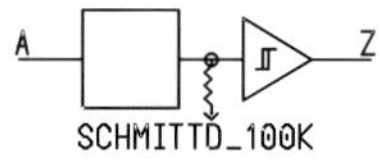
Tri-State Output Pad	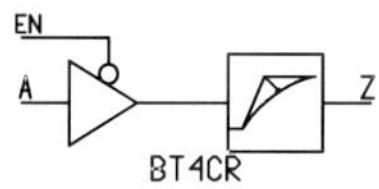
Bidirectional Input/Output Pad	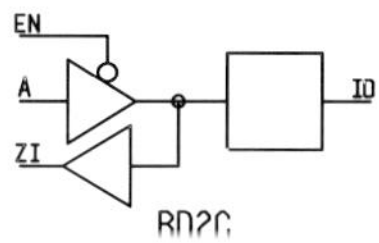
Output Pad, 8 mA	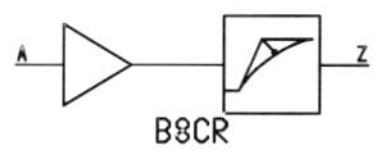
Two Pads for 32 kHz Crystal Oscillator	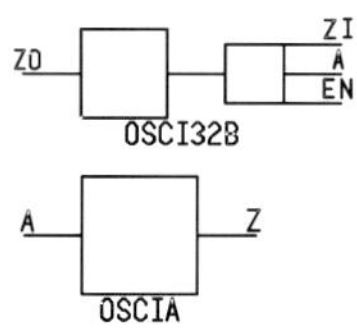
Positive Power Supply Pad	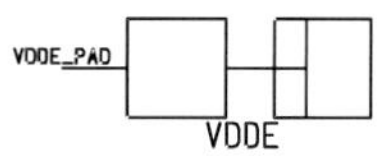
Negative Power Supply Pad	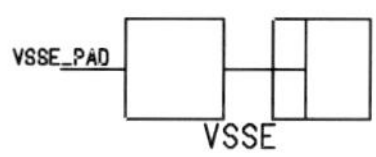
1-Stage CMOS AOAI Gate	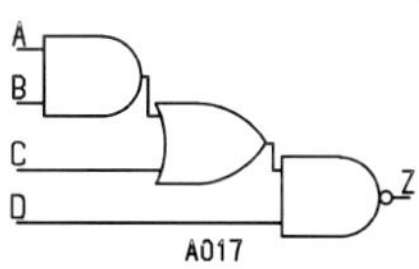
1-Stage CMOS OAOI Gate	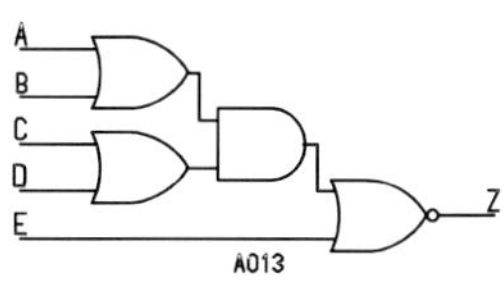
1-Stage CMOS AOI Gate	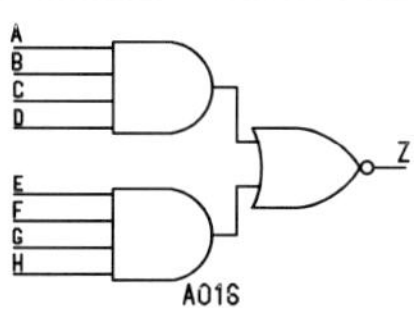

A

Table A.5 Some symbols for CBIC designs (MIETEC library on MENTOR) (continued)

Function	Symbol
Internal logical 0	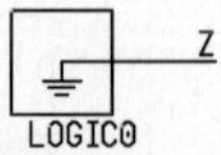
Internal logical 1	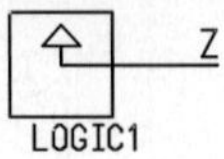
Bus-Keeper	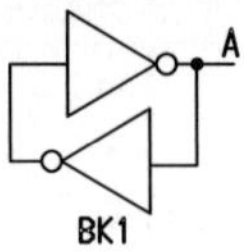

Note: There is a difference between IEC and ANSI symbols regarding active-low inputs and negated outputs: ANSI symbols usually have a negation circle and the pin name is marked as active-low by a negating bar or possibly a trailing "n". For example, Qn is used for the negated output of a flip flop.

The IEC symbols do not have pin names inside the symbol because inner characters are qualifying symbols. Thus, the pin has only a negation circle and may carry the non-negated pin name!

Table A.5 represents some symbols of functions for Standard Cell designs (also: Cell Based Integrated Circuits, CBICs). They are not subject to any standards [A.7]. LOGIC0 and LOGIC1 are used to tie unused internal input signals low or high. The Bus-Keeper will keep a Tri-State bus at the last active logical state as long as all drivers are high Z.

A.2.3 IEC Symbols for Printed Circuit Boards

Any type of electronic component may potentially be placed on a Printed Circuit Board (PCB). The standardization of such devices is especially important, although the implemented symbols are not necessarily portable between design systems.

A large symbol library for the Schematic Editor of the PSPICE design system (Cadence/OrCAD) is, for example, commercially available with over 10000 standardized IEC symbols [A.8].

Note: It should be noted that ANSI and IEC symbols are not necessarily pin compatible. An existing design may have to be updated by hand when switching from ANSI to IEC symbols or vice versa. It may be necessary to move connecting wires to new pin locations.

A.3 IEC Symbols for Analogue Elements

The passive and active elements of table A.6 are described in different standardization publications [A.1].

- **IEC 60617-2**: Symbol elements, qualifying symbols and other symbols having general application;
- **IEC 60617-4**: Basic passive components;
- **IEC 60617-5**: Semiconductors and electron tubes;
- **IEC 60617-13**: Analogue elements.

The symbols may be used for the following applications:

- Circuit simulation of general electric or electronic problems,
- Circuit schematics for the design of Printed Circuit Boards,
- Design of digital or analogue cells for Integrated Circuits (IC).

Table A.6 A selection of IEC symbols for analog designs

Function	Symbol
Voltage Supply	
General Ground	
Equipotential	
Resistor	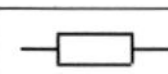
Inductor (Coil)	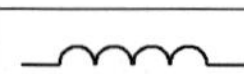
Capacitor	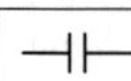
Electrolytic Capacitor	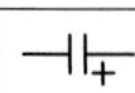
Diode	
Zener Diode	

Table A.6 A selection of IEC symbols for analog designs (continued)

Function	Symbol
Bipolar Transistor (npn)	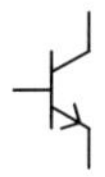
Bipolar Transistor (pnp)	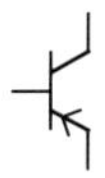
Junction Field Effect Transistor (JFET, n channel)	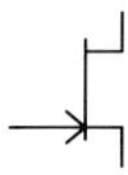
Junction Field Effect Transistor (JFET, p channel)	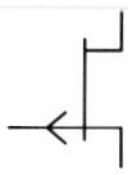
MOS Transistor (n channel, accumulation type)	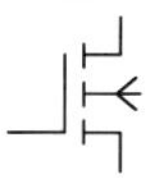
MOS Transistor (p channel, accumulation type)	
Operational Amplifier with Offset Control	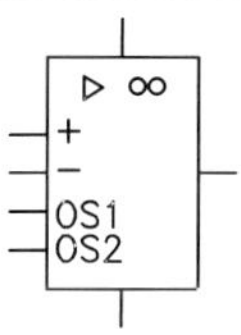

Table A.6 A selection of IEC symbols for analog designs (continued)

Function	Symbol
Voltage Comparator	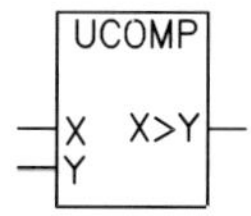
Voltage Regulator	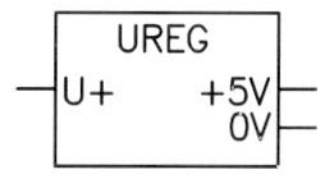
Sample&Hold Amplifier	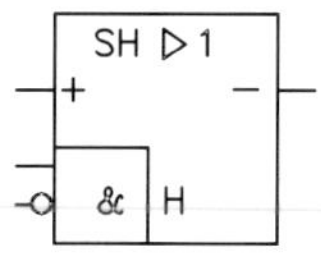
Voltage to Frequency Converter	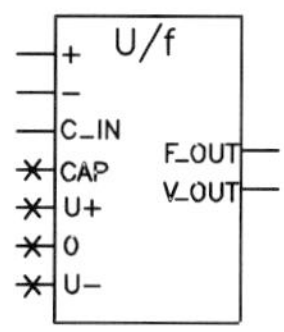
Analog to Digital Converter (ADC)	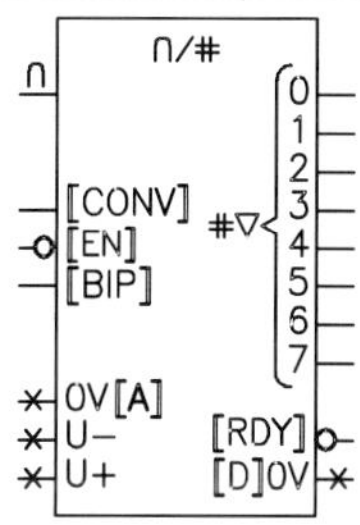

A.4 References

[A.1] http://www.iec.ch
IEC Central Office; 3, rue de Varembé, CH-1211 GENEVA 20, Switzerland

[A.2] http://www.ansi.org
ANSI Headquarters; 1819 L Street, NW, Suite 600, Washington, DC 20036, USA

[A.3] http://www.ieee.com
IEEE-USA; 1828 L Street, NW, Suite 1202, Washington, DC 20036, USA

[A.4] IEEE Standard No.: 91A/91; ISBN 1-5593-7135-8;
Product No.: SH14456-TBR; IEEE 1991

[A.5] http://www.vde-verlag.de
VDE Verlag; Bismarckstr. 33, D 10625 Berlin, Germany

[A.6] http://www2.din.de
DIN Deutsches Institut für Normung e. V.; Burggrafenstraße 6, D 10787 Berlin, Germany

[A.7] Standard Cell Library 0.5 µm, Mietec MTC 35, EUROPRACTICE, 1998

[A.8] Design Center/MicroSim PSpice DIN-Bibliothek. – Karlsruhe: HOSCHAR Systemelektronik GmbH, 1996 (Diskette and Handbook)

A

Appendix B VHDL Syntax

FRIEDEMANN STOCKMAYER, HANS KREUTZER

AGGREGATES: List of Elements, put in parenthesis, separated by comma.

Syntax: (value1, value2, ...) or (element1 => value1, element2 => value2, ...)

Example:
```
SIGNAL X : BIT;
SIGNAL Y : BIT_VECTOR (7 DOWNTO 0);
Y <= (7 => '1', 6 DOWNTO 3 => '0', 2 => X,
OTHERS => '0');
```

ALIAS DECLARATION: Alternate name of an object, it is not a new object.

Syntax: ALIAS aliasname : aliastyp IS objectname ;

Example:
```
ALIAS parity :  BIT IS data(7);
```

ARCHITECTURE: Defines the body of a design entity.

Syntax:
```
ARCHITECTURE architecture_name OF entity_name IS
    declarations
BEGIN
    concurrent statements
END architecture_name ;
```

Example:
```
ARCHITECTURE arch_my OF entity_my IS
    SIGNAL A, B, C, D, Y INTEGER;
BEGIN
    A <= B + C;
    Y <= A - D;
END arch_my ;
```

ARRAY: Consists of elements which have the same type. The declaration of an object of the type array must reference objects of an already specified type.

Syntax: TYPE typename IS ARRAY (range) OF element_type;

Example:
```
TYPE word IS ARRAY (15 DOWNTO 0) OF BIT;
SIGNAL w:  word;
```
... The objects 'string', 'bit_vector' and 'std_logic_vector' are, for example, defined as unconstrained array types:
```
TYPE bit_vector IS ARRAY(NATURAL RANGE <>)OF BIT;
SIGNAL Addr:  bit_vector(15 DOWNTO 0);
```

ASSERTION STATEMENT: Checks if a boolean expression is true. If it is not, an error will be reported and the simulator can be controlled with the severity level (note, warning, error and failure).

Syntax: ASSERT condition REPORT string SEVERITY severity_level;

Example:
```
ASSERT DATA <= X"00FF" REPORT "DATA is higher than 00FF!" SEVERITY
note;
```

ATTRIBUTES: Gives additional information about an object.

Syntax: object'attribute_name

Example: Predefined attributes for scalar types and array types:

X'HIGH	The upper bound of X
X'LOW	The lower bound of X
X'LEFT	The left bound of X
X'RIGHT	The right bound of X

Predefined attributes for array types und array subtypes:

X'RANGE	The Range of X
X'REVERSE_RANGE	The reversed range of X
X'LENGTH	X'HIGH – X'LOW + 1 taken as an integer

Predefined attributes for all signals X:

X'EVENT	True (Type Boolean), if an event on X has just occured
X'ACTIVE	True (Type Boolean), if an assignment to X was made
X'LAST_EVENT	Time since the last event on signal X
X'LAST_ACTIVE	Time since the last assignment to X
X'LAST_VALUE	The previous value of X, before the last change of X
X'DELAYED(T)	X delayed T units of time
X'STABLE(T)	True (Type Boolean), if no event has occured within T
X'QUIET(T)	True (Type Boolean), if X has no assignment within T
X'TRANSACTION	Change of value of BIT, if assignment to X

BLOCK: Concurrent assignment, which contains concurrent statements.

Syntax:
```
label: BLOCK (optional_guard_condition)
    declarations
BEGIN
    concurrent statements
END BLOCK label;
```

Example:
```
expl:  BLOCK (en = '1')
BEGIN
    Y <= GUARDED X;
END BLOCK expl;
```

CASE: Selects one statement out of a number of alternative sequential statements.

Syntax:
```
[case_label] CASE expression IS
    WHEN choice =>
        sequential statements
    {WHEN choice =>
        sequential statements}
END CASE [case_label];
```

Example:
```
CASE address IS
    WHEN "000" => decode <= X"01";
    WHEN "011" => decode <= X"F3";
    WHEN "100" => decode <= X"33";
    WHEN OTHERS => decode <= X"00";
END CASE;
```

COMPONENT DECLARATION: Declaration of a virtual design entity interface that may be used in a component instantiation.

Syntax:
```
COMPONENT component_name [IS]
    GENERIC (List);
    PORT (List);
END COMPONENT;
```

A

Example:
```
COMPONENT parity
    GENERIC (n:  INTEGER);
    PORT (w:  IN STD_ULOGIC_VECTOR (n-1 DOWNTO 0);
          p:  OUT STD_ULOGIC);
END COMPONENT;
```

COMPONENT INSTANTIATION: Defines a subcomponent of a design entity in a hierarchical design.

Syntax: instance_label: [COMPONENT] component_name
GENERIC MAP (generic_association_list)
PORT MAP (port_association_list);

Example:
```
U1:  parity
    GENERIC MAP (n => 8)
    PORT MAP (w => DATA_BYTE, p => PARYITY_BIT);
```

CONFIGURATION DECLARATION: Declaration, which describes the hierarchy of a design.

Syntax:
```
CONFIGURATION config_name OF entity_name IS
  FOR architecture_name
    FOR instance_label: component_name
      USE ENTITY
      library_name.entity_name(arch_name);
        FOR arch_name
          ... lower-level configuration specifications ...
          END FOR;
    END FOR;
  END FOR;
END config_name;
```

Example:
```
LIBRARY DIGPARTS;
CONFIGURATION struct_con OF struct_entity IS
  FOR struct_arch
    FOR U1,U2:  REG
      USE ENTITY WORK.UNIVREG;
    END FOR;
    FOR ALL: FULLADD
      USE ENTITY DIGPARTS.OPTFULLADD;
        FOR OPTIMIZED
          FOR ALL: AND2
            USE ENTITY DIGPARTS.AND2;
          END FOR;
          FOR ALL: OR2
            USE ENTITY DIGPARTS.OR2;
          END FOR;
        END FOR;
    END FOR;
  END FOR;
END;
```

CONFIGURATION SPECIFICATION: Associates binding information with component labels representing instances of a given component.

Syntax: FOR instance_label: component_name
USE ENTITY library_name.entity_name(arch_name);

Example:
```
FOR U1:  fulladd USE ENTITY WORK.fadd(arch_fadd);
```

CONSTANT DECLARATION: Name and value declaration of a CONSTANT.
Syntax: CONSTANT constant_name : type [:= expression];
Example:
```
CONSTANT address_width:  INTEGER := 8;
```

ENTITY: Describes the inputs and outputs of a design entity.
Syntax:
```
ENTITY entity_name IS
    GENERIC (generic_list);
    PORT (port_list);
END [ENTITY] entity_name;
```
Example:
```
ENTITY fulladd IS
    GENERIC (DELAY: time := 5ns);
    PORT (a,b,c_in:  IN BIT;
    sum, c_out:  OUT BIT);
END fulladd;
```

EXIT: Used to complete the execution of an enclosing loop statement.
Syntax: EXIT [loop_label] [WHEN condition];
Example:
```
loop1:  FOR i IN loop_var DOWNTO 0 LOOP
    EXIT loop1 WHEN i > 15;
    mem(i) <= (OTHERS => '0');
END LOOP;
```

EXPRESSION: Is a formula, that defines the computation of a value.
Syntax: relation {logical_operator relation} ;
logical_operator ::= AND | OR | XOR | NAND | NOR | XNOR.
relation ::= shift_expression [relational_operator shift_expression]
relational_operator ::= = | /= | < | <= | > | >=
shift_expression ::= simple_expression [shift_operator simple_expression]
shift_operator ::= sll | srl | sla | sra | rol | ror
simple_expression ::= [sign] term {adding_operator term}
sign ::= + | - , adding_operator ::= + | - | &
term ::= factor {multiplying_operator factor}
multiplying_operator ::= * | / | mod | rem
factor ::= primary [** primary] | ABS primary | NOT primary
primary ::= name | literal | aggregate | function_call | qualified_expression | type_conversion | allocator | (expression)

FILE DECLARATION: Declares a file of the specified type and links it to an existing file.
Syntax: FILE logical_name : file_type IS mode 'file_name';
Example:
```
FILE intext:  TEXT IS IN "stimuli.txt";
FILE erg_file:  TEXT IS OUT "results.txt";
```

FOR LOOP: Will be used to implement a sequence of statements to be executed repeatedly.
Syntax:
```
[loop_label :] FOR parameter IN range LOOP
    sequential statements
END LOOP [loop_label];
```
Example:
```
FOR i IN 7 DOWNTO 0 LOOP
    IF i = 2 THEN
        NEXT;
    ELSE
        fifo(i) <= (OTHERS => '0');
    END IF;
END LOOP;
```

A

FUNCTIONS: Subprogram, that returns a value. The function call is an expression.

Syntax:
```
FUNCTION function_name (parameter_list) RETURN type IS
    declarations
BEGIN
    sequential statements
END function_name;
```

Example:
```
FUNCTION boolean_to_bit(a:  BOOLEAN) RETURN BIT IS
    BEGIN
        IF a THEN
            RETURN '1';
        ELSE
            RETURN '0';
        END IF;
    END boolean_to_bit;
```

GENERATE: Used for iterative or conditional elaboration of a part of a description.

Syntax:
```
[gen_label] : FOR parameter IN range GENERATE
    [{declarative_item}
BEGIN]
    concurrent statements
END GENERATE [gen_label];
```

Example:
```
gen_bsp:  FOR k IN 4 DOWNTO 1 GENERATE
    y(k) <= a(k) AND b(k);
END GENERATE gen_bsp;
```

GENERICS: Defines generic constants. Its values may be determined by the environment.

Syntax: see Entity

Example: see Entity

IF STATEMENT: Statement, that selects one or none of the enclosed sequences of statements for execution, depending on the value of one or more corresponding conditions. Used in processes, functions or procedures.

Syntax:
```
[if_label:] IF condition_1 THEN
    sequential_statements
{ELSIF condition_n THEN
    sequential_statements}
[ELSE
    sequential_statements]
END IF [if_label] ;
```

Example:
```
ARCHITECTURE archmux4to1 OF mux4to1 IS
BEGIN
mux:  PROCESS (a, b, c, d, s)
    BEGIN
        IF s = "00" THEN x <= a;
    ELSIF s = "01" THEN x <= b;
    ELSIF s = "10" THEN x <= c;
    ELSE s = "11" THEN x <= d;
    END IF;
END PROCESS mux;
END archmux4to1;
```

LIBRARY CLAUSE: Defines logical names for design libraries in the host environment.

Syntax:
```
LIBRARY library_name_1, library_name_2, .. ;
```

Example:
```
LIBRARY IEEE;
USE IEEE.STD_LOGIC_1164.ALL;
```

LITERALS: is the value of a type.
Syntax: see Example.
Example:
```
CONSTANT count:  INTEGER:= 32;     -- Integer literal
CONSTANT end:  real:= 15.0;        -- Real literal
int1 <= 16#FA#                     -- Literal of base 16
int2 <= 2#1111_1010#               -- Literal of base 2
great_number := 2.8E6              -- Literal with exponent
```

NAMES: Used as identifier.
Syntax: May contain letters, digits and underline. The first position of a name must be a letter.
Example:
```
RTL_Unit
Bus_16Bit
```

NEXT STATEMENT: Used to complete the execution of one of the iterations of an enclosing loop statement.
Syntax: [label:] NEXT [loop_label] [WHEN condition];
Example: see FOR LOOP.

NULL: Performs no action.
Syntax: [label :] NULL;
Example:
```
CASE state IS
    WHEN s1 | s2 | s4 => next_state <= s0
    WHEN OTHERS => NULL;
END CASE;
```

PACKAGE BODY: Defines the bodies of subprograms and the values of deferred constants declared in the interface to the package.
Syntax:
```
PACKAGE BODY package_name IS
    declarations
    deferred constant declarations
    subprogram bodies
END [PACKAGE BODY] [package_name];
```
Example:
```
PACKAGE BODY my_package IS
    FUNCTION resolved (drivers:  my_logic_vector)
        RETURN my_logic IS
    BEGIN
    - code for the function
    END resolved;
END my_package;
```

PACKAGE DECLARATIONS: Defines the interface to a package.
Syntax:
```
PACKAGE package_name IS
    declarations
END [PACKAGE] [package_name];
```
Example:
```
PACKAGE my_package IS
    TYPE my_logic IS ('0', '1', 'X', 'Z');
    TYPE my_logic_vector IS ARRAY (NATURAL RANGE <>) OF my_logic;
    FUNCTION resolved (drivers:  my_logic_vector)
        RETURN my_logic;
    CONSTANT propagation_delay:  time;
END my_package;
```

A

VARIABLE ASSIGNMENT: Replaces the current value with a new value specified by an expression.

Syntax: [label :] variable_name := expression;

Example:
```
y := a XOR b;
y := '1';
```

VARIABLE DECLARATION: Declares variables and optional an initialization value.

Syntax: [SHARED] VARIABLE variable_name_list : type [:=expression];

Example:
```
VARIABLE a, b, c :  BIT;
VARIABLE count :  INTEGER RANGE 0 TO 256;
```

WAIT STATEMENT: Is a sequential statement and suspends a process or a procedure.

Syntax: [label :] WAIT [ON sensitivity_list][UNTIL condition][FOR time_exp.];

Example:
```
WAIT UNTIL CLK'EVENT and CLK = '1';
WAIT ON S1 FOR 5 ns;
```

WHILE LOOP: Used to execute a sequence of statements repeatedly, zero or more times while a condition is true.

Syntax:
```
[loop_label :] WHILE condition LOOP
    sequential statements
END LOOP [loop_label];
```

Example:
```
WHILE (i <= 7) LOOP
    IF i = 2 THEN
        NEXT;
    ELSE
        fifo(i) <= (OTHERS => '0');
    END IF;
    i := i + 1;
END LOOP;
```

Appendix C Packages

FRIEDEMANN STOCKMAYER, HANS KREUTZER

C.1 Package STD.STANDARD

```
package STANDARD is

  type BOOLEAN is (FALSE, TRUE);
  type BIT is ('0', '1');
  type CHARACTER IS (
    NUL,    SOH,    STX,    ETX,    EOT,    ENQ,    ACK,    BEL,
    BS,     HT,     LF,     VT,     FF,     CR,     SO,     SI,
    DLE,    DC1,    DC2,    DC3,    DC4,    NAK,    SYN,    ETB,
    CAN,    EM,     SUB,    ESC,    FSP,    GSP,    RSP,    USP,
    ' ',    '!',    '"',    '#',    '$',    '%',    '&',    ''',
    '(',    ')',    '*',    '+',    ',',    '-',    '.',    '/',
    '0',    '1',    '2',    '3',    '4',    '5',    '6',    '7',
    '8',    '9',    ':',    ';',    '<',    '=',    '>',    '?',
    '@',    'A',    'B',    'C',    'D',    'E',    'F',    'G',
    'H',    'I',    'J',    'K',    'L',    'M',    'N',    'O',
    'P',    'Q',    'R',    'S',    'T',    'U',    'V',    'W',
    'X',    'Y',    'Z',    '[',    '\',    ']',    '^',    '_',
    '`',    'a',    'b',    'c',    'd',    'e',    'f',    'g',
    'h',    'i',    'j',    'k',    'l',    'm',    'n',    'o',
    'p',    'q',    'r',    's',    't',    'u',    'v',    'w',
    'x',    'y',    'z',    '{',    '|',    '}',    '~',    DEL);

  type SEVERITY_LEVEL is (NOTE, WARNING, ERROR, FAILURE);
  type INTEGER is range -2147483648 to 2147483647;
  type REAL is range -1.7e38 to 1.7e38;

  type TIME IS range -9223372036854775808 to 9223372036854775807
  units
    fs;             - femtosecond
    ps = 1000 fs;   - picosecond
    ns = 1000 ps;   - nanosecond
    us = 1000 ns;   - microsecond
    ms = 1000 us;   - millisecond
    sec = 1000 ms;  - second
    min = 60 sec;   - minute
    hr = 60 min;    - hour
    end units;

  subtype DELAY_LENGTH is TIME range 0 fs to TIME'HIGH;

  impure function NOW return DELAY_LENGTH;

  subtype NATURAL is INTEGER range 0 to INTEGER'HIGH;
  subtype POSITIVE is INTEGER range 1 to INTEGER'HIGH;

  type STRING is array (POSITIVE range <>) of CHARACTER;
  type BIT_VECTOR is array (NATURAL range <>) of BIT;
```

```
  type FILE_OPEN_KIND is (READ_MODE, WRITE_MODE, APPEND_MODE);
  type FILE_OPEN_STATUS is (OPEN_OK, STATUS_ERROR, .NAME_ERROR, MODE_ERROR);

attribute FOREIGN: STRING;

end STANDARD;
```

C.2 Package STD.TEXTIO

```
package TEXTIO is

  type LINE is access STRING;
  type TEXT is file of STRING;
  type SIDE is (RIGHT, LEFT);

  subtype WIDTH is NATURAL;
  file INPUT: TEXT open READ_MODE is "STD_INPUT";
  file OUTPUT: TEXT open WRITE_MODE is "STD_OUTPUT";

  procedure READLINE (file F: TEXT; L: inout LINE);
  procedure READ (L: inout LINE; VALUE: out BIT; GOOD: out BOOLEAN);
  procedure READ (L: inout LINE; VALUE: out BIT);
  procedure READ (L: inout LINE; VALUE: out BIT_VECTOR; GOOD: out BOOLEAN);
  procedure READ (L: inout LINE; VALUE: out BIT_VECTOR);
  procedure READ (L: inout LINE; VALUE: out BOOLEAN; GOOD: out BOOLEAN);
  procedure READ (L: inout LINE; VALUE: out BOOLEAN);
  procedure READ (L: inout LINE; VALUE: out CHARACTER; GOOD: out BOOLEAN);
  procedure READ (L: inout LINE; VALUE: out CHARACTER);
  procedure READ (L: inout LINE; VALUE: out INTEGER; GOOD: out BOOLEAN);
  procedure READ (L: inout LINE; VALUE: out INTEGER);
  procedure READ (L: inout LINE; VALUE: out REAL; GOOD: out BOOLEAN);
  procedure READ (L: inout LINE; VALUE: out REAL);
  procedure READ (L: inout LINE; VALUE: out STRING; GOOD: out BOOLEAN);
  procedure READ (L: inout LINE; VALUE: out STRING);
  procedure READ (L: inout LINE; VALUE: out TIME; GOOD: out BOOLEAN);
  procedure READ (L: inout LINE; VALUE: out TIME);

  procedure WRITELINE (FILE F: TEXT; L: inout LINE);

  procedure WRITE (L: inout LINE; VALUE: in BIT;
    JUSTIFIED: in SIDE:= RIGHT; FIELD: in WIDTH := 0);
  procedure WRITE (L: inout LINE; VALUE: in BIT_VECTOR;
    JUSTIFIED: in SIDE:= RIGHT; FIELD: in WIDTH := 0);
  procedure WRITE (L: inout LINE; VALUE: in BOOLEAN;
    JUSTIFIED: in SIDE:= RIGHT; FIELD: in WIDTH := 0);
  procedure WRITE (L: inout LINE; VALUE: in CHARACTER;
    JUSTIFIED: in SIDE:= RIGHT; FIELD: in WIDTH := 0);
  procedure WRITE (L: inout LINE; VALUE: in INTEGER;
    JUSTIFIED: in SIDE:= RIGHT; FIELD: in WIDTH := 0);
  procedure WRITE (L: inout LINE; VALUE: in REAL;
    JUSTIFIED: in SIDE:= RIGHT; FIELD: in WIDTH := 0; DIGITS: in NATURAL:= 0);
  procedure WRITE (L: inout LINE; VALUE: in STRING;
    JUSTIFIED: in SIDE:= RIGHT; FIELD: in WIDTH := 0);
  procedure WRITE (L: inout LINE; VALUE: in TIME;
    JUSTIFIED: in SIDE:= RIGHT; FIELD: in WIDTH := 0; UNIT: in TIME:= ns);

- function ENDFILE (file F: TEXT) return BOOLEAN;

end TEXTIO;
```

C.3 Package IEEE.STD_LOGIC_1164

(Standard Multivalue Logic System IEEE Std 1164-1993)

```
PACKAGE std_logic_1164 IS
  TYPE std_ulogic IS (
  'U',   - Uninitialized
  'X',   - Forcing Unknown
  '0',   - Forcing 0
  '1',   - Forcing 1
  'Z',   - High Impedance
  'W',   - Weak Unknown
  'L',   - Weak 0
  'H',   - Weak 1
  '-'    - Don't care
  );

TYPE std_ulogic_vector IS ARRAY ( NATURAL RANGE <> ) OF std_ulogic;
FUNCTION resolved ( s :  std_ulogic_vector ) RETURN std_ulogic;
SUBTYPE std_logic IS resolved std_ulogic;
TYPE std_logic_vector IS ARRAY ( NATURAL RANGE <> ) OF std_logic;

SUBTYPE X01 IS resolved std_ulogic RANGE 'X' TO '1';
SUBTYPE X01Z IS resolved std_ulogic RANGE 'X' TO 'Z';
SUBTYPE UX01 IS resolved std_ulogic RANGE 'U' TO '1';
SUBTYPE UX01Z IS resolved std_ulogic RANGE 'U' TO 'Z';

FUNCTION "and" ( l :  std_ulogic; r :  std_ulogic ) RETURN UX01;
FUNCTION "nand" ( l :  std_ulogic; r :  std_ulogic ) RETURN UX01;
FUNCTION "or" ( l :  std_ulogic; r :  std_ulogic ) RETURN UX01;
FUNCTION "nor" ( l :  std_ulogic; r :  std_ulogic ) RETURN UX01;
FUNCTION "xor" ( l :  std_ulogic; r :  std_ulogic ) RETURN UX01;
FUNCTION "xnor" ( l :  std_ulogic; r :  std_ulogic ) RETURN UX01;
FUNCTION "not" ( l :  std_ulogic ) RETURN UX01;

FUNCTION "and" ( l, r :  std_logic_vector ) RETURN std_logic_vector;
FUNCTION "and" ( l, r :  std_ulogic_vector ) RETURN std_ulogic_vector;
FUNCTION "nand" ( l, r :  std_logic_vector ) RETURN std_logic_vector;
FUNCTION "nand" ( l, r :  std_ulogic_vector ) RETURN std_ulogic_vector;
FUNCTION "or" ( l, r :  std_logic_vector ) RETURN std_logic_vector;
FUNCTION "or" ( l, r :  std_ulogic_vector ) RETURN std_ulogic_vector;
FUNCTION "nor" ( l, r :  std_logic_vector ) RETURN std_logic_vector;
FUNCTION "nor" ( l, r :  std_ulogic_vector ) RETURN std_ulogic_vector;
FUNCTION "xor" ( l, r :  std_logic_vector ) RETURN std_logic_vector;
FUNCTION "xor" ( l, r :  std_ulogic_vector ) RETURN std_ulogic_vector;
FUNCTION "xnor" ( l, r :  std_logic_vector ) RETURN std_logic_vector;
FUNCTION "xnor" ( l, r :  std_ulogic_vector ) RETURN std_ulogic_vector;
FUNCTION "not" ( l :  std_logic_vector ) RETURN std_logic_vector;
FUNCTION "not" ( l :  std_ulogic_vector ) RETURN std_ulogic_vector;

FUNCTION To_bit ( s :  std_ulogic; xmap :  BIT := '0') RETURN BIT;
FUNCTION To_bitvector ( s :  std_logic_vector ; xmap :  BIT := '0') RETURN BIT_VECTOR;
FUNCTION To_bitvector ( s :  std_ulogic_vector; xmap :  BIT := '0') RETURN BIT_VECTOR;
FUNCTION To_StdULogic ( b :  BIT ) RETURN std_ulogic;
FUNCTION To_StdLogicVector ( b :  BIT_VECTOR) RETURN std_logic_vector;
FUNCTION To_StdLogicVector ( s :  std_ulogic_vector ) RETURN std_logic_vector;
FUNCTION To_StdULogicVector ( b :  BIT_VECTOR) RETURN std_ulogic_vector;
FUNCTION To_StdULogicVector ( s :  std_logic_vector ) RETURN std_ulogic_vector;
FUNCTION To_X01 ( s :  std_logic_vector ) RETURN std_logic_vector;
FUNCTION To_X01 ( s :  std_ulogic_vector ) RETURN std_ulogic_vector;
```

A

```
FUNCTION To_X01 ( s :  std_ulogic) RETURN X01;
FUNCTION To_X01 ( b :  BIT_VECTOR) RETURN std_logic_vector;
FUNCTION To_X01 ( b :  BIT_VECTOR) RETURN std_ulogic_vector;
FUNCTION To_X01 ( b :  BIT ) RETURN X01;

FUNCTION To_X01Z ( s :  std_logic_vector ) RETURN std_logic_vector;
FUNCTION To_X01Z ( s :  std_ulogic_vector ) RETURN std_ulogic_vector;
FUNCTION To_X01Z ( s :  std_ulogic) RETURN X01Z;
FUNCTION To_X01Z ( b :  BIT_VECTOR) RETURN std_logic_vector;
FUNCTION To_X01Z ( b :  BIT_VECTOR) RETURN std_ulogic_vector;
FUNCTION To_X01Z ( b :  BIT ) RETURN X01Z;
FUNCTION To_UX01 ( s :  std_logic_vector ) RETURN std_logic_vector;

FUNCTION To_UX01 ( s :  std_ulogic_vector ) RETURN std_ulogic_vector;
FUNCTION To_UX01 ( s :  std_ulogic) RETURN UX01;
FUNCTION To_UX01 ( b :  BIT_VECTOR) RETURN std_logic_vector;
FUNCTION To_UX01 ( b :  BIT_VECTOR) RETURN std_ulogic_vector;
FUNCTION To_UX01 ( b :  BIT ) RETURN UX01;

FUNCTION rising_edge (SIGNAL s :  std_ulogic) RETURN BOOLEAN;
FUNCTION falling_edge (SIGNAL s :  std_ulogic) RETURN BOOLEAN;

FUNCTION Is_X ( s :  std_ulogic_vector ) RETURN BOOLEAN;
FUNCTION Is_X ( s :  std_logic_vector ) RETURN BOOLEAN;
FUNCTION Is_X ( s :  std_ulogic) RETURN BOOLEAN;

END std_logic_1164;
```

C.4 Package IEEE.NUMERIC_STD

(Synthesis Package IEEE Std 1076.3-1997)

```
type UNSIGNED is array (NATURAL range <>) of STD_LOGIC;
type SIGNED is array (NATURAL range <>) of STD_LOGIC;
```

Arithmetic Operators

Operator	Type of operand		Type of result
	ARG		
abs(ARG)	SIGNED		SIGNED(ARG'LENGTH-1 downto 0)
Operator (unary)	Type of argument		Type of result
	ARG		
ARG	SIGNED		SIGNED(ARG'LENGTH-1 downto 0)
Operators (binary)	Type of operands		Type of result
	L	R	
L + R	UNSIGNED	UNSIGNED	UNSIGNED(MAX(L'LENGTH,R'LENGTH)-1 downto 0)
L − R	SIGNED	SIGNED	SIGNED(MAX(L'LENGTH,R'LENGTH)-1 downto 0)
L ∗ R	UNSIGNED	NATURAL	UNSIGNED(L'LENGTH-1 downto 0)
L / R [1)]	NATURAL	UNSIGNED	UNSIGNED(R'LENGTH-1 downto 0)
L rem R [1)]	INTEGER	SIGNED	SIGNED(R'LENGTH-1 downto 0)
L mod R [1)]	SIGNED	INTEGER	SIGNED(L'LENGTH-1 downto 0)

[1)] Restrictions for synthesis: R is constant and a literal of base 2.

Relational Operators

Operators	Type of operands		Type of result
	L	R	
L > R	UNSIGNED	UNSIGNED	BOOLEAN
L < R	SIGNED	SIGNED	BOOLEAN
L >= R	UNSIGNED	NATURAL	BOOLEAN
L <= R	NATURAL	UNSIGNED	BOOLEAN
L = R	INTEGER	SIGNED	BOOLEAN
L /= R	SIGNED	INTEGER	BOOLEAN

Logical Operators

Operator	Type of operand	Type of result
	L	
not L	UNSIGNED	UNSIGNED(L'LENGTH-1 downto 0)
	SIGNED	SIGNED(L'LENGTH-1 downto 0)

Operators	Type of operands		Type of result
	L	R	
L and R L or R L nand R	UNSIGNED	UNSIGNED	UNSIGNED(L'LENGTH-1 downto 0)
L nor R L xor R L xnor R [1)]	SIGNED	SIGNED	SIGNED(L'LENGTH-1 downto 0)

[1)] xnor is not compatible with IEEE Std 1076-1987.

Rotation Operators

Operators	Type of operands		Type of result
	ARG	COUNT	
ARG sll COUNT ARG srl COUNT	UNSIGNED	INTEGER	UNSIGNED(ARG'LENGTH-1 downto 0)
ARG rol COUNT ARG ror COUNT	SIGNED	INTEGER	SIGNED(ARG'LENGTH-1 downto 0)

Rotation Functions

Function	Type of arguments		Type of return value
	ARG	COUNT	
ROTATE_LEFT (ARG, COUNT)	UNSIGNED	NATURAL	UNSIGNED(ARG'LENGTH-1 downto 0)
ROTATE_RIGHT (ARG, COUNT)	SIGNED	NATURAL	SIGNED(ARG'LENGTH-1 downto 0)

Shift Functions

Function	Type of arguments		Type of return value
	ARG	COUNT	
SHIFT_LEFT (ARG, COUNT)	UNSIGNED	NATURAL	UNSIGNED(ARG'LENGTH-1 downto 0)
SHIFT_RIGHT (ARG, COUNT)	SIGNED	NATURAL	SIGNED(ARG'LENGTH-1 downto 0)

Sizing of Arrays

Function	Type of arguments		Type of return value
	ARG	NEW_SIZE	
RESIZE (ARG, NEW_SIZE)	UNSIGNED	NATURAL	UNSIGNED(NEW_SIZE-1 downto 0)
	SIGNED	NATURAL	SIGNED(NEW_SIZE-1 downto 0)

Conversion Functions

Function	Type of argument	Type of return value
	ARG	
TO_INTEGER (ARG)	UNSIGNED	NATURAL
	SIGNED	INTEGER

Function	Type of arguments		Type of return value
	ARG	SIZE	
TO_UNSIGNED (ARG, SIZE)	NATURAL	NATURAL	UNSIGNED(SIZE-1 downto 0)

Function	Type of arguments		Type of return value
	ARG	SIZE	
TO_SIGNED (ARG, SIZE)	INTEGER	NATURAL	SIGNED(SIZE-1 downto 0)

Element compare (with don't cares, '-')

Function	Type of arguments	Type of return value
	L, R	
STD_MATCH (L, R)	STD_ULOGIC	BOOLEAN
	STD_LOGIC_VECTOR	BOOLEAN
	STD_ULOGIC_VECTOR	BOOLEAN
	UNSIGNED	BOOLEAN
	SIGNED	BOOLEAN

Translation Functions

Function	Type of arguments		Subtype of return value
	S	XMAP	
TO_01 (S, XMAP)	UNSIGNED	STD_LOGIC	UNSIGNED(S'RANGE)
	SIGNED	STD_LOGIC	SIGNED(S'RANGE)

Appendix D Standardization in Electronic Design Automation

DIRK JANSEN

This table follows the listing of the Silicon Integration Initiative (SI2), taken from http://www.si2.org/roadmap.

UCB SPICE

SPICE is an analog simulator that nowadays is a quasi-standard. The program includes analytical models of nearly all basic circuit components and is able to handle complex, non-linear circuits. The original software was developed at the University of California, Berkeley. SPICE is still implemented in many actual analog simulation tools and commercial simulators, the netlist description and transistor models are basics of many applications. The program is still distributed and maintained by UCB.

IEC IEC 61360: Standard Data Element Types

This standard defines requirements on data types (parameters) that are used for the description of electronic components in form and function. The standard summarizes amongst others, a library of definitions, which comply with the standard, as well as procedures for maintenance, and extension of this library.

ISO 13584: Parts Library Standard

This standard provides rules and guidelines for setting up a hierarchy of component families. The rules of these standards are used by standardization groups, which are now able to set up reference hierarchies. The rules have to be obeyed by providers of libraries. An important chapter is the 'methods for structuring of component families'.

EIA 567 A: VHDL Model Component and Hardware Interface Standard

This standard defines requirements on VHDL models at the component level, which have to obey a common signal convention, will show unified behaviour in simulation, and which will be re-usable in many different designs. With this included the models may be distributed to different users, and by this support a dissemination independent of technology.

EIA IS 103: Library of Parameterized Modules

LPM is a standard describing parameterizable function modules (macros) with the intention of simplified usage, placement and routing of complex, re-usable basic blocks such as adders, counters, shift registers, multipliers, etc. They are often used in combination with FPGAs, they may show a common symbol, and are well suited to symbolic input. The technological representation is the task of the technology provider. Common to all LPMs are generators, which allow one to specify the bit length, the architecture, and other features of a, in general, regular structure.

IEEE 1076 VHDL: VHSIC Hardware Description Language

VHDL is a formal hardware description language, suited to all phases of a design process. VHDL can be read by humans as well as by computers, so it supports in development, verification, and test phase of a hardware design. VHDL may be used for the transfer of design data, for the maintenance, modification, and programming of hardware.

IEEE 1164 MV: Multi-Value Logic

This standard describes the std_logic_1164 VHDL Package, the included declarations and functions, and how they have to be used. The package defines a multi-value logic system, suited for port signal declarations. With these functions and operators VHDL models can be built up which may be transferred to other EDA systems, giving an unified description of the behaviour on different kinds of simulators.

IEEE P1076.2: Maths Package

The VHDL maths package is a standardized package of mathematical functions, and arithmetic functions, as well as trigonometric functions in VHDL with guaranteed accuracy and unified error handling.

IEEE P1076.3: Standard VHDL Synthesis Package

This package defines standard procedures and functions for synthesis of binary digital electronics from VHDL code. It contains the hardware interpretation of values which belongs to the data types *bit, Boolean*, and *std_ulogic*; some functions to handle '*don't cares*'; functions to detect rising and falling edges; declaration of vectors and operators on these vectors, which may be interpreted as *signed* or *unsigned* numbers. The unified definitions simplify the automatic synthesis of circuits from VHDL code.

IEEE P1076.4 VITAL: VHDL Initiative Towards ASIC Libraries

VITAL is an accepted industrial standard, founded on VHDL, for the specification of ASIC libraries in a 'sign off' quality. VITAL is a set of standard timing/test routines, logic primitives, and programmable truth tables. The standard includes further methods for model development and back annotation of data that describes timing (SDF).

IEEE MHDL: MIMIC Hardware Description Language

This is a standard in development, intended to extend the VHDL language in such a way that special requirements in microwave technology can be handled in simulation and analysis.

IEEE P1076.1 VHDL-A: VHDL Analog/VHDL-AMS

This standard is an extension of the VHDL language to allow the description of mixed analog/digital systems (AMS = Analog Mixed Signal). The language supports all application levels of VHDL. Using these new constructs, non-electrical, mechanical, thermal, as well as mixed analog/digital systems may be described and simulated.

IEEE 1029.1 WAVES: Waveform and Vector Exchange Standard

WAVES is an application of the VHDL language. The standard covers format, header, architecture, and processing model of a data structures (file) for the transfer of wave forms (traces) for test purposes.

IEEE 1364-2001: Verilog Standard

This standard describes the Verilog Hardware description language. Verilog is a formal hardware description language suited to all phases of a design process. Verilog can be read by humans as well as by computers, so it supports the development, verification, and test phases of a hardware design. Verilog may be used for the transfer of design data, for maintenance, modification, and programming of hardware.

OVI Verilog-A: Verilog HDL for Analog

This standard will define the analog extension for the Verilog hardware description language, similar to VHDL-AMS.

IEEE OMF: Open Modelling Forum

These documents will improve the re-usability of models by defining an open procedural interface between the model and the simulator. The proprietary data of the model provider is encapsulated.

SII: Open Access

OpenAccess is a community effort to provide true inter-operability, not just data exchange, amongst IC design tools through an open standard data API and a reference database supporting that API for IC design. The OpenAccess Coalition is a neutral organization of industry leaders that are leading this effort operating under Si2 bylaws.

SII OLA: Open Library Architecture

Increased requirements for accuracy and consistency of electrical analysis within deep sub-micron design flows have led to the development of a new open library architecture for EDA. OLA is a comprehensive Application Procedural Interface (API) that can be used by EDA tools for the determination of circuit and inter-connect characteristics (such as delay or power). It has been endorsed and approved by the ASIC Council and Si2 membership. OLA uses the IEEE 1481 (Delay and Power Calculation System) Standard as the fundamental architecture.

IEEE P1029.2 FDL: Fault Dictionary Language

FDL is a diagnosis language. The standard describes syntax, semantics and format of diagnosis data in test systems.

IEEE P1029.3 TRSL: Test Requirements Specification Language

TRSL defines syntax, semantics, and format of a language for writing test requirements in test systems.

IEEE 716 ATLAS: Abbreviated Test Language for All Systems

ATLAS is a test language and is used for specification of tests as well as for test programming. ATLAS is closely linked to ARINC 626, an aeronautical standard. In the future, both standards will be combined in some way.

ISO 10164 Parts 12 & 14 (Test Management)

ISO 10164 standard describes communication protocols for system level test management functions in an open client server architecture. Part 14 defines the related terms.

IEEE P1149.5 BSCN: Boundary Scan

This standard describes a serial scan through special I/O cells of chips or even modules, MCMs or boards, allowing all the inputs and outputs of the chip to be observed. By using *boundary scan* one may avoid using needle adapters or special test pins on outputs (see JTAG).

IEEE P1226 ABBET: A Broad Based Environment for Test

P1226.10 (SCC20) Standard for Software Interface for Runtime Services for a Broad Based Environment for Test (ABBET). This is a set of standards, which cover the test problem, beginning with product specification, and up to the final product. These standards are linked to many more standards in the test topic field.

IEEE P1232 AIESTATE: Artificial Intelligence and Expert System Tie to ATE

This standard describes the interface methodology and encapsulation of a self-acting computer reasoner for test purposes. The architecture of a client server environment is defined with separation between test and diagnosis strategy.

IEEE P1389 TMIMS: Test Management and Information Maintenance Services

TMIMS includes definitions, formats, and services for a collection of data and for analysis in the framework of system maintenance.

IEEE P1029.1 DTIF: Digital Test Information Format

This standard is founded on the LASARTAP format for digital test. It covers test vectors, errors lists, and timing data for use in digital test equipment.

EIA EDIF: Electronic Design Interchange Format

EDIF 3.0 defines the syntax and language constructs for electronic design data exchange, mainly the netlist data of the connections, and related attributes, and includes symbolic representations. EDIF supports further architectural levels of design as well as the system and PCB board design level. EDIF 3.5 is an interim standard, suited to the transfer of data for design, manufacturing, and assembly of printed circuit boards and electronic assemblies. EDIF 4.0 will extend the features further to MC Modules and will contain technology rules and assembly drawings.

ANSI/EIA 656 IBIS: I/O Buffer Information

IBIS describes the behaviour of electronic circuits IC at the interface, the analog characteristics at the pins (I/O). The standard is founded on a tabular description of the voltage-current characteristics of each terminal, of the time domain behaviour, and gives information on the package. IBIS simulations run fast and with high accuracy. IBIS models are used for delay time calculations and for the control of signal integrity at package or board level.

ISO AP210: Electronic PCA Design and Manufacture

This STEP application protocol covers requirements and data needed to manufacture printed circuit assemblies PCA. The data should contain all information needed, from development to manufacturing of the PCAs.

ISO AP211: Product Application Protocol – Electrical

PAP-E is still worked on. It is intended to set up an information model for timing requirements in test and diagnosis.

ISO AP220: PCA Manufacturing Planning

This is a STEP application protocol for blank PCBs as well as mounted PCBs. The model should describe both views and will be used in development as well as in manufacturing.

CFI DR: CAD Framework Initiative Design Representation

DR is a programming interface reference for the design model, it includes, for example, netlist formats.

CFI DCL: Delay Calculation Language

Contains definitions and methods for delay time calculations.

EIA DIE-T: Die Information Exchange for Timing

The DIE-T format is a readable exchange format for delay data of digital ICs. It includes all information that is needed to carry out a static or dynamic timing analysis.

OVI PDEF: Physical Delay Extraction Format

PDEF is a method of describing floor plan data. The format sets up a critical link between logical and physical sub-micron design. It includes all necessary information for the interchange of floor plan data between synthesis tools and layout tools.

OVI SDF: Standard Delay Format

This standard allows a file-based exchange of component-specific timing data and timing constraints between the ASIC delay calculation tools and the simulation tools, thus allowing a high precision timing calculation.

OVI SPF: Standard Parasitic Format

The SPF format includes in a readable form the extracted *parasitics* (resistances, capacities), which are supplied to the synthesis and simulation tools.

NIST Model Validation

This standard defines methods for analysis and assessment of circuit models. Requirements for testing circuits are set up which minimize parasitics. It covers further methods of translating data sheet values of manufacturers into circuit parameters.

IEEE P1450 (STIL): Standard Test Interface Language

The STIL is a language definition, allowing the exchange of large amounts of test data between EDA equipment and digital testers (ATE). It defines pattern formats and timing data required for the 'device under test' (DUT). Several test philosophies are supported.

Addresses of the Organisations

EIA
c/o Patti Rusher
2500 Wilson Boulevard #203
Arlington, VA 22201
USA
(703) 907-7545 phone
(703) 907-7501 fax
http://www.edif.org/edif/eia.html
prusher@eia.org

IEC International Electrotechnical Commission
3, rue de Varembé
P.O. Box 131
CH- 1211 Genève 20
Switzerland
+41 22 919 02 11 phone
+41 22 919 03 00 fax
E-mail: info@iec.ch

IEEE
Service Center
445 Hoes Lane
P.O. Box 1331
Piscataway, NJ 08855-1331,
USA
(908) 981-0060 phone
(908) 981-9667 fax
http://stdsbbs.ieee/org/
stds.info@ieee.org

ISO International Standards Organization
1 Rue de Yarembe
Case Postale 56
CH-1211, Genève 20
Switzerland/Suisse
44 22 749-0111 phone
44 22 733-3430 fax
http://www.iso.ch/welcome.html
sales@isocs.iso.ch

CFI
4030 W. Braker Lane #550
Austin, TX 78759
USA
(512) 342-2244 phone
(512) 342-2037 fax
http://www.si2.org/specs.html
cfi@cfi.org

OVI
15466 Los Gatos Boulevard #109-071
Los Gatos, CA 95032
USA
(408) 353-8899 phone
(408) 353-8869 fax
www.chronologic.com/ovi/membership.html
ovi@netcom.com

NIST
Publicity Affairs Director
Administration A903
National Institute of Standards & Technology
Gaithersburg, MD 20899-0001
USA
(301) 975-2000 phone
http://www.nist.gov
aeshop@micf.nist.gov

UCB

University of California Berkeley
http://www.berkeley.edu
Berkeley, CA 94720
USA

Si2

Silicon Integration Initiative
4030 West Braker Lane, Suite 550
AUSTIN, Texas 78759
USA
+1 (512) 342 2037 phone
http://www.si2.org

Appendix E Symbols

DIRK JANSEN

Symbols derived from basic symbols by indices are described in the related text.

a	distance between wires
a	scaling factor
a	acceleration
A	area
A	transconductance matrix
A	forward current amplification
A	amplification
A_{vt}, A_β	technology constants
$\boldsymbol{b}$	current vector
B	width of wires
B_F	current amplification (DC)
b_L	width of wire
c	phase-velocity of light
c_0	velocity of light in vacuum
C	capacity
C_0	coefficient
C_{ox}	gate capacity
d	diameter
D	damping coefficient (mechanical)
D_v	density of interconnection
$\mathrm{d}t$	differentiation in time domain
E_g	band gap energy
f	frequency
F	force (mechanical)
f_E	technology constant
Δf	bandwidth
$f(x)$	function
g_m	transconductance (small signal, AC)
G_m	transconductance
i	current (small signal, AC)
I	current (DC)
I_F	forward current (Diode)
I_s	saturation current
j	complex operator
k	Boltzmann's constant
K	spring constant (mechanical)
K_1, K_2	process dependent noise constants
K_p	process factor with MOSFETs
L	L(ow) Matrix, elements below the diagonal
L	channel length
L_{eff}	effective channel length
m	gate effectivity
M	mass
n	number
n	turns per minute
n	dividing factor
n_F	emission coefficient
n_i	intrinsic density
N_s	substrate doping
p	possibility
q	elementary charge
Q	charge
r	factor
r	resistance (dynamical)
R	resistor
R_s	sheet resistance
s	complex frequency, Laplace parameter
S_i	binary coeffizient i
S_R	slew rate
S_{vt}, S_β	technology constants
SNR	signal to noise ratio (in dB)
t	time
T	absolute temperature
T	time constant
t_n	discrete time
t_r, t_f	rise time, fall time
u	deviation

Symbol	Meaning
v	velocity
v	voltage (small signal, AC)
V	voltage (DC)
V	V(p)-Matrix, elements above the diagonal
V_{ref}	reference voltage
V_T	temperature voltage (about 26 mV at 300 K)
V_{Th}	threshold voltage
V_{To}	threshold voltage
W	width, channel width
$\boldsymbol{x}$	voltage vector
X	decimal equivalent
x_1	element of the vectors $\boldsymbol{x}$
$x(k)$	value, discrete at time index k
$y(k)$	deviated value, discrete at time index k
z	operator of the z-transformation
Z	complex resistance

Symbol	Meaning
α	stuck at error
α_T	temperature coefficient of a resistance
β	process gain-factor
β	current amplification (AC)
γ	gamma, substrate threshold parameter
ε_0	permeability of vacuum
ε_r	relative permeability
λ	Lambda, channel length modulation factor
λ	wavelength
μ	mobility of the charge carrier
μ_r	relative permeability
ξ	damping factor
σ	sigma, statistical scattering
τ	delay, time constant
τ_B	intrinsic (basic) delay
Φ	Phi, surface potential

Authors

Professor Dr.-Ing. Gerhard Albert has acted as a Professor in the Department of Information Technology at the University of Applied Sciences, Mannheim, since 1987. He is engaged in teaching and research in the *Design of Analog Integrated Circuits*. In addition he is head of the *Transfer Centre for Sensor Technology, Microelectronics, and Software Technology* at the UAS/Mannheim. He is the author of several publications and has been a member of the IEEE and MPC for many years.

Prof Dr.-Ing. Hermann Clauss has taught *Electronic Circuit Integration* and *Electronic Components* at the University of Applied Sciences, Heilbronn, since 1987. He is director of the laboratory for VLSI design, with his main focus on analog and bipolar circuit design. Before that he had many years of experience at Temic/Germany in IC manufacturing. He is the author of several publications and has been a member of MPC from the beginning.

Prof. Dr. rer. nat. Andreas Foitzik teaches *Materials and Manufacturing in Micro Mechanics* at the University of Applied Sciences/Esslingen and is a member of the mechatronics faculty. Before that he worked as a scientist at the Max Planck Institute for Materials Research, and at the Fraunhofer Institute for Material Mechanics. He twice received a grant as a Feodor Lynen student of the Alexander von Humboldt Stiftung to study at the Case Western Reserve University in Cleveland, Ohio, USA. His focus in research and teaching is on materials, micro-systems (MEMS), physical measurements, printed circuit manufacturing, and assembly and connection technology. He is the author/co-author of about 40 publications.

Prof. Dr.-Ing. Dirk Jansen has taught *Digital Circuit Design* and *Microelectronics* at the University of Applied Sciences, Offenburg, since 1986. After several years of industrial experience in the aircraft industry, he set up the laboratories for electronic components at UAS, optoelectronics (course book), and electronic circuit design. In 1989 he founded the *ASIC Design Centre* at the UAS, Offenburg, which has now successfully designed more than 25 ASICs. He is the director of this centre as well as the director since 1995 of the Institute of Applied Research of the UAS, Offenburg. He is the speaker for the MPC-Group, a member of the IEEE, VDE, EUROCHIP, and EUROPRACTICE, and author/editor of several publications (EDA Handbuch, Hanser). His research focus is ASIC design, system on a chip design, hardware/software codesign, integrated processor cores, and inductive data transmission.

Prof. Dr.-Ing. Bernd Kohlhammer teaches *Semiconductor Technology*, and *Electronic Circuit Design* at the University of Applied Sciences, Aalen. He is director of the *EDA Centre* and is engaged in chip design for several industrial applications. His focuses in research are computer aided circuit simulation and realization, with emphasis on printed circuits and complex mixed signal designs. He has been a member of the MPC group from the beginning.

Prof. Dr.-Ing. Hans Kreutzer has taught for more than 10 years *Digital Electronics*, *Computer Aided Engineering*, and *Signals and Systems* at the University of Applied Sciences, Reutlingen, Before that he was responsible in industry for the development of measuring equipment for protocols in data transmission. He has been a member of the MPC Group since many years.

Prof. Dr.-Ing. Dr. h.c. Horst Nielinger, now retired, taught *Communication Technology* and *Analog Techniques* at the UAS, Furtwangen, until 1998. He is the author of several publications in the field of SPICE applications, and one of the authors of the first books in German language about SPICE. His continuing interest is the application and development of simulation techniques for electronic circuits, nowadays with VHDL-AMS. He is one of the co-founders of the MPC group and is still very active.

Prof. Dr.-Ing. Gerd-Uwe Paul lectures at the FH Mannheim/ University of Applied Sciences in the institute for communicational engineering. He teaches the *Application of Programmable Logic Components* and the *Design of Integrated Digital Circuits/Devices*. He also lectures on the *Fundamental Principles of Electrical Engineering* and on *Error Detecting and Error Correcting Codes*. He is a member of the IEEE and belongs to the MPC group.

Prof. Dr.-Ing. Ermenfried Prochaska teaches *Microelectronics* and *Digital Circuit Design* at the mechanical production faculty of the University of Applied Sciences, Heilbronn. His focus is on programmable logic devices and digital signal processors. He is the author/co-author of several books on digital design, signal processor applications, and logic analysers. He is responsible for didactic affairs at UAS, Heilbronn, and a member of the commission for didactics in university studies. He is a member of the IEEE and MPC.

Prof. Dr.-Ing. Martin Rieger has taught *Electronic Circuit Design*, *Microprocessor Technology*, and *Computer Aided Engineering* at the University of Applied Sciences, Albstadt-Sigmaringen, since 1993. The main focus of his interest is microelectronics, VLSI design and the application of microprocessors for the automations of buildings. In industry he was responsible for a department for fast analog IC design, with emphasis on signal processing ICs for TV and RF applications. He participated in the European JESSI projects DAB and HDTV. He is a member of the IEEE solid state society, the IEEE computer society, the IEEE educational society, VDE, VDI, and MPC.

Prof. Dr. Wolfgang Rülling teaches *Computer Science* and *Microelectronics* at the University of Applied Sciences, Furtwangen. His areas of interest are ASIC design, test methods, algorithms, and data structures. Before that he was a leading scientist at the University of Saarbrücken, working on the topic of VLSI design methods and concepts in parallel computing. He is a member of the MPC since many years and author of several publications.

Prof. Dr.-Ing. Günther Schuster teaches *Foundations of Electronics*, *Measurement Techniques*, *Technology of Assembly and Interconnections*, and *Mechanical Microsystems* at the University of Applied Sciences, Esslingen. He is the director of the laboratory of test and measurement in microelectronics in Göppingen,, with emphasis on mechanical production measurements, quality assurance, and electromagnetic interference. The focus of his work is the technology of production and assembly technology for microelectronics (MEMS).

Dr. techn. Alfred Schütz works for Infineon, München, Germany. He is responsible for the definition, development, and application of DRAM modules that are integrated into complex logic chips. Before that he hold a management function in ASIC-design as well as in the application of EDA systems. The focus of his activity lies in the establishing of ASIC libraries and the definition of design flows. He studied electro-techniques, and gained his PhD in Vienna, Austria. He is the author of several publications.

Dr.-Ing. habil. Peter Schwarz is the head of the Department Modeling and Simulation of the Fraunhofer Institute for Integrated Circuits, Branch Lab Design Automation EAS Dresden. After his study at the Dresden University of Technology he worked for 14 years in the company *Meßelektronik Dresden*. He was responsible for research and development in a CAD department. Later he was the leader of a research team at the *Akademie der Wissenschaften* of the former GDR engaged in the development of the multi-level, mixed-signal simulator KOSIM, which was used in industry and research institutions at that time. Today, his focus is on simulation methods and model generation for *IC Electronics, Communications, Mechatronics, Automation* and *MEMS*. He is the author of many publications and a member of the IEEE and VDE.

Dipl.-Ing. Friedemann Stockmayer worked with Wavetek Wandell Goltermann in Germany. For several years he designed measurement equipment for digital communication techniques, ATM, and SDH, and is responsible as a project leader for ASIC and FPGA designs. Because of his long-term experience in digital design and modern synthesis methods he teaches *Computer Aided Engineering* at the University of Applied Sciences, Reutlingen.

Prof. Dr.-Ing. Harald Töpfer has taught *Electronic Circuit Design*, *Simulation* and *Microelectronics* at the University of Applied Sciences, Esslingen, since 1991. Before that he developed ASICs at Temic, Germany and Mikroelektronik Erfurt, Germany with an emphasis on the full custom design of analog and digital VLSI. He is the co-author/author of several publications and is a member of the IEEE and MPC.

Prof. Dr. Gert Voland received his MSc in Experimental Physics at Darmstadt, Germany, and received his PhD for research in the field of surface states on silicon MOS structures. He has worked at the semiconductor branch of Siemens AG in Munich on the

development of microprocessors and as a CAD manager. After that he was Director of Design Engineering at European Silicon Structures (ES2) in Munich and Aix-en-Provence. He currently teaches *Digital Electronics, Analog Electronics,* and *ASIC Methodology* in the Computer Science Department of the University of Applied Sciences/Konstanz, Germany.

D

J

K

L

M

N

O

P

Q

U

V